BIOLOGY OF
TETRAHYMENA

BIOLOGY OF *TETRAHYMENA*

Edited by **Alfred M. Elliott**

University of Michigan at Ann Arbor

Dowden, Hutchinson & Ross, Inc.

Stroudsburg, Pennsylvania

Library of Congress Cataloging in Publication Data

Elliott, Alfred Marlyn, 1905-   comp.
    Biology of Tetrahymena.

    Bibliography: p.
    1. Tetrahymena. I. Title. [DNLM: 1. Tetra-
hymena. 2. Tetrahymena pyriformis. QX151 E46b
1973]
QL368.H87E44   593'.172   73-12911
ISBN 0-87933-013-9

Manufactured in the United States of America
Exclusive distributor outside the United States and Canada:
John Wiley & Sons, Inc.

Contributors

SALLEY ALLEN   *Departments of Botany and Zoology, University of Michigan, Ann Arbor*

HERMAN BAKER   *Departments of Medicine and Preventive Medicine, College of Medicine and Dentistry of New Jersey, East Orange*

IVAN L. CAMERON   *Department of Anatomy, The University of Texas Medical School at San Antonio, San Antonio*

ROBERT L. CONNER   *Department of Biology, Bryn Mawr College, Bryn Mawr, Pennsylvania*

JOHN O. CORLISS   *Department of Zoology, University of Maryland, College Park*

DUDLEY COX   *Haskins Laboratories at Pace College, and Department of Biology, Pace College, New York*

PHILLIP DUNHAM   *Department of Biology, Syracuse University, Syracuse, New York*

ALFRED M. ELLIOTT   *Department of Zoology, University of Michigan, Ann Arbor*

OSCAR FRANK   *Department of Medicine, College of Medicine and Dentistry of New Jersey, East Orange*

JOSEPH FRANKEL   *Department of Zoology, University of Iowa, Iowa City*

IAN GIBSON   *School of Biological Sciences, University of East Anglia, Norwich, Norfolk England*

GEORGE G. HOLZ, JR.   *Department of Microbiology, State University of New York Upstate Medical Center, Syracuse*

S. H. HUTNER   *Haskins Laboratories at Pace College, Pace College, New York*

JOHN R. KENNEDY   *Department of Zoology, University of Tennessee, Knoxville*

MARY J. KOROLY   *Department of Biology, Bryn Mawr College, Bryn Mawr, Pennsylvania*

DONNA L. KROPP   *Department of Physiology, Yale University of Medicine, New Haven, Connecticut*

MICHAEL R. LEVY   *Department of Biological Sciences, Southern Illinois University, Edwardsville*

BARBARA B. McDONALD *Department of Biology, Dickinson College, Carlisle, Pennsylvania*

NORMAN E. WILLIAMS *Department of Zoology, University of Iowa, Iowa City*

Preface

Through time biologists have sought out organisms for experimentation that could aid them answer the questions they were asking. In the field of microbiology, where axenic clonal cultures were essential to obtain meaningful results, the bacteria were the first to be isolated—followed by the fungi, algae, and finally the protozoa. Among representatives of the last group, the blood flagellates were the first to be cultured axenically. Other protozoa, the amoebae, sporozoa and ciliates, did not lend themselves so well to axenic culture. Among the ciliates, members of the genus *Tetrahymena* proved to be the most easily grown under axenic conditions and *T. pyriformis* became the species of choice. It was the first isolated from natural waters in France by Andre Lwoff in 1923. At the time it was known as *Glaucoma pyriformis*. This discovery created little excitement at the time and for the next decade almost nothing was done with it. Starting in the early 1930's American workers became interested in members of the genus *Tetrahymena* and papers appeared in the literature at an ever-increasing rate. This has continued and at present the literature is vast and scattered through numerous publications, from journals of molecular biology to veterinary medicine. The growing use of these cells in experimental biology stems from their ease of handling, fantastic growth rate, genetic stability, sexual processes and animal characteristics. They have been used in ultrastructural, developmental, genetical, biochemical, molecular and other studies. Many problems in biology can be investigated to advantage using these ciliates as experimental tools.

Investigators initiating a study of *Tetrahymena* find it extremely difficult to survey the literature because the publications appear in such diverse journals, many of which are found only in professional libraries. Whereas there have been a number of excellent reviews on various aspects of protozoan morphology, physiology, and biochemistry where research on *Tetrahymena* has been included, there is only one book* on the genus alone and this is limited to the biochemistry and physiology of the cell. It seemed that a more comprehensive coverage of the genus was needed.

*Hill, D. L., 1972, *The Biochemistry and Physiology of* Tetrahymena, Academic Press, New York and London. 230 pp.

This book was planned at the Toronto meeting of the Society of Protozoologists in 1967 by Ivan L. Cameron, Robert L. Conner, Seymour H. Hutner and myself. Most of the tentative topics and authors were selected at that time. Later other topics as well as authors were added. In order that the book be as useful as possible it was decided to include a general bibliography with full citations. This appears at the back of the book.

Unfortunately all of the topics originally selected have not been covered. However, most of the material that would have appeared in these missing chapters has been included in relevant chapters.

I wish to express my appreciation to those who helped me plan the book during its initial stages and all of the authors who have contributed chapters. I especially want to thank Robert L. Conner, John O. Corliss, George G. Holz, and Seymour H. Hutner for their continued encouragement during the difficult periods of bringing the book through to publication. Thanks are also due Alex Tompa for his help in updating the general bibliography. I think I speak for all of us when I express the hope that this book will be useful to present workers in the field and future investigators who decide to use these delightful ciliates to help clarify some of the mysteries of cells.

A.M.E.
1973

Contents

Preface   v

CHAPTER 1
History, Taxonomy, Ecology, and Evolution of Species of *Tetrahymena*
    JOHN O. CORLISS   1

CHAPTER 2
Morphology of *Tetrahymena*
    ALFRED M. ELLIOTT and JOHN R. KENNEDY   57

CHAPTER 3
The Nutrition of *Tetrahymena*: Essential Nutrients, Feeding and Digestion
    GEORGE G. HOLZ, JR.   89

CHAPTER 4
The Composition, Metabolism, and Roles of Lipids in *Tetrahymena*
    GEORGE G. HOLZ, JR. and ROBERT L. CONNER   99

CHAPTER 5
Chemistry and Metabolism of Nucleic Acids in *Tetrahymena*
    ROBERT L. CONNER and MARY J. KOROLY   123

CHAPTER 6
Regulation of Solutes and Water in *Tetrahymena*
    PHILIP B. DUNHAM and DONNA L. KROPP   165

CHAPTER 7
Growth Characteristics of *Tetrahymena*
    IVAN L. CAMERON   199

CHAPTER 8
Effects of Some Environmental Factors on the Biochemistry, Physiology, and Metabolism
        of *Tetrahymena*
    MICHAEL R. LEVY   227

CHAPTER 9

Life Cycle and Distribution of *Tetrahymena*
  ALFRED M. ELLIOTT   259

CHAPTER 10

Nucleic Acids in *Tetrahymena* during Vegetative Growth and Conjugation
  BARBARA B. McDONALD   289

CHAPTER 11

Genetics of *Tetrahymena*
  SALLY ALLEN and IAN GIBSON   307

CHAPTER 12

Cortical Development in *Tetrahymena*
  JOSEPH FRANKEL and NORMAN E. WILLIAMS   375

CHAPTER 13

Tetrahymena as a Nutritional Pharmacological Tool
  S. H. HUTNER, HERMAN BAKER, OSCAR FRANK, and DUDLEY
    COX   411

General Bibliography   435
Subject Index   501

CHAPTER 1

History, Taxonomy, Ecology, and Evolution of Species of *Tetrahymena*

John O. Corliss

Department of Zoology
University of Maryland
College Park, Maryland

I. INTRODUCTION

The bulk of the literature on species of ciliates belonging to the single hymenostome holotrich genus *Tetrahymena* has been published within the past 25 years; most of it is of a physiological or biochemical nature, and a majority of the articles are concerned with only one species, *Tetrahymena pyriformis*. These three facts are verifiable by an inspection of the General Bibliography section of this volume; careful counts have been published showing such trends (19,22,24,27,34,45).

Tetrahymena has been the best known and most "plastic" of a "pivotal group" of ciliates (27), the "amenable ciliate" (139), the "*Escherichia coli* of the non-photosynthetic eukaryotes" (84), and "the star performer" of the phylum Protozoa (Kidder, in ref. 73). This volume undertakes the responsibility of justifying such accolades, with presentation, analysis, and synthesis of data gleaned from the publications of scores of workers.

Specific chapters following this one are concerned with subjects in the overall field of "tetrahymenology," principally, *T. pyriformis*. Yet there is a need to clarify and explain the taxonomic status of all presently known species of the genus. Furthermore, there is a value in revealing the research attractiveness of various of the species; many problems of the future may demand as the experimental organism tetrahymenas with characteristics differing from those possessed by the presently better known strains of *T. pyriformis*.

In the following discussion problems of *Tetrahymena* taxonomy are treated with emphasis on the comparative approach. Ten species are recognized as distinct, and a detailed characterization is offered for each of them. A brief discussion of the ecology of species of the genus is presented, noting that six of them exhibit the parasitic habit to some degree. Finally, a few speculations on the phylogeny and evolution of the group are indulged in.

II. HISTORICAL

A. Biological History

The commonness of small freshwater hymenostome ciliates of the genus *Tetrahymena*, or at least of the family Tetrahymenidae (which includes the genera *Colpidium, Glaucoma*, and others not as well known; see ref. 42), allows one to suspect that Antony van Leeuwenhoek, "father of protozoology," observed them on more than one occasion three centuries ago (e.g., see his 1676 letter, as published in ref. 48). The Danish zoologist Müller (123,124) saw them in the eighteenth century. Many workers of the 1800s reported these organisms under various labels, although such articles are scattered or else the data are frequently hidden in other, larger works. Great protozoologists such as Bütschli, Claparède, Dujardin, Ehrenberg, Lachmann, Maupas, Perty, Roux, Schewiakoff, Stein, and Stokes (for references see ref. 27) had at least passing familiarity with these small, undifferentiated ciliates.

In the twentieth century, *before* the year 1940, original observations on the morphology, systematics, or general biology of *Tetrahymena* seem to have slackened off, although the work of Kahl (see appropriate sections in refs. 88 and 89) deserves particular citation. In the field of physiology, however, the groundwork for nearly all subsequent experimental investigation was laid in the publications of Lwoff (111, published exactly 50 years ago; and see ref. 112a,b) in France, and of Hall and his students (for citations see ref. 71), and then shortly later of Kidder and associates (92,93) in America.

After 1940, the tremendous growth of the literature on the physiology and biochemistry of these axenically culturable ciliates set in. At about the same time, an outburst of works on the morphology and taxonomy of *Tetrahymena* (and relatives) took place, directly stimulated by the now classic major publication of Furgason (66). Researchers included, at first, primarily Corliss, Fauré-Fremiet, Kozloff, and Mugard. Later these investigators were joined by their students, associates, and other workers such as Brooks, Czapik, Dragesco, Dysart, Elliott, Frankel, Holz, Jankowski, Kazubski, Loefer, McArdle, Roque, Savoie, Stout, Thompson, Williams, and Windsor. The advent of techniques of silver impregnation meant as much to cytological studies of *Tetrahymena* species as did attainment of axenic culture methods to the physiological investigations; cortical "fingerprinting" (as coined by J. C. Thompson) became the rage.

Finally, notably in the past decade and a half, species of *Tetrahymena* have come to be used in important research in genetics, cytogenetics, morphogenesis, and ultrastructure, fields often involving refined biochemical and biophysical approaches. Principal workers in these areas include R. D. Allen, S. L. Allen, Buhse, Corliss, Elliott, Fauré-Fremiet, Mandel, Nanney, Nilsson, Pitelka, Scherbaum, Small, Williams, and Zeuthen.

B. Taxonomic History

Although the *nomenclature* ascribed to various species of *Tetrahymena* (and closely related genera) is unfortunately intertwined with the *systematics* of the group, I wish to attempt a brief historical account of the taxonomy of the group apart from nomenclatural involvements, reserving treatment of the latter for the following section. The matter is further complicated by the fact that the genus (now) allegedly contains about 10 to 14 species with differing degrees of taxonomic affinity. It is, however, primarily the type species, *T. pyriformis,* that has had such an amazingly mixed-up and colorful history.

Descriptions of tetrahymenid ciliates by early workers are of extremely limited value. Although they may not contravene more complete recent accounts, they are often entirely worthless when it comes to recognition of differences among closely related species. This is understandable when one considers both the small size and relatively undifferentiated state of these organisms and the lack of adequate techniques of microscopy and of staining. Very possibly Ehrenberg (54, and earlier) studied species of *Tetrahymena* now known as *T. pyriformis* and *T. patula;* other workers in the midnineteenth century also must have seen these two important species. Yet the earliest accounts of lasting value are the remarkably accurate ones of Maupas (114,115). In the first of these two articles, he described *T. pyriformis* (unfortunately calling it *Glaucoma pyriformis*); in the second, *T. patula* (under a separate generic name, *Leucophrys*). Maupas recognized affinities between these two organisms. Had he only (1) placed them taxonomically in a single genus, and (2) not usurped the name of a perfectly good and separate genus, *Glaucoma,* for one of them, posterity would have been spared considerable polemics on the subject.

Not until about 50 years later did the second good (and even better) morphological descriptions of *T. pyriformis* and *T. patula* appear—in an important article by Furgason (66) in which the Klein silver technique was employed to great advantage (the species, assigned to separate genera, were identified as *Tetrahymena geleii* and *Leucophrys patula*). Furgason created the generic name *Tetrahymena* in this article and clearly recognized differences between his new genus and the neighboring genera *Glaucoma* and *Colpidium*. In the same year a third species of major importance—*T. vorax* (as *Glaucoma vorax*)—was discovered and described by Kidder et al. (95).

Generic characteristics of *Tetrahymena,* particularly revealed in the oral area by use of silver techniques, subsequently became easily recognizable, and several other species—separable on a variety of more-or-less clear-cut characters—became established, mostly through the work of Corliss and close associates. Some of these species had already been described, under different names, by earlier workers: for example, *T. rostrata, T. limacis,* and *T. stegomyiae.* Others were described as new: for example, *T. paravorax, T. setifera, T. chironomi,* and *T. corlissi.* To date, 14 species have been recognized, including the recent *T. bergeri* (25,26,34,40,47,142); 10 are accepted here.

Taxonomically, it is important to note that practically all of the *amicronucleate* hymenostome ciliates grown in "pure" culture in the 1930s and 1940s were strains of the now very well-known species *T. pyriformis*. This must still be kept in mind today when reading the older literature. Unfortunately, many—although not all—the names employed for such strains were (and are) the names of other perfectly valid species of ciliates. Furgason (66), 33 years ago, was keenly aware of the increasing and unnecessary confusion arising through taxonomic-nomenclatural "misjudgments" of his generally physiologically oriented contemporaries; and he predicted that the diverse amicronucleate strains under study (some of which he was not able at that time to examine personally) would prove to be members of a single species. The validity of this prediction was confirmed beyond doubt about 10 years later (19,27), when it became possible to carry out a large-scale comparative investigation. When *micronucleate* strains were isolated, little confusion ensued; the name *T. pyriformis* was by then generally well established.

C. Nomenclatural History

This subject has been treated quite exhaustively in a recent article published in the *Bulletin of Zoological Nomenclature* (45); even more recently, the International Commission on Zoological Nomenclature (136) has handed down legal decisions accepting the recommendations embodied in the petition referred to above, recommendations that validate the following names (with authors and dates as shown): *Tetrahymena* Furgason, 1940; *T. pyriformis* (Ehrenberg, 1830) Lwoff, 1947; *T. patula* (Ehrenberg, 1830) Corliss, 1951; Tetrahymenidae Corliss, 1952; and others of less importance to the present account.

About 27 nominal genera and at least 43 specific names have been involved in the history of *Tetrahymena*. Generic names that have been used in the literature in association with species now known to be members of the genus *Tetrahymena* generally fall into two major categories: (1) those that represent perfectly valid genera, with "good" species of their own, and which therefore should *not* be submerged as synonyms of *Tetrahymena:* for example, *Glaucoma* Ehrenberg, 1830; *Colpidium* Stein, 1860; *Balantidium* Claparède and Lachmann, 1858; *Colpoda* Müller, 1773; and (2) names that *should* be synonyms (in spite of their earlier dates of origin) of *Tetrahymena*, since their species are now included within *Tetrahymena* (itself now accepted as a *nomen conservandum*); for example, *Leucophrys* Ehrenberg, 1830; *Leucophrydium* Roux, 1899; *Paraglaucoma* Kahl, 1926; *Lambornella* Keilin, 1921; *Leptoglena* Grassé and de Boissezon, 1929; *Turchiniella* Grassé and de Boissezon, 1929; *Protobalantidium* Abé, 1927. (Under rejected generic names, incidentally, must also be included the misspellings of *Tetrahymena*, for example, *Tetrahymen* and *Tetrahymenia*, which are still occasionally seen in the literature.)

Of the many specific names involved, only a few need be mentioned here. Such epithets as *pura, pirum, pyrum, piriforme,* and *piriformis* are either rejected or are synonyms of *pyriformis*, in the combination *T. pyriformis*. *Tetrahymena geleii* Furgason, 1940, also is a synonym of the earlier, now accepted name *T. pyriformis* (Ehrenberg, 1830). It should also be emphasized that *Leucophrys patula*, a rather popular name in the literature, now is a synonym of *T. patula*, with the recognition that it represents a species congeneric with *T. pyriformis*.

In following sections of this chapter, very brief nomenclatural histories are indicated for each of the species presently comprising the genus by listing principal synonyms. Reference is also made to type specimens, in recognition of the growing importance of such material in the alpha taxonomy of ciliates today (29,30,33,36,41,43).

Incidentally, many minor nomenclatural errors in the literature on *Tetrahymena* have stemmed from ignorance or abuse of simple provisions set forth in the *International Code of Zoological Nomenclature* (85). A convenient review and interpretation of the *Code* appeared several years ago (30).

III. TAXONOMIC CONSIDERATIONS

A. Basic Problems Involved

There are many problems, some yet to be resolved, in the recognition and separation of species within the genus *Tetrahymena*. Most of these are related to the small size of many of the forms; the relatively undifferentiated nature of the cortex; the cosmopolitan distribution of many strains; the overlapping ranges in body sizes and shapes, and in numbers of rows of cilia; the facile adaptability of many strains of several of the species to a variety of natural and laboratory conditions; the differences that exist between strains within a single species; and the frequent lack of truly unique specific characters. The existence of clear-cut *generic* characteristics, easily separating tetrahymenas as a group from members of other species of hymenostome ciliates, is of no help of course. This point is mentioned because, in the older (and occasionally the newer) literature, such generic features have often been presented as characters usable at the *specific* level, *within* the genus. Finally, the shortage of meaningful, statistically treatable data (morphological *and* physiological) is glaringly evident for most of the forms that have been assigned to the genus.

On the positive side, however, several facts may also be noted. Techniques are now available for more careful comparative studies (32,37); these include not only cytological methods of silver impregnation (20,49,59,65,67,157) and electron microscopy (scanning and high-resolution: 1,2,13,15a,b,138,139,146,147), but also biochemical, genetic, and molecular approaches (see subsequent chapters of this volume). Experimental and physiological considerations are beginning to be given weight, whenever possible, comparable to that formerly reserved primarily for morphological data. New concepts, directly or indirectly, have played an important role, too: for example, the ideas of cytotaxis (self-perpetuation of cortical patterns) and, earlier, of syngens, both proposed by Sonneborn (148,149); the standardization of corticotypes by Nanney (127–131a,b); and the emphasis on morphogenesis [by Corliss and others (37–39,59,145; Chapter 12)], serology (e.g., 104–107,144), statistics and "numerical taxonomy" (86), and chemical genetics (3; Chapter 11).

It is important to emphasize that it is only by consideration of an *assemblage* or *combination* of characteristics (and with data for each derived from the study of entire populations, not isolated, individual specimens) that separate species of *Tetrahymena* may be differentiated with a high degree of reliability. Loefer (105) and Corliss (40) have pointed this out briefly in recent articles.

Once I (and others) thought that the number of ciliary meridians might serve as an isolated major criterion in recognition of the different species. Both the great range in number of rows within a single species and the heavy overlap in ranges among separate species have made such a character differentially unreliable *if used alone*. If considered sensibly, however, and in combination with other characteristics, the corticotype *is* of considerable value, both in quick identificatioin and in refined taxonomic assignment of the organism.

B. Sources of Differential Characteristics

Corliss has recently proposed that there are four major categories of features that are consistently of value in research on the comparative systematics and evolution of *Tetrahymena* species (40). As sources of criteria utilized in the following sections of this chapter, these descriptive categories, admittedly overlapping in nature, merit brief attention here. (It is also to be hoped that they will be noted and followed by future workers on tetrahymenid and related ciliates.)

1. *Ciliature, infraciliature, and cortical features.* Under this heading may be considered such characteristics as: number and distribution of ciliary rows, including postoral meridians (POMs), during the entire life cycle and under varying natural and experimental conditions; presence or absence of a caudal cilium and its polar basal body (PBB) complex; number, distribution, and location of the contractile vacuole pores (CVPs or EVPs, if one today recognizes the vacuole as a water "expulsion vesicle"); and other, minor details of the "silverline system" (e.g., the cytoproct or CYP) and of the infraciliature of both the body proper and the oral area. Silver techniques are nearly indispensable for gathering many of these data; phase and electron microscopy are also useful. The scanning electron microscope will prove to be of help.

2. *Life cycle characteristics.* To list major ones: amount, degree, and kind of polymorphism exhibited under both natural and experimental conditions; body size and form at different stages and under differing environmental situations, natural and experimental, quite apart from clear-cut polymorphism; occurrence of doublets; presence or absence of encysted stages, and kind(s) of cyst(s); kind(s) of stomatogenesis; features of the nuclei and of (macro)nuclear division; presence or absence of sexuality: conjugation, autogamy, or cytogamy.

3. *Ecological data.* Included here is detailed information on both natural and experimental habitats. Examples are food preferences in nature; culturability and growth characteristics under experimental laboratory conditions; degree of parasitism (obligate, facultative, or none); hosts involved, if any, and effect(s) on such hosts and their parasites under varied conditions of the environment.

4. *Physiological-biochemical properties* (other than those already covered above). Included here are data resulting primarily from laboratory study and experimentation: for example, growth and nutritional studies, metabolic pathway investigations, and molecular approaches; also, comparative serological and immunological data and refined genetic information. Although this category has been least used to date, it may provide the source of the most critical and thus the most valuable differential data in future

taxonomic studies. Yet, while recognizing that utilization of nonmorphological characteristics in protozoan systematics is a highly desirable goal, one must appreciate major inherent difficulties: the frequent variability of properties *within* single strains or sublines; also, the frequent similarity or overlapping of patterns among *different* strains, species, or even genera; the importance of such factors as temperature control in growth, nutritional, serological, and genetic experiments; and even the general universality of many chemical reactions at the cellular level. Still, on the positive side, it should be realized that whole constellations of physiological properties often do support the species distinctiveness originally determined on purely morphological grounds.

Even casual inspection of the comparative taxonomic accounts, including the key to species, reveals that uneven usage has been made of characteristics from the four major categories considered briefly above. Future workers must "lean over backward" to improve the situation.

C. Current Composition of Genus

Although 14 species of *Tetrahymena* have been recognized* (25,26,34,40,47,141), three are badly in need of redescription and, indeed, must now be considered doubtful species; and our knowledge of four others is rather limited. Three of the latter, however—*T. paravorax, T. setifera,* and *T. stegomyiae*—are treated below with the seven better known species; the fourth, *T. bergeri,* which still needs to be studied further from a comparative point of view, is considered in the following paragraph. However, the three forms really requiring rediscovery—*T. faurei, T. glaucomaeformis,* and *T. parasitica*—are dispensed with immediately following the description of *T. bergeri,* with brief comment as to their possible proper taxonomic placement, and are not referred to again in this chapter.

Tetrahymena bergeri Roque, de Puytorac, and Savoie, 1970. Very briefly described as a new species in an abstract (141); later a fuller treatment was published (142; the correct year is 1971 not 1970) in which the ciliate was again (erroneously) described as new. Much like *T. rostrata,* the organism is obviously a member of the *rostrata* complex. According to its discoverers, *T. bergeri* is a free-living species with the following characteristics: size (theront) 50–70 × 25–30 μm; single caudal cilium; two CVPs in the vicinity of meridians 5 and 6 (but is this correct?); 26 to 33 ciliary meridans, with two POMs; ovoid micronucleus; reproductive, but no resting, cyst; infraciliary bases of buccal organelles possibly differing in detail from those of other species in the complex; parasitized by a curious, large, flagellated, gram-negative bacterium; possibly exhibits

---

*Jankowski (87) erected a new genus, *Dexiostoma* Jankowski, 1967, to contain an organism which he believed at the time to be *Colpidium campylum* of the literature. In 1969 he suggested that his *Dexiostoma campylum* should perhaps be transferred to the genus *Tetrahymena,* as *T. campyla.* The taxonomic problem is quite complicated, however, and too complex to consider further here. From my own studies I believe that *Colpidium campylum* is a good species of *Colpidium,* and that Jankowski possibly had either an aberrant strain of *T. pyriformis* or an entirely different ciliate, perhaps a new and undescribed species of either *Colpidium* or *Tetrahymena.* Thus if his organism is a fifteenth member of the genus *Tetrahymena,* only further study would establish the fact.

cytogamy during a part of its life cycle; histophagous in its food preference. (For illustrations see ref. 142).

Tetrahymena faurei Corliss, 1952. Isolated from spinal canal and tail musculature of injured embryonic trout (18). Ovoid body form with elongate anterior tip; average body size 44 × 27 μm; prominent micronucleus ca. 2.5 μm in diameter; 32 to 37 rows of cilia, with two to three POMs; two to four CVPs in meridians 8 to 11. No caudal cilium; neither cysts nor conjugation observed. Although the meridian number falls nicely in the range of that of *T. limacis,* other characteristics suggest close affinity to *T. corlissi.* The single strain found cannot be positively identified unless what can be determined to be an identical organism is relocated and described in considerably more detail.

Tetrahymena glaucomaeformis (von Gelei, 1935) Furgason, 1940. Found free-living in Hungary (68) and France (21,26). *Tetrahymena pyriformis*-like in almost all respects studied, although minor silverline differences are purported to exist; most striking characteristic is a remarkable constancy in the number of rows of cilia: 22 to 23, most often 23, reported from studies of large populations of the organism—thus the ciliate shows an amazingly restricted range in kinetal number. Small micronucleus. Body darkens noticeably when the organism is fed on insect intestinal tissue. No caudal cilium; neither cysts nor conjugation observed. In spite of one or two seemingly unique characteristics, this ciliate should probably be considered only an atypical micronucleate strain of *T. pyriformis,* as Czapik (47) and others have also concluded.

Tetrahymena parasitica (Penard, 1922) Corliss, 1952. Originally found by Penard (137) in or on the gills of gammarids from Geneva, Switzerland, and named *Glaucoma parasiticum.* In a brief description Penard stated that its ciliation was more dense than that of *T. pyriformis* (which he called *Glaucoma pyriforme*). Until rediscovery and redescription, however, this ciliate cannot be positively identified, since few reliable data concerning it exist. If it should prove to be a strain of either *T. limacis* or *T. chironomi,* which is possible, then its name—having been proposed at an earlier date—would supersede either of those names as the correct designation of the species in question. However, it may have been a strain of the ubiquitous *T. pyriformis,* the name of which would not be affected by such an identification.

Before turning to characterizations of the 10 "good" species of *Tetrahymena* described to date (there are doubtless more awaiting discovery), it should be restressed that *generic* characteristics must be (and are) possessed by *all species* within the genus. Thus, in at least one stage of the life cycle, some strains of practically all tetrahymenas show several structures or characters often associated with only the type species, *T. pyriformis.* As examples, one may cite: the typical tetrahymenal buccal apparatus; the shape and often the size of the buccal overture; the general overall shape of the body; the general arrangement and distribution and appearance of the ciliary meridians; the appearance and location of the CYP (always in meridian 1) and the CVPs; the type of stomatogenesis evoked during binary fission; the shape, position, size (relative to body size), and composition of the macronucleus; the mode of conjugation, when it occurs; the widespread distribution in nature, often with exhibition of some degree of parasitism; and the relative ease of axenic cultivation, even on defined media. Many of the features just cited are illustrated in drawings and photomicrographs distributed throughout this chapter; note particularly Figs. 1–17.

The 10 species to be considered individually below are sortable into three species complexes designated (40) as: A, the *pyriformis* complex, containing *T. pyriformis, T. setifera,* and *T. chironomi;* B, the *rostrata* complex, comprised of *T. rostrata, T. limacis, T. corlissi,* and *T. stegomyiae;* and C, the *patula* complex, including *T. patula, T. vorax,* and *T. paravorax.* (*T. bergeri,* see above, if ultimately fully established as an independent species, would fall into B, the *rostrata* complex.)

There is a clear-cut distinction between group C and groups A + B, for the species (although perhaps occasionally not all their strains) comprising C all exhibit in their free-swimming stages a striking dimorphic life cycle involving reversible microstome-macrostome stages and the presence of a huge cytopharyngeal pouch in the carnivorous macrostome form. Generally larger species, no members of the C complex exhibit parasitism to any degree. Furthermore, in exhibition of macronuclear chromatin extrusion (MCE) (which occurs at nearly all nuclear fissions in all species of *Tetrahymena;* 53), a heavy, dark band across the predivision macronucleus indicates the future MCE body in species of the *patula* complex, whereas there is no such band in ciliates of the other two groups.

Jankowski (87) not only noted the distinction between groups A + B and C but attached subgeneric names to the two subdivisions of the genus: the nominate subgenus *Tetrahymena* Furgason, 1940, and *Roquea* Jankowski, 1967, the latter to contain the *patula* complex and the former the other two complexes. I am not convinced of the need, value, or wisdom in formal attachment of subgeneric names to the several groups of ciliates under discussion; but legalistically Jankowski's names were properly presented and defended and remain available in the literature.

Groups A and B may be separated from each other on the basis of: the strongly histophagous tendencies of the species of the B complex (e.g., 72); their larger size (especially in the form of trophonts filled with food vacuoles of host tissue); their loss of a simple pyriform shape; their greater number of kineties; the ovoid (rather than spherical) shape of their micronucleus; and, commonly, their production of cysts (of one or more kinds) in their polymorphic life cycles.

Another way to divide the 10 species would be to consider the presence or absence of a caudal cilium as a major difference, as Czapik (47), influenced by earlier work of Corliss (e.g., 77), has done. However, it seems more natural now to treat the origin of such a structure as having occurred independently within each of the three groups described above. It is interesting to note that one species of the *pyriformis* complex (*T. setifera*), two of the *rostrata* group (*T. rostrata* and *T. corlissi*), and one of the *patula* complex (*T. paravorax*) possess this distinctive organelle of such value as a character in practical keys. The four species named (except for the two members from the same group) have little else in common, however.

Still another way to subdivide the genus would be to consider their ranges in number of ciliary meridians with respect to their life cycles; Loefer and colleagues (105,109) have in effect suggested this. Three groupings emerge: species with monophasic meridional ranges (including *T. pyriformis, T. setifera, T. chironomi, T. stegomyiae,* and *T. corlissi*); species with diphasic or bimodal ranges (*T. limacis and T. rostrata*), depending on their life-style of the moment (parasitic or free-living); and polymorphic species with meridional ranges of microstome and macrostome forms showing considerable overlapping (*T. patula, T. vorax,* and *T. paravorax*). Such an arrangement is not far from being equivalent to the three species complexes endorsed here. The principal difference is

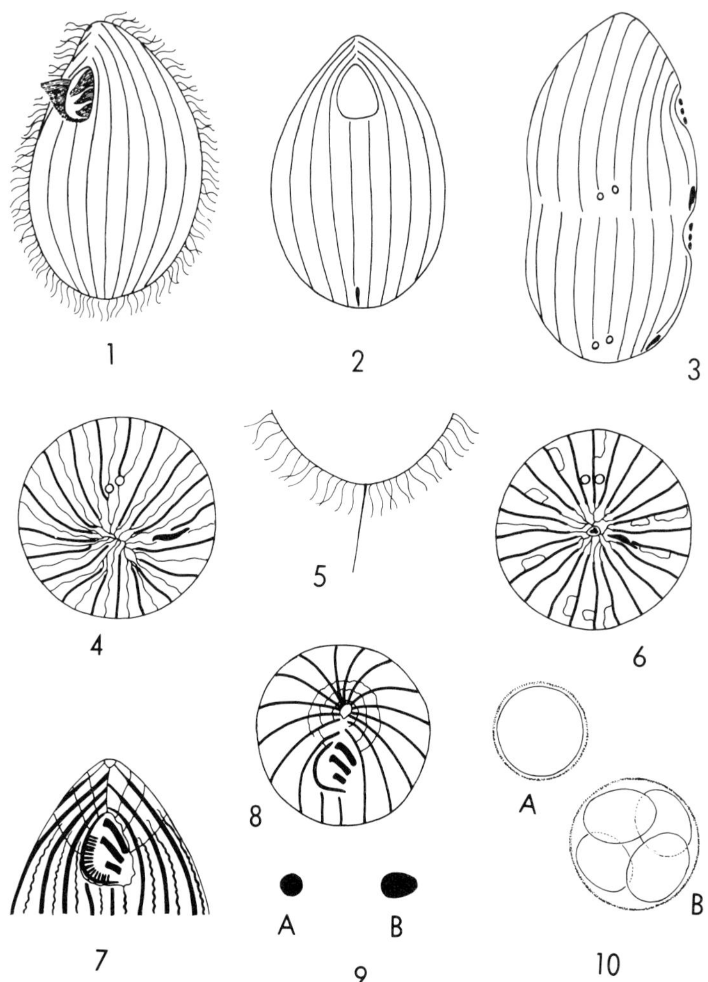

Figures in illustration, semidiagrammatically, of characteristics of a primarily generic nature.

Fig. 1.

Tetrahymena pyriformis, depicting general body shape and somatic and buccal ciliature (AZM + UM), as in life. Left ventral view.

Fig. 2.

Tetrahymena pyriformis, showing body shape, course of the rows of cilia or kineties, buccal overture, and cytoproct. Ventral view.

Fig. 3.

Tetrahymena pyriformis in early stage of binary fission. Stomatogenesis is already essentially complete; new CYP and CVPs have formed; but division of the (parental) body has just commenced. Right lateral view.

Fig. 4.

Tetrahymena pyriformis, posterior polar aspect, showing CYP in meridian 1, CVPs in meridians 5 and 6, and confluence of 1° and 2° meridians in a pattern typical of all species except those with a caudal cilum (cf. Fig. 6).

the removal, in my scheme, of the facultatively parasitic forms, *T. corlissi* and *T. stegomyiae,* from association with *T. pyriformis,* including them instead in the *rostrata* complex.

A brief remark should be made concerning the location of the genus in the overall hierarchical scheme of ciliate classification. It is the type genus of the family Tetrahymenidae Corliss, 1952, in which it is associated with five other supposedly bonafide genera (*Colpidium, Deltopylum, Paratetrahymena, Platynematum,* and *Stegochilum*). The Tetrahymenidae and two other families (Curimostomatidae and Glaucomidae) comprise the suborder Tetrahymenina, one of five that are considered today (42) to make up the order Hymenostomatida of the subclass Holotrichia in the class Ciliatea and the subphylum Ciliophora of the phylum Protozoa.

D. Key to Ten Species

Although the ciliate specialist will require the more detailed characterizations appearing on subsequent pages, the following simple dichotomous key to the 10 species currently fully accepted by me is offered for use in rapid sorting or in reasonably accurate identification of these members of the genus.

Fig. 5.

Tetrahymena setifera, posterior end of body in side view, showing a few ordinary somatic cilia and the caudal cilium.

Fig. 6.

Tetrahymena setifera, posterior polar aspect, showing features also depicted in Fig. 4; but note particularly the small yet distinct PBB complex, a characteristic found in the silverline system of all four species of *Tetrahymena* possessing a caudal cilium.

Fig. 7.

Tetrahymena pyriformis, anterior end of body, depicting schematically principal features of the infraciliature and the silverline system found in the area of the oral apparatus; certain additional, minute "fibrils," of limited taxonomic significance, have been omitted. The tripartite AZM (on the organism's left, the viewer's right) and the UM (on the organism's right)—giving the genus its name: tetra-hymena—hardly require labeling, they are now so well known (see also Fig. 1, in which the membranous structures are depicted as in life). Ventral, and slightly apical, view.

Fig. 8.

Tetrahymena pyriformis, apical polar view, revealing particularly the anterior convergence of kineties; the total number (here 17) of meridians (counted, by convention, from number 1, the rightmost of the postorals, clockwise around apex of organism); and the location of the (infraciliary bases of the) buccal organelles.

Fig. 9.

Shapes of micronuclei known in *Tetrahymena.* (A) Spherical, relatively small diameter, as found in members of the *pyriformis* and *patula* complexes. (B) Ovoid, often quite large, as known in the *rostrata* complex.

Fig. 10.

Cysts known in *Tetrahymena.* (A) Resting or "resistant" cyst, known only for *T. rostrata.* (B) Larger reproductive cyst, known for *T. rostrata, T. corlissi, T. patula,* and *T. vorax.* In both kinds of cysts, the cystic membrane is highly transparent. (See Fig. 29 for illustration of the unique cuticular cyst of *T. stegomyiae.*)

Key to Species of *Tetrahymena*

1. Both microstome (small-mouthed) and macrostome (large-mouthed) free-swimming stages found in life cycle; macrostome form relatively large (e.g., 85–250 μm), and microstome seldom under 60–70 μm; huge cytopharyngeal pouch in macrostome form; commonly either two or four POMs and more than two CVPs; all species completely free-living (*patula* complex)   2

1.′ Only microstome forms found, even in polymorphic life cycles; body size small to medium (typically less than 70 μm, several under 50 μm, although one parasitic species to 100 μm); no cytopharyngeal pouch; typically only two POMs and only two CVPs; some species parasitic (*pyriformis* and *rostrata* complexes)   4

2. Without caudal cilium (and thus no PBB complex); microstome stage pyriform or tailed; reproductive cyst common or relatively uncommon   3

2.′ With caudal cilium; microstome stage common and elongate, but not tailed; no reproductive cyst known; 22 to 30 meridians, with typically two POMs . . . . *T. paravorax* (p. 41)

3. Meridional number high (32 to 45); and characteristically four POMs; microstome form uncommon, pyriform, and not tailed; reproductive cyst common . . . . *T. patula* (p. 32)

3.′ Meridional number considerably lower (17 to 28), and only two POMs; microstome form common and typically tailed; reproduction cyst uncommon . . . . *T. vorax* (p. 36)

4. Small species (length often only 40–50 μm, although one species to 80 μm); generally found free-living (except one species); not particularly histophagous; cysts unknown or unconfirmed in monomorphic life cycle; micronucleus spherical (*pyriformis* complex)   5

4.′ Generally larger species (length often 60 μm or greater, except one species only 40–55 μm); commonly found as parasites (although one species often free-living in edaphic habitats); strongly histophagous; encysted stage(s) common in polymorphic life cycle (except in one species); micronucleus typically ovoid (*rostrata* complex)   7

5. Without caudal cilium; typically parasitic in chironomid larvae; small body size (e.g., 40 μm); commonly 24 to 26 meridians . . *T. chironomi* (p. 21)

5.′ With or without caudal cilium; typically free-living, although one species occasionally a facultative parasite; small body size (although range 30–80 μm); rarely over 24 meridians   6

6. Without caudal cilium; free-living or facultatively parasitic in variety of hosts; variable body size, considering many strains (e.g., 30–80 μm); commonly 17 to 21 ciliary meridians; both micronucleate and amicronucleate populations widely distributed in nature . . . . *T. pyriformis* (p. 13)

6.′ With caudal cilium; found only free-living; small body size (e.g., 40 μm);

commonly 23 to 24 meridians .... *T. setifera* (p. 19)

7.   Without caudal cilium; obligate parasites in nature of either mosquito larvae or snails and slugs 8

7.'   With caudal cilium; capable of free-living existence in nature or found as parasites of either molluscs, annelids, and other edaphic invertebrates or wounded aquatic vertebrates 9

8.   Parasite of mosquito larvae, with presence fatal to host; peculiar cyst on cuticle of host probably provides entry into coelomic cavity; body size 60–100 μm; 25 to 30 meridians .... *T. stegomyiae* (p. 30)

8.'   Nonpathogenic parasite of snails and slugs; ingestion by host provides entry into digestive tract and liver; no cystic stage; body size small (50–55 μm); meridional number generally drops in laboratory culture (e.g., from 35 to 26) .... *T. limacis* (p. 27)

9.   Edaphic forms also found as pathogenic parasite in slugs, snails, and enchytraeid worms; entry not through host's mouth; renal organ primary site of infection (in slugs); both reproductive and resting cysts; meridional number of parasitic forms generally drops in laboratory culture (e.g., from 48 to 32) .... *T. rostrata* (p. 23)

9.   Often found as parasite in bruised or wounded fishes or amphibians (especially in embryonic or larval stages of such aquatic vertebrates); reproductive cysts; 25 to 31 meridians, with no change in number in laboratory culture .... *T. corlissi* (p. 29)

E. Detailed Characterization of Species

In the following accounts a dozen characters are described, as concisely as possible, for each species. Such usage of a combination of taxonomically valuable characteristics should provide the reader with enough data to cause him to agree that the several species are indeed distinguishable and to enable him to place an unknown form in the proper category.

Considerably more space is devoted to *T. pyriformis* than to the species following it for several reasons: there is much more information on it; it is of far greater interest to most nontaxonomic workers; it is most apt to be encountered in collections from nature; it is the "prototype" of the whole genus and is obviously the most representative of the *pyriformis* complex; and, coming first, it permits back-reference when similar structures, and so on, are being discussed under subsequent descriptions of the other species. In the *rostrata* complex the characterization of *T. rostrata* occupies more space than any of the other three members, reasonably enough. Similarly, *T. patula* requires more space than the other species in the *patula* group, although *T. vorax*, better known to more workers, is also purposely favored.

1. Tetrahymena pyriformis (Ehrenberg, 1830) Lwoff, 1947
(Figs. 1–4, 7–9, 11–17, and 20)

Type species of the genus, by designation (66; see ref. 45).

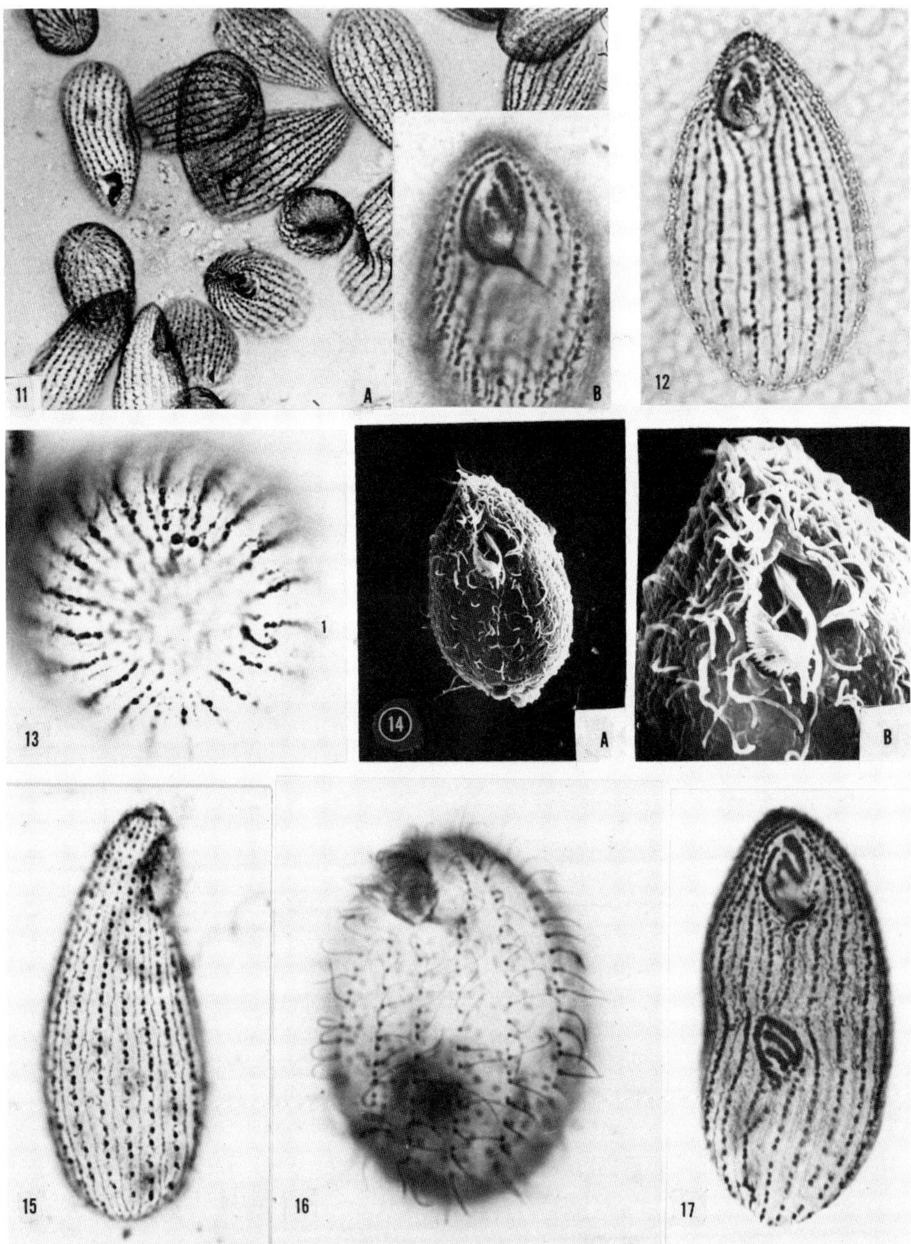

Further figures illustrating typically generic *characteristics, although all are of* T. pyriformis.

Fig. 11.

(A) Group of silver-impregnated specimens, under low power. (B) Anterior end of body, in ventral view. Note buccal area in which some of the deeper "elements" are revealed with particular clarity: not only the bases of the UM and of the three membranelles (AZM) but also the oral fibers associated with the UM and the deep fiber coursing down into the cytopharynx. Chatton-Lwoff technique.

Type material (slide of silver-impregnated specimens of strain E) deposited in the United States National Museum, at the Smithsonian Institution (41,43): USNM no. 24111.

Principal synonyms: *Acomia inflata, A. ovata, Balantidium knowlesii, Colpidium glaucomaeforme, C. putrinum, Glaucoma piriformis, G. pyriforme, G. pyriformis, Kolpoda pirum, Leptoglena knowlesii, Leucophrys piriformis, L. pyriformis, Protobalantidium knowlesii, Tetrahymena geleii, T. glaucomaeformia, T. glaucomaeformis, Trichoda carnium, T. pura, T. pyrum, Turchiniella culicis* (for original literature references, see refs. 19,21,26,45).

Ciliary Rows. Range in number quite wide, 15 to 25; but not only is the "usual" number considerably restricted—17 to 21 (26), and the mode for 36 strains is 19 $\pm$ 1 (108,110)—but there is no bimodal curve. That is, there are no separate or distinct stages or phases in the life cycle (natural or experimental) that have differing ranges (99). Even strains originating from widely separated sources, especially if cultured under similar conditions, generally show overlapping in the number of meridians (21,26,103,110). Nanney's (127–131a,b) usage of the number of kinetics in his corticotype analyses has found "stability sinks" for various strains or lines within syngens (also in amicronucleate lines). Incidentally, there do not seem to be statistically significant differences in the number of ciliary rows between micronucleate and amicronucleate strains.

Fig. 12.

Another specimen impregnated with silver according to the Chatton–Lwoff method. Ventral view.

Fig. 13.

Posterior polar view showing the same silverline features emphasized in the drawing in Fig. 4. Note especially the CYP and the CVPs.

Fig. 14.

Whole-mount, ventral views, as seen with scanning electron microscopy. (A) Note both somatic and buccal ciliature, neatly preserved. (B) Higher magnification of anterior end of body, same specimen, showing clearly the UM (on organism's right) and some of the membranellar ciliature which protrudes from the left wall of the buccal cavity.

Fig. 15.

A specimen showing a body form commonly observed in cultures, although it is more cucumber- than pear- ("pyriform") shaped. Chatton–Lwoff technique. Right lateral view.

Fig. 16.

A specimen impregnated with silver according to the protargol technique. Note major differences in comparison with organisms treated by the Chatton–Lwoff method (cf. Figs. 11–13, 15, and 17). Basal bodies and cilia are clearly stained; many features of the so-called silverline system are refractive; buccal area is filled with cilia of UM and AZM (although resolvable individually with careful focusing); nuclear apparatus (here out of focus) is stained. In this particular specimen the body shape of the organism has not been perfectly preserved.

Fig. 17.

A dividing form, impregnated according to the Chatton–Lwoff technique. Note that the phenomenon of stomatogenesis clearly precedes the actual process of binary fission, as also depicted in Fig. 3. Ventral view.

Acknowledgments with gratitude to J. Frankel for Figs. 11B and 17, to E. B. Small for Figs. 14A and B, and to N. E. Williams for Figs. 15 and 16.

In spite of the great total range of meridians reported for *T. pyriformis*—even larger than 15 to 25 if various rare, freak, or anomalous specimens are included (105)—it is important to stress the overall *stability* of the number found over many years (representing literally hundreds of thousands of generations) in cultures of numerous strains, maintained under a variety of "normal" conditions of growth in several centers of ciliate research in the United States and Europe.

The silverline and ultrastructural nature of the meridians, both 1° and 2°, are not reviewed here (but see refs. 1,2,19,21,26,66,68,138,139). The number of postoral kineties (POMs) is typically two. This feature, however, is of limited taxonomic significance within the genus, since all other species (except *T. patula*) also often show the same number.

Caudal Cilium. None.

Contractile Vacuole Pores. Typically two, occasionally from one to several, commonly located in meridians 5 and 6, not far from the posterior pole of the organism (19,21,110). The pore in the lower-numbered kinety is always anterior to the second one. No cilia present on the pore-bearing meridians posterior to the level of the CVPs.

Other Silverline Features. Several articles (21,26,66,68,97,99–102) have offered detailed descriptions of the so-called silverline system in *Tetrahymena* species; there is little value in repeating such specialized data here, for it appears that most of the features are common to all species of the genus. Somatic elements include the 2° meridians, the CVPs, the CYP or cytopyge or cell anus (normally located in meridian 1, near the posterior end of the body), the apical loop, the preoral suture, the circumoral ring, the intermeridional connectives, the intrameridional cross-fibrils, and the PBB complex (in the four species, *not* including *T. pyriformis,* that have a caudal cilium). Buccal elements include the infraciliary bases of the undulating membrane (UM), and three membranelles (the latter comprising the adoral zone of membranelles, or AZM), the oral fibers (formerly recognized as "riblike striations of the UM"), the buccal overture (pyriform-shaped and oriented in the longitudinal body axis), and silver-impregnable portions of the cytostome–cytopharyngeal complex. Several of these features, both somatic and oral, are illustrated semidiagrammatically in Figs. 1–4,7,8, and 20, and in photomicrographs in Figs. 11–17. Study of the ultrastructure of many of them has underlined their universality not only among species of *Tetrahymena* but in other ciliate genera as well (1,2,13,15b,63,134,138,139,169; and other chapters in the present volume).

Body Shape and Size. Normally a small, pear-shaped organism; but the extreme plasticity of the *Tetrahymena* pellicle permits considerable temporary distortion. Starved individuals, too, may lose the usual pyriform shape. Populations cultivated in axenic media often contain specimens showing highly asymmetrical patterns of body ciliature. The body size ranges in length from 10–11 μm to over 90 μm. A common mean figure is ca. 50 μm; such forms typically show a width of ca. 30 μm (21,103). The average size of the pyriform buccal overture in such specimens is 10.5 × 6 μm; this size is generally characteristic of all species in the genus, except for the macrostome stage of members of the *patula* complex.

No true polymorphism is exhibited under any conditions. Thus the life cycle is best described as being basically monomorphic.

It is appropriate to make brief mention of the occurrence of "doublet" organisms, long known for this species in both naturally occurring cases and in experimental work.

Homopolar doublets, with two sets of oral structures typically located 180° from one another and both at the same (anterior) pole of the body, show a normal body shape but an increase of ca. 50% in number of ciliary rows; and certain silverline structures are doubled (e.g., there is a second set of CVPs and a second CYP related to the two sets of POMs and two stomatogenous kineties). Doublets divide, producing daughter doublets, stomatogenesis usually occurring synchronously. There is a high degree of stability in growing populations of homopolar doublets, although eventual reversion to singlets appears to be inevitable. Doublets must not be mistaken for the "normal" organisms when making counts of kineties, CVPs, and so on. Also worthy of note is the occurrence of chain formation in this species (64).

Cysts. With the exception of the still-baffling reports of Watson (159) and Hurst (82), cyst formation has never been described for this species. We are obliged to conclude that, if it occurs at all, it is an extremely abnormal phenomenon in the life cycle of this ciliate, in contrast to the situation in certain other species of the genus.

Stomatogenesis. New mouth formation is now well known, at least descriptively, for many tetrahymenid ciliates (e.g., see articles by Buhse, Corliss, Frankel, Furgason, von Gelei, Nanney, Williams, and others; Chapter 12). To describe the phenomenon briefly in *Tetrahymena* species, in the normal situation, first an anarchic (?) field of basal bodies, seemingly proliferated from the "stomatogenous" postoral meridian number 1, develops midway down the ventral surface of the organism prior to the onset of cytokinesis. From this field arise the infraciliary bases of the new tetrahymenal apparatus destined to appear in the buccal cavity of the posterior daughter organism (opisthe). The oral ciliature of the anterior organism (the proter) is either that of the original parental animal or represents a "replacement mouth" derived, apparently, from contributions at the infraciliary level from the "old" UM or the anterior end of meridian number 1 or both. Since the production of mouthparts (for the opisthe) involves the kinetosomes of a somatic meridian, this mode of stomatogenesis is called *somatic-meridional* (38). Scanning EM work: Refs. 13, 15b,143b.

Nuclear Characteristics. Size, shape, and location of the nuclei, indeed, even the presence or absence of the micronucleus, are generally of little use at the infrageneric level in ciliate taxonomy. In earlier years of *Tetrahymena* systematics, more weight was sometimes given such features than we now know was justified. Early strains of *T. pyriformis* appear to have been nearly all amicronucleate races. Later, micronucleate strains were identified, many of these becoming important in cytogenetic and genetic studies. (It is possible of course that some of the amicronucleate tetrahymenas were micronucleate at the time of original isolation, losing their micronucleus during subsequent laboratory culture. In most cases there is no way of knowing now what the original nuclear condition of such strains may have been.)

The macronucleus, in both micronucleate and amicronucleate forms, is typically an ovoid to irregularly spherical body, measuring ca. 9.5×11 μm and located in the central portion of the ciliate's body, often a little posterior to the midline (Fig. 20B). It is assumed to be a polyenergid structure; its DNA ploidy level has been determined to be ca. 50. Macronuclear chromatin extrusion (MCE-) bodies are regularly noted in dividing organisms (21,51–53) and are not to be confused with micronuclei. Their mode of production in species of the *pyriformis* and *rostrata* complexes is strikingly different from that in members of the *patula* group, as mentioned above. The micronucleus, when present at all, is singular in number, spherical in shape (Fig. 9A), with a diameter of ca. 1.5–2 μm, and is located nearby, or even in a cleft or notch of, the macronucleus. There are 10 chromosomes in *T. pyriformis* in the diploid state (140).

Sexuality. Whether or not most of the strains of *T. pyriformis* isolated in axenic culture years ago were amicronucleate, any exhibition or, at least, "carrying-through" of sexuality was precluded once the amicronucleate condition was reached. (It is not known whether "mating competence" of old amicronucleate strains has ever been rigorously tested against today's known, micronucleate, mating types.) Following the breakthrough discovery of conjugation in micronucleate *T. pyriformis*, by Elliott and Nanney (58) [actually conjugation, of the selfing kind, had been reported a year earlier by Horn (81), but in an unpublished Master's thesis], however, the phenomenon of conjugation could be and has been studied in some detail, especially from a genetic point of view. Mating types and syngens ("varieties") have become recognized as clearly among strains of this species as they are in *Paramecium aurelia*. A dozen or more syngens are now known (see Chapter 11), and others doubtless await discovery. Although mating type systems are established, the phenomenon of selfing is well known among certain strains of *T. pyriformis*. Autogamy has never been reported.

Except for the fact that some biologists may wish to recognize syngens (148) as "physiological (or biological) species," there is no other direct taxonomic (*sensu* classificational) significance to be attached to the occurrence of conjugation in ciliates. Exclusive of genetic criteria, combinations of other features—physiological as well as morphological and ecological—do not support systematic separation of syngens at any level above the "strain" (which handy designation, incidentally, is purposely very imprecisely defined by ciliate protozoologists) or some other infraspecific category.

Habitat. *Tetrahymena pyriformis* is very widely distributed throughout the world (see Chapter 9) and is capable of adjusting to many differing ecological niches. Commonly found in freshwater habitats, such as ponds, pools, and streams, it has been reported from thermal springs, salt marshes, and soil (21,22,34,55,56). At the same time it has been discovered as a coprophilic ciliate and as a facultative parasite (26,44a,96,122,160). Presumably a microphagous organism (bacteria feeder) in nature, it also can survive saprozoically on a liquid diet, the substrate that might be found in the body cavity of larval insects. It is also facultatively histophagous; that is, it can ingest tissues of various invertebrates and vertebrates in particulate form, although it does not do so with the speed, thoroughness, or avariciousness of certain other species of *Tetrahymena* (72), nor does the body appreciably change shape under such conditions.

Experimentally, strains of *T. pyriformis* have been established in axenic culture, on nonparticulate media (standard proteose-peptone or chemically defined synthetic medium). Also, *T. pyriformis* has been found to survive and multiply well as an induced "facultative parasite" of several kinds of host organisms, particularly insect larvae (see references in ref. 26). The host generally succumbs to the invasion or infection, which rapidly becomes a heavy one even if few ciliates were originally introduced into the body cavity (26,154). [Incidentally, the discovery of ciliates living within the body of some metazoan organism must be viewed, taxonomically, with considerable caution. There are dangers inherent in allowing the pendulum to swing too far in either direction; that is, such protozoa are not necessarily different species when found in different hosts; nor is necessarily the same species found in all hosts. In the history of the genus *Tetrahy-*

mena, one may find commission of the error in both extremes, ranging from the trend to assigning forms even to new genera when found in new hosts to the fashion of concluding (with no other evidence) that any and every ciliate exhibiting facultative parasitism must ipso facto be a strain of the well-known species *T. pyriformis.* The taxonomic problem in the case of *Tetrahymena* is further complicated by the occurrences of two different species simultaneously in the same host.⌋

Physiological-Biochemical Traits. As pointed out briefly above and elsewhere (37,86,105), criteria derived from comparative physiological and biochemical studies have to date yielded data of rather limited usefulness in the classification of ciliated protozoa. Strains of *T. pyriformis* have served as favorite experimental organisms in hundreds of such investigations (e.g., 73,75,84,94; and the remainder of this book). Groupings *within* the species have sometimes resulted, or been proposed, from the studies, but they have been of little taxonomic significance (78,105,107). Separation of species *has* been supported by some physiological findings, but such data have rarely contradicted the conclusions already drawn from consideration of existing morphological or life cycle information.

Genetics. As indicated earlier, genetic evidence for taxonomic interrelationships with respect to species of *Tetrahymena* is limited. Twelve syngens, with multiple mating type systems, have been recognized to date for *T. pyriformis;* by the criteria of some biologists, these represent separate species (Refs. in 148). However, few other characteristics support such taxonomic separation—at this stage of our knowledge of ciliate systematics. The recent observations of S. L. Allen, Mandel, Nanney, and others (see Chapter 11) attempting molecular approaches (base ratio analyses, nucleic hybridization studies, immunological comparisons) to ciliate genetics may—to date—have posed more taxonomic problems than they have solved; nevertheless, there are fascinating possibilities in such approaches and they must be continued.

2. *Tetrahymena setifera* Holz and Corliss, 1956
(Figs. 5, 6, and 18)

Type material (slide of silver-impregnated specimens, strain HZ-1): USNM no. 24112.
Synonym: Possibly *Saprophilus* (=*Sathrophilus*) *ovatus,* a questionable species described by Kahl (89).
Ciliary Rows. "Normal" range reported as 22 to 26, with 23 and 24 being most common (77). A curious subline established by Holz (74) not only became amicronucleate but underwent a loss in number of meridians to 17 to 21; the organisms retained the distinctive caudal cilium, however; and the subline did not differ in any physiological reactions from the parental strain. Furthermore, and perhaps most significantly, there occurred over an extended period of months a gradual drift or regulatory return to the characteristic range of 22 to 26 kineties (Holz, personal communication). There are two, rarely one or three, POMs.
Caudal Cilium. The single most striking property of *T. setifera* is its caudal cilium arising from a PBB complex which is revealed in silver-impregnated organisms (Figs. 5, 6, and 18C). The organelle is inconspicuous but measures 13–17 μm in length, whereas body cilia average 6–7 μm. The PBB complex appears to be of a tripartite

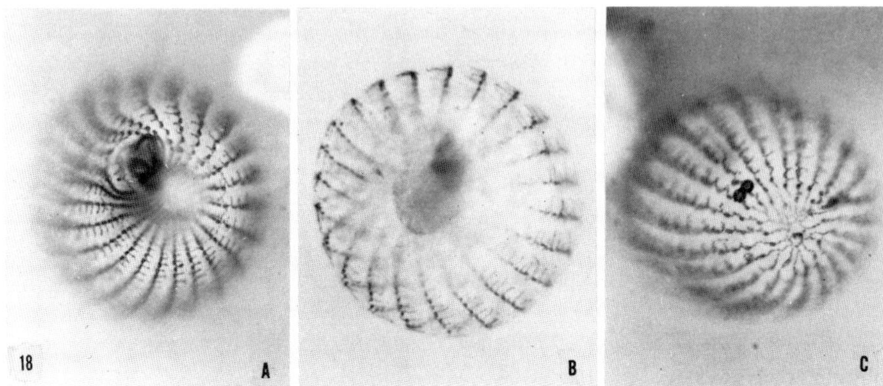

Fig. 18.

Three optical levels through a single specimen of *Tetrahymena setifera* impregnated with silver by the Chatton–Lwoff technique. Features of the silverline system are clearly evident and need little explanation. (A) Apical pole. (B) Subequatorial plane. (C) Posterior pole (see also Fig. 6).

nature; it is encircled by a delicate argentophilic fibril(?) separating it from the posterior terminations of the 1° and 2° ciliary meridians (Figs. 6 and 18C).

Contractile Vacuole Pores. Typically two, although one to four have been noted, commonly located in ciliary meridians 8 and 9, although rows 6, 7, and 10 are occasionally involved. The two pores are always arranged parallel to one another; they appear to have larger diameters than those in *T. pyriformis;* and one or more cilia are frequently associated with the pore-bearing kineties posterior to the level of the CVPs (77).

Other Silverline Features. Minor details of the silverline system are of limited taxonomic value and need not be described here.

Body Shape and Size. Essentially indistinguishable from certain specimens of *T. pyriformis*. The range in sizes reported is less, as might be expected, since many more strains of the latter species have been investigated. The mean size, 40 × 25 μm, is significantly smaller, with the difference in body volume being even more impressive. No polymorphism has been observed. Homopolar doublets have been noted.

Cysts. None.

Stomatogenesis. As in *T. pyriformis*.

Nuclear Characteristics. Essentially as found in *T. pyriformis*. Both micronucleate and amicronucleate strains are known.

Sexuality. No sexual phenomena noted to date in nature, in laboratory clonal cultures, in populations of mixed strains, or in mixtures with sexually reactive clones of various syngens of *T. pyriformis*.

Habitat. Geographical and ecological distribution unknown. The only strains studied were isolated from a single freshwater pond in New York State. The organism has been established in standard axenic and monoxenic (with *Aerobacter* spp.) culture and in chemically defined media (75,77). Facultative parasitism is unknown; it has been shown, however, to be mildly facultatively histophagous.

Physiological-Biochemical Traits. Little studied. Nutritional and biochemical findings known from the investigations of Holz and associates (75) are not particularly valuable in taxonomic considerations, yet the requirement of *T. setifera* for a sterol for best growth differentiates it from all strains of *T. pyriformis* and represents a point of similarity with two other caudal cilium-bearing species, *T. paravorax* and *T. corlissi* (80).

Genetics. No information available.

3. *Tetrahymena chironomi* Corliss, 1960
(Fig. 19)

Type material (slide of silver-impregnated specimens, strain N-12): USNM no. 24113.

Synonyms: *Glaucoma piriformis* of Treillard and Lwoff (156); *T. parasitica* (Penard, 1922) Corliss, 1952, subdivision "a," of Corliss (18); *T. parasitica* of Barthelmes (4); possibly *Glaucoma parasiticum* of Penard (137).

Ciliary Rows. Reported as 23 to 28, with 24 to 26 most commonly found (26). There are two POMs. The apparent lack of definitive 2° meridians, in specimens impregnated with silver according to the Chatton–Lwoff technique, represents a curious difference, admittedly of unknown and possibly minor significance, between this species and several of the others, including *T. pyriformis*.

Caudal cilium. None.

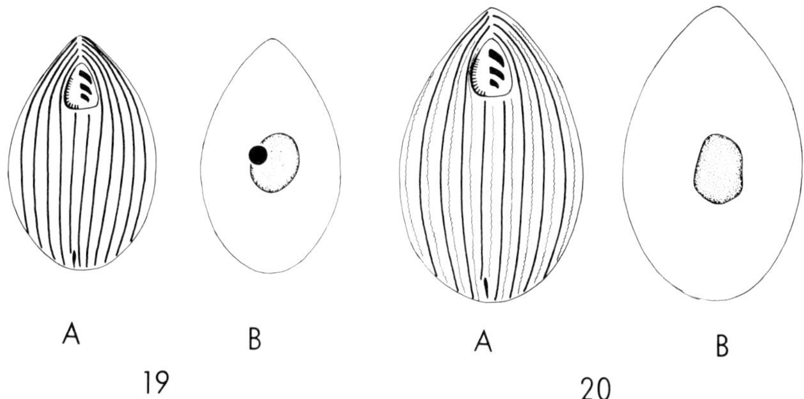

Semidiagrammatic drawings of specimens taken from a single larval chironomid host with a mixed infection of ciliates.

Fig. 19.

Tetrahymena chironomi. (A) Based on a silver-impregnated specimen. (B) Based on a Feulgen-stained specimen, revealing micro- and macronuclei. Ventral view.

Fig. 20.

Tetrahymena pyriformis. (A) From a silver-impregnated specimen. (B) From a Feulgen specimen; note absence of micronucleus (thus an amicronucleate strain). Ventral view. Drawn to same scale as Fig. 19; size and other differences between the two species are immediately apparent.

Contractile Vacuole Pores. Typically three, although two to four have been seen. The kineties involved are numbers 6 to 9, usually 6 to 8 with one in each. Generally, no cilia occur posterior to the level of the pores.

Other Silverline Features. There seem to be no significant differences in the silverline picture from that known for *Tetrahymena* species in general.

Body Shape and Size. Superficially indistinguishable from certain specimens of *T. pyriformis*. However, with its average body size only 40 × 23 μm and its possession of more rows of cilia, *T. chironomi* more closely resembles *T. setifera*. Preconjugants, conjugants, and exconjugants are generally small, lean, and often mouthless; but this is not to be considered true polymorphism. No homopolar doublets noted. Conjugating pairs are sometimes joined by another individual, making triplets for a period of time while conjugation is occurring.

Cysts. None.

Stomatogenesis. As in *T. pyriformis*.

Nuclear Characteristics. Essentially as in micronucleate strains of *T. pyriformis*, except that the spherical micronucleus, especially in specimens of the parasitic phase taken directly from the body of the host, is relatively much larger, its diameter ranging from 3.0 to 3.7 μm.

Sexuality. Conjugation noted repeatedly by Treillard and Lwoff (156) 49 years ago; by Barthelmes (4); by Corliss (26). There are several curious features about the phenomenon in this organism: its occurrence within the body cavity of its host; its high frequency; the occurrence of triplets as well as pairs; its apparent selfing (intraclonal) nature; the mouthlessness and seemingly inevitable death (an unknown "influence" of the host?) of the exconjugants.

Habitat. That *T. chironomi* may be found widespread in larval midges (species *Chironomus plumosus*) is attested to by its discovery in these insects in collections made near Paris (26,156), some taken 27 years apart; near Berlin (4); near London (26); and from locations in Michigan (26). Corliss also found, years ago, what may have been free-living members of the species from freshwater ponds near Paris, although they were never studied intensively. The ciliate probably should be categorized as a parasite, with a predilection for chironomid larvae, with a facultatively free-living stage of unknown duration (44a). The exact mode of entry into the body cavity of the larval midge is not known, but it may involve invasion through "natural" breaks or weaknesses present during the molting process of the host. Heavily infected fourth-instar larvae may contain as many as a quarter of a million tetrahymenas in their hemocoel. Most curious is the still-perplexing fact that, having enjoyed rapid multiplication at the expense of various nutritive substances in the body cavity of their host and having caused its (eventual) death, the ciliates inevitably undergo, at the time of or just preceding the death of the larval midge, a fatal kind of conjugation; thus they themselves perish by the tens of thousands not long after demise of their host.

Physiological-Biochemical Traits. Attempts to establish *T. chironomi* in axenic culture have failed. It is quite possible, since only populations taken from midge larvae were so treated, that the ciliates represented only exconjugants, organisms destined to die in due time anyway. Thus refined physiological and biochemical studies remain to be carried out. Multiplying very rapidly in the host hemocoel, *T. chironomi* is mildly histophagous, probably nourishing itself on such tissues as fat bodies, as well as liquid nutrients such as hemoglobin. Strains isolated from nature as free-living forms (kept for short lengths of time only before fixation) were probably bacteria feeders (26).

Genetics. The widespread occurrence of conjugation, even if of a lethal (and very likely selfing) kind, and the possibility that its larger micronucleus might contain a greater number of chromosomes than 10 (the number found in *T. pyriformis*), make work on the cytogenetics of this species highly desirable.

4. *Tetrahymena rostrata* (Kahl, 1926) Corliss, 1952
(Figs. 9 and 21–24)

Type material (slide of silver-impregnated specimens, strain A): USNM no. 24114.
Synonym: *Paraglaucoma rostrata* of Kahl (88).

Ciliary Rows. Since this species exhibits a shift in total number of kineties depending on phases in its life history, it would be misleading to cite the entire range of meridians without relating the numbers involved to the conditions under which the organisms may be found. Specimens taken directly from edaphic habitats or established for some time in "normal" culture in the laboratory generally show a range of 27 to 35, based on observations of Corliss (17b,25,26,35,77, and unpublished) and of McArdle (118). Occasionally the number dips to 26 or is extended as high as 39. Stout's (152) forms, found in enchytraeid worms as well as in soil and litter samples, showed a similar range, 29 to 35. However, in its parasitic phase, in the renal organ of certain slugs and snails, the number of kineties ranges from 36 to 58 (5,90,102,155). After ciliates of the parasitic phase have been established for a while in laboratory (free-living) culture, the range drops to 28 to 37, according to the careful investigations of Kozloff (102) and Brooks (5). When isolated in axenic culture, strains of *T. rostrata* have usually shown a range not greater than 27 to 37; in one study, however, the limit went as high as 48 (155).

McArdle (118–120) discovered several unusual edaphic strains which bear special mention for, with passage of time in bacterized laboratory culture, the total meridional counts in these "abnormal" forms dropped from that known for about 40 "normal" strains, falling eventually to a range of only 17 to 22. The taxonomic significance of these *T. pyriformis*-like strains is not altogether clear; they appeared to be "irreversible" forms, small in size, often amicronucleate, incapable of forming cysts, but retaining a caudal cilium. It is unwise, at least at the present stage of our knowledge, to consider that these data destroy our concept of the "normal" species as having the ranges in number of meridians given in the preceding paragraph. Nevertheless, they do suggest great caution in attempting either to recognize a given tetrahymena from only a few individuals, or to depend solely on the criterion of meridional number for specific identification; and they certainly underscore the need for knowing the past history, the whole life cycle, and the current environmental conditions in which the organism is or has been living.

In spite of the relatively large number of somatic kineties, there are typically only two, rarely one, three, or four, POMs.

Caudal Cilium. Present; noted by Kahl (88) 47 years ago; missed in redescriptions in the early 1950s (e.g., 152); rediscovered by Corliss (77) and Kozloff (102). Inconspicu-

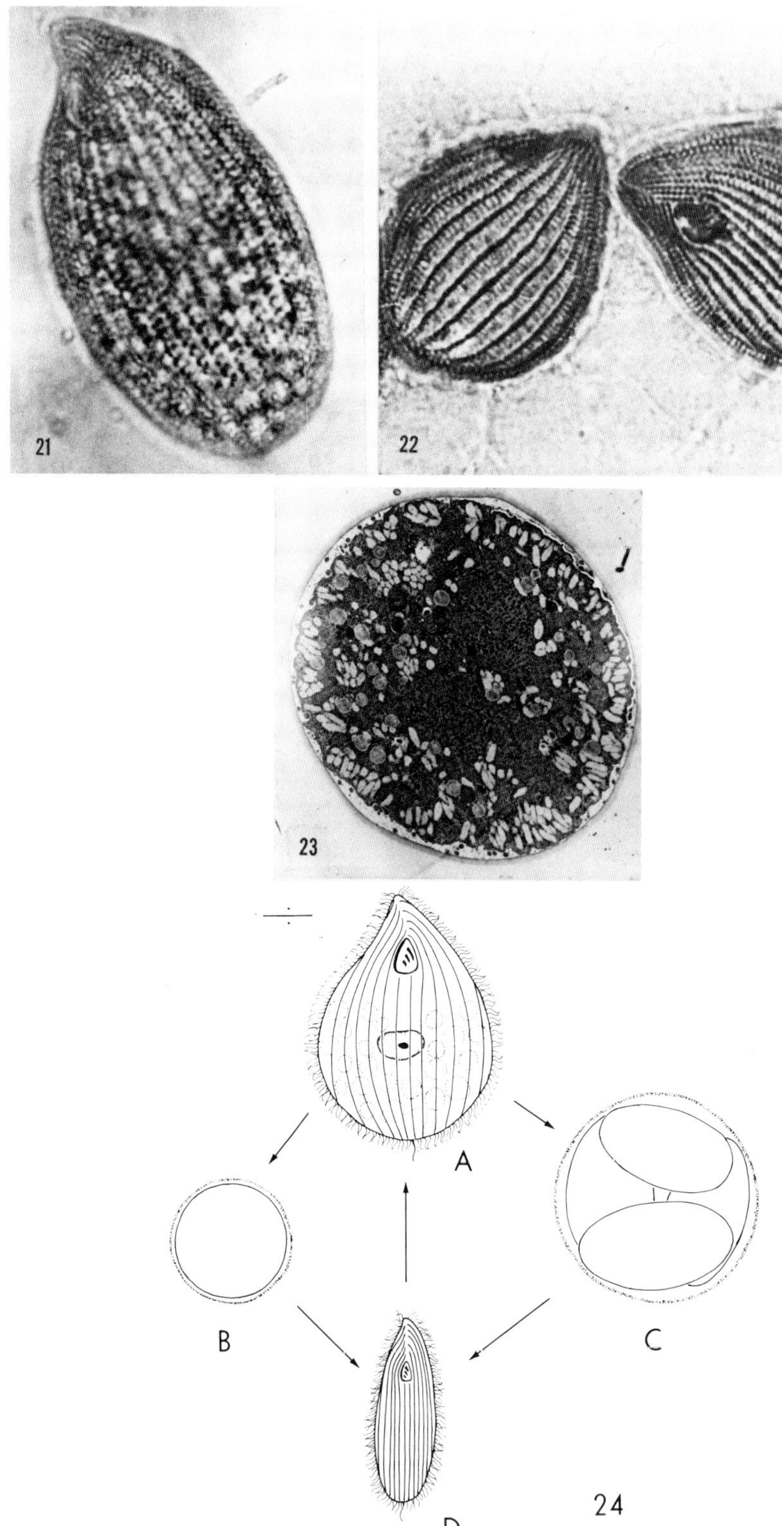

ous in living ciliates (and only slightly longer than regular somatic cilia), it is most easily proved to be present by revelation of its PBB complex in silver-impregnated specimens.

Contractile Vacuole Pores. There may be one to six, occurring in kineties 6 through 12; commonly found numbers, however, are three to four, occupying ciliary rows 9 to 11. Pore diameter, as measured in silver-impregnated individuals, averages ca. 1.5 μm.

Other Silverline Features. Minor details of the silverline system do not warrant special consideration.

Body Shape and Size. A polymorphic species, the body form and size depend on the stage or phase in the life history (Fig. 24).

In the trophont (or tissue-feeding) stage, the species exhibits a more-or-less pyriform shape with rounded posterior end and pointed, tapered, or occasionally rostrate, anterior end. Its size in "normal" strains—as found in nature or as cultured on bacterized media in the laboratory—ranges in length from over 80 μm to less than 20 μm; a common length is 50–60 μm, although some strains average well over that. The width is typically two-thirds the length. The theront or "hunter" form is much leaner.

Two additional stages are possible, even in the normal life cycle. Instead of undergoing simple binary fission while freely feeding or swimming, the trophont may round up and form a tomont, a form that immediately encysts and then commences to undergo one or two, or occasionally more, fissions in quick succession. Tomites emerging from the reproductive cyst (35–40 μm in diameter) feed, and are soon again recognizable as trophonts. Generally in the absence of food, a trophont may form a resting, resistant, or protective cyst (usually 18–20 μm in diameter). Such a stage may last a long time; division seldom occurs until excystment.

Superimposed on this polymorphism may be a diphasic life history: that is, the trophont may be found in both a parasitic phase, for example, living inside slugs and snails,

---

Figures of the edaphic, histophagous species Tetrahymena rostrata, *based on forms found in stages of the polymorphic life cycle.*

Fig. 21.

A growing trophont, becoming engorged on tissue; impregnated with silver by the Chatton–Lwoff technique. Left ventral view.

Fig. 22.

Two young, rotund products of the recent division of a tissue-filled trophont; stained as in Fig. 21. Left lateral and ventral views.

Fig. 23.

An electron micrograph of a specimen within a resting cyst. Note thin cystic membrane, the great number of mucocysts in the peripheral cytoplasm, and two macronuclear anlagen (evidence that the sexual phenomenon of autogamy has occurred in the encysted organism). Thin section, relatively low magnification.

Fig. 24.

Basic life cycle of *T. rostrata*. (A) Trophont. Division to produce daughter trophonts may occur directly (indicated by division sign; and see Fig. 22). (B) Resting cyst. (C) Reproductive cyst, not obligatory in life cycle. (D) Tomite, generally the product of excystment (B or C), does not divide.

Acknowledgment with gratitude to E. W. McArdle for Figs. 21, 22, and the basic idea of Fig. 24.

and a free-living phase, for example, growing in laboratory cultures. The significance of the diphasic life cycle for this species lies not in change of form (for essentially there is none in the trophont) but in striking differences in number of kineties, as discussed above.

Cysts. Reproductive and resting cysts; but neither is obligatory in the life cycle. Both have been described briefly above. The nature of the cystic membrane(s), its possible relationship to the organism's rather prominent and numerous mucocysts, and so on, are important problems still requiring solution (167; Fig. 23).

Stomatogenesis. As in *T. pyriformis.*

Nuclear Characteristics. Whereas the macronucleus is essentially identical in shape and appearance to that of *T. pyriformis* and other species, although very likely possessing a distinctive ploidy number, the micronucleus is significantly different from that of members of the *pyriformis* complex. It is ovoid in shape, commonly measuring 3–3.5 μm (but up to 4.5 μm) in its long axis and 2–2.3 μm (up to 3.8 μm) in its short axis, a clear contrast to the smaller, spherical micronucleus of *T. pyriformis* (Figs. 9A and B). There appear to be about 40 to 50 chromosomes in *T. rostrata,* contrasted with 10 in the typespecies (35). The mode of formation of MCE bodies, as described under *T. pyriformis,* separates members of the *pyriformis* and *rostrata* groups from species of the *patula* complex (51,53).

Sexuality. In the same year in which Elliott and Nanney (58) announced the breakthrough news of sexuality (conjugation) in *T. pyriformis,* Corliss (17b) published his discovery of sex (autogamy) in *T. rostrata.* Full details, cytological and otherwise, are yet to be published, although a comprehensive preliminary report describing the process generally and discussing its possible implications from the point of view of senescence and rejuvenescence has appeared (35). No mating types, and thus no conjugation, have been reported to date; thus the species appears to be highly inbred. Autogamy in this species occurs in the resting cyst stage, so the organism holds considerable potential value in experimental studies in genetics and cytogenetics—once conjugation can be found in nature or induced in the laboratory. [The possibility that *T. bergeri* (142), itself perhaps a strain of *T. rostrata,* exhibits cytogamy is an exciting discovery highly worthy of confirmation.]

Habitat. Readily found free-living in edaphic habitats in nature (moss, sphagnum, forest litter, soil, and so on); but it is also frequently discovered as a facultative parasite (or sometimes histophagous scavenger) in such widely diverse hosts as enchytraeid obligochaetes, rotifers, tardigrades, snails, and slugs (26). Geographical distribution is wide; it has been reported, for example, from Canada, France, Germany, Hungary, Poland, New Zealand, and the United States (5,21,26,35,47,69,88,90,102,118,153,155). The organism is easily grown in both crude (e.g., on bacterized beefsteak) and refined (e.g., on axenic, nonparticulate media) laboratory culture. Although only a facultative parasite, it is truly pathogenic. In slugs the kidney or renal organ is the primary site of infection, but other tissues and organs may also be entered; when albumen gland and oviduct are invaded, transovum transmission to a fresh host can occur; but the usual portal of entry into an adult slug is apparently through a little-known dorsal integumentary pouch (5,44a).

Physiological-Biochemical Traits. Little studied to date; this is unfortunate, since the species has considerable potential as an experimental organism in several areas of cellular research. It is hardly to be expected, however, that nutritional or biochemical findings would be of much taxonomic significance; more than likely they would, at best, support distinctiveness already established by mode of life, characteristics of polymorphism, cortical features, and so on.

Genetics. No information available to date because of lack of mating types, and so on. Yet the discovery of autogamy in the life cycle, indicating a high degree of inbreeding, makes the species potentially valuable for (future) genetic work, as pointed out above. Cytogenetic studies have been started (35); and the mean DNA base composition (expressed in percent guanine plus cytosine) has been determined to be 24, while that of strains of *T. pyriformis* covers the range 24 to 32 (113).

5. *Tetrahymena limacis* (Warren, 1932) Kozloff, 1946
(Fig. 25)

Type material (slide of silver-impregnated specimens, strain D.r.IV): USNM no. 24115.
Synonym: *Paraglaucoma limacis* of Warren (158).

Ciliary Rows. This species also exhibits distinct diphasic meridional ranges; thus, as pointed out in the case of *T. rostrata,* figures on numbers of kineties should not be presented without statements concerning the conditions under which the organisms were living at the time of study. In its free-living phase, as occasionally found in nature or as often raised in laboratory culture, the meridional range is 24 to 32, with 26 common. For the parasitic phase, the entire range is 26 to 46, although generally restricted to 32 to 41, with 35 a commonly found number (foregoing data derived from analysis of information in refs. 5,26,46,90,97,100,101,121,158,166). Thus there is always a significant drop in kineties when laboratory populations of this species are established from originally parasitic populations.

There are one to four, typically two or three, POMs.

Caudal Cilium. None.

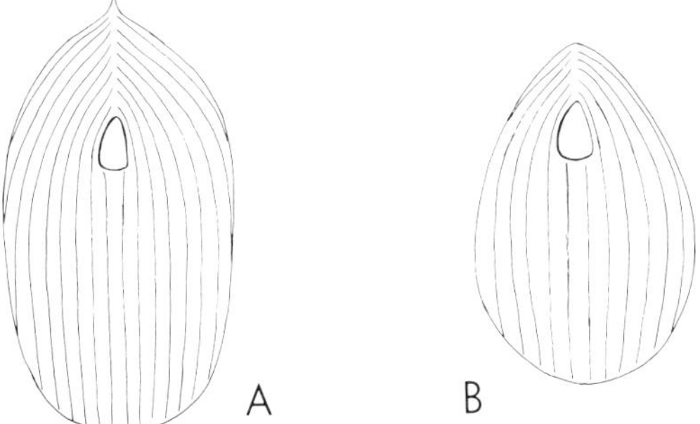

Fig. 25.

Semidiagrammatic drawings of *Tetrahymena limacis,* a diphasic species, in ventral view. (A) Obligate parasitic form; note particularly the shape of the body. (B) Cultured form, drawn to same scale; note differences in body size and shape, number of meridians, and relative size of buccal overture.

Contractile Vacuole Pores. There are generally two to three, occurring in kineties 5 to 8; often one (or two) is (or are) associated with row 6 or 7, the other two (or one) with 7 or 8.

Other Silverline Features. Minor details are of limited taxonomic significance, but the reader is referred to the excellent article on this species by Kozloff (97), one of the pioneering works in comparative systematics of tetrahymenas employing the technique of silver impregnation.

Body Shape and Size. In its diphasic life cycle, this species, unlike *T. rostrata*, exhibits changes referable to the two differing situations. However, the dimorphism is also unlike that found in members of the *patula* complex in that only microstome forms are involved. As many studies have shown (e.g., 5,26,46,97,100,101,121,158, 166), living ciliates from slug or snail hosts are generally larger than those from laboratory cultures, for example, average lengths of 50–55 μm as contrasted with 40–45 μm. The parasitic-phase forms are cucumber-shaped and apiculate at the anterior end; cultured forms are generally pyriform and pointed anteriorly but not apiculate (cf. Fig. 25A and B). Indeed, except for the greater number of rows of cilia (and the shape of the micronucleus; see below), the laboratory *T. limacis* is practically indistinguishable from strains of *T. pyriformis*. Homopolar doublets have been noted in domesticated cultures of various strains (101; J. O. Corliss, unpublished).

Cysts. None. [Michelson (121) reported resting and reproductive cysts, but there is presumptive evidence that his article in part confused *T. limacis* with *T. rostrata* and that the cysts were of the latter species. As is now well known, both of these species can inhabit the same slug host at the same time (5,44a).]

Stomatogenesis. As in *T. pyriformis*.

Nuclear Characteristics. Like *T. rostrata* and unlike members of the *pyriformis* and *patula* complexes, *T. limacis* possesses an ovoid micronucleus. Small, its longer diameter never exceeds 2.5 μm (whereas in *T. rostrata* it reaches a length of 4.5 μ with, therefore, a tremendous difference in volume).

Sexuality. A presumably selfing type of conjugation seems to be common, having been reported by various investigators in strains isolated from widely separated parts of the world, for example, South Africa, Illinois, and the West Coast of the United States (5,26,100,101,158,165,166). Unfortunately, no one to date has undertaken a careful study of it.

Habitat. *Tetrahymena limacis*, similar to *T. rostrata*, is widely distributed as a parasite (5,26,46,97,101,158,166); unlike *T. rostrata*, it is more host-restricted, occurring primarily in slugs and snails and not other invertebrates (although reported from freshwater mussels, ref. 90); it occurs in the liver (digestive gland) rather than the renal organ; it is an obligate rather than facultative parasite, yet only slightly pathogenic for its host; it is seldom found in nature free of its host. Thus *T. limacis* is edaphic only in the sense that it is occasionally found free-living in its host's environment. Transmission is very likely through ingestion of fecal-contaminated material (5,44a). It is easily cultured in the laboratory, in both bacterized and axenic media.

Physiological-Biochemical Traits. Little studied with any sophistication. Scattered facts accumulated, but few of comparative taxonomic value.

Genetics. In spite of the known existence of conjugation (presumably a selfing type) in strains of this species, and some work (53; M. P. Dysart, unpublished) on DNA/RNA ratios in their nuclei, no truly genetic study has yet been undertaken.

6. *Tetrahymena corlissi* Thompson, 1955
(Figs. 26 and 27)

Type material (slide of silver-impregnated specimens, strain W): USNM no. 24116. Synonyms: None.

Ciliary Rows. Range of 25 to 31, based on data from the few studies (published and unpublished) made to date, with no significant differences between parasitic and free-living phases. Typically, there are two POMs.

Caudal Cilium. Present, with a clearly detectable PBB complex (26,72,77,117,153).

Contractile Vacuole Pores. There are two to three, located in kineties 6 to 10; often the number is three, found in meridians 8 and 9.

Other Silverline Features. Minor differences may exist, but they are of limited taxonomic significance.

Body Shape and Size. Monomorphic; body elongate, pointed anteriorly and rounded posteriorly, differing principally from *T. pyriformis* in being thinner at its "waistline."

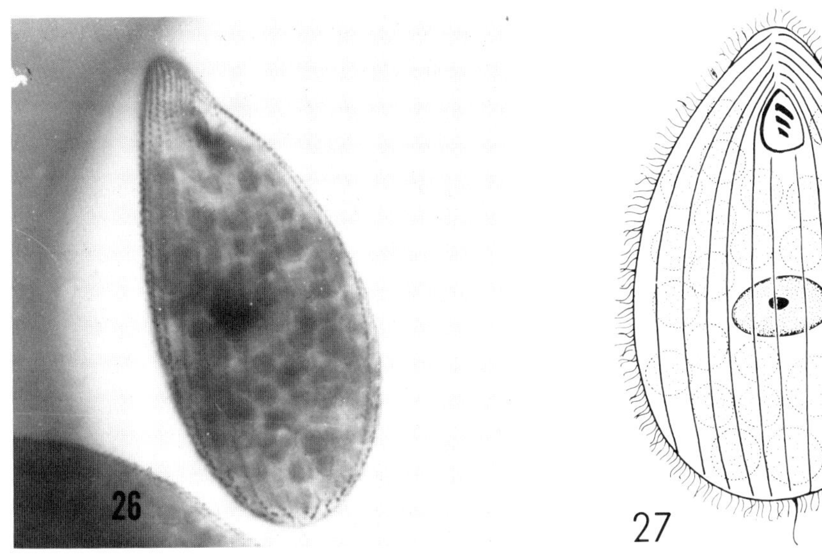

Tetrahymena corlissi, *histophagous species which is the only regular invader of vertebrate hosts among all species in the genus.*

Fig. 26.

Silver-impregnated specimen in early trophont stage, very recently derived from the marauding theront ("hunter") stage. Chatton-Lwoff technique. Right ventral view.

Fig. 27.

Semidiagrammatic drawing of engorged trophont raised on tissue in bacterized laboratory culture; note many dark food vacuoles, hugeness of body, presence of caudal cilium. Ventral view.

Acknowledgment with gratitude to R. M. Jones for Fig. 26.

An avid histophage, the organism's size may vary considerably depending on the culture medium and its own physiological state. Thompson (153) reported mean sizes of ca. 47 × 31 μm for bacteria feeders, and ca. 52 × 32 μm for forms in axenic culture. McArdle (117) recorded an average size of 68 × 21 μm for ciliates feeding on *Drosophila* gut tissue. In bacterized meat-fed strains, lengths of over 80 μm have frequently been observed, the relatively huge and body-darkened (with food vacuoles) histophages completely dwarfing such species as *T. pyriformis* when the two are placed together (unpublished observations of R. M. Jones, M. Albertson, J. C. Thompson, Jr., and J. O. Corliss; and see ref. 72).

Cysts. Reproductive, but no resting, cysts. They may contain two to eight tomites which, on emergence, quickly attain the mature trophont stage in the presence of adequate food.

Stomatogenesis. As in *T. pyriformis*.

Nuclear Characteristics. The species possesses an ovoid micronucleus with typical dimensions of 2.1 × 2.8 μm. Amicronucleate strains have been reported (150; and other articles not cited), produced in the laboratory by x-radiation.

Sexuality. None known.

Habitat. The strain that served as the basis for description of the organism as a new species (153) was discovered as a parasite in the circulatory system of a *Pseudotriton* larva, although it should be noted that Thompson (personal communication) also found the organism often as a free-living bacteria feeder; and it thrives in laboratory culture. It apparently occurs widely in a variety of lower vertebrates, fishes through amphibians, generally in (freshwater) larval forms or very young forms or in bruised or wounded specimens. Speidel and associates (e.g., 150) discovered it repeatedly in experimental studies on poorly nourished larvae of *Bufo, Rana, Pseudacris,* and *Pseudotriton* spp. and R. M. Jones and colleagues noted it over and over again in the guppy, *Lebistes reticulatus,* in battered shipments from a commercial source. Nigrelli's (e.g., 133) ciliates from skin or surface wounds of several species of fishes from aquaria in New York City may well have been strains of *T. corlissi.* It is reasonable to consider the species as a natural parasite, with a facultatively free-living stage, whose invasion is frequently fatal to its host (40,44a). McArdle's (117) isolated detection of the species in a sample of dried moss spotlights our ignorance of its distributional range in nature.

Physiological-Biochemical Traits. Little studied; of limited value from a taxonomic point of view.

Genetics. No information available.

7. *Tetrahymena stegomyiae* (Keilin, 1921) Corliss, 1960
(Figs. 28 and 29)

Type material (slide of silver-impregnated specimens from material fixed by J. Muspratt in South Africa): USNM no. 24117.

Synonym: *Lambornella stegomyiae* of Keilin (91).

Ciliary Rows. Range of 25 to 30; not known whether or not the number varies (but I predict not) during the complete life cycle which itself has been insufficiently studied (26). Two POMs.

Caudal Cilium. None.

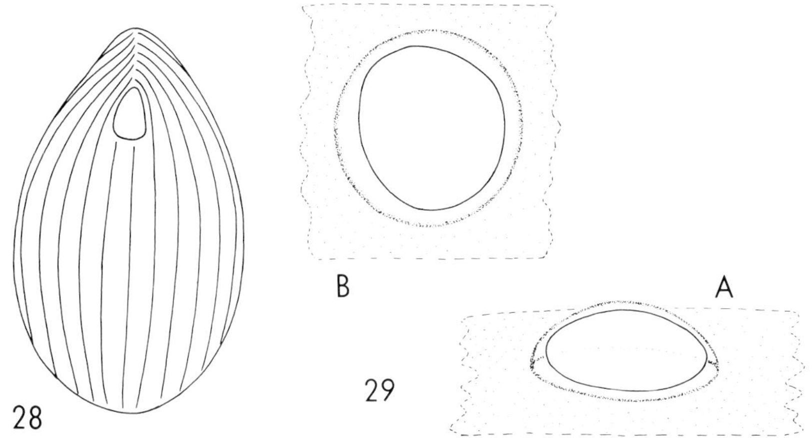

Semidiagrammatic drawings of Tetrahymena stegomyiae, *curious parasite of tropical mosquito larvae.*

Fig. 28.

Pyriform-shaped culture form. Ventral view.

Fig. 29.

Cuticular cyst, unique in the genus, found on outside of the host (mosquito larva). (A) Side view. (B) Top view. (Other *Tetrahymena* cysts are shown in Fig. 10A and B.)

Contractile Vacuole Pores. Not described.

Other Silverline Features. Typical of the genus, to the extent studied.

Body Shape and Size. Pear-shaped, large, densely ciliated. Length shows a range of ca. 60–100 μm.

Cysts. Possibly an obligate stage in the parasitic phase of this species is a peculiar cyst (Fig. 29). Relatively large and rounded, ca. 35 μm in diameter, it occurs only on the outside of parasitized mosquito larvae, on their body cuticle, and appears to be neither a reproductive nor a resting cyst. On the assumption that the ciliate itself enters the host's body cavity by penetration through the cuticle, is it then an "invasion" cyst? The term "cuticular cyst" is perhaps the best to use, in view of the limited data available concerning its nature and exact role. [Muspratt (126) has also reported briefly on the existence of a true resting cyst of some sort, a structure formed *within* the cuticular cyst in instances in which penetration into the body cavity of the host presumably has been unsuccessful.]

Stomatogenesis. Not studied, but safely assumed to be as in *T. pyriformis.*

Nuclear Characteristics. Micronucleus, although studied in relatively few specimens through lack of proper material, appears to be a prominent, near-spherical structure, at least 2.5–3.0 μm in diameter.

Sexuality. None known.

Habitat. Association of the organism with the larvae of certain tropical tree-hole-breeding culicine mosquitoes—*Aedes and Culex* species from Northern Rhodesia—suggested an obligate host-parasite relationship to earlier workers (91,125,126). Still the ciliate is able to live outside its host, in nature as well as in the laboratory. The infection is ultimately fatal to the host. Development of the cuticular cyst, the mode of entry into the host by boring through the cuticular layers to the hemocoel, indicates a long if curious association with certain mosquito larvae and supports the conclusion that *T. stegomyiae* should be considered an obligate parasite with a free-living, free-swimming stage between hosts (26,40,44a).

Physiological-Biochemical Traits. No information available.

Genetics. No information available.

8. *Tetrahymena patula* (Ehrenberg, 1830) Corliss, 1951
(Figs. 30–35)

Type material (slide of silver-impregnated specimens, strain L-FF): USNM no. 24118.

Synonyms: *Leucophrys patula* of the literature (see refs. 17a,21,40,45). *Leucophrydium putrinum* of Roux (see ref. 143a).

Ciliary Rows. Although the species exhibits striking dimorphism in free-swimming stages of its life cycle, there is strong overlapping in the number of meridians for the two forms, the microstome and macrostome. Macrostomes have 33 to 45, with the upper extreme rare; the modal number has varied from 36 (see ref. 161) to 40 (see ref. 21). Microstomes show a range of 32 to 41, commonly 36 to 38, according to Corliss' studies (21); but Williams (161), studying rapidly growing forms in a specific axenic medium, arrived at the same modal number, 36, which he found for macrostomes cultured under similar laboratory conditions.

The species is unique in the genus in its possession, regularly, of more than two POMs. It generally has four, in both microstome and macrostome; occasionally three or five are found. Reports of a greater number are not accurate; the crowding of kineties in the oral area, especially in the macrostome, may give such an illusion.

Caudal Cilium. None.

Contractile Vacuole Pores. There are two to six, occurring in kineties 9 to 16. Generally the number is two to three, in meridians 9 to 11, in microstomes; and four to five, in rows 10 to 12, in macrostomes. Diameter of a single pore is commonly 0.9 μm in microstome forms but as large as 1.5 μm in macrostomes.

Other Silverline Features. Strikingly similar to those of *T. pyriformis* and other members of the genus in details, in spite of differences in body size, life cycle, and total number of meridians. Secondary meridians, it might be mentioned, often appear to be discontinuous in this species, whereas they are typically heavy and "complete" in *T. pyriformis* (at least in amicronucleate strains) and in various of the other species.

Body Shape and Size. *Tetrahymena patula,* and the other two members of the complex, differ from all other species in the genus in demonstration of a polymorphic life cycle (Fig. 35) highlighted by striking *di*morphism in its free-swimming stages: a small form with a small oral aperture (microstome form) and a large one with a greatly enlarged buccal area (macrostome). A third stage, known in *T. patula* and *T. vorax* but not *T. paravorax,* is the reproductive cyst, a form unknown in other species of *Tetrahymena* except *T. rostrata* and *T. corlissi.* Under appropriate environmental conditions (as determined in laboratory culture; see refs. 151,161), interconversions between microstome, macrostome, and cyst occur freely. It should be stressed that binary fission can occur in populations of free-swimming microstomes or macrostomes, however, without intervention of the reproductive cystic stage (161).

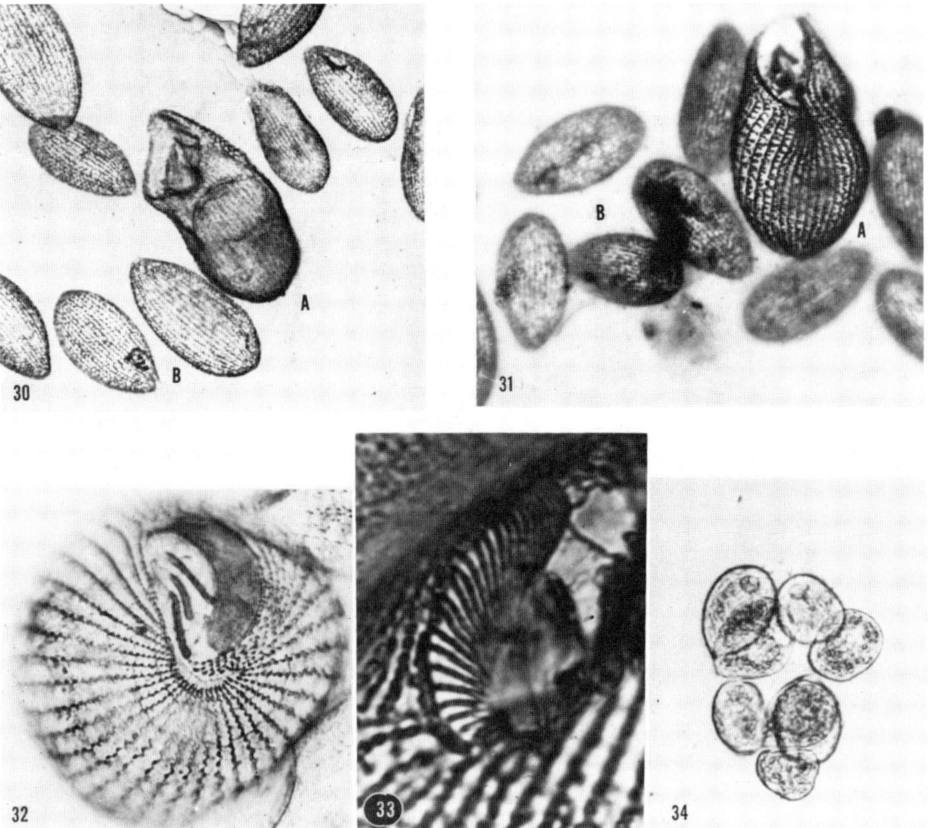

Figures of Tetrahymena patula *polymorphic species with striking dimorphism in free-swimming stage and reproductive cyst in full life cycle.*

Figs. 30 and 31.

Silver-impregnated specimens from a mixed culture, showing two macrostomes (A; at slightly different magnifications and in slightly different views in the two photomicrographs) and several microstomes (B). Note that the microstomes are hardly distinguishable from specimens of *T. pyriformis* (Fig. 11 and others), except for more closely packed ciliary rows. Chatton-Lwoff technique.

Fig. 32.

Apical polar view of a recently produced daughter macrostome, silver-impregnated specimen, showing details of the silverline system and buccal infraciliature typical of the species which, however, are generally paralleled in other species of the genus as well (see preceding figures). Chatton-Lwoff technique.

Fig. 33.

Close-up of UM area of a macrostome form, showing especially clearly the riblike oral fibers of that portion of the buccal cavity. Chatton-Lwoff technique.

Fig. 34.

Photomicrograph of unstained, living tomites, eight in number, squirming within the invisible gelatinous envelope of a reproductive cyst.

Acknowledgments, with gratitude to E. Fauré-Fremiet for Fig. 32 and to N. E. Williams for Figs. 33 and 34.

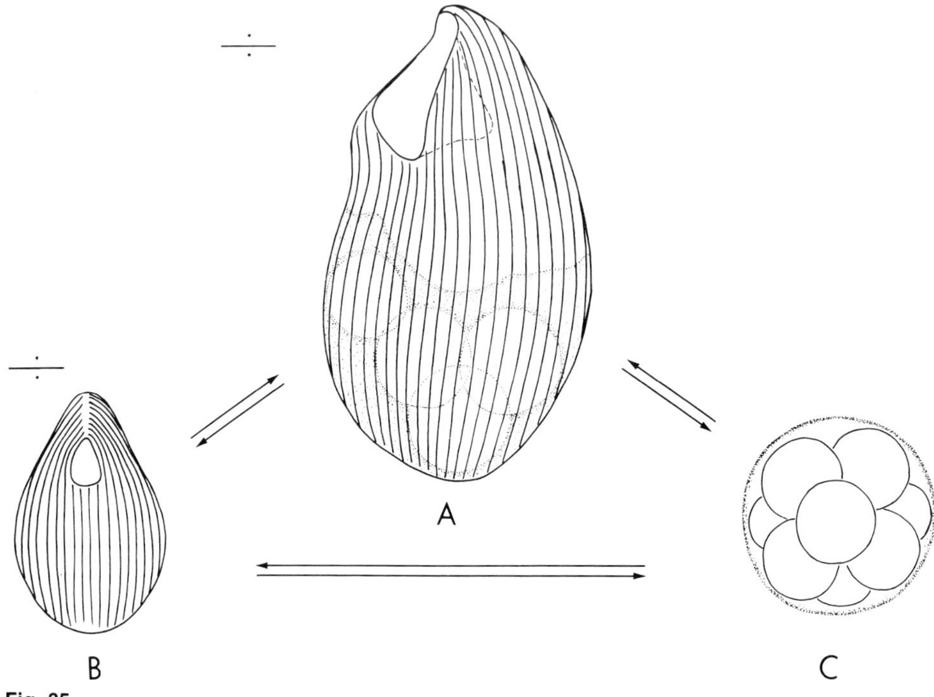

Fig. 35.

Basic life cycle of *T. patula*. (A) Macrostome (carnivorous) form, capable of producing daughter macrostomes (as indicated by the division sign), may also transform to microstome (B) or to a reproductive cyst (C). (B) Microstome form, similarly capable of dividing, transforming into macrostome (A), or producing a cyst (C). (C) Reproductive cyst, within which several divisions may occur to produce tomites capable of maturing into either macrostome (A) or microstome (B) forms. [Essentially adapted from Williams (161).]

In nature, and generally in the laboratory in bacterized or axenic culture media, the most frequently encountered form of *T. patula* is the voracious, carnivorous, sometimes cannibalistic, macrostome. It is broadly and irregularly pyriform, bag-shaped, the anterior third of the body being scooplike, and the posterior end plump and hemispherical. Body size seems to vary with the strain under consideration, or else (unknown) environmental factors can affect the range, especially the upper limit, markedly. The largest sizes (lengths) in the literature range from 110 μm (21), to 150 μm (60), to 160 μm (66). From accumulated observations, Corliss (21) reported an average size of 95 × 54 μm; studying the same strain some years later, Williams (161) gave 85 × 55 μm as a mean from an axenic culture and 87 × 65 μm from a culture containing *T. pyriformis* as prey. Since Fauré-Fremiet, who reported greater lengths (60), employed the same strain subsequently examined by Corliss and Williams, it is possible that long years in laboratory culture cause a gradual diminishing in size; the fact that the organism had become amicronucleate (and thus senescent?) by the time of Williams' study (the most recent of the three) may also be related to its decrease in maximal size.

Microstome forms occur regularly under certain conditions of culture (161). Their strong resemblance to *T. pyriformis* has been universally recognized, but reports on size ranges vary considerably: for example, range in length 32–55 μm, with average size 45 × 28 μm (21), as opposed to a general size of 60 × 40 μm (161). Again nutritional conditions, as well as age of strains, are probably the principal factors responsible for such variation. Also, there are different kinds of microstomes: free-swimming, freely dividing cultures, for example, as opposed to the small-mouthed tomites that emerge from the reproductive cyst.

Hompolar doublets have been observed (21,60) and studied experimentally (61).

The buccal area in *T. patula* is characteristic of that of the genus; thus the description given under *T. pyriformis* suffices. Even the dimensions of the entire as well as separate parts of the oral apparatus are essentially the same in microstomes as they are in *T. pyriformis*. In the macrostome, however, the buccal cavity is, relative to the body size, much bigger (e.g., 30 × 22 μm in a body 95 × 55 μm, contrasted with 10 × 5 μm in a microstome body 45 × 28 μm), and the undulating membrane, outstanding among the oral ciliary organelles, is far more extensive. Also, the cytopharyngeal area becomes tremendously expanded and enlarged; the resulting cytopharyngeal pouch acts as a "preparatory vacuole" or "receiving vacuole" in the organism's carnivorous state and is one of the distinctive features of the whole *patula* complex. Its origin is very likely the same in all three species; Buhse (8) made a thorough study of its development (it starts as an independent vacuole, subsequently fusing within the distal portion of the inconspicuous cytopharynx, enlarging then as a single cavity) in *T. vorax* (Figs. 41 and 43). The morphogenetics of the change in the oral area, whether from microstome to macrostome (by oral replacement) or from macrostome to microstome (involving binary fission of the organism and different processes for the proter as compared with the opisthe), is beyond the scope of the present account (but see refs. 7–10,151, and Chapter 12). Another topic beyond discussion is that of stimulation of (the formation of) the macrostome forms; the presence of living, potential prey (e.g., *T. pyriformis* or *Colpidium* spp.) is known to invoke transformation of microstomes to macrostomes. However, Buhse (9; and see refs. 14,132) isolated a soluble factor, stomatin, from *T. pyriformis* which induces the process under chemically controllable conditions in the neighboring species *T. vorax*.

Cysts. The reproductive cyst should be recognized as a third important stage in the life history of *T. patula* although, as emphasized above, it does not represent the only—or even the most common—way of increasing the population; typical binary fission of microstomes and macrostomes occurs frequently under many cultural conditions. Williams (161) found that either microstome or macrostome forms could become encysted; earlier observers, including Corliss, had seen reproductive cysts only in populations of macrostomes. The number of divisions within the thin, transparent cystic membrane ranges from one to four, thus producing 2 to 16 (commonly 8) tomites which, no doubt under the influence of some stimulus not yet understood, break out of the membrane and grow or transform into either (new) microstomes or macrostomes. No resting or resistant cyst is known for the species.

Stomatogenesis. Essentially as in *T. pyriformis*, during regular division of both microstomes and macrostomes. In transformation of microstomes into macrostomes, however, the process of oral replacement is invoked; and in the production of microstomes from macrostomes, the proter's oral apparatus arises by a remodeling, involving partial resorption and regression, of the original macrostome buccal complex.

Nuclear Characteristics. Macronucleus, except for expected differences in size directly correlatable to body size, is essentially identical to that known for other species in the genus. Micronucleus, single, spherical, located in an indentation of the macronucleus, has been seen by some investigators; the classic strain L-FF, originally reported to possess such a micronucleus (21), was later found to be amicronucleate (161). MCE bodies are regularly produced during macronuclear division in macrostome trophonts (51,53), and now they are known in the fission of tomites in the reproductive cyst (161). The mode of production of MCE bodies in members of the *patula* group is strikingly different from that in the *pyriformis* and *rostrata* complexes, as noted in Section III,C, and thus the phenomenon is of taxonomic significance. Only in the *patula* complex does the MCE body commence (i.e., is recognizable at a very early stage of nuclear fission) as a heavy band of chromatin across the center of the elongating, predivision macronucleus.

Sexuality. Maupas (115,116) noted conjugation in *T. patula* about 65 years before it was discovered and described in *T. pyriformis*. He observed that only microstome forms seemed to be involved, and that they had to be derived from diverse stocks. Unfortunately, no worker since that time has found sexuality among his populations of the species.

Habitat. Fresh water; easily found in the vicinity of watercress, according to Maupas (116). Apparently, the species is not a common ciliate, however, since it has been so relatively infrequently reported in the literature (21,22). It is easily grown in the laboratory, in bacterized and axenic media, both particulate and nonparticulate, and in monoxenic culture with some other ciliate as prey.

Physiological-Biochemical Traits. *Tetrahymena patula* has been little investigated for such properties.

Genetics. There are no data of genetic significance.

9. *Tetrahymena vorax* (Kidder, Lilly, and Claff, 1940) Kidder, 1941
(Figs. 36–43)

Type material (slide of silver-impregnated specimens, strain V_2): USNM no. 24119. Synonyms: *Glaucoma vorax* of Kidder et al. (95), *Leucophrys vorax* of Fauré-Fremiet (61).

Ciliary Rows. Number changes little during the whole life cycle, although the size and volume of the body change enormously in the tranformation from microstome to macrostome form. Studying strain V_2, years ago, Corliss (21) reported a range of 18 to 23 kineties for microstomes, with or without tails, with a modal number of ca. 20; and a range of 19 to 26, with a mode of ca. 22, for the macrostomes. Williams (162), using V_2 some years later and under conditions of more carefully controlled growth parameters, found no significant difference in number for microstomes and macrostomes; but he reported means for three separate sublines as ca. 24.5, 20.0, and 18.5. The number of meridians in the strain used in describing the species as new (95) was given as 19 to 21; later workers (105,109) have given 17 to 28 or 18 to 28 as the range for the whole species.

The number of true POMs is two at all stages, although occasionally meridian numbers n and $n-1$ are so encroached on by the huge buccal cavity of the macrostome form that their continuity to the apical end of the body is disrupted; and the inexperienced observer often fails to note that several of the other ventral kineties, although crowded, do maintain their full-length integrity and are not terminated at the posterior edge of the oral opening.

Caudal Cilium. None.

Contractile Vacuole Pores. There may be two to six, occurring in meridians 5 to 10. From data collected on strain V_2, there are generally two to three, in kineties 5 to 8, in microstomes; and five, in rows 6 to 9, in macrostomes (21; J. O. Corliss, unpublished).

Other Silverline Features. Differences in details, when compared with *T. pyriformis*, are generally of minor taxonomic significance. However, it may be worthwhile pointing out that 2° and even 3° meridians seem to be present more constantly and "stain" more heavily in *T. vorax* than in any other species (e.g., see Fig. 39). Yet electron microscopy has revealed that these "lines" are in effect artifacts; so their significance is slight.

Body Shape and Size. The species exhibits a polymorphic life cycle: microstome and macrostome forms and a reproductive cyst (Fig. 42). Interconversions among microstome, macrostome, and cyst occur freely (7–9,144,162), but division of free-swimming microstomes and macrostomes can and does often occur; that is, the reproductive cyst is definitely not an obligatory stage in the life cycle.

In nature, and commonly in routine nondefined media (both bacterized and axenic) in laboratory cultures, the most frequently encountered form of *T. vorax* is the microstome form, but in a shape thought to be unique among species of the entire genus. [However, Williams (161) reported as a rare variant microstome form of *T. patula* a tailed organism reminiscent of the form of *T. vorax* described here.] The posterior end of the organism is drawn out into a tail, more or less attenuated depending on cultural conditions (Figs. 36 and 42A). The CVPs are not carried posteriorly down into the tapered part of the posterior quarter of the ciliate's body. This tailed microstome may persist in culture indefinitely, with periodic binary fission, under conditions of appropriate substrate (21,162,163).

Another microstome form was described by Corliss (21), although generally not found by subsequent workers: a typical *T. pyriformis*-like stage, with rounded posterior end. [*Tetrahymena vorax* Tur has been reported as tailless, incidentally, but the microstome form remains larger and more elongate than the typical *T. pyriformis*. In two earlier special cases, causing some controversy in the literature, Shaw and Williams (144) concluded, verifying earlier evidence obtained by Loefer (104), that two *T. pyriformis*-like strains long labeled *T. vorax* are best considered to belong to the species *T. pyriformis*, since none of their present-day characteristics seems to relate them to typical polymorphic strains of *T. vorax*. This conclusion has continued to receive support (105,110).] In some instances, old, bacterized cultures seemed to reach this stage *irreversibly*. Morphologically such populations (all amicronucleate) became completely indistinguishable from members of the monomorphic species *T. pyriformis*, recalling a similar situation for certain allegedly aberrant sublines of *T. rostrata*.

The size (especially length) of the tailed microstome varies considerably under differing conditions of culture. It can generally be distinguished from *T. pyriformis*—which it otherwise resembles in so many ways—not only by its tapered posterior end but also by its greater size. Lengths have been given as follows: range 31–115 μm, depending

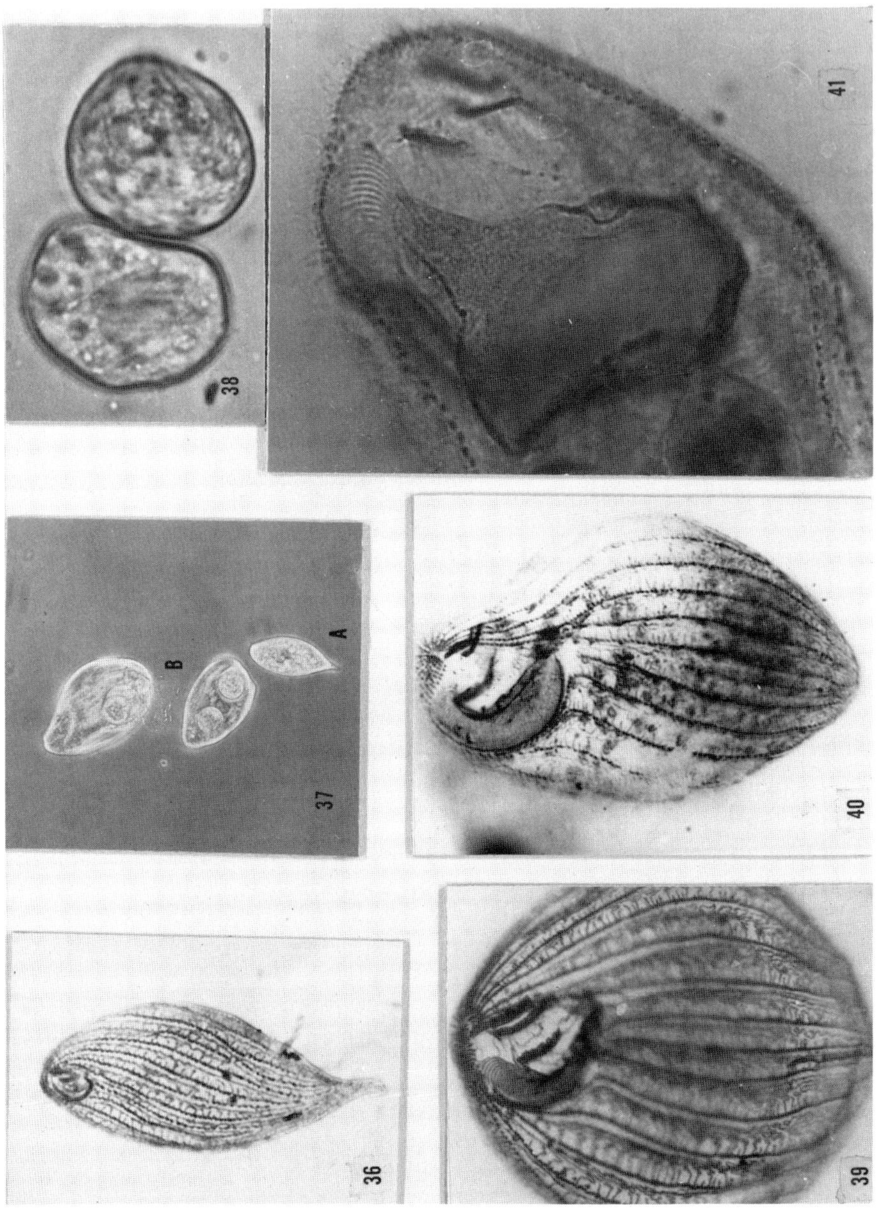

Figures of Tetrahymena vorax, *a second polymorphic species in the* patula *complex with general characteristics like those of* T. patula.

Fig. 36.

Tailed microstome form, the commonly found stage in the life cycle of the species, strikingly similar to *T. pyriformis* except for the attenuated posterior end of the body. Chatton-Lwoff technique. Ventral view.

Fig. 37.

Microstome (A) and macrostome (B) forms, as seen under phase microscopy.

on culture medium (95); 108, 86, and 78 μm, for three different sublines of strain V₂ (162).

Although the macrostome form may not be the naturally persistent form for *T. vorax*, it is easily caused to occur—in nature as well as in laboratory culture. On being fed other ciliates (e.g., *T. pyriformis* or species of *Colpidium*), or occasionally arising as a cannibal, the carnivorous form of *T. vorax* is quickly developed. It is characterized by its relatively great size, its tremendously expanded buccal area, and its huge cytopharyngeal pouch (e.g., see Figs. 41 and 43). The body size can apparently reach dimensions as large as 250 × 150 μm (95), although no investigators in more recent times, and working with different strains, have recorded lengths of over 150 μm; an average size is close to 110 × 75 μm (e.g., 162). On appropriate substrate, axenic cultures of macrostomes may persist indefinitely, undergoing regular binary fission periodically with resort to neither microstome nor cyst formation (162).

The fate of the old mouth (and associated buccal organelles), origin of the new mouth, and development of the cytopharyngeal pouch—in the transformation of microstome to macrostome form in *T. vorax*—have recently been under intense investigation by Buhse and associates (e.g., 7–14,70,132), with emphasis on aspects of biochemical morphogenesis and ultrastructure (using both transmission and scanning electron microscopy). The intriguing chemical factor stomatin, an RNAase-sensitive principle released by *T. pyriformis*, induces transformation from microstome to macrostome form in the absence of prey organisms.

Cysts. Reproductive cysts, although not an obligate stage in the laboratory life cycle, can be produced from either tailed microstome or carnivorous macrostome forms (70,162). Within an entirely transparent gelatinous cyst membrane, binary fission occurs to produce 2 to 16, typically 2 to 4, tomites. Rounded at first, with a diameter in the range of 20–30 μm, they quickly assume the shape of either tailed microstomes or typical macrostomes. No resting cyst is known.

Stomatogenesis. As in *T. patula*.

Fig. 38.

High-power photomicrograph of unstained, living tomites, two in number, about ready to emerge through invisible membrane of a reproductive cyst.

Fig. 39.

Ventral aspect of recently produced daughter macrostome, silver-impregnated specimen, showing rotundity, immature state of development of buccal cavity, and details of the silverline system (even including bits of 3° ciliary meridians). Chatton-Lwoff technique.

Fig. 40.

Mature macrostome, silver-impregnated according to the Chatton-Lwoff technique. Ventral view.

Fig. 41.

Buccal area of macrostome with focus purposely on the huge, internal cytopharyngeal pouch characteristic of members of the *patula* complex. Chatton-Lwoff technique.

Acknowledgments with gratitude to H. E. Buhse, Jr., for Fig. 40, to K. R. Groh for Fig. 38 and to N. E. Williams for Figs. 36, 37, and 39.

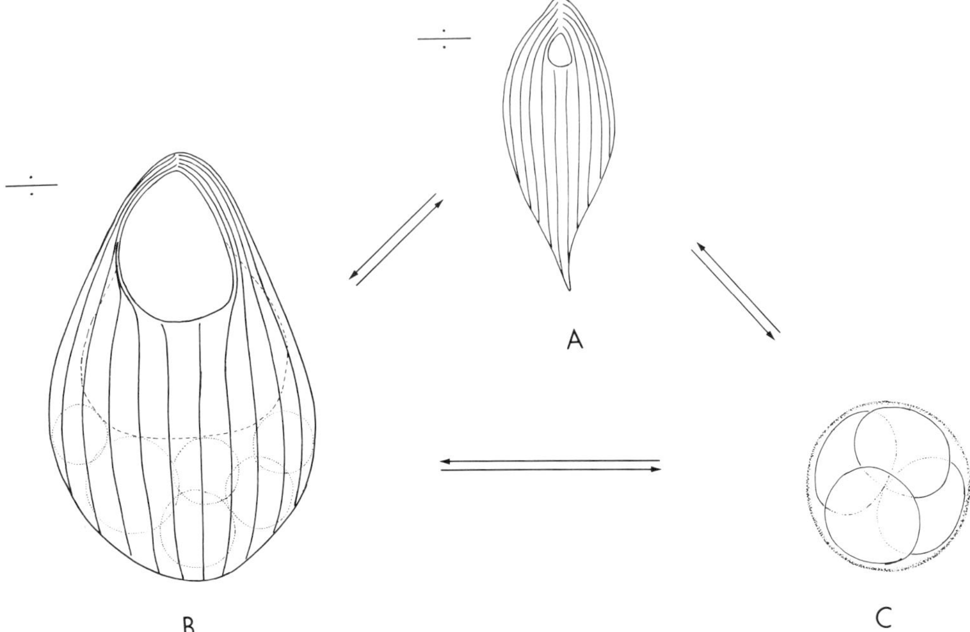

Fig. 42.

Basic life cycle of *T. vorax*. See legend for Fig. 35, and text for details. (A) Tailed microstome form. (B) Carnivorous macrostome form. (C) Reproductive cyst. Note that all stages are interconvertible. Drawn to same scale as used in Fig. 35. [Essentially adapted from Williams (162).]

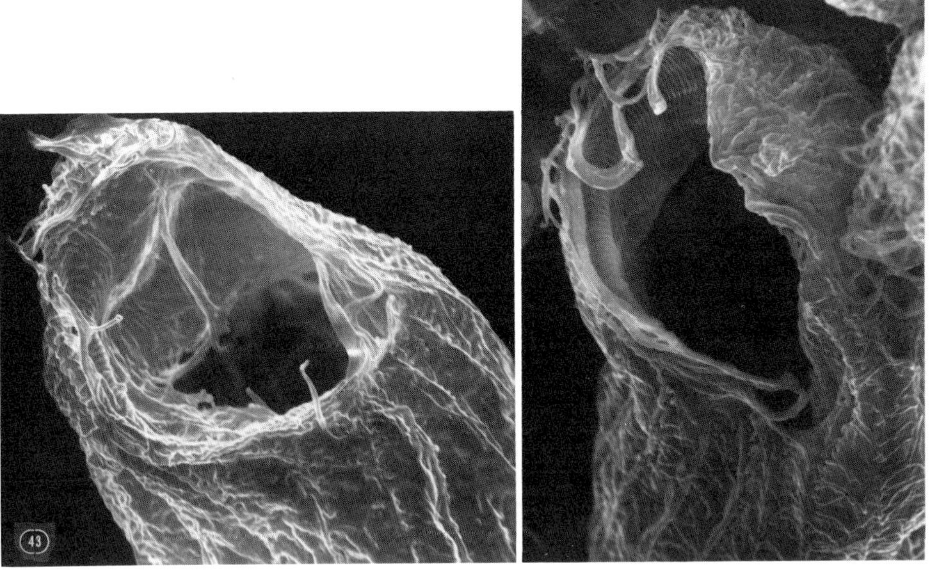

Fig. 43.

Ventral views of anterior end of mature macrostome form (in whole mounts) of *Tetrahymena vorax*, as observed with the scanning electron microscope, with emphasis solely on the "yawning cavern" which represents the confluence of the enlarged buccal cavity with the tremendous cytopharyngeal pouch.

Acknowledgment with gratitude to H. E. Buhse, Jr., for direct aid in production of Fig. 43.

Nuclear Characteristics. All strains studied to date, except the one (now extinct) originally described by Kidder et al. (95), have been reported as being amicronucleate. MCE bodies have been noted in dividing organisms (51,53).

Sexuality. None known; but this is not at all surprising, since essentially only amicronucleate strains have been studied to date.

Habitat. Fresh water; little else is known about the ecology of the species, although it is easily cultured in the laboratory in a variety of media and under a wide range of conditions.

Physiological-Biochemical Traits. Data on such properties have often been related to the species' exhibition of polymorphism, described above. The biochemical information available is seldom of much significance in the comparative taxonomy and classification of tetrahymenas, although serological data of Loefer (104) and nutritional and verifying serological data of Shaw and Williams (144) were presented in partial evidence of the identity of "*T. vorax*" strains V_1 and PP with strains of *T. pyriformis.*

Genetics. No information available.

10. Tetrahymena paravorax Corliss, 1957
(Figs. 44–46)

Type material (slide of silver impregnated specimens, strain RP): USNM no. 24120. Synonym: *T. vorax* (strain RP) of Elliott and Hayes (57).

Ciliary Rows. Range in total number 22 to 30, with a fairly distinct difference in modal numbers for the microstome as opposed to the macrostome form in strain RP: 24 and 28, respectively (6,23). There are only two POMs in material studied by Corliss (23). The illusion of a greater number, especially in the macrostome (see drawings in ref. 50), is caused by the crowding of meridians in the buccal area.

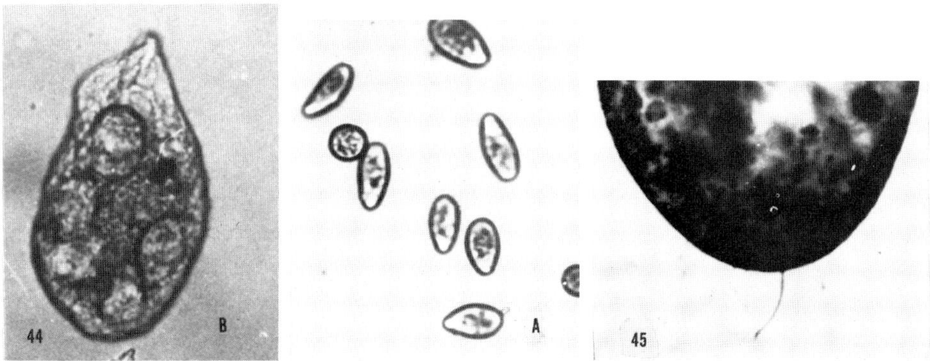

Figures of Tetrahymena paravorax, *third member of the* patula *complex.*

Fig. 44.

Macrostome (A) and microstome (B) forms. Silver-impregnated specimens; not at identical magnifications. Chatton-Lwoff technique.

Fig. 45.

Protargol silver preparation of posterior end of a specimen, unretouched, showing caudal cilium.

Acknowledgment with gratitude to H. E. Buhse, Jr., for Figs. 44A and B and 45.

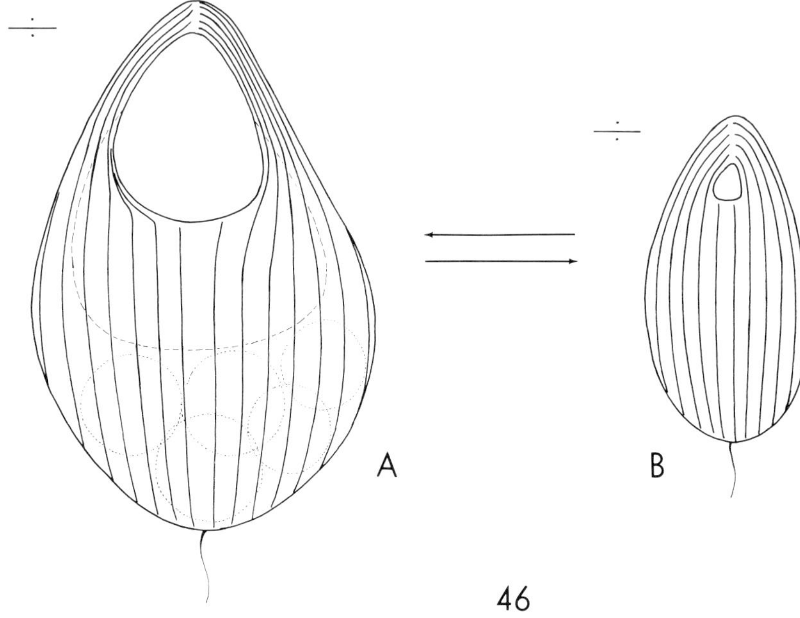

Fig. 46.

Basic life cycle of *T. paravorax,* as determinable to date. Macrostome (A) and microstome (B) forms are interconvertible, although also capable of independent growth and division for indefinite (?) periods of time. No cystic stages known. Drawn to same scale as used in Figs. 35 and 42.

Caudal Cilium. Present; thicker than ordinary somatic cilia, and at least twice their length (Fig. 45). In specimens impregnated with silver by the Chatton–Lwoff technique, the PBB complex is clearly seen.

Contractile Vacuole Pores. There may be one to six, often two, CVPs, occurring in ciliary meridians number 6 to 9, often 8 (see ref. 23). However, Dragesco and Njiné (50) very recently described a strain from the Cameroons which, in the macrostome stage, possesses as many as six entire vacuoles, *each* with two to four CVPs.

Other Silverline Features. Basically as in *T. pyriformis.*

Body Shape and Size. Tetrahymena paravorax can exhibit a striking dimorphism in its life cycle, with interconversions between microstome and macrostome forms (Fig. 46). No reproductive cysts have been found (6,23,50). The form commonly encountered in strain RP is the microstome (6,23). In outline it resembles *T. pyriformis,* although generally considerably larger in size and relatively thinner at the "waistline." Size of the body may reach 115 μm in length. An average of 70–90 μm has been reported for organisms from the wild (50) and in axenic culture (6). Microstomes, with periodic fissions, may persist indefinitely in culture. Macrostomes are typical of members of the *patula* complex. Buhse (6) found their production and maintenance most difficult; but I have noted macrostomes, in bacterized cultures of the same strain, both in the presence of introduced prey (e.g., *T. pyriformis*) and in the absence of other ciliates (i.e., via cannibalism). Corliss (23) has seen them up to 200 μm in length; Dragesco's (50) African strain averaged 85–140 μm; and Buhse (6) reported a mean size of 107 $\times$ 56 μm. Macrostomes in culture frequently revert to microstomes after only a few fissions. Homopolar doublets have been seen.

Without doubt the same kinds of morphogenetic phenomena known for the two preceding species, *T. patula* and *T. vorax,* occur during transformation of microstome ↔ macrostome in *T. paravorax,* although they have been little studied in this third member of the complex. Corliss (23) has briefly reported persistence of the cytopharyngeal pouch in the microstome; if confirmable in other strains, it would represent a unique characteristic.

Cysts. Neither reproductive nor resting cysts have been reported (6,23,50). Considerable effort was made to induce the "missing" stage (reproductive cyst), using techniques known to work with other members of the *patula* complex, but without success (6).

Stomatogenesis. There is no reason to believe that "ordinary" new mouth formation and such phenomena as oral replacement would not be the same in this species as in the two preceding ones, although detailed studies are yet to appear. [T. Njiné is currently engaged in such research, personal communication (see citation, under his name, in the General Bibliography, Chapter 14).]

Nuclear Characteristics. Micronucleus is spherical, with a diameter of ca. 2.5 μm. Corliss (23) reported a variable number, one to four, unusual for species of *Tetrahymena.* Dragesco and Njiné (50) mentioned only one, with a diameter of 3–5 μm. Elliott and Hayes (57) stated that there was "usually one micronucleus or rarely two," and reported the existence of "a higher number of chromosomes" than known for *T. pyriformis.*

Sexuality. A nonlethal selfing kind of conjugation has been noted by all workers who have studied strain RP (6,23,57; and M. P. Dysart, G. G. Holz, Jr., unpublished).

Habitat. Fresh water. Strain RP was recovered from a source in the Republic of Panama (57); the recent African strain from a polysaprobic pond in Nkolbisson (50).

Physiological-Biochemical Traits. Little studied, although some potentially important observations have been made (6,75,79). For example, optimal temperature for laboratory growth of strain RP in axenic culture is 33°C, which is much higher than that determined for *T. patula* and *T. vorax.*

Genetics. No data; this is understandable, in spite of the availability of conjugation (although presumably of a selfing, intraclonal type), since only two strains have been collected from nature to date.

IV. ECOLOGICAL CONSIDERATIONS

A. Generalities and Free-Living Forms

The modern, biochemically oriented ecologist of "higher" organisms is often disappointed in the apparent lack of significant progress in elucidation of problems of potential ecological importance in studies of such microbial forms as the protozoa. With rare exception, protozoologists have seldom managed to accumulate more than scattered facts about distribution of species: their tolerance, in a general way, to variation in such obvious physicochemical factors as light, temperature, salinity, pH, and oxygen content; and their reaction to degrees of availability of essential nutrients (in dissolved form or otherwise). The fact that protozoa exhibit wide ranges of adaptability to their environment explains, as Fauré-Fremiet (62) and Noland and Gojdics (135) have recently pointed out, their very wide geographical distribution. *Tetrahymena* stands as no exception to this general condition; in fact, some of its species may have abilities in ecological adaptability greater than those known for many species of even the most widespread plants and animals. More needs to be done (44c).

The following accounts represent only summaries—or analyses from a comparative point of view—of data available (usually under the heading "Habitat") in preceding pages treating the species one by one.

The type species, *T. pyriformis,* appears to be the most cosmopolitan. Known to protozoologists for the longest time, it has been found in the greatest number and in different kinds of places (22,40,56) and has been determined (generally in laboratory culture) to be the most adaptable to changes in physical and chemical (including nutritional) factors. Of course, to date, it has been the most studied species, its ease of cultivation under refined conditions understandably maintaining its popularity as an experimental organism (16,34,73,83,84,94).

The "test-tube ecology" of *T. pyriformis* is not my main concern, especially since such biochemical data are treated most adequately in other chapters of this volume and/or in earlier easily available reviews. Its ubiquitousness in nature is the most important consideration here—the fact that it has been isolated not only from the usual freshwater habitats but also from thermal springs and soil and the body cavities of various aquatic or edaphic invertebrates. Its gustatory proclivities are broad; it can be satisfied, as a microphage, on bacteria; as a macrophage or histophage, on tissues of both invertebrate and vertebrate organisms; as a facultative parasite, on the liquid diet afforded by fluids of the host (saprozoic nutrition); as an experimental laboratory organism, on dissolved nutrients even of totally known chemical composition and in the complete absence of other organisms (defined axenic media).

The geographical and ecological distribution of the other nine species treated in this chapter is relatively little known, unfortunately. Such gaps in our knowledge of these ciliates need very much to be filled.

We have practically no data on *T. setifera.* The ecology of *T. chironomi* is poorly known. There is no doubt concerning its wide distribution as a parasite of chironomid larvae, but much remains to be learned about its occurrence in the free-living stage.

Members of the *rostrata* complex are considered for the most part in the following section (in a discussion of parasitism), but they too are capable of life outside their hosts. *Tetrahymena rostrata* itself, for example, is properly classified as an edaphic ciliate, obtainable in field collections of moss, sphagnum, forest litter, and soil. Its resting cyst apparently serves as a successful adaptive measure when a particular habitat becomes temporarily desiccated. Its histophagous tendencies are attested to by its frequent discovery, within the edaphic niche, in the moribund body or empty carcass of some small invertebrate occupying the same habitat. Its geographical distribution is wide.

Tetrahymena limacis appears to be an obligate parasite (primarily of slugs and snails), although widely distributed and surely possessing at least a temporary stage outside the host. As with all species of the genus known to date, *T. limacis* can be cultured axenically. The species *T. corlissi*, frequently noted as a parasite in the tissues of bruised aquatic vertebrates (generally larvae), is certainly facultatively free-living; and Thompson (personal communication) recalls finding it in nature as a bacteria feeder, not to mention its isolation from dried moss (117). It is an avid histophage in bacterized culture but also thrives on a liquid diet. *Tetrahymena stegomyiae*, with its unique association (via a cuticular cyst) with the larvae of certain species of mosquito, is perhaps the most enigmatic tetrahymenid species of all with respect to our knowledge of other ecological aspects in its full life cycle.

Members of the *patula* complex have also been far too little investigated in nature. Reports on their distribution are so spotty that we have no idea how truly widespread the three species may be. For example, only very recently our knowledge of the distribution of *T. paravorax* was doubled—by its discovery in Africa (50); until then, the only other collection had been made in Panama (57).

Tetrahymena patula, T. vorax, and *T. paravorax* are all freshwater forms; in view of their body size (in the macrophage stage), and the striking picture presented by the carnivorous macrostome form, it seems strange that these species have not been detected more often in the field. It may be that the macrostomes are relatively rare in nature, and that the microstomes cannot compete well when captured with populations of their rapidly multiplying "look-alike" congener, *T. pyriformis*. It should be recalled, however, that ordinarily when laboratory cultures of the two species are mixed, the *T. patula* or *T. vorax* or *T. paravorax* immediately becomes carnivorous and, feeding on its prey (the *T. pyriformis* present), transforms into the readily recognizable macrostome form. So the mystery remains.

With the exception of *T. setifera*, only members of the *patula* complex are totally free from involvement in parasitism. Their microstomes are presumably microphages (bacteria feeders) in nature; their macrostomes are macrophages, carnivorous on other ciliates or occasionally cannibalistic. Neither form is particularly histophagous; and attempts to induce parasitism experimentally in such species as *T. patula* and *T. vorax* have met with extremely limited success. [It is important to note that the strains of "*T. vorax*," V_1 and PP, that were used in Thompson's (154) experimental infection work and which quite readily took to a facultatively parasitic way of life, are now considered to have been strains of *T. pyriformis* (104,105,110,114; and p. 37).]

B. Species Exhibiting the Parasitic Habit

Parasitism plays a major role in the bionomics or ecology of many organisms. The species belonging to the genus *Tetrahymena* represent a rather unusual group of forms in that they, or stages in their full life cycles, represent a wider spread of degrees of parasitism than is perhaps known for any other genus of the same (i.e., small) size in the entire animal kingdom (26,34,44a). With respect to the condition obtaining in nature, *Tetrahymena* may be placed into four major categories with respect to parasitism, although two factors must be kept clearly in mind: (1) that it is practically impossible to define the four groupings without recognizing degrees of overlapping, and (2) that some species may fit as well, or nearly as well, in one category as in another. Also, we are concerned here primarily with *natural* conditions; but the fact that under certain *experimental* conditions some species may show a shift in category cannot be ignored.

The categories of parasitism (only *endo*parasitism is under consideration here) may be defined briefly (see also ref. 44a) as:

1. *Free-living forms.* (Admittedly such forms show *zero* degree of parasitism; but it is convenient to include them here in order to present the whole gamut from no to obligate parasitism.) Typically found as nonparasitic in nature; thus essentially obligate free-living forms, although capable of exhibiting "parasitism" in a limited way under experimental conditions (e.g., when inoculated into the body cavity of certain insect larvae).

2. *Facultatively parasitic forms.* Forms generally or often free-living in nature but for which parasitism is (also) a natural and perhaps widespread occurrence in the bionomics of the species; thus the ciliates are capable of living, for no particularly limited period of time, a life of endoparasitism (experimentally as well as naturally).

3. *Facultatively free-living forms.* Forms typically or most often found in association with a host but also capable, in nature as well as experimentally, of thriving for indefinite periods free of the host's body.

4. *Parasitic forms.* Typically obligate parasites, passing at least a major portion of their life cycle within the body of a host, although capable of existence (experimentally) free of the host and surely possessing a stage of the natural life cycle outside the host, although possibly one of short duration.

Briefly, the 10 species of *Tetrahymena* may be distributed among the four categories as indicated below. The accounts are concise, because most of the data presented are already available to the reader in preceding pages of this chapter.

1. Free-living, four species: *T. setifera, T. patula, T. vorax,* and *T. paravorax,* all essentially totally free-living forms in nature, to the best of our present knowledge. It is interesting to note that the last three have a carnivorous stage in their life cycle, and that *T. patula* and *T. vorax* have an encysted stage, although of a reproductive and not resistant kind.

2. Facultatively parasitic, two species: *T. pyriformis,* often found within the body cavity of aquatic invertebrates, such as insect larvae, the infection usually being fatal. Yet the invasion is, with rare exception, purely an accidental, even though naturally occurring, one, and probably takes place via breaks or lesions in the host's integument or, possibly on occasion, via the oral opening. *Tetrahymena rostrata,* an edaphic, soil-moss-litter ciliate, a scavenger, a histophage, (yet also) found parasitic, *not* by accident, in such hosts as land gastropods and enchytraeid worms, via definite portals of entry (none oral), infection being highly pathogenic. The second species exhibits both reproductive and resting cysts, the latter surely related to its ecological niche when free-living.

3. Facultatively free-living, two species: *T. chironomi,* a frequent parasite of chironomid larvae, with entrance into the hemocoel apparently through breaks available during molting, inevitably causing death of the host; little really known about its free-living stages beyond knowledge of viability of the ciliate in culture free of the host. *Tetrahymena corlissi,* the only species found in regular association with (aquatic) vertebrates; in body

lesions or within the circulatory system, especially of injured larval forms, entering via breaks or wounds plus actual attack on exposed tissues, invasion eventually being fatal to the host. Again, there is little knowledge of the ciliate's nonparasitic life, although it is known to be a voracious histophage in laboratory culture and has been found as a bacteria feeder in nature. The only cyst known for either species is the reproductive cyst of *T. corlissi*.

4. Parasitic, two species: *T. stegomyiae*, apparently an obligate parasite of the larvae of certain mosquitoes, with entrance directly from the outside via a peculiar cuticular cyst, an invasion fatal to the host; little known about the free-living stage in the water outside the host, although long-lived in laboratory culture. *Tetrahymena limacis*, very likely an obligate parasite of terrestrial gastropods, with entrance through the mouth, and infection mildly or nonpathogenic; free-living stage, in nature, possibly of short duration, although the organism thrives well in laboratory culture as a histophage or even on a nonparticulate substrate. No cystic stage has ever been discovered (discounting the recent report in ref. 121; see p. 28 for the second species; this is curious, for a resting cyst of some kind would be helpful for preservation during any extended period of time between hosts, especially in more-or-less desiccated habitats.

Further consideration of host–parasite interrelationships involving tetrahymenid ciliates is beyond the scope of this abbreviated account. However, the interesting cases of mixed infections should be recalled; most significant are *T. pyriformis* and *T. chironomi*, often found together (although *T. chironomi* predominates) in the hemocoel of chironomid larvae (26), infection being fatal whether one or both species present; and *T. rostrata* and *T. limacis* in slugs (5), in different locations and with different effects on the host (*T. rostrata*, kidney primary site of infection, highly pathogenic; *T. limacis*, digestive gland, only slightly pathogenic when the sole parasite). The latter case can be even more complicated by the simultaneous presence of a third guest in the slug, the unrelated (trichostome) ciliate *Colpoda steinii*, a harmless accidental transient of the intestinal tract only (5,26,165). Mosquitoes (larval stages) serve as hosts to more than one species of *Tetrahymena*, but mixed ciliate infections, that is, more than one species at the same time, have not yet actually been reported from such insects (26,28).

Corliss (44a) has recently proposed a new hypothesis on the origin of obligate endoparasitism, suggesting that this way of life may, for many organisms, have had its evolutionary origin as the end result of a process in which facultative parasitism played a transitional role between the strictly free-living and the parasitic state. As documented above, the extant species of *Tetrahymena* may be considered to serve as a "living example" of a series of otherwise "missing links" in demonstration of the way in which this may have actually taken place, step by step, in past eons.

V. PHYLOGENY AND EVOLUTION

In spite of a considerable body of information about the comparative morphology, physiology, behavior, and (for *T. pyriformis*) genetics and cytogenetics of these ciliates, we still have little basis for postulating phylogenetic interrelationships within the genus *Tetrahymena*, or for suggesting the evolutionary origin of any given species. Although mere speculation itself would be out of order, there is perhaps value in a brief consideration of some factors that are of probable or potential importance in attempts to determine the evolution of such organisms.

The molecular approach, although it may be the most exciting and eventually the most profitable, can be disposed of first and with the greatest brevity here—simply because we do not yet have enough data to indicate even its potential value in evolutionary considerations at the protozoan level. Base ratio analyses, nucleic hybridization studies, and immunological comparisons all should be exploited, for obvious reasons; yet the meager evidence we have to date appears to reveal that there are tremendous difficulties in obtaining truly meaningful data from such research. There must ultimately be a genetic basis for discontinuities, but discovery and recognition are made very difficult because of limitations in technique and in materials to study, and—perhaps most of all—because of "biological noise." Thus "molecular taxonomy" needs further development before it can support any phylogenetic speculation of significance with respect to the evolution of ciliates, especially at levels *above* that of the species.

Thus, for the time being at least, it appears that one is obliged to fall back on comparative analyses of morphological, ecological, general physiological,* and morphogenetic characteristics in any attempt to "arrange" the species of *Tetrahymena* on a "genealogical tree" or dendrogram. Decisions have to be made concerning the nature of the characters that are best considered "basic," "primitive," or most conservative, and which are secondarily derived, more "advanced," more specialized, less stable, and so on. The age-old problems of parallel evolution, convergence, false homologies, possible loss of "key" characters, necessity of dealing purely with extant forms, and especially *paucity of information* render all "conclusions" reached most tentative indeed (39,44b). In the present situation the possession in common of very strong *generic* properties presents a further difficulty in evaluating the limited diversity observable *within* the genus.

Five groups of characteristics appear to lend themselves to phylogenetic considerations, characters already described for each of the 10 species of *Tetrahymena* treated in previous sections of this chapter:

1. Polymorphism of free-swimming stages in the life cycle: presence, absence, kind.
2. Cyst(s): presence, absence, kind(s).
3. Feeding habits (and ecological preferences): microphagous, macrophagous (carnivorous, histophagous), (endo)parasitic (saprozoic), or some combination of these.

---

*Very little experimental work has been done that might throw light on problems of evolution of these ciliates, although relevant biochemical data have been treated, very briefly, by such workers as Holz (76) and Hill (73). The carefully controlled studies of Williams (162) and Shaw and Williams (144), often overlooked, might be cited as important investigations pertinent to certain aspects of the overall question posed here. Combining physiological and serological approaches, these workers gathered data which clearly suggest, among other implications, that recognition of interrelationships among strains and species of *Tetrahymena* can be very difficult and that conclusions should not be drawn without considerable caution. In the case of the polymorphic *T. vorax*, Williams discovered that three markedly different types of sublines (differing, for example, in macronuclear size, body size, and number of rows of cilia) could be obtained from a single parental clone. He postulated that these subline types were not the result of genic mutation, but rather that they represented alternative epigenetic stable states of a single cellular system. Space does not permit further discussion of this significant research; the reader is urged to consult the original literature.

4. Host–parasite relationships: kind and degree of parasitism (natural and/or experimental), if any exhibited; nature and diversity of host; entry into and effect on host.

5. Selected morphological features, best used in combination: typical number of rows of cilia (in all stages of life cycle, if variable), body sizes and shapes, presence or absence of caudal cilium, micronuclear size and shape.

Although it is always possible that "simpler" forms have been spun off from species with more complex, polymorphic life cycles (such as *T. pyriformis* from *T. vorax* or from *T. rostrata*), there is some logic in concluding that such has not been the case in *Tetrahymena*. Not only is it reasonable, without any evidence to the contrary, to suppose that "simpler" gave rise to "more complex" (i.e., ancestors of present-day *T. pyriformis* preceded both extant *T. pyriformis* and the other species currently with us) but, had the reverse been the case, then a different "simpler type" would have had to exist at one time in order to give rise to the "more complex," which then gave rise to the "simpler" again.

Also, the appearance of *pyriformis*-like forms in the life cycles of species in the "higher" groups might well be interpreted as a kind of ontogenetic "recapitulation" of the past evolutionary histories of the more complex tetrahymenas, following the same sort of reasoning which I have attempted to apply to suprafamilial levels among the Ciliophora (39) and among the Protozoa in general (44b).

Thus it seems not unlikely that the *pyriformis* complex appeared first, with subsequent origin, along quite diverse paths, of the *rostrata* and *patula* groups (probably via *T. vorax* in the latter case). With reference to the five groups of characteristics outlined above, it may be noted that there is generally either absence or simplicity of features in *T. pyriformis* (and presumed close relatives, *T. setifera* and *T. chironomi*) in comparison with the situations generally occurring in the other, more "advanced" complexes. Space permits, unfortunately, only the briefest development of this idea here.

For example, *T. pyriformis* is monomorphic, without cysts, microphagous and ubiquitous, and only accidentally facultatively parasitic (with infection fatal for the host, frequently an insect larva). It has a relatively low number of ciliary rows and no caudal cilium (although one is present in the neighboring species *T. setifera*). The micronucleus, if present at all, is small in diameter, spherical in shape, and possesses a low number of chromosomes.

However, using mostly *T. rostrata* as reasonably representative, one finds in the *rostrata* complex greater changes in body form and size during the full life cycle, both reproductive and resting cysts (although not in all members of the complex), edaphic forms (with emphasis on histophagous feeding tendencies), and both strongly facultative and even obligatory (e.g., *T. limacis*) parasitism (with varying degrees of pathogenicity, differing routes of entry, and diverse and widely distributed host species). There is a higher number of ciliary rows, and a caudal cilium is found in two species of the complex. The micronucleus is generally large and definitely ovoid and, at least in *T. rostrata,* contains a relatively large number of chromosomes.

Finally, in the *patula* complex, there is pronounced dimorphism in the free-swimming stages, with a microphagous microstome form and a carnivorous macrostome form, the latter exhibiting a specially developed cytopharyngeal pouch; there are reproductive

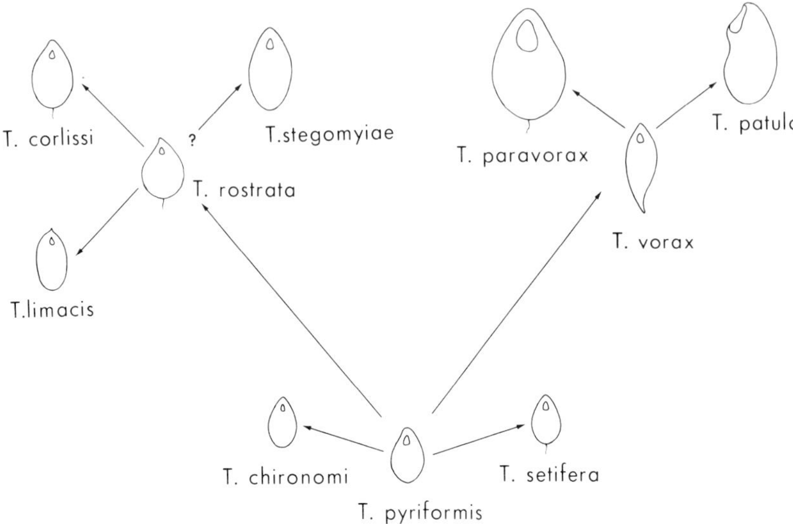

Fig. 47.

Graphic illustration of possible evolutionary pathways within the genus *Tetrahymena*. As discussed in the text, it seems reasonable, until contradictory evidence or hypotheses are available for consideration, that the *pyriformis* complex is the most ancient and that from it arose the *rostrata* and *patula* complexes, with persistence of course of "representative" species of the *pyriformis* complex. Although such conclusions are speculative, they may serve heuristic purposes. As the dendrogram is arranged, the forms on the (reader's) left represent species that engage more or less heavily in the parasitic habit; those on the right are entirely nonparasitic species. Because we are dealing purely with extant forms, it should be clearly noted that any vertical axis in the figure does *not* indicate the geological time scale. Since every line indicated in the figure is highly questionable, it seems gratuitous and superfluous to obscure the drawing with numerous interrogation marks, so only what is probably the *very most* worrisome spot is so indicated.

cysts (only) in the life cycle; and the species are never found as parasites. An increased number of ciliary meridians is evident, although the body size is (also) considerably larger; a caudal cilium exists in one species *(T. paravorax)*. The micronucleus, when present, is spherical.

With reference to the possible evolution of the parasitic habit in the genus, it may be observed that a great deal of plasticity still seems to exist. The degree of parasitism exhibited at the present time by various of the species within the three complexes must have been reached quite independently (44a) and is therefore of limited taxonomic significance. Although all members of the *patula* complex are obligately free-living forms, the species comprising the other two groups show wide ranges in their degrees of parasitism; in the *pyriformis* complex one finds a totally free-living species, a facultatively parasitic one, and a facultatively free-living one; in the *rostrata* complex, a facultatively parasitic species, a facultatively free-living one, and two obligately parasitic forms.

Perhaps some of the forms with a strong predilection for parasitism will evolve into species similar to those which, among ciliates of the family Tetrahymenidae, appear to have reached the ultimate in adaptation to such a life: for example, the mouthless *Curimostoma renalis,* in the renal organ of certain freshwater clams, so ably described and boldly classified by Kozloff (98) nearly 20 years ago.

In Fig. 47, a dendrogram is presented suggesting possible lines of evolution within the genus *Tetrahymena*. I am fully aware of its weaknesses and of many of the unanswered problems involved. Yet the "tree" may at least serve to stimulate more data gathering and to represent a point of departure for the surely more elegant *and* more defensible diagrams of the future.

ACKNOWLEDGMENTS

Support of National Science Foundation grants is gratefully acknowledged. I wish also to thank many "tetrahymenological" colleagues, too numerous to cite, for their critical suggestions, loan of figure material (not all of which could be included), and unpublished data and ideas generously offered over the years. I am particularly indebted to Drs. Howard E. Buhse, Jr., Frank E. G. Cox, Joseph Frankel, George G. Holz, Jr., Eugene N. Kozloff, John B. Loefer, Eugene W. McArdle, and Norman E. Williams for most substantial aid in such respects, including critical reading of preliminary drafts of this chapter. Persons supplying original photomicrographs for my use are directly acknowledged at the end of the appropriate figure explanation. Finally, I am grateful to my daughters Susan and Margaret, and to my good colleague Dr. Jiří Lom, for indispensable help in preparation of the line drawings, and to Mrs. Carol Knox and Mrs. Dorothy Shaffer for their patient typing and retyping of the manuscript in its several forms and revisions.

REFERENCES

1. Allen, R. D. 1967. J. Protozool. 14:553–565.
2. Allen, R. D. 1969. J. Cell Biol. 40:716–733.
3. Allen, S. L. 1967. In M. Florkin and B. J. Scheer, eds., Chemical zoology, vol. I, Protozoa (G. W. Kidder, ed.), Academic Press, New York, pp. 617–694.
4. Barthelmes, D. 1960. Z. Fisch. Hilfswiss. 9:273–280.
5. Brooks, W. M. 1958. Hilgardia 39:205–276.
6. Buhse, H. E., Jr. 1960. J. Protozool. 7(suppl.):9–10.
7. Buhse, H. E., Jr. 1966. J. Protozool. 13:429–435.
8. Buhse, H. E., Jr. 1966. Trans. Amer. Microsc. Soc. 85:305–313.
9. Buhse, H. E., Jr. 1967. J. Protozool. 14:608–613.
10. Buhse, H. E., Jr. and Cameron, I. L. 1968. J. Exp. Zool. 169:229–236.
11. Buhse, H. E., Jr. and Corliss, J. O. 1968. J. Protozool. 15 (suppl.):10
12. Buhse, H. E., Jr. and Corliss, J. O. 1969. Proc. 2nd Ann. Scanning Electron Microsc. Symp., Illinois Institute of Technology Research Institute, Chicago, 169–176.
13. Buhse, H. E., Jr., Corliss, J. O., and Holsen, R. C., Jr. 1970. Trans. Amer. Microsc. Soc. 89:328–336.
14. Buhse, H. E., Jr., and Nicolette, J. A. 1970. J. Protozool. 17(suppl.):12.
15a. Buhse, H. E., Jr., Stamler, S. J., and Corliss, J. O. 1970. Proc. 3rd Ann. Scanning Electron Microsc. Symp., Illinois Institute of Technology Research Institute, Chicago, 171–176.
15b. Buhse, H. E., Jr., Stamler, S. J., and Corliss, J. O. 1973. Trans. Amer. Micros. Soc. 92:95-105.
16. Chen, T.-T., ed. 1967-1972. Research in protozoology, vols. 1–4, Pergamon Press, New York.
17a. Corliss, J. O. 1951. Proc. Amer. Soc. Protozool. 2:11.
17b. Corliss, J. O. 1952. C. R. Acad. Sci. Paris 235:399–402.
18. Corliss, J. O. 1952. Proc. Soc. Protozool. 3:3.
19. Corliss, J. O. 1952. Trans. Amer. Microsc. Soc. 71:159–184.
20. Corliss, J. O. 1953. Stain Technol. 28:97–100.
21. Corliss, J. O. 1953. Parasitology 43:49–87.
22. Corliss, J. O. 1954. J. Protozool. 1:156–169.
23. Corliss, J. O. 1957. J. Protozool. 4(suppl.):13.
24. Corliss, J. O. 1957. J. Protozool. 4(suppl.):15–16.
25. Corliss, J. O. 1959. J. Protozool. 6(suppl.):24

26. Corliss, J. O. 1960. Parasitology 50:111–153.
27. Corliss, J. O. 1961.The ciliated protozoa: Characterization, classification, and guide to the literature. Pergamon Press, New York.
28. Corliss, J. O. 1961. Trans. Roy. Soc. Trop. Med. Hyg. 55:149–152.
29. Corliss, J. O. 1962. Symp. Soc. Gen. Microbiol. 12:37–67. .
30. Corliss, J. O. 1962. J. Protozool. 9:307–324.
31. Corliss, J. O. 1962. Parasitology 52:10P.
32. Corliss, J. O. 1963. Proc. XVI Int. Cong. Zool., Washington, D.C. 4:97–102.
33. Corliss, J. O. 1963. J. Protozool. 10:247–249.
34. Corliss, J. O. 1965. Acta Protozool. 3:1–20. .
35. Corliss, J. O. 1965. Ann. Biol. 4:49–69.
36. Corliss, J. O. 1966. Protistologica 2:5–14.
37. Corliss, J. O. 1967. Bull. Natl. Inst. Sci. India, No. 34:26–33.
38. Corliss, J. O. 1967. J. Protozool. 14:1–8.
39. Corliss, J. O. 1968. Trans. Amer. Microsc. Soc. 87:1–20.
40. Corliss, J. O. 1970. J. Protozool. 17:198–209.
41. Corliss, J. O. 1971. Trans. Amer. Microsc. Soc. 90:243–252.
42. Corliss, J. O. 1971. Trans. Amer. Microsc. Soc. 90:344–362.
43. Corliss, J. O. 1972. Trans. Amer. Microsc. Soc. 91:221–235.
44a. Corliss, J. O. 1972. Trans. Amer. Microsc. Soc. 91:566–573.
44b. Corliss, J. O. 1972. Amer. Zool. 12:739–753.
44c. Corliss, J. O. 1973. Amer. Zool. 13:145-148.
45. Corliss, J. O. and Dougherty, E. C. 1967. Bull. Zool. Nomencl. 24:155–186.
46. Corliss, J. O., Smith, A. C., and Foulkes, J. 1962. Nature 196:1008–1009.
47. Czapik, A. 1968. Acta Protozool. 5:315–357.
48. Dobell, C. 1932. Antony van Leeuwenhoek and his "little animals." Swets and Zeitlinger, Amsterdam.
49. Dragesco, J. 1962. Bull. Microsc. Appl. 11:49–58.
50. Dragesco, J. and Njiné, T. 1971. Ann. Fac. Sci. Cameroun, No. 7–8:97–140.
51. Dysart, M. P. 1959. J. Protozool. 6(suppl.):17–18.
52. Dysart, M. P. 1963. J. Protozool. 10(suppl.):8–9.
53. Dysart, M. P. 1964. Diss. Abstr. 24:3892.
54. Ehrenberg, C. G. 1838. Die Infusionsthierchen als vollkommene Organismen. Leipzig.
55. Elliott, A. M. 1959. Ann. Rev. Microbiol. 13:79–96.
56. Elliott, A. M. 1970. J. Protozool. 17:162–168.
57. Elliott, A. M. and Hayes, R. E. 1954. Biol. Bull. 107:309.
58. Elliott, A. M. and Nanney, D. L. 1952. Science 116:33–34.
59. Evans, F. R. and Corliss, J. O. 1964. J. Protozool. 11:353–370.
60. Fauré-Fremiet, E. 1948. Arch. Anat. Microsc. Morphol. Exp. 37:183–203.
61. Fauré-Fremiet, E. 1949. Belg.-Nederl. Cyto-embryol. Dagen., Gent, 100–102.
62. Fauré-Fremiet, E. 1967. In M. Florkin and B. J. Scheer, eds., Chemical zoology, vol. I, Protozoa (G. W. Kidder, ed.), Academic Press, New York, pp. 21–54.
63. Forer, A., Nilsson, J. R., and Zeuthen, E. 1970. C. R. Trav. Lab. Carlsberg 38:67–86.
64. Frankel, J. 1964. J. Protozool. 11:514–526.
65. Frankel, J. and Heckmann, K. 1968. Trans. Amer. Microsc. Soc. 87:317–321.
66. Furgason, W. H. 1940. Arch. Protistenk. 94:224–266.
67. Galigher, A. E. and Kozloff, E. N. 1971. Essentials of practical microtechnique. 2nd ed. Lea and Febiger, Philadelphia.
68. Gelei, J. von. 1935. Arch. Protistenk. 85:289–302.
69. Gellért, J. 1956. Acta Biol. Acad. Sci. Hung. 6:337–359.
70. Groh, K. R. and Buhse, H. E., Jr. 1968. J. Protozool. 15(suppl.):9.
71. Hall, R. P. 1941. In G. N. Calkins and F. M. Summers, eds., Protozoa in biological research, Columbia University Press, New York, pp. 475–516.
72. Harman, W. J. and Corliss, J. O. 1956. Trans. Amer. Microsc. Soc. 75:332–333.
73. Hill, D. L. 1972. The biochemistry and physiology of *Tetrahymena.* Academic Press, New York.

74. Holz, G. G., Jr. 1955. Biol. Bull. 109:306.
75. Holz, G. G., Jr. 1964. In S. H. Hutner, ed., Biochemistry and physiology of Protozoa, vol. III, Academic Press, New York, pp. 199–242.
76. Holz, G. G., Jr., 1966. J. Protozool. 13:1–4.
77. Holz, G. G., Jr., and Corliss, J. O. 1956. J. Protozool. 3:112–118.
78. Holz, G. G., Jr., Erwin, J. A., and Davis, R. J. 1959. J. Protozool. 6:149–156.
79. Holz, G. G., Jr., Erwin, J. A., and Wagner, B. 1961. J. Protozool. 8:297–300.
80. Holz, G. G., Jr., Erwin, J. A., Wagner, B., and Rosenbaum, N. 1962. J. Protozool. 9:359–363.
81. Horn, G. F. 1951. M. S. Thesis, University of Washington.
82. Hurst, D. D. 1957. J. Protozool. 4(suppl.):18.
83. Hutner, S. H., ed. 1964. Biochemistry and physiology of Protozoa, vol. III, Academic Press, New York.
84. Hutner, S. H., Baker, H., Frank, O., and Cox, D. 1972. In R. N. Fiennes, ed., Biology of nutrition, Pergamon Press, New York, pp. 85–177.
85. International code of zoological nomenclature adopted by the XV International Congress of Zoology, London, July 1958. 1964. 2nd ed. London.
86. Jahn, T. L. and Bovee, E. C. 1967. Bull. Natl. Inst. Sci. India, No. 34:1–12.
87. Jankowski, A. W. 1967. Zool. Zh. 46:17–23.
88. Kahl, A. 1926. Arch. Protistenk. 55:197–438.
89. Kahl, A. 1930–1935. In F. Dahl, ed., Die Tierwelt Deutschlands, G. Fischer, Jena, pp. 1–886.
90. Kazubski, S. L. 1958. Bull. Acad. Polon. Sci. 6:247–252.
91. Keilin, D. 1921. Parasitiology 13:216–224.
92. Kidder, G. W. 1941. In G. N. Calkins and F. M. Summers, ed., Protozoa in biological research, Columbia University Press, New York, pp. 448–474.
93. Kidder, G. W. 1951. Ann. Rev. Microbiol. 5:139–156.
94. Kidder, G. W., ed. 1967. Protozoa, vol. I of Chemical zoology (M. Florkin and B. J. Scheer, eds.), Academic Press, New York.
95. Kidder, G. W., Lilly, D. M., and Claff, C. L. 1940. Biol. Bull. 78:9–23.
96. Kirby, H. 1941. In G. N. Calkins and F. M. Summers, eds., Protozoa in biological research, Columbia University Press, New York, pp. 1009–1113.
97. Kozloff, E. N. 1946. J. Morphol. 79:445–465.
98. Kozloff, E. N. 1954. J. Protozool. 1:200–206.
99. Kozloff, E. N. 1956. J. Protozool. 3:17–19.
100. Kozloff, E. N. 1956. J. Protozool. 3:20–28.
101. Kozloff, E. N. 1956. J. Protozool. 3:204–208.
102. Kozloff, E. N. 1957. J. Protozool. 4:75–79.
103. Loefer, J. B. 1952. J. Morphol. 90:407–414.
104. Loefer, J. B. 1958. Proc. XV Int. Congr. Zool., London, 126–127.
105. Loefer, J. B. 1967. Bull. Natl. Inst. Sci. India, No. 34:34–47.
106. Loefer, J. B., Owen, R. D., and Christensen, E. 1958. J. Protozool. 5:209–217.
107. Loefer, J. B. and Scherbaum, O. H. 1963. Syst. Zool. 12:175–177.
108. Loefer, J. B., Small, E. B., and Furgason, W. H. 1960. C. R. X Cong. Int. Biol. Cell., Paris, 229–230.
109. Loefer, J. B., Small, E. B., and Furgason, W. H. In Progress in protozoology, International Congress Series No. 91, 2nd International Conference on Protozoology, Excerpta Medica Foundation, London, p. 218.
110. Loefer, J. B., Small, E. B., and Furgason, W. H. 1966. J. Protozool. 13:90–102.
111. Lwoff, A. 1923. C. R. Acad. Sci., Paris 176:928–930.
112a. Lwoff, A. 1932. Recherches biochimiques sur la nutrition des protozoaires. Le pouvoir synthèse. Monogr. Institut Pasteur. Masson, Paris.
112b. Lwoff, A. 1943. L. Évolution physiologique: Étude des pertes de fonctions chez les microorganismes. Hermann, Paris.
113. Mandel, M. 1967. In M. Florkin and B. J. Scheer, eds., Chemical zoology, vol. I (G. W. Kidder, ed.), Academic Press, New York, pp. 541–572.
114. Maupas, E. 1883. Arch. Zool. Exp. Gén., Sér. 2, 1:427–664.

115. Maupas, E. 1888. Arch. Zool. Exp. Gén.,Sér. 2, 6:165–277.
116. Maupas, E. 1889. Arch. Zool. Exp. Gén.,Sér. 2, 7:149–517.
117. McArdle, E. W. 1959. J. Protozool. 6(suppl.): 27.
118. McArdle, E. W. 1960. Diss. Abstr. 20:4760.
119. McArdle, E. W. 1960. J. Protozool. 7(suppl.):17.
120. McArdle, E. W. 1960. J. Protozool. 7(suppl.):23.
121. Michelson, E. H. 1971. Parasitology 62:125–131.
122. Mugard, H. 1949. Ann. Sci. Nat. Zool.,Sér. 11, 10:171–268.
123. Müller, O. F. 1773. Vermium terrestrium et fluviatilium, seu animalium infusorium, helminthicorum et testaceorum, non marinorum, succincta historia. Havniae et Lipsiae.
124. Müller, O. F. 1786. Animalcula infusoria fluviatilia et marina. Havniae et Lipsiae.
125. Muspratt, J. 1945. J. Entomol. Soc. S. Afr. 8:13–20.
126. Muspratt, J. 1947. Parasitology 38:107–110.
127. Nanney, D. L. 1966. Amer. Nat. 100:303–318.
128. Nanney, D. L. 1966. J. Protozool. 13:483–490.
129. Nanney, D. L. 1968. J. Protozool. 15:109–113.
130. Nanney, D. L. 1968. Science 160:496–502.
131a. Nanney, D. L. 1971. J. Protozool. 18:33–37.
131b. Nanney, D. L. 1972. Ann. N.Y. Acad. Sci. 193:14–28.
132. Nicolette, J. A., Buhse, H. E., Jr., and Robin, M. S. 1971. J. Protozool. 18:87–90.
133. Nigrelli, R. F., Jakowska, S., and Padnos, M. 1956. J. Protozool. 3(suppl.):10.
134. Nilsson, J. R. and Williams, N. E. 1966. C. R. Trav. Lab. Carlsberg 35:119–141.
135. Noland, L. E. and Gojdics, M. 1967. In. T.-T. Chen, ed., Research in protozoology, vol. 2, Pergamon Press, New York, pp. 215–266.
136. Opinion 915 (of the ICZN) 1970. Bull. Zool. Nomencl. 27:33–38.
137. Penard, E. 1922. Études sur les infusoires d'eau douce. Georg, Geneva.
138. Pitelka, D. R. 1963. Electron-microscopic structure of protozoa. Pergamon Press, New York.
139. Pitelka, D. R. 1969. In T-T. Chen, ed., Research in protozoology, vol. 3, Pergamon Press, New York, pp. 279–388.
140. Ray, C., Jr. 1956. J. Protozool. 3:88–96.
141. Roque, M., de Puytorac, P., and Savoie, A. 1970. J. Protozool. 17(suppl.):37.
142. Roque, M., de Puytorac, P., and Savoie, A. 1971. Protistologica 6 (year 1970):343–351.
143a. Roux, J. 1901. Fauna infusorienne des eaux stagnantes des environs de Genève. Kündig, Geneva.
143b. Ruffolo, J. J., Jr. and Frankel, J. 1972. J. Protozool. 19 (suppl.):21.
144. Shaw, R. F. and Williams, N. E. 1963. J. Protozool. 10:486–491.
145. Small, E. B. 1967. Trans. Amer. Micros. Soc. 86:345–370.
146. Small, E. B. and Marszalek, D. S. 1969. Science 163:1064–1065.
147. Small, E. B., Marszalek, D. S., and Antipa, G. A. 1971. Trans. Amer. Microsc. Soc. 90:283–294.
148. Sonneborn, T. M. 1957. In E. Mayr, ed., The species problem, AAAS, Washington, D.C., pp. 155–324.
149. Sonneborn, T. M. 1964. Proc. Natl. Acad. Sci. 51:915–929.
150. Speidel, C. C. 1958. Biol. Bull. 115:366.
151. Stone, G. E. 1963. J. Protozool. 10:74–80.
152. Stout, J. D. 1954. J. Protozool. 1:211–215.
153. Thompson, J. C., Jr. 1955. J. Protozool. 2(suppl.):12.
154. Thompson, J. C., Jr. 1958. J. Protozool. 5:203–205.
155. Thompson, J. C., Jr. 1958. Va. J. Sci. 9:315–318.
156. Treillard, M. and Lwoff, A. 1924. C. R. Acad. Sci., Paris 178:1761–1764.
157. Tuffrau, M. 1967. Protistologica 3:91–98.
158. Warren, E. 1932. Ann. Natal Mus. 7:1–53.
159. Watson, J. M. 1946. J. Trop. Med. Hyg. 49:44–46.
160. Wenyon, C. M. 1926. Protozoology, vols. I and II. Baillière, Tindall and Cox, London.
161. Williams, N. E. 1960. J. Protozool. 7:10–17.
162. Williams, N. E. 1961. J. Protozool. 8:403–410.

163. Williams, N. E. 1964. J. Protozool. 11:566–572.
164. Williams, N. E. and Luft, J. H. 1968. J. Ultrastruct. Res. 25:271–292.
165. Windsor, D. A. 1959. J. Protozool. 6(suppl.):33.
166. Windsor, D. A. 1960. J. Protozool. 7(suppl.):27.
167. Zebrun, W., Corliss, J. O., and Lom, J. 1967. Trans. Amer. Microsc. Soc. 86:28–36.

CHAPTER 2

Morphology of *Tetrahymena*

Alfred M. Elliott

Department of Zoology
University of Michigan
Ann Arbor, Michigan

John R. Kennedy

Department of Zoology
University of Tennessee
Knoxville, Tennessee

I. INTRODUCTION

Most of the research on the morphology of *Tetrahymena* is based on growing, probably logarithmic-phase, cells. The earlier light microscope observations revealed the general morphology, but it was not until ultrastructural studies were possible that the details became known. Even with this information, the identity and function of several structures remains a mystery. In this chapter we attempt to describe our present knowledge, drawn from both light and electron microscopy, of the morphology of *Tetrahymena*, which is largely that of *T. pyriformis*.

Tetrahymena is a pyriform-shaped and uniformly ciliated cell (Fig. 1). Ciliary rows (primary meridians) number from 17 to 42 and alternate with secondary meridians of mucocyst pores. The oral area is located close to the anterior pole of the cell (Figs. 1 and 2). A centrally positioned oval macronucleus (Fig. 2) is always present, but the micronucleus (Fig. 57) may be absent from some strains. A contractile vacuole is found in the posterior part of the cell (Fig. 27). *Tetrahymena* ranges in length from 28 to 250 μm depending on the species (for size variation in *T. pyriformis*, see Chapter 9).

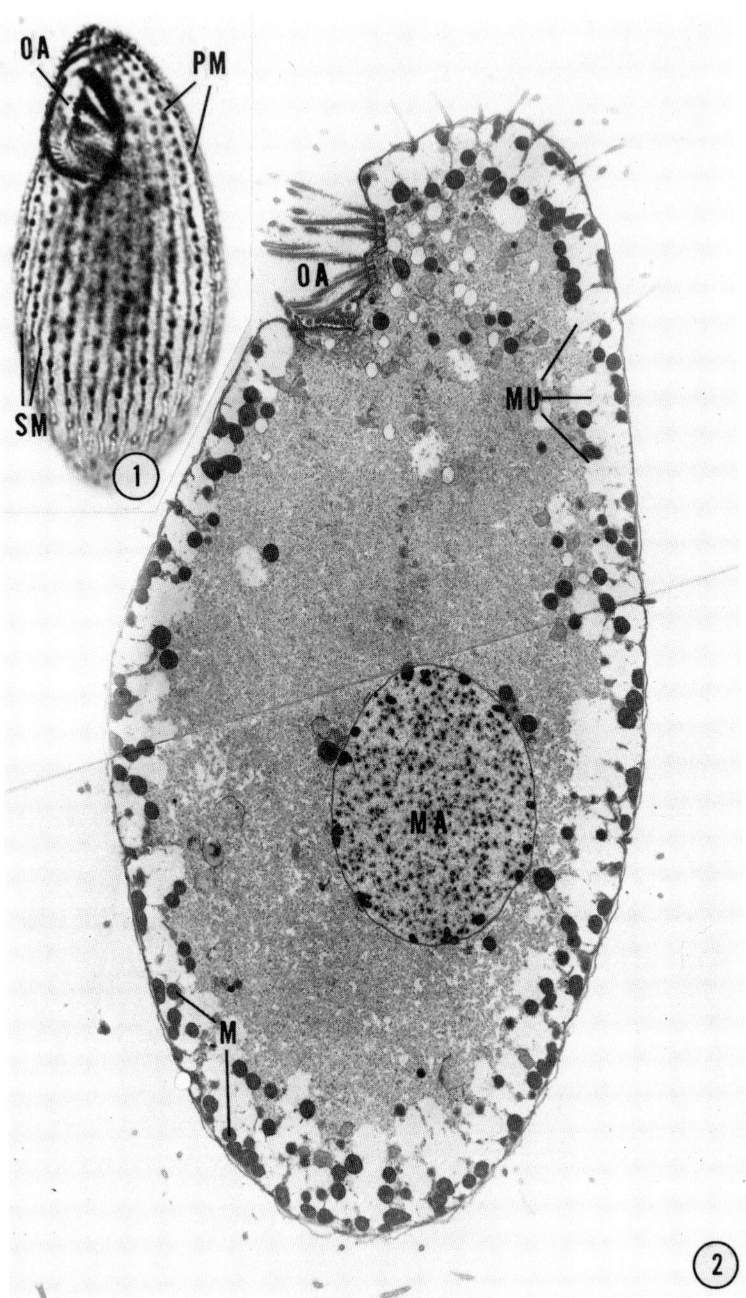

Fig. 1.

A Chatton–Lwoff silver stain of a cell from a logarithmic culture. The oral area (OA) is in the anterior quarter of the cell. The dark granules of the primary meridians (PM) represent sites of cilia. Alternating with them are secondary meridians (SM) composed of mucocyst pores. ×480.

Fig. 2.

A longitudinal section through a stationary-phase cell (strain GL). The oral area (OA) is in the anterior of the cell. The macronucleus (MA) is centrally located. Peripherally, the mucocysts (MU) and mitochondria (M) are indicated. ×1500.

II. CORTICAL STRUCTURES

A. The Cortex

The cortical pattern, its chemistry, and its development have been areas of major interest in *Tetrahymena*. Thus there have been several in-depth studies of this region's fine structure (1,3,56,74,89,94). In considering the cortical architecture, it is evident that the pattern is based on the repetition of a single unit. This repeating unit is the ciliary–kinetosome complex (Figs. 1, 3, and 4a) and associated microtubules. While the structure of its components has been presented in many studies, the entire unit has only recently been examined in detail (64). Here we attempt to bring together the information on the organization of the ciliary–kinetosome (basal body) complex.

The ciliary filaments of *Tetrahymena* arise from the kinetosome in conjunction with a proliferation of the outer cell membrane. Thus the nine pairs of peripheral ciliary filaments are continuous with components of the kinetosome wall (Fig. 4). The central pair terminates at the axosome near the cell surface (Fig. 4). In addition to these components common to nearly all cilia, Gibbons (36,37) observed substructure in the cilia of *Tetrahymena*. Use of the Markham rotation technique has further clarified this substructure (2). A sheath encloses the central pair of ciliary filaments and an adjacent midfiber (Fig. 5). Also present are secondary fibers connected with subfibril A of the peripheral filament (the unit closest to the center of the cilium) by radially placed dense material. These connections are more evident in longitudinal sections (Fig. 4). Extending from subfibril A are paired "arms" recently demonstrated by Gibbons (36–38) to be sites of ciliary ATPase.

At the junction of the cilium and kinetosome, certain structural changes are evident. The central pair of ciliary filaments terminates on the axosome, an amorphous dense-staining granule. This granule rests upon an apparently unstructured thin membrane (Fig. 4). At this level the peripheral ciliary filaments are still paired and are joined by an amorphous band. Immediately below the axosome a dense-staining septum, at the level of the circumciliary ring (64), has the structure of two series of pentagons around a central ring and is surrounded by the peripheral filaments of the kinetosome (Fig. 6). At this point the beginning of formation of the nine triplet fibers is first apparent with subfibril C arising from subfibril B as an incomplete microtubule open to the inner portion of the kinetosome (Fig. 6). Each triplet shows an amorphous connection with the cell surface, which probably represents close binding of the kinetosome to the cell membrane. This no doubt accounts for Hoffman's (41) failure to obtain membrane-free kinetosomes. As subfibril C progresses inward, its inner edge makes contact with subfibril B to form the nine sets of triplets characteristic of kinetosomes and centrioles. Connection of subfibril C with subfibril A of the adjacent triplet can also be seen at this level (Figs. 7 and 8). The presence of subfibril C on only a third of the basal body as reported by Wolfe (102) may reflect a loss through disruption of that portion of subfibril C not closely bound to the adjacent subfibrils. The central portion of the kinetosome contains granular material extending from the terminal plate to near the proximal end of the kinetosome. The kinetosome terminates proximally in a cartwheel-shaped structure. Radiating from subfibril A to a central hub are spokelike fibers. Just internal to subfibril A, a secondary granule or filament is also evident (Fig. 8). Critical

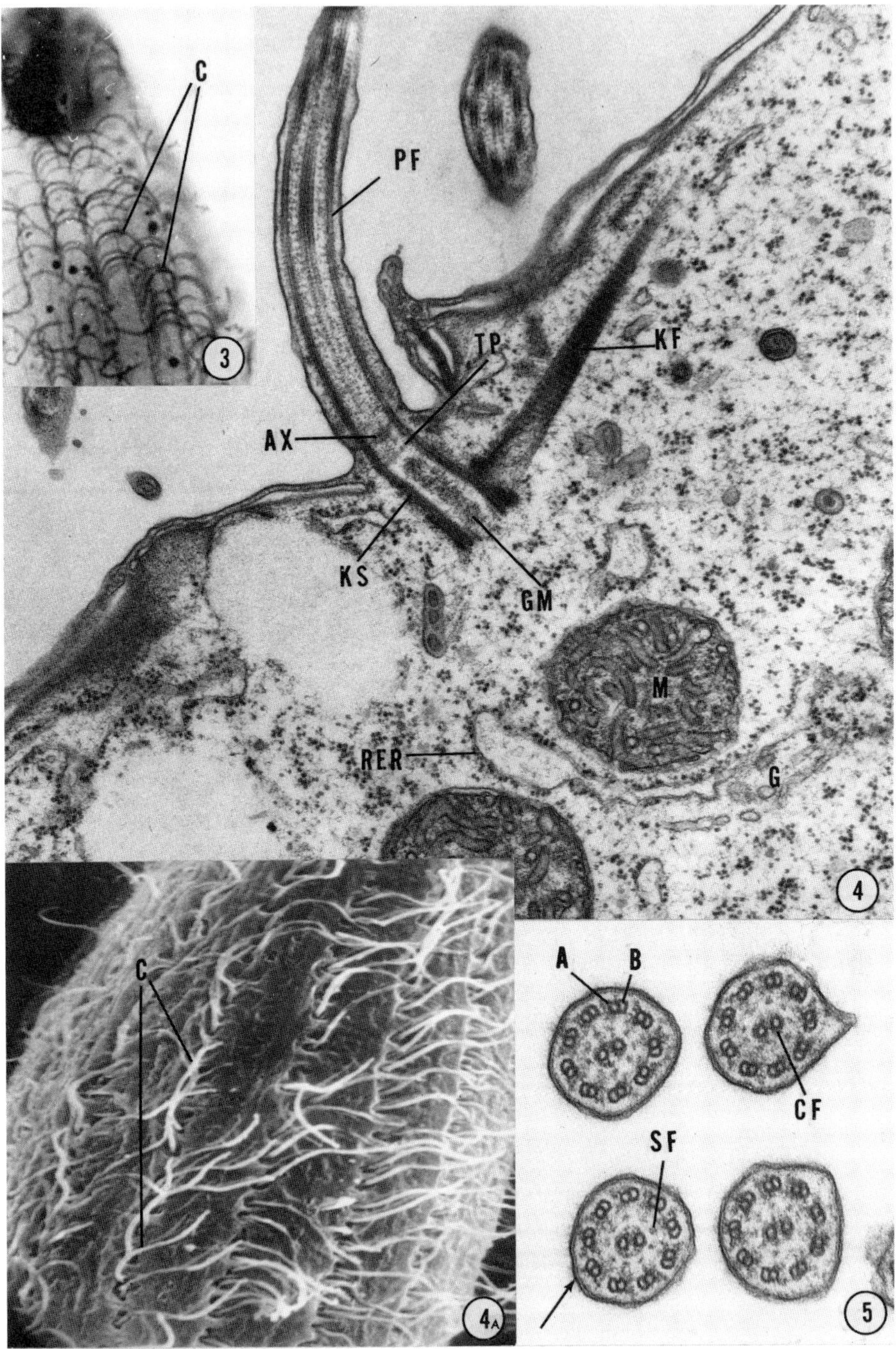

Fig. 3.

A segment of the cell (Grebecki stain) shows the pattern of cilia (C) as they project from the surface. ×1200.

measurements of the components of basal bodies in *Tetrahymena* have recently been made (64).

Various fibers and tubules are associated with each kinetosome. The kinetodesmal fiber extends from the upper right quadrant in the lower third of the kinetosome (Figs. 4,7, and 9). It arises from the kinetosome through three or four square or slightly rhomboid structures and short fibrils attached directly to the kinetosome triplets (1). The major periodicity of the kinetodesmal fiber has been estimated to be from 26 nm (1) in thin-section material to 32 nm in negative-stained whole mounts (64). The fiber extends anteriorly for two or three kinetosomes without making direct contact with adjacent fibers (Fig. 9). A ring of amorphous material is continuous with the kinetodesmal fibers and encloses the lower portion of the kinetosome. Arising from the kinetosome's anterior portion adjacent to Allen's (1) triplet fiber 6, four to ten transverse microtubules arch toward the surface of the cell and extend toward the kinety on the left (Figs. 9–11). Arising from the kinetosome's posterior portion between triplet fibers 1 and 2, five to eight postciliary microtubules (Figs. 7 and 8) arch posteriorly to the surface. Both sets of microtubules measure about 24 nm in diameter. This system of fibers, microtubules, and associated kinetosome with its cilium constitute the functional unit of the cortex. Its organization into a continuous series with other cortical components comprises the kineties of the organism, which make up the primary meridians of the cortical pattern.

Tetrahymena is enclosed in a system of membranes closely corresponding to those described for *Paramecium* (76); and, as in the latter, all show the unit membrane structure (94,100). The outer limiting membrane [outer pellicular membrane of Tokuyasu and Scherbaum (94)] corresponds to the cell membrane of *Paramecium* (76). It is continuous with the ciliary membrane as described above. Immediately beneath the cell membrane and separated by 120–125 Å, a second membrane [inner pellicular membrane (94)] corresponds to the outer alveolar membrane of *Paramecium* (76). These two membranes are joined by a series of cross bridges (94), no doubt accounting for their constant spacing. The outer alveolar membrane is continuous with a third or inner alveolar membrane [cytoplasmic membrane (94)]. The space formed by the junction of these two (Figs. 10 and 19) has been termed the alveolar sac (76). In well-fixed material secondary

---

Fig. 4.

A longitudinal section through a cilium shows the peripheral ciliary filaments (PF) as they extend into the cell to form the kinetosome (KS). Also visible are the axosome (AX), terminal plate (TP), kinetodesmal fiber (KF), and granular matrix (GM) of the kinetosome. In the cytoplasm are mitochondria (M), Golgi (G), and rough ER (RER). ×38,000.

Fig. 4a.

The distribution of cilia (C) in the primary meridians on the surface of a cell can be seen in this scanning electron micrograph. Also evident are depressions in the cell surface from which the cilia project. A section through the depression can be seen in Fig. 4. ×3800.

Fig. 5.

This cross section through a group of cilia shows the unit membrane nature of the ciliary sheath (arrow). Also evident are the paired peripheral subfibrils (A, B), central ciliary filaments (CF), and secondary fibers (SF). ×75,000.

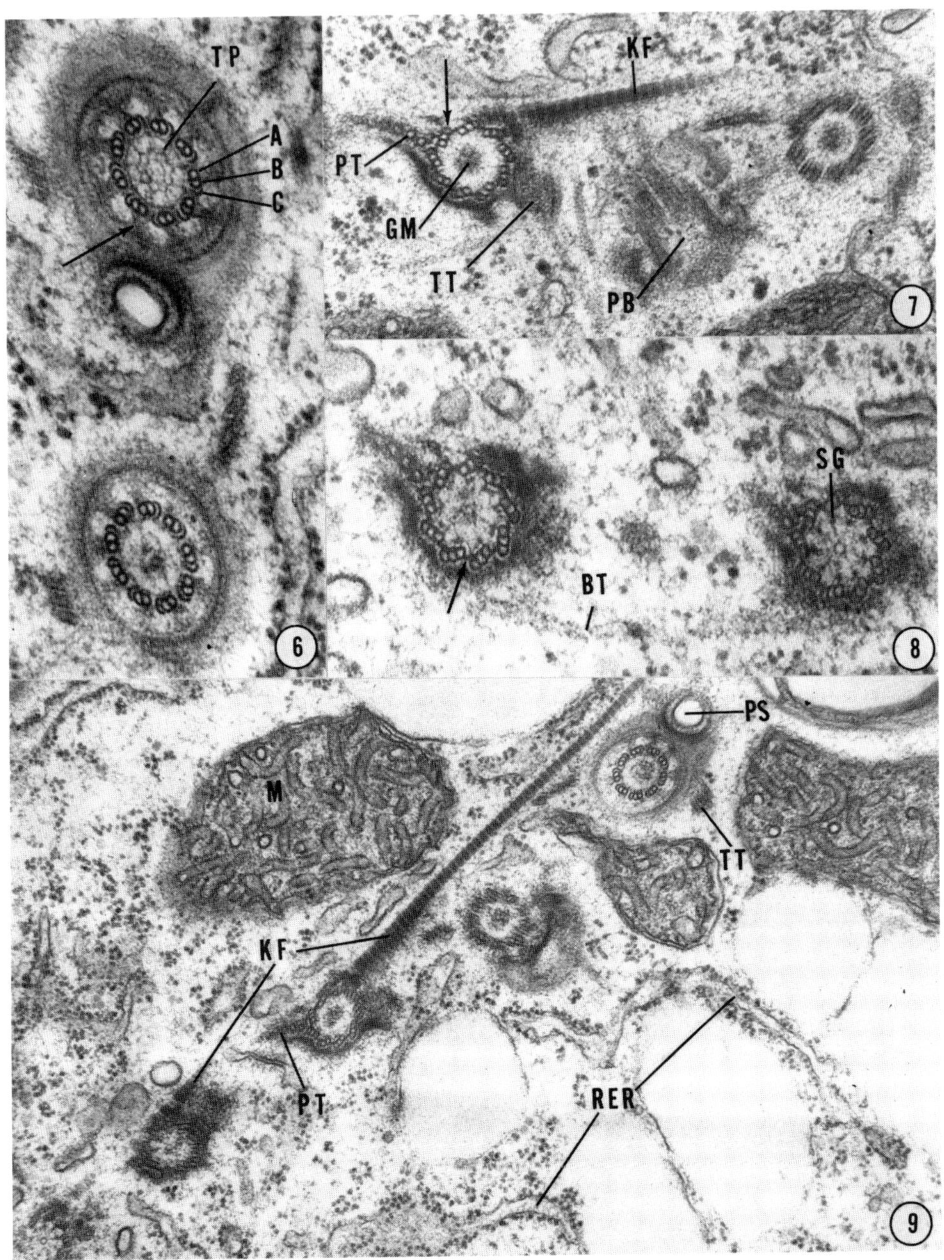

Fig. 6.

The structure of the terminal plate (TP) and peripheral filaments of the kinetosomes is shown in this section. Subfibrils A (A), B (B), incomplete subfibril C (C), and the amorphous connections to the cell surface (arrow) are also evident. ×73,000.

Fig. 7.

In the lower third of the kinetosome, subfibril C is complete (arrow), and the kinetodesmal fiber (KF) extends anteriorly. Both the transverse (TT) and posterior (PT) tubules are evident as is the granular material (GM) in the center of the kinetosome. A probasal body (PB) can also be seen. ×54,000.

cross-connections between the outer and inner alveolar membranes can be seen to bind the cell membrane closely to the cortical cytoplasm (Fig. 10). Thus alveolar size is probably considerably reduced over that previously described for both *Tetrahymena* (1,94) and *Paramecium* (76). The distribution of the alveolar sacs can be seen in Fig. 12 of the surface of *T. pyriformis* after freeze-etching. An amorphous (ectoplasmic) layer, the granular layer of *Paramecium* (76), underlies the inner alveolar membrane (Fig. 19).

Although there is substantial variation in the number of primary meridians in various strains of *T. pyriformis* (55), the organization of the kineties remains fairly constant at the ultrastructural level. Several investigators have considered this organization (1,56,74,75), the most recent detailed description being that of Allen (1).

The kineties are rows of repeating ciliary–kinetosome units arranged in anterior-posterior direction over the animal's surface (Figs. 1, 3, 4a, 9, and 10). To the right of each kinety, a layer of longitudinal microtubules runs anterior-posteriorly between the inner alveolar membrane and the amorphous ectoplasmic layer (Fig. 10). They have been described as numbering from 7 to 12 (1) and 8 to 14 (74). In strain GL of *T. pyriformis*, however, we have observed from as few as 4 to more than 20 microtubules in one layer. This range variation may represent a strain characteristic. Variation within a given strain of *T. pyriformis*, however, probably represents a difference in tubule number at different sites along the cell surface. The greatest number of longitudinal microtubules may be found in the midregion of the cell. They seem to terminate at varying intervals as they approach the poles of the cell. Thus, at a given level adjacent ciliary rows seem to contain comparable numbers of longitudinal microtubules (Fig. 10). The transverse microtubules extend under the ectoplasmic layer and terminate beneath the longitudinal tubules of the adjacent kinety. However, the posterior microtubules terminate beneath the longitudinal fibers of the corresponding row (1). The kinetodesmal fibers extend anteriorly without making direct contact with each other. Another longitudinal series of two or three microtubules, the basal microtubules have been described as "lying against the dense material at the proximal end of the basal body (kinetosome)" (1) with at least one microtubule continuous with two kinetosomes (Fig. 8).

Allen (1) considers the possibility that the basal microtubules may be the conducting mechanism with regard to ciliary coordination and discusses its merits and disadvantages. We find that under optimal conditions of fixation other longitudinal microtubules can be seen in the cytoplasm of *Tetrahymena*, which are not associated with the kinetosomes.

Fig. 8.

The nine triplets of the kinetosomes are surrounded by an amorphous ring of material. A fiber connects subfibril A to subfibril C (arrow). The cartwheel-shaped structure at the base of the kinetosome is also evident, as are the secondary granules (SG) adjacent to subfibril A. A segment of the basal tubule (BT) is also present. ×73,000.

Fig. 9.

A parasomal sac (PS) and kinetosomes of a single primary meridian are seen here at different levels. A kinetodesmal fiber (KF) extends anteriorly for several kinetosomes. Its periodicity is distinct. Also present are mitochondria (M), rough ER (RER), and posterior (PT) and transverse (TT) tubules. ×38,000.

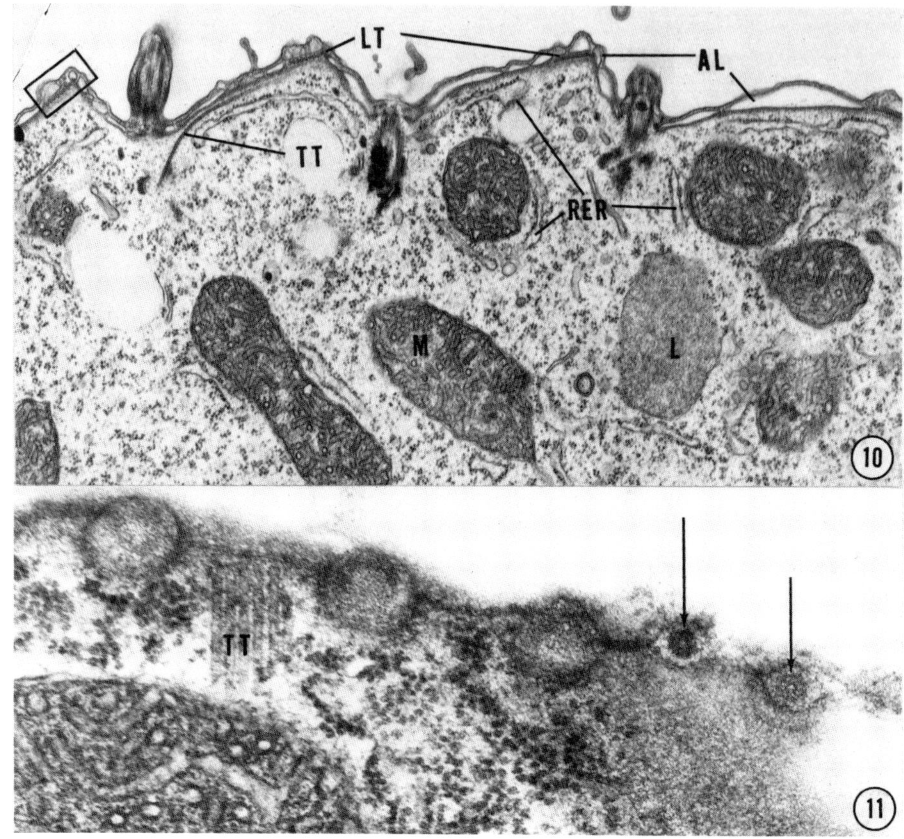

Fig. 10.

This section through a portion of the cell surface shows the organization of the alveolar sac (AL) around three primary meridians. A transverse tubule (TT) and four groups of longitudinal tubules (LT) are evident, as are numerous mitochondria (M), rough ER (RER) and a lysosome (L). ×16,000.

Fig. 11.

A tangential section through a secondary meridian showing the ring structure of the mucocyst pore (arrows). Also present are five transverse tubules (TT). ×54,000.

In addition, a close association between the other cortical fibers and the ciliary–kinetosome complex is indicated by the fact that, when cells disrupt, the cortical network separates as a unit (72,74). Therefore it does not seem likely that the basal microtubules function in ciliary coordination more than other microtubules of the cortical system. In fact, in reviewing the material on cortical function in ciliates, Jahn and Bovee (44) indicate that little basis exists for attributing a neuromotor function to fibrillar systems in general. Kennedy and Zimmerman (49) found that exposure of *Tetrahymena* to high hydrostatic pressure (7500 and 10,000 psi) causes the longitudinal microtubules to break down (depolymerize). This depolymerization appears to begin in the posterior of the cell and progress anteriorly. Associated with this disruption of microtubules is a rounding of the cell in the posterior region. This study is in agreement with the hypothesis of Allen

Fig. 12.

The distribution of the alveolar sacs over a portion of the surface of *T. pyriformis* can be seen in this freeze-etched preparation. Mucocyst pores (arrows) are indicated. ×16,000.

(1) that the fibers and microtubules of *Tetrahymena* have a role in cellular support and morphogenesis (form stabilization).

Certain openings are visible in the cortex of *Tetrahymena*. Anterior to each kinetosome is the small frequently observed parasomal sac (Figs. 6 and 9) of unknown function. The outside openings of the mucocysts are located between the cilia along the primary meridians and also make up the secondary meridians (Fig. 11). They are circular (0.1 μm in diameter) structures with "an outer light ring and a smaller inner dense ring with a tiny electron transparent point in the center" (1). The contractile vacuole pores, depressions in the cortical meridians and usually two in number (Fig. 27), are located in the posterior region of *Tetrahymena*. These are described in Section III,d. Variation in position and number of these pores is discussed by Nanney (65).

The nature of the chemical composition of cilia and kinetosomes has been more elusive than their morphology. With the development of ciliary isolation techniques (11,12,96) for *Tetrahymena*, however, information is rapidly accumulating. Analysis of cilia have indicated that the bulk of the isolated material is protein (11). This protein has been described as ethanol-soluble and having a distinctive acid composition. It accounts for 80% of the weight of the fraction and has a large content of aspartic and glutamic acids and alanine. There is also a small content of proline and methionine (97). Much of the amino acid content is of a very general type (16,95), however. Gibbons (37) found that about one-half of the ciliary protein came from the axoneme, and the membrane and soluble matrix accounted for about one-fourth each. Several investigators (36,97)

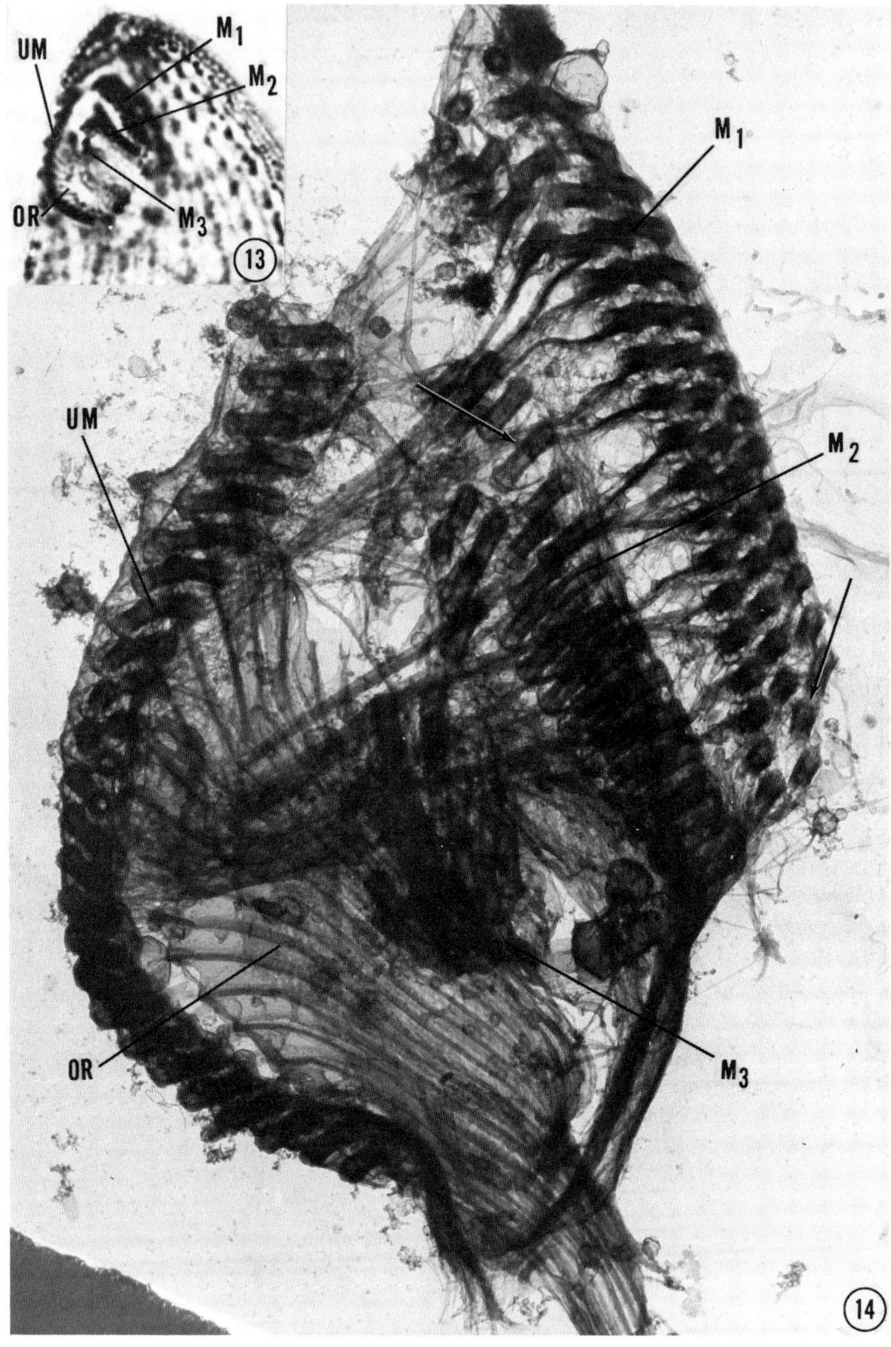
UM
M1
M2
OR
M3
13
UM
M1
M2
M3
OR
14

have described multiple boundaries on ultracentrifugation of ciliary extracts, suggesting the existence of several ciliary proteins. One of these, dynen, has been ascribed to the arms of subfibril A and is high in ATPase activity (36–38). Renaud et al. (78) demonstrated that the outer ciliary fiber protein in *Tetrahymena* has a sedimentation coefficient of 6 S. This 6 S particle is probably a dimer formed by two 55,000-molecular-weight polypeptide chains. The amino acid composition of this protein resembles that of actin, with variation being comparable to variation between actin from different species. By employing antiserum to ciliary microtubule protein, Tamura (92) demonstrated that common proteins are shared by the microtubules of cilia and the oral apparatus. He also suggests that macronuclear microtubules may contain some of this protein. Carbohydrate content has been reported to vary from 0.6–1.2% (95) to 2–6% (16). In addition, Child (11) reported 2.7% nucleic acid content in cilia in the form of adenine and uracil nucleotides. Watson and Hopkins (95) and Culbertson (16) found nucleic acid content only in the range of 0.2–0.6%, however. Thus, although considerable work has been performed on this chemically complex organelle, much more must be done before it is possible to fully relate structure to function.

B. The Mouth

The oral area of *Tetrahymena* (Fig. 13) is located in the anterior quarter of the animal (5,6,14,35). Its position denotes the ventral surface of the cell. Along the right oral margin is the undulating membrane from which a series of oral ridges pass inward to the cytostome–cytopharyngeal complex. The left side of the mouth contains three ciliary membranelles (M_1, M_2, and M_3) of varying length (Figs. 13 and 14). The cortical repeating unit, the ciliary–kinetosome complex and associated tubules described above, is the basic component of the oral area in *Tetrahymena*. Thus primary attention is given to their organization and modification to form the mature mouth.

The undulating membrane (Fig. 15) is composed of a ciliated and a nonciliated row (23,57,71), each with 27 to 30 kinetosomes (30). In Fig. 14, 25 kinetosomes can be counted in a single row. In both *T. patula* (57) and *T. pyriformis* (71), the outer kinetosome row is ciliated. Fibrous and tubular interconnections, similar to those described in other ciliates, are evident between adjacent rows of kinetosomes in both the undulating membrane and membranelles (Figs. 14, 15, and 17). Internal to each nonciliated kinetosome is

---

Fig. 13.

The three membranelles (M_1, M_2, M_3), the undulating membrane (UM), and oral ridges (OR) are demonstrated in this Chatton-Lwoff silver stain of the oral area. ×525.

Fig. 14.

This whole-mount preparation of the oral area of *Tetrahymena* was prepared by spreading homogenized cells on a Langmuir trough and drying by the critical-point method. Kinetosomes and interconnecting microtubules are clearly visible, as is the organization of the three membranelles (M_1, M_2, M_3), the undulating membrane (UM), and oral ridges (OR). Detail can frequently be seen in individual kinetosomes (arrows). ×7500. (Micrograph courtesy of J. N. Grim.)

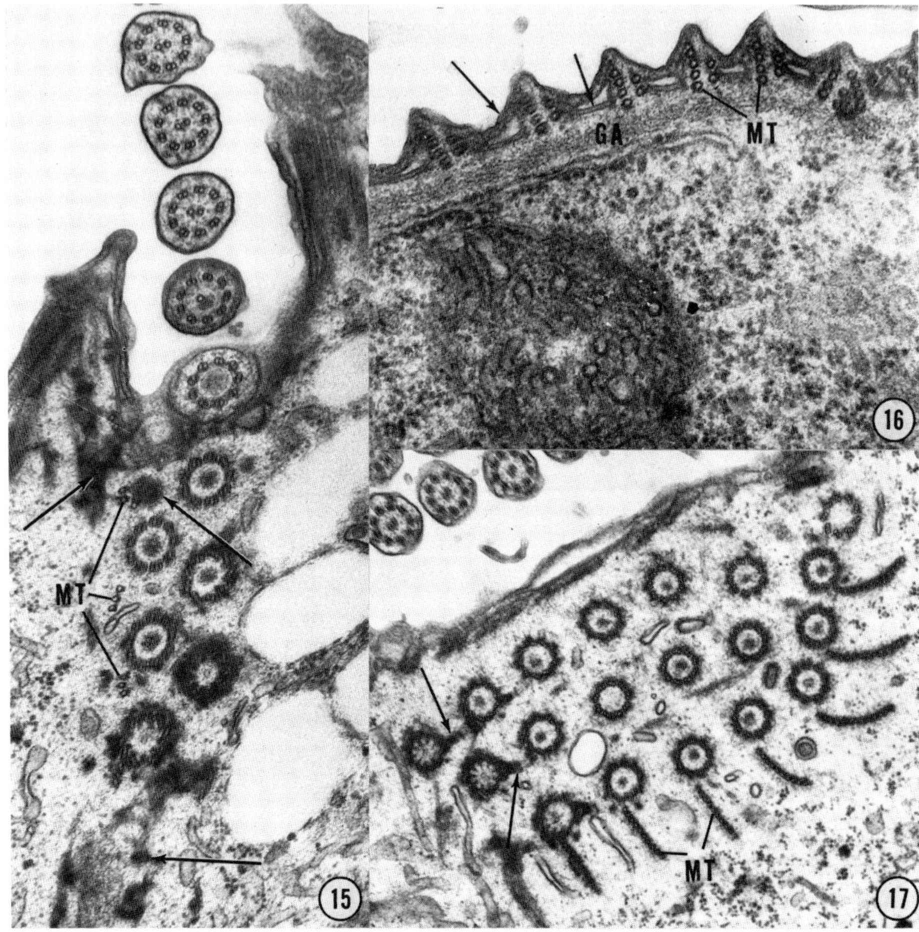

Fig. 15.

The undulating membrane is composed of a ciliated and a nonciliated row of kinetosomes. Fibrous (arrows) and tubular (MT) connectives are also evident. Ciliary–kinetosome structure is comparable to that of the somatic area of the cell. ×39,000.

Fig. 16.

The oral ridges seem to be composed of modified alveoli with unit membrane structure (arrows). Microtubules (MT) of the ridges are evident, as is the underlying granular or fine fibrous area (GA). ×57,000.

Fig. 17.

In a membranelle the basic kinetosome structure and the fibrous interconnections (arrows) are visible. Microtubules (MT) arise from the posterior row of kinetosomes. ×25,000.

a group of two or three microtubules. Arising from the proximal portion of the kinetosome and frequently associated with the microtubules is some fibrous material (Fig. 15), which Nilsson and Williams (71) suggest is continuous with the oral ridge. This material may be the counterpart of the cortical kinetodesmal fiber.

The oral ridges of *Tetrahymena* (Figs. 13, 14, 16, and 18) are closely aligned modifications of the basic cortical alveoli (Fig. 16) and extend from the undulating membrane into the cytopharynx of the cell. Underlying and composing a part of each ridge are two sets of four and two microtubules each (1). A thick granular ectoplasmic layer (Fig. 16), characteristic of the cortex in general, underlies this entire area. This layer has recently been described as a banded, "fine fibrous reticulum" (100).

The three membranelles are each composed of three rows of cilia (Figs. 14 and 17). It has been estimated that M_1 contains about 19 cilia per row, M_2 about 13 cilia per row, and M_3 (Fig. 17) only about 6 cilia per row (71), although other investigators have reported minor variations from these numbers (30). As in the undulating membrane, these cilia and kinetosomes show the basic structural characteristics observed for the cortical units. The outer ciliary row of the first membranelle, however, contains fine projections from one side of their cilia in the tip region (100). Cross-connections between kinetosomes of adjacent ciliary rows are evident (Fig. 17). Here, however, as in the undulating membrane, modifications of tubular and fibrous components occur. Arising from the posterior row of kinetosomes in each membranelle, a network of fine fibrils encloses a series of microtubules and both pass internally to the cytoplasm to form connections with various parts of the mouth (Fig. 17). These connections have been described in detail by several investigators (30,71). M_1 and M_2 and M_3 are joined by "membranellar connections" originating from the inner row of each membranelle and continuing with the outer row of the membranelle immediately posterior. In addition, cross-connections are described as joining the right margins of membranelles M_1, M_2, and M_3 with the undulating membrane, the left margins of M_1 and M_2 being connected to the basal portion of the undulating membrane by a "peripheral connective." In conjunction with fibers from M_3, the oral rib fibers extend into the cytoplasm of the cell to form the "deep fibers" of the oral area. Kennedy (46) showed that formation of microtubules in the developing oral apparatus can be blocked by addition of 0.005 M colchicine. Associated with this block is a delay in oral development and cell division.

The cytostome–cytopharyngeal complex is an area worthy of further examination. The right wall of the cytostome is composed of the oral ribs. Although oral ribs can be seen adjacent to both the undulating membrane and the cytopharynx (Figs. 14 and 18), the relationship between them is unclear. Separating the cytostome from the cytopharynx is a projection, relatively devoid of ribosomes and with finely granular cytoplasm (Fig. 18), called the cytostomal lip (71) or valve (23). Microtubules of the cytostomal lip project into and seem to surround at least a portion of the cytopharynx and possibly connect with the oral ribs (Fig. 18). These tubules seem to be projections from the membranelles which together with the oral rib fibers may form the deep fiber bundle. Below the cytostomal lip the cytopharynx is a small cavity from which food vacuoles arise. Numerous dense-membraned, flattened saccules found along the base of the cytopharynx may coalesce to form food vacuoles during ingestion, or a type of Golgi apparatus during mating (27). The series of fibers that constitute the oral ribs, surround the cytopharynx, and are continuous with the fibers of the cytostomal lip no doubt are supportive in nature. In fact, these tubules in conjunction with the interconnections of the membranelles and the undulating membrane probably give form and stability to the oral area. Their apparent spiral organization (Fig. 14) may also function in directing currents of food into the cytostome. Certainly, further clarification of the ramifications of this system of tubules is necessary.

Techniques have been established for isolation of the oral area (30,71,98). Although

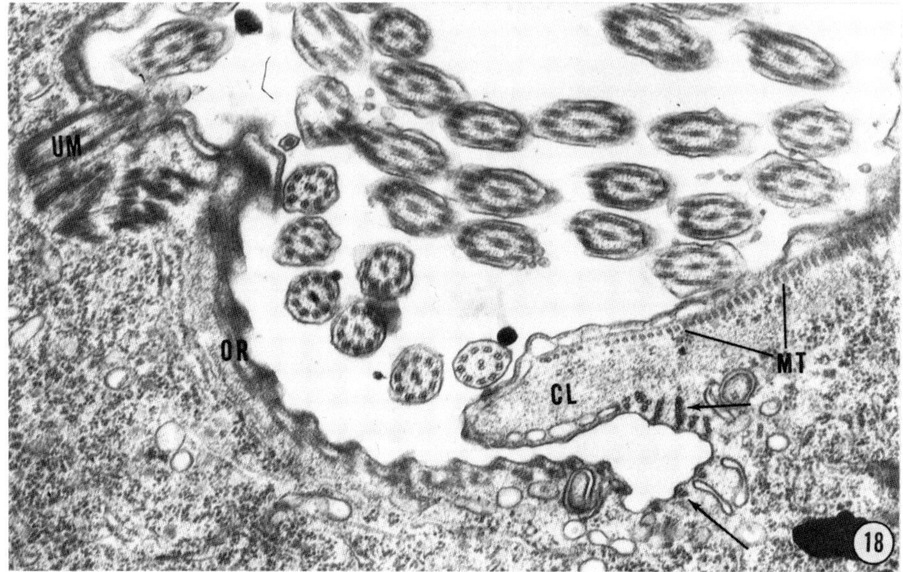

Fig. 18.

The cytostomal lip (CL) contains microtubules (MT) from the membranelles along one margin. Other microtubules (arrows), possibly from the undulating membrane (UM), also extend into this area from the oral ridges (OR). ×24,000.

isolated developing oral areas have been examined with the electron microscope (30,101), biochemical studies of the oral area have not been conducted. This is an area that certainly needs further careful examination.

III. CYTOPLASMIC COMPONENTS

A. Endoplasmic Reticulum

Two basic types of membrane systems have been described in *Tetrahymena*. One, a system devoid of ribosomes and including the Golgi apparatus (described below) is the smooth (agranular) endoplasmic reticulum (SER). The other is the ribosome-containing rough (granular) endoplasmic reticulum (RER).

Early studies reported that SER was associated with stomatogenic kinetosomes during cilia formation in *T. pyriformis* GL (99). Recent examination of developing oral areas of this strain tend to confirm this (47). Small segments of SER, however, are also found surrounding the kinetosomes in the mature mouth (Fig. 17) and body regions and are scattered throughout the cytoplasm of the cell (Fig. 19). These segments or fragments of SER do not appear to form a continuous system as with the contractile vacuole nephridial tubules (see below).

The RER, presently considered a common feature of the cytoplasm of *Tetrahymena*, was rarely referred to prior to 1961. At that time, Roth and Minick (80) briefly mentioned the presence of RER near the nuclear envelope and suggested that its complexity increased

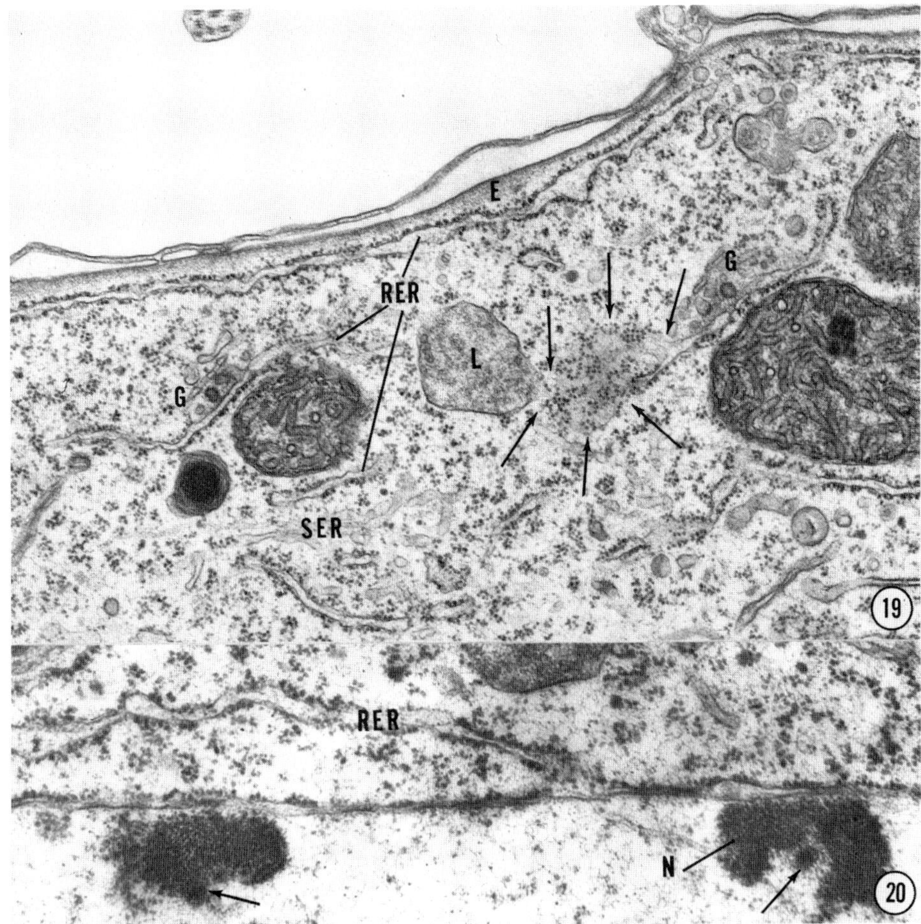

Fig. 19.

Underlying the cell surface just beneath the ectoplasmic layer (E) is rough ER (RER). Golgi (G) networks arising from RER are also present, as are mitochondria and some smooth ER (SER). At one point there is a suggestion of continuity between a lysosome (L) and rough ER (arrows). ×25,000.

Fig. 20.

The nuclear membrane is frequently found to be continuous with rough ER (RER). A dense body (arrows) is often associated with the nucleolus (N). ×39,000.

with approaching cell division. While describing RER in variety 9, we (24) suggested that the extensive membrane system might be a strain characteristic since in strain WH$_6$ membranes were inconspicuous and granules (ribosomes) were evenly dispersed (26). With the advent of improved fixation techniques, the number of references to RER increased (22,38,41,44,94); and only recently has it received specific attention (32,33).

Many biosynthetic processes are carried out by the RER. Thus its distribution in the cell probably reflects these different processes. Profiles of individual RER cisternae are generally observed underlying the cell membrane (Figs. 10 and 19) and in association with the kinetosomes where they may be responsible for the synthesis of fibrous material of the cortical area. Other branching cisternae are scattered throughout the cytoplasm and in association with other organelles. During logarithmic growth cisternae are often adjacent to and surround mitochondria (Figs. 4, 10, and 19), probably representing rapid enzyme synthesis (22). The cisternal membrane proximal to the mitochondrion and most of the distal membrane contains ribosomes. However, a region of the distal membrane lacks granules where it is closely associated with a third organelle, the Golgi apparatus (Figs. 4 and 19). A similar relationship between the RER and the Golgi apparatus has been described in *Campanella umbellaria,* in which a transfer of acid phosphatase from the ergastoplasm to the Golgi was suggested (39), and in *Frontonia leucas* (45). The mitochondria-associated RER cisternae in *Tetrahymena* have also been shown to contain acid phosphatase reaction products (Fig. 22), suggesting that hydrolases may be synthesized by the ribosomes (23). Although unequivocal connections have not been observed, morphological evidence suggests the possibility of a continuity between RER and primary lysosomes (20), another organelle high in phosphatase activity. In some sections (Fig. 19), portions of the membrane of a lysosomelike body appears to taper off to become continuous with RER and/or areas of high ribosome concentration. Whether this is an artifact of the plane of section, or whether it does in fact represent the formation of lysosomes within the cisternae of certain segments of RER, is a question requiring further and more critical investigation.

Tokuyasu and Scherbaum (94) suggest that the RER is responsible for the formation of mucocysts. They hypothesize that the RER forms a bleb which increases in size outside the cisternae of the reticulum, eventually giving rise to an immature mucocyst. The dense bodies reported by them to be immature mucocysts seem to us to be comparable to the structures described above as primary lysosomes. Although mucocysts may be formed by RER, we feel that available morphological evidence does not justify their conclusion.

The outer unit of the nuclear membrane contains abundant ribosomes (see Section IV,A). It has been described as branching and extending into the cell cytoplasm (22) as part of the RER network (Fig. 20). In cells synchronized for division by multiple heat shocks, light, amorphous material can be found in the space between the inner and outer nuclear membrane at points of origin of the RER (47). Thus it seems possible that this network serves as a channel for distribution of nuclear information to the cytoplasm.

In addition to the RER, the cell cytoplasm is filled with ribosomes often appearing as polyribosomes (Figs. 19 and 30) in growing cells and not directly associated with membranes. No doubt, this system of granules also synthesizes some of the cell's protein.

B. Mitochondria

Early electron microscope studies of the mitochondria of *Tetrahymena* (81) describe the internal processes as fingerlike microvilli. Since then, little attention has been given to their basic structure, although some examination of mitochondrial fractions with freeze-etch techniques has been undertaken (86). Major interest rather has been concerned with the behavior of mitochondria under different growth conditions.

In *Tetrahymena*, as in most ciliates, mitochondria are elongate or oval and are composed of two unit membranes. The outer mitochondrial membrane frequently appears discontinuous (Figs. 4 and 9), although this is no doubt due to the plane of section. The inner membrane is modified to form the microvilli projecting internally into the matrix of the mitochondria. More recently, these fingerlike projections of the inner membrane, referred to as mitochondrial tubules, have been found to branch (86). Continuity of the inner tubular matrix with the space between the two membranes is frequently evident (Figs. 9 and 19). Connection of the outer mitochondrial membrane with profiles of endoplasmic reticulum (ER) have also been observed with a frequency of about two to three per mitochondria in negatively stained preparations. Thus a direct continuity exists between the cisternae of some segments of ER and inner tubular matrix of the mitochondria, which could provide "a route for transfer of special proteins . . . from the rough ER into the mitochondria" (34). Elementary particles about 75 Å in diameter have been identified on the membrane of *Tetrahymena* mitochondria, but they are not necessarily confined to the matrix side of the tubules (86).

The structure of the internal mitochondrial matrix varies with different conditions of fixation. Generally, it appears to contain only scattered granules, but with good preservation the matrix is densely granular. These granules may be ribosomes which are consistent with the idea of extensive enzyme content. The existence of mitochondrial DNA in *Tetrahymena* has been clearly established (72,91). It is apparently native and is synthesized when nuclear DNA turnover is not occurring (72). During macronuclear DNA synthesis, however, there is a significant increase in uptake of [^{3}H]thymidine by mitochondria (73). Furthermore, this label remains randomly distributed among the mitochondria for four generations. Little else is known about its composition or activity.

Intramitochondrial masses are common components of the mitochondrial matrix. Having been described as varying from spindle-shaped to spherical in organization (19,22,26,80), they often appear as rodlike structures. Roth and Minick (80) suggest a relationship between the appearance of intramitochondrial structures and the onset of macronuclear division. We have been able to find no basis for their conclusion. In fact, intramitochondrial masses have been observed in both logarithmic- and stationary-phase cells (22,26). Although they increase on about the fifth day of culture growth, suggesting a relationship between the onset of their appearance and cell aging (22), they have also been observed to increase markedly in logarithmic cultures subjected to heat synchrony, high pressure (50), or high levels of carbon dioxide (21). Thus it might be concluded that the intramitochondrial mass is a constant feature of mitochondria in *Tetrahymena*, which under various conditions of physiological stress (stationary growth, high carbon dioxide, temperature, or pressure) becomes enlarged and globular.

The chemical composition of these bodies is unknown. They are certainly not RNA since exposure to RNase does not affect them while removing nuclear RNA (25), nor do they appear related to mitochondrial DNA when the distribution of [^{3}H]thymidine label uptake by mitochondria is examined. In fact, any label association with the intramitochondrial mass appears to be coincidental (91). Loading the medium with various concentrations of cations (Ca^{2+}, Ba^{2+}) has had no effect in increasing or decreasing their number (21); thus it is unlikely that they are inorganic in nature as in metazoan cells.

Changes in mitochondrial shape and position have been correlated with changes in cell age and thus with cellular activity. In young, rapidly growing (logarithmic) cultures, elongate mitochondria are distributed just beneath the cell membrane (22). As the cells become older (stationary phase) and cell activity decreases, some of the mitochondria move internally, becoming more spherical with the number of intramitochondrial masses apparently increasing. Mitochondrial degeneration also seems to increase at this time (22). Internal alterations of mitochondrial structure can be induced by exposure of cells to cigarette smoke residue (48). The tubular inner membranes become shelflike and are eventually degraded. This change correlates closely with reduced respiration of the cell.

The origin of mitochondria at cell division is still somewhat of an enigma. Sato (83) reported that the mitochondria of *T. geleii* gradually increased from 600 to 800 in daughter cells to 1200 to 1500 just before division. By following changes in mitochondrial number as estimated by counting the number of mitochondria in 75 typical cross sections, Elliott and Bak (22) observed a correlation between the number of mitochondria and the rate of cell growth. More mitochondria were present in rapidly dividing logarithmic cultures than in the more slowly dividing early logarithmic cultures. Their belief that increase is accomplished through direct mitochondrial division seems more plausible than the many other proposed methods (for review, see ref. 4). This view has been supported by the studies of Parsons and Rustad (73).

There are many features of the basic mitochondrial structure that are still unclear. Questions to be answered involve the nature and location of mitochondrial DNA and its relation to the intramitochondrial mass. What is the general composition of this structure, and what is the distribution and organization of mitochondrial enzymes? What is evident is that the behavior of mitochondria in the cell directly reflects the physiological state of the cell at a given time.

C. Golgi Apparatus

Only recently have well-defined structures corresponding to the classic Golgi apparatus of most eucells (74) been seen in *Tetrahymena* (32,33). These Golgi-like bodies (dictyosome equivalents) scattered throughout the cytoplasm of *T. pyriformis* (Figs. 4 and 19) may number as many as several hundred per cell (33). The Golgi bodies are composed of one or usually two parallelly arranged smooth-membraned cisternae and associated vesicles (Figs. 4 and 19). The vesicles appear to arise from a smooth-membraned portion of RER closely associated with a mitochondrion. These bleblike structures separate from the membrane, accumulate adjacent to the Golgi, and eventually may coalesce to form cisternae. The possibility of transfer of material from the RER to the Golgi by way of these vesicles has been suggested for *Epistylis anastatica* (29). RER in close association with mitochondria has been demonstrated to be the site of acid phosphatase reaction (23), although resolution was inadequate to demonstrate blebs on the membrane. Gold-fischer et al. (39) demonstrated acid phosphatase activity in these blebs in *C. umbellaria* and suggest that they may function in the transfer of this enzyme to the Golgi apparatus. A similar situation may exist in *Tetrahymena*. The close association and possible formation of the Golgi apparatus from the endoplasmic reticulum was thoroughly examined by Franke and his co-workers (32,33).

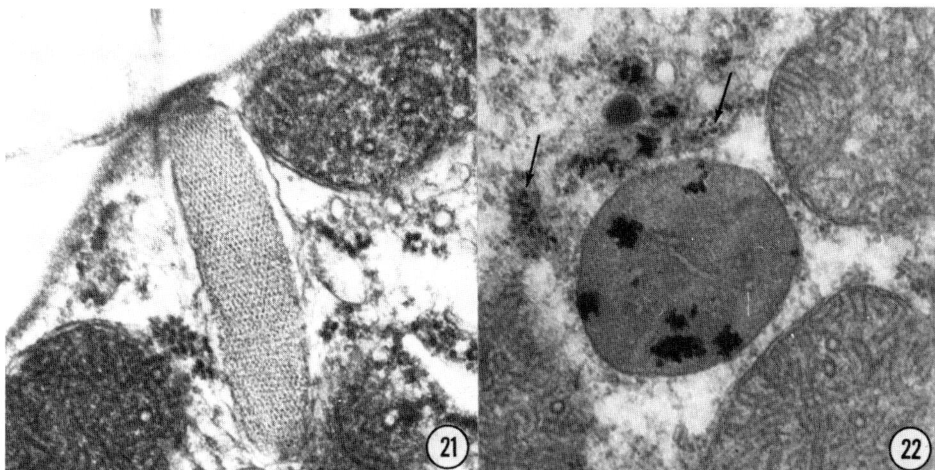

Fig. 21.

A mature mucocyst showing its location prior to discharge. The transverse periodicity is visible. ×40,000.

Fig. 22.

A primary lysosome and RER with dilated cisternae (arrows), both showing acid phosphatase reaction products. ×28,000.

Whereas these Golgi-like structures may function in the segregation and packaging of hydrolases, another typical Golgi apparatus has been observed prior to and during mating which probably is related to the fusion process (see Chapter 9).

D. Lysosomes

According to DeDuve and Wattiaux (for review, see ref. 17), lysosomes are subcellular particles enveloped by a unit membrane and containing acid hydrolases. Lysosomes have been identified in several protozoa, including *T. pyriformis* strains GL (60,67,68) and E (20,23), and several sexually active strains (21). The localization of hydrolases, principally acid phosphatase, has been considered positive evidence for the presence of lysosomes. There is no doubt about the existence of secondary lysosomes (food vacuoles, autophagic vacuoles, and so on) in *T. pyriformis,* but evidence for primary lysosomes is not so easily obtained. Opinions vary on the existence of these structures in metazoan cells, from those who claim that they are too small to visualize to those who believe that they are hydrolase-containing sacs which fuse with phagosomes (food vacuoles) and other vacuoles (for discussion, see ref. 17). It is reasonable to assume that hydrolases are synthesized at the ribosomes and make their way into the cisternae of the RER. Acid phosphatase has been localized in these spaces in the peritrich *C. umbellaria* (39). Nearby vesicles also contain the enzymes thought to arise from the food vacuoles without involvement of the Golgi apparatus. Acid phosphatase has also been observed in the cisternae of the RER of *T. pyriformis* (Fig. 22) and, as pointed out above, primary lysosomes may arise from terminal dilations of the RER (20). This evidence is from ultrastructural observations and is equivocal, owing to the failure to obtain clear-cut connections between the primary lysosomes and the RER in stained sections. There seems to be no question about the existence of single-membraned vacuoles, slightly smaller than mitochondria, which are acid phosphatase-positive. Staining is not always consistent, however, which throws some doubt on the uniformity in function of these vacuoles. There may be more than one kind, or they may react differently to the acid phosphatase stain. Moreover, some contain sparsely scattered tubules, resembling fragments of mitochondrial tubules, and may be peroxisomes (Fig. 22). These bodies have not yet been identified morphologically in *T. pyriformis*, although their enzymes have been identified (Chapter 8).

E. Mucocysts

Small, argentophilic granules lying beneath the pellicle along the meridians have been observed in several tetrahymenids. These have been called protrichocysts (74), secretory ampules (10), mucigenic bodies (75) and, more recently, mucocysts (94). The last-mentioned term seems to describe the probable function of these structures, hence we use it in this discussion.

Tokuyasu and Scherbaum (94) studied mucocysts and their development in *T. pyriformis* strain GL at the ultrastructural level. The mucocysts may arise from invaginations of the RER, which form small vacuoles. These undergo maturation resulting in the mature mucocyst which becomes attached to the pellicle by its pointed end. Its content is at first crystalline (Fig. 21) but, when stimulated, becomes amorphous mucoid material which is discharged through a pore in the pellicle. The pore is formed by a fusion of the outer mucocyst and pellicular membranes (Fig. 11). Before discharge the crystalline material has a longitudinal periodicity of 150 Å and a lateral one of about 100 Å. We have seen what appears to be the later stages of the development of mucocysts (Fig. 21) but have been unable to confirm the above observations on their origin from RER in strain E or several sexually active strains (21).

The function of mucocysts in *T. pyriformis* is unknown. They could coat the cell with a mucoid substance that might protect it against osmotic shock. We have noted a spectacular increase in mucocysts when cells are subjected to a medium of low osmotic pressure and also when they are mating. Zebrun et al. (107) suggest that mucocysts function during cyst formation in *T. rostrata*, a process not found in *T. pyriformis*.

F. Food Vacuoles

Under natural conditions the various species of *Tetrahymena* utilize different food sources. Most are bacteria feeders but some, such as *T. chironimi*, *T. rostrata* (15), and *T. limacis* (53), are facultative parasites, living on the body tissues and fluids of their hosts. Most species can utilize soluble natural products such as proteose–peptone, or even synthetic media (chemically defined). Particulate matter is not essential to induce food vacuoles. The process of food vacuole formation may be the same in all species of *Tetrahymena*.

Is it possible to assign functions to the various parts of the elaborate mouth structures? Obviously, the nutrients, bacteria or fluids, are brought into the buccal cavity by the strong beating of the membranelles and undulating membrane. Just what function the oral ribs play, other than support, is not known. The same is true of the "valve" or "flap." It seems that such a unique structure as the latter must play some role in food intake, either in the selection of food or for regulating the size of the vacuole. Before the food vacuole forms, a double unit membrane closes the internal cytopharynx. Beneath this are numerous double-membraned vesicles (Fig. 18). Although the source of the food vacuole membrane is obscure, the numerous double-membraned vesicles in the cytopharyngeal region tempt one to suggest that they are involved in its formation. These vesicles, however, together with other larger vesicles or saccules concentrated in the oral region, may have other functions (Chapter 9). As the vacuole increases in size, it slides along the adjacent fibers until, when fully formed, it extends beyond the fibers and lies deep in the cytoplasm. The fibers may direct the vacuole to its central location in the ciliate. Dense vesicles adhere to the vacuole membrane when it is completely filled. Recently, we have observed that these vacuoles contain whorls of membranes, suggesting that they may be residual vacuoles (54). They are acid phosphatase-negative,indicating that they do not contain hydrolases. It may be that they are the final stages of autophagic vacuoles which are discharged from the cell by recycling through the food vacuoles.

Apparently, many factors in the environment influence the rate of food vacuole formation. Long ago, Mills (58) studied the effects of pH on the rate of food vacuole formation in *Colpidium* which may have been a species of *Tetrahymena*. She showed that vacuoles formed at the rate of two to three per minute at pH 6.0–6.5, whereas the number dropped to less than two per minute at pH 8.0. The viscosity of the medium, as well as other factors, influenced the rate the vacuoles formed.

Tetrahymena pyriformis is quite selective regarding the kinds of materials it ingests. It readily accepts carbon particles (India ink) alone or when mixed with bacteria. It also ingests colloidal gold (13), but if mixed with bacteria it selects only the bacteria (23). Their ability to discriminate between these different particles is quite remarkable. Food vacuoles form readily in particulate media and in soluble natural products (87). They form more slowly in chemically defined media, as has been observed by several workers (4,42). We have observed that in many strains starved cells form vacuoles (water) when suspended in distilled water prior to mating or in starvation experiments (54).

The first account of feeding, at the electron microscope level, was that of Muller and Rohlich (59) in *T. corlissi*. They briefly described changes in the mouth region as the food vacuole formed and showed several stages in digestion. The food source was rat or mouse spleen. They concluded that much of digestion occurred extracellularly since none of the original particles could be identified in the newly formed vacuoles. They also stated that the vacuole membrane was derived from mouth structures, but how this happened was not determined.

The events associated with cellular digestion have been the object of much research in mammalian cells (17) and to some extent in protozoa (8,23,51,52,59–63,85). In *T. pyriformis,* with which most of the work has been done, at least two strains have been studied. Muller et al. (60), using strain GL, fed the cells inert polystyrene latex particles (PLP) and followed the food vacuoles throughout digestion. They found that

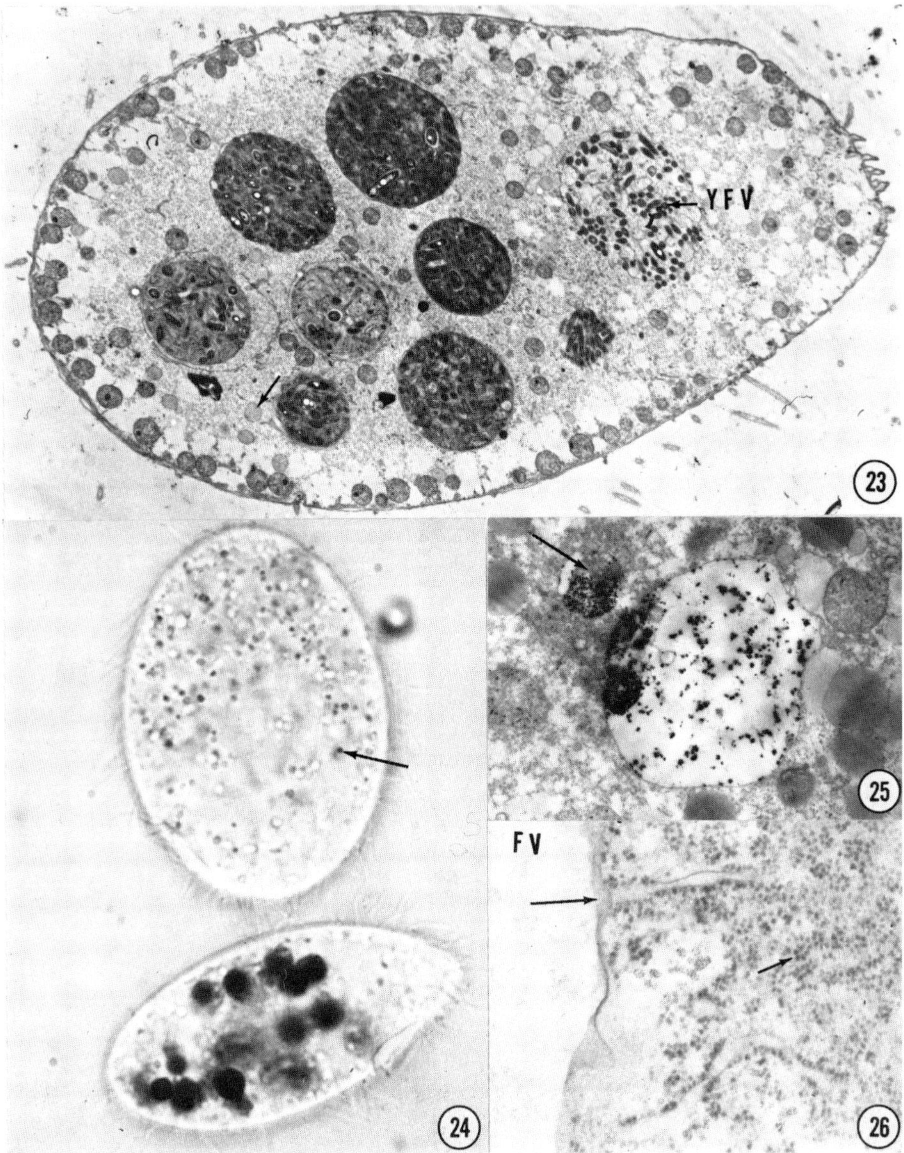

Fig. 23.

A longitudinal section of a whole cell fed bacteria, showing food vacuoles in various stages of digestion. Bacteria in the newly formed vacuole (YFV) are loosely packed. When digestion starts, the vacuole contents become more compact and dense. Primary lysosomes (arrow) are numerous during feeding. ×2300.

Fig. 24.

A light microscope micrograph showing two ciliates stained with neutral red. The upper one has not been fed and contains numerous neutral red granules (arrow). The lower ciliate has been fed (India ink mixed with bacteria); it shows many food vacuoles but very few neutral red granules, which suggests that the granules have fused with the food vacuoles. ×1000.

the ciliates accepted the PLP as readily as natural food (bacteria) and that acid phosphatase normally appeared in the vacuoles. They concluded that induction of food vacuoles was due to the mechanical action of the particles rather than the food itself.

After strain E is starved for 24 h and then fed a mixture of bacteria *(Aerobacter aerogenes)* and India ink, ingestion begins at once and within 5 min dense vacuoles can be seen (23). They pass through several stages during digestion, probably similar to those described for *Paramecium multimicronucleatum* (61). Food vacuoles can be seen in all stages in a low-magnification micrograph (Fig. 23). The newty formed vacuoles (stage 1) contain evenly dispersed bacteria and when tested for acid phosphatase show no reaction products. After 30 min the bacteria condense into a compact mass, leaving a clear space between it and the membrane. This has been referred to as the "halo" stage (stage II). Such food vacuoles contain acid phosphatase (Fig. 25), suggesting that they have acquired hydrolytic enzymes and that digestion is occurring. Muller et al. (60) have suggested, and using the Gomori technique we have confirmed at the electron microscope level (23), that these enzymes arise from vesicles external to the food vacuole. We believe that the vesicles that contribute acid phosphatase to the food vacuole are primary lysosomes. The fusion of primary lysosomes with food vacuoles has also been recently reported in *Blepharisma* (18) and in mouse strain L fibroblasts by Gordon et al. (40). Later the vacuoles become small and dense (stages III and IV). Presumably, they contain only undigested material and carbon particles. Acid phosphatase stains reveal that at this time reaction products are diminished and usually appear at the periphery of the vacuoles. Occasionally, fingerlike evaginations of the vacuole membrane occur, which could be one of the methods by which end products move into the cytoplasm.

The contents of the residual vacuoles are rapidly discharged through the cytopyge. Such fecal pellets are without a membrane, suggesting that the membrane either disintegrates upon discharge or never leaves the ciliate but forms the membrane that closes the cytopyge.

Some recent light microscope observations in our laboratory (21) also confirm the fusion of primary lysosomes with newly formed food vacuoles. By employing neutral red staining of starved (24–72 hr) cells, numerous neutral red granules quickly appear throughout the cytoplasm. When the cells are then fed bacteria and food vacuoles form, all the granules disappear and the food vacuoles take on a pinkish color, suggesting that the neutral red granules fuse with the food vacuoles (Fig. 24). This implies that the neutral red granules about which a great deal of controversy has existed for many years, are primary lysosomes. Other investigators have recently supported this idea (42,79).

Fig. 25.

A small acid phosphatase–positive food vacuole lies near a larger one. They may fuse. Such fusion is observed in living cells. ×7500.

Fig. 26.

Rough endoplasmic reticulum often concentrates near a food vacuole (FV) membrane. Occasionally, cisternae can be seen to open directly into the vacuole (arrow), suggesting that hydrolases may flow into the food vacuole without forming primary lysosomes. Polyribosomes (arrows) are observed in rapidly growing cells. ×38,000.

There may be additional methods by which food vacuoles acquire hydrolases. We have often seen whorls of RER surrounding food vacuoles in growing cells. In some regions the cisternae appear to enter the vacuole (Fig. 26). This suggests that the enzymes are deposited directly without intervening primary lysosomes.

G. Pinocytic Vacuoles

It is well known that ciliates, similar to most protozoa, take in water at their surfaces by osmosis (51). Much of it may also enter, along with soluble nutrients, by pinocytosis. In strain E small clear vacuoles lying just beneath the plasma membrane may be pinocytic vacuoles (23). Their origin is unknown, but they are often seen in the vicinity of breaks in the pellicle which have been described as regions through which the contents of mucocysts have been discharged (94). These vacuoles are acid phosphatase-negative when lying near the plasma membrane but become positive as they migrate deeper into the cytoplasm (23). They may acquire hydrolases by fusing with primary lysosomes similar to food vacuoles. Under the phase-contrast microscope, fusion of food vacuoles with one another and with smaller vacuoles can be observed, but one can never be sure of the identity of the smaller vacuoles. Physiological evidence (87) seems to support the contention that the ciliate does take in small particles through the surface, but whether this is by diffusion or pinocytosis, or both, has not been established.

H. Autophagic Vacuoles

In normal stationary-phase ciliates, but particularly under periods of stress (starvation, old age, high carbon dioxide levels, and so on), cytoplasmic particles such as mitochondria, ER, ribosomes, and nuclear material become sequestered into autophagic vacuoles which undergo digestive processes similar to food vacuoles. They become acid phosphatase-positive and degradation of their contents occurs (22,23). The end products are apparently utilized by the cell, but the ultimate fate of the vacuoles is questionable. They may finally disintegrate in the cytoplasm or they may fuse with residual vacuoles prior to expulsion of the latter.

I. Contractile Vacuole

All species of *Tetrahymena* have a contractile vacuole which, similar to that of *Paramecium* (84), is a complex structure. The following account applies only to *T. pyriformis* strain E (22). It is unlikely that the contractile vacuole of other species is radically different.

The nephridial system consists of two pores (Fig. 27), a branching network of tubules, and a large vacuole which fills (diastole) and empties (systole) at regular intervals (Fig. 28). The entire system is located in the posterior half of the cell. The vacuole and tubules are bounded by a unit membrane which is continuous with the ER. Thus the entire organelle is a morphologically integrated structure.

The contractile vacuole pores, located on meridians 5 and 6 (Fig. 27), are composed of thick walls from which over 200 microtubules (250 Å) originate and radiate anteriorly

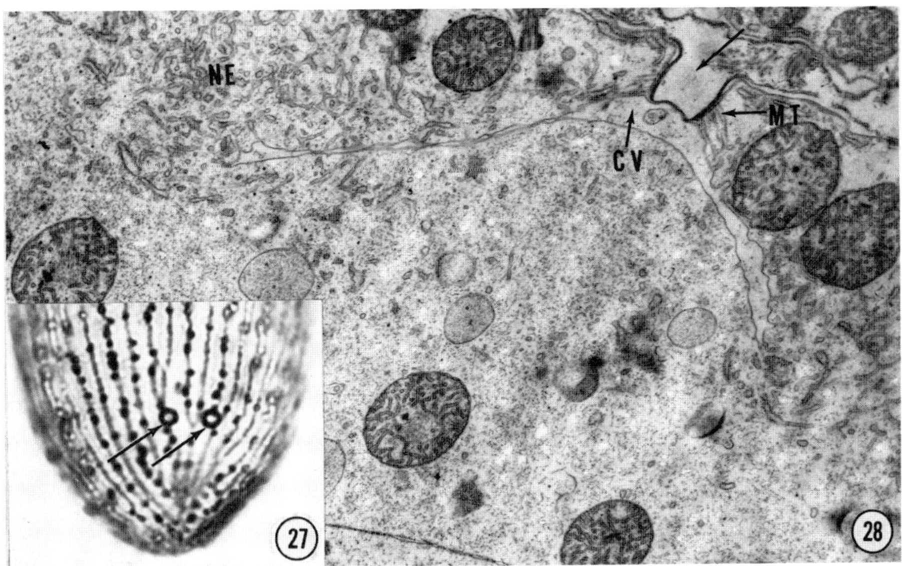

Fig. 27.

Light microscope preparation (Chatton-Lwoff stain) showing paired contractile vacuole pores (arrows). ×720.

Fig. 28.

A low-power micrograph of a collapsed contractile vacuole (CV) with its nephridial tubules (NE) and a pore (long arrow). Microtubules (MT) originate from the pore wall and extend toward the vacuole. ×5600.

some distance into the cytoplasm (59,60). Thus far, resolution has not been sufficient to determine if they actually attach to the membrane. The pore is closed internally by a dense double membrane (400 Å) which ruptures when the vacuole contents are expelled. The inner membrane is continuous with the vacuole membrane, and the outer one is a continuation of the inner part of the pore wall (Fig. 28). The diameter of the pore seems to vary during the vacuolar cycle, being more distended prior to discharge than following the expulsion of the vacuolar contents. The contractile vacuole in most cells opens into both pores. Sometimes two vacuoles have been observed, each with its own pore.

The nephridial tubules have a more-or-less uniform diameter of approximately 200 nm with a matrix slightly more dense than that of the contractile vacuole (Fig. 28), suggesting that the tubules contain material in addition to water. The tubules remain attached to the contractile vacuole throughout its cycle, thus providing a continuous system of channels.

The contractile vacuole fills and empties at regular intervals; the rate depends on the osmolarity, and perhaps other factors, of the environment (see Chapter 6). The entire cycle may be passive, that is, the filling results from the influx of water until the intracellular pressure initiates expulsion. However, microtubules do extend from the pore walls (Fig. 28) close to the contractile vacuole membrane. They may function only in support, or possibly they are involved in the dilation of the pores at the time

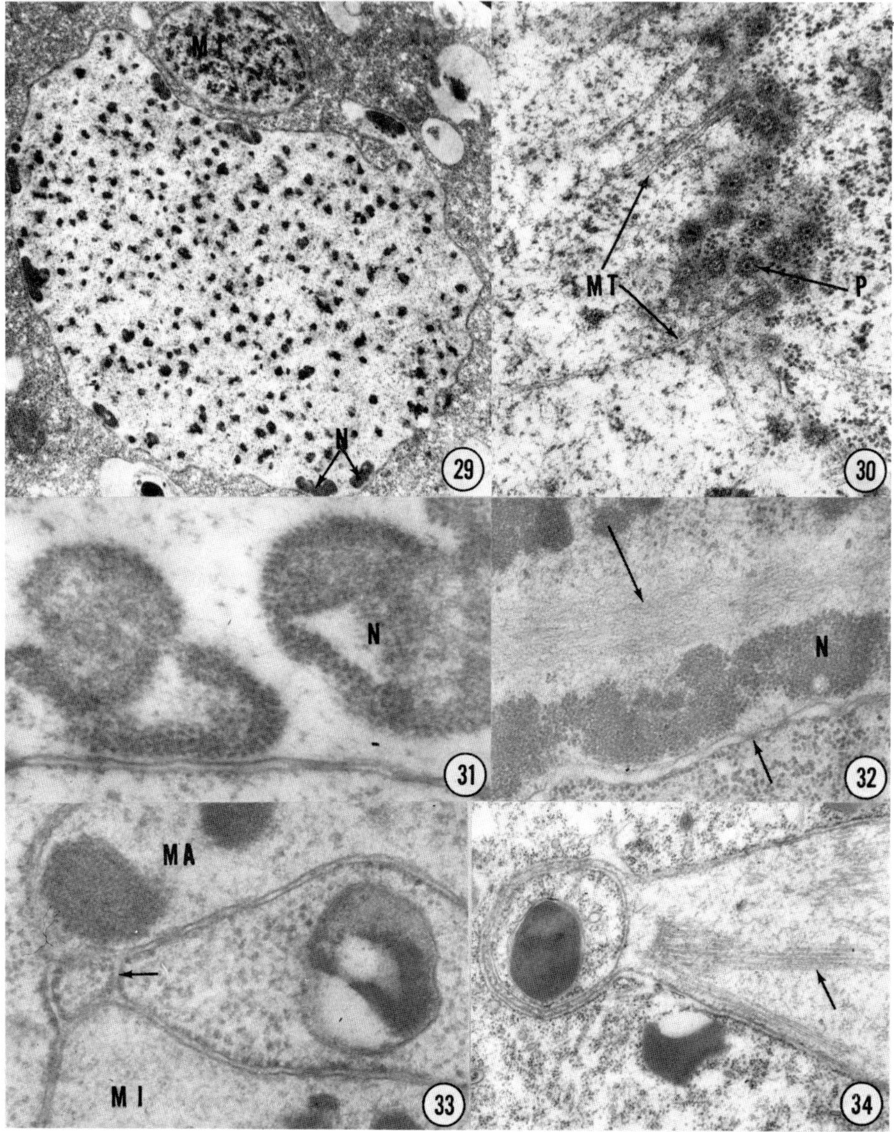

Fig. 29.

The micronucleus (MI) lies in an indentation of the macronucleus during interphase. The nucleoli (N) lie just under the macronuclear envelope, and the chromatin bodies are randomly distributed in the nucleoplasm. ×6800.

Fig. 30.

The pores (P) in the macronuclear envelope are shown in tangential section. Macronuclear microtubules (MT) present during division are also evident. ×34,000.

Fig. 31.

The nucleolus consists of outer ribosomes, an inner mass of less discrete granules, and a central less dense region. ×5500. (See Chapter 10 for a more detailed discussion of nucleoli.)

of discharge. They seem to be under tension when the vacuole reaches its greatest diameter. The intravacuolar pressure probably increases as the vacuole fills, resulting in tension of the membrane, microtubules, and pore walls. When the intravacuolar pressure is highest (completely filled), the contractile vacuole pore membranes apparently rupture and mend following discharge. Since the membranes have not been observed at the time of rupture it is not possible to state where the break occurs or how the membranes are reestablished. Expulsion appears to be a result of the combined action of the microtubules, intracellular turgor, and the tensile strength of the pore membranes. At the termination of expulsion, the contractile vacuole is completely collapsed; the membrane forms narrow channels which retain their continuity with the nephridial tubules. Filling starts at once and the cycle is repeated.

J. Storage Products

Tetrahymena pyriformis, as well as other species (42), has a remarkable capacity to store lipid and carbohydrate under varying environmental conditions (see Chapters 4, 5, and 8). Lipid droplets are usually without membranes and are more-or-less uniform in size. They stain intensely with osmium and less densely when fixed with glutaraldehyde. Glycogen appears in dense granular clusters when fixed with glutaraldehyde but is usually lost in osmium-fixed cells. When glycogen is abundant, it frequently forms islands, some of which are enveloped in a membrane (21). It may reach such high concentrations that all cytoplasmic structures are crowded into small irregular regions of the ciliate.

IV. NUCLEI

A. Macronucleus

The macronuclei of the various species of *Tetrahymena* seem to be very much alike. They resemble those of the different species of *Paramecium* (75). The first account of the ultrastructure of the macronucleus of *T. pyriformis* was that of Sedar and Rudzinska in 1956 (88). It has been described by numerous workers since (7,25,80, and others).

The hyperpolyploid structure is bounded by an envelope consisting of two distinct unit membranes with a low-density space between them (Figs. 20 and 31–33). The

---

Fig. 32.

Occasionally, the macronucleus contains fine fibrous material (long arrow) of unknown function. During aging or other stressed conditions, the nucleoli (N) fuse as shown here. The short arrow indicates a pore in transverse section. ×38,000.

Fig. 33.

Nuclear envelope continuity (arrow) exists between the micronucleus (MI) and the macronucleus (MA). ×49,000.

Fig. 34.

During meiosis spindle fibers (microtubules) come to an abrupt end adjacent to a dense body enveloped in RER. The nature or function of this body is obscure. ×17,000.

overall thickness of the envelope is approximately 200 nm, although there is some variation which may be caused by the fixative. Each membrane is about 70 Å thick. The envelope is interrupted by "pores" (95 to 135 pores per square micron, ref. 31) which result from a fusion of the outer and inner membranes. Examination of nuclear envelope fragments during differing conditions of growth suggested that the smaller the pore diameter the higher the pore frequency (104). This observation has not been confirmed with freeze-etch studies (90), however. There seems to be no actual opening in the pores but rather a modified septum surrounded by ribosomes with a dense particle at the center. Wunderlich (103) states that alterations in the frequency of the central granule support the hypotheses that this particle is ribonucleoprotein migrating from the nucleus to the cytoplasm. The structure of nuclear pores can best be seen in tangential sections of the envelope (Fig. 30). Two types of pores have been demonstrated from whole-mount preparations of the macronuclear envelope of logarithmically growing *Tetrahymena* (31). Type-B pores have well-developed annuli surrounding the pores, while type-A pores seem to lack an annulus as revealed by negative staining. Macronuclear envelopes from stationary-phase cells show only type-A pores, type-B pores being entirely absent (105). Franke (31) suggests that "these types represent different structural states of the pore complex relating to changes in the pore's functional role as a system regulating the nucleocytoplasmic interactions." The appearance of an annulus on type-B pores may be due to the presence of ribosomes on the outer segment of the nuclear membrane. The outer membrane of the envelope is contiguous with the ER, thus providing a continuous channel·from the space in the nuclear envelope to the cisternae of the ER as mentioned above (Fig. 20).

The contents of the interphase macronucleus, as observed in light microscopy, consist of Feulgen-positive, small, granular, evenly distributed bodies and peripherally located, RNA-positive, larger bodies. It is generally agreed that the former are chromatin (chromosomes) and the latter nucleoli.

The chromatin bodies are 100–200 nm across and are composed of numerous granules. Structural changes in chromatin have been reported to occur during the cell cycle and in response to environmental conditions (66,69). The nucleoli lie close to the inner membrane of the envelope; indeed, they show a definite orientation to it (Figs. 29, 31, and 32). They consist of numerous dense granules resembling cytoplasmic ribosomes arranged around a larger more dense region (Figs. 20 and 32) which has been identified as the site of nucleolar DNA (9). Most of their structure is destroyed with RNase. Under certain environmental (see Chapter 8) and growth (70) conditions, the granules take on a specific pattern, suggesting that some sort of maturation of the ribosomes is occurring. The peripheral particles lie in rows which seem to be leaving the parent body (Fig. 31). Beneath the rows of granules, fibrous material is interspersed with less discrete granules which make up the remaining dense portion of the nucleolus. Recently (82), changes in the structure of the nucleolus associated with cell culture cycle have been described.

Fine fibers (Fig. 32) and microtubules (Figs. 20 and 30) have been repeatedly observed in the nucleoplasm (10,24,43,80,93,106) of *Tetrahymena*. These macronuclear microtubules have been described as both persisting throughout the cell cycle (43) and being confined only to the period of macronuclear division (28,93,106). They appear singly or in bundles (10,28,43) and may be connected with the inner nuclear membrane (106). It has been suggested by Ito et al. (43) that macronuclear microtubules function in maintaining the individuality of diploid subunits. Other studies indicate that disruption of macronuclear microtubules with colchicine inhibits cell division in heat-synchronized populations (46,106). Addition of colchicine after the physiological transition point in synchronized cells, however, inhibits macronuclear microtubule formation and amitosis but does not block cytokinesis (93). Thus it seems that microtubule-directed macronuclear elongation is not essential for cytokinesis in *Tetrahymena*.

B. Micronucleus

Most sexually active strains possess a diploid micronucleus lying in an indentation of the macronucleus (Fig. 29). It is approximately one-fifth the diameter of the macronucleus. The envelope resembles that of the macronucleus in dimensions, but no pores have been seen. Occasionally, the outer membranes of both nuclei are continuous (Fig. 33), which may constitute a means of communication between them during interphase. Obviously, the connection is severed during mitosis. At interphase the chromatin bodies are dispersed and resemble those of the macronucleus. No other structures are seen in the interphase nucleus. However, during mitosis spindle fibers, resembling microtubules, are conspicuous in glutaraldehyde-fixed material (Fig. 34). Chromosomal behavior during meiosis at the light microscope level has been carefully analyzed by Ray (77). Nothing new has been added from ultrastructural studies of the process.

ACKNOWLEDGMENT

Original research reported here has been supported by grants AI 01416 and GM 13386 from the National Institutes of Health.

REFERENCES

1. Allen, R. D. 1967. J. Protozool. 14:553–565.
2. Allen, R. D. 1968. J. Cell Biol. 37:825–831.
3. Allen, R. D. 1969. J. Cell Biol. 40:716–733.
4. Allen, S. L., Misch, M. S., and Morrison, B. M. 1963. J. Histochlem. Cytochem. 11:706–719.
5. Buhse, H. E. and Corliss, J. O. 1969. Proc. 2nd Ann. Scanning Electron Micros. Symp., 171–176.
6. Buhse, H. E., Corliss, J. O., and Holsen, R. C. 1970. Trans. Amer. Microsc. Soc. 89:328–336.
7. Cameron, I. L. and Guile, E. E. 1965. J. Cell Biol. 26:845–855.
8. Carasso, N., Favard, P., and Goldfischer, S. 1964. J. Microsc. 3:297–322.
9. Charret, R. 1969. Exp. Cell Res. 54:353–361
10. Chessin, E. M. and Mosevich, T. N. 1962. Arch. Protistenk. 106:181.
11. Child, F. M. 1959. Exp. Cell Res. 18:258–267.
12. Child, F. M. and Mazia, D. 1956. Experimentia 12:161–162.
13. Cohen, A. I. 1959. Ann. N. Y. Acad. Sci. 78:609–622.
14. Corliss, J. O. 1953. Parasitology 43:49–87.
15. Corliss, J. O. 1962. Parasitology (Abstr.), 52:10.
16. Culbertson, J. R. 1966. J. Protozool. 13:397–406.
17. DeDuve, C., and Wattiaux, R. 1966. Ann. Rev. Physiol. 28:435–492.
18. Dembitzer, H. M. 1968. J. Cell Biol. 37:329–344.
19. Elliott, A. M. 1963. In L. Levine, ed., The cell in mitosis, pp. 107–121, Academic Press, New York.

20. Elliott, A. M. 1965. Science, 194:640–641.
21. Elliott, A. M. unpublished.
22. Elliott, A. M. and Bak, I. J. 1964. J. Protozool. 11(2):250–261.
23. Elliott, A. M. and Clemmons, G. L. 1966. J. Protozool. 13:(2):311–323.
24. Elliott, A. M. and Kennedy, J. R. 1962. Trans. Amer. Microsc. Soc. 81:300–308.
25. Elliott, A. M., Kennedy, J. R., and Bak, I. J. 1964. J. Cell Biol. 12:515–531.
26. Elliott, A. M., Travis, D. M., and Work, J. A. 1966. J. Exp. Zool. 161:177–192.
27. Elliott, A. M. and Zieg, R. G. 1968. J. Cell Biol. 36:391–398.
28. Falk, H., Wunderlich, F., and Franke, W. W. 1968. J. Protozool. 15:776–780.
29. Faure-Fremiet, E., Fayard, P., and Carasso, N. 1962. J. Microsc. 1:287–312.
30. Forer, A., Nilsson, J. R., and Zeuthen, E. 1970. C. R. Trav. Lab. Carlsberg 38:67–86.
31. Franke, W. W. 1967. Z, Zellforsch. 80:585–593.
32. Franke, W. W. and Eckert, W. A. 1971. Z. Zellforsch. 122:244–253.
33. Franke, W. W., Eckert, W. A., and Krien, S. 1971. Z. Zellforsch. 119:577–604.
34. Franke, W. W. and Kartenbeck, J. 1971. Protoplasma 73:35–41.
35. Furgason, W. H., 1940, Arch. Protistenk. 94:224–266.
36. Gibbons, I. R. 1963. Proc. Natl. Acad. Sci. 50:1002–1010.
37. Gibbons, I. R. 1965. Arch. Biol. (Liege), 76:317–352.
38. Gibbons, I. R. and Rowe, A. J. 1965. Science 149:424–426.
39. Goldfischer, S., Carasso, N., and Favard, P. 1963. J. Microsc. 2:621–628.
40. Gordon, G. B., Miller, L. R., and Bensch, K. G. 1965. J. Cell Biol. 25:(pt. 2):41–55.
41. Hoffman, E. J. 1965. J. Cell Biol. 25:217–228.
42. Holz, G. G., 1964, In S. H. Hutner, ed., Biochemistry and Physiology of protozoa, vol. III, Academic Press, New York. pp. 199–242.
43. Ito, J., Lee, Y. C., and Scherbaum, O. H. 1968. Exp. Cell Res. 53:85–93.
44. Jahn, T. L., and Bovee, E. C. 1967. In T. T. Chen, ed., Research in protozoology, vol. I, Pergamon Press, New York, pp. 41–200.
45. Kennedy, J. R. 1967. J. Protozool. 14:(suppl.):25.
46. Kennedy, J. R. 1969. In The cell cycle – Gene, enzyme interaction, Academic Press, New York pp. 227–248.
47. Kennedy, J. R., unpublished.
48. Kennedy, J. R. and Elliott, A. M. 1970. Science 168:1097–1098.
49. Kennedy, J. R. and Zimmerman, A. M. 1970. J. Cell Biol. 47:568–576.
50. Kennedy, J. R. and Zimmerman, A. M. Unpublished.
51. Kitching, J. A. 1956. Protoplasmatologia, 3(D,3,a):1–45.
52. Kitching, J. A. 1967. In T. T. Chen, ed., Research in protozoology, Pergamon Press, New York. pp. 307–337.
53. Kozloff, E. N. 1956. J. Protozool. 3:204–208.
54. Levy, M. R. and Elliott, A. M. 1968. J. Protozool. 15:208–222.
55. Loefer, J. R., Small, E. B., and Furgason, W. H. 1966. J. Protozool. 13:90–103.
56. Metz, C. B. and Westfall, J. A. 1954. Biol. Bull. 107:106–122.
57. Miller, O. L. and Stone, G. E. 1963. J. Protozool., 10:280–288.
58. Mills, S. M. 1931. J. Exp. Biol 8:17–29.
59. Muller, M. and Rohlich, P. 1961. Acta Morphol. Acad. Sci. Hung. 10:297–305.
60. Muller, M., Rohlich, P., and Törö, I. 1965. J. Protozool. 12:27–34.
61. Muller, M. and Törö, I. 1962. J. Protozool. 9:98–102.
62. Muller, M., Rohlich, P., Toth, J., and Törö, I. 1963. In Ciba Foundation symposium on lysosomes, Churchill, London, pp. 210–216.
63. Muller, M., Törö, I., Polgar, M., and Druga, A. 1963. Acta Biol. Acad. Sci. Hung. 14:209–213.
64. Munn, E. A. 1970. Tissue Cell 2:499–512.
65. Nanney, D. L. 1966. J. Exp. Zool. 161:307–318.
66. Nilsson, J. R. 1970. C. R. Trav. Lab. Carlsberg 37:285–300.
67. Nilsson, J. R. 1970. C. R. Trav. Lab. Carlsberg 38:87–106.
68. Nilsson, J. R. 1970. C. R. Trav. Lab. Carlsberg 38:107–121.

69. Nilsson, J. R. 1970. J. Protozool. 17:539–548.
70. Nilsson, J. R. and Leick, V. 1970. Exp. Cell Res. 60:361–372.
71. Nilsson, J. R. and Williams, N. E. 1966. C. R. Trav. Lab. Carlsberg 35:119–141.
72. Parsons, J. A. 1965. J. Cell Biol. 25:641–646.
73. Parsons, J. A. and Rustad, R. C. 1968. J. Cell Biol. 37:683–693.
74. Pitelka, D. R. 1961. J. Protozool. 8:75–89.
75. Pitelka, D. R. 1963. Electron microscopic structure of protozoa, Pergamon Press, New York.
76. Pitelka, D. R. 1965. J. Microsc. 4:373–394.
77. Ray, C., Jr. 1956. J. Protozool. 3:88–97.
78. Renaud, F. L., Rowe, A. J., and Gibbons, I. R. 1968. J. Cell Biol. 36:79–90.
79. Rosenbaum, R. M. and Wittner, M. 1962. Arch. Protistenk. 106:223–240.
80. Roth, L. E. and Minick, O. T. 1961. J. Protozool. 8:12–21.
81. Rudzinska, M. A. and Porter, K. R. 1954. Trans. N. Y. Acad. Sci. 16:408–411.
82. Satir, B. and Dirksen, E. R. 1971. J. Cell Biol. 48:143–154.
83. Sato, H. 1960. Anat. Rec. 138:381.
84. Schneider, L. 1961. J. Protozool. 7:75–90.
85. Schneider, L. 1964. Z. Zellforsch. 62:225–245.
86. Schwab–Stey, H., Schwab, D., and Krebs, W. 1971. J. Ultrastruct. Res. 37:82–93.
87. Seaman, G. R. 1961. J. Protozool. 8:204–212.
88. Sedar, A. W. and Rudzinska, M. A. 1956. J. Biophys. Biochem. Cytol. 2:331–336.
89. Small, E. B., Maiszalek, D. S., and Antipa, G. A. 1971. Trans. Amer. Micros. Soc. 90:283–294.
90. Speth, V. and Wunderlich, F. 1970. J. Cell Biol. 47:772–776.
91. Stone, G. E. and Miller, O. L. 1965. J. Exp. Zool. 159:33–38.
92. Tamura, S. 1971. Exp. Cell Res. 68:169–179.
93. Tamura, S., Tsuruhara, T., and Watanabe, Y. 1969. Exp. Cell Res. 55:351-358.
94. Tokuyasu, K. and Scherbaum, O. 1965. J. Cell Biol. 27:67–81.
95. Watson, M. R. and Hopkins, J. M. 1962. Exp. Cell Res. 28:280–295.
96. Watson, M. R., Hopkins, J. M., and Randall, J. T. 1961. Exp. Cell Res. 23:629–631.
97. Watson, M. R., Alexander, J. B., and Silvester, N. R. 1964. Exp. Cell Res. 33:112–129.
98. Whitson, G. L., Padilla, G. M., Canning, R. E., Cameron, I. L., Anderson, N. G., and Elrod, L. H. 1966. Natl. Cancer Inst. Monogr. 21:317–321.
99. Williams, N. E., Anderson, E., Kessel, R. G., and Beams, H. W. 1960. J. Protozool. 7(suppl.):27.
100. Williams, N. E. and Luft, J. 1968. J. Ultrastruct. Res. 25:271–292.
101. Williams, N. E. and Zeuthen, E. 1966. C. R. Trav. Lab. Carlsberg 35:101–118.
102. Wolfe, J. 1970. J. Cell Sci. 6:679–700.
103. Wunderlich, F. 1969. Z. Zellforsch. 101:581–587.
104. Wunderlich, F. 1969. Exp. Cell Res. 56:369–374.
105. Wunderlich, F. and Franke, W. W. 1968. J. Cell Biol. 38:458–462.
106. Wunderlich, F. and Speth, V. 1970. Protoplasma 70:139–152.
107. Zebrun, W., Corliss, J. O., and Lom, J. 1967. Trans. Amer. Miscrosc. Soc. 86:28–36.

CHAPTER 3

The Nutrition of *Tetrahymena:* Essential Nutrients, Feeding, and Digestion

George G. Holz, Jr.

Department of Microbiology
State University of New York
Upstate Medical Center
Syracuse, New York

It has been almost a half-century (or 73,000 6-h cell generations) since Lwoff (57) reported the successful axenic cultivation of *Tetrahymena pyriformis,* a pioneering accomplishment which led to definition of the ciliate's nutritional requirements and to the development of a chemically defined culture medium. This work was summarized in 1951 by Kidder and Dewey (51), the investigators who were the most responsible. Since that time the major investigational preoccupation has been with the metabolic bases of the nutritional requirements; this research has been reviewed by Hutner et al. (44), Kidder (48), and Holz (35). Now we are coming full-circle with the return to a question raised by Lwoff in 1923. What are the forms and routes of nutrient entry into *Tetrahymena?*

To describe the current state of our knowledge, we *summarize* here the often reviewed information on nutritional requirements, including those of species other than *T. pyriformis,* and then discuss recent work on feeding and digestion.

I. NUTRITIONAL REQUIREMENTS

Tetrahymena species have absolute nutritional requirements for amino acids, B-complex vitamins, a purine, a pyrimidine, and elements supplied in inorganic salts.

Table 1.

A Chemically Defined Medium Designed for the Cultivation of *T. pyriformis* Syngen 1, Mating Type II, at 35°C, but Suitable for Most Other Strains of the Species (42)

Component	Final concentration per 100 ml of medium	Component	Final concentration per 100 ml of medium
DL-Alanine	30 mg	Nicotinic Acid	90 μg
L-Arginine·HCl	30 mg	Calcium *d*-pantothenate	75 μg
L-Asparagine·H₂O	20 mg	Thiamine·HCl	50 μg
L-Glutamic acid	40 mg	Sodium riboflavin-5'-phosphate	45 μg
L-Glutamine	10 mg		
Glycine	40 mg	Pyridoxamine·2HCl	10 μg
L-Histidine·HCl·H₂O	20 mg	Biotin	0.1 μg
L-Isoleucine	20 mg	DL-Thioctic acid	1 μg
L-Leucine	20 mg	Folic acid	1 μg
L-Lysine·HCl	20 mg	K₂HPO₄·3H₂O	100 mg
DL-Methionine	30 mg	MgSO₄·7H₂O	50 mg
DL-Phenylalanine	30 mg	CaCO₃	7.5 mg
L-Proline	20 mg	Citric acid·H₂O	60 mg
DL-Serine	30 mg	Iron	270 μg
DL-Threonine	40 mg	Manganese	40 μg
L-Tryptophan	15 mg	Zinc	100 μg
DL-Valine	20 mg	Copper	8 μg
Guanosine-3'(2')-phosphoric acid·2Na·H₂O	9 mg	Cobalt	10 μg
		Molybdenum	1.2 μg
Uracil	4 mg	Glucose	1 g

An example of a chemically defined medium for *T. pyriformis,* containing the required nutrients and some stimulatory ones, is shown in Table 1.

The L-amino acids needed are arginine, histidine, isoleucine, leucine, lysine, methionine, phenylalanine, threonine, tryptophan, and valine. Most species also require serine, and a few need glycine, proline, alanine, aspartic acid, glutamic acid, and tyrosine (Table 2). Nonessential amino acids are often stimulatory to growth and are included in chemically defined media (Table 1).

Amino acids are used by *Tetrahymena* for protein synthesis, as sources of energy, and as precursors of other essential metabolites (16,77).

A particular amino acid requirement often depends upon the levels of other nutrients in the medium. Serine precursors and serine-dependent metabolites spared or eliminated the serine requirement of *T. pyriformis,* and the substitution of glycine or threonine for serine was folic acid-dependent (14,15). High levels of serine inhibited growth when dextrin was supplied in the medium, and the inhibition was reversed by glucose (71). Proline, ornithine, and citrulline spared the requirement for arginine (17). Proline is a product of arginine catabolism, while citrulline and ornithine are intermediates. Arginine sparing by citrulline and ornithine has been attributed to ornithine inhibition of arginine deiminase (34).

Table 2.
L-Amino Acid Requirements of *Tetrahymena*

Arginine, histidine, isoleucine, leucine, lysine, methionine, phenylalanine, serine, threonine, tryptophan, valine

Above required by:	Amino acids also required	Ref.
Tetrahymena corlissi Th-X	—	(37)
Tetrahymena paravorax RP	Glycine	(38)
Tetrahymena patula L-FF	—	(43)
Tetrahymena patula LI	Glycine	(47)
Tetrahymena pyriformis, all strains examined; serine not required by "serine mutants" and some other strains	Glycine, proline, alanine, aspartic acid, glutamic acid, and tyrosine by a few strains	(14,22,25,26,51)
Tetrahymena setifera HZ-1	—	(41)
Tetrahymena vorax Tur	—	(47,81)
Tetrahymena vorax V$_2$S, V$_2$D, and V$_2$M	—	(81)

Amino acid requirements may be lost by *Tetrahymena.* Aspartic acid-dependent strains of *T. pyriformis* grew without the amino acid after a period of laboratory cultivation (26).

The need for amino acids may be met by the free acids themselves or by their polymers: peptides, polypeptides, and proteins. The polymers, particulate or in solution, can be digested extra- and intracellularly (18,52,58,64,89).

Tetrahymena species require two nitrogenous bases: a purine and a pyrimidine. Guanine and uracil and their nucleosides and nucleotides are preferred. The guanine requirement can be substituted for by cytidine (and by some of its derivatives), but cytosine and thymine (and its derivatives) either are inactive or only spare the requirement (Table 3). The need for a purine or pyrimidine base also can be met with a nucleic acid, since extra- and intracellular nucleases are produced by *T. pyriformis* (19,30,53,55, 64).

Most *Tetrahymena* species and strains require the B-complex vitamins nicotinic acid, thiamine, riboflavin, pantothenic acid, and folic acid. They also require thioctic acid but not the fat-soluble vitamins needed by vertebrates (Table 4). Although biotin is only rarely required, it is often included in chemically defined media for *Tetrahymena* species (Table 1). Some clones of *T. pyriformis* collected in the Pacific area did not require niacin or thioctic acid when they were isolated but developed requirements for these vitamins during laboratory cultivation (26). Some species and strains of species require a B$_6$ vitamin. Pyridoxine is relatively inactive in meeting this need, but pyridoxal and pyridoxamine are highly active and equivalent. A strain of a species that does not require B$_6$, *T. corlissi* 21, was assayed with *Saccharomyces carlsbergensis* and found to synthesize the vitamin (47).

Table 3.

Purine and Pyrimidine Requirements of *Tetrahymena*

Guanine or some derivative required. Requirement usually spared by adenine and some of its derivatives.

Uracil or a derivative, or cytidine or a derivative, required.

Cytosine, and thymine and its derivatives, show modest sparing activity or are inactive.

Above required by:	Ref.
Tetrahymena corlissi, three strains	(27,47)
Tetrahymena limacis Mf. 1	(47)
Tetrahymena paravorax RP	(27,38)
Tetrahymena patula LI	(47)
Tetrahymena pyriformis W	(51)
Tetrahymena pyriformis syngen 1, MT II	(36)
Tetrahymena setifera HZ-1	(41)
Tetrahymena vorax Tur	(47)

Table 4.

Vitamin Requirements of *Tetrahymena*

Nicotinic acid, thiamine, riboflavin, pantothenic acid, folic acid, and thioctic acid required by:	Ref.
Tetrahymena corlissi, four strains	(37,47)
Tetrahymena paravorax RP	(38)
Tetrahymena patula LI and L-FF	(43,47)
Tetrahymena pyriformis, all strains examined except clones of syngen 10 which do not require thiamine, and clones of syngen 4 which also require biotin and choline	(25,51,86)
Tetrahymena setifera HZ-1	(41)
Tetrahymena vorax Tur	(47)

B_6 vitamin required by:	Ref.
Tetrahymena patula LI and L-FF	(43,47)
Tetrahymena pyriformis, most strains	(25,51)
Tetrahymena vorax Tur	(47)

B_6 vitamin not required by:	Ref.
Tetrahymena corlissi, four strains	(37,47)
Tetrahymena paravorax RP	(38)
Tetrahymena pyriformis, "pyridoxine mutants"	(21)
Tetrahymena setifera HZ-1	(41)

Tetrahymena species can utilize some conjugates of vitamins as readily as the free forms. For example, riboflavin monophosphate can substitute for riboflavin (Table 1). *Tetrahymena pyriformis* W, however, responded to pantothenate better than to panthetheine and its phosphate, or to coenzyme A (13).

Attempts to identify the requirements of *Tetrahymena* for the inorganic elements have encountered the difficult problem of contamination of components of the culture medium with the essential nutrient being tested for (31,32,51,83,84). Two principal approaches were followed: (1) the element was deleted from the medium, and in some cases the medium was subsequently treated with a complexing agent and any complex removed, or (2) the element was deleted and a complexing agent was incorporated into the medium. Population growth response was the index of essentiality of the element. No unequivocal demonstration of an absolute nutritional requirement for an element has resulted from these efforts. The elements routinely included as inorganic salts in chemically defined media for *Tetrahymena* (Table 1) reflect both the responses of investigators to these studies and their knowledge of the essentiality of many of the elements for enzymic activities known to be carried on by *Tetrahymena*.

Tetrahymena species grow best in chemically defined media when there is a source of carbohydrate [*T. corlissi*, (37,47), *T. patula* (47), *T. limacis* (47), *T. setifera* (41), and *T. paravorax* (38)]. Glucose was favored but dextrin was also acceptable; in fact, Reynolds and Wragg (73) reported that *T. pyriformis* grew better and utilized nitrogen more effectively with dextrin as the carbohydrate source than with glucose. The nature of the carbohydrate supplied in the medium also influenced amino acid utilization, synthesis, pool size and composition, and excretion (71,72).

Cox et al. (10) have described a carbohydrate-free medium for *T. pyriformis* enriched with amino acids (serine, proline, glutamic acid) and with C_2 and C_3 compounds (acetate, glycerophosphate, glycerol).

Glucose, mannose, maltose, fructose, dextrin, and starch all supported fermentation by strains of *T. pyriformis* (24,49,56,81).

β-Glucosidase, β-*N*-acetylglucosaminidase, α-glucosidase, and amylase are secreted into the environment by *T. pyriformis* (55,64,85).

Organic acids and alcohols alone have relatively little nutritional value for *Tetrahymena* (36,51,78). Acetate, however, stimulated growth of *T. pyriformis*, particularly in the absence of thioctic acid and glucose (51). Ethanol or methanol, but not a wide variety of other alcohols, organic acids, or aldehydes, was required by *T. setifera* HZ-1 in a defined medium with a sterol, but not in a peptone-based medium (41).

Three species of *Tetrahymena* required a sterol for best growth in chemically defined media: *T. corlissi* Th-X, *T. paravorax* RP, and *T. setifera* HZ-1 (38,39,41). The requirement in all cases was satisfied by growth factor amounts (1 μg/ml) of desmosterol, Δ^7-cholestenol, 7-dehydrocholesterol, cholesterol, 24-methylene cholesterol, fucosterol, campesterol, β-sitosterol, clionasterol, brassicasterol, stigmasterol, poriferasterol, ergosterol, or neospongesterol. All these sterols have a β-hydroxyl group at position 3 and may have double bonds in the ring system or side chain at positions 5, 7, 22, or 24; α- or β-methyl or ethyl substitutions at position 24; or methylene or ethylene substitutions at position 24. Zymosterol was slightly active for two of the ciliates *(T. corlissi, T. paravorax)* and lophenol for one *(T. corlissi)*. Other sterols considered potential precursors of cholesterol, or green plant sterols, or ergosterol (for example, lanosterol and products of its demethylation) were inactive, as were precursors of lanosterol (squalene, mevalonic acid). Combinations of synthetic phosphoglycerides, oleic acid, and glycerophosphate spared the sterol requirement of the three ciliates (39,41,90). The metabolic bases of the nutritional activity of the sterols, and of the sterol-sparing activity of the other lipids, are unknown. Paradoxically, *T. pyriformis,* which does not have a nutritional requirement for a sterol, synthesizes tetrahymanol, a pentacyclic triterpenoid alcohol (59). Analytical methods of great sensitivity have failed to detect sterols, however. Tetrahymanol was also synthesized by the ciliates that required sterols (3). This subject is discussed in greater detail by Holz (35) and in Chapter 4.

Phosphoglycerides have been useful in the cultivation of some species of *Tetrahymena* [*T. patula* L-FF and LI, *T. limacis* Mf. 1, and *T. corlissi* 24 (47)]. The metabolic rationale is unknown (40).

Long-chain fatty acids are not required by *Tetrahymena* species. Even under conditions in which they are taken up and incorporated into complex lipids, or are converted into other fatty acids, they do not stimulate population growth. For example, palmitic, stearic, oleic, linoleic, and γ-linolenic acids, all natural constitutents of *T. pyriformis* lipids, were taken up and metabolized readily by the ciliate (28,54) but had no positive nutritional effect (37,51).

II. FEEDING AND DIGESTION

Nutrient entry into *Tetrahymena* may follow a variety of routes and employ a variety of mechanisms: (1) by diffusion (free, restricted, exchange, facilitated) through the cell membrane, uniformly over the surface or at specific loci, and through the food vacuole membrane; and (2) by active transport, pinocytosis, or phagocytosis at any of these locations. The paths and modes of entry depend upon the nature of the nutrient under consideration, the environmental circumstances, and the physiological state of the ciliate. Water, ions, and small nonpolar molecules may traverse the surface by diffusion, active transport, and pinocytosis. Macromolecules in solution and particulates can enter by pinocytosis and phagocytosis. Size, charge, lipoid solubility and the metabolic activities of the organism influence membrane permeation. Size and charge influence the engulfment processes (pinocytosis and phagocytosis), and a charge-dependent concentration at the surface may trigger engulfment.

This subject has been reviewed exhaustively: Chapter 6, Conner (8), Müller (61), Holz (35). Only food vacuole formation and function are dealt with here. Uhlig et al. (87) characterized the stages for the ciliate *Metafolliculina* as ingestion, digestion, resorption, and egestion.

Tetrahymena can ingest particulate material, animate or inanimate, digestible or not, with great efficiency, and shows considerable powers of selection.

Many bacteria (33) and yeast (74) serve as food, dead or alive. Curds and Cockburn (12) showed that the rate of feeding on bacteria by an individual *T. pyriformis* was dependent both on the ciliate population density and the bacterial population density. Among the inanimate particulate materials that have been ingested under appropriate conditions are colloidal gold (23), latex (66,75), carmine (7), and carbon (10). The suspension medium is taken into the food vacuole with the particles (76). Edible particulates may be selectively ingested from mixtures containing inedible ones. For example, bacteria were eaten in preference to colloidal gold when *T. pyriformis* was presented with a mixture under conditions in which the gold was taken in when no bacteria were present (23). Ricketts (75), however, found that, when mixtures of latex particles (of various sizes) and yeast cells were fed to *Tetrahymena* in a nutritive medium, preferential uptake of a particle depended upon the proportions of the two particles present and not upon their nutritive value or size.

It had been thought that *Tetrahymena* could be cultured in media containing only soluble components, and that under these conditions fewer food vacuoles were formed than in the presence of particulates. Rasmussen and Kludt (70), however, showed that the success of peptone-based and of chemically defined media depends upon their content of small particles arising from precipitation of metal salts, particularly iron. A proteose–peptone medium that normally supported growth when it was sterilized by autoclaving became unsuitable for *Tetrahymena* when it was filter-sterilized. Addition of particulate material restored the growth-supporting properties of the medium. In contrast to the peptone medium, a chemically defined medium supported growth whether it was sterilized by autoclaving *or* by filtration. The medium apparently restored its particle content by new precipitation.

The activity of iron in enhancing growth, respiration, and numerous metabolic capacities (9,50,82) may be attributed to its contribution of particles to the culture medium, as well as to its association with heme synthesis.

We are faced with the question, What are the effective triggers of food vacuole formation; solutes, particles, combinations of both? Studies of pinocytosis in *Amoeba* (5) showed that relatively low-molecular-weight solutes of appropriate charge initiated food vacuole formation as readily as did macromolecules. This has yet to be established for *Tetrahymena*. For example, Ricketts (75) reported that salt solutions used to stimulate pinocytosis in *Amoeba* failed to induce food vacuole formation in *Tetrahymena*. The concentration of dyes at the cytopharyngeal membrane during vacuole formation, however, is well-documented for *Tetrahymena* (79) and for *Paramecium* (2,29). Solutes are concentrated within pinocytic vacuoles of *Amoeba* by binding to the layer of mucopolysaccharide on the surface (6). A similar mechanism for soluble nutrient acquisition may operate in ciliates. Jahn et al. (45) observed the secretion of a mucoid material associated with the oral apparatus of *Tetrahymena* and *Paramecium,* and Nilsson (69) described the incorporation into food vacuoles of alcian blue-stainable material (a mucocyst product?) which was abundant in the oral area of *T. pyriformis*. Secretion of alcian blue-stainable material has also been reported in *T. rostrata* (60).

Food vacuoles appear to be formed by inpocketing of the cell membrane at the base of the cytopharynx. Thus the wall of the vacuole is a membrane derived in part from this portion of the cell membrane (23). Since the area of membrane of the completed vacuole greatly exceeds that of the membrane at the base of the cytopharynx at the time of vacuole formation, membrane synthesis must occur during vacuole formation. Chapman-Andresen and Nilsson (7) estimated that a log-phase *Tetrahymena* that had been fed carmine particles and contained 30 food vacuoles had formed membrane equivalent to 30% of its body surface area and had taken in 7% of its own volume. Ricketts (75) arrived at somewhat higher values in experiments with latex particles. It is not known if the organization, chemical composition, and physiological characteristics of the vacuole membrane are homologous with those of the cytopharyngeal membrane, nor if that membrane is homologous with the membrane investing the general body surface. The isolated food vacuole membrane of latex-fed *Acanthamoeba castellani* had the same phospholipid, sterol, and enzyme content as the cell surface membrane (88).

Food vacuoles detach from the point of their formation and move rapidly into the cytoplasm toward the posterior of the cell. Under some conditions they may be formed by *Tetrahymena* and detached at a rate of four per 10 min (7,75). The rapid movement to the posterior of the cell appears under the light microscope to be directed and assisted by some means (Holz, unpublished), perhaps by the fibrils that lie next to the forming vacuole (23).

As digestion is initiated, the wall of the food vacuole may be thrown into folds, perhaps providing thereby an increased surface area for permeation of soluble products of digestion into the cytoplasm (87). Some inpocketings give the appearance of being pinched off as secondary food vacuoles. They also contain acid phosphatase. In addition to these secondary vacuoles, smaller, pinocytic vacuoles have been observed in stages of formation at the food vacuole membrane of *Tetrahymena* (23).

Food vacuoles void their indigestible contents through the cytopyge. No systematic investigation of this process has been made with the electron microscope, but Elliott and Clemmons (23) reported that the voided contents were not enclosed in a membrane. We have observed defecation in *T. setifera*. A papilla approximately 5 μm in diameter appeared at the cytopyge and particulate material was discharged from food vacuoles through the tip of the papilla. Vacuolar discharges occurred every 3.3 min. If this was a near-maximum rate, it could be compared with the food vacuole formation rate of one per 2.5 min reported for *T. pyriformis* (7,75). Ricketts (75) observed that when *T. pyriformis* was fed latex particles (1.95 μm diam) *ad libitum* it voided approximately 6.3 particles per minute over a 150-min experimental period. Nothing is known about how food vacuoles "find" the cytopyge or how discharge is carried out.

REFERENCES

1. Allen, S. L., Misch. M. S., and Morrison, B. M. 1963. J. Histochem. Cytochem. 11:706–719.
2. Ball, G. H. 1927. Biol. Bull. 52:68–78.
3. Beach, D., Erwin, J. A. and Holz, G. G., Jr. 1969. In Progress in protozoology, A. A. Strelkov, K. M. Sukhanova, and I. B. Raikov, eds., Abstracts, 3rd International Congress on Protozoology, Nauka, Leningrad, pp. 133–134.
4. Berger, J. D. and Kimball, R. F. 1964. J. Protozool. 11:534–537.
5. Chapman-Andresen, C. 1962. C. R. Trav. Lab. Carlsberg 33:73–264.
6. Chapman-Andresen, C. and Holter, H. 1964. C. R. Trav. Lab. Carlsberg 34:211–226.
7. Chapman-Andresen, C., and Nilsson, J. R. 1968. C. R. Trav. Lab. Carlsberg 36:405–432.
8. Conner, R. L. 1967. In M. Florkin and B. T. Scheer, eds. Chemical zoology, vol. 1, Protozoa (G. W. Kidder, ed.), Academic Press, New York, pp. 309–350.
9. Conner, R. L. and Cline, S. G. 1964. J. Protozool. 11:486–491.
10. Cox, D., Frank, O., Hutner, S. H. and Baker, H. 1969. J. Protozool 15:713–716.
11. Cox, F. E. G. 1967. Trans. Amer. Microsc. Soc. 86:261–267.
12. Curds, C. R. and Cockburn, A. 1969. J. Gen. Microbiol. 54:343–358.

13. Dewey, V. C. and Kidder, G. W. 1954. Proc. Soc. Exp. Biol. Med. 87:198–199.
14. Dewey, V. C. and Kidder, G. W. 1960. J. Gen. Microbiol. 22:72–78.
15. Dewey, V. C. and Kidder, G. W. 1960. J. Gen Microbiol. 22:79–92.
16. Dewey, V. C. and Kidder, G. W. 1972. J. Protozool. 19:50–53.
17. Dewey, V. C., Heinrich, M. R. and Kidder, G. W. 1957. J. Protozool. 4:211–219.
18. Dickie, N. and Liener, I. E. 1962. Biochim. Biophys. Acta 64:41–51.
19. Eichel, H. J., Conger, N. and Figueroa, E. 1963. J. Protozool. 10(suppl.):6.
20. Elliott, A. M. 1965. Science 149:640–641.
21. Elliott, A. M. and Clark, G. M. 1958. J. Protozool. 5:235–240.
22. Elliott, A. M. and Clark, G. M. 1958. J. Protozool. 5:240–246.
23. Elliott, A. M. and Clemmons, G. L. 1966. J. Protozool. 13:311–323.
24. Elliott, A. M. and Outka, D. E. 1956. Biol. Bull. 111:301–302.
25. Elliott, A. M., Addison, A. M. and Carey, S. E. 1962. J. Protozool. 9:135–141.
26. Elliott, A. M., Studier, M. A. and Work, J. A. 1964. J. Protozool. 11:370–378.
27. Erwin, J. 1960. Ph.D. Thesis, Syracuse University.
28. Erwin, J. and Bloch, K. 1963. J. Biol. Chem. 238:1618–1624.
29. Grebecki, A. 1963. Protoplasma 56:89–98.
30. Haessler, H. A. and Cunningham, L. 1957. Exp. Cell Res. 13:304–311.
31. Hall, R. P. 1954. J. Protozool. 1:74–79.
32. Hall, R. P. 1954. Trans. N. Y. Acad. Sci. 16:418–419.
33. Harris, J. O. 1967. J. Protozool. 14:600–602.
34. Hill, D. L. and Chambers, P. 1967. Biochim. Biophys. Acta 148:435–447.
35. Holz, G. G., Jr. 1964. In S. H. Hutner, ed., Biochemistry and physiology of Protozoa, vol. 3, Academic Press, New York, pp. 201–240.
36. Holz, G. G., Jr., Erwin, J., and Davis, R. J. 1959. J. Protozool. 6:149–156.
37. Holz, G. G., Jr., Erwin, J., and Wagner, B. 1960. Unpublished.
38. Holz, G. E., Jr., Erwin, J. A., and Wagner, B. 1961. J. Protozool. 8:297–300.
39. Holz, G. G., Jr., Wagner, B., Erwin, J., Britt, J. J., and Bloch, K. 1961. Comp. Biochem. Physiol. 2:202–217.
40. Holz, G. G., Jr., Wagner, B. Erwin, J. and Kessler, D. 1961. J. Protozool. 8:192–199.
41. Holz, G. G., Jr., Erwin, J., Wagner, B., and Rosenbaum, N. 1962. J. Protozool. 9:359–363
42. Holz, G. G., Jr., Erwin, J., Rosenbaum, N., and Aaronson, S. 1962. Arch. Biochem. Biophys. 98:313–322.
43. Holz, G. G., Jr., Patterson, K., and Watts, T. 1962. Unpublished.
44. Hutner, S. H., Baker, H., Frank, O., and Cox, D. 1972. In Biology of nutrition (R. N. Fiennes, ed., International encyclopedia of nutrition, vol. 18, Pergamon Press, New York, pp. 85–177.
45. Jahn, T. L., Bovee, E. C., Dauber, M., Winet, H., and Brown, M. 1965. Ann. N. Y. Acad. Sci. 118:912–920.
46. Kaudewitz, F. 1958. Arch. Protistenk. 102:332–448.
47. Kessler, D. 1961. M. S. Thesis, Syracuse University.
48. Kidder, G. W. 1967. In M. Florkin and B. T. Scheer, eds., Chemical zoology, vol. 1, Protozoa (G. W. Kidder, ed.), Academic Press, New York, pp. 93–159.
49. Kidder, G. W. and Dewey, V. C. 1945. Physiol. Zool. 18:136–157.
50. Kidder, G. W. and Dewey, V. C. 1949. Arch. Biochem. Biophys. 20:433–443.
51. Kidder, G. W. and Dewey, V. C. 1951. In A. Lwoff, ed., Biochemistry and physiology of Protozoa, vol. 1, Academic Press, New York, pp. 323–400.
52. Lantos, T., Muller, M., Toro, I., Druga, A., and Vargha, I. 1964. Acta Biol. Acad. Sci. Hung. Suppl. 6:29.
53. Lazarus, L. H. and Scherbaum, O. H. 1967. Biochim. Biophys. Acta 142:368–384.
54. Lees, A. M. and Korn, E. D. 1966. Biochemistry 5:1475–1481.
55. Lloyd, D., Brightwell, R., Venables, S. E., Roach, G. I., and Turner, G. 1971. J. Gen. Microbiol. 65:209–223.
56. Loefer, J. B. and McDaniel, M. R. 1950. Proc. Am. Soc. Protozool. 1:8.
57. Lwoff, A. 1923. C. R. Acad. Sci., Paris, Ser. D 176:928–930.

58. Lwoff, A. and Roukelman, N. 1926. C. R. Acad. Sci., Paris, Ser. D. 183:156–158.
59. Mallory, F. B., Gordon, J. T., and Conner, R. L. 1963. J. Amer. Chem. Soc. 85:1362–1363.
60. McArdle, E. W. 1969. J. Protozool. 16(suppl.):24.
61. Muller, M. 1967. In M. Florkin and B. T. Scheer, eds., Chemical zoology, vol. 1, Protozoa (G. W. Kidder, ed.), Academic Press, New York, pp. 351–380.
62. Muller, M. 1971. Acta Biol. Acad. Sci. Hung. 22:179–186.
63. Muller, M. 1971. J. Protozool. 18(suppl.):10.
64. Muller, M. 1972. J. Cell Biol. 52:478–487.
65. Muller, M., Baudhuin, P., and de Duve, C. 1966. J. Cell. Physiol. 68:165–175.
66. Muller, M., Rohlich, P., and Toro, I. 1965. J. Protozool. 12:27–34.
67. Muller, M. Rohlich, P., Toth, J., and Toro, I. 1963. Ciba Foundation Symposium on Lysosomes, Churchill, London, pp. 201–216.
68. Muller, M., Toth, J., and Toro, I. 1960. Nature 187:65.
69. Nilsson, J. R. 1971. J. Protozool. 18(suppl.):33.
70. Rasmussen, L. and Kludt, T. 1970. Exp. Cell Res. 59:457–463.
71. Reynolds, H. 1969. J. Protozool. 16:204–210.
72. Reynolds, H. 1970. J. Bacteriol. 104:719–727.
73. Reynolds, H. and Wragg, J. B. 1962. J. Protozool. 9:214–22.
74. Ricketts, T. R. 1970. Protoplasma 71:127–137.
75. Ricketts, T. R. 1971. Exp. Cell Res. 66:49–58.
76. Ricketts, T. R. 1972. Arch. Mikrobiol. 81:344–349.
77. Roth, J. S., Eichel, H. J., and Ginter, E. 1954. Arch. Biochem. Biophys. 48:112–119.
78. Ryley, J. G. 1952. Biochem. J. 52:483–492.
79. Seaman, G. R. 1961. J. Protozool. 8:204–212.
80. Seaman, G. R. 1961. J. Cell Biol. 1:243–245.
81. Shaw, R. F. and Williams, N. E. 1963. J. Protozool. 10: 486–491.
82. Shug, A. L., Elson, C., and Shrago, E. 1969. J. Nutr. 99:379–386.
83. Slater, J. V. 1952. Physiol. Zool. 25:283–287.
84. Slater, J. V. 1952. Physiol. Zool. 25:323–332.
85. Smith, I. 1961. Ph.D. Thesis. Columbia University.
86. Stokstad, E. L. R., Hoffmann, C. E., Regan, M. A., Fordham, D., and Jukes, T. H. 1949. Arch. Biochem. Biophys. 20:75–82.
87. Uhlig, G., Komnick, H., and Wohlfarth-Bottermann, K. 1965. Helgolaender Wiss. Meeresunters 12:61–77.
88. Ulsamer, A. G., Wright, P. L., Wetzel, M. G., and Korn, E. D. 1971. J. Cell Biol. 51:193–215.
89. Viswanatha, T. and Liener, I. E. 1956. Arch. Biochem. Biophys. 61:410–421.
90. Wagner, B. and Erwin, J. 1961. Personal communication.

CHAPTER 4

The Composition, Metabolism, and Roles of Lipids in *Tetrahymena*

George G. Holz, Jr.

Department of Microbiology
State University of New York
Upstate Medical Center
Syracuse, New York

and

Robert L. Conner

Department of Biology
Bryn Mawr College
Bryn Mawr, Pennsylvania

I. INTRODUCTION

Interest in the lipids of *Tetrahymena* has increased in recent years with the recognition that *T. pyriformis* is a particularly favorable organism for studying general problems in lipid structure, function, and metabolism. This ciliate grows rapidly to high population densities in axenic culture in chemically defined media, biosynthesizes a variety of unusual lipids de novo, can take up lipids from the environment and incorporate them unaltered or convert them into other lipids, and in cell-free preparations is active in the biosynthesis and biodegradation of lipids. Further, there is a growing literature on the isolation and subfractionation of its membrane-encompassed organelles. Finally, the metabolism of *T. pyriformis* has been studied in greater detail than that of any other protozoan.

In this chapter we have been selective in our choice of subjects, citations, and degree of detail in discussion. For a more exhaustive account of the lipid composition, nutrition, and metabolism of protozoa, Dewey (27) should be consulted. The reader is cautioned not to place absolute faith in the quantitative analytical information cited below. *Tetrahymena pyriformis* contains lipases which alter complex lipids soon after any disturbance in the organism's integrity, and the lipids are rich in polyunsaturated fatty acids. *Tetrahymena pyriformis* has been reported to contain phospholipase A, lysophospholipase, and phospholipase C (128). Lipolytic enzymes have bedeviled enzymologists working with cell-free preparations (30,103,127). Measures taken to guard against lipolysis and oxidation include low temperature, nitrogen atmosphere, antioxidants, and lipase inhibitors. Application of these safeguards has not been uniform. In addition, there may be strain differences in lipid composition within a species. Most of the work reviewed here was done with *T. pyriformis,* and we have included strain designations where it seems appropriate. Finally, environmental conditions can influence lipid content and composition.

II. TOTAL LIPIDS

The lipid content of *T. pyriformis* has been estimated variously as 8–25% of the dry weight (1,9,20,72,93,94,126). There is a rise in total lipids that parallels population growth through the log and stationary phases (32,37,76). Allison and Ronkin (2) followed this increase by cytochemical methods and correlated the lipid content with changes in the number and distribution of lipid deposits in the cells (GL, S, W, WH-6). Koch and Scherbaum (76) observed that the ratio of total lipid to protein rose when log-phase cells were exposed to low oxygen tension, or received heat shocks to induce synchronous cell division (GL). Kidder et al. (72) increased the total lipids in proteose–peptone-grown cells to 44.8% of the dry weight by adding Tween 80 (polyoxyethylene sorbitan monooleate) to the medium (W).

Among the total lipids are neutral lipids (fatty acids, alcohols, aldehydes, ethers and esters, glycerides, hydrocarbons, squalene, triterpenoid alcohols, and a terpenoid quinone) and polar lipids (phospholipids, phosphonolipids, and a sulfolipid). The neutral components account for 8–76% of the total lipids, and the polar ones for 24–92% (6,9,20,32,117,126). The ratio of neutral to polar lipids is dependent upon population age and oxygen tension. The increase in the total lipids of cells during the transition from log to stationary phase has been attributed almost exclusively to increases in triglycerides (90% of neutral lipids) and unesterified fatty acids (32,37,76,127). Acetyl-CoA and long-chain acyl-CoA become depleted in the cells at this time (GL) (12).

III. TOTAL FATTY ACIDS

A. The Composition of Fatty Acids

Table 1 lists the fatty acids of *T. pyriformis* WH-14 found by Erwin and Bloch (32) after direct saponification of log-phase cells grown in a chemically defined, fatty acid-free medium. The acids were mostly even-numbered, straight-chain, saturated and unsaturated compounds. Lauric, myristic, palmitic, and stearic comprised ~15% of the total, the monoenoic acids palmitoleic and oleic ~20%, and the polyunsaturates linoleic and γ-linolenic ~55%. The remaining 10% was made up of small amounts of odd-numbered straight- and branched-chain acids, arachidic acid, and eicosenoic acids. *Tetrahymena*

Table 1.
Fatty Acid Composition of *T. pyriformis* WH-14 (32)

Fatty acid	Chain length (number of carbon atoms)	Double bond positions (number of carbon atoms from carboxyl group)	Total fatty acids (%)
Lauric	12	—	1.9
Myristic	14	—	6.5
13-Methyltetradecanoic	15	—	2.5
Pentadecanoic	15	—	1.0
14-Methylpentadecanoic	16	—	1.0
Palmitic	16	—	4.8
Palmitoleic	16	9	11.5
15-Methylhexadecanoic	17	—	tr
Margaric	17	—	1.0
Heptadecenoic	17	?	1.0
Stearic	18	—	1.0
Oleic	18	9	8.7
Linoleic	18	9,12[a]	17.9
γ-Linolenic	18	6,9,12	37.7
Arachidic	20	—	1.0
Eicosenoic	20	?	1.0

[a]The 6,9-isomer also has been reported (6,38,101).

does not contain notable amounts of the eicosenoic and docosenoic acids characteristic of plants and animals, which are products of the chain elongation and desaturation of α-linolenic and γ-linolenic acids.

Cells grown in chemically defined, fatty acid-free media demonstrate directly their capacity for de novo fatty acid biosynthesis. *Tetrahymena* can, however, incorporate exogenous long-chain fatty acids of great variety into its lipids directly, or after metabolic conversion by chain elongation and/or desaturation (82). Also, the configurations of the carbon chains of biosynthesized long-chain fatty acids can be influenced by the types of short-chain fatty acids supplied to the organism (117). Descriptions of the fatty acids of *T. pyriformis* have often been based on analyses of cells grown in mixtures of extracts and digests of natural materials (proteose–peptone, and liver or yeast extract). Palmitic, stearic, and oleic are the major fatty acids found in such culture media (76). These fatty acids are readily converted by the ciliate to its characteristic polyunsaturates: linoleic and γ-linolenic acids (32).

No striking differences in fatty acid composition have been reported for the various strains [WH-14 (32), W (82,117), and GL (76)]. Lees and Korn (82) did not report palmitoleic acid, but other investigators found that it accounted for up to 15% of the total. Peng and Elson (101), Ferguson et al. (38), and Berger et al. (6) recently found the $\Delta^{6,9}$ isomer of linoleic acid. It was concentrated in the phosphorus-containing lipids, particularly in a glyceryl ether aminoethylphosphonate, while linoleic acid ($\Delta^{9,12}$) was the major fatty acid of cardiolipin (6).

Erwin and Bloch (32) also described the fatty acid compositions of *T. corlissi* Th-X, *T. setifera* HZ-1, and *T. paravorax* RP. All were similar to that of *T. pyriformis* except that *T. setifera* HZ-1 and *T. paravorax* RP contained considerable amounts of odd-numbered, branched-chain acids (12 and 23% of the total fatty acids, respectively).

The fatty acids of the various species of *Tetrahymena* constitute ~5% of the dry weight of log-phase cells and ~10% of the dry weight of stationary-phase cells. Analyses of qualitative and quantitative distributions of the fatty acids in the complex lipids of *T. pyriformis* revealed that the neutral lipids (mostly triglycerides) were rich in the saturated (myristic, palmitic, and stearic) and monounsaturated (palmitoleic and oleic) acids. The polar lipids contained the bulk of the polyunsaturated acids (WH-14, W) (6,32,82).

Fatty acid composition is a function of culture age. The ratio of saturates to unsaturates (S/U) increases in the interval between early and late log phase. Erwin and Bloch (32) found an increase in the ratio of saturates to monounsaturates with increasing culture age, without an accompanying change in the relative contribution of polyunsaturates to the total fatty acids (WH-14). In contrast, Koch and Scherbaum (76) found that the ratio of saturates to monounsaturates did not change and that there was a decrease in the contribution of polyunsaturates to the total (GL). An interesting feature of similar work on other species was the finding that the amounts of C_{17} and C_{19} iso acids in *T. paravorax* decrease with culture age, from 28% of the total fatty acids in early log phase to only 3% in stationary phase (32). It was suggested that the basis for this decrease might be the utilization of leucine for protein synthesis, and its unavailability for iso fatty acid synthesis late in culture life.

The temperature dependence of the fatty acid composition of *T. pyriformis* has been studied with exponentially multiplying (WH-14) and division-synchronized (GL) populations (32,76). Constant exposure during growth to an incubation temperature higher than that used for control cells or alternating exposures to optimal and just sublethal temperatures to induce synchrony caused an increase in the S/U ratio. The S/U ratio change in division-synchronized cells was associated with a reduction in polyunsaturates. Upon release from cyclic heat treatments, the γ-linolenic acid content of these cells rose sharply and the S/U ratio fell.

Oxygen tension has a profound effect on fatty acid composition. The S/U ratio in late log-phase cells was reported to be 2.6 times greater for cells grown under conditions of limited oxygen tension than for control cells (76).

The fatty acid compositional changes associated with changes in culture age, temperature, and oxygen tension may have as a common basis the availability of oxygen to the enzyme systems responsible for unsaturated fatty acid biosynthesis. As populations grow in size, oxygen becomes limiting in the standing cultures normally used, even when the ratio of culture medium surface area to culture medium volume is large (84). This situation is aggravated by elevating the temperature, thus lowering oxygen solubility while at the same time stimulating oxygen consumption. Heat-induced division synchronization is carried out with relatively low population densities (5×10^4–1×10^5), but much higher temperatures are used than in population growth experiments.

B. Metabolism of Fatty Acids

The species of *Tetrahymena* that can be grown in a chemically defined medium and that do not require fatty acids include all strains of *T. pyriformis, T. corlissi* Th-X, *T. setifera* HZ-1, and *T. paravorax* RP (52). The characteristic fatty acids were biosynthesized de novo when the cells were grown in a basal medium containing glucose and amino acids as sources of carbon and energy. When fatty acids were added, the effect on cell fatty acid composition varied with the nature of the supplement. Acetate was incorporated readily into fatty acids (32,127), and the fatty acid composition of the cells was the same as that observed when the cells were grown in the basal medium. When propionate was supplied, however, the fatty acid composition was altered qualitatively and quantitatively (117). Odd-numbered acids (13:0, 15:0, 17:0, 17:1) became conspicuous. When isobutyrate was added, 30% of the total fatty acids were even-numbered, saturated, iso acids (iso 14:0, 16:0, 18:0). Supplementation with α-methyl-*n*-butyrate resulted in 20% anteiso, odd-numbered, saturated acids (anteiso 13:0, 15:0, 17:0, 19:0) and a variety of unknown saturated and unsaturated acids. In all these cases the proportion of unsaturates decreased. None of the short-chain carboxylic acid supplements influenced the ratio of neutral lipids to polar lipids (W).

All the species and strains of *Tetrahymena* that have been examined take up palmitic, stearic, oleic, and linoleic acids and elongate and/or desaturate these compounds to the characteristic polyunsaturate, γ-linolenic acid (32). When $[1\text{-}^{14}C]$palmitic acid was used, as much as 30% of the ^{14}C radioactivity supplied was recovered from the cells. The relative specific activities of the individual fatty acids indicated a sequence of events in which palmitic acid is chain-elongated to stearic, which is desaturated to oleic, to linoleic, and finally to γ-linolenic acid. The metabolism of added ^{14}C-labeled stearic, oleic, and linoleic acids followed the same pathway (Fig. 1). The $\Delta^{6,9}$ isomer of 18:2 was not observed in this study.

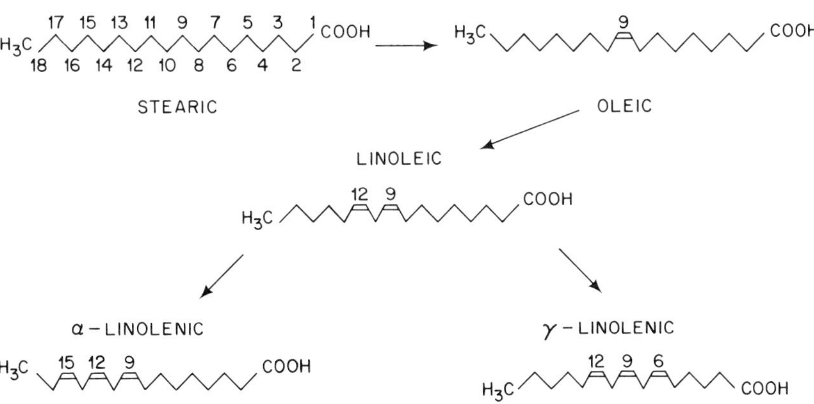

Fig. 1.

Polyunsaturated fatty acid biosynthesis.

The first double bond inserted into stearic acid was at the 9-position. The second double bond was introduced at the 12-position, in a divinyl methane relationship to the existing double bond, toward the methyl group end of the molecule. The third double bond was introduced at the 6-position, toward the carboxyl group end of the molecule. Randomization of ^{14}C by degradation of the labeled acids and recycling of fragments was limited, even when exposure of the cells to the labeled acids lasted for as long as 5 days. However, the capacity for both α and β oxidation of stearic acid has been reported for *T. pyriformis* W (3).

The sequence of reactions up to linoleic acid is characteristic of eucaryotic cells in general. The desaturation of linoleic to γ-linolenic acid is the step in vertebrate polyunsaturate synthesis that is followed by chain elongation, and a second, carboxyl group-oriented desaturation, to yield the characteristic 20-carbon, tetraenoic arachidonic acid. γ-Linolenic acid and arachidonic acid are also conspicuous in some eucaryotic microbial groups: chrysomonads (49), an ameba (77), phycomycetes (115), trypanosomatids (78,95), nonphotosynthetic or dark-grown euglenids and phytomonads (33,62), red (80) and brown algae (13,75), and mosses and ferns (111). Photosynthesizing euglenids, unicellular and filamentous green algae, and higher plants emphasize the conversion of linoleic acid to α-linolenic acid, the ω3 isomer of octadecatrienoic acid, instead of the ω6 isomer γ-linolenic acid. With few exceptions, for example, yeasts and molds (116), α-linolenic acid is found in eucaryotes as a component of chloroplast glycolipids, and γ-linolenic acid is found as a component of mitochondrial phospholipids. *Tetrahymena* and the other forms that synthesize γ-linolenic acid differ from vertebrates in being able to execute de novo biosynthesis of this polyene from acetate. Vertebrates are unable to desaturate oleic to linoleic acid, but when given a dietary source of linoleic acid (essential fatty acid) can perform the chain elongation and carboxyl group-oriented desaturations necessary to form arachidonic acid.

Tetrahymena also can incorporate long-chain fatty acids, unaltered, into its complex lipids. Thompson (127) found that log-phase *T. pyriformis* W introduced palmitic acid mainly into phospholipids. Stationary-phase cells, however, incorporated 30% of the palmitic acid into phospholipids, and the remainder into glycerides and other lipids that do not contain phosphorus. *Tetrahymena* can also incorporate fatty acids that it cannot biosynthesize. For example, exogenous α-linolenic acid was taken up and partitioned about equally between neutral and phospholipids. Relatively little was found in the unesterified fatty acid fraction (82).

The bias of *T. paravorax* for iso acid biosynthesis was examined by Erwin and Bloch (32). They supplied the ciliate with [*methyl*-^{14}C] methionine, [1-^{14}C] palmitic acid, or [U-^{14}C]leucine throughout early log phase and analyzed the distribution of radioactivity in individual fatty acids extracted from the cells. Little radioactivity from the methyl group of methionine was incorporated into the total fatty acids; therefore iso acids were not synthesized by methylation of straight-chain analogs. A small amount of radioactivity from palmitic acid was incorporated into the C_{17} and C_{19} iso acids, but the relative specific activities were low. Twenty percent of the radioactivity supplied as [U-^{14}C] leucine was recovered in the total fatty acids. All the fatty acids were labeled to some extent; however, the relative specific activities of the iso acids (C_{15}, C_{17}, and C_{19}) were two to four times greater than those of the straight-chain acids. This indicated that the iso acids were synthesized by chain elongation of isovaleric acid derived from leucine. The radioactivity observed in the straight-chain acids was believed to have originated from randomization of label caused by β oxidation of isovaleric acid.

Lees and Korn (81,82) attempted to define the specificities of the enzymes responsible for polyunsaturate synthesis, and of the sites where polyunsaturates play their structural and functional roles. *Tetrahymena pyriformis* W was supplied during growth with various C_{18} and C_{20} unsaturated fatty acids and the cell lipids were analyzed. An eicosamonoenoic acid (20:1 ω9) was taken up; part was converted to an eicosadienoic acid (20:2 ω6) and the remainder was incorporated, mainly into the phospholipids. When the eicosadienoic acid was supplied in the medium, some was converted into an eicosatrienoic acid (20:3 ω6), which was not further altered, and both unsaturates were also incorporated into the phospholipids. Added eicosatetraenoic acid, arachidonic acid (20:4 ω6), was incorporated without conversion. Unlike the other C_{20} polyunsaturates tested, it depressed population growth and the cells were abnormal in appearance. In tests of C_{18} unsaturates, the octadecamonoenoic acid vaccenic acid (18:1 ω7) was desaturated to an octadecadienoic acid (18:2 ω7), and a small amount was chain-elongated to an eicosamonoenoic acid (20:1 ω7). The octadecamonoenoic acid petroselinic acid (18:1 ω12) was toxic, but surviving cells were rich in the isomer. The octadecadienoic acid linoleic acid (18:2 ω6) was tolerated well, and its normal intracellular concentration was increased. The octadecatrienoic acid isomers, α-linolenic (18:3 ω3) and γ-linolenic (18:3 ω6), were toxic but were incorporated when supplied in small amounts. The γ-linolenic acid concentration could be raised to as high as 61% of the total fatty acids by this means. Some of the α-linolenic acid was desaturated to an octadecatetraenoic acid (18:4 ω3). Incorporation of "foreign" polyunsaturates was usually associated with a reduction in the content of linoleic and γ-linolenic acids.

The experiments of Shorb (117), Lees and Korn (82), and Erwin and Bloch (32) illustrate the ability of *T. pyriformis* to take up and incorporate, and/or convert and incorporate, a considerable variety of short-chain and long-chain fatty acids, and by so doing to modify greatly its fatty acid composition.

C. Temperature Effects on Fatty Acid Metabolism

Rosenbaum et al. (109) found that growing *T. pyriformis* WH-14 at 40°C (optimal temperature ~35°C) induced cell surface abnormalities visible with the light microscope. Levy et al. (86) observed that *T. pyriformis* E grown at 33°C (optimal temperature ~28° C) underwent similar changes, and they examined the ultrastructure of the deformed cells. Extensive derangement of the membranous interfaces of cell organelles was evident. Rosenbaum et al. (109) achieved some amelioration of the abnormalities by supplementing a chemically defined culture medium with soybean phospholipids rich in unsaturated fatty acids. Less effective were a combination of dipalmitoylphosphatidylcholine and dipalmitoylphosphatidylethanolamine, or the fatty acids of cells grown at 35°C. These findings suggested that high temperature caused a derangement of the normal metabolism and function of membrane lipids. Erwin (31) analyzed the lipids of cells grown at 35°C and at 40°C with and without phospholipid nutritional supplements. At 35°C, ~70%

of the phospholipid fatty acids of the cells grown with the dipalmitoyl phospholipids used by Rosenbaum et al. (109) were unsaturated, while at 40°C only ~15% were unsaturated (83% was palmitic acid). Free palmitic acid (and its di- and triglycerides), choline, and ethanolamine were inactive in supporting growth at 40°C in the absence of a phospholipid supplement. These observations were believed to indicate direct incorporation of the synthetic dipalmitoyl phospholipids into the membranes of cells grown at 40°C, where they spared a loss of capacity for normal phospholipid biosynthesis and function. The low concentration of unsaturated fatty acids in these cells, and in cells grown in unsupplemented medium at 40°C, suggested that high temperature might impair the biosynthesis of unsaturated fatty acids. Accordingly, the fatty acids of cells grown at 35°C were isolated and fed to cells in the basal medium at 40°C. Normal population growth and normal morphology resulted. Erwin (31) concluded that the primary lesion in cells grown at 40°C was in fatty acid desaturation, and that this was responsible for a derangement of phospholipid metabolism and function which led to the cell surface abnormalities observed. This conclusion is in harmony with the observations of Chou and Scherbaum (11) that acid-soluble phosphorus complexes, containing ethanolamine and aminoethylphosphonic acid, accumulated in *T. pyriformis* GL at 34°C (optimal temperature ~28°).

Reid et al. (105) and Cox (24) have described abnormalities of the surface of *T. pyriformis* WH-14 kept at low temperatures (0–5°C). In this case survival was prolonged by phytosterols and other lipids supplied in crude soybean lecithin.

IV. NEUTRAL LIPIDS

A. Composition

The neutral lipid fraction of *Tetrahymena* contains triglycerides, hydrocarbons, fatty alcohols, aldehydes, ethers, and esters. Ubiquinone (20,25,133), squalene (20,50,118), two pentacyclic alcohols (tetrahymanol and diplopterol) (90,92), free fatty acids (94), and waxy esters (97,135) are found; however, no tetrahymanol fatty acyl esters have been detected (20,135; but see ref. 64). It is reasonable to assume that farnesol (112, 113), chimyl alcohol, and *n*-hexadecanol are present (127).

The triglycerides are the major fraction of the neutral lipids, and it has been suggested that as in other eucaryotic cells they function as a carbon reserve (32). The mechanisms of regulation of triglyceride synthesis and utilization are unknown. Glycogen synthesis via glyconeogenesis, involving glyoxylate cycle induction by partial anaerobiosis or entrance into stationary phase, is believed to occur at the expense of triglycerides (84,85). Starvation is accompanied by a sequential use of lipids. First, the triglyceride reserves are utilized, and then the membrane lipids (83). The triglycerides were found to be rich in saturated (14:0, 16:0, 18:0—49% of the total) and monounsaturated (16:1, 18:1—32%) fatty acids and poor in polyunsaturates (18:2, 18:3—9%) (WH-14) (32).

The unsaponifiable lipid fraction from *T. pyriformis* W shows 10 discrete materials which can be resolved by silicic acid chromatography, and even more may be present (20). It is important to remember that some of the unsaponifiable components may arise from more complex materials during alkaline hydrolysis, and thus represent low-polarity compounds not found as such in the cells (20).

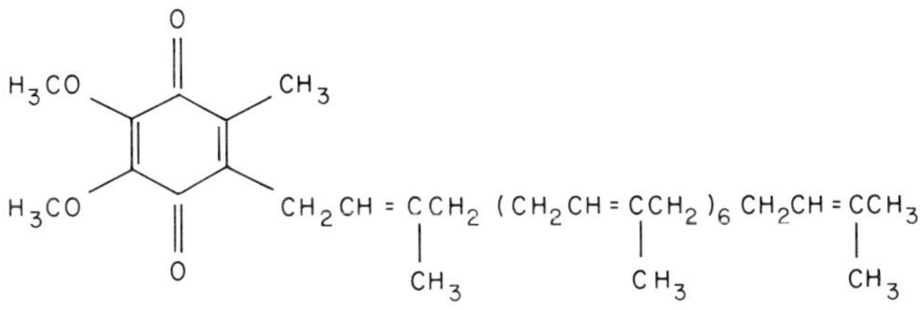

H_3CO ... CH_3 ... $CH_2CH = CCH_2 (CH_2CH = CCH_2)_6 CH_2CH = CCH_3$ (with CH_3 substituents and a benzoquinone ring bearing two $=O$ groups)

Fig. 2.

Ubiquinone$_8$.

All the unsaponifiables are labeled when the cells are grown with $[2\text{-}^{14}C]$ acetate. $[2\text{-}^{14}C]$Mevalonate serves as a precursor for the four identified constituents as well as at least three others. $[^{14}C]$Squalene incubations give rise to only two derivatives, tetrahymanol and diplopterol (20). No true sterols were detected by using sensitive isotopic techniques (4,20). These investigations extend and confirm the observations of Aaronson and Baker (1) who reported a negative Liebermann–Burchard test for sterols, and of Conner and Ungar (19) and Williams et al. (136) who could not detect sterols by gas-liquid chromatographic procedures. Dewey (27) noted small amounts of a highly polar lipid material which gave an unusual Liebermann–Burchard reaction and formed a digitonin precipitate soluble in pyridine. The nature of this trace component remains to be resolved.

B. Ubiquinone$_8$ (Coenzyme Q$_8$)

Ubiquinone$_8$ (coenzyme Q$_8$) (Fig. 2) has been reported in *T. pyriformis* W (25,133). Relatively large amounts (0.66 μmole/g dry weight) are found. The biosynthesis of the polyisoprenoid side chain is thought to be via the usual acetate $\rightarrow$ mevalonate $\rightarrow$ Δ^3-isopentenyl pyrophosphate $\rightarrow$ farnesyl pyrophosphate pathway (8,20). The benzoquinone ring arises from glucose (8), p-hydroxybenzoic acid, and shikimic acid (96). The surprising finding of a functional shikimic acid pathway in *Tetrahymena* is of interest since tryptophan and phenylalanine are absolute nutritional requirements. It may be that this metabolic pathway has been preserved solely for ubiquinone synthesis and that the large amount of this compound found in the cells is due to a lack of utilization of the shikimic acid for aromatic amino acid biosynthesis.

C. Tetrahymanol and Diplopterol

The major solid alcohol of *T. pyriformis* was originally isolated by McKee et al. (94) and identified as a compound isomeric with cholesterol; however, Mallory et al. (90) established that it is a pentacyclic triterpenoid alcohol. An unambiguous chemical synthesis by Tsuda et al. (132) and the x-ray crystallographic studies by Gordon and Doyne (45) have confirmed the structure to be gammaceran-3β-ol (Fig. 3).

The occurrence of a pentacyclic triterpenoid alcohol in this ciliate is unusual since this class of compounds has not been found in other "animal" cells. Pentacyclic triterpenoids have in no other case been found to be the dominant solid alcohol in an organism, although tetrahymanol has been found in a plant (140). This fatty alcohol comprises 0.14% of the dry weight (3.25 μmole/g) of cells (90) and constitutes approximately

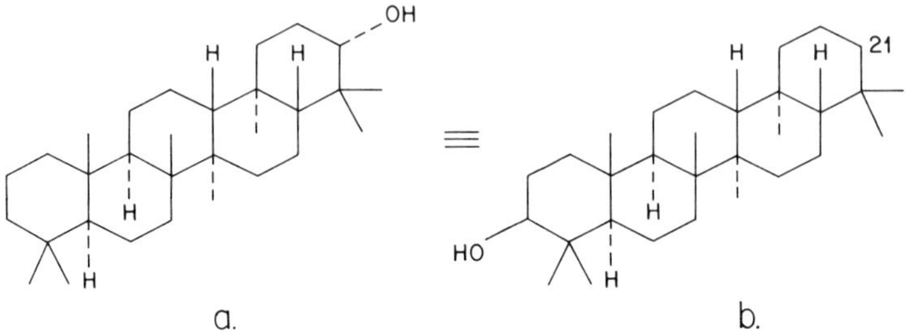

Fig. 3.
Tetrahymanol.

50% of the unsaponifiable fraction (17,118). A second minor component has been identified as diplopterol on the basis of thin-layer and gas-liquid chromatographic comparison
with an authentic sample (16), and by a mass spectral analysis (92).

Beach et al. (4) found tetrahymanol in *T. pyriformis* WH-14, *T. setifera* HZ-1, *T.
vorax* Tur and V$_2$S, *T patula* LI and L-FF, *T. corlissi* Th-X, *T. paravorax* RP, and
T. limacis Mf.1. Even though all of these species synthesize tetrahymanol, only *T. pyriformis* grows well in the absence of lipid supplements (52).

The biosynthesis of tetrahymanol and diplopterol follows the acetate → mevalonate
→ squalene sequence (10,17,20,92); however, a nonoxidative, possibly "proton-
induced," mechanism for squalene cyclization originally proposed by Tsuda et al. (132)
has gained experimental support (10,139) (Fig. 4).

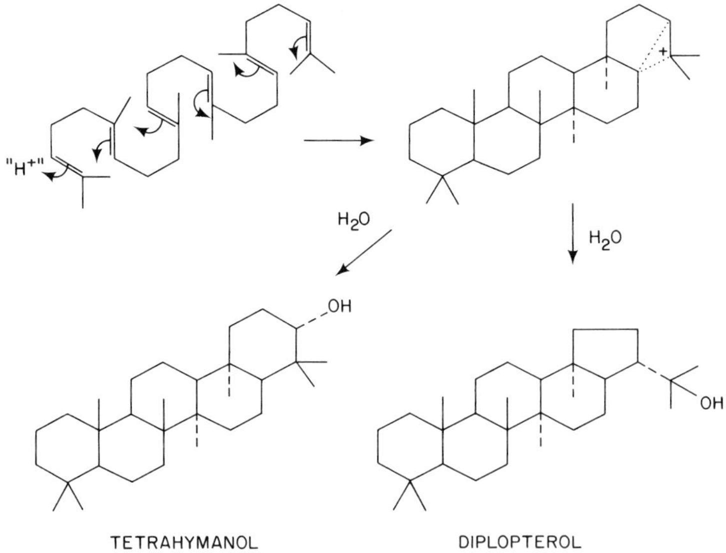

Fig. 4.
Biosynthesis of tetrahymanol.

The biosynthetic studies on tetrahymanol and diplopterol represent the first direct experimental evidence for a pathway not involving squalene 2,3-oxide as an intermediate in squalene cyclization. Figure 3 shows that tetrahymanol can be drawn either as the 21α-ol (A) or the 3β-ol (B); however, the biosynthetic mechanism probably proceeds through the 21α-ol, water serving as the source for the oxygen function (139) as shown in Fig. 4. Molecular oxygen is not required for the squalene cyclization. Diplopterol would also be anticipated as a product of this reaction mechanism. The squalene-tetrahymanol cyclase has been localized in "microsomal" particulates (22).

D. Sterol Effects on Lipid Biosynthesis

Although sterol synthesis does not appear to occur in *T. pyriformis,* these compounds are accumulated by the ciliate (17,19,20,21,91,99). Three separate phases of accumulation occur at 13–19°C. There is a rapid initial, temperature-independent phase which probably involves surface adsorption, a lag phase, and a temperature-dependent, hence metabolic, phase. When cholesterol is added to culture fluid, approximately 60% is taken up by the cells regardless of the external concentration employed. This has led to the speculation that the food vacuole may represent the major site of cholesterol entry (17). Many exogenous sterols are accumulated and transformed by the ciliates to a common derivative. Exogenous sterols also inhibit tetrahymanol and diplopterol biosynthesis from acetate, mevalonate, and squalene (20,21). Cholesterol or a metabolic derivative appears to block the cyclization of squalene to tetrahymanol and diplopterol, as well as the incorporation of mevalonic acid into squalene, although acetate and mevalonate incorporation into other terpene derivatives continues.

It can be concluded from these observations that the major blocks induced by cholesterol in the metabolic sequence leading to tetrahymanol occur after isoprene formation and before squalene synthesis, as well as at the squalene cyclization step. This pattern of inhibition seems to differ from that reported in mammalian liver and other tissues. Siperstein and Guest (119) suggested that the primary point of feedback inhibition induced by exogenous cholesterol in liver is the reduction of a hydroxymethylglutaryl derivative to mevalonic acid. Several groups (29,46) have also reported a lowering of the rate of incorporation of mevalonic acid and of squalene into sterols when rats are fed an exogenous source of cholesterol, but these blocks seem to be secondary in the inhibition pattern. These additional sites of inhibition of sterol synthesis in rat liver by dietary cholesterol have been confirmed and extended by SubbaRao and Olson (124) and by Krishnaiah et al. (79), who have noted that ubiquinone (coenzyme Q) biosynthesis is not reduced by low levels of cholesterol feeding, and who also found an inhibition of squalene cyclization to lanosterol. The situation in *Tetrahymena* appears to be similar to these later reports with an additional control point occurring after isoprene formation, since cholesterol does not block incorporation of the mevalonate into derivatives other than squalene, tetrahymanol, and diplopterol.

Some species of *Tetrahymena* require growth factor amounts of a sterol when they are grown in a chemically defined medium (53,55,57). This requirement may be spared or bypassed by proper supplementation with other lipids. In the case of *T. setifera* HZ-1, cholesterol can be replaced by the addition of synthetic L-α-(dipalmitoyl)-

phosphatidylethanolamine or "natural" phosphatidylethanolamines containing saturated and unsaturated fatty acids. The sterol bypassing action of the phospholipids is enhanced by free monounsaturated fatty acids, for example, oleic acid (57). Phosphatidylethanolamine rich in linoleic and γ-linolenic acids is the main phospholipid fraction of *T. setifera* grown in the presence of cholesterol, and a phosphatidylethanolamine which contains 95% palmitic acid is found in the ciliate when synthetic L-α-(dipalmitoyl)phosphatidylethanolamine is added to the culture fluid (36). *Tetrahymena setifera* grown in a chemically defined medium supplemented with cholesterol and with L-α-(dipalmitoyl)-phosphatidylethanolamine exhibited a depression in incorporation of [1-^{14}C]acetate or [1-^{14}C]stearate into the major polyunsaturated fatty acids of the organism, when compared to cells grown with cholesterol alone (35). In other growth experiments purified fatty acids from *T. setifera* partially replaced both cholesterol and phospholipids. It was concluded from these observations that the sterol (or some conversion product) is involved in the stepwise desaturation of the C_{18} fatty acids and that an impairment of these reactions interferes with normal phospholipid biosynthesis.

E. Sterol Transformations

Tetrahymena pyriformis W converts cholesterol to 7,22-bisdehydrocholesterol ($\Delta^{5,7,22}$-cholestatrien-3β-ol) by the introduction of double bonds at the C-7:8 and C-22:23 positions (see Fig. 5) (21). The stereospecificity of the proton removal at C-5, C-7, and C-22 has been elucidated (7,98,137,138).

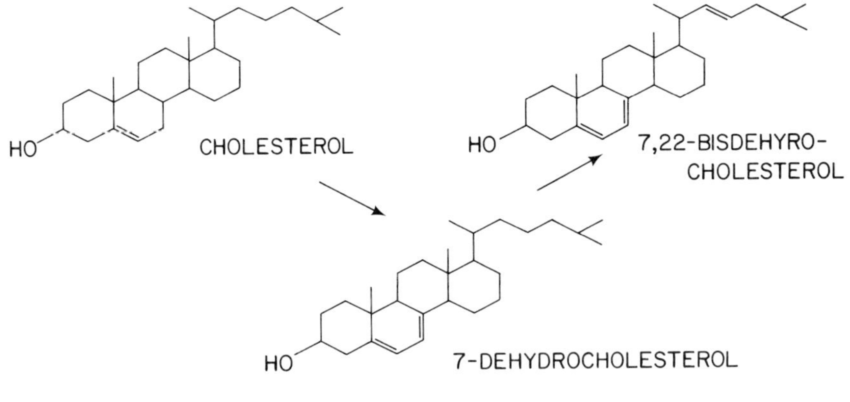

Fig. 5.
Cholesterol metabolism.

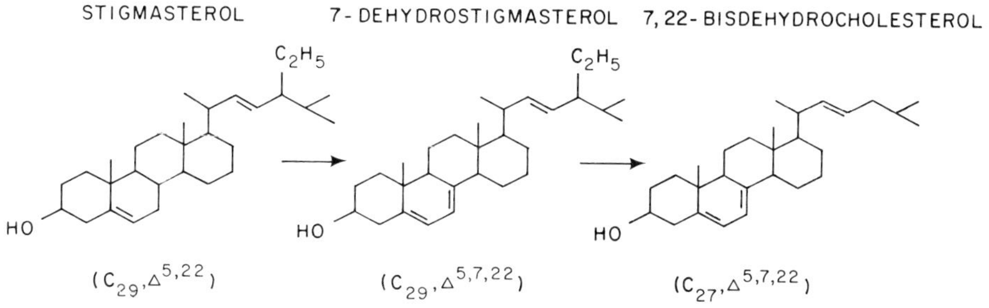

Fig. 6.

Stigmasterol metabolism.

22-Dehydrocholesterol has been identified in the sterol mixtures from cholesterol incubations, and reasonable evidence has been offered demonstrating the presence of 7-dehydrocholesterol (17,21,91). Conner and Ungar (19) had earlier noted the appearance of polar material in $[4\text{-}^{14}C]$ cholesterol incubations and concluded that a metabolic transformation had occurred. It now appears likely that the highly polar derivatives observed by these investigators were, in reality, degradation products of the labile 7,22-bisdehydrocholesterol.

Several sterol incubations have been carried out with *T. pyriformis* that indicate a rather remarkable ability to introduce isolated double bonds at the Δ^5, Δ^7, and Δ^{22} positions (89). 22-Dehydrocholesterol is readily converted to 7,22-bisdehydrocholesterol by the introduction of the Δ^7 (Fig. 5). No other sterols were detected. 7-Dehydrocholesterol is converted to 7,22-bisdehydrocholesterol (Fig. 5), and again no other sterols are found in the cellular lipids. Lathosterol, possessing a Δ^7, is converted to the cholestatrienol by the introduction of the Δ^5 and Δ^{22} double bonds, while cholestanol which contains no double bonds is metabolized to give a good yield of the 7,22-bisdehydrocholesterol. 22-Dehydrocholestanol has been characterized in this latter incubation, and there is reasonable evidence that suggests the presence of a Δ^5 (cholesterol) and Δ^7 (lathosterol) component. If verified this would indicate that each double bond can be introduced independently. This conclusion is reinforced by the detection of what appears to be three dienes, $\Delta^{5,7}$ (7-dehydrocholesterol), $\Delta^{7,22}$ (22-dehydrolathosterol), and $\Delta^{5,22}$ (22-dehydrocholesterol).

Not only can *Tetrahymena* desaturate sterols, but these cells can also dealkylate certain phytosterols, namely, 7-dehydrostigmasterol, stigmasterol, and β-sitosterol. Stigmasterol incubations give rise to 7-dehydrostigmasterol and 7,22-bisdehydrocholesterol (Fig. 6).

It appears that the dealkylation process is more specific with regard to substrate than the oxidation reactions, since no 22-dehydrocholesterol, the direct dealkylation product of stigmasterol, is detected in the sterol mixtures. β-Sitosterol is converted to 7,22-bisdehydrocholesterol, and two intermediates, $\Delta^{5,7}$-stigmastadiene-3β-ol and 7-dehydrostigmasterol are probably present in the sterol fraction. Little if any stigmasterol is detected. This suggests that introduction of the Δ^{22} double bond is hindered by the bulky ethyl substitution at C-24. This conclusion seems to be substantiated by the recent work of Nes et al. (99) who demonstrated that the conformation of the sterol side chain determines whether or not deethylation occurs.

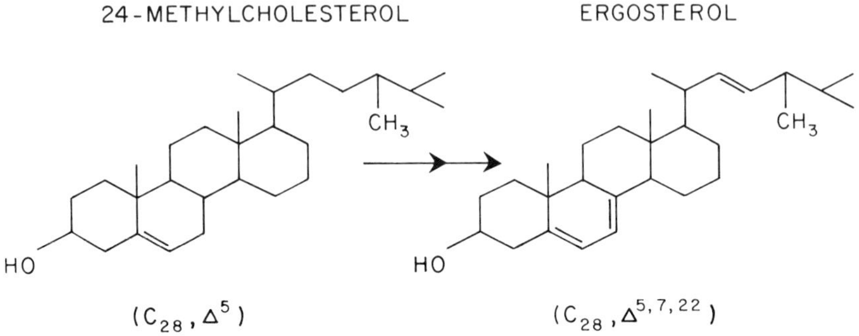

24-METHYLCHOLESTEROL   ERGOSTEROL

Fig. 7.

24-Methylcholesterol metabolism.

Ergosterol is incorporated into the cells without change, while 24-methylcholesterol is converted to ergosterol (Fig. 7). Dealkylation of the C-24 methyl substituent in these compounds does not occur.

V. POLAR LIPIDS

A. Composition of Phosphorus-Containing Lipids

Phospholipids were estimated as 24–92% of the total lipids of *T. pyriformis* W and WH-14 when cells were grown in a proteose–peptone-based medium (6,20,121,126) or in a chemically defined one (32,64,119). In early experiments chromatography of the total phospholipids had revealed that phosphatidylethanolamine and phosphatidylcholine were the major fractions (32,47,76,87,107,117,126). Upon further analysis each proved to be a mixture of structurally related compounds (Table 2). The phosphat-

Table 2.

Phosphorus-Containing Lipids of *T. pyriformis*

	Relative abundance[a]
Diacylglyceryl phosphorylethanolamine	45
Monoether monoacylglyceryl aminoethylphosphonate	14
Monoether monoacylglyceryl phosphorylcholine	13
Diacylglyceryl phosphorylcholine	9
Diacylglyceryl aminoethylphosphonate	5
Cardiolipin	5
Ceramide aminoethylphosphonate and other sphingolipids	—
Phosphatidylinositol	—
Lysophosphatidylethanolamine, lysophosphatidylcholine and lysoglyceryl aminoethylphosphonate	—

[a]Approximate proportions from averages of published values of percent total lipid phosphorus in each component (6,64,100).

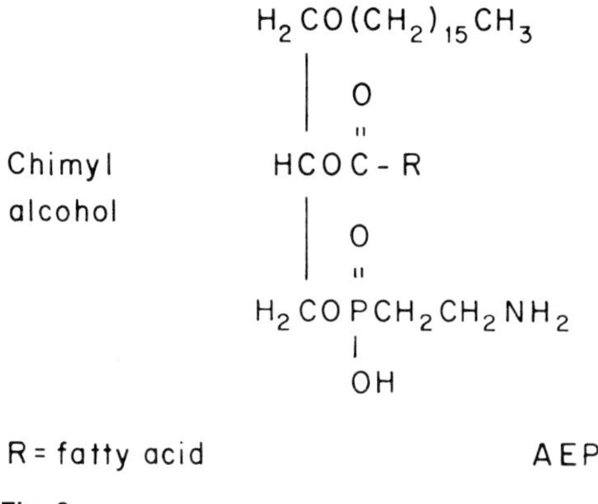

Fig. 8.

Glyceryl ether aminoethylphosphonate.

idylethanolamine fraction contained three components: diacylglyceryl phosphorylethanolamine, diacylglyceryl aminoethylphosphonate, and a monoether monoacylglyceryl aminoethylphosphonate (W, WH-14, GL) (5,6,87,121,125,127). The phosphatidylcholine fraction contained diacylglyceryl phosphorylcholine and a monoether monoacylglyceryl phosphorylcholine (W) (6,127).

Kandatsu and Horiguchi (65) were the first to find aminoethylphosphonic acid (AEP) in *T. pyriformis* W, where it was reported to be free, bound in lipids, and associated with proteins (WH-14) (110). It is a phosphonic acid analog of phosphorylethanolamine, discovered first in sheep rumen ciliates (60). A second phosphonic acid, 2-amino-3-phosphonopropionic acid (APP) was found by Kittredge and Hughes (73) in *T. pyriformis* WH-14. It is a phosphonic acid analog of phosphorylserine. The biological occurrence of the phosphonic acid analogs of the phosphate esters of all the common bases of phospholipids (ethanolamine, serine, choline) has been established with the finding of the *N*-methyl derivatives of AEP [*N*-methyl-AEP, *N,N*-dimethyl-AEP, *N,N,N*-trimethyl-AEP in an anemone (74), and ceramide *N*-methyl-AEP in molluscs (58)]. The methylated AEPs are phosphonic acid analogs of monomethylethanolamine, dimethylethanolamine, and choline phosphates. They have not been found in *Tetrahymena* (5,6,121,129; but see ref. 101).

The diacylglyceryl phosphorylethanolamine of *T. pyriformis* W was unusually rich in medium-chain length fatty acids (12:0 and 14:0 comprised 50% of the total saturates) and relatively poor in polyunsaturates, while the monoether monoacylglyceryl aminoethylphosphonate (Fig. 8) contained almost exclusively the polyunsaturates 18:2 $\Delta^{6,9}$ and 18:3 $\Delta^{6,9,12}$ acylating the C-2 position (6). The diacylglyceryl aminoethylphosphonate contained 16:0 (13%), 18:0 (53%), and 18:1 (22%) (GL) (125). The two components of the phosphatidylcholine fraction have not been separated and purified; therefore their fatty acid compositions are unknown. The entire fraction has a composition biased toward the unsaturates (W, WH-14) (6,64).

Other phospholipids that have been identified include cardiolipin (5,6,47,64, 100,107,122,126,127), sphingolipids (5,6,9,39,64,76,117,126,127), phosphatidylin-

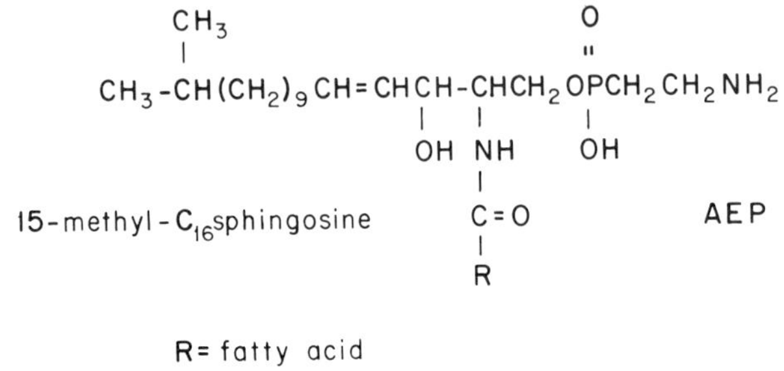

$$CH_3-CH(CH_2)_9 CH=CHCH-CHCH_2OPCH_2CH_2NH_2$$

15-methyl-C_{16}sphingosine   C=O   AEP

R= fatty acid

Fig. 9.

Ceramide aminoethylphosphonate.

ositol (47,64,107,117,127), and lysolipids (lysophosphatidylcholine, lysophosphatidyleth-anolamine, and lysoaminoethylphosphonolipid (32,64,100,107,117,122,127).

The cardiolipin had a fatty acid composition dominated by the monounsaturates 16:1 and 18:1 and the polyunsaturates 18:2 (mostly $\Delta^{9,12}$) and 18:3 $\Delta^{6,9,12}$ (W, WH-14) (6,64).

There is a mixture of sphingolipids in *T. pyriformis* W. Carter and Gaver (9) isolated a ceramide, a ceramide aminoethylphosphonate (Fig. 9), and a sphingomyelinlike lipid. Both ceramides contained branched, long-chain bases (15-methyl-C_{16} sphingosine and 17-methyl-C_{18} sphingosine) and small amounts of other bases, and were acylated with a branched-chain C_{17} acid (66%), stearic acid (19%) and smaller amounts of palmitic acid, and a branched-chain C_{19} acid. Berger et al. (6) obtained similar results. Ferguson et al. (39) examined the fatty acids of the sphingolipids in greater detail and found, in addition to the three major saturated acids (16:0, iso 17:0, and 18:0), the corresponding α-hydroxy forms.

B. Metabolism of Phosphorus-Containing Lipids

The incorporation of ^{14}C-labeled precursors into the phospholipids of *T. pyriformis* was studied first by Thompson (W) (127). [1-^{14}C]Acetate was an excellent source of the carbon skeletons. [U-^{14}C]glucose, [6-^{14}C]glucose, and [1,3-^{14}C]glycerol were poor sources. [1,2-^{14}C]Ethanolamine was slightly more active than glucose or glycerol and labeled both the ethanolamine and choline phospholipids. [1-^{14}C]Palmitate was rapidly incorporated. Cells from early log-phase populations incorporated palmitate mainly into phospholipids; however, in stationary-phase cells only 30% of the palmitate was found in the phospholipids. Turnover of [1-^{14}C]palmitate in phospholipids was low in cells multiplying rapidly. An attempt to demonstrate direct incorporation of [^{14}C]palmitate-labeled phosphatidylethanolamine into the membrane lipids failed. The distribution of radioactivity among the lipid fractions suggested that the phospholipid was degraded.

Dennis and Kennedy (26) presented evidence for the occurrence in *T. pyriformis* of the phosphatidylserine "decarboxylation cycle" operative in animal tissues; an exchange reaction of phosphatidylethanolamine with serine, followed by decarboxylation of phosphatidylserine to yield phosphatidylethanolamine and carbon dioxide. The exchange enzyme was found in the mitochondrial fraction (W). It occurs in a microsomal fraction in rat liver.

Smith and Law (121) studied phosphatidylcholine biosynthesis in *T. pyriformis* WH-14. The radioactivity of [*methyl*-^{14}C]methionine and of [*methyl*-^{14}C]choline was incorporated by intact cells into the choline of phosphatidylcholine and into no other base or phospholipid. Enzyme assays of subcellular fractions disclosed a microsomal phospholipid *N*-methyltransferase system which introduced the radioactivity of [*methyl*-^{14}C]-*S*-adenosyl-L-methionine into phosphatidylcholine. Characteristic of the system suggested that the biosynthesis of phosphatidylcholine followed the pathway described for many other organisms (vertebrates, eucaryotic microbes, and the bacterial genus *Agrobacterium*): phosphatidylethanolamine → phosphatidylmonomethylethanolamine → phosphatidyldimethylethanolamine → phosphatidylcholine. Smith and Law (121) also found a mitochondrial CDP-choline:diglyceride phosphocholine transferase which introduced the radioactivity of [1,2-^{14}C]CDP-choline into the choline of phosphatidylcholine and its glyceryl ether analog. They were unable to obtain incorporation of choline into sphingolipids by intact cells or by cell-free systems.

Interest in the metabolism of the phosphonolipids has centered on the biosynthesis of the C—P bond. To date, no mechanism postulated has received general acceptance; therefore we only attempt to summarize the relevant literature in approximate chronological order. The initial observations upon which subsequent work was based were mainly those of Liang and Rosenberg (87). They found that log-phase *T. pyriformis* WH-14 in a proteose–peptone medium, and washed organisms suspended for short periods in a chemically defined medium, incorporated ^{32}P-labeled orthophosphate most rapidly and to highest specific activity into lipid-bound AEP. Free AEP was labeled more slowly. This suggested that synthesis of lipid-bound AEP did not follow a pathway involving free AEP. The lipid fraction in which the AEP was bound to saponifiables was labeled twice as rapidly as the fraction in which AEP was bound to unsaponifiables. This suggested that AEP phosphorus might originate from the phosphatidic acid ester, and that the C—P bond was formed by the reaction of a carbon precursor with lipid-bound phosphorus.

Segal (114) proposed that AEP could be derived from phosphatidylethanolamine by conversion of ethanolamine phosphate to a nitrogenous analog of phosphoenolpyruvic acid, followed by an intramolecular phosphoramidic rearrangement to form the C—P bond. He also considered that APP synthesis could follow an analogous course in which phosphatidylserine took the place of phosphatidylethanolamine. Thompson (127), however, exposed log-phase ciliates to [1,2-^{14}C]ethanolamine and failed to obtain ^{14}C]AEP (W). Ethanolamine phospholipids and choline phospholipids were labeled. Liang and Rosenberg (88) and Horiguchi et al. (61) also found that ^{14}C-labeled ethanolamine *and* serine were poor precursors of the AEP skeleton (WH-14, GL). Warren (134) confirmed the negative finding for serine (W). All these observations argued against AEP synthesis via phosphatidylethanolamine.

The search for the carbon precursor of AEP has centered on intermediates in glycolysis. When whole cells were exposed to glucose, to pyruvate, and to acetate with specific carbon atoms labeled with ^{14}C, the distributions of radioactivity in phosphonic acids were consistent with the proposition that a single glycolytic intermediate was the precursor of both the AEP and APP carbon skeletons (61,88,131,134). [1-^{14}C]Glucose labeled primarily the phosphono carbon of AEP, and [2-^{14}C]glucose labeled the amino carbon.

There was no randomization of ^{14}C from specifically labeled glucose before incorporation into the phosphonic acids. $[3\text{-}^{14}C]$ Pyruvate (but not $[1\text{-}^{14}C]$pyruvate or $[2\text{-}^{14}C]$ pyruvate) and $[2\text{-}^{14}C]$acetate labeled AEP, as did $[2\text{-}^{14}C]$- and $[3\text{-}^{14}C]$succinate, -fumarate, and -malate (but not the $[1\text{-}^{14}C]$ forms), and $[2\text{-}^{14}C]$butyrate, $[3\text{-}^{14}C]$- or $[4\text{-}^{14}C]$glutamate, and $[2\text{-}^{14}C]$glycerol (GL) (61). Degradation of the ^{14}C-labeled AEP and examination of the ratio of radioactivity in the two carbon atoms showed equal labeling by $[2\text{-}^{14}C]$-acetate, $[3\text{-}^{14}C]$pyruvate, $[2\text{-}^{14}C]$- or $[3\text{-}^{14}C]$succinate, $[3\text{-}^{14}C]$- or $[4\text{-}^{14}C]$-glutamate, and unequal labeling (the amino carbon was much more heavily labeled than the phosphonocarbon) by $[2\text{-}^{14}C]$glucose, -glycerol, and -butyrate. These results suggested that the phosphono carbon of AEP came from the C-3 of phosphoenolpyruvate (PEP) or enoloxalacetate (EOA), and the amino carbon from the C-2 of PEP or EOA. The latter two compounds are interconvertible, and AEP could be produced by transamination of 3-phosphonopyruvate formed either directly by phosphonolation of PEP or by phosphonolation of the C-3 of EOA followed by decarboxylation.

Liang and Rosenberg (88) postulated that PEP may react with phosphatidic acid ester to form a glyceride ester of 3-phosphonopyruvate which upon transamination would yield an APP phosphonolipid. Decarboxylation of the APP would then give AEP. Phosphoenolpyruvate could react as well with a glyceryl ether derivative to form glyceryl ether AEP (127). When *T. pyriformis* was incubated with $[^{14}C]$palmitate or $[^{3}H]$chimyl alcohol, the choline glycerophosphatides were labeled rapidly and the ethanolamine glycerophosphatides slowly. There was a shift of radioactivity with time to the AEP glycerophosphatides. Thompson (129) interpreted these observations as an indication that choline glycerophosphatide was the "lipid carrier" of the AEP precursors. In support of this suggestion, Berger et al. (6) pointed out the fact that only aminoethylphosphonate and phosphorylcholine were linked to the glyceryl ether, chimyl alcohol. Dennis and Kennedy (26) showed that PEP and oxaloacetate (OAA) do not act as substrates for the *T. pyriformis* enzyme that exchanges L-serine for the ethanolamine of phosphatidylethanolamine.

Trebst and Geike (131) and Warren (134) favored the intramolecular rearrangement of PEP to 3-phosphonopyruvic acid, followed by transamination to APP and decarboxylation to AEP. In this circumstance both the phosphorus and the carbon of PEP would be incorporated into AEP. To harmonize the intramolecular rearrangement hypothesis with the bimolecular reaction hypothesis of Liang and Rosenberg (88), it was suggested that the rearrangement occurs in phosphatidylenolpyruvate (134). This compound has not yet been demonstrated.

Smith and Law (120) approached the question of C—P bond synthesis by examining the ability of intact cells to take up and metabolize AEP and APP (WH-14). Radioactivity of $[1,2\text{-}^{14}C]$AEP was incorporated ($\sim$10%) into the glyceryl ether–AEP fraction of the polar lipids. There was no evidence for in vivo degradation of the labeled AEP. Radioactivity of $[2\text{-}^{14}C]$APP appeared in glyceryl ether–AEP ($\sim$19%) and in triglycerides, fatty acids, and other uncharacterized materials. This incorporation was taken as evidence that there is a direct conversion of APP to AEP and that this conversion occurs with the free phosphonates. In support of this claim, APP was not found in phospholipids in the experiments nor in the original isolation of APP from *T. pyriformis* (73). Also, APP may be considered an analog of serine phosphate, and no mechanism is known for the incorporation of serine phosphate into phospholipids.

Attempts have been made to resolve the knotty problem of C—P bond synthesis with cell-free preparations of *Tetrahymena*. Liang and Rosenberg found (87) that diacylglyceryl AEP could be formed in vitro by the "Kennedy reaction," with AEP transferred to a diglyceride (dipalmitin) from cytidine monophosphate AEP. Warren (134) reported a limited conversion of ^{32}P-labeled PEP into AEP and APP (equivalent specific activity), and of ^{32}P-labeled APP into AEP, by a disrupted-cell preparation of *T. pyriformis* W. Geike (43) described a sulfhydryl-dependent cell-free system which labeled AEP when supplied with $[\text{U-}^{14}\text{C}]$ glucose (WH-14). Finally, Horiguchi (59) has recently been successful in obtaining incorporation of $[^{32}\text{P}]$ orthophosphate and $[\text{3-}^{14}\text{C}]$ PEP into AEP by a crude homogenate of *T. pyriformis* GL. The ^{14}C preferentially labeled the phosphono carbon. The rate of incorporation of ^{32}P and ^{14}C was greatest into free AEP and was more susceptible to inhibition by phosphonoacetaldehyde than by APP. These results were believed to support a major pathway to AEP via amination of phosphonoacetaldehyde. The decarboxylation of APP to AEP was relegated to the status of a bypass. Roberts et al. (106) earlier had suggested the formation of AEP by transamination of glutamate with phosphonoacetaldehyde.

The biodegradation of AEP and APP has been examined with dialyzed homogenates of *T. pyriformis* (106). Under anaerobic conditions, transaminations occurred between AEP and APP and α-ketoglutaric acid, which probably yielded phosphonoacetaldehyde and phosphonopyruvic acid, respectively. Introduction of a carbonyl group alpha or beta to the C—P bond is believed to increase the lability of the bond so that the products of transamination decompose, either spontaneously or by enzymic action, to orthophosphate and the corresponding carbonyl compound (acetaldehyde in both cases).

Improved techniques for the detection, isolation, and determination of phosphonates, including chromatographic and ^{31}P nuclear magnetic resonance methods, have been instrumental in accelerating research on the phosphonolipids of *Tetrahymena* (5,44,67,108,123,125).

C. Glyceryl Ether Metabolism

The AEP and choline phospholipids of *T. pyriformis* contain a glyceryl ether: chimyl alcohol (6,47,126,127). Biosynthesis of the ether bond has been achieved in microsomal preparations of *T. pyriformis* W and HSM (40,68). The complete system required CoA, Mg^{2+}, ATP, and NADPH. Dihydroxyacetone phosphate has been shown to be the precursor of the glycerol backbone, and hexadecanol (cetyl alcohol) the precursor of the alkyl side chain (41,42).

Catabolism of chimyl alcohol has been studied with intact cells, homogenates, and subcellular fractions (W) (69,70). A tracer amount of $[^{3}\text{H}]$ chimyl alcohol was incorporated by whole cells, rapidly and without degradation, into the phospho- and phosphonolipids. When a substrate amount was used, radioactivity appeared in triglycerides as well, and the intracellular pool of free chimyl alcohol increased. The major product of glyceryl ether degradation was palmitic acid, plus some fatty aldehyde and fatty alcohol. The responsible enzyme did not appear to be inducible or population age-dependent and was not associated with a cell organelle. The reaction required O_2, NAD^+ and an NADPH-generating system, but not a pteridine.

D. Sulfolipid

Haines (48) has found a long-chain alkyl sulfate in *T. pyriformis:* 1,14-docosyl disulfate. It is also found in *Ochromonas, Chorella, Chlamydomonas, Pseudomonas,* and *Streptomyces* and is probably an omnipresent microbial membrane component.

VI. DRUGS AND LIPID METABOLISM

Triparanol, 1-[*p*-(β-diethylaminoethoxy)phenyl]-1-(*p*-tolyl)-2-(*p*-chlorophenyl)ethanol, impaired population growth of *T. pyriformis,* delayed heat-induced synchronous division, blocked tetrahymanol biosynthesis with the accumulation of squalene, and inhibited the formation of unsaturated fatty acids (WH-14, GL, W) (54,56,63,102,118). The population growth inhibition was antagonized by certain sterols and by oleic acid. The dual nature of the effects of triparanol on the biosynthesis of tetrahymanol and unsaturated fatty acids was particularly apparent with synchronously dividing ciliates. The inhibition of division was antagonized by oleic acid but not by sterols.

A number of other substances elicit the need for a sterol supplement for growth. These inhibitors include certain substituted purines (28), colchicine (18), the oxidative phosphorylation uncoupling agent, 2,4-dinitrophenol (14), pregnane steroids (15,18), and 3β-(β-dimethylaminoethoxy)androst-5-en-17-one (DMAE-DHA) (50,51).

Holmlund (50) showed that 2,4-dinitrophenol, progesterone, and DMAE-DHA each act in a different fashion on respiration and carbon dioxide release, as well as on protein, RNA, and lipid synthesis. DMAE-DHA seems to act directly on squalene-tetrahymanol cyclase, while the modes of action of the other inhibitors are less direct and at present remain obscure. The relationship of sterols to these diverse inhibitors has not been established and will be resolved only when the cellular function(s) of tetrahymanol and the sterols are known.

Polyene antibiotics are relatively inactive as inhibitors of the growth of *T. pyriformis* WH-14 (34), and do not cause lysis of strain W unless the cells have been grown with a sterol supplement (23).

VII. LIPID DISTRIBUTION

A marked differential occurs in the chemical content of the various membranous cellular organelles of *Tetrahymena.* The most striking pattern is seen in the cilium in which phosphonolipids are found in greater abundance (64,71,122) than in the whole cell. The same pattern is true for glyceryl ethers (122,130). Berger et al. (6) showed that a high proportion of the phosphonolipid consists of 1-hexadecyl-2-acyl-*sn*-glyceryl-3-aminoethylphosphonate, which argues that this is the compound that is enriched in the cilium. In addition, tetrahymanol has been found to be enriched fivefold in the cilium (23,130). This finding has been extended to the ciliary membrane (23). Several investigators have assumed that the lipid of the isolated cilium is localized exclusively in the membrane portion of the organelle. Until the membranes have been isolated, analyzed, and compared to the intact cilium, the assumption is not warranted. The work reported to date, with one exception, has been with the membrane-bound organelle and not with the membrane per se.

Ceramide AEP is enriched in the cilia of strain W (104), as well as in the ciliary membrane (66). Cardiolipin, ceramide, and phosphatidylcholine are present in lesser amounts in the cilium than in the remainder of the cells (64,71,122). Mitochondria are enriched in cardiolipin (64,130).

The unique lipid class composition of each organelle examined thus far (64,130) is but one facet of the chemical specificity involved. There is also evidence that the fatty acid composition of a given phospholipid class may vary as a function of the organelle from which the phospholipid was isolated (64).

It is evident that additional chemical investigation coupled with physical measurements is necessary to interpret the descriptions currently available. The use of selected mutants does not seem to be a realistic approach at the present time; however, nutritional supplements which *Tetrahymena* so readily incorporates may provide considerable insight into the question of how the lipid composition of these cells relates to the biochemical-physiological functions.

REFERENCES

1. Aaronson, S. and Baker, H. 1961. J. Protozool. 8:274–277.
2. Allison, B. M. and Ronkin, R. R. 1967. J. Protozool. 14:313–319.
3. Avins, L. R. 1968. Biochem. Biophys. Res. Commun. 32:138–142.
4. Beach, D., Erwin, J. A., and Holz, G. G., Jr. 1969. *In* A. A. Strelkov, K. M. Sukhanova, and I. B. Raikov, eds. Progress in protozoology, Abstracts of the 3rd International Congress on Protozoology, Nauka, Leningrad, pp. 133–134.
5. Berger, H. and Hanahan, D. J. 1971. Biochim. Biophys. Acta 231: 584–587.
6. Berger, H., Jones, P., and Hanahan, D. J. 1972. Biochim. Biophys. Acta 260:617–629.
7. Bimpson, T., Goad, L. J., and Goodwin, T. W. 1967. Biochem, J. 115:857–858.
8. Braun, R., Dewey, V.C., and Kidder, G. W. 1963. Biochemistry 2:1070–1072.
9. Carter, H. E. and Gaver, R. C. 1967. Biochem. Biophys. Res. Commun. 29:886–891.
10. Caspi, E., Zander, J. M., Greig, J. B., Mallory, F. B., Conner, R. L., and Landrey, J. R. 1968. J. Am. Chem. Soc. 90:3563–3564.
11. Chou, S. C. and Scherbaum, O. H. 1967. Exp. Cell Res. 45:31–38.
12. Chua, A. S. and Ronkin, R. R. 1967. Comp. Biochem. Physiol. 21:425–429.
13. Chuecas, L. and Riley, J. P. 1966. J. Mar. Biol. Ass. U. K. 46:153–159.
14. Conner, R. L. 1957. Science 126:698.
15. Conner, R. L. 1959. J. Gen. Microbiol. 21:180–185.
16. Conner, R. L. Unpublished.
17. Conner, R. L. and Mallory, F. B. 1969. In K. Schubert, ed., Symposium uber biochemische Aspekte der Steroidforschung, Akademie-Verlag, Berlin, pp. 5–16.
18. Conner, R. L. and Nakatani, M. 1958. Arch. Biochem. Biophys. 74:175–181.
19. Conner, R. L. and Ungar, F. 1964. Exp. Cell Res. 36:134–144.
20. Conner, R. L., Landrey, J. R., Burns, C. H., and Mallory, F. B. 1968. J. Protozool. 15:600–605.
21. Conner, R. L., Mallory, F. B., Landrey, J. R., and Iyengar, C. -W. L. 1969. J. Biol. Chem. 244:2325–2333.
22. Conner, R. L., Mallory, F. B., Landrey, J. R. and Kritchevsky, D. Unpublished.
23. Conner, R. L., Mallory, F. B., Landrey, J. R., Ferguson, K. A., Kaneshiro, E. S., and Ray, E. 1971. Biochem. Biophys. Res. Commun. 44:995–1000.
24. Cox, D. 1970. J. Protozool. 17:150–152.
25. Crane, F. L. 1962. Biochemistry 1:510–517.
26. Dennis, E. A., and Kennedy, E. P. 1970. J. Lipid Res. 11:394–403.

27. Dewey, V. C. 1967. In M. Florkin and B. T. Scheer, eds., Chemical zoology, vol. 1, Protozoa (G. W. Kidder, ed.), Academic Press, New York, pp. 161–274.
28. Dewey, V. C., Kidder, G. W., and Markees, D. G. 1959. Proc. Soc. Exp. Biol. Med. 102:306–308.
29. Dietschy, J. M., and Siperstein, M. D. 1967. J. Lipid Res. 8:97–104.
30. Eichel, H. J. 1961. Symp. Genet. Biol. Ital. 8:381–416.
31. Erwin, J. A. 1970. Biochim. Biophys. Acta 202:21–34.
32. Erwin, J. and Bloch, K. 1963. J. Biol. Chem. 238:1618–1624.
33. Erwin, J. and Bloch, K. 1963. Biochem. Z. 338:496–511.
34. Erwin, J. and Shatz, G. W. Unpublished, cited in ref. 35.
35. Erwin, J. A., Beach, D. and Holz, G. G. Jr. 1966. Biochim. Biophys. Acta 125:614–616.
36. Erwin, J., Beach, D., and Holz, G. G., Jr. Unpublished, cited in ref. 35.
37. Everhart, L. P., Jr. and Ronkin, R. R. 1966. J. Protozool. 13:646–650.
38. Ferguson, K. A., Conner, R. L., and Mallory, F. B. 1971. Arch. Biochem. Biophys. 144:448–450.
39. Ferguson, K. A., Conner, R. L., Mallory, F. B., and Mallory, C. W. 1972. Biochim. Biophys. Acta 270:111–116.
40. Friedberg, S. J. and Greene, R. C. 1968. Biochim. Biophys. Acta 164:602–603.
41. Friedberg, S. J., Heifetz, A. and Greene, R. C. 1971. J. Biol. Chem. 246:5822–5827.
42. Friedberg, S. J., Heifetz, A., and Greene, R. C. 1972. Biochemistry 11:297–301.
43. Geike, F. 1969. Naturwissenschaften 56:462.
44. Glonek, T., Henderson, T. O., Hilderbrand, R. L., and Myers, T. C. 1970. Science 169:192–194.
45. Gordon, J. T. and Doyne, T. H. 1966. Acta Crystallogr. 21(suppl.):A113.
46. Gould, R. G. and Swyryd, E. A. 1966. J. Lipid Res. 7:698–707.
47. Hack, M. H., Yaeger, R. G. and McCaffery, T. D. 1962. Comp. Biochem. Physiol. 6:247–252.
48. Haines, T. H. 1965. J. Protozool. 12:655–659.
49. Haines, T. H., Aaronson, S., Gellerman, J. L., and Schenk, H. 1962. Nature 194:1282–1283.
50. Holmlund, C. E. 1971. Biochim. Biophys. Acta 248:363–378.
51. Holmlund, C. E. and Bohonos, N. 1966. Life Sci. 5:2133–2139.
52. Holz, G. G., Jr. 1964. In S. H. Hutner, ed., Biochemistry and physiology of protozoa, vol. 3, Academic Press, New York, pp. 199–242.
53. Holz, G. G., Jr., Erwin, J. A., and Wagner, B. 1961. J. Protozool. 8:297–300.
54. Holz, G. G., Jr., Rasmussen, L., and Zeuthen, E. 1963. C. R. Trav. Lab. Carlsberg 33:289–300.
55. Holz, G. G., Jr., Wagner, B., Erwin, J., Britt, J. J., and Bloch, K. 1961. Comp. Biochem. Physiol. 2:202–217.
56. Holz, G. G., Jr., Erwin, J., Rosenbaum, N., and Aaronson, S. 1962. Arch. Biochem. Biophys. 98:312–322.
57. Holz, G. G., Jr., Erwin, J., Wagner, B., and Rosenbaum, N. 1962. J. Protozool. 9:359–363.
58. Hori, T., Sugita, M., and Itasaka, O. 1969. J. Biochem. (Tokyo) 65:451–457.
59. Horiguchi, M. 1972. Biochim. Biophys. Acta 261:102–113.
60. Horiguchi, M. and Kandatsu, M. 1960. Nature 184:901–902.
61. Horiguchi, M., Kittredge, J. S., and Roberts, E. 1968. Biochim. Biophys. Acta 165:164–166.
62. Hulanicka, D., Erwin, J., and Bloch, K. 1964. J. Biol. Chem. 239:2778–2787.
63. Johnson, W. J. and Jasmin, R. 1961. Fed. Proc. 20:281.
64. Jonah, M. and Erwin, J. A. 1971. Biochim. Biophys. Acta 231:80–92.
65. Kandatsu, M. and Horiguchi, M. 1961. Agr. Biol. Chem. 26:721–722.
66. Kaneshiro, E. and Conner, R. L. 1972. In press.
67. Kapoulas, V. M. 1969. Biochim. Biophys. Acta 176:324–329.
68. Kapoulas, V. M. and Thompson, G. A., Jr. 1969. Biochim. Biophys. Acta 187:594–597.
69. Kapoulas, V. M., Thompson, G. A., Jr., and Hanahan, D. J. 1969. Biochim. Biophys. Acta 176:237–249.
70. Kapoulas, V. M., Thompson, G. A., Jr., and Hanahan, D. J. 1969. Biochim. Biophys. Acta 176:250–264.
71. Kennedy, K. E. and Thompson, G. A., Jr. 1970. Science 168:989–991.
72. Kidder, G. W., Dewey, V. C., and Heinrich, M. R. 1954. Exp. Cell Res. 7:256–264.
73. Kittredge, J. S. and Hughes, R. R. 1964. Biochemistry 3:991–996.

74. Kittredge, J. S., Isbell, A. F., and Hughes, R. R. 1967. Biochemistry 6:289–295.

75. Klenk, E., Knipprath, W., Eberhagen, D., and Koof, H. P. 1963. Hoppe-Seyler's Z. Physiol. Chem. 334:44–59.

76. Koch, E. G. and Scherbaum, O. H. 1967. Z. Allg. Mikrobiol. 7:349–361.

77. Korn, E. D. 1963. J. Biol. Chem. 238:3584–3587.

78. Korn, E. D., Greenblatt, C. L., and Lees, A. M. 1965. J. Lipid Res. 6:43–50.

79. Krishnaiah, K. V., Joshi, V. C., and Ramasarma, T. 1967. Arch. Biochem. Biophys. 121:147–153.

80. Laur, M. H. 1961. Theses, Faculte des Sciences de L'Universite de Paris, pp.1–93.

81. Lees, A. M. and Korn, E. D. 1966. Biochim. Biophys. Acta 116:403–406.

82. Lees, A. M. and Korn, E. D. 1966. Biochemistry 5:1475–1481.

83. Levy, M. R. and Elliott, A. M. 1968. J. Protozool. 15:208–222.

84. Levy, M. R. and Scherbaum, O. H. 1965. J. Gen. Microbiol. 38:221–230.

85. Levy, M. R. and Scherbaum, O. H. 1965. Arch. Biochem. Biophys. 109:116–121.

86. Levy, M. R., Gollon, C. E., and Elliott, A. M. 1969. Exp. Cell Res. 55:295–305.

87. Liang, C. R. and Rosenberg, H. 1966. Biochim. Biophys. Acta 125:548–562.

88. Liang, C. R. and Rosenberg, H. 1968. Biochim. Biophys. Acta 156:437–439.

89. Mallory, F. B. and Conner, R. L. 1971. Lipids 6:149–153.

90. Mallory, F. B., Gordon, J. T., and Conner, R. L. 1963. J. Am. Chem. Soc. 85:1362–1363.

91. Mallory, F. B., Conner, R. L., Landrey, J. R., and Iyengar, C.-W. L. 1968. Tetrahedron Lett. 58:6103–6106.

92. Mallory, F. B., Conner, R. L., Landrey, J. R., Zander, J. M., Greig, J. B., and Caspi, E. 1968. J. Am. Chem. Soc. 90:3564–3566.

93. McCashland, B. W. and Steinacher, R. H. 1962. Proc. Soc. Exp. Biol. Med. 111:789–793.

94. McKee, C. M., Dutcher, J. D., Groupe, V. and Moore, M. 1947. Proc. Soc. Exp. Biol. Med. 65:326–332.

95. Meyer, H. and Holz, G. G., Jr. 1966. J. Biol. Chem. 241:5000–5007.

96. Miller, J. E. 1965. Biochem. Biophys. Res. Commun. 19:335–339.

97. Ming Chu, I., Wheeler, M. A., and Holmlund, C. E. 1972. Biochim. Biophys. Acta 270:18–22.

98. Mulheirn, L. J., Aberhart, D. J. and Caspi, E. 1971. J. Biol. Chem. 246:6556–6559.

99. Nes, W. R., Malya, P. A. G., Mallory, F. B., Ferguson, K. A., Landrey, J. R., and Conner, R. L. 1971. J. Biol. Chem. 246:561–568.

100. Nozawa, Y. and Thompson, G. A., Jr. 1971. J. Cell Biol. 49:712–721.

101. Peng, Y. -M. and Elson, C. E. 1971. J. Nutr. 101:1177–1184.

102. Pollard, W. O., Shorb, M. S., Lund, P. G., and Vasaitis, V. 1964. Proc. Soc. Exp. Biol. Med. 116:539–543.

103. Rahat, M., Judd, J., and van Eys, J. 1964. J. Biol. Chem. 239:3537–3545.

104. Ray, E. 1972. Ph.D. Thesis, Bryn Mawr College, Bryn Mawr, Pennsylvania.

105. Reid, R., Cox, D., Baker, H., and Frank, O. 1969. J. Protozool. 16:231–235.

106. Roberts, E., Simonsen, D. G., Horiguchi, M., and Kittredge, J. S. 1968. Science 159:886–888.

107. Rogers, C. G. 1968. Can. J. Biochem. 46:331–339.

108. Roop, W. E., Tan, S. A., and Roop, B. L. 1971. Anal. Biochem. 44:77–80.

109. Rosenbaum, N., Erwin, J., Beach, D., and Holz, G. G., Jr. 1966. J. Protozool. 13:535–546.

110. Rosenberg, H. 1964. Nature 203:299–300.

111. Schlenk, H. and Gellerman, J. L. 1965. J. Amer. Oil Chem. Soc. 42:504–511.

112. Schneiderman, H. A. and Gilbert, L. I. 1964. Science 143:325–333.

113. Schneiderman, H. A., Gilbert, L. I., and Weinstein, M. J. 1960. Nature 188:1041–1042.

114. Segal, W. 1965. Nature 208:1284–1286.

115. Shaw, R. 1965. Biochim. Biophys. Acta 98:230–237.

116. Shaw, R. 1966. Adv. Lipid Res. 4:107–174.

117. Shorb, M. S. 1963. In J. Ludvik, J. Lom, and J. Vavra, eds., Progress in protozoology, Proceedings of the 1st International Conference on Protozoology, Czechoslovak Academy of Sciences, pp. 153–158.

118. Shorb, M. S., Dunlap, B. E., and Pollard, W. O. 1965. Proc. Soc. Exp. Biol. Med. 118:1140–1145.
119. Siperstein, M. D. and Guest, M. J. 1960. J. Clin. Invest. 39:642–652.
120. Smith, J. D. and Law, J. H. 1970. Biochim. Biophys. Acta 202:141–152.
121. Smith, J. D. and Law, J. H. 1970. Biochemistry 9:2152–2157.
122. Smith, J. D., Snyder, W. R., and Law, J. H. 1970. Biochem. Biophys. Res. Commun. 39:1163–1169.
123. Snyder, W. R. and Law, J. H. 1970. Lipids 5:800–802.
124. SubbaRao, K. and Olson, R. E. 1967. Biochem. Biophys. Res. Commun. 26:668–673.
125. Sugita, M. and Hori, T. 1971. J. Biochem. (Tokyo) 69:1149–1150.
126. Taketomi, T. 1961. Z. Allg. Mikrobiol. 1:331–340.
127. Thompson, G. A., Jr. 1967. Biochemistry 6:2015–2022.
128. Thompson, G. A., Jr. 1969. J. Protozool. 16:397–400.
129. Thompson, G. A., Jr. 1969. Biochim. Biophys. Acta 176:330–338.
130. Thompson, G. A., Jr. 1972. J. Protozool. 19:231–236.
131. Trebst, A. and Geike, F. 1967. Z. Naturforsch. B22:989–991.
132. Tsuda, Y., Morimoto, A., Sano, T., Inubushi, Y., Mallory, F. B., and Gordon, J. T. 1965. Tetrahedron Lett. 19:1427–1431.
133. Vakirtzi-Lemonias, C., Kidder, G. W., and Dewey, V. C. 1963. Comp. Biochem. Physiol. 8:331–334.
134. Warren, W. A. 1968. Biochim. Biophys. Acta 156:340–346.
135. Wheeler, M. A. and Holmlund, C. E. 1972. Fed. Proc. 31:477.
136. Williams, B. L., Goodwin, T. W., and Ryley, J. F. 1966. J. Protozool. 13:227–230.
137. Wilton, D. C. and Akhtar, M. 1970. Biochem. J. 116:337–339.
138. Zander, J. M. and Caspi, E. 1970. J. Biol. Chem. 245:1682–1687.
139. Zander, J. M., Greig, J. B. and Caspi, E. 1970. J. Biol. Chem. 245:1247–1254.
140. Zander, J. M., Caspi, E., Pandey, G. N., and Mitra, C. R. 1969. Phytochemistry 8:2265–2267.

Chemistry and Metabolism of Nucleic Acids in *Tetrahymena**

Robert L. Conner and Mary J. Koroly

Department of Biology
Bryn Mawr College
Bryn Mawr, Pennsylvania

I. INTRODUCTION

Studies related to the biochemistry and physiology of nucleic acids in *Tetrahymena* are numerous and encompass many areas of investigation. The use of radiolabeled compounds has allowed cellular biologists to follow the synthesis of RNA and DNA during the transition from the lag phase to log growth, during log phase, and from log to stationary phase. The description of cell cycle activities is quite extensive; the course of events during and after induction of synchrony has been described; and the meiotic cycle has been explored in a sophisticated fashion. In addition, our knowledge about the enzymic activities is reaching a stage equivalent to that of the most widely used bacterial and mammalian tissues.

*The following abbreviations are used in this chapter. C, cytosine; T, thymine; U, uracil; G, guanine; A, adenine; 2′(3′)-CMP, 2′(3′)-UMP, 2′(3′)-GMP, 2′(3′)-AMP, mixtures of the 2′- and 3′-ribonucleoside monophosphate isomers of cytidine, uridine, guanosine, and adenosine, respectively; 5′-CMP, 5′-UMP, 5′-GMP, and 5′-AMP, the 5′-ribonucleoside monophosphate isomers; dCMP, dUMP, dGMP, dAMP, TMP, the 5′-deoxyribonucleoside monophosphate isomers of deoxycytidine, deoxyuridine, deoxyguanosine, deoxyadenosine, and thymidine, respectively; CDP, UDP, GDP, ADP, the 5′-ribonucleoside diphosphate derivatives; CTP, UTP, GTP, ATP, the 5′-nucleoside triphosphate derivatives; dCTP, dGTP, dATP, TTP, the 5′-deoxyribonucleoside triphosphate derivatives; CMP!, UMP!, GMP!, AMP!, the 2′,3′-cyclic ribonucleoside monophosphate derivative of cytidine, uridine, guanosine and adenosine, respectively; P_i, inorganic orthophosphate; PP_i, inorganic pyrophosphate; PRPP, 5-phospho-α-D-ribosyl pyrophosphate; tRNA, transfer RNA; sRNA, soluble RNA; mRNA, messenger RNA; rRNA, ribosomal RNA.

The reasons for this rather extensive knowledge are straightforward. Lwoff (126) successfully established *Tetrahymena* in axenic culture many years ago, and the development of a chemically defined medium by Kidder and Dewey (95) set the stage for definitive chemical studies. Control of the nutritional environment allowed unambiguous interpretation of data. Experimental design was unencumbered by the presence of bacteria, other microorganisms, or by variations encountered when using a complex medium of a poorly defined nature. The information available stems primarily from investigations with several strains of *T. pyriformis*, although some data are present in the literature for *T. setifera, T. vorax, T. paravorax,* and for the tetrahymenid ciliates *Glaucoma* and *Colpidium.*

While our current state of knowledge is impressive, several poorly resolved aspects concerning nucleic acid metabolism remain and, unfortunately, some conflicting literature reports exist that seemingly cannot be resolved satisfactorily. We attempt in this chapter to point out what in our opinion seem to be the best documented pathways.

The comparative aspect of the biochemistry of nucleic acids is one of the most important considerations in these studies; however, it is important to note that knowledge of nutritional requirements for a preformed purine and for a pyrimidine, the biosynthesis of pseudouridine, and the development of purine and pyrimidine analogs for experimental purposes, as well as for the treatment of certain human diseases, first originated from studies with *Tetrahymena.*

More recently, the study of extranuclear DNA has been facilitated by the use of *Tetrahymena.* Cellular organelles can be isolated in quantity, which in turn allows adequate physicochemical measurements to be made on purified material. Such investigations in time will allow determination of the means of cellular propagation of the organelles, of the unique features of metabolism of the cellular particulates, of the contribution of each organelle to the economy of the whole cell, and of how these various cellular inclusions interact with one another in the regulation of cellular activity. As a model system, *Tetrahymena* holds a unique position in the eucaryotes world.

This chapter is limited for the most part to the chemistry and metabolism of nucleic acids. The other aspects of nucleic acid involvement in cellular physiology are thoroughly covered in other chapters of this volume.

II. NUTRITION

A. Purines

All the ciliated protozoa have an obligatory purine need. *Tetrahymena pyriformis* was the first organism noted to require a purine, guanine, for growth (91). The capacity to synthesize the heterocyclic ring structure is lacking (97). Glycine and formate are not incorporated into purine bases (74). Guanine, guanosine, and the guanine nucleotides (or the deoxy derivatives) can serve as the sole purine source, while adenine (and derivatives) and hypoxanthine (and derivatives) are able to spare, but not replace, guanine. This appears to be true for all tetrahymenids (62,63,74,85,90,92,98). Table 1 gives a partial list of the purines that are capable of supporting growth, of sparing guanine, or are nutritionally inert.

Table 1.
Activity of Purines in Promoting the Growth of *T. pyriformis*[a]

Sole source[b]	Sparing only	Inert
Guanine, 100	Adenine	Xanthine
Guanosine, 200	Adenosine	Uric acid
Guanylic acid (2′,3′,5′) 200	Adenylic acid (2′,3′,5′)	
1-Methylguanine, 75	Hypoxanthine	
1-Methylxanthine, 15	Inosine	
2,5-Diaminopurine, 2.3	Inosinic acid	
8-Methylguanine, 4	(deoxy derivatives)	

[a]Based on information from Kidder and Dewey (94,95).
[b]Molar basis where guanine equals 100.

Nutritional studies have been of great value in allowing an adequate projection of the cellular enzymic complement. With few exceptions the predicted reactions have been found. The structural formulas of some of the more commonly discussed purines and pyrimidines are given in Fig. 1.

B. Pyrimidines

Tetrahymena pyriformis was the first organism shown to have a pyrimidine nutritional requirement (91,92). Uracil can serve as the sole pyrimidine source, while cytosine and thymine cannot (93). Uridine and cytidine are equally effective, while thymidine is inactive. 2′(3′)-CMP, 5′-CMP, 2′(3′)-UMP, and 5′-UMP are all active nutritionally (90), as are the deoxy derivatives (206).

Pyrimidine ring biosynthesis does not occur. 2,6-Diamino-4-hydroxy-5-formylamino pyrimidine does not satisfy the pyrimidine requirement but can be used for pteridine synthesis (98). Orotic acid cannot be utilized by intact cells (90). While permeability considerations may account for the inertness of this pyrimidine precursor, this is deemed unlikely since compounds such as 5′-AMP seem to enter the cells in unaltered form (79). Thymine and cytosine derivatives have been demonstrated to arise from uracil (75).

Thymine and thymidine are reported to be incorporated specifically into DNA (1,13,206), which indicates that no demethylation occurs; however, a recent investigation calls for reinterpretation of many of the earlier studies. Lanzetta and Berech (103) have demonstrated not only that *Tetrahymena* degrade thymidine to β-aminoisobutyric acid but also metabolize the latter compound to methylmalonate and subsequently to precursors of various macromolecules other than DNA. These investigators found that incubation with [14C]thymidine yielded radiolabeled glucose (glycogen), ribose (RNA) and alanine, aspartic acid and glutamic acid (protein). Clearly, these results must be used to reevaluate much of the autoradiographic work in which the unfortunate assumption was made that the pyrimidine ring was not degraded and hence that only DNA could serve as a repository for the radiocarbon.

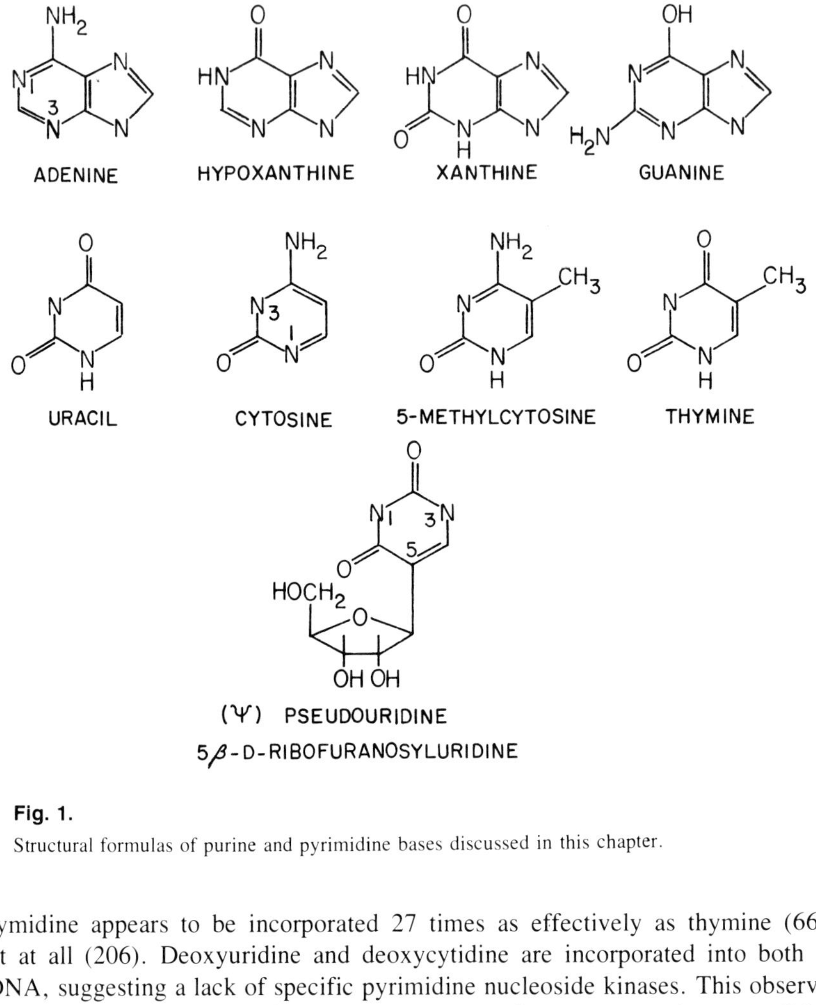

ADENINE   HYPOXANTHINE   XANTHINE   GUANINE

URACIL   CYTOSINE   5-METHYLCYTOSINE   THYMINE

(Ψ) PSEUDOURIDINE

5β-D-RIBOFURANOSYLURIDINE

Fig. 1.

Structural formulas of purine and pyrimidine bases discussed in this chapter.

Thymidine appears to be incorporated 27 times as effectively as thymine (66,74), or not at all (206). Deoxyuridine and deoxycytidine are incorporated into both RNA and DNA, suggesting a lack of specific pyrimidine nucleoside kinases. This observation also demonstrates the ability of the cells to convert these compounds to uracil before incorporation into the polymers occurs. Cells rendered deficient in folic acid in order to reduce methylation can be grown, although slowly, with 5-methyldeoxycytidine, thymidine, 5-methyl-dCMP, and TMP, but not with thymine, 5-methylcytosine, or 5-methyluridine (206).

The most interesting difference among the *Tetrahymena* species is the ability of *T. setifera* to use cytosine or thymidine as a sole pyrimidine source, while neither of these compounds is effective for *T. pyriformis* (86).

C. Analogs

A large number of purine and pyrimidine analogs have been tested for growth inhibition (80,90), however, reports on the exact mode of action or intermediary metabolism are rare. The purine analog 8-azaguanine has been shown to be incorporated into RNA

(76), and hence must be "activated" to give a nucleotide. Dewey and Kidder (46) showed that 8-azaguanine growth inhibition is a complex phenomenon (46,90). 6-Mercaptopurine is converted to the nucleotide by the purine pyrophosphate-phosphoribosyl transferase (79) but does not inhibit growth (45). Several analogs were tested in the *Tetrahymena* pyrophosphorylase system with negative results. These include 6-chloropurine, 2,6-diaminopurine, orotic acid, 2-fluoroorotic acid, and thiouracil (79); however, several of these are potent growth inhibitors (45). Small amounts of a "nucleotidelike" material were found after prolonged incubations when 4-aminopyrazolo-(3-4-*d*)-pyrimidine was used as the substrate (79).

III. CELLULAR CONTENT OF NUCLEIC ACIDS

A. Deoxyribonucleic Acid

Deoxyribonucleic acid comprises approximately 5–10% of the total nucleic acid in *T. pyriformis* S, GL, H, and B (7,61,76,168). The major portion of this nucleic acid exists in the polygenomic macronucleus, while smaller amounts are found in the diploid micronucleus. There is evidence for the existence of extranuclear DNA within mitochondria, and possibly even within kinetosomes. The studies attempting to localize cytoplasmic DNA, however, have encountered serious problems because of the difficulties involved in separating cell organelles from the nuclei in a way that excludes the possibility of contamination from nuclear DNA.

Various physical and chemical assays have been employed to determine the quantity of DNA per cell in the several strains of *Tetrahymena*. The results of these investigations are presented in Table 2. Cellular DNA content appears to vary with *Tetrahymena* strain (21,194), state of ploidy (160), and culture age of cells (12,130).

The DNA content does not increase appreciably during the lag phase of growth (19). Ratios of DNA to protein remain constant throughout the log phase, regardless of the rate of growth, whereas DNA/RNA ratios vary considerably (115,116), a pattern similar to that in bacteria (139). McDonald (130) demonstrated that the macronucleus of a stationary-phase cell (strain H) contains twice as much DNA as the macronucleus of a log-phase cell.

Debault and Ringertz (43) state that in *T. pyriformis* GL the division of DNA between two daughter cells is a "remarkably sloppy process" in which one of the fission products has approximately 4% more DNA than the other. Cleffman (26) used strain HSM to show that the DNA content is regulated after unequal division by elimination of excess DNA from the macronucleus of one daughter cell and by additional replication of the DNA in the macronucleus of the second. This most interesting mechanism for the maintenance of a constant DNA amount per cell deserves further investigation.

The DNA base compositions for several strains of *Tetrahymena* are shown in Table 3. The techniques used and problems involved in determining base composition for protozoa have been reviewed by Mandel (128). The data listed indicate that the quanine-cytosine content of all strains examined is remarkably low (between 22 and 32%). Thus, adenine (A) and thymine (T) make up about 70% of the DNA. Although A and T

Table 2.
Tetrahymena DNA Content

Strain	DNA per cell (pg)	Culture age	Method	Ref.
H	5.0	Stationary	Indole	(61)
W	3.0–30.3	Log	Indole Diphenylamine	(88)
S	~25.0	Log	Feulgen microspectrophotometry	(157)
HSM	9–14	Lag	Spectrophotometry	(19,20)
A,T	~11	Log	Diphenylamine	(60)
GL	13.5–17.5	Log	Diphenylamine, spectrophotometry	(168,169)
GL	14.9	Newly formed	Staining electron microscopy	(43)
GL	29.2	Before division	Staining electron microscopy	(43)

Strain	DNA per macronucleus	Culture Age	Method	Ref.
GL	~20	Log	Microspectrophotometry, interferometry, autoradiography	(160)
GL	~13.7	Log	Diphenylamine	(114)
Syngen 1	10.2	Newly formed	Photographic plate	(202)
Syngen 1	21.3	Before division	Microdensitometry	(202)

Strain	DNA per micronucleus	Culture age	Method	Ref.
Syngen 1	0.86	Log	Photographic plate microdensitometry	(202)
A,T	0.13	Log	Diphenylamine	(60)

Strain	DNA per mitochondrion	Culture age	Method	Ref.
ST	3.7×10^{-4}	—	Modified indole	(187)

are present in equimolar amounts, the G/C ratio appears to be in favor of cytosine (168), the one exception being strain EU 6000 (194). The base composition of DNA remains constant throughout the cell cycle (168). The base ratio of mitochondrial DNA in the one strain examined (strain 4) also has a low GC content—approximately 26% (182).

Other characteristics of tetrahymenid DNA are presented in Table 9.

Table 3.
Tetrahymena DNA Base Composition

Species	Thymine[a]	Guanine[a]	Cytosine[a]	Adenine[a]	GC (%)	Purine/ pyrimidine	Ref.
Tetrahymena pyriformis							
EU 6002	38.94	9.45	14.51	37.10	24	0.87	(194)
HSM	36.60	11.91	14.89	36.60	26	0.94	(194)
1A	—	—	—	—	26	—	(179)
1WH-52	—	—	—	—	26	—	(179)
1-IL-12	—	—	—	—	26	—	(179)
2-1	—	—	—	—	26	—	(179)
5-1	—	—	—	—	26	—	(179)
6-1	—	—	—	—	26	—	(179)
8-2	—	—	—	—	26	—	(179)
GL	36.9	11.5	15.0	36.6	27	0.86–0.93	(168)
3-1	—	—	—	—	28	—	(179)
7-1	—	—	—	—	28	—	(179)
9-1	—	—	—	—	28	—	(179)
EU 6000	35.80	14.50	14.50	35.20	30	0.99	(194)
W	—	—	—	—	30	—	(170)
GL	—	—	—	—	30	—	(179)
E	—	—	—	—	32	—	(179)
Tetrahymena patula	—	—	—	—	22	—	(170)
Tetrahymena rostrata	—	—	—	—	24	—	(179)

[a]Values given in mole fractions $\times$ 100.

B. Ribonucleic Acid

1. Ribonucleic Acid Content

The RNA content of *T. pyriformis* is a function of the growth phase (Table 4). Figure 2 shows the typical growth pattern for strain W grown at 28.5°C in a 2% proteose–peptone medium supplemented with 0.1% yeast extract and 90 μM iron EDTA (31). A lag phase of about 3 h was noted when cells from a 48-h stationary culture were used as the inoculum. Following a short acceleration period, the cells entered the log phase of growth, with a generation time of 2.8 h from the 4th through the 16th hour after inoculation. The culture began a prolonged deceleration phase lasting from the 16th to about the 25th hour. The stationary phase commenced at about the 25th hour, while the decline period began on the 3rd day.

Satir (166) cultured strain W in 2% proteose–peptone without supplements at 25°C; the growth rate observed was lower, and the maximum population density was ~50% of the value given in Fig. 2. Differences in the culture fluid composition and other environmental factors would be anticipated to influence both the growth rate and the final cellular densities. The variation in iron content noted in various lots of proteose–peptone makes the use of such culture fluid without an iron supplement undesirable (31,148,175).

Table 4.
The RNA Content of Several Strains of *T. pyriformis*

Strain	Age	RNA per cell (ng)	Ref.
W	Early log (maximum RNA)	~0.28	(166)
W	Early log (maximum RNA)	1.80	(100)
W	Log (unknown point)	1.28	(88)
W	Deceleration	0.59–0.64	(100)
W	Early stationary	0.58	(100)
W	Late stationary	0.50	(100)
W	Early stationary	0.40	(29)
W	Stationary	~0.14	(166)
GL	Log	0.25	(168)
GL	Log	0.11–0.28 (dependent on culture fluid)	(115)
HSM	Log	~0.2	(19)

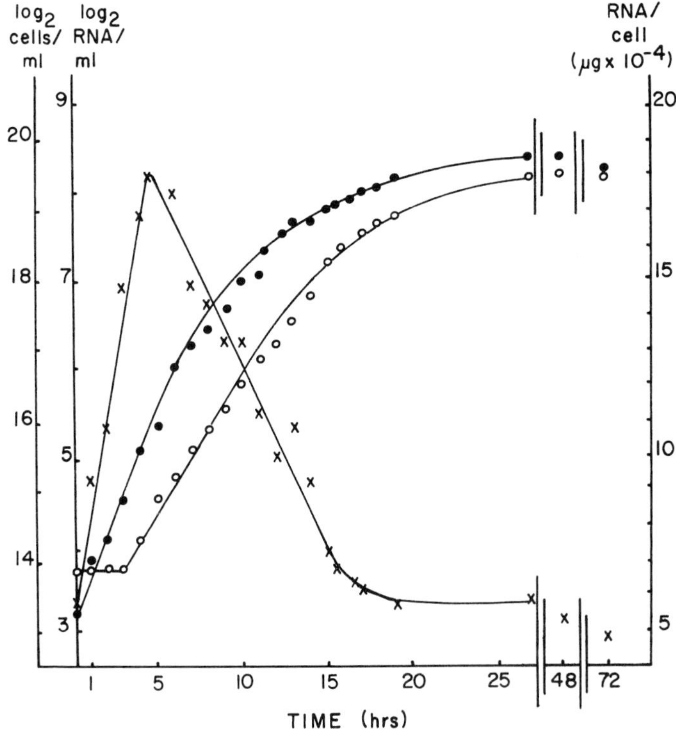

Fig. 2.

A growth curve for *T. pyriformis* W in a 2% (w/v) proteose–peptone medium supplemented with 0.1% (w/v) yeast extract and an iron chelate. Cell numbers (open circles) and RNA (closed circles) are expressed in $\log_2$, while the RNA content per cell (x) is given in micrograms $\times$ 10^{-4}.

The length of the lag phase in *T. pyriformis* is, within limits, independent of the size of the initial inoculum, but is affected by the age of the cells (144,151,154,166). The relative durations of the log and deceleration phases, however, can vary with inoculum size.

The RNA content of the ciliate population was followed during the phases of culture growth. There was a rapid buildup of RNA during the lag period (Fig. 2). The maximum level per cell was reached after 5 h, at which time the organisms had just entered the log phase of growth. The total amount of RNA in the culture continued to rise with the increase in cell number; however, the rate of synthesis of this macromolecule was not as great as the rate of cellular division. This lack of correspondence resulted in a steady decline in the RNA content of the individual cell throughout log and decelerating growth. A similar qualitative pattern has been reported for strain HSM (19) and for strain W grown under the conditions described above (166). The difference in the growth rate of the cells in the two studies may explain the dissimilarity in the maximum amounts of RNA observed in strain W (Table 4). A gradual loss of RNA in the culture during the stationary phase was noted (100).

It is important to remember that the characteristics of culture growth and individual cell growth need not be the same. Studies made with cells in synchronous growth or with single cell cycles are not comparable in many aspects to investigations on cell populations. For example, the ultrastructure of the nucleoli changes throughout the growth phases regardless of individual cell cycle and may correspond to the differing levels of RNA produced per cell during the life history of a culture (19,167).

2. Base Composition: RNA

The base composition of the bulk RNA for several strains is reported in Table 5. These ciliates have a purine/pyrimidine ratio of 0.96:1.22 and contain RNA of the AU type. Mandel (128) has pointed out that, generally, protozoa with DNA of low GC content likewise have a lower amount of GC in their bulk RNA.

Table 5.
RNA Base Composition

Strain	Cytosine[a]	Guanine[a]	Uracil[a]	Adenine[a]	GC (%)	Purine/ pyrimidine	Ref.
GL	19.3	19.8	30.0	30.9	39.1	1.00–1.10	(168)
W	22.2	20.6	26.7	30.3	42.8	1.04	(41)
EU 6000	19.21	24.71	28.88	27.20	43.9	1.08	(194)
EU 6002	18.41	26.43	26.83	28.33	44.8	1.22	(194)
HSM	17.99	27.43	26.87	27.72	45.4	1.20	(194)
WH 14	20.8	22.8	30.1	26.3	43.6	0.96	(101)
GL (MN)[b]	16.0	14.0	33.0	36.8	30.0	1.00	(112)

[a]Values are given in mole fractions × 100.
[b]MN, macronucleus.

IV. NUCLEIC ACID CONTENT OF CELLULAR ORGANELLES

A. Macronucleus

The large polyploid macronucleus is the functional nucleus during normal vegetative growth. The macronucleus contains 80 to 90 "subnuclei" or large chromatin aggregates at the time of division, and about 45 in each daughter. These subnuclei appear to be haploid, although the ploidy level of the macronucleus is still to be resolved (60,141,202).

Several methods have been developed for the isolation of macronuclei (69,113,132,160). This organelle comprises about 5% of the total cellular dry weight in log-phase cells (160). When examined in terms of whole-cell mass, most of the DNA, 2% of the RNA, and 5% of the protein are contained within the macronuclear envelope (117). Deoxyribonucleic acid constitutes 15–23% of the dry weight of the macronucleus of strain GL; RNA, 2–3%; and protein, 75–85% (114,160). The relative percentages change during the growth of the culture, since the macronucleus size varies with age (114,122,160,180).

The average macronucleus of a cell in the log phase of growth contains 500 to 1000 nucleoli, electron-dense bodies located on the inner surface of the nuclear envelope (142). These nucleoli begin to fuse in the deceleration phase of culture growth and continue to aggregate throughout stationary phase, changing in appearance as well as in size (19,167). Disaggregation appears to occur when the nongrowing cells are inoculated into fresh culture fluid. Further, these nucleolar changes can be induced artificially by manipulation of the cellular environment (19,122,142,167). A correlation can be made between the size and number of the nucleoli in cells at different stages of population growth, or under various experimental conditions, and the amount of RNA (rRNA) synthesized in these cells (19,122,142,167).

Each of the nuclear nucleoli contains a small amount of DNA which presumably has the active genes for rRNA synthesis (22,57). Engberg and Pearlman (57) calculate about 340 rRNA genes per haploid genome in log phase and show a 30–40% lower homology between rRNA and nuclear DNA in stationary-phase cells. Ribonucleic acid synthesis has been demonstrated both in vivo and in isolated macronuclei (112). A DNA-dependent RNA polymerase has been isolated from the macronucleus, which has as its product "DNA-like" RNA, probably messenger (112). Most of the nuclear RNA, however, is of the ribosomal precursor variety (102,118) and is discussed later.

B. Ribosomes

Elliott et al. (56) examined the distribution of ribosomes in strains WH-6 and E during the growth phases. In the log phase, rough endoplasmic reticulum is abundant 'nd is randomly distributed throughout the cytoplasm. As the cells increase in age, less and less rough endoplasmic reticulum is observed, and that which is present is located primarily near mitochondria. Free ribosomes are plentiful (18).

Ribosomes isolated from *T. pyriformis* HSM have a mean diameter of 210 Å (18). The estimated molecular weight of an average ribosome from strain GL is about 4 $\times$ 10^6 (193). The sedimentation coefficients for ribosomes from various strains are reported in Table 6. Weller et al. (193) noted that the values obtained from measurements on

Table 6.
Sedimentation Values for Ribosomes, Ribosomal Aggregates, and Polysomes Obtained from *T. pyriformis*

| Strain | Growth stage | Sedimentation Value (S) | | Ref. |
		Ribosomes	Aggregates or polysomes	
GL	Stationary (?)	72–78	120, 150, 170 aggregates	(127)
GL	Stationary	80	123, 125, 160, 190, 220 aggregates	(193)
GL	Log	80	—	(115)
GL	Log	80	—	(120)
GL	Heat-synchronized	70, 80, 100	—	(152)
GL	Heat-synchronized	84	120	(209)
HSM	Lag	73	125, 178, 186 polysomes	(18)
HSM	Cold-synchronized	82	123, 148, 174 polysomes dimers, trimers, tetramers	(199)
ST	Stationary	70–80	—	(23)

sucrose density gradients vary from experiment to experiment; therefore the range of values may be due in part to technical considerations. The sedimentation values may also vary with cell cycle (152), Mg^{2+} concentration (23,127,152), cold synchrony, or when inhibitors such as actinomycin D are used (198). Ribosomal aggregation occurs in the presence of high salt concentrations, calcium, magnesium, and EDTA (73,127,193). Ribosomes disaggregate in cells maintained at supraoptimal temperatures (15).

The possible action of nucleases and proteases must be taken into account in evaluating the sedimentation coefficients in all these studies. These enzymes are quite active in disrupted *Tetrahymena* preparations, even at low temperatures (38,135). It is worthy of note that *Tetrahymena* ribosomes are stable in low magnesium concentrations. This behavior pattern is similar to that observed in mammalian systems and unlike that in bacteria (120,127,152,193). The stability noted at low concentrations of this cation may be due to the high polyamine content of these organelles, in which the polyamine/RNA ratio is 0.24 (193). Spermidine and putrescine are present in equivalent quantities, and together make up about 95% of the total polyamine bound to the ribosome. A small amount of spermine is also found.

Polyribosomes have been isolated by indole treatment and by intracytoplasmic cavitation with nitrogen followed by zonal ultracentrifugation (18,73,198). Electron micrographs show monomers, dimers, trimers, and tetramers which correspond to each distinct sedimentation peak (18) (Table 6). The polysome pattern varies during the cell cycle in a fashion that correlates well with the extent of protein synthesis, which implies the involvement of the polyribosomes in this process. Actinomycin D and cold synchrony appear to depress the polysome content.

Table 7.
Sedimentation Values for Ribosomes, RNP Particles, Derived Ribosomal Subunits, and RNA[a]

Strain	Ribosome	RNP particles	Derived subunits[b]	Subunit RNA	Ref.
GL	80	60	50	→ 25 + 5	(121,207)
		40	30	→ 17	
GL	80	—	—	26, 14, 4	(193)
GL	Whole-cell RNA	—	—	24.5, 16.8, 4.5 (?)	(131)
HSM	82	61 (?)	60	—	(199)
			30	—	
WH 14	75	—	50	→ 28	(101)
			30	→ 18	
ST	80	—	60	→ 26 + 4–5	(23)
			40	→ 17	
ST	80 (mitochondrial)	—	55	→ 21 + 14 + 5	(23)

[a]Values are given in Svedberg units.
[b]Subunits derived by dialysis.

Ribosomes are composed of approximately equal amounts of protein and RNA (115,193). The proteins are heterogeneous. Weller et al. (193) report 13 prominent bands on disk gel electrophoresis at pH 3.8. Few if any bands are resolved at pH 9.4. The electrophoretic behavior with regard to pH is similar to that observed in mammalian ribosomal protein separations (124). Cytoplasmic ribosomes can be dissociated into two subunits of dissimilar size by dialysis. These derived subunits and the specific class(es) of RNA found in each are characterized by sedimentation coefficients as listed in Table 7.

The large ribosomal subunit contains 25–28 S RNA and 5 S RNA, while the smaller subunit possesses 14–18 S RNA. These values can be compared and contrasted with the 70 S bacterial ribosome containing 23 and 16 S RNA (139), mammalian ribosomes of 80 S with 28 and 19 S RNA (150), and tetrahymenid mitochondrial ribosomes consisting of 21 and 14 S RNA (Table 7).

Base ratios of the RNA classes are listed in Table 8. In general, these values indicate a higher proportion of purines than pyrimidines, and a relatively low quantity of GC, which is similar to whole-cell RNA (Table 5). This is not surprising, since particulate RNA comprises 85–90% of the total cellular RNA in strain GL (115) and at least 85% of the total RNA in strain W (38).

Ribosomal RNA is synthesized in the macronucleus (16,70,102,112,120,121,155) within the nucleoli (22,57,142). Leick and Andersen (118) calculated that eight rRNA molecules are made per nucleolus per second at 28°C. The 25 and 17 S RNA are produced in stoichiometric amounts. When labeled RNA precursors are employed for short time periods and the appearance of the label in cytoplasmic ribosomal particulates is followed, a chronological sequence of the synthetic processes can be examined. Leick and Plesner (120,121) and Kumar (102) noted that 17 S RNA associated with protein appears in the cytoplasm within a short period of time, whereas labeled 25 S RNA accumulates in the nucleus before being released. The macronucleus is estimated to

Table 8.
Base Composition of RNA

RNA species	Cytosine[a]	Guanine[a]	Uracil[a]	Adenine[a]	GC (%)	Purine/ pyrimidine	Ref.
72 S ribosomal	16.7[b]	26.8[b]	25.1[b]	31.4[b]	43.5	1.39	(127)
60 S RNP	21.9[b]	14.0[b]	30.7[b]	33.3[b]	36	0.90	(101)
45 S ribosomal	18.1 ± 1.4	20.9 ± 1.4	26.3 ± 0.9	34.7 ± 1.9	39	1.25	(117)
25 S + 17 S	25.6 ± 0.5	25.5 ± 1.3	25.2 ± 0.9	31.7 ± 0.7	51	1.13	(117)
25.4 S	16.2[b]	16.8[b]	32.1[b]	34.8[b]	33	1.07	(101)
17 S	15.8[b]	20.6[b]	31.8[b]	31.9[b]	36	1.10	(101)
26 S	17.9	25.3	24.3	32.3	43	1.36	(23)
17 S	15.8	33.4	24.8	26.0	49	1.46	(23)
Mitochondrial 21 S ribosomal	11.7	16.2	35.2	36.9	28	1.13	(23)
Mitochondrial 14 S ribosomal	11.8	18.8	35.7	33.8	31	1.11	(23)

[a]Values given in mole percent.
[b]Average of reported values.

contain 0.5% of the cytoplasmic 17 S RNA and 1.6% of the cytoplasmic 25 S RNA, corresponding to turnover rate constants (movement from macronucleus to cytoplasm) of 0.92 min^{-1} and 0.29 min^{-1}, respectively (118).

This differential appearance of the two RNA classes in the cytoplasm is not due to a different rate of transcription from DNA (120), but may be dependent on a different rate of completion of the larger subunit in the macronucleus or to a different rate of transport across the macronuclear membrane. The dissimilar rate of movement of radiolabeled 25 S and 17 S RNA from the macronucleus leads to an unequal labeling pattern in the two ribosomal subunits which are derived by dialysis from the 80 S ribosome. This is analogous to the situation in mammalian tissue culture cells (149), although the processing of the ribosomal particles proceeds more than 10 times faster in tetrahymenid cells (118).

It has been suggested that ''the large number of nucleoli and the high ribosome assembling efficiency of each nucleolus may both account for the high ribosome forming capacity of *Tetrahymena*. The high efficiency of each nucleolus may be related to its location near the nuclear envelope which is likely to offer easy access for precursor material and favorable conditions for export of nascent ribosomal particles to the cytoplasm'' (142).

The formation of rRNA appears to be initiated by the synthesis of 45 S RNA (117) or 35 S RNA (102, 156, 207) in the nucleoli. When ribosome formation is experimentally inhibited in some manner such as by introducing the culture into a ''step-down'' medium, a large proportion of the radioactivity is found associated with the high-molecular-weight RNA in a short period of time. This short-lived compound can be methylated (102,117,207), suggesting a similarity to the 45 S rRNA precursor in mammalian cell nucleoli. The labeling kinetics suggest that this large nuclear RNA may be a parent compound of the 25 and/or 17 S rRNA (102,117). A second possible nuclear precursor of rRNA has been described; 30-32 S RNA, possibly derived from the 45 S (35 S) RNA, also incorporates label rapidly, can be methylated, and may well be an intermediate in the formation of 25 and/or 17 S RNA (102,119). Apparent "clustering" of rRNA 17 and 25 S genes lend support to the proposed high-molecular-weight intermediate (57).

The newly formed rRNA becomes associated with protein shortly after synthesis in the nucleus (119). The 17 S RNA plus protein forms a 40 S ribonucleoprotein (RNP) particle, while the 25 S RNA complexes with protein to form a 60 S RNP unit. These particles then pass into the cytoplasm at different rates—the larger of the two is retained in the nucleus longer. There are few if any 40 S RNP particles free in the cytoplasm, and these smaller units cannot be isolated from the polysome region by sucrose density centrifugation. The 60 S RNP particle, however, can be isolated both from the cell sap and from the polysome fraction (102,117), and both 40 and 60 S RNP particles are rapidly labeled by $[^{14}C]$amino acids (119). It is of interest that the synthesis of RNP ribosomal protein and RNA are parallel (119).

Labeling kinetics suggest that the 40 and 60 S RNP particles are the biological precursors of the 80 S ribosome (117), and ultimately of the 30 and 50 S subunits, respectively. The differential in sedimentation values between the ribosomal precursor, or subribosomal particles and the 30 and 50 S subunits derived from ribosomes may be due to the loss of some lower-molecular-weight material (119,152,198). Leick et al. (119) present evidence based on buoyant densities of the ribosome, the RNP particles, and the derived subunits, and on double-label experiments to show that the 40 and 60 S particles contain more protein than the corresponding subunits. These investigations suggest that the extra protein is removed when the precursor particles are converted to the ribosomal subunits. The evidence obtained suggests that the subunits are not recycled through the 40 and 60 S particles, nor through the polysomes.

The following scheme can be suggested for the pathway of ribosome synthesis:

Nucleus:

45 (35) S RNA $\xrightarrow{\text{Rapid}}$ 32 (30) S RNA

32 (30) S RNA $\longrightarrow$ 26 and 17 S rRNA

or   26 S rRNA $\longrightarrow$ 17 S rRNA

26 S rRNA + protein + 5 S rRNA $\xrightarrow{\text{Slow}}$ 60 S RNP particle

17 S rRNA + protein $\xrightarrow{\text{Rapid}}$ 40–45 S RNP particle

Cytoplasm:

60 S RNP + 40–45 S RNP $\longrightarrow$ 70–80 S monosome $\rightleftharpoons$ polysome

Derived by isolation procedures:

70–80 S monosome $\longrightarrow$ 50 S subunit + 30 S subunit + protein

C. Micronucleus

The micronucleus, absent in many strains, differs markedly from the macronucleus in structure and ploidy level. The micronucleus is diploid and is reported to contain anywhere from 12- to 85-fold less DNA than the macronucleus (60,70,141,202) (Table 2). This organelle is enclosed by a nuclear envelope but has no apparent nucleoli nor RNA particles (69). The presence of RNA is questionable (69,138).

D. Mitochondria

1. Deoxyribonucleic Acid

Convincing evidence has been presented for the existence of DNA in the mitochondria of various organisms (71). The first reports of mitochondrial DNA in *Tetrahymena* were published by Stone and Prescott (178) and Parsons (145). This work was extended using high-resolution autoradiography coupled with DNase digestion in cellular homogenates to show that DNA is within the mitochondria, that $[^3\text{H}]$thymidine is incorporated into this organelle and, further, that the radiolabel is incorporated into DNase-sensitive material (177). The long-term stability of mitochondrial DNA was suggested since the radioactive label is maintained within a population of mitochondria through at least four generations of log growth (145,147). Further, it was shown that labeling of mitochondria occurs throughout most if not all of the cell cycle, even in the absence of nuclear DNA synthesis (17,147), which suggests the possibility that DNA is not only associated with mitochondria but is synthesized in this organelle as well.

Suyama and colleagues (182,186,187) attempted to characterize *Tetrahymena* mitochondrial DNA. Cesium chloride density gradient equilibrium centrifugation analyses were carried out on DNA isolated from whole-cell and mitochondrial preparations from six strains (see Table 9). The ratio of satellite DNA to major DNA was greatly enhanced in DNA isolated from a mitochondrial preparation in three of these strains (182). The density of this satellite DNA was reported to be similar to, but not identical with, that of the respective nuclear DNA. Parsons and Dickson (146) also reported nuclear and satellite DNA bands for two strains.

Brunk and Hanawalt (13), using a cesium chloride gradient and a double-labeling technique, however, were unable to detect satellite DNA in any of the strains examined by Suyama. The method of Brunk and Hanawalt shows nuclear and mitochondrial DNA to be of identical buoyant density within an uncertainty of 0.0005 g/cc. These investigators found, however, that mitochondrial DNA metabolism is unlike that of nuclear DNA. A difference in the synthesis of mitochondrial and nuclear DNA was demonstrated by the use of $[^{14}\text{C}]$thymidine and $[^3\text{H}]$-5-bromodeoxyuridine. The latter compound is incorporated into nuclear DNA in greater amounts, nearly fourfold, than into mitochondrial DNA. This differential allows a ready separation of the two DNA species by analytical ultracentrifugation techniques. Thus Brunk and Hanawalt claim that mitochondrial and nuclear DNA can be resolved only when one species is more heavily labeled with

Table 9.
Buoyant Density and Thermal Denaturation Temperatures for *Tetrahymena* DNA

| Species | Buoyant density (g/cc) | | | Thermal denaturation (°C) | | Ref. |
| | Major | Satellite | | Whole-cell | Mito. | |
	Whole-cell	Mito.	Unknown			
Tetrahymena pyriformis						
W	1.690	—	—	81.2	—	(170)
HSM	1.685	1.671[a]	—	—	—	(146)
6-III	1.685	—	—	—	—	(146)
GL	1.688	1.684[a]	1.693	—	—	(182)
1	1.685	1.685	—	—	—	(182)
4 (ST?)	1.692	1.686[a]	1.700	—	79.5	(182)
5	1.687	1.686	—	—	—	(182)
6	1.685	1.685	1.698	—	—	(182)
9	1.690	1.684[a]	—	—	—	(182)
GL	1.691	1.684	—	—	—	(182)
Tetrahymena patula	1.684	—	—	—	77	(170)

[a]Densities of mitochondrial (mito.) DNA differ from that of corresponding nuclear DNA according to refs. 146 and 182; however, others dispute this claim (13). See text for details.

$[^3H]$-5-bromodeoxyuridine. It is clear that additional investigation is necessary to clarify the conflicting claims of the two groups.

Suyama (182) noted that whole-cell DNA preparations of strains 4, 6, and GL exhibit additional minor bands in cesium chloride density gradients. These bands are less than one-half the actual satellite band of mitochondrial DNA. These unknown bands were eliminated with DNase and therefore were regarded as DNA in nature (182). Brunk and Hanawalt (13) suggest this extra band may be glycogen, a possibility that is not likely if this material is DNase-sensitive.

Melting analyses were made to determine the hyperchromicity resulting from thermal denaturation of DNA (182). The melting temperature of mitochondrial DNA is 79.5°C. At 75°C only 3–4% hyperchromicity is observed. The sharp increase in optical density on heating is characteristic of doubled-stranded, native DNA. Melting temperatures for nuclear DNA are reported by Schildkraut (170) (Table 9).

Suyama (182) also noted that heat-denatured mitochondrial DNA has a buoyant density that differs from that of denatured nuclear DNA (1.703 g/ml versus 1.711 g/cc). This has been interpreted as suggesting that isolated mitochondrial DNA is distinct from nuclear DNA, and that the mitochondrial material does not seem to be an RNA–DNA complex.

Purified mitochondrial DNA shows a major relatively uniform sedimenting component of $S^0_{20,w} = 41.68$, indicating a molecular weight of approximately 4×10^7 daltons (60, 182). Brunk and Hanawalt (13) report a similar value of $3–4 \times 10^7$ daltons.

Suyama and Miura (186) postulated a linear structure for mitochondrial DNA and described the mean length of the linear fibers to be about 17.6 μm $\pm$ 0.077. Electron micrographs suggest seven or more strands of DNA per mitochondrion. Brunk and Hanawalt (13) also could not detect circular DNA.

2. Ribonucleic Acid

Ribonucleic acid appears to be present in the mitochondria of *Tetrahymena* (183,189). The sedimentation coefficients have been determined as 18, 14, and 4 S. The age of the cells, the addition of bentonite or magnesium, or change in the isolation medium does not change this sedimentation pattern (185). Suyama (183) suggests that the low sedimentation values may be due to nuclease activity during the isolation procedure, although there is some evidence that these RNA fractions are not merely degradation products (23).

The 4 S fraction exhibits aminoacyl acceptor activity (183). Mitochondrial leucyl tRNA can be separated on methylated albumin kieselguhr columns from cytoplasmic leucyl tRNA. The presence of a mitochondrial leucyl tRNA synthetase has also been shown (184). It is of interest that, while tRNA is found in the mitochondria in large quantities, this class of RNA does not appear to be synthesized in situ. No hybridization with mitochondrial DNA was detectable (183), and no synthesis of tRNA was observed when isolated mitochondria were examined (185). Radiolabeled ATP is incorporated into this fraction while [³H]UTP is not, which suggests that a turnover of the terminal three nucleotides (pCpCpA) of tRNA occurs.

The pellet obtained by high-speed centrifugation of disrupted mitochondria contains 14–18 S RNA. This fraction contains 75–80% of the total mitochondrial RNA and appears to be ribosomal in nature.

Chi and Suyama (23) recently compared the physicochemical properties of mitochondrial and cytoplasmic ribosomes isolated from strain ST. Although the ribosomes from both sources sediment at 80 S in 10 mM Mg^{2+}, the cytoplasmic ribosomes split into 60 and 40 S bands with lower amounts of this cation or in the presence of EDTA, whereas the mitochondrial ribosomes exhibit only a 55 S band. Further, although 26 and 17 S RNA are found in the cytoplasmic ribosomes (Table 7), the most recent evidence indicates that their mitochondrial counterparts contain 21 and 14 S RNA (23). The base compositions of the RNA from the two sources seem to be distinct (Table 6). Density and electrophoretic differences are also evident. Finally, no significant complementarity is found between mitochondrial DNA and cytoplasmic rRNA, while some hybridization between the two mitochondrial nucleic acids was demonstrated.

When in vitro mitochondrial RNA synthesis was followed, a large percentage of label was incorporated into the 14–18 S RNA. This newly synthesized fraction, however, was soluble, and Suyama and Eyer (185) suggest, therefore, that synthesis of mRNA is occurring. Protein synthesis has been demonstrated using mitochondrial ribosomes from *Tetrahymena* (2).

E. Kinetosomes

Several investigators have reported the presence of nucleic acids in kinetosomes. The data obtained by Randall and Disbrey (157), using fluorescent microscopy with acridine orange and autoradiography, suggest the existence of approximately 2 × 10^{-16} g of DNA per cell in the kinetosomes. Seaman (171,172) reported relatively large quantities of nucleic acids in kinetosomes; however, these results have been disputed because of the lack of controls and inability to insure the removal of nuclear DNA. Argetsinger (7) (strain B) and Hoffman (81) (strain S) report dry weight values of 0.1% DNA in a kinetosomal fraction, but these investigators, as well as Randall and Disbrey (157), agree that detection of such small amounts is seriously hampered by the sensitivity of the methods used, and that the possibility of nuclear contamination cannot be overlooked.

Both Hoffman (81) and Argetsinger (7) also report the presence of RNA in the kinetosomal fraction.

F. Cilia

Cilia have been examined for nucleic acids. Culbertson (39), using spectrophotometric and phosphate analysis of cilia obtained from strain W, observed that about 0.4% of the dry weight was nucleic acid. Watson and Hopkins (192) also reported a nucleic acid content of 0.2–0.4% in cilia isolated from strain S. However, the studies of Child (24) and Gibbons (67) were unable to confirm the previous claims. Child (24), Hoffman (81), and Culbertson (39,40) presented evidence for nucleotides in the cilia. Renaud et al. (159) isolated a structural protein in cilia having a bound guanine nucleotide component.

V. METABOLISM OF NUCLEIC ACIDS

A. Deoxyribonuclease

1. Deoxyribonuclease (Deoxyribonuclease II, Deoxyribonucleate 3'-Nucleotidohydrolase, EC 3.1.4.6)

The depolymerization of DNA appears to follow the sequence:

$$\text{DNA} + \text{H}_2\text{O} \rightarrow 3'\text{-deoxyribonucleates}$$

Deoxyribonuclease activity is due to an acid hydrolase(s) which may be associated with the digestive process (134,135). The intracellular and extracellular DNases are type-II enzymes which are characterized by an acid pH optimum and no Mg^{2+} requirement (72,84). Holm (84) reports the absence of a DNase I (EC 3.1.4) in strain GL. Deoxyribonuclease activity has been noted in strains W and S (51,163,205). No rigorous attempt to purify either the intracellular or extracellular enzyme has been reported.

2. 3-Deoxynucleotidase (3'-Nucleotidase, 3'-Ribonucleotide [3'-Deoxyribonucleotide?] Phosphohydrolase, EC 3.1.3.6)

The 3'-deoxyribonucleates presumably are metabolized to the corresponding deoxynucleosides by phosphatase activity, possibly a 3'-nucleotidase (37), although no enzymological studies on the 3'-deoxynucleotides have been reported.

3. Deoxyribonucleic Acid Polymerases

Two DNA polymerases, one associated with the mitochondrial fraction and one with the postmitochondrial supernate have been reported (196,197). The former, which may be a repair enzyme, appears to be induced approximately 40-fold by thymidine deprivation, ethidium bromide, and ultraviolet or high-energy electron irradiation (89). The two enzymes have been partially purified by using molecular sieve techniques (195).

4. Others

Purine and pyrimidine nucleoside deaminases and phosphorylases are thought to metabolize deoxynucleosides to the free bases in much the same way as the ribonucleosides. These reactions are discussed in Section V.B.

B. Ribonucleic Acid

The following represents a proposed scheme for the depolymerization of RNA in *Tetrahymena*.

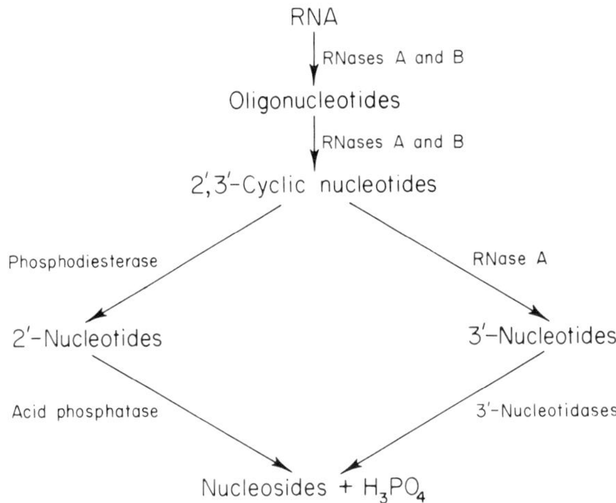

Purine nucleoside metabolism:

Pyrimidine nucleoside metabolism:

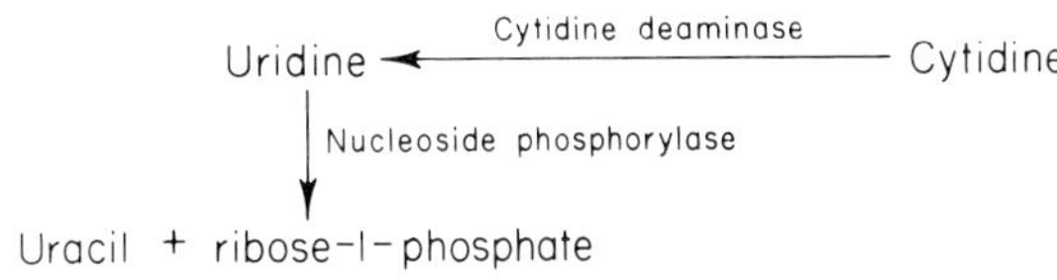

Table 10.
Comparison of the Literature Values concerning RNase Solubility in *T. pyriformis*

Investigator and strain	Harvesting solution	Homogenate solution	Disruption technique	RNase sedimentability (%)
Roth (162), strain W	Distilled water	Distilled water	Colloid mill	10
Lazarus and Scherbaum (105), strain GL	0.4% NaCl	50 mM tris, pH 7.45	Aminco French press	8–14
Müller et al. (135), variety I, MT IV	Prescott solution[a]	0.2 M sucrose	Filtration through sintered glass or fritted stainless-steel filters	84
Eichel et al. (50) strains W, S, and GL	Distilled water	Distilled water	Filtration through sintered glass or stainless-steel filters	75
Wu-Tcheng (205), strain W	70 mM tris-succinate, pH 6.5	70 mM tris-succinate, pH 6.5	Ultrasound	52

[a]KCl, 6mg; CaHPO$_4$, 4 mg; MgSO$_4$ · 7H$_2$O, 2 mg in 1-liter glass distilled water.

1. Ribonucleases and Nucleotidases

Ribonuclease activity [RNase, ribonucleate nucleotido-2′-transferase (cyclizing), EC 2.7.7.17] was first found in strain W (51,163). Both intracellular and extracellular enzymes are present, and Eichel et al. (50) has described the distribution of activity between the cells and their environment in strains GL, S, E, H, WH-6, HS, HSM, and Y of *T. pyriformis*, *T. vorax* PP, *T. rostrata* I$_0$C$_2$ and 1P9, and in *Glaucoma chattoni* A. The extracellular enzyme has a pH optimum of 6.0 and is considerably more heat-stable than intracellular RNases. The intracellular RNase activity is divided between particulate and soluble forms (Table 10). A great diversity of values for the RNase distribution appears in the literature, with a range from 84% particulate to 90% soluble (50,105,135,161,162,205). It appears that the culture technique, age of culture, method of harvesting the cells, tonicity and pH of the solutions used to wash the cells free of culture fluid, and cellular disruption method all may play a part in determining the amount of soluble RNase.

Müller et al. (135) suggested a lysosomal origin for at least part of the RNase activity, and also the presence of a latent form of the enzyme. The activation process is poorly understood at this time; therefore it must remain an open question as to the physiological significance of the RNase activity observed in disrupted cellular preparations. Koroly (100) provided evidence that the cellular activity of the RNases is dependent upon the energy state of the cell.

The intracellular RNases from two strains of *T. pyriformis* have been examined in some detail. Strain GL (105,106) contains a microsomal bound phosphodiesterase [as characterized by the hydrolysis of calcium bis(*p*-nitrophenyl phosphate)], as does strain W (205). In addition, a soluble phosphodiesterase (devoid of RNase activity) which cleaves 2′,3′-cyclic nucleotides to 2′-derivatives is found in strain W (30). Three soluble components from strain GL, RNases I, II, and III, can be resolved on a sulfoethoxy Sephadex column (SE-50) when a rather steep sodium chloride gradient is applied (105). The same column can be employed to separate the two enzymes (RNase A and RNase B) found in strain W (205), although a shallower gradient must be employed to obtain satisfactory resolution. The same two RNase components from strain W can be resolved on Sephadex G-200, diethylaminoethylcellulose, or carboxymethylcellulose columns (205).

The pH optima for strain-W enzymes are ~5.5 for RNase A and ~5.0 for RNase B (205). Eichel et al. (50) reported the intracellular activity to be optimum at pH 5.85. All three components from strain GL show maximum activity at pH 5.0 (105).

A further differential between strain-GL and -W enzymes appears in heat stability. Eichel et al. (50) reported that the intracellular RNase activity from strain W was inactivated completely after heating for 5 min at 100°C, while strain GL retained 20% of the original activity after a 20-min period at 100°C (104).

A final difference in the enzymes from the two strains appears in the products formed when 2′-3′-cyclic nucleotides are degraded. All three GL components appear to form 3′-nucleotides (205); RNase A from strain W was shown to give 3′-AMP when degrading 2′-3′-cyclic AMP (205), while RNase B did not attack this substrate.

Roth (161) found strain-W RNases to be sensitive to lilac leaf inhibitor (9) and to copper ions, which may explain the stimulatory effects noted with EDTA (105). No sulfhydryl dependency was noted by Roth (161), although cysteine stimulates activity (38). The strain-W enzymes are not sensitive to 10 m*M* sodium fluoride or sodium cyanide (205).

The RNases in *Tetrahymena* are endonucleolytic and show no base specificity (50,105,162,205).

Ribonuclease preparations from strains W (161) and GL (105) degrade AMP! and UMP!. These results have been confirmed and extended to CMP! and GMP! (205). Further, polyadenylic acid has been shown to give rise to AMP! and polyuridylic acid to UMP! (205).

The nature of the extracellular RNase of strain W, first observed by Eichel et al. (50), is of interest in terms of the digestive processes of this cell. Further work is required to establish if the extracellular enzyme is related to either of the intracellular RNases or represents, as seems likely, still another RNase species.

The acid pH optimum of RNases, as well as the apparent lack of specificity with regard to the internucleotide bond attacked, are properties shared mainly by the same class of enzymes in plants (5,83,158,200). The RNases of the animal kingdom, however, usually have a specificity for either a purine or pyrimidine base.

Only 17% of the RNase in homogenates was active when assayed in 0.25 *M* sucrose at 0°C, but after exposure to lowered osmotic pressure, freezing and thawing, and incubation at temperatures over 0°C, the latency was lost (135). This observation suggests that little of the RNase in these cells is normally active in the cytoplasm, and that this enzyme may be present in lysosomes or their protozoal equivalent. Much remains to be done with the RNases. The absolute purity of the enzymes isolated needs to be proved by disk polyacrylamide electrophoresis, by analytical ultracentrifugation (which will also give the molecular weights), and by crystallization. Not only are these physical measurements necessary to show the absolute number of RNases, but they are also of prime importance in establishing the relationship of the enzymes to cellular metabolism. The high degree of latency and particulate localization make an analysis of the activation steps critical in an attempt to understand the role of these enzymes in cellular function.

In strain W the specificities of RNase A and B and the localization of the two nucleotide phosphatases, namely, 3'-nucleotidase (EC 3.1.3.6) and acid phosphatase (EC 3.1.3.2), lead to an interesting speculation about the possible roles these two RNases might play in vivo. Since 3'-nucleotides are the substrates for the specific 3'-nucleotidase(s), and 2'-nucleotides are degraded mainly by acid phosphatase activity, a coupling could be present between the activities of the two RNases, a phosphodiesterase(s), and the two phosphatases. Conner and Macdonald (37) reported that the major part of the 3'-nucleotidase was soluble. This enzyme therefore might function mainly in the cytoplasm and its substrate, 3'-nucleotides, could be the degradative products of endogenous RNA degraded by RNase A. The nucleosides after conversion to bases, and orthophosphate produced from the 3'-nucleotides, are excreted (29,107) or returned to the common pool for RNA synthesis (100). These reactions form a reasonable scheme for cytoplasmic RNA turnover. However, acid phosphatase activity has long been recognized to be associated with particles in cells (6,8,11,44,109,111,136,137). In *Tetrahymena* several types of membrane-bound structures with acid phosphatase activity have been described by various workers (3,54,55). Food vacuoles, after the attachment of acid phosphatase-containing granules, have been considered as a secondary lysosome and termed phagosomes (135). During starvation autophagic vacuoles form de novo and often contain mitochondria and other cytoplasmic inclusions in various stages of lysis (55). Hydrolytic esterase activities are associated with these "internal" food vacuoles (134). Conner and Macdonald (37) observed that a large proportion of the acid phosphatase in *Tetrahymena* retained a particulate nature even after a variety of severe disruptive treatments, and suggested that this activity is localized in granules of the type associated with food vacuoles rather than in the more general type of lysosome. If acid phosphatase degrades 2'-nucleotides produced by RNase B and phosphodiesterase activity, it is a reasonable assumption that RNase B is present in the food vacuoles also. The role of RNase B, the phosphodiesterase, and of the acid phosphatase would be the digestion of foreign polyribonucleotides present in engulfed food or cytoplasmic RNA in the case of the starvation vacuoles. The speculation on the roles of RNase A and RNase B is summarized in Fig. 3.

2. Metabolism of Nucleosides Arising from RNA

Purine and pyrimidine nucleoside metabolism seems to follow a similar pattern. Deamination of adenosine (48) and cytidine (49) occurs to yield inosine and uridine, respectively. Nucleoside phosphorylases are present, which result in the formation of purine or pyrimidine bases and, presumably, ribose-1-phosphate. Inosine (33,48,79), guanosine, and uridine phosphorylase have been reported in cell-free preparations (79).

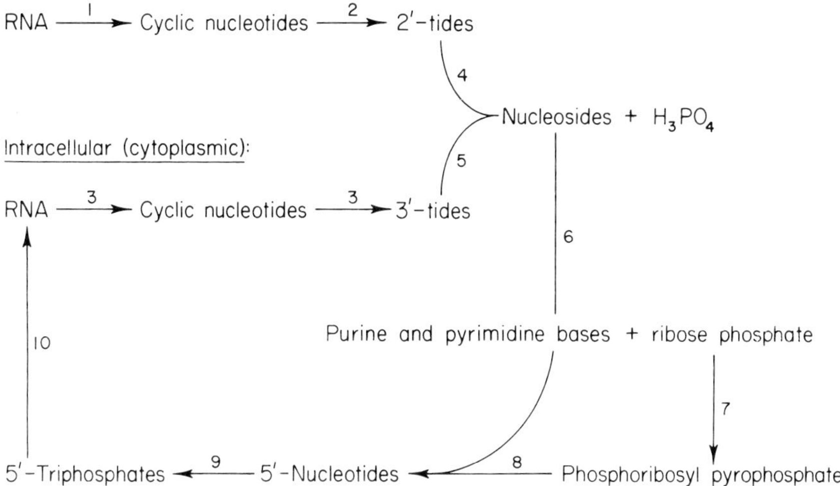

Fig. 3.

Schematic diagram of the proposed roles for RNase A and RNase B in RNA metabolism in vivo. 1, RNase B (EC 2.7.7.17); 2, 2′-3′-cyclic phosphodiesterase; 3, RNase A (EC 2.7.7.17); 4, acid phosphomonoesterase (EC 3.1.3.2); 5, 3′-nucleotidase (EC 3.1.3.6); 6, nucleoside phosphorylase; 7, ATP:D-ribose-5-phosphate pyrophosphotransferase (EC 2.7.6.1); 8, 5′-nucleotide:pyrophosphate phosphoribosyl transferase (EC 2.4.2.7, 2.4.2.8, or 2.4.2.9); 9, nucleotide kinases; 10, RNA polymerase (EC 2.7.7.6).

Under starvation conditions 25–30% of the RNA present in *T. pyriformis* W is degraded and converted to perchloric acid-soluble products. Hypoxanthine (29,107), uracil (29), and orthophosphate (107) are excreted by the ciliates.

C. Purine Metabolism

Numerous reactions involving purine and pyrimidine bases, nucleosides, deoxynucleosides, nucleotides, deoxynucleotides, and analogs have been reported. In some cases there is disagreement as to the presence or absence of a given enzymic activity, or as to the reaction mechanism involved. It is clear, however, that de novo synthesis of purine and pyrimidine ring structures does not occur.

The listing that follows gives the known reactions, the suspected conversions, and the negative assays that appear in the literature (see Fig. 4).

1. Purine Bases

a. Adenine

i. Adenase (adenine deaminase, adenine aminohydrolase, EC 3.5.4.2)

$$\text{Adenine} \xrightarrow{\;\;/\!\!/\;\;} \text{hypoxanthine}$$

The presence of this enzyme was claimed by Seaman (173) in strain S; however, this finding was not substantiated by Eichel (48) for strains S and W or Leboy et al. (107) for strain W. Heinrich et al. (74) inferred the presence of this enzyme, but Kidder

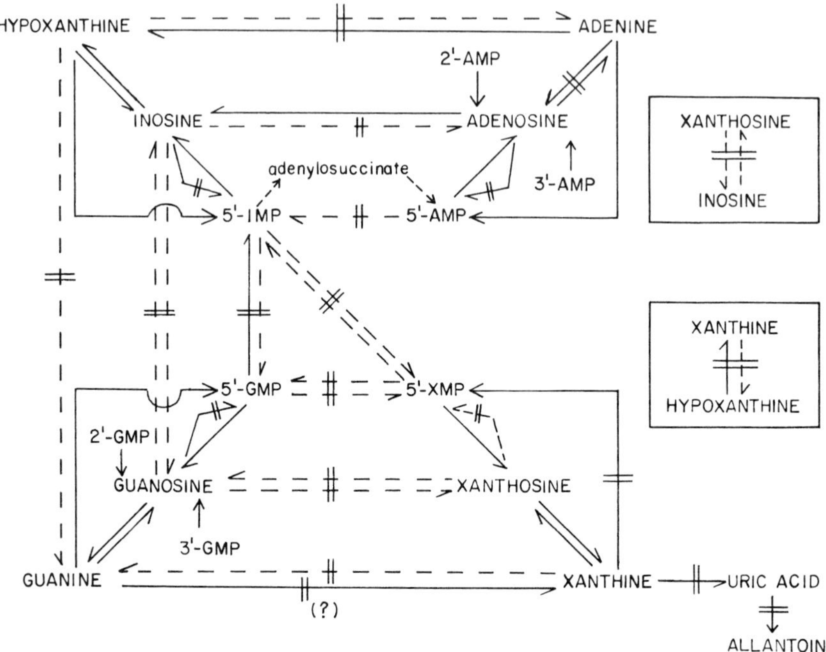

Fig. 4.

A summary of purine metabolism in *T. pyriformis*. Solid lines indicate those reactions in which enzymic assays have been reported to be positive. Solid lines with crossbars indicate negative results. Dashed lines represent those reactions that are assumed to occur, and dashed lines with crossbars are cases in which activity is believed to be absent, although no enzymic assays have been reported.

(90) reinterpreted this work in the light of more definitive enzymic evidence that became available.

ii. AMP pyrophosphorylase (AMP:pyrophosphate phosphoribosyltransferase, EC 2.4.2.7)

$$\text{Adenine} + \text{PRPP} \longrightarrow 5'\text{-AMP} + \text{PP}_i$$

The occurrence of the pyrophosphorylase activity has been demonstrated by Hill and Chambers (79). This reaction would explain the ability of adenine to spare but not replace guanine as a growth requirement. The activation of adenine via 5'-AMP formation would allow incorporation of this compound into nucleic acid derivatives and coenzymes containing adenine and hypoxanthine.

iii. Nucleoside phosphorylase (purine nucleoside phosphorylase, purine-nucleoside: orthophosphate ribosyltransferase, EC 2.4.2.1)

$$\text{Adenine} + \text{ribose-1-phosphate} \;\not\rightleftharpoons\; \text{adenosine} + H_3PO_4$$

The absence of this enzyme is indicated by the work of Leboy et al. (107) and of Hill and Chambers (79).

b. Hypoxanthine

i. Xanthine oxidase (hypoxanthine oxidase, xanthine:oxygen oxidoreductase, EC 1.2.3.2)

$$\text{Hypoxanthine} \xrightarrow{\;/\!/\;} \text{xanthine}$$

Extensive investigation has failed to detect this enzyme (48,79,107,203). The unconfirmed report of Seaman (173) claiming the occurrence of this enzyme must remain in question. Villela (190) agrees there is a lack of oxidases but claims the presence of the highly labile anaerobic dehydrogenases for guanine, isoguanine, hypoxanthine, and xanthine, which can couple to triphenyltetrazolium chloride and methylene blue. The role of these enzymes in cellular metabolism is not clear at this time, since Koroly (100) has established a stoichiometric relationship between the loss of RNA from the cells and the appearance of orthophosphate, pyrimidines, and purines as degradation products. Further, the physiological significance of the purine dehydrogenases can be questioned, since Hill and Chambers (79) found that intact cells under anaerobic conditions did not give rise to xanthine or uric acid when incubated with $[8\text{-}^{14}\text{C}]$hypoxanthine.

ii. Inosine monophosphate (IMP) pyrophosphorylase (hypoxanthine phosphoribosyltransferase, IMP:pyrophosphate phosphoribosyltransferase, EC 2.4.2.8)

$$\text{Hypoxanthine} + \text{PRPP} \rightarrow 5'\text{-IMP} + \text{PP}_i$$

This enzyme, described by Hill and Chambers (79), provides a means for the activation of hypoxanthine which, similar to adenine, spares but does not replace the guanine nutritional requirement.

The nucleoside phosphorylase reaction is probably reversible (see Section V,C,1,a, iii); however, since inosine kinase appears to be absent (79), formation of the nucleoside would represent a dead-end pathway of no physiological significance in the production of $5'$-IMP.

c. *Guanine*

i. Guanase (guanine deaminase, guanine aminohydrolase, EC 3.5.4.3)

$$\text{Guanine} + \text{H}_2\text{O} \xrightarrow{\;/\!/\;} \text{xanthine} + \text{NH}_3$$

Eichel (48) and Hill and Chambers (79) were unable to detect guanase; however, Seaman (173) claims the presence of this activity. Villela (190) found only dehydrogenase activity. Leboy and Conner (108), using more sensitive assays, noted the conversion of small amounts of guanine by cell-free extracts to a material that can be converted by xanthine oxidase to uric acid as judged by the usual spectrophotometric assay. Since guanine can serve as the sole purine source in these cells, conversion to adenine (presumably via hypoxanthine) must occur. The "guanase" activity may reflect conversion of guanine or a derivative in the incubation mixtures to hypoxanthine or a derivative which can then react with added xanthine oxidase to give uric acid. It must be assumed that, if a hypoxanthine derivative is involved, it must be metabolized back to the base in the extracts in order to react with xanthine oxidase.

ii. Guanosine phosphorylase (purine nucleoside:orthophosphate ribosyltransferase, EC 2.4.2.1)

$$\text{Guanine} + \text{ribose-1-phosphate} \rightleftharpoons \text{guanosine} + \text{H}_3\text{PO}_4$$

The enzymological studies by Hill and Chambers (79) provide excellent evidence for the existence of this enzyme.

iii. GMP:pyrophosphorylase (GMP:pyrophosphate phosphoribosyltransferase, EC 2.4.2.8)

$$\text{Guanine + PRPP} \rightarrow \text{5'-GMP + PP}_i$$

This reaction is catalyzed by an enzyme found in the soluble fraction of cell-free preparations obtained from *T. pyriformis* W (79).

d. Xanthine

i. Xanthine oxidase (see section V,C,1,b,i)

ii. Xanthine monophosphate (XMP):pyrophosphate phosphoribosyltransferase (xanthine pyrophosphorylase)

$$\text{Xanthine + PRPP} \nrightarrow \text{5'-XMP + PP}_i$$

Hill and Chambers (79) reported that no detectable activity could be measured in *Tetrahymena* preparations.

e. Uric Acid

i. Uricase (urate oxidase, urate:oxygen oxidoreductase, EC 1.7.3.3)

$$\text{Uric acid + O}_2 \nrightarrow \text{unidentified products}$$

The absence of activity in cell-free preparations has been noted by Eichel (48), Villela (190), and Conner and Hershkowitz (33), although Seaman (173) claims to have found this enzyme.

2. Purine Nucleoside-Metabolizing Enzymes

a. Purine Kinases (Adenosine Kinase, ATP:Adenosine-5'-Phosphotransferase, EC 2.7.1.20; Guanosine Kinase; Inosine Kinase)

$$\text{Adenosine + ATP} \nrightarrow \text{5'-AMP + ADP}$$
$$\text{Guanosine + ATP} \nrightarrow \text{5'-GMP + ADP}$$
$$\text{Inosine + ATP} \nrightarrow \text{5'-IMP + ADP}$$

No kinase activity toward purine nucleosides was detected by Hill and Chambers (79) in *T. pyriformis*.

b. Nucleoside Hydrolase (Inosinase, Inosine Ribohydrolase, EC 3.2.2.2; Adenosine, Guanosine, Inosine, or Xanthosine Hydrolases)

$$\text{Inosine + H}_2\text{O} \rightarrow \text{hypoxanthine + ribose}$$

No direct evidence for the presence of purine nucleoside hydrolases exists. These enzymes have been reported in *Paramecium aurelia* (176); however, in this case nucleoside phosphorylases are absent. *Tetrahymena pyriformis* W incubated with $[8\text{-}^{14}\text{C}]$AMP appear to give rise to a small amount of adenine (<1%) (79). This interesting observation, if not the result of a chromatographic artifact since rigorous proof of adenine occurrence was not reported, suggests, in the absence of detectable adenosine phosphorylase, the following scheme:

$$5'\text{-AMP} \xrightarrow[\text{phosphatase}]{\text{acid}} \text{adenosine} \xrightarrow{\text{hydrolase}} \text{adenine}$$

c. *Xanthosine Phosphorylase* (Hydrolase)

$$\text{Xanthosine} + H_3PO_4 \rightleftarrows \text{xanthine} + \text{ribose-1-phosphate}$$

or

$$\text{Xanthosine} + H_2O \rightleftarrows \text{xanthine} + \text{ribose}$$

Xanthosine phosphorylase or xanthine ribosylhydrolase activity can be demonstrated in homogenates of *T. pyriformis* W (33). This reaction may have no physiological significance, since neither xanthosine nor xanthine can serve in a nutritional capacity.

3. *Purine Nucleotide-Metabolizing Enzymes*

a. *5'-AMP*

i. Acid phosphatase (orthophosphoric monoester phosphohydrolase, EC 3.1.3.2)

$$5'\text{-AMP} + H_2O \rightarrow \text{adenosine} + H_3PO_4$$

Evidence for acid phosphatase activity toward 5'-AMP was first provided by Elliott and Hunter (53) with intact cells, and later by enzymological assays by Allen et al. (3) and Conner and Macdonald (37). There is, however, no definitive evidence that would bar the existence of a 5'-nucleotidase (5'-ribonucleotide phosphohydrolase, EC 3.1.3.5). The acid phosphatase(s) also dephosphorylates 5'-dAMP, 2'-AMP (but not 3'-AMP), 5'-GMP, 5'-dGMP, 5'-IMP, 5'-CMP, 5'-dCMP, 5'-UMP, and 5'-TMP (35).

ii. 5'-AMP deaminase (AMP aminohydrolase, EC 3.5.4.6)

$$5'\text{-AMP} + H_2O \not\rightarrow 5'\text{-IMP} + NH_3$$

The claim has been made that an unusual soluble 5'-AMP deaminase can be observed in cell-free preparations of strain S (173). Other investigators (33,48), however, have provided evidence that casts doubt on the existence of such an enzyme. The presence of nucleotide phosphatases that give rise to nucleosides known to be deaminated, prevents a clear-cut interpretation of the earlier nucleotide deaminase studies. Since deaminase activity previously was followed only by spectrophotometric techniques or by ammonia release, it was not possible to determine whether the nucleotide or the nucleoside is the substrate for the deaminase. To circumvent this technical limitation, radiolabeled substrates were employed. Neither of the expected deamination derivatives of ^{14}C-labeled 5'-AMP or 5'-CMP, that is, 5'-IMP or 5'-UMP, respectively, could be detected after incubation of the labeled substrates with cell-free preparations and subsequent chromatographic isolation of the nucleotides. Thus no evidence could be found for the presence of 5'-AMP and 5'-CMP deaminases in strain W (35).

iii. Other 5'-nucleotide reactions. A series of 5'-nucleotide reactions known to occur in several organisms is important in the interconversion of guanine and adenine derivatives. Since guanine can serve as the sole purine source, while adenine and hypoxanthine can only spare the guanine requirement, it is obvious that a full complement of enzymes is not present in *Tetrahymena*. Xanthine shows no sparing action in growth studies, hence has been assumed not to be involved in any of the interconversions. Hill and Chambers (79) showed that xanthine cannot be activated by the 1-pyrophosphoryl-ribosyl-5-phosphate system (nucleotide pyrophosphorylase), while adenine and hypoxanthine can. This observation may explain the lack of effect of xanthine and its derivatives, since xanthosine phosphorylase activity is present which would give rise to the inactive base xanthine. Further, since purine kinases are not found, it would not be possible to "activate" xanthosine by direct phosphorylation. Exogenous xanthosine-5-phosphate would have to run the gauntlet of the acid phosphatases and thus probably could not enter the metabolism of the cell unscathed. Even though xanthine and its derivatives are not active nutritionally, these compounds have not been eliminated completely as intermediates in the conversion of guanine to hypoxanthine and adenine derivatives. The most likely reaction (90) for bypassing xanthine in the conversion of 5'-GMP to 5'-IMP is GMP reductase [reduced NADP:GMP oxidoreductase (deaminating), EC 1.6.6.8]; however, Hill and Chambers (79) were unable to demonstrate this enzyme in cell-free preparations of strain W.

$$5'\text{-GMP} + \text{NADPH}^+ + \text{H}^+ \rightleftharpoons\!\!\!\!/ \ \text{NADP}^+ + 5'\text{-IMP} + \text{NH}_3$$

These investigators also were unable to detect xanthine in cells incubated with [^{14}C]-8-guanine. Thus the question concerning what reactions or what compounds are involved in guanine transformation to adenine remains to be answered.

The equivalence of hypoxanthine and adenine in sparing guanine implies that adenine derivatives can arise from hypoxanthine nucleoside or, more likely, 5'-mono-, di-, or triphosphonucleotides.

On the basis of the above discussion, the following enzymes are deemed not to be present.

    1. 5'-GMP synthetase (xanthosine-5-phosphate:ammonia ligase (AMP), EC 6.3.4.1)

$$5'\text{-GMP} + \text{PP}_i + 5'\text{-AMP} \rightleftharpoons\!\!\!\!/ \ \text{xanthosine-5-phosphate} + \text{NH}_3 + \text{ATP}$$

    2. 5'-GMP synthetase (xanthosine-5'-phosphate:*l*-glutamine amido-ligase (AMP), EC 6.3.5.2)

$$5'\text{-GMP} + l\text{-glutamate} + \text{PP}_i + \text{AMP} \rightleftharpoons\!\!\!\!/ \ \text{xanthosine-5'-phosphate} + l\text{-glutamine} + \text{ATP} + \text{H}_2\text{O}$$

    3. IMP dehydrogenase (IMP:NAD oxidoreductase, EC 1.2.1.4)

$$5'\text{-IMP} + \text{NAD}^+ + \text{H}_2\text{O} \rightleftharpoons\!\!\!\!/ \ \text{xanthosine-5'-phosphate} + \text{NADH} + \text{H}^+$$

The following reactions are presumed to be responsible for the conversion of 5'-IMP to 5'-AMP.

    1. 5'-IMP + GTP + *l*-aspartate $\rightleftharpoons$ GDP + H$_3$PO$_4$ + adenylosuccinate (adenylosuccinate synthetase, IMP:*l*-aspartate ligase (GDP), EC 6.3.4.4)

    2. Adenylosuccinate $\rightleftharpoons$ fumarate + AMP (adenylosuccinate lyase, adenylosuccinate AMP lyase, EC 4.3.2.2)

In summary:

1. Phosphatases have been demonstrated to degrade 2'- and 3'-AMP to adenosine. Mixtures of 2'(3')-GMP are dephosphorylated. Further, acid phosphatases release orthophosphate from 5'-AMP, 5'-dAMP, 5'-IMP, 5'-XMP, 5'-GMP, and 5'-dGMP.

2. Adenosine-deoxyadenosine deaminase has been established; 5'-AMP-5'-dAMP deaminase is absent.

3. Nucleoside kinase activity is lacking toward adenosine, guanosine, inosine and, presumably, xanthosine.

4. Nucleoside phosphorylase metabolizes inosine (but not adenosine), guanosine, and xanthosine to the free base.

5. Nucleotide pyrophosphorylase converts adenine, guanine, inosine, and probably 8-azaguanine to the corresponding 5'-nucleotides.

The metabolic scheme for purine metabolism is reasonably complete (see Fig. 4), although there remain some disputed areas and a need in the case of a few reactions for a more complete description of the mechanism by employing techniques that have become available since the original investigations. There are still some observations that cannot be explained by the scheme presented. The most serious of these concerns the interaction of azapurines, especially 8-azaguanine and guanine and its derivatives on growth. Dewey and Kidder (46) found a bimodal curve for 8-azaguanine inhibition when guanine was used in the growth medium, but only a unimodal curve when guanosine was employed. The inhibition could be partially overcome by increasing the guanine concentration; however, guanosine was ineffective. This suggests a block at the guanosine phosphorylase level. The surprising finding that 2'(3')-mixtures of GMP annulled the azaguanine inhibition is difficult to explain. Both guanylic isomers are known to undergo dephosphorylation, and it would be anticipated that the 2'(3')-GMP mixture would give rise to guanosine which had been shown to lack the ability to counteract the inhibitory agent. These data can be explained in several ways: (1) the 2'(3')-guanylic acid mixture contained a sizable quantity of guanine (or guanine arises during the preparation of the medium); (2) there is a mechanism for bypassing guanosine in the intermediary metabolic pathway; or (3) the 2'(3')-isomeric mixture interferes with 8-azaguanine entry into the cells, or with the conversion of 8-azaguanine to the "active" form—presumably (5')-8-azaguanosine mononucleotide. Clearly, additional investigation is necessary to clarify this important point.

D. Pyrimidine Metabolism

1. Cytosine

a. Cytosine Deaminase (Cytosine Aminohydrolase, EC 3.5.4.1)

$$\text{Cytosine} + H_2O \xrightarrow{\quad\not\quad} \text{uracil} + NH_3$$

Cytosine deaminase activity is not found in cell-free preparations (49), nor is CMP pyrophosphorylase (79):

$$\text{Cytosine} + \text{5-phosphoribosyl-1-pyrophosphate} \;\not\rightleftharpoons\; \text{5'-CMP} + PP_i$$

b. Cytidine Phosphorylase

This enzyme appears lacking, since cytosine is inert nutritionally (95). The presence of a nucleoside phosphorylase would allow the conversion of cytosine to cytidine which can fulfill the pyrimidine nutritional requirement.

$$\text{Cytosine} + \text{ribose-1-phosphate} \not\rightleftharpoons \text{cytidine} + H_3PO_4$$

2. Cytidine

a. Cytidine kinase is not detected by using conventional assays (79,206).

$$\text{Cytidine} + \text{ATP} \not\rightleftharpoons 5'\text{-CMP} + \text{ADP}$$

b. Cytidine-deoxycytidine deaminase (cytidine aminohydrolase, EC 3.5.4.5)

The deamination of cytidine (deoxycytidine) has been demonstrated by Eichel (49), Winicur and Roth (201), and Hill and Chambers (79).

$$\text{Cytidine (deoxycytidine)} + H_2O \rightarrow \text{uridine (deoxyuridine)} + NH_3$$

3. 2'-,3'-,5'-CMP

a. Phosphatases

The dephosphorylation of 2'(3')-CMP mixtures, 5'-CMP, and 5'-dCMP has been demonstrated (35,37). Phosphatase activity toward 2'(3')-UMP mixtures, 5'-UMP, and 5'-TMP has also been noted. It is believed that phosphoric acid removal from 2'- and 5'-nucleotides and 5'-deoxynucleotides is due to acid phosphatase (EC 3.1.3.2). The pH optimum in all cases is from 4.5 to 5.0, and the reaction is blocked by fluoride. 3'-Nucleotides and, presumably, 3'-deoxynucleotides are dephosphorylated by 3'-nucleotidase (EC 3.1.3.6) which has a pH optimum at 6.8 and is not inhibited by fluoride but is partially affected by cysteine and cyanide (35,37).

b. 5'-CMP Deaminase

$$5'\text{-CMP} + H_2O \not\rightarrow 5'\text{-UMP} + NH_3$$

No definitive evidence has been presented concerning a 5'-CMP deaminase; however, some data suggest the absence of this enzyme (79,201). This is reinforced by the finding that sodium fluoride, which prevents dephosphorylation of 5'-CMP, also prevents deamination (35).

The formation of cytidine derivatives from uracil must occur, since uracil can serve as the sole pyrimidine source. This may take place at the 5'-nucleotide level but more probably occurs in the nucleotide triphosphates.

4. Uracil

a. UMP Pyrophosphorylase (Uracil Phosphoribosyltransferase, UMP:Pyrophosphate Phosphoribosyltransferase, EC 2.4.2.9)

$$\text{Uracil} + \text{5-phospho-}\alpha\text{-D-ribosylpyrophosphate} \rightleftharpoons 5'\text{-UMP} + PP_i$$

This is one of two enzymes in *Tetrahymena* known to utilize uracil as a substrate (68,77–79). This type of scavenger mechanism appears to be necessary in the formation

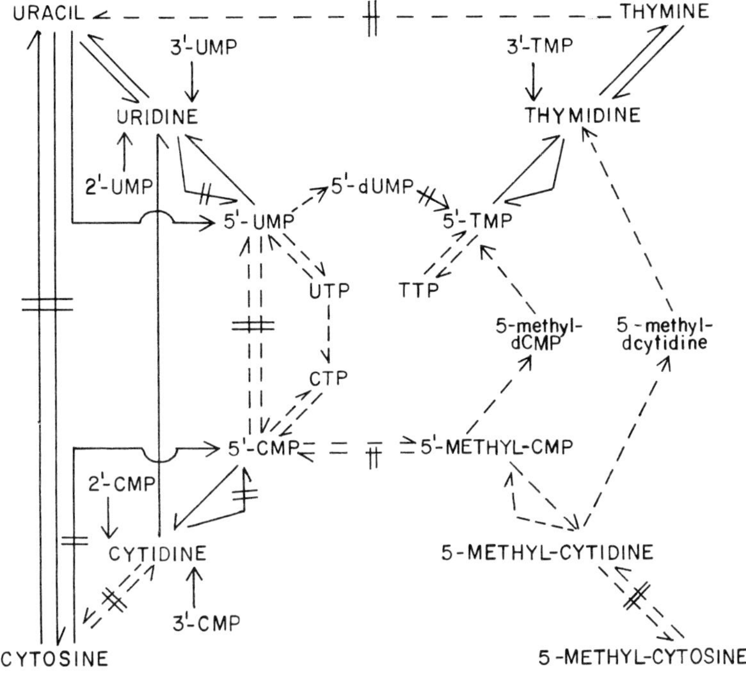

Fig. 5.

A summary of pyrimidine metabolism in *T. pyriformis*. Solid lines indicate those reactions in which enzymic assays have been reported to be positive. Solid lines with crossbars indicate negative results. Dashed lines represent those reactions that are assumed to occur, and dashed lines with crossbars are cases in which activity is believed to be absent, although no enzymic assays have been reported.

of all the precursors for RNA synthesis. These enzymes may represent important agents in the regulation of RNA synthesis, since the control mechanism noted thus far in other cellular systems appears to block the de novo synthesis of the purine and pyrimidine heterocyclic ring structures. Since de novo synthesis does not occur in *Tetrahymena*, the availability of the purine and pyrimidine bases and of 5'-phospho-α-D-ribosyl-1-pyrophosphate must be limiting under given conditions, hence regulate the overall rate of growth. An examination of the factors influencing these and related enzymes may provide important clues to the maintenance of balanced growth.

b. Pseudouridine (ψ) Synthetase

The first demonstration of the synthetase of ψ-uridine was carried out by Heinrikson and Goldwasser (68,77,78) by using *Tetrahymena* extracts.

$$\text{Uracil} + \text{ribose-5-phosphate} \rightarrow \psi\text{-uridine monophosphate}$$

The formation of this pyrimidine derivative, found as one of the unusual bases in tRNA, has been shown to proceed via a reaction mechanism differing from the UMP pyrophosphorylase. The formation of the C—C bond rather than the N—C linkage represents a unique type of chemistry in the nucleic acids (see Fig. 5). Since ψ-UMP is found in tRNA, it can be assumed that kinases convert the mononucleotide to di- and triphosphates.

5. *Uridine*

a. Uridine Kinase (ATP:Uridine-5′-Phosphotransferase, EC 2.7.1.8)
Some apparently conflicting reports concerning uridine kinase appear in the literature. This enzymic activity has been described as absent (79) or weak (78). Plunkett and Moner (153) found uridine was phosphorylated rapidly to give the triphosphate derivative. ATP and GTP were effective phosphorylating agents. UTP and CTP probably act as feedback inhibitors on the uridine kinase and thus regulate the imput of precursors for RNA and DNA synthesis. Such a feedback system could account for the lack of or limited activity noted previously.

$$\text{ATP (GTP)} + \text{uridine} \rightarrow \text{ADP (GDP)} + 5′\text{-UMP}$$

b. Uridine Phosphorylase (Uridine-Orthophosphate Ribosyltransferase, EC 2.4.2.3)
Uridine-deoxyuridine phosphorylase was originally reported by Eichel (49) and has been confirmed by Hill and Chambers (79).

$$\text{Uridine} + H_3PO_4 \rightleftarrows \text{uracil} + \text{ribose-1-phosphate}$$

6. *2′-, 3′-, and 5′-UMP*

The dephosphorylation of 2′(3′)-UMP mixtures appears to be due to acid phosphatase (EC 3.1.3.2) and 3′-nucleotidase (EC 3.1.3.6), while acid phosphatase is active toward 5′-UMP (35). There is no information about a specific 5′-nucleotidase (EC 3.1.3.5) available at this time.

$$\text{UMP} + H_2O \rightarrow \text{uridine} + H_3PO_4$$

b. Nucleoside-deoxynucleosidomonophosphate (ATP: Nucleosidomonophosphate Phosphotransferase, EC 2.7.4.4, 2.7.4.5, and 2.7.4.9) and ***Nucleosidodiphosphate Kinases*** (ATP:Nucleosidediphosphate Phosphotransferase, EC 2.7.4.6)
5′-UMP kinase has been purified 300-fold. The preparation also phosphorylates dUMP and 5′-CMP. Some evidence indicates cooperation and two 5′-UMP sites on the enzyme. Nucleoside di- and triphosphates in high concentrations are inhibitory (164).

$$5′\text{-UMP} + \text{ATP} \rightarrow 5′\text{-UDP} + \text{ADP}$$
$$5′\text{-UDP} + \text{ATP} \rightarrow 5′\text{-UTP} + \text{ADP}$$

7. *Thymidine*

a. Thymidine Phosphorylase (Thymidine:Orthophosphate Deoxyribosyltransferase, EC 2.4.2.4)

$$\text{Thymidine} + H_3PO_4 \rightleftarrows \text{thymine} + \text{deoxyribose-1-phosphate}$$

This enzyme was demonstrated by Friedkin (65).
b. Thymidine Kinase (ATP:Thymidine-5′-Phosphotransferase, EC 2.7.1.21)

This enzyme appears to be a constitutive enzyme in *Tetrahymena* (174) and is not inhibited by TTP, TDP, dCTP, or dCMP.

$$\text{Thymidine} + \text{ATP} \rightarrow \text{TMP} + \text{ADP}$$

Uracil can serve as the sole pyrimidine source and therefore can give rise to both cytidine (cytidylate) and thymidine (thymidylate). The pathways are unknown, and only fragmentary evidence is currently available. The report that labeled thymidine appears only in DNA (1,13,206) has been challenged (103). Deoxycytidine and deoxyuridine label both RNA and DNA, which indicates that both compounds pass through uracil before incorporation. These two observations provide strong evidence that direct demethylation of the pyrimidine ring does not occur in *T. pyriformis*, although the metabolism of the pyrimidine ring in other ways may lead to radiolabel incorporation into compounds other than those found in nucleic acids. The lack of direct demethylation is further suggested by the inability of thymidine or 5′-methyluridine to support growth (90,206). A thymidine deficiency can be induced in *Tetrahymena* grown in a chemically defined medium with 4-amino-N^{10}-methylpteroylglutamic acid (methotrexate, A-methopterin), a folic acid antagonist (47). Zeuthen (208) noted that a combination of methotrexate and uridine results in growth inhibition in proteose–peptone culture fluid, which can be reversed by thymidine addition. This interesting observation suggests that uridine inhibits the uptake of exogenous thymidine, while the folate analog appears to block the methylation step in the conversion of a uracil or cytosine derivative to TMP. This conclusion is reinforced by the finding that 5-methyldeoxycytidine is as effective as thymidine in releasing the inhibition (206,208). Thymidine, 5-methyldeoxycytidine, 5-methyl-dCMP, or TMP also partially release the growth block induced by a folic acid deficiency; however, 5-methyluridine is inactive (206). These findings indicate that 5′-methylcytidine is not deaminated to 5′-methyluridine but do not answer the question concerning the biosynthetic pathway of thymidine monophosphate. 5′-dMethylcytidine may be converted to 5′-dCMP and then deaminated to give TMP; however, this remains to be established as the de novo biosynthetic pathway in *Tetrahymena*. TMP synthetase catalyzing the reaction dUMP $-\!/\!/\!\!\rightarrow$ TMP appears to be lacking (206).

Tetrahymena pyriformis is quite insensitive to 5-fluorodeoxyuridine (64,206,208) which is a powerful inhibitor of DNA synthesis in bacteria (87) and mammalian cells (52) as a result of the inhibition of TMP synthetase. Wykes and Prescott (206) suggest that the absence of this enzyme could account for the lack of sensitivity in these ciliates.

In summary:

1. Phosphatases have been demonstrated to degrade 2′(3′) mixtures of UMP and CMP to uridine and cytidine, respectively. Further, 5′-UMP, 5′-CMP, 5′-dCMP, and 5′-TMP are dephosphorylated by cell-free extracts.

2. Deaminases for cytidine-deoxycytidine have been established. 5′-CMP and 5′-dCMP deaminase are absent, and cytosine deaminase assays are negative.

3. Nucleoside kinase activity toward uridine-deoxyuridine is present, while cytidine-deoxycytidine kinase appears to be absent; thymidine kinase is found; and a 5-methylcytidine kinase is suspected.

4. A nucleoside phosphorylase is active toward uridine, thymidine, and probably 5-fluorouridine (206), although the last-mentioned reaction has not been verified by an enzymic assay.

5. Nucleotide pyrophosphorylase activity seems to be restricted to the conversion of uracil to 5′-UMP and probably 5-fluorouracil to the nucleotide derivative.

6. Nucleotide kinases are unquestionably present, as is an aminating system which converts a uracil to a cytosine derivative, and a methylating mechanism which converts a uridine or cytidine derivative to 5′-TMP.

E. Ribonucleic Acid Polymerase

A macronuclear DNA-dependent RNA polymerase has been described by Lee and Byfield (112). Expression of this activity is dependent on DNA, Mg^{2+}, and all four riboside triphosphates. Puromycin and pyrophosphate are inhibitory, while rifampicin and orthophosphate are not. Thus the enzyme(s) resembles those of mammalian cells rather than bacteria. Activity is lost if the nucleus is ruptured.

Ribonucleic Acid-Adenylate (-Cytidylate) Pyrophosphorylase [*ATP:sRNA Adenyltransferase* (sRNA Adenyltransferase, EC 2.7.7.20); *CTP:sRNA Cytidyltransferase* (sRNA Cytidyltransferase, EC 2.7.7.21)]

A soluble extract prepared from the particulate (macronuclear?) fraction of *T. pyriformis* S appeared to contain RNA-adenylate (-cytidylate) pyrophosphorylase activity (59):

$$sRNA_n + 2\,CTP + ATP \rightleftarrows sRNA_{n+3}(pCpCpA) + PP_i$$

The enzyme(s) was quite unstable and had to be prepared in a hypertonic, buffered 50% glycerol solution. The incorporation of labeled ATP and CTP into sRNA could be followed. The large amounts of contaminating nucleic acid, nucleases, and proteinases hinder purification. Fitt (59) reports a pH optimum of 8.2 and a magnesium requirement of 10–16 mM. This investigation represents a major step toward isolating the labile enzymes involved in RNA synthesis.

F. Regulation of Nucleic Acid Synthesis and Degradation

The description of RNA amounts in the various phases of growth shows clearly that the cellular content of this nucleic acid does not remain at any one level for an extended period of time (Fig. 2). This failure of RNA to reach a steady-sate level in a culture, even when cells are reproducing at a constant rate, may be due to the rapidly changing cellular environment. The RNA levels observed in *Tetrahymena* follow a pattern that is similar in all types of cells. Bacteria (133,139,191), yeast (125), *Tetrahymena* (19,166), *Euglena* (14), and mammalian tissue culture cells (165,188) all show an increase in cellular RNA throughout the lag and acceleration phases of growth until a maximum is reached at a point specific for each type of culture. Once the maximum level is attained, the cellular content decreases steadily for the remainder of the log phase of growth.

The problem of the regulation of RNA levels by environmental factors has been approached in bacteria by resuspending the cells in either enriched or minimal media and examining the amount and synthesis of RNA in these organisms for a period of time. Neidhardt and Magasanik (140) showed that enrichment of a minimal growth medium results in an immediate increase in cellular RNA synthesis, followed by an increase in protein content and an elevation of the growth rate in *Aerobacter aerogenes*. These investigators, as well as Mandelstam and Halvorson (129), Kjeldgaard and Kurland (99), and Niwa, et al. (143), showed that various log phase bacterial cells placed in medium deficient in one or more essential nutrients, such as a required nucleic acid precursor, an amino acid, or an energy source, respond to the "shift-down" conditions by a rapid and marked depression of RNA synthesis followed temporally by a decline in protein production. Thus the level of RNA is closely coupled to the growth rate of the cells.

Tetrahymena pyriformis also responds to a shift-down situation by a series of rapid readjustments in ultrastructure (122) and metabolism. When these protozoa are suspended in a nonnutrient buffered medium, cell division ceases in a short time and there is an initial, rapid reduction followed by a more gradual decline in the rate of respiration (32), a loss of protein (203), a loss of RNA (107), and an excretion of the products of RNA degradation into the suspension fluid (29).

The terminal products of RNA degradation excreted from the ciliates suspended in a buffered solution have been established as uracil, hypoxanthine, and orthophosphate (29,107). Ribose does not appear in the excretion products in stoichiometric amounts and is presumed to be metabolized as an energy source under these adverse conditions. Cline and Conner (29) established that after a 3-h incubation in the buffer, the appearance of orthophosphate in the external environment was equivalent to the cellular RNA decrement.

Koroly (100) extended these quantitative estimates to the purines and pyrimidines. She showed that a stoichiometric relationship can be established between the RNA degraded under starvation conditions and the formation and excretion of the purine and pyrimidine bases. This type of straightforward quantitative purine and pyrimidine analysis is based on a number of assumptions. Such studies are possible only when de novo synthesis and extensive degradation of the purine and pyrimidine heterocyclic ring structures do not occur. Continued de novo synthesis of purine and pyrimidine bases throughout the incubation period would add the difficult technical complication of correcting for a pool of increasing size. The same consideration holds if complete breakdown of the bases is a part of the metabolic capacity of the cell employed. If both facets occur simultaneously, the problem cannot be approached in a satisfactory manner.

Tetrahymena pyriformis meets both of these qualifications as an experimental organism. Purines and pyrimidines cannot be synthesized de novo (90) and are not degraded (29,107), although thymidine metabolism may occur (103).

Two other conditions are important for satisfactory quantitation in the purine-pyrimidine assay; first, the base ratios in RNA must be known; and second, the intermediary metabolism of purines and pyrimidines must be established. Ribonucleic acid base ratios for strain W have been reported by Cummins and Plaut (41). The occurrence of the unusual bases in tRNA does not present a difficulty, since these compounds are present only in trace amounts. The purine-pyrimidine end products of RNA metabolism have been established as hypoxanthine, probably guanine, and uracil (29). Using this information, Koroly (100) demonstrated that the cellular pool of purines and pyrimidines remains constant when the cells are incubated in the absence of an exogenous source of these compounds, even though large quantities of RNA may be degraded during starvation.

Several agents are known to alter the qualitative and quantitative composition of the excretion products of RNA from *Tetrahymena*. These agents include glucose (36), Na$^+$ and Mg^{2+} (27), sterols and steroids (36), dinitrophenol (27,36), and iodoacetate (32). The influence of these materials on the loss of catabolic products could be due to any one of several possibilities. First, RNA degradation per se may be altered; second, RNA synthesis may be affected; or last, certain of the RNA breakdown products may be selectively retained by the cells or assimilated into non-RNA materials.

Mg^{2+}, cholesterol, and glucose do not influence RNA synthesis or degradation; however, each appears to reduce orthophosphate and/or purine and pyrimidine efflux from *Tetrahymena* in a distinctly different way. The primary influence of Mg^{2+} is to increase the amount of phosphorus arising from RNA breakdown in the cellular acid-soluble pool (100). It can be assumed this is due to the formation of an insoluble magnesium phosphate complex which may be organic or inorganic in nature (27,28). These two possibilities can be distinguished readily by a determination of the cellular levels of orthophosphate in the presence and absence of the cation. These results appear to negate the speculations offered by Cline (27) who suggested that Mg^{2+} may stabilize ribosomes and thus reduce RNA degradation. Koroly (100), however, found little or no influence of Mg^{2+} on RNA levels.

Cholesterol was found to have no influence on RNA degradation. The sterol led, however, to the selective assimilation of phosphate into one of the non-RNA acid-insoluble cellular constituents (100). The lack of influence on purines and pyrimidines appears to eliminate DNA from consideration and indicates incorporation of the anion into phospholipid or phosphoproteins. Day et al. (42) found that the addition of cholesterol to a macrophage suspension increased the rate of incorporation of radiophosphate into phospholipids. Cholesterol has also been noted to influence the type and amount of fatty acid synthesized and incorporated into *T. setifera* phospholipds (58).

Glucose addition to the shift-down suspensions had little effect on RNA loss, and only a small differential was noted when these suspensions were compared to the controls (100). This hexose leads to the assimilation into an acid-insoluble component of an equivalent amount of purine and pyrimidine bases, and orthophosphate arising from RNA. These results strongly suggest that glucose, in some fashion, induces DNA synthesis. This speculation can be tested experimentally by following the size of the DNA fraction and by the use of radiothymidine as an indicator of DNA synthesis provided the DNA is isolated in reasonably pure form (103).

Glucose and acetate, excellent carbon sources for these cells, had little or no effect on the extent of RNA degradation or synthesis in starved cells; therefore energy does not appear to be the limiting factor (100). That energy is important in determining the minimum rate of RNA degradation was established by the use of the oxidative phosphorylation uncoupling agent, 2,4-dinitrophenol (DNP), and the glycolytic inhibitor iodoacetic acid (IAA). The addition of DNP to the cellular suspensions increases RNA breakdown and blocks synthesis (100). Glucose, which would be anticipated to lead to ATP formation even in the presence of DNP, effectively annuls the increment of loss brought on by the inhibitor but does not restore RNA synthesis. It may be that the availability of the chemical energy source for RNA synthesis is limited because of the nature of the metabolism of glucose. This compound is rapidly incorporated into glycogen (4,82,123), even in the presence of DNP (34), although a portion also is respired (32). Cirillo (25) found that glucose enters *Tetrahymena* by a facilitated diffusion mechanism and that only traces of free glucose are present in the cells. This indicates rapid phosphorylation of the hexose, which requires the expenditure of ATP. This reaction, coupled with the glycogen deposition (which does not yield energy), may account for the lack of influence of glucose in counteracting the DNP inhibition of RNA synthesis. Glycolytic ATP production may be the limiting factor, and the competition for this compound by other reactions may not allow RNA synthesis in the presence of the inhibitor.

The cooperative nature of the inhibition produced by the combination of glucose and iodoacetate has been established in *Tetrahymena* (32) and in tumor cells (204). An increase in RNA degradation and a decrease in RNA synthesis was observed in the presence of the glycolytic inhibitor and glycolytic substrate. This argues that energy availability is an important facet for both of these processes (100). Acetate addition in the presence of IAA and glucose restores the levels to the control values. The impact of this carbon source, which bypasses the glycolytic block by direct entry into the citric acid cycle and thus augments oxidative phosphorylation, indicates that energy yields are involved.

In summary, energy availability does not appear to be limiting in the regulation of RNA degradation nor for synthesis in "starved" cells. The addition of agents that lower the production of ATP or increase consumption of this compound, however, do show the importance of this parameter in the regulation of the extent of RNA depletion and of RNA synthesis in *Tetrahymena*. Bertles and Beck (10) reached a similar conclusion in their studies on reticulocytes.

Levy and Elliott (122) noted the appearance of acid phosphatase "autophagic vacuoles" which appeared to contain portions of the cytoplasm in starved cells. Further, these investigators noted that large nucleoli form in the macronucleus and that a portion of the macronucleus is pinched off. The terminal portion of the "bleb" usually was connected to what appeared to be an autophagic vacuole. This sequence of events could account for the smaller size of the macronucleus noted under starvation conditions.

A similar series of observations has been reported for cells in stationary phase. Cameron and Guile (19) found large nucleoli in stationary-phase cells. Elliott and Bak (54) and Elliott and Clemmons (55) found autophagic vacuoles containing mitochondria and other cytoplasmic inclusions. Summers et al. (181) and McDonald (130) showed that the macronuclei in stationary-phase cells are smaller than those found in log phase. These morphological similarities in stationary and starved cells are in good agreement with the RNA degradation pattern observed in each case. The sequence of cytological events might have a direct bearing on the general observation that no agent tested reverses the RNA loss observed in nonnutrient suspensions of the ciliate. If a series of irreversible events, such as autophagic vacuole formation, is initiated by the removal of the cells from the culture fluid, it is difficult to envision how any agent could alter the process.

Our knowledge about the factors regulating RNA levels is, at present, too scanty to determine the interplay between nucleic acid degradation and synthesis in *Tetrahymena*. The established enzymology and more extensive information about environmental parameters, however, will allow definitive studies concerning RNA synthesis and turnover to be carried out in the near future. The intriguing studies of Lee and Byfield, mentioned earlier, provide an important beginning of an understanding of the process of RNA synthesis in isolated macronuclei and should provide a basis for the study of cellular regulation.

ACKNOWLEDGMENTS

The original work reported here was supported by grants from the National Science Foundation (GB-4605, GB-12133). M.J.K. was supported by a USPHS training grant (5-T1-AI-148). R.L.C. was supported in part by a USPHS special fellowship (1-F3-AM-39).

REFERENCES

1. Albach, R. A. and Hull, R. W. 1963. J. Protozool. 10(suppl.):20.
2. Allen, N. E. and Suyama, Y. 1972. Biochim. Biophys. Acta 259:369–377.
3. Allen, S. L., Misch, M. S., and Morrison, B. M. 1963. J. Histochem. Cytochem. 11:706–719.
4. Allison, B. M. and Ronkin, R. R. 1967. J. Protozool. 14:313–319.
5. Anfinsen, C. B. and White, F. H. 1961. In P. D. Boyer, H. Lardy, and K. Myrback, eds., The enzymes, 2nd ed., vol. V, Academic Press, New York, pp. 95–122.
6. Appelmans, F., Wattiaux, R., and De Duve, C. 1955. Biochem. J. 59:438–445.
7. Argetsinger, J. 1965. J. Cell Biol. 24:154–157.
8. Arsenis, C. and Touster, O. 1967. J. Biol. Chem. 242:3399–3401.
9. Bernheimer, A. W. and Steele, J. M., Jr. 1955. Proc. Soc. Exp. Biol. Med. 89:123–126.
10. Bertles, J. F. and Beck, W. S. 1962. J. Biol. Chem. 237:3770–3777.
11. Birns, M. 1960. Exp. Cell Res. 20:202–205.
12. Bolund, L. and Ringertz, N. R. 1966. Exp. Cell Res. 44:606–613.
13. Brunk, C. F. and Hanawalt, P. C. 1969. Exp. Cell Res. 54:143–149.
14. Buetow, D. E. and Levedahl, B. H. 1962. J. Gen. Microbiol. 28:579–584.
15. Byfield, J. E., Young, C. L., and Bennett, L. R. 1969. Biochem. Biophys. Res. Commun. 37:806–812.
16. Cameron, I. L. 1965. J. Cell Biol. 25:9–18.
17. Cameron, I. L. 1966. Nature 209:630–631.
18. Cameron, I. L., Cline, G. B., Padilla, G. M., Miller, O. L., and van Dreal, P. A. 1966. Natl. Cancer Inst. Mongr. 21:361–371.
19. Cameron, I. L. and Guile, E. E. 1965. J. Cell Biol. 26:845–855.
20. Cameron, I. L. and Jeter, J. R. 1970. J. Protozool. 17:429–431.
21. Cameron, I. L. and Stone, G. E. 1964. Exp. Cell Res. 36:510–514.
22. Charret, R. 1969. Exp. Cell Res. 54:353–361.
23. Chi, J. C. H. and Suyama, Y. 1970. J. Mol. Biol. 53:531–556.
24. Child, F. M. 1967. In M. Florkin and B. J. Scheer, eds., Chemical zoology, vol. 1 (G. W. Kidder, ed.), Academic Press, New York, pp. 381–393.
25. Cirillo, V. P. 1962. J. Bacteriol. 84:754–758.
26. Cleffman, G. 1968. Exp. Cell Res. 50:193–207.
27. Cline, S. G. 1965. Ph.D. Thesis, Bryn Mawr College, Bryn Mawr, Pennsylvania.
28. Cline, S. G. 1966. J. Cell. Physiol. 68:157–164.
29. Cline, S. G. and Conner, R. L. 1966. J. Cell. Physiol. 68:149–156.
30. Conner, R. L. Unpublished.
31. Conner, R. L. and Cline, S. G. 1964. J. Protozool. 11:486–491.

32. Conner, R. L. and Cline, S. G. 1967. J. Protozool. 14:22–26.
33. Conner, R. L. and Hershkowitz, R. Unpublished.
34. Conner, R. L. and Koroly, M. J. Unpublished.
35. Conner, R. L. and Linden, C. 1970. J. Protozool. 17:659–662.
36. Conner, R. L. and Longobardi, A. E. 1963. J. Protozool. 10(suppl.):8.
37. Conner, R. L. and Macdonald, L. A. 1964. J. Cell. Comp. Physiol. 64:257–263.
38. Conner, R. L. and Travis, B. Unpublished.
39. Culbertson, J. R. 1966. J. Protozool. 13:297–406.
40. Culbertson, J. R. and Hull, R. W. 1963. J. Protozool. 10(suppl.):8.
41. Cummins, J. E. and Plaut, W. 1962. Biochim. Biophys. Acta 55:418–420.
42. Day, A. J., Fidge, N. H., and Wilkinson, G. N. 1966. J. Lipid Res. 7:132–140.
43. Debault, L. E. and Ringertz, N. R. 1967. Exp. Cell Res. 45:509–518.
44. De Duve, C., Pressman, B. C., Gianetto, R., Wattiaux, R., and Appelmans, F. 1955. Biochem. J. 60:604–617.
45. Dewey, V. C., Heinrich, M. R., Markees, D. G., and Kidder, G. W. 1960. Biochem. Pharmacol. 3:173–180.
46. Dewey, V. C. and Kidder, G. W. 1960. Z. Allg. Mikrobiol. 1:1–12.
47. Dewey, V. C., Kidder, G. W., and Parks, R. E., Jr. 1951. Proc. Soc. Exp. Biol. Med. 78:91–95.
48. Eichel, H. J. 1956. J. Biol. Chem. 220:209–220.
49. Eichel, H. J. 1957. J. Protozool. 4(suppl.):16.
50. Eichel, H. J., Conger, N. F., and Figueroa, E. 1963. J. Protozool. 10(suppl.):6.
51. Eichel, H. J. and Roth, J. S. 1953. Biol. Bull. 104:351–358.
52. Eidinoff, M. L. and Rich, M. A. 1959. Cancer Res. 19:521–527.
53. Elliott, A. M. and Hunter, R. L. 1951. Biol. Bull. 100:165–172.
54. Elliott, A. M. and Bak, I. J. 1964. J. Cell Biol. 20:113–129.
55. Elliott, A. M. and Clemmons, G. L. 1966. J. Protozool. 13:311–323.
56. Elliott, A. M., Travis, D. M., and Work, J. A. 1966. J. Exp. Zool. 161:177–192.
57. Engberg, J. and Pearlman, R. E. 1972. Eur. J. Biochem. 26:393–400.
58. Erwin, J. A., Beach, D., and Holz, G. G., Jr. 1966. Biochim. Biophys. Acta 125:614–616.
59. Fitt, P. S. 1966. J. Protozool. 13:507–509.
60. Flavell, R. A. and Jones, I. G. 1970. Biochem. J. 116:155–157.
61. Flavin, M. and Engelman, M. 1953. J. Biol. Chem. 200:59–68.
62. Flavin, M. and Graff, S. 1951. J. Biol. Chem. 191:55–61.
63. Flavin, M. and Graff, S. 1951. J. Biol. Chem. 192:485–488.
64. Frankel, J. J. 1965. J. Exp. Zool. 159:113–148.
65. Friedkin, M. 1953. J. Cell. Comp. Physiol. 41(suppl.):261–282.
66. Friedkin, M. and Wood, H., IV. 1956. J. Biol. Chem. 220:639–651.
67. Gibbons, I. R. 1963. Proc. Natl. Acad. Sci. 50:1002–1010.
68. Goldwasser, E. and Heinrikson, R. L. 1966. Progr. N. A. Res. Mol. Biol. 5:399–415.
69. Gorovsky, M. A. 1970. J. Cell Biol. 47:619–630.
70. Gorovsky, M. A. and Woodard, J. 1969. J. Cell Biol. 42:673–682.
71. Granick, S. and Gibor, A. 1967. Prog. Nucleic Acid Res. Mol. Biol. 6:143–186.
72. Haessler, H. A. and Cunningham, L. 1957. Exp. Cell Res. 13:304–311.
73. Hartman, H. and Dowben, R. M. 1970. J. Cell Biol. 44:676–678.
74. Heinrich, M. R., Dewey, V. C., and Kidder, G. W. 1953. J. Amer. Chem. Soc. 75:1741–1742.
75. Heinrich, M. R., Dewey, V. C., and Kidder, G. W. 1957. Biochim. Biophys. Acta 25:199–200.
76. Heinrich, M. R., Dewey, V. C., Parks, R. E., and Kidder, G. W. 1952. J. Biol. Chem. 197:199–204.
77. Heinrikson, R. L. and Goldwasser, E. 1963. J. Biol. Chem. 238:485–486.
78. Heinrikson, R. L. and Goldwasser, E. 1964. J. Biol. Chem. 239:1177–1187.
79. Hill, D. L. and Chambers, P. 1967. J. Cell. Physiol. 69:321–330.
80. Hill, D. L., Straight, S., and Allan, P. W. 1970. J. Protozool. 17:619–623.
81. Hoffman, E. J. 1965. J. Cell Biol. 25:217–228.
82. Hogg, J. F. and Wagner, C. 1956. Fed. Proc. 15:275–276.
83. Holden, M. and Pirie, N. W. 1955. Biochem. J. 60:53–62.

84. Holm, B. 1966. Exp. Cell Res. 41:12–19.
85. Holz, G. G., Jr. 1964. In S. H. Hutner, ed., Biochemistry and physiology of Protozoa, vol. 3, Academic Press, New York, pp. 199–242.
86. Holz, G. G., Jr., Erwin, J., Wagner, B., and Rosenbaum, N. 1962. J. Protozool. 9:359–363.
87. Horowitz, J. and Chargaff, E. 1959. Nature 184:1213–1214.
88. Iverson, R. M. and Giese, A. C. 1957. Exp. Cell Res. 13:213–223.
89. Keiding, J. and Westergaard, O. 1971. Exp. Cell Res. 64:317–322.
90. Kidder, G. W. 1967. In M. Florkin and B. J. Scheer, eds., Chemical zoology, vol. 1 (G. W. Kidder, ed.), Academic Press, New York, pp. 93–159.
91. Kidder, G. W. and Dewey, V. C. 1945. Arch. Biochem. 8:293–301.
92. Kidder, G. W. and Dewey, V. C. 1948. Proc. Natl. Acad. Sci. 34:566–574.
93. Kidder, G. W. and Dewey, V. C. 1949. J. Biol. Chem. 178:383–387.
94. Kidder, G. W. and Dewey, V. C. 1949. J. Biol. Chem. 179:181–187.
95. Kidder, G. W. and Dewey, V. C. 1951. In A. Lwoff, ed., Biochemistry and physiology of Protozoa, vol. 1, Academic Press, New York, pp. 323–400.
96. Kidder, G. W. and Dewey, V. C. 1955. Arch. Biochem. Biophys. 55:126–129.
97. Kidder, G. W. and Dewey, V. C. 1957. Arch. Biochem. Biophys. 66:486–492.
98. Kidder, G. W., Dewey, V. C., Parks, R. E., and Heinrich, M. R. 1950. Proc. Natl. Acad. Sci. 36:431–439.
99. Kjeldgaard, N. O. and Kurland, C. G. 1963. J. Mol. Biol. 6:341–348.
100. Koroly, M. J. 1969. Ph.D. Thesis, Bryn Mawr College, Bryn Mawr, Pennsylvania.
101. Kumar, A. 1969. Biochim. Biophys. Acta 186:326–331.
102. Kumar, A. 1970. J. Cell Biol. 45:623–634.
103. Lanzetta, P. A. and Berech, J. 1972. Personal communication.
104. Lazarus, L. H. and Scherbaum, O. H. 1966. J. Cell Physiol. 68:95–98.
105. Lazarus, L. H. and Scherbaum, O. H. 1967. Biochim. Biophys. Acta 142:368–384.
106. Lazarus, L. H. and Scherbaum, O. H. 1967. Nature 213:887–888.
107. Leboy, P. S., Cline, S. G., and Conner, R. L. 1964. J. Protozool. 11:217–222.
108. Leboy, P. S. and Conner, R. L. Unpublished.
109. Lee, D. 1970. Biochim. Biophys. Acta 211:550–554.
110. Lee, D. 1970. J. Cell Physiol. 76:17–22.
111. Lee, D. 1971. Biochim. Biophys. Acta 225:108–112.
112. Lee, Y. C. and Byfield, J. E. 1970. Biochem. 9:3947–3959.
113. Lee, Y. C. and Scherbaum, O. H. 1965. Nature 208:1350–1351.
114. Lee, Y. C. and Scherbaum, O. H. 1966. Biochem. 5:2067–2075.
115. Leick, V. 1967. C. R. Trav. Lab. Carlsberg 36:113–126.
116. Leick, V. 1968. Nature 217:1153–1155.
117. Leick, V. 1969. Eur. J. Biochem. 8:221–228.
118. Leick, V. and Andersen, S. B. 1970. Eur. J. Biochem. 14:460–464.
119. Leick, V., Engberg, J., and Emmersen, J. 1970. Eur. J. Biochem. 13:238–246.
120. Leick, V. and Plesner, P. 1968. Biochim. Biophys. Acta 169:398–408.
121. Leick, V. and Plesner, P. 1968. Biochim. Biophys. Acta 169:409–415.
122. Levy, M. R. and Elliott, A. M. 1968. J. Protozool. 15:208–222.
123. Levy, M. R. and Scherbaum, O. H. 1965. J. Gen. Microbiol. 38:221–230.
124. Low, R. B. and Wool, I. G. 1967. Science 155:330–332.
125. Lucas, J. M., Schuurs, A. H. W. M., and Simpson, M. V. 1964. Biochemistry 3:959–967.
126. Lwoff, A. 1923. C. R. Acad. Sci., Paris 176:928.
127. Lyttleton, J. W. 1963. Exp. Cell Res. 31:385–389.
128. Mandel, M. 1967. In M. Florkin and B. J. Scheer, eds., Chemical zoology, vol. 1 (G. W. Kidder, ed.), Academic Press, New York, pp. 541–572.
129. Mandelstam, J. and Halvorson, H. 1960. Biochim. Biophys. Acta 40:43–49.
130. McDonald, B. B. 1958. Biol. Bull. 114:71–94.
131. Mita, T. 1965. Biochim. Biophys. Acta 103:182–185.
132. Mita, T., Shiomi, H., and Iwai, K. 1966. Exp. Cell Res. 43:696–698.

133. Morse, M. L. and Carter, C. E. 1949. J. Bacteriol. 58:317–326.
134. Muller, M. 1967. In M. Florkin and B. J. Scheer, eds., Chemical zoology, vol. 1 (G. W. Kidder, ed.), Academic Press, New York, pp. 351–380.
135. Muller, M., Baudhuin, P., and De Duve, C. 1966. J. Cell Physiol. 68:165–176.
136. Muller, M., Hogg, J., and De Duve, C. 1968. J. Biol. Chem. 243:5385–5395.
137. Muller, M. and Toro, I. 1962. J. Protozool. 9:98–102.
138. Murti, K. G. and Prescott, D. M. 1970. J. Cell Biol. 47:460–467.
139. Neidhardt, F. C. 1964. Prog. Nucleic Acid Res. Mol. Biol. 3:145–181.
140. Neidhardt, F. C. and Magasanik, B. 1960. Biochim. Biophys. Acta 42:99–116.
141. Nilsson, J. R. 1970. J. Protozool. 17:539–548.
142. Nilsson, J. R. and Leick, V. 1970. Exp. Cell Res. 60:361–372.
143. Niwa, M., Yamadeya, Y., and Kuwajima, Y. 1964. J. Bacteriol. 88:809–810.
144. Ormsbee, R. A. 1942. Biol. Bull. 82:423–437.
145. Parsons, J. A. 1965. J. Cell Biol. 25:641–646.
146. Parsons, J. A. and Dickson, R. C. 1965. J. Cell Biol. 27:77A.
147. Parsons, J. A. and Rustad, R. C. 1968. J. Cell Biol. 37:683–693.
148. Peng, Y. M. and Elson, C. E. 1971. J. Nutr. 101:1177–1184.
149. Penman, S. 1966. J. Mol. Biol. 17:117–130.
150. Perry, R. P. 1967. Progr. Nucleic Acid Res. Mol. Biol. 6:219–257.
151. Phelps, A. 1935. J. Exp. Zool. 70:109–130.
152. Plesner, P. 1961. Cold Spring Harbor Symp. Quant. Biol. 26:159–162.
153. Plunkett, W. and Moner, J. G. 1969. Amer. Zool. 9:1106.
154. Prescott, D. M. 1957. Exp. Cell Res. 12:126–134.
155. Prescott, D. M. 1964. Progr. Nucleic Acid Res. Mol. Biol. 3:33–58.
156. Prescott, D. M., Bostock, C., Gamow, E., and Lauth, M. 1971. Exp. Cell Res. 67:124–128.
157. Randall, J. and Disbrey, C. 1965. Proc. Roy. Soc., Ser. B. 162:473–491.
158. Reddi, K. K. 1959. Biochim. Biophys. Acta 33:164–169.
159. Renaud, F. L., Rowe, A. J., and Gibbons, I. R. 1968. J. Cell Biol. 36:79–90.
160. Ringertz, N. R., Bolund, L., and Debault, L. E. 1967. Exp. Cell Res. 45:519–532.
161. Roth, J. S. 1959. Ann. N. Y. Acad. Sci. 81:611–618.
162. Roth, J. S. 1963. J. Protozool. 10(suppl.):24–25.
163. Roth, J. S. and Eichel, H. J. 1955. Biol. Bull. 108:308–317.
164. Ruffner, B. W. and Anderson, E. P. 1969. J. Biol. Chem. 244:5994–6002.
165. Salzman, N. P. 1959. Biochim. Biophys. Acta 31:158–163.
166. Satir, B. 1967. Exp. Cell Res. 48:253–262.
167. Satir, B. and Dirksen, E. R. 1971. J. Cell Biol. 48:143–154.
168. Scherbaum, O. 1957. Exp. Cell Res. 13:24–30.
169. Scherbaum, O., Lauderback, A. L., and Jahn, T. 1959. Exp. Cell Res. 18:150–166.
170. Schildkraut, C. L., Mandel, M., Levisohn, S., Smith-Sonneborn, J. E., and Marmur, J. 1962. Nature 196:795–796.
171. Seaman, G. R. 1960. Exp. Cell Res. 21:292–302.
172. Seaman, G. R. 1962. Biochim. Biophys. Acta 55:889–899.
173. Seaman, G. R. 1963. J. Protozool. 10:87–91.
174. Shoup, G. D., Prescott, D. M. and Wykes, J. R. 1966. J. Cell Biol. 31:295–300.
175. Shug, A. L., Elson, C., and Shrago, E. 1970. J. Nutr. 99:379–386.
176. Soldo, A. T. and van Wagtendonk, W. J. 1961. J. Protozool. 8:41–55.
177. Stone, G. E. and Miller, O. L., Jr. 1965. J. Exp. Zool. 159:33–37.
178. Stone, G. E. and Prescott, D. M. 1964. J. Protozool. 11(suppl.):24.
179. Sueoka, N. 1961. Cold Spring Harbor Symp. Quant. Biol. 26:35–43.
180. Summers, L. G. 1963. J. Protozool. 10:288–293.
181. Summers, L., Bernstein, E., and James, T. W. 1957. Exp. Cell Res. 13:436–437.
182. Suyama, Y. 1966. Biochemistry 5:2214–2221.
183. Suyama, Y. 1967. Biochemistry 6:2829–2839.
184. Suyama, Y. and Eyer, J. 1967. Biochem. Biophys. Res. Commun. 28:746–751.

185. Suyama, Y. and Eyer, J. 1968. J. Biol. Chem. 243:320–328.
186. Suyama, Y. and Miura, K. 1968. Proc. Natl. Acad. Sci. 60:235–242.
187. Suyama, Y. and Preer, J. R. 1965. Genetics 52:1051–1058.
188. Swaffield, M. N. and Foley, G. E. 1960. Arch. Biochem. Biophys. 86:219–224.
189. Swift, H., Adams, B. J., and Larsen, K. 1964. J. Royal Microsc. Soc. 83:161–167.
190. Villela, G. G. 1965. Proc. Soc. Exp. Biol. Med. 118:834–838.
191. Wade, H. E. and Morgan, D. M. 1957. Biochem. J. 65:321–331.
192. Watson, M. R. and Hopkins, J. M. 1962. Exp. Cell Res. 28:280–295.
193. Weller, D. L., Raina, A., and Johnstone, D. B. 1968. Biochim. Biophys. Acta 157:558–565.
194. Wells, C. 1962. Amer. Zool. 2:457–458.
195. Westergaard, O. 1970. Biochim. Biophys. Acta 213:36–44.
196. Westergaard, O., Marcker, K. A., and Keiding, J. 1970. Nature 227:708–710.
197. Westergaard, O. and Pearlman, R. E. 1969. Exp. Cell Res. 54:309–313.
198. Whitson, G. L., Padilla, G. M., and Fisher, W. D. 1966. In Cell synchrony, Academic Press, New York, pp. 289–306.
199. Whitson, G. L., Padilla, G. M., and Fisher, W. D. 1966. Exp. Cell Res. 42:438–446.
200. Wilson, C. M. 1967. J. Biol. Chem. 242:2260–2263.
201. Winicur, S. and Roth, J. S. 1963. Fed. Proc. 22:292.
202. Woodward, J., Kaneshiro, E., and Gorovsky, M. A. 1972. Genetics 70:251–260.
203. Wu, C. and Hogg, J. F. 1952. J. Biol. Chem. 198:753–764.
204. Wu, R. and Racker, E. 1959. J. Biol. Chem. 234:1029–1035.
205. Wu-Tcheng, J. J. 1969. Ph.D. Thesis, Bryn Mawr College, Bryn Mawr, Pennsylvania.
206. Wykes, J. R. and Prescott, D. M. 1968. J. Cell Physiol. 72:173–184.
207. Yuyama, S. and Zimmerman, A. M. 1972. Exp. Cell Res. 71:193–203.
208. Zeuthen, E. 1968. Exp. Cell Res. 50:37–46.
209. Zimmerman, A. M. 1969. In G. M. Padilla, G. L. Whitson, and I. L. Cameron, eds., The cell cycle, Academic Press, New York, pp. 203–225.

CHAPTER 6

Regulation of Solutes and Water in *Tetrahymena*

Philip B. Dunham

Department of Biology
Syracuse University
Syracuse, New York

and

Donna L. Kropp

Department of Physiology
Yale University School of Medicine
New Haven, Connecticut

I. INTRODUCTION

Tetrahymena pyriformis can inhabit aquatic environments ranging from fresh water to brackish water as concentrated as 65% seawater (63). In fresh water it maintains osmolarity and concentrations of Na and K higher than those of the environment. Under slightly brackish conditions, however, intracellular Na concentration is lower than that of the environment.

Freshwater protozoa contain concentrations of ions much lower than the concentrations in cells of most metazoans (except for a few freshwater metazoans such as hydroids and lamellibranchs) (14,33). Protozoa (marine and freshwater) resemble metazoan cells in that K is the predominant intracellular inorganic substance (30,55). A striking feature in protozoa with regard to osmotic and ionic regulation is the presence in virtually all freshwater forms and many marine forms of contractile vacuoles.

This chapter reviews the evidence on the mechanisms involved in ionic and osmotic regulation in *T. pyriformis*. The functions of the contractile vacuole and other structures in the regulation of cell volume and solutes are discussed.

Ionic regulation and the structure and function of the contractile vacuole in protozoa in general have been reviewed (14,57). Also available are recent general reviews of active transport (98) and osmoregulation (73).

II. INTRACELLULAR CONCENTRATIONS OF POTASSIUM, SODIUM, AND CHLORIDE

The intracellular concentrations of Na, K, and Cl in *T. pyriformis* W in media of various compositions are shown in Table 1. Na and K concentrations were determined by flame photometry and Cl concentrations by electrometric titration. All the intracellular concentrations have been corrected for extracellular spaces of the packed pellets of cells on which the determinations were made.

The K concentration in *T. pyriformis* is typical for protozoa, in which K is generally between 20 and 35 mmoles/kg (29,33). Na concentrations are similar to those in *Paramecium* (29,103) but are considerably higher than those in freshwater amebas (7,57;

Table 1.

Concentrations of K, Na, and Cl in *T. pyriformis* W Cultured under Various Conditions[a]

Condition	Concentration in cells			Concentration in media			Ref.
	K_i	Na_i	Cl_i	K_0	Na_0	Cl_0	
1. Cultured in P medium (for many years)	32	13	6	5	37	29	(33)
2. Cultured in B medium + 5% seawater (1 year +)	22	2	—	1	24	—	(2)
3. Cultured in B medium + 10% seawater (1 year+)	25	4	—	3	47	—	(2)
4. Cultured in B medium + 30% seawater (1 year +)	44	10	—	6	144	—	(2)
5. Cultured in P medium + 1% NaCl (2 years)	33	21	10	5	223	215	(30)
6. Cultured in P medium; equilibrated 30 min. in P medium + 1% NaCl	39	105	53	5	223	215	(33)
7. Cultured in P medium; equilibrated 45 min. in P medium diluted 13-fold	24	5	2	0.4	3	2	(33)

[a]The culture media were 2% proteose–peptone (P medium) or 1% bactotryptone (B medium). Concentrations in the media are in mmoles/liter; concentrations in cells are in mmoles/liter cells, corrected for extracellular spaces. Data in lines 6 and 7 are from cells equilibrated for short times under conditions other than the culture medium. (Data in lines 2, 3, and 4 are not precise values but were read from a diagram in the original article.)

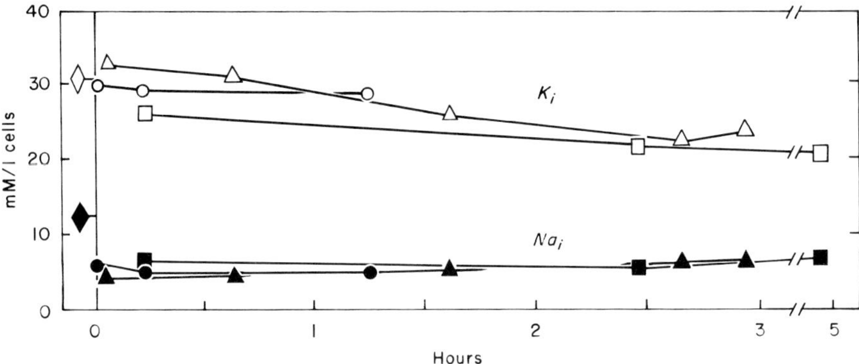

Fig. 1.

Changes in cellular K and Na concentrations after dilution of the medium (2% proteose–peptone). Open symbols, K_i; solid symbols, Na_i. Initial K_i and Na_i (diamonds) are mean values for cells in 2% proteose–peptone. The dilutions of the culture medium were: triangles, 6-fold; circles, 8-fold; squares, 13-fold. [From Dunham and Child (33); reproduced by permission of the Managing Editor, *The Biological Bulletin*).

R. D. Prusch and P. B. Dunham, unpublished). However, Cl concentrations in *T. pyriformis* and other ciliates are considerably lower than in a freshwater amebas (73).

The intracellular concentrations of Na (Na_i) in *T. pyriformis* cultured for protracted periods in high-NaCl media (Table 1, lines 4 and 5) were not much greater than in cells cultured in a slightly brackish medium (Table 1, lines 1, 2, and 3), even though the initial effect in high-NaCl medium (Table 1, line 6) is a large increase in Na_i. The nature of the adaptation of *T. pyriformis* to high-NaCl medium is discussed further in Section IV,A.

The ability of *T. pyriformis* to maintain intracellular K in a very dilute medium (external K concentration, K_o, = 0.4 mM) is shown in Table 1, line 7. In the diluted medium Na_i was much lower in cells in 2% proteose–peptone but was greater than the external Na concentration (Na_o). The Na efflux occurred rapidly upon dilution of the 2% proteose–peptone and was completed in 2 min; K efflux was considerably slower and was not completed in 2 h (Fig. 1).

III. REGULATION OF POTASSIUM

Andrus and Giese (2) have provided several lines of evidence indicating that maintenance of the relatively high intracellular K concentration (K_i) depends in part on active accumulation of K. Exposure of cells to low temperature (6°C), hypoxia, or metabolic inhibitors [iodoacetate, 2,4-dinitrophenol (DNP), or azide] reduced K_i by 30–50%, indicating a dependence of the level of K_i on metabolism. After a return to normal conditions from the cold or from hypoxia, K was reaccumulated by the cells to the normal concentration. This net flux was against a sizable concentration gradient, 5-fold at the outset and nearly 10-fold at the end of recovery. Figure 2 shows this reversible effect of low temperature on K_i (and on Na_i as well). It was not reported whether the effects of the metabolic inhibitors on K_i were reversible. The loss of 50% of intracellular K in the cold has been confirmed (95).

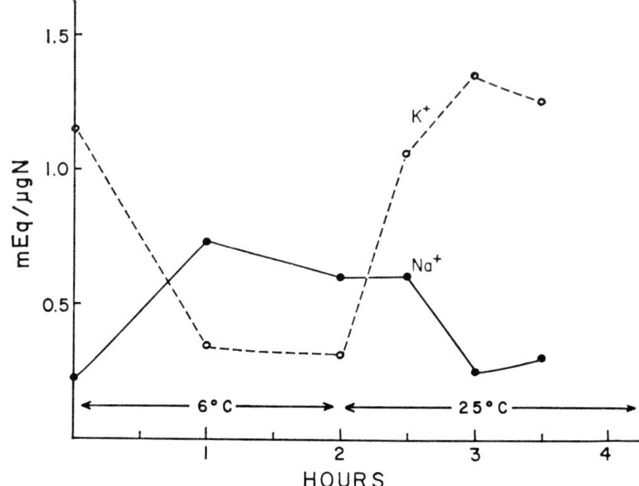

Fig. 2.

Changes in cellular K and Na content (in 1% bactotryptone plus 10% seawater) upon exposure to low temperature (6°C) and subsequent return to the original temperature (25°C). Cellular contents are in milliequivalents per microgram of cell nitrogen [From Andrus and Giese (2).]

The relation between unidirectional influx of K and K_o has been determined by using ^{42}K (31; P. B. Dunham, unpublished). The results (Fig. 3) show that the rate of unidirectional influx is not a linear function of K_o. The driving forces for passive unidirectional influx are the external concentration and the transsurface electrical potential difference. The decrease in slope of the curve in Fig. 3 above 5 mM K_o could be due to saturation of the active influx process. If so, the slope above 5 mM K_o would indicate the relationship between unidirectional influx and the passive driving forces alone. It is possible, however, that this decrease in slope was due in part to a decreased driving force as a result of the depolarizing effect of K_o. A rigorous treatment of the data is not possible in the absence of information on the relationship between K_o and the potential difference. Nevertheless, the data in Fig. 3 are consistent with active accumulation of K.

Dunham and Child (33) determined the steady-state relationship between K_i and K_o over a wide range of external K concentrations. Figure 4 shows the concentrations of K in cells equilibrated for 45 min or more in various concentrations of K. The relationship between K_i and K_o resembles an adsorption isotherm, suggesting again that K_i is not a passive function of K_o. Such relationships have been construed as evidence for retention of K by specific cytoplasmic fixed-charge sites (62). The change in slope of the K_i/K_o curve at 15 mM K_o would in this view indicate the saturation level of the K retention sites. The extrapolation of the linear portion of the curve (above 15 mM K_o) to zero K_o gives an estimate of the capacity of these sites, 43 mmoles/liter cells. Thus intracellular K would be in two states, one associated with specific sites and another proportional to K_o. Although maintenance of K_i might be due to fixed-charge sites, such a system does not exclude membrane transport.

Andrus and Giese provided evidence for a different scheme for comparmentation of K_i. Exposure of *T. pyriformis* to various metabolic inhibitors caused a loss of 30%

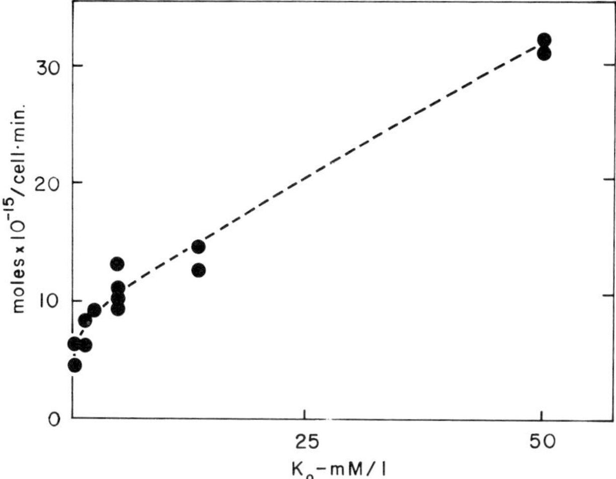

Fig. 3.

Unidirectional influxes of K in cells in steady state in various external concentrations of K, determined from ^{42}K influxes and expressed as moles $\times$ 10^{-15}/cell per minute. (P. B. Dunham, unpublished.)

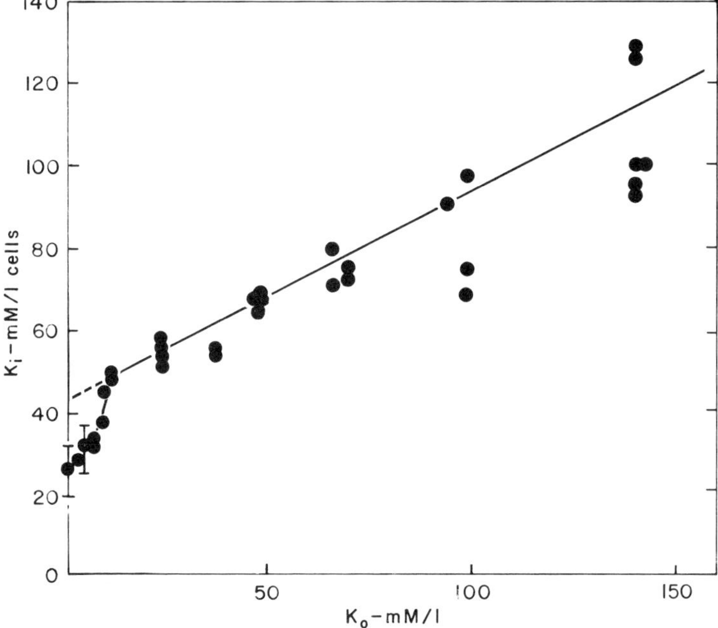

Fig. 4.

Cellular K concentrations in cells equilibrated for 30–120 min in media with various K concentrations. Points and brackets at 5 mM and 1 mM K_o show means and total range of 44 and 31 determinations, respectively. Point at 2 mM K_o is a mean of four determinations. All other points represent single determinations. The dashed line extrapolated to the ordinate (at 43 mmoles/liter cells) shows the postulated maximum capacity of intracellular binding sites (see text for explanation). [Modified from Dunham and Child (33); reproduced by permission of the Managing Editor, *The Biological Bulletin.*]

of K_i. Exposure to low temperature caused the loss of 50% of K_i. Simultaneous exposure to inhibitors and the cold caused no further loss of K. These data suggest two mobile compartments of K_i: one compartment, 30% of K_i, is sensitive to metabolic inhibitors and to low temperature; and another compartment, 20% of K_i, is sensitive to low temperature only. A third compartment, 50% of K_i, is unaffected by any of the treatments. Evidence was presented suggesting that the maintenance of the temperature-sensitive compartment (20% of K_i) is coupled to Na extrusion, while the more labile compartment (30% of K_i) is independent of Na movements. Coupling of Na and K transports is discussed in Section V. The presence of temperature-sensitive and -insensitive compartments of K in *T. pyriformis* has been confirmed (95).

There is no clear relationship between the above two schemes for compartmentation of K. They are based on different kinds of evidence and need not be mutually exclusive. Both kinds of evidence show the heterogeneous distribution of intracellular K, which is not surprising considering the structural and functional complexity of the cells. Neither the K in the temperature-insensitive compartment nor the K associated with selective fixed-charge sites can be tightly bound, since over 90% of the intracellular K is exchangeable with ^{42}K (33). The morphological basis of K compartmentation is not known for *T. pyriformis* or for any other cell type (cf. 92).

IV. REGULATION OF SODIUM

A. Active Extrusion of Sodium

The relationship between Na_i and Na_o in cells equilibrated in various concentrations of Na_o is shown in Fig. 5. Above 5 mM Na_o, Na_i is less than Na_o. Between 2 and 20 mM Na_o, Na_i is constant at about 5 mmoles/liter cells but increases linearly with increasing Na_o above 20 mM Na_o. These results suggest that regulation of Na_i depends upon active extrusion of Na by a mechanism which becomes saturated, or is operating maximally, in 20 mM Na_o (33).

Further evidence for active extrusion of Na was obtained by Andrus and Giese (2) who showed that, in cells transferred from 25 to 6°C, Na_i increased more than twofold (Fig. 2), perhaps as a result of inhibition of a metabolically dependent process. Upon return to 25°C Na_i decreased to the original concentration. The efflux of Na was against a concentration gradient, and against an electrical gradient as well (assuming a potential difference oriented in the usual manner, with the cytoplasm negative to the cell's environment).

B. Adaptation to High-NaCl Medium

Cells adapted to a medium containing a high NaCl concentration demonstrate an enhanced ability to maintain low intracellular Na concentration (compare lines 5 and 6 in Table 1). Cells equilibrated briefly in the high-NaCl medium had a Na concentration of 105 mmoles/liter cells, whereas cells cultured for 2 years in the high-NaCl medium had a much lower Na_i (21 mmoles/liter cells). Some of the details of the process of adaptation and physiological characteristics of the adapted cells have been determined

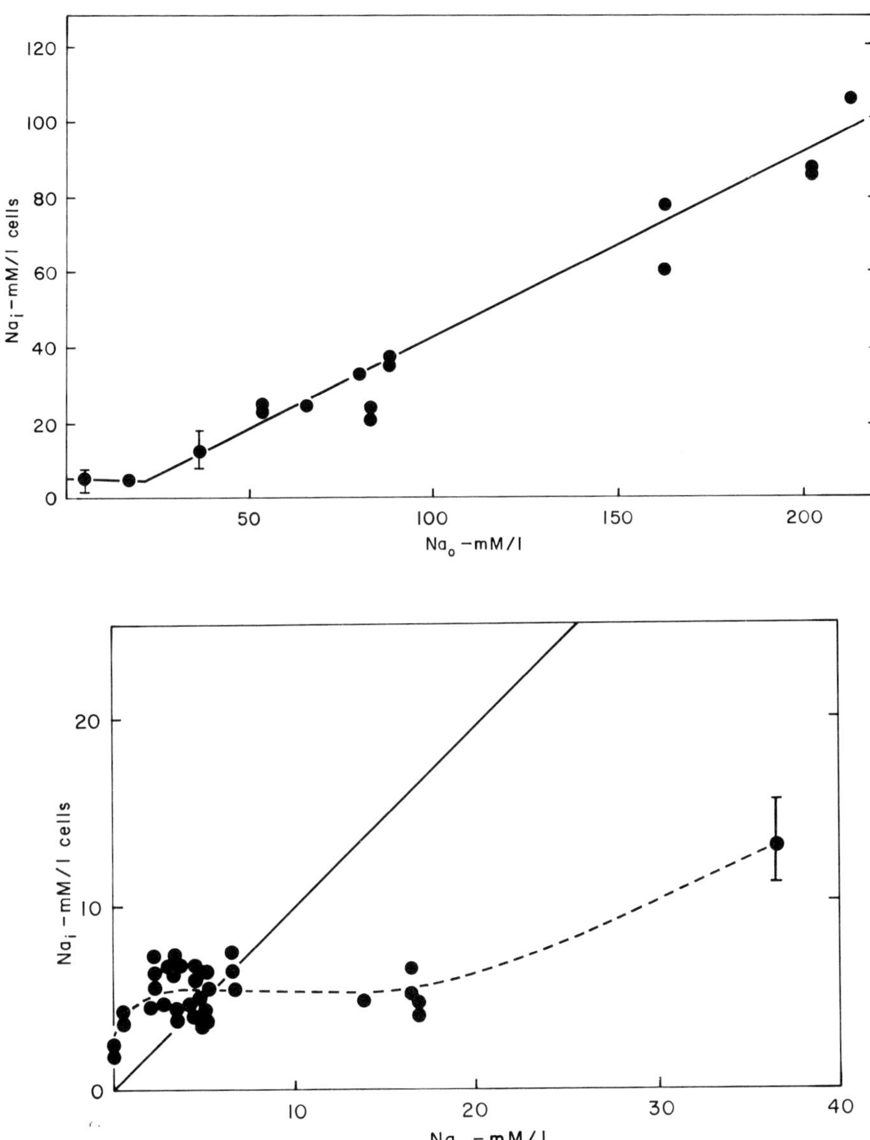

Fig. 5A and B.

Cellular Na concentrations in cells equilibrated for 30–120 min in media with various Na concentrations. The and brackets at 37 mM Na$_o$ (Fig. 5A and B) show the mean and total range of 44 determinations. The symbols at 5 mM Na$_o$ in (A) shows the mean and total range of 28 determinations between 0 and 10 mM Na$_o$; the point at 17 mM Na$_o$ in (A) shows the mean of five determinations between 10 and 20 mM Na$_o$. All other points represent single determinations. (B) shows the individual values below 20 mM Na$_o$ that are summarized in (A). The solid line in (B) connects points of equal cellular and external concentrations [Modified from Dunham and Child (33); reproduced by permission of the Managing Editor, *The Biological Bulletin*].

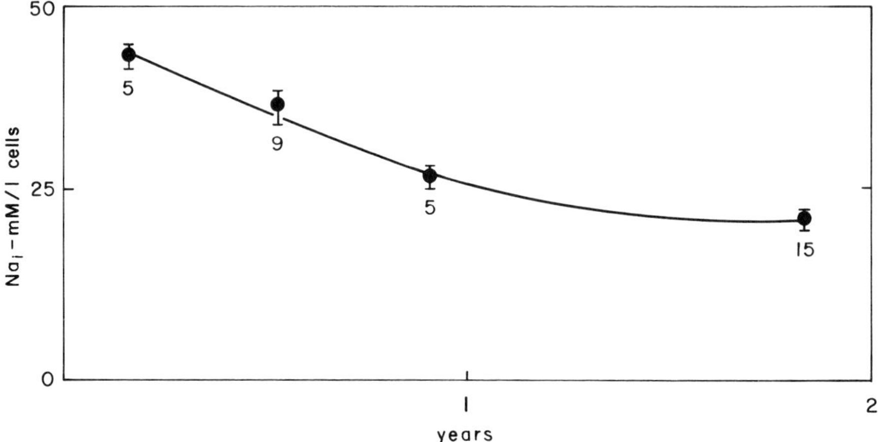

Fig. 6.

Cellular Na concentrations at various times after initiation of adaptation to a high-NaCl medium. Means, standard errors of the means, and numbers of determinations are shown. [Data from Dunham (30).]

(30). Only 2% of a population of *T. pyriformis* cultured in 2% proteose–peptone medium (35 m*M* NaCl) survived transfer to the high-NaCl medium (200 m*M* NaCl). After equilibration for a short time in an intermediate concentration of NaCl, however, 72% of the cells survived to divide in the high-NaCl medium. Therefore tolerance to the stress of the transfer directly to high-NaCl medium, and not the ability to acclimate to high NaCl, characterized the 2% of the population surviving the direct transfer.

It was shown further that the stress tolerance was a heritable characteristic which preadapted a few of the cells to the high-NaCl environment. After eight generations in high NaCl, descendants of the 2% of cells surviving the original transfer were equilibrated for 2 h in 35 m*M* NaCl and then returned to 200 m*M* NaCl. Almost all (33 of 35) survived the stress. The same test was made of cells after eight generations in 200 m*M* NaCl, which originally had been introduced into the 200 m*M* medium without stress of direct transfer by prior equilibration in an intermediate concentration of NaCl. Only 36% (13 of 36) of these cells survived the retransfer to 200 m*M* NaCl. These results suggest that the tolerance of stress on transfer to high NaCl is a heritable characteristic, and that the population of cells after direct transfer to high NaCl were selected for stress tolerance.

Chemical analysis of cells cultured in the high-NaCl medium for 1–2 months showed Na$_i$ of 43 mmoles/liter cells. In the course of nearly 2 years, Na$_i$ in the cells in high-NaCl medium continued to decline, reaching 21 mmoles/liter cells in 22 mo. The time course of the change in Na$_i$ in the adapted cells is shown in Fig. 6. It is likely that the *T. pyriformis* cultured by Andrus and Giese (2; see Table 1, line 4) in medium with 30% seawater, and by Loefer (63) in 65% seawater, underwent adaptation of a similar nature to the high concentrations of NaCl. Adaptation of *T. pyriformis* (= *Glaucoma pyriformis*) (cf. 21) to high concentrations of NaCl (up to 310 m*M*) had also been shown by Chatton and Tellier in 1927 (9). Holz et al. (52) demonstrated that tolerance to high concentrations of NaCl is a varietal characteristic in *Tetrahymena*.

The results suggested two steps in the process of adaptation to the high-NaCl environment: (1) selection for tolerance to the stress of transfer to the high concentration of
NaCl; (2) selection, over the course of 22 mo, or 1500 generations, for the ability
to maintain low intracellular Na.

The increasing ability to maintain a low intracellular Na concentration was due in
part to an increased rate of active Na extrusion. In cells in normal medium, the Na
extrusion mechanism was saturated at 20 mM Na$_o$ (33). After 4 mo in the high-NaCl
medium, the saturation level for Na extrusion in NaCl-adapted cells was about 45 mM;
after 22 mo the saturation level was 125 mM (30).

C. The Contractile Vacuole as the Site of Sodium Extrusion

Na extrusion by the contractile vacuole was postulated several years ago (33,74).
Schmidt-Neilsen and Schrauger (84) proposed that the contractile vacuole in *Amoeba
proteus* reabsorbs K in addition to extruding Na. Chapman-Andresen and Dick (7) provided
preliminary evidence suggesting that some Na is expelled by the contractile vacuole
of *Chaos chaos,* but there was no evidence for active extrusion. Dick (28, p. 68) suggested,
on the basis of the evidence from *T. pyriformis* (33) and *C. chaos* (7), that the vacuolar
fluid is formed by passive filtration, followed by active reabsorption of K into the cytoplasm. Na would remain in the vacuole to be expelled with the fluid. Electrogenic
transport of cations into the contractile vacuole of *A. proteus* may account for the potential
difference observed between the cytoplasm and the vacuole, the vacuole contents being
electrically positive to the cytoplasm (76).

A preliminary report has been published presenting the first clear evidence for Na
transport by the contractile vacuole in *T. pyriformis* (34). In one series of experiments,
the relationship between rate of expulsion of fluid by the contractile vacuole and rate
of net efflux of Na was determined (Fig. 7). Upon reduction of Na$_o$ from 20 to 2
mM, Na$_i$ decreased from 10 to 4 mmoles/kg cells. The efflux was complete in 3 min.
The rate of fluid output by the contractile vacuole was 0.8 pl/min per cell. The loss

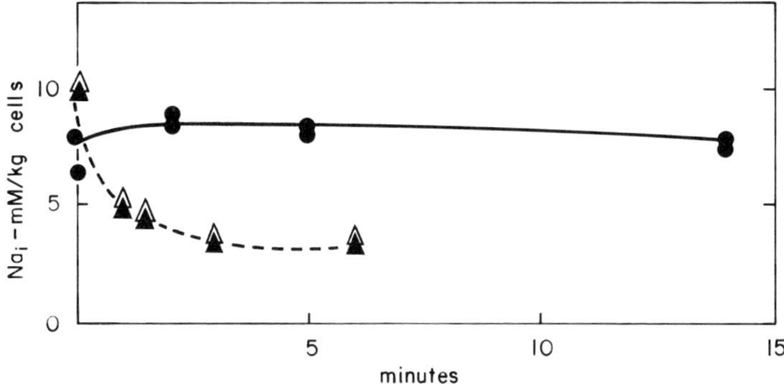

Fig. 7.

Cellular Na concentrations after reduction of external Na concentration at zero time from 20 to 2 mM. Circles,
sucrose added to the medium at zero time to a concentration of 100 mM; triangles, no sucrose added. (P.
B. Dunham, unpublished.)

of intracellular Na as a result of reduction in Na_o could be prevented by increasing osmotic pressure with sucrose (100 mM) at the same time that Na_o was reduced. The slight, transient increase in Na_i after addition of sucrose (Fig. 7) can be accounted for by the transient decrease in cell volume (see also Fig. 13). The addition of sucrose also inhibited the activity of the contractile vacuole nearly completely (by more than 20-fold) for at least 40 min (Fig. 12). In an attempt to find further support for this conclusion, ^{24}Na efflux was measured in cells both in the presence and absence of sucrose. The cells were loaded with ^{24}Na for 1 h in medium with 20 mM Na, and a sample was taken for initial radioactivity of the cells. Then the cells were washed three times in unlabeled medium containing either 20 mM Na or 20 mM Na and 100 mM sucrose. When the cells were virtually free of extracellular ^{24}Na, samples were taken at intervals for determination of ^{24}Na content. As shown in Fig. 8, in the presence of sucrose, Na efflux is reduced fourfold. These results confirm that the contractile vacuole is the major avenue of Na efflux.

Preliminary experiments have shown that the cells are effectively impermeable to [^{14}C]sucrose, although sucrose may reduce the permeability of the cells to Na (cf. 3,4). This possibility has not been ruled out conclusively. It was shown that sucrose has no effect on unidirectional influx of Na (P. B. Dunham, unpublished), but influx of Na may be largely active and may not be an indication of permeability to Na (see Section IV,D).

In another series of experiments, Na_o was kept constant as the determination was made of the effect on Na_i produced by decreasing fluid output by the contractile vacuole. The addition of sucrose (to a concentration of 100 mM) resulted in an increase in Na_i from 9 to 27 mmoles/kg cells (Fig. 9), suggesting that the contractile vacuole contributes to the maintenance of a low Na_i by active secretion of Na. Whereas the effect of sucrose on the net efflux of Na might be complicated by an effect on Na permeability, the interpretation of the latter experiment could be complicated only by an effect of sucrose on Na extrusion. More likely, Na is transported into the vacuolar fluid, and the reduced Na extrusion is a result of the reduced rate of fluid discharge.

With knowledge of net Na fluxes, vacuolar outputs, and unidirectional fluxes of Na, it was possible to calculate concentrations of Na in the fluid expelled by the contractile vacuole for cells with various intracellular Na concentrations. Under the conditions for which sufficient data were available for the calculations, Na_i ranged from 2 to 30 mmoles/kg cells. Calculated Na concentrations in the vacuolar fluid were always higher than Na_i, and appeared to reach a maximum at 100 mM. A complete account of these methods, results, and conclusions is in preparation and will be published elsewhere.

Intracellular Cl increased only slightly when vacuolar output was reduced (Fig. 9), suggesting either that an anion in addition to Cl accompanies Na in the vacuolar fluid, or that the cell surface is permeable to Cl, and Cl diffuses from the cell rather than accumulating when vacuolar function is stopped. *Tetrahymena pyriformis* contains an appreciable concentration of orthophosphate (about 5 mmoles/kg cells, ref. 13), which is excreted from the cell and may accompany Na in the vacuole.

Finally, preliminary evidence suggests that the K concentration in the vacuolar fluid is lower than in the cytoplasm (D. L. Kropp and P. B. Dunham, unpublished). If transfer of K to the vacuolar fluid accompanies formation of the fluid, either by secretion of

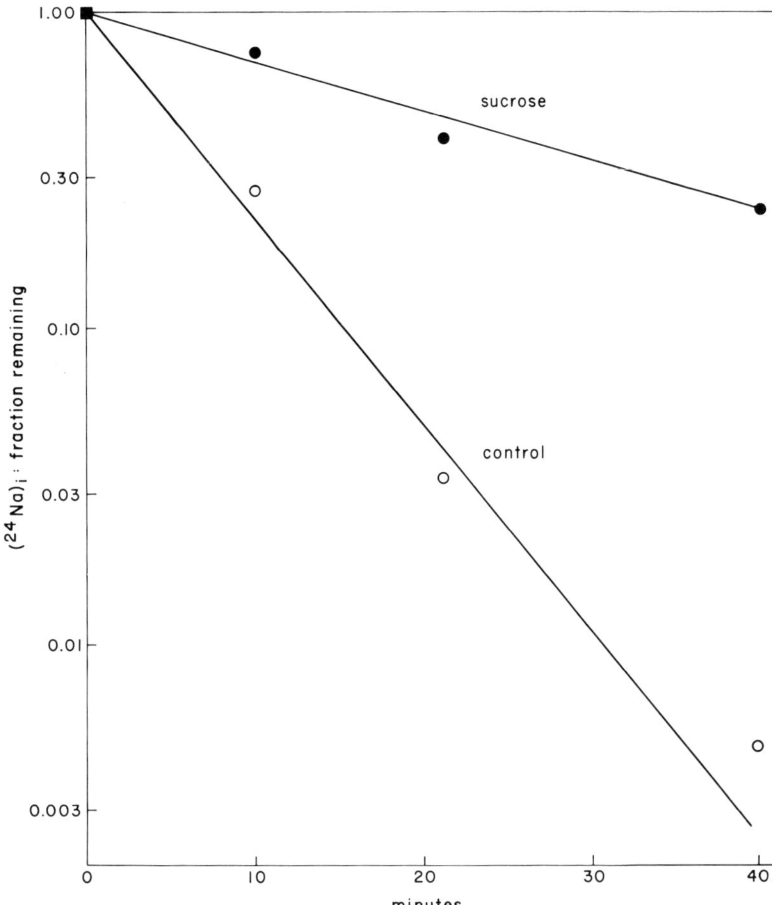

Fig. 8.

Effect of sucrose on unidirectional efflux of ^{24}Na. The cells were loaded with ^{24}Na to isotopic equilibrium (1 h) in medium with 20 mM Na. Then the cells were washed three times by centrifugation in isotope-free medium containing either 20 mM Na (control medium) or 20 mM Na and 100 mM sucrose (sucrose medium). When the medium was virtually ^{24}Na-free, samples of cells were taken for determination of radioactivity. Washing was repeated during the efflux to maintain the medium isotope-free. The rate constants of the efflux were 0.16/min in the control medium and 0.036/min in the sucrose medium. (P. B. Dunham, unpublished.)

K or by filtration, then K must be actively reabsorbed from the vacuolar fluid, as suggested by Schmidt-Nielsen and Schrauger (84) and by Dick (28). The low concentration of K in the vacuolar fluid and secretion of Na into it are probably responsible for the slow net efflux of K relative to the Na efflux when the medium is diluted (Fig. 1). Otherwise, a high permeability of the cell to Na relative to K would have to be proposed, which is not true of cells in general. Measurements have been made of the relative effects of altered extracellular concentrations of Na and K on the membrane potential of *Paramecium* (102). The results, while not conclusive, showed that the permeability to Na was no greater than to K.

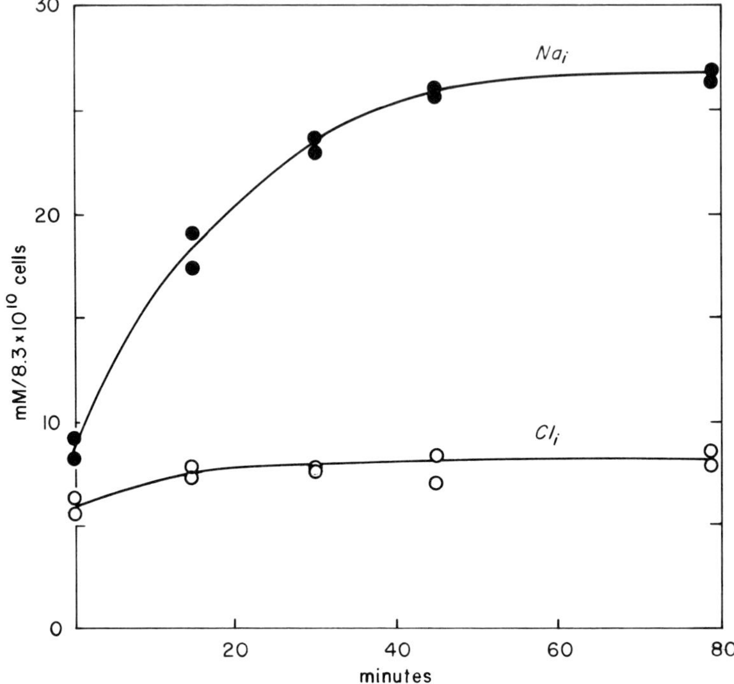

Fig. 9.

Cellular Na and Cl contents in cells in 20 mM Na after addition of sucrose to the medium at zero time to a concentration of 100 mM. Cellular ion contents are expressed as mmoles/8.3 × 10^{10} cells, the number of cells in a liter at time zero. (P. B. Dunham, unpublished.)

As stated above, the results in Fig. 7 suggest that the general surface of *T. pyriformis* has a relatively low permeability to Na. This may reflect an important adaptation for hyperosmotic regulation in fresh water.

D. Accumulation of Sodium

Cells in media with less than 5 mM Na$_o$ have an intracellular Na concentration greater than Na$_o$ (Fig. 5). Internal compartmentalization of Na may enable *Tetrahymena* to retain Na against its chemical concentration gradient, since 25% of the total intracellular Na is unable to exchange with extracellular ^{24}Na. However, even the exchangeable compartment was greater than Na$_o$ below 2.5 mM. Therefore in *T. pyriformis* there may be a mechanism for accumulation of Na in addition to one for extrusion.

Observations of net influxes of Na against its concentration gradient suggest that *Tetrahymena* may accumulate Na by a carrier-mediated process. A net influx of Na, resulting in an increase in its intracellular concentration, may be a result of either increasing unidirectional Na influx (e.g., by increasing Na$_o$), or decreasing unidirectional efflux (e.g., by reducing the activity of the contractile vacuole; see Section IV, C). In one kind of experiment cells were equilibrated in a medium with a very low Na concentration

(0.02 mM). In these cells Na_i was 1.9 mmoles/kg cells. When Na_o was increased to 1.1 mM, Na was accumulated to a final intracellular concentration of 6.5 mmoles/kg cells. In another type of experiment, unidirectional Na efflux was reduced by the addition of sucrose to an extracellular concentration of 100 mM. Na_o was 20 mM. In these cells Na_i incresed to a level above Na_o, despite the fact that Na_o was kept constant (Fig. 9). This is evidence for a mechanism for Na accumulation, since if Na is both accumulated and extruded, Na_i should exceed Na_o when extrusion is inhibited.

Further support for a mechanism for Na accumulation was obtained from measurements of unidirectional Na influx by using ^{24}Na (59,60). Cells were allowed to reach steady state in media containing various concentrations of Na. Trace amounts of ^{24}Na were then added to the external media such that the concentrations of extracellular Na were not significantly changed. The unidirectional influx was calculated by following the increase in intracellular ^{24}Na with time (the total Na_i remained constant during this time). A plot of unidirectional Na influx over a range of Na_o from 0.02 to 8.0 mM yielded a curve with two components. An initial portion at low Na_o, had a steep slope which saturated at an Na_o of 0.8 mM. This saturating curve relating Na influx to Na_o probably represents carrier-mediated Na influx. The second component of the curve was a portion which increased linearly with increasing Na_o and probably represents a passive unidirectional influx of Na.

When unidirectional Na influx was determined in the presence of 5.0 mM K_o, the saturable component of unidirectional Na influx was eliminated. Furthermore, a net influx of Na could be seen upon reduction of K_o, while an increase in K_o resulted in a net efflux of Na. This net efflux was shown to be the result of a decrease in unidirectional Na influx. These effects of K on Na influx indicate competition between K and Na for transport sites.

In addition, it was found that extracellular Na inhibited K influx. Since the apparent maximum active K influx is much greater than that for Na influx, while the apparent affinity of the transport mechanism for Na is greater than for K, it is suggested that there are two distinct sets of transport sites, one with a high affinity for Na but with a low maximum velocity, and another which selectively transports K. However, Na and K are mutually competitive for the two transport systems.

Figure 5 shows that the Na-accumulating mechanism has a higher affinity for Na than does the extrusion mechanism, since in very low Na_o, Na_i exceeds Na_o. The maximal rate of the accumulating mechanism is lower than that of the mechanism for extrusion, however, since Na_i can be maintained lower than Na_o at higher levels of Na_o.

There are several aspects of the probable mechanisms for both extrusion and accumulation. A requirement for Na for physiological functions might necessitate a mechanism for accumulation from a Na-poor environment. There is evidence (discussed in Section VIII) for coupling between Na accumulation and the influx of other solutes, such as sugars and amino acids, as there is in vertebrate intestine and other tissues (23). Extrusion of Na into the contractile vacuolar fluid by a solute–solvent coupled transport system may be necessary for the formation of the fluid. Homeostatic regulation of Na_i at very low Na_o would enable *Tetrahymena* to maintain a cytoplasm hyperosmotic to the external medium, ensuring a passive entry of water for extrusion through the vacuole and permitting the removal of metabolic wastes.

V. COUPLING OF THE TRANSPORT OF SODIUM AND POTASSIUM

It was stated that, because large net fluxes of either Na or K were not accompanied by reciprocal fluxes of the other ion, the transports of Na and K were not coupled (33). However, more recent evidence suggests partial coupling of Na extrusion and K accumulation.

Andrus and Giese (2) described three compartments of K_i (discussed above). One of the compartments, containing 20% of K_i, was sensitive to low temperature but not to metabolic inhibitors. This compartment was lost from cells only under those circumstances in which Na extrusion was also inhibited. The metabolic inhibitors appeared to have no effect on Na extrusion and did not cause the loss of the 20% compartment of K_i. Furthermore, cells in Na-free medium subjected to low temperature lost only the 30% compartment but not the 20% compartment of K_i. This suggests that the net efflux of the 20% compartment must be coupled to Na entry. In another experiment exposure of cells to the cold in low-K medium resulted in loss of both labile compartments of K_i (Na was present in the medium for exchange with the 20% compartment). The recovery after return to 23°C in low-K medium was characterized by very slow extrusion of Na, as well as slow reaccumulation of K. These three observations of Andrus and Giese (2) suggest that, since net fluxes of about 50% of Na_i and 20% of K_i are interdependent, active transport of Na and of K are partially coupled.

Further evidence for coupled net fluxes of Na and K in *T. pyriformis* comes from experiments in which cells were equilibrated in media with various concentrations of K but with Na_o kept constant at 30 mM (Fig. 10). The expected decrease in K_i was observed. A concomitant increase in Na_i was also observed, despite the constant Na_o.

If Na extrusion is associated with the contractile vacuole, then K accumulation may take place from the vacuolar fluid. Partial coupling would be due to a separate mechanism for accumulation of K directly from the external medium.

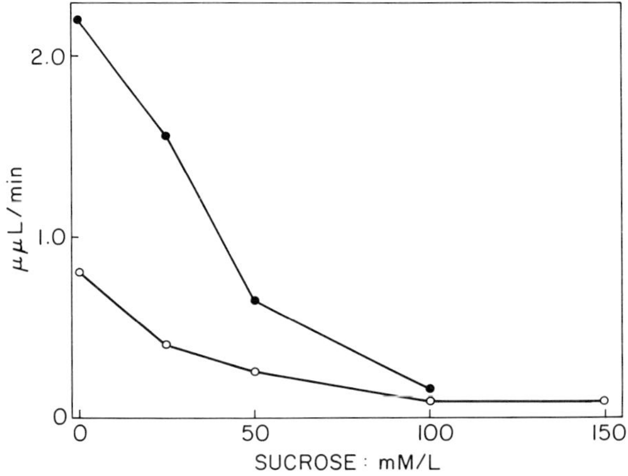

Fig. 10.

Cellular K and Na concentrations in cells equilibrated 45 min in media with 30 mM Na_o and various K · concentrations. Open symbols, K_i; solid symbols, Na_i. (P. B. Dunham, unpublished.)

Adenosinetriphosphatase (ATPase), requiring Mg and activated in vitro by Na and K, has been implicated in coupled transport of Na and K in many kinds of cells (98). Conner et al. (15) found a Mg-dependent ATPase in homogenates of *T. pyriformis*, but the activity of this enzyme was enhanced only slightly (10–15%) by Na and K and was not inhibited by cardiac glycosides. These workers also found pyrophosphatase activity which was Mg-dependent and which was stimulated greatly by addition of Na and K together. The presence of a coupled transport of Na and K in *T. pyriformis*, which is dependent on pyrophosphatase rather than ATPase, might explain the failure of the cardiac glycoside ouabain (in concentrations as high as $10^{-3}\ M$) to inhibit transport of Na and K (2).

VI. OTHER INORGANIC SUBSTANCES

A. Chloride

Chloride concentrations in *T. pyriformis* are generally less than Na_i (Table 1) and less than the external concentrations of Cl. If there is a mechanism for Cl extrusion, it is probably not associated with the contractile vacuole as Na extrusion appears to be, since the increase in Na_i after inhibition of the contractile vacuole is accompanied by only a slight increase in Cl_i (Fig. 9).

Cells in low-Cl media appear to contain a Cl_i lower than Cl_o (30). If so, there is no mechanism for accumulation of Cl as there is for Na. In the absence of evidence for active extrusion or accumulation of Cl in *T. pyriformis*, it is assumed that Cl is distributed passively.

B. Regulation of Orthophosphate

Regulation of intracellular orthophosphate involves not only unidirectional fluxes but also the balance between fluxes and the incorporation of phosphate into (or release from) pyrophosphate and organic molecules such as nucleic acids, phospholipids, and organic esters. Correspondingly, the regulation of orthophosphate is important in the regulation of various aspects of metabolism.

A remarkable ability to regulate intracellular orthophosphate in *T. pyriformis* has been demonstrated by Cline and Conner (13). Strain-W cells in log phase culture were transferred from nutrient medium to 70 mM Tris–HCl containing no phosphate. During 5 h, intracellular orthophosphate remained constant at about 4.9 mmoles/liter cells, while 21% of total intracellular P was released to the medium. Phosphate from the degradation of about 30% of the RNA in the cells, shown to occur after transfer to the nonnutrient medium, could account for all the loss. The phosphate concentration in the medium increased during the experiment from 0 to 0.3 mM, well below the concentration maintained in the cells. The acid-soluble organic phosphate, presumably mostly mononucleotide esters, also remained constant in the cells during RNA degradation and loss of phosphate. The intracellular concentration of acid-soluble phosphate esters is about 24 mmoles/liter cells [P. O. Pruett (1965) cited in ref. 14], making these compounds important anions in *T. pyriformis* in terms of total solutes. The total concentration of both orthophosphate and Cl is less than 15 mmoles/liter.

Earlier work had shown that [^{32}P]orthophosphate was accumulated from very dilute phosphate solutions (17,18,91). Leboy et al. (61) showed an accumulation of 0.02 μmole/h per 10^6 cells from a Tris–HCl medium containing 0.1 or 0.2 mM phosphate. The cells were not in steady state, having been transferred from the culture medium to the Tris–HCl at the beginning of the experiments. The accumulation of [^{32}P]orthophosphate reflected a turnover of intracellular phosphate rather than net influx, since concurrent net effluxes of phosphate were observed. Nevertheless, several characteristics of [^{32}P]orthophosphate entry suggest that carrier-mediated transport is involved.

The extent of [^{32}P]orthophosphate accumulation after a 2-h exposure showed a maximum at an external pH of 6.5; accumulation was considerably reduced at both higher and lower pH values. The rate of accumulation may be influenced by the ionic state of orthophosphate, but the maximum suggests that, in addition, pH effects the cells directly, perhaps at specific sites responsible for the accumulation.

The relationship between the rate of influx and external orthophosphate concentration showed saturation kinetics with a concentration at half-maximal influx of $8.4 \times 10^{-5}\,M$ at pH 7.5, and a maximal influx of about 0.12 μmole/h per 10^6 cells (75). Preliminary evidence for inhibition of orthophosphate entry by arsenate has been reported (40).

The Q_{10} for influx of [^{32}P]orthophosphate was 1.7 between 18 and 28°C, lower than the Q_{10} values for accumulation of arabinose and phenylalanine (see Section VIII) but still somewhat higher than expected for simple, passive diffusion (75).

These points of evidence suggest carrier-mediated accumulation of orthophosphate, although Pruett et al. (75) pointed out that it is difficult to distinguish between incorporation and transport. Net influx of orthophosphate has not been observed, but cells allowed to approach steady state in a phosphate-free medium would probably accumulate phosphate against a concentration gradient after addition of a small amount of phosphate to the medium.

Several studies have shown relationships between regulation of orthophosphate and various aspects of the physiology of *T. pyriformis*. Hamburger and Zeuthen (44) showed a correlation between [^{32}P]orthophosphate accumulation and the cell division cycle in strain GL-R. Populations of cells were caused to divide synchronously by a series of exposures to high temperature (cf. 104). The first synchronous division occurred about 70 min after the end of the heat treatment (EH); the second division, 170 min after EH. A maximum rate of accumulation of [^{32}P]orthophosphate, three- to fourfold greater than at EH, occurred 20–30 min before the first division. At the end of the first division, the rate of accumulation had declined to about two-thirds of the maximum. Another maximum was observed 30–40 min before the second division, followed by another decline in rate of accumulation.

Accumulation of phosphate is influenced by strain and mating type differences (91), and by composition of the culture medium (16,19).

Orthophosphate entry seems to be associated with carbohydrate metabolism. Under some conditions glucose enhances accumulation (19). Iodoacetate and DNP inhibit the influx and increase the efflux of orthophosphate. Glucose partially reverses the effects of DNP (14). The loss of orthophosphate induced by iodoacetate is not reversed by glucose but is partially reversed by acetate; however, acetate does not reverse the inhibition of entry of orthophosphate. The processes responsible for the unidirectional influx and efflux of orthophosphate have different characteristics and therefore, similar to accumulation and extrusion of Na, probably occur at different sites on the cell.

The contractile vacuole may be the site of orthophosphate efflux. Cl is probably not the only anion accompanying Na in the fluid of the contractile vacuole (Fig. 9), and orthophosphate may be the principal anion in the vacuolar fluid. If the pellicle is relatively impermeable to solutes, the contractile vacuole is likely to be responsible for all excretion (cf. 41). Cline (12) showed a complex relationship between phosphate efflux and the ionic and osmotic composition of the medium. Proof of the association between phosphate efflux and the contractile vacuole is not yet available, however.

As shown in the next section, there is a close relationship between orthophosphate and the regulation of Ca, Mg, and pyrophosphate.

C. Calcium, Magnesium, and Pyrophosphate

Under certain culture conditions *T. pyriformis* contains numerous granules, 0.2–2 μm in diameter, consisting almost entirely of a hydrated, insoluble, complex salt of Ca, Mg, and pyrophosphate (79). The granules are surrounded by a membrane which appears to contain a pyrophosphatase. This enzyme is activated by Mg and inhibited by Ca. Thus the control of Ca and Mg concentrations is involved in the regulation of the utilization of the granules, which in turn must play a role in regulation of overall phosphorus metabolism.

The deposition of the granules is a highly controlled process, since it requires both Ca and Mg as well as orthophosphate, and since Ca and Mg are always in equimolar amounts in the granules, independent of the relative concentrations of Ca and Mg in the medium (68). There is about 1 mole each of Ca and Mg per mole of pyrophosphate in the granules. There is no indication as to the mechanism of deposition and its relationship to the pyrophosphatase.

The accumulation of Ca and Mg during deposition of the granules (after phosphate deprivation) was studied by Rosenberg and Munk (80). Ca influx (measured by using ^{45}Ca) proceeded relatively slowly and was stimulated only slightly by the addition of Mg alone or of orthophosphate alone; Ca influx was very greatly enhanced by the addition of both Mg and orthophosphate. The same pattern was observed for Mg influx (measured by chemical analysis); influx was slow except in the presence of both Ca and orthophosphate. The rapid accumulation of Ca and Mg was correlated with the appearance of the granules.

Under conditions of phosphate deprivation, there was a loss of granules, presumably through the action of the pyrophosphatase in the membranes of the granules. Concomitant with the loss of granules there was an efflux of Ca from the cells. The Ca efflux ceased after most of the granules were gone, and only then did an efflux of Mg commence. This pattern of regulation of Ca and Mg probably facilitates the mobilization of phosphate from the granules in the face of phosphate deprivation, since Ca inhibits the pyrophosphatase and Mg is an activator of the enzyme. Thus the regulation of Ca and Mg concentrations may be vital to the regulation of phosphorus metabolism. There are no indications as to the mechanisms of Ca and Mg release. The contention of Rosenberg and Munk that Ca efflux is an active process is not substantiated.

It is difficult to determine from the data given in Rosenberg's publications what the concentrations of Ca, Mg, and pyrophosphate are in the cells, since the amounts are given per 10^6 cells and mean cell volume is not given. If a cell volume of 20 pl is assumed, some approximations can be made. The uptake of 40 ng-atoms of Mg per 10^6 cells during deposition of granules would give a concentration of 2 mmoles/liter cells. An equivalent amount of Ca is accumulated. The amounts of pyrophosphate, from 0.2 to 3 μmoles per 10^6 cells, would give concentrations of 10–150 mmoles/liter cells. The low concentration is for cells before starving and the high concentration for cells after deposition of granules. Similar levels of orthophosphate were found. If the technique for extracting the phosphates is valid, then only a small fraction of the pyrophosphate is in the granules.

The total concentration of Ca in *T. pyriformis,* measured by flame photometry, is between 7 and 12 mmoles/liter cells for external Ca concentrations of 0.2–0.5 m*M* (31; P. B. Dunham, unpublished). If 2 mmoles/liter cells are in the granules, as calculated above, then the granules account for only about 20% of the intracellular Ca. There have been no analyses of total Mg concentrations.

In general, Ca and probably Mg are maintained at concentrations greater than the external concentration. There may be mechanisms for active uptake, but there is little direct evidence. In addition, Ca and Mg can be excreted, depending upon the metabolic state of the cells. Ca and Mg are undoubtedly important in many aspects of cellular structure and function. One clear example is in regulation of phosphorus metabolism.

D. Trace Metals

Two trace metals, Zn and Co, have been studied and both appear to be accumulated. Radioactive zinc (^{65}Zn) was accumulated to a saturation level in 4 h (90). The ^{65}Zn became firmly bound and was not removed in 26 h in isotope-free medium. Strain E took up radioactive cobalt (^{60}Co) during the log phase of culture growth (89). When stationary phase was reached, the cells abruptly began to release the Co. The rate of influx was slow, reaching half-maximal saturation in about 12 h. Upon return to isotope-free medium, an efflux of ^{60}Co was observed with a half-time of about 20 h, indicating that the Co was not tightly bound. Assuming complete exchange of intracellular Co with ^{60}Co, the maximum concentration of intracellular Co observed in these experiments was about 0.3 mmole/liter cells (calculated from ref. 89).

VII. REGULATION OF CELL VOLUME AND OSMOLARITY

Tetrahymena pyriformis, similar to other freshwater protozoa, is hyperosmotic to fresh water. Volume changes in response to osmotic challenges show that the cells are permeable to water. Continual passive entry of water necessitates expenditure of energy to remove this water in order to maintain a constant cell volume. The possible avenues of water entry are: (1) formation of food vacuoles; (2) osmosis across the cytopharyngeal membrane; (3) osmosis across the pellicle; and (4) pinocytosis at the parasomal sacs. According to Kitching (57), metabolic water is a minor contribution to water balance in freshwater protozoa. The normal water content of *T. pyriformis,* about 80% of cell volume (33), is typical of animal cells.

A. Function of the Contractile Vacuole

Hartog (45) in 1888 was the first to propose that the contractile vacuole in protozoa serves an osmoregulatory function by removing fluid at a rate equal to passive entry. No serious suggestion has been made for any other means of fluid removal for the purpose of regulating volume. Although the contractile vacuole has been the subject of considerable interest before and since Hartog's proposal, almost no information has been published on the function of the contractile vacuole in *Tetrahymena.* Degen (25) in 1905 published results of experiments on the function of the contractile vacuole in *Colpidium colpidium (= Glaucoma colpidium)* (cf. 22), a form closely related to *Tetrahymena.* The frequency of contractions of the vacuole was reduced by increased concentrations of various electrolytes and sugars. Degen's results were interpreted as a demonstration of the osmoregulatory function of the contractile vacuole and were among the first and clearest in support of Hartog's hypothesis.

The physiological data on *T. pyriformis* in this section are observations from our laboratory. A preliminary report of some of the results has been published (36). Kitching's article (57) should be consulted for information and references on the contractile vacuole in other protozoa.

The influence of external osmolarity on the rate of fluid output by the contractile vacuole was determined for cells of strain W equilibrated in various media (Fig. 11). As in other protozoa, the rate of fluid output decreases with increasing osmolarity. The effect of altered Na_o on vacuolar output is consistent with the conclusion drawn above, namely, that *T. pyriformis* regulates intracellular Na.

If intracellular osmolarity were constant, the rate of vacuolar output would decrease linearly with external osmolarity (57). In *Tetrahymena,* as shown in Fig. 11, this was not the case. The data suggest that, as external osmolarity increases, intracellular osmolarity also increases. Direct evidence demonstrating this relationship is presented in Section VII,B.

Intracellular osmolarity is a function not only of the rate of fluid output by the contractile vacuole but also of the concentration of solutes in the vacuolar fluid. Schmidt-Nielsen and Schrauger (84) showed that the fluid in the contractile vacuole of *A. proteus* is hyposmotic to the cytoplasm. They suggested that reabsorption of K may be involved in the production of the hyposmotic fluid. Transport of Na and K associated with the contractile vacuole in *T. pyriformis* is certainly involved in regulation of intracellular osmolarity. More recently, Riddick (78), in an elegant study on *Pelomyxa carolinensis,* showed that the concentration of Na in fluid in the contractile vacuole was 19.9 m*M*, compared to 0.6 m*M* in the cytoplasm, while that of K was 4.6 m*M* in the vacuolar fluid and 31 m*M* in the cytoplasm. These results, based on direct chemical analysis of vacuolar fluid, support the above suggestions, based on indirect evidence, regarding the transport of Na into the vacuolar fluid, and some mechanism for maintaining low K in vacuolar fluid.

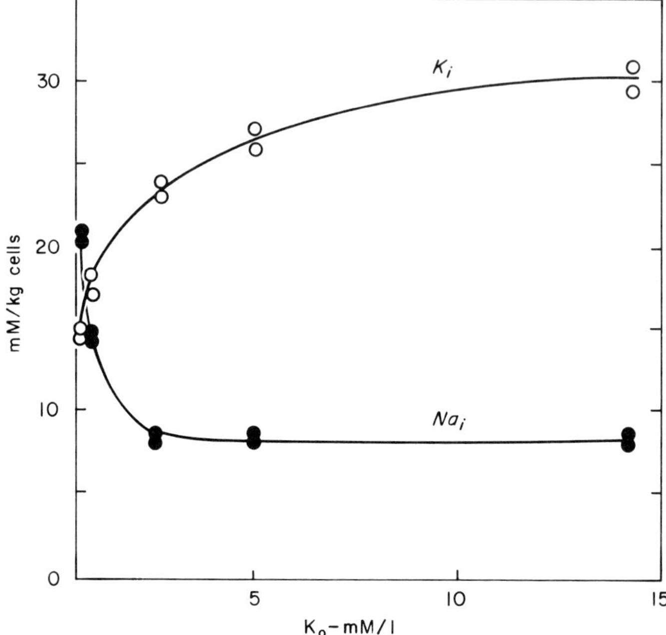

Fig. 11.

Rates of fluid output by the contractile vacuole (in picoliters per minute per cell) in cells equilibrated in media of various concentrations of sucrose and Na (solid symbols, 2 mM Na$_o$; open symbols, 20 mM Na$_o$). Rates of fluid output were determined on individual cells with the aid of a camera lucida. Each point is the mean of at least 40 determinations. (L. C. Stoner and P. B. Dunham, unpublished.)

In addition to maintaining volume and osmolarity of the cell under constant external conditions, the contractile vacuole regulates cell volume under nonsteady-state conditions. The passive volume change after osmotic challenge is a function of the delay between the change in the medium and the achievement of the new rate of fluid output (Fig. 12). Rapid adjustment to a new output rate minimizes the passive volume changes.

Furthermore, *T. pyriformis* is capable of restoring cell volume to the original volume after passive volume changes (Fig. 13). Although changes in the water permeability coefficient have not been ruled out, this form of regulation seems more likely to be a function of the contractile vacuole. Accordingly, during recovery after passive shrinkage, the output of fluid by the contractile vacuole must be slower than passive influx until the original volume is attained; after passive swelling output would temporarily be faster than passive entry. Ciliates have not generally been described as capable of volume recovery (57,74). Both the regulation of volume against a constant osmotic gradient and recovery after passive volume changes indicate a feedback system which relates cell volume to vacuolar output. In no way could the return of volume after passive swelling be accounted for as a passive process. Experiments involving [14C]sucrose have shown that the recovery of cell volume after passive shrinkage cannot be accounted for by the passive entry of sucrose (94). The initial sucrose influx from 100 mM sucrose

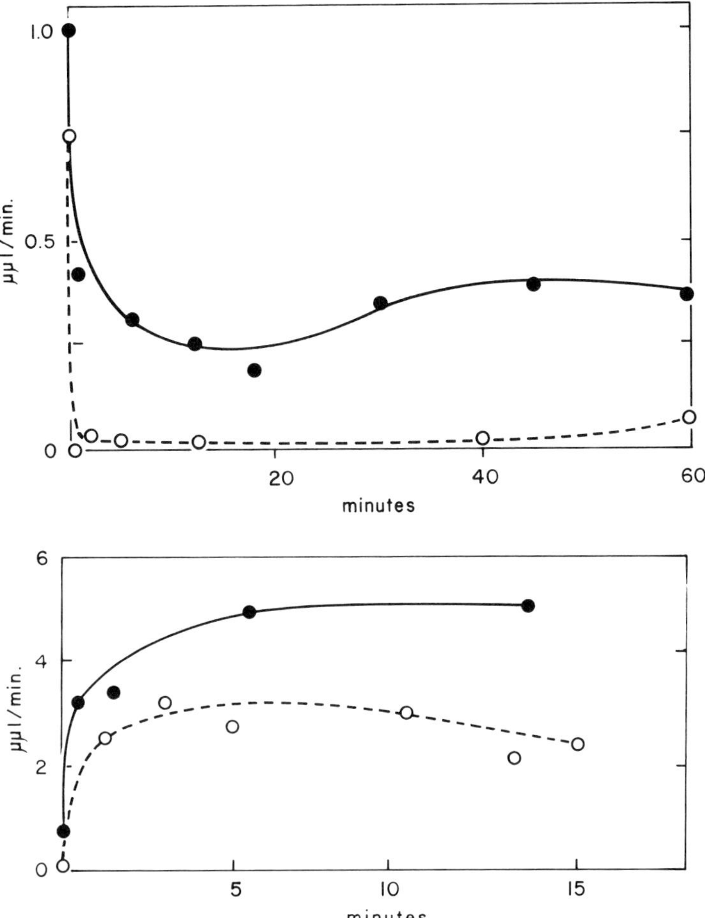

Fig. 12A and B.

Time course of changes of fluid output (in picoliters per minute per cell) by the contractile vacuole after osmotic challenges. Outputs were determined with the aid of a camera lucida. Each point is the mean of at least four determinations. (A.) Increased external osmolarity. Cells equilibrated in 20 mM NaCl were transferred at time zero to 20 mM NaCl plus 25 mM sucrose (solid line) or 100 mM sucrose (broken line). (B.) Decreased external osmolarity. Cells equilibrated in 20 mM NaCl plus 100 mM sucrose were transferred at time zero to sucrose-free medium containing 20 mM NaCl (broken line) or 2 mM NaCl (solid line) (L. C. Stoner and P. B. Dunham, unpublished.)

is 7 mmoles/(kg cells × hours), a rate of entry that can just be accounted for by the rate of formation of food vacuoles (cf. 8).

The variables determining the rate of fluid output by the contractile vacuole are frequency of contraction and volume of the vacuole just before discharge. Both are altered in response to osmotic challenge (56) and may vary independently. For example, after increasing external osmolarity the duration between contractions increases, and the maximum volume of the vacuole decreases. However, the maximum duration of the interval

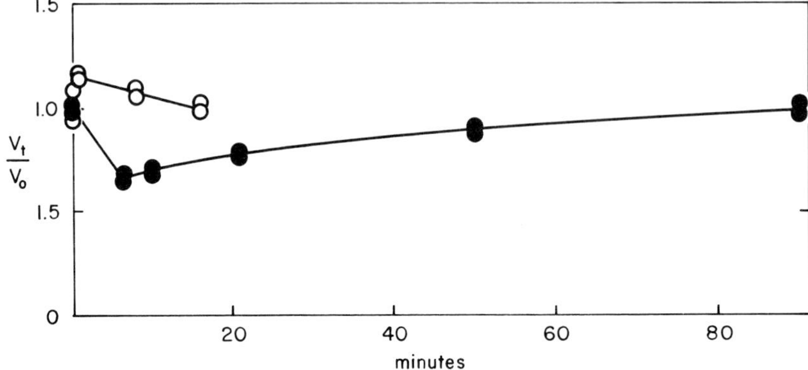

Fig. 13.

Time course of changes in cell volume after osmotic challenges. Volumes were determined from volumes of packed pellets of cells, and are expressed as fractions of initial volume (V_t/V_o). Solid symbols, increased external osmolarity (cells transferred at zero time from 30 mM NaCl to 100 mM sucrose); open symbols, decreased external osmolarity (cells transferred at zero time from 20 mM NaCl plus 100 mM sucrose to 100 mM sucrose). (P. B. Dunham, unpublished.)

between discharges is attained immediately, whereas the minimum volume of the vacuole just before discharge is not attained for five or six vacuolar cycles (Stoner and Dunham, unpublished).

Studies of the ultrastructure of *T. pyriformis* (38) have revealed a system of "nephridial tubules 20 nm in diameter which are continuous with and located over much of the surface of the contractile vacuole (see Chapter 2). These tubules are continuous with the rough endoplasmic reticulum and with some clear vesicles called water vacuoles. The nephridial tubules may provide the membrane surface by which and ions are separated from the cytoplasm. The force responsible for the separation of the vacuolar fluid is unknown. Filtration, active transport of water, and passive movement of water coupled with transport of solutes have been proposed (56,57). The geometry of the nephridial tubules is suited for a coupled water-solute transport system. Na extrusion or K extrusion followed by K reabsorption could play a role. (See ref. 27 for a theoretical treatment of coupled water and solute transport.)

If the formation of the fluid and determination of its composition takes place in the nephridial tubules, the function of the vacuole may be primarily temporary storage. It is uncertain whether or not the vacuole itself generates the force for discharge of the fluid. In *T. pyriformis* (several strains) there are fibrils between the wall of the channel leading to the vacuole and the surface of the vacuole itself (5,38). If these fibrils are contractile, they are more likely associated with opening of the pore than with expulsion of fluid from the vacuole. An indentation of the pellicle occurs around the contractile vacuole pore just prior to expulsion of the vacuolar fluid (36). The indentation indicates a periodic triggering mechanism and may be due to shortening of the fibrils just described. In a cell in which the expulsions are temporarily arrested, the periodic indentations persist, indicating the presence of a mechanism controlling indentation, which is separate from the onset of systole. The vacuole is bounded by a single membrane which does not contain fibrils such as were observed in the wall of the vacuole in *Paramecium* (85). The presence of these fibrils suggests that the vacuole in *Paramecium* may be contractile, although Organ, et al. (70) have suggested that the vacuole in *Paramecium* collapses from force generated by the cytoplasm adjacent to the vacuole. The same mechanism for discharge of the contractile vacuole has been suggested for *T. pyriformis* (71) and *A. proteus* (99). Isolated contractile vacuoles from *A. proteus* contract when exposed to 15 m*M* ATP (76), in contradiction to the proposal above (99).

B. Maintenance of Osmolarity

Indirect evidence has indicated that *T. pyriformis* is hyperosmotic to the external medium. The concentration of all measured intracellular solutes exceeds the total external concentration. The functioning of the contractile vacuole and the relation between vacuolar

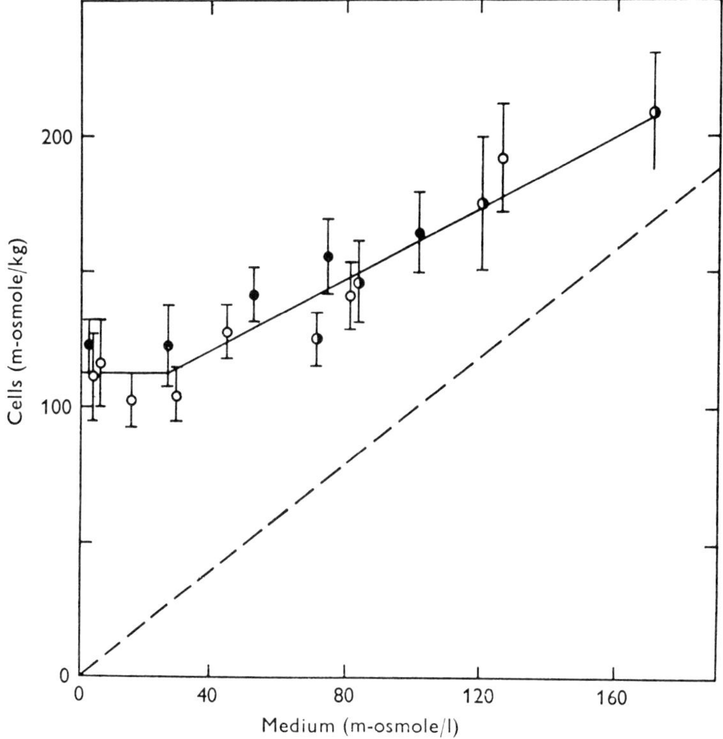

Fig. 14.

The intracellular osmolarity of *Tetrahymena* in media of various osmolarities. Both intracellular and external osmolarities were calculated from freezing-point depressions. The results are expressed as milliosmoles per kilogram of cells for cells that had been exposed to: dilutions of culture medium for 4–5 h (open circles); low ionic strength solutions of various sucrose concentrations for 4–5 h (solid circles); inorganic medium for 12 h prior to a 2-h exposure to inorganic medium plus various concentrations of sucrose (half-open circles). Each symbol indicates the mean of 10 determinations (five experiments) ±2 SEM. The dashed lines connect points of equal cellular and external osmolarities. [From Stoner and Dunham (96), reproduced by permission of the Editor, Journal of Experimental Biology.]

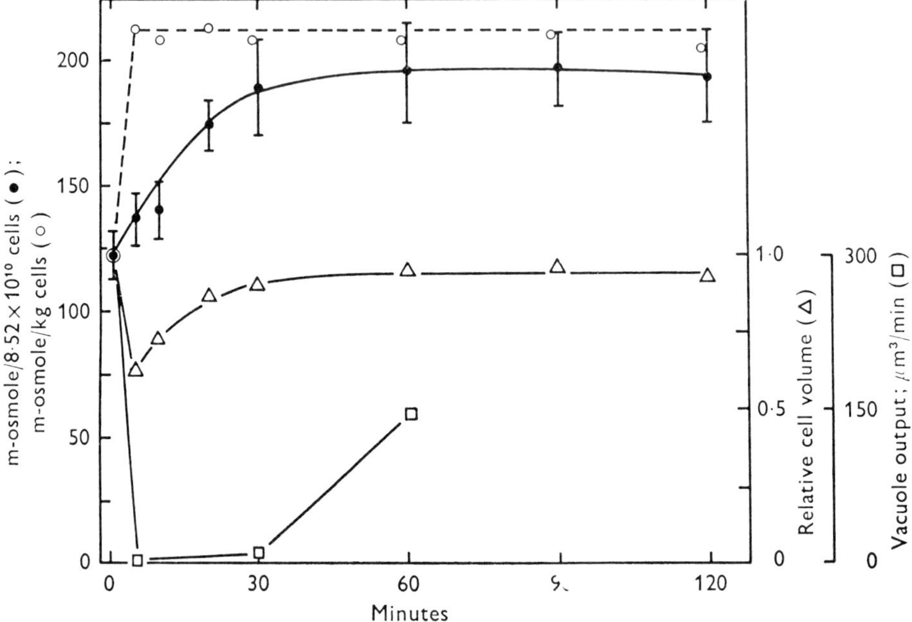

Fig. 15.

The regulation of intracellular osmolarity and cell volume by *Tetrahymena* exposed to an osmotic stress, exposure to 100 m*M* sucrose. Cell suspensions were in inorganic medium for 12 h prior to the addition of sucrose. Open circles, intracellular osmolarity (from freezing points); triangles, relative cell volume; solid circles, amount of solute in millimoles per 8.52 × 10^{10} cells (calculated from osmolarity and relative cell volume); the brackets show ±2 SEM; squares, rate of fluid elimination by the contractile vacuole in cubic microns per minute per cell. Each symbol indicates the mean of 10 determinations from five experiments. The symbols for all measurements coincide at time zero, as indicated by the large circle. [From Stoner and Dunham (96), reproduced by permission of the Editor, Journal of Experimental Biology.]

output and external osmolarity are consistent with maintenance of intracellular hyperosmolarity in *Tetrahymena.*

Recently, intracellular osmolarity was measured directly (95). Freezing-point depressions were determined on extracts of cells equilibrated in media of various osmolarities. The results are shown in Fig. 14. The cells maintained an intracellular osmolarity greater than the external osmolarity throughout the range of external osmolarities. Up to about 40 mosmoles/liter in the external medium, intracellular osmolarity was approximately constant. At higher external osmolarities intracellular osmolarity increased.

The response of cells to altered external osmolarity was investigated by measuring cell volume, output of the contractile vacuole, and intracellular osmolarity at various times after the osmotic challenge. The results are shown in Fig. 15. Upon exposure to 100 m*M* sucrose, the cells shrink immediately and the contractile vacuole ceases its activity. The intracellular osmolarity rises because of the shrinkage. Then cell volume begins to increase concomitant with an increase in the amount of intracellular solutes. The cells return to within 90% of their initial volume, at which time the contractile vacuole resumes functioning and the increase in total solutes ceases. Thus the cells respond to the hyperosmotic stress by increasing water influx, thus increasing total solutes, until cell volume is restored. A regulatory mechanism that determines a particular volume is implied. In addition, the cells seek to maintain an internal hyperosmolarity, which is interpreted as reflecting the requirement for a small, constant water influx in order to operate the contractile vacuole. This structure is involved not only in volume regulation but in the regulation of Na concentration as described above, and perhaps of other solutes as well.

When the cells are subjected to a hypoosmotic challenge, the reverse responses are observed: an initial swelling, a transient increase in rate of vacuolar output, and an initial decrease in osmolarity due to the swelling. These events are followed by a reduction in the amount of free solutes per cell, which is paralleled by a reduction in cell volume to near the initial volume. Following both hyper- and hypoosmotic challenges, after the initial change in intracellular osmolarity due to the passive volume change, intracellular osmolarity undergoes no further change as amount of osmotically active solutes and cell volume change. This indicates the causal relationship between the regulated change in cell volume and the change in amounts of intracellular free solute.

Further evidence for the active nature of the return of cell volume to the initial level after osmotic challenges comes from experiments in which cells were exposed to 0.1 mM DNP for 30 min prior to the osmotic challenge. If the cells are then exposed to 100 mM sucrose, they shrink and the osmolarity rises, but the volume remains at 50% of initial volume and the amount of solute per cell remains constant (94). Thus a metabolic inhibitor prevents the regulated recovery of cell volume and the change in the amount of solute per cell. DNP also prevented cessation of the activity of the contractile vacuole. Thus this response, under normal conditions, also reflects a regulatory system that depends on metabolism.

In a medium of 80 mosmoles/liter, intracellular osmolarity was 123 mosmoles. Of this, about 114 mosmoles were accounted for from the measured solutes: Na, K, Cl (values given in Table 2), and 32 mosmoles/liter of free amino acids (94), an earlier published estimate of a total concentration of free amino acids of 57 mosmoles/liter (96) has been found to be in error). After exposure to 100 mM sucrose for 20 min total intracellular osmolarity increased to 212 mosmoles/liter, a change of 89 mosmoles/liter. Of this, increases in Na and K accounted for 24 mosmoles, and amino acids, principally glutamic

Table 2.
Approximate Concentrations of the Major Known Solutes in *T. pyriformis* cultured in 2% proteose–peptone or Similar Slightly Brackish media

Cation	Concentration (mmoles/kg cells)	Anion	Concentration (mmoles/kg cells)
K	30	Cl	6
Na	10	P[a]	5
Ca	7	R—P[b]	24
		AA[c]	32

[a]Orthophosphate.
[b]Organic phosphate esters.
[c]Total free amino acids (not all of which are anionic).

acid, alanine, and glycine, increased by 38 mosmoles/liter. Thus more than two-thirds of the additional solutes were accounted for, and amino acids were the most important of these solutes. From redistribution of $[^{14}C]$ amino acids between protein and the free pool of amino acids, it was concluded that the increase in free amino acid content was through release from a bound compartment, perhaps from a protein with the specific function of participation in osmoregulation.

C. Role of the Cytopharynx in Water Balance; Other Avenues of Water Entry

In freshwater ciliates the rate of ingestion of water into food vacuoles is only a fraction of the rate of fluid discharge by the contractile vacuole (56). Estimates range from less than 10% (a peritrich, ref. 57) to 30% (*Paramecium,* ref. 37). Food vacuoles in *Paramecium* undergo a significant loss of volume after becoming free from the cytopharynx, but subsequently increase in size to nearly the initial volume such that the volume of fluid eliminated from a food vacuole at the cytoproct is a large fraction of the volume ingested (64). It has never been shown, however, that all food vacuoles are emptied at the cytoproct. Some food vacuoles may deliver their contents into the cytoplasm by pinocytosis (54).

Seaman (87) attempted to estimate the rate of fluid ingestion in *T. pyriformis* S from the rate of uptake of a dye, trypan blue, and the uptake of killed bacterial cells. He found a rate of 0.037 ml/h per 10^6 cells, or 0.62 pl/min per cell. This rate of fluid ingestion is nearly the same as the rate of fluid output by the contractile vacuole of strain W in a 20 mM NaCl medium (0.7 pl/min per cell). However, photomicrographs in Seaman's paper (87) show that in cells ingesting trypan blue, the dye becomes highly concentrated in the food vacuoles as they are formed, precluding valid calculations of the rate of fluid ingestion based on uptake of the dye. In amebas inducers of pinocytosis are taken into pinocytic vacuoles in concentrations much higher than in the external medium (6). Nor can fluid ingestion be calculated from the rate of ingestion of bacteria, since in the phagotrophy of particulate matter in *T. pyriformis* water is excluded to a considerable extent (67).

Measurements have been made of the rate of fluid ingestion in strain W based on frequency of formation of food vacuoles and diameters of the vacuoles as they were formed (94). A typical value for cells in 2% proteose–peptone was 0.06 pl/min per cell, 10% of the output of the contractile vacuole measured under the same conditions. No measurements have been made of the rate of elimination of fluid at the cytoproct, but it is certainly a significant fraction of the fluid ingested.

Several early workers (37,41) postulated that the cytopharynx was the principal site of passive water entry in ciliates (see refs. 64 and 74 for further discussion and references). Kitching (57) holds the opposite view, namely, that most of the water expelled by the contractile vacuole enters passively over the general cell surface.

The low permeability of *T. pyriformis* to Na is probably a function of the pellicle, a complex structure consisting of three membranes underlaid by the ectoplasmic layer (1,72,97; Chapter 2). The pellicle covers most of the surface of the cell and is interrupted only by discharged mucocysts (97), parasomal sacs (1), the cytopharynx, the cytopyge, and contractile vacuole pores. The pellicle extends through the cytostome into the cytopharynx (39,66,69; Chapter 2), but a portion of the cytopharynx is bounded by a single membrane, the cytopharyngeal membrane (65,66,69). Restriction of surfaces across which passive exchanges of water and solutes take place to the cytopharyngeal membrane and parasomal sacs would be of obvious advantage to an osmoregulating cell.

On one side of the cytostome is a lip or valve (39,69) which may regulate entry of materials into the cytopharynx and thus play a role in regulation of entry of water and solutes into the cytoplasm (cf. 41). It is conceivable that passive exchange via the parasomal sacs is also subject to regulation. It has been suggested that pinocytosis takes place through the pellicle at structures resembling discharged mucocysts (39) and at the parasomal sacs (1). If this is true, pinocytosis through the pellicle may be an important means of water entry, but more data must be obtained to confirm these proposals.

VIII. ACCUMULATION OF ORGANIC SOLUTES

The regulation of organic solutes bears directly on the general problems of nutrition and metabolism. Treated here are the influx of organic solutes in *T. pyriformis,* and the question of the relative importance of phagotrophy and active transport in the acquisition of organic solutes. Related problems are treated in earlier chapters of this volume.

Tetrahymena pyriformis (=Glaucoma ficaria) (cf. 22) can acquire nutrients from particulate matter alone (53), or from a medium containing only dissolved materials (51). Particulate matter or fluids may be taken into food vacuoles by a process analogous to phagocytosis in amebas. The process of formation of food vacuoles has been held to be the same whether particulate matter or fluid alone is ingested (51,87).

Recent experiments have suggested, however, that the presence of particulate matter in the medium is necessary for "normal" rates of formation of food vacuoles and for rapid multiplication in strain GL (77). In proteose–peptone medium sterilized by filtration, growth of cultures was suppressed severalfold. Addition of various kinds of particulate matter, including powdered quartz, potassium aluminum sulfate (an insoluble salt), and clay, resulted in stimulation of growth of the cultures. The chemical nature of the particles did not seem to be critical. These results indicate that the acquisition of at least some nutrients in *Tetrahymena* depends in large part on the formation of food vacuoles.

There were earlier attempts to evaluate the relative roles of phagotrophy and membrane transport in the acquisition of dissolved nutrients (43,51). In proteose–peptone liquid medium, cells of strain S may contain 15 to 30 food vacuoles, each with a turnover time of 1 h or less (87), suggesting that food vacuole formation may be important in the acquisition of dissolved nutrients. However, Seaman (86) calculated that uptake of acetate by formation of food vacuoles could account only for about 2% of the total rate of utilization of acetate. The calculation was based on the rate at which fluid was engulfed by food vacuoles, and on the assumption that the concentration of acetate in the food vacuoles was the same as in the medium. No account was taken of possible selective binding of acetate at the cytopharyngeal membrane prior to formation of food vacuoles, however, which would result in a higher rate of accumulation of acetate by phagotrophy than Seaman calculated. Such selective binding has been observed with trypan blue in *T. pyriformis* (87), with several dyes in *Paramecium* (42), and with proteins and other inducers of pinocytosis in amebas (6). There have been no studies on active transport of acetate in *Tetrahymena*.

Conner (14) calculated that, in order to account for the rate of uptake of $[^{32}P]$orthophosphate in *T. pyriformis* by phagotrophy (or pinocytosis), a volume of fluid 15 times the cell volume would have to be engulfed in 1 h. This appears unlikely, but again selective binding in the cytopharynx, resulting in initial concentration of orthophosphate, has not been ruled out.

Since cells cultured in chemically defined medium form food vacuoles at a much lower rate than do cells in solutions of proteinaceous materials such as proteose–peptone, membrane transport may be important in the nutrition of *T. pyriformis* (51). However, the importance of solutes in the nutrition of *T. pyriformis* in a natural environment is unclear, since it and related ciliates are generally believed to be bacteria feeders in nature (39).

If active transport accounts in part for the entry of solutes into the cytoplasm from food vacuoles, transport may also take place across the cytopharyngeal membrane, the structure that gives rise to food vacuoles. In this view, dissolved nutrients enter the cytoplasm by active transport whether or not food vacuoles are formed, and the two points of view Seaman and others have attempted to distinguish are not mutually exclusive. Phagotrophy, however, would increase the surface area available for transport, and would allow some control of the environment from which transport takes place. The parasomal sacs might also be involved in active accumulation of solutes.

Formation of pinocytotic vesicles from food vacuoles, suggested for *T. pyriformis* (39) and *P. auerlia* (54), could further increase the surface area for transport. The ultimate fate of these pinocytotic vesicles is not known.

It is possible, however, that the properties of the cytopharyngeal membrane are altered after it becomes a food vacuole, and that transport might take place from food vacuoles, but not at the cytopharyngeal membrane.

The entry of $[^{32}P]$orthophosphate into *Stentor* takes place at the anterior end of the cell (26). Stentors were dissected into anterior and posterior ends and exposed to labeled medium. In eight experiments (three cell fragments per sample), the mean ratio of ^{32}P accumulated by anterior ends/posterior ends was 92. It was concluded that orthophosphate accumulation is accomplished either by formation of food vacuoles or at a specialized locus for transport in the oral region. It might be further concluded that the general cell surface of *Stentor* is impermeable to orthophosphate.

From several studies on the influx of organic solutes in *T. pyriformis*, evidence was obtained for membrane transport as the mechanism of accumulation. Cirillo (11) showed that arabinose, a pentose not metabolized by *T. pyriformis*, was not accumulated against a concentration gradient. The intracellular concentration never exceeded 60% of the external concentration. The external concentrations used in the experiments, however, about 50 mM, were relatively high; accumulation against a gradient might have occurred if the external concentrations had been lower. Nevertheless, the evidence obtained by Cirillo suggests that entry of arabinose was by a carrier-mediated transport process. First, the Q_{10} of arabinose entry between 10 and 30°C was 2.5, higher than expected for simple, passive diffusion. Second, the presence of glucose in the medium markedly inhibited arabinose entry, suggesting competition between glucose and arabinose for a common carrier. When glucose was added to the medium after arabinose entry had reached a steady state, an outward counterflow of arabinose was elicited. That is, the influx of arabinose was inhibited by glucose, but the efflux was not, giving a net efflux of arabinose against a concentration gradient. A counterflow of this sort is predicted for a carrier-mediated transport process when a large concentration gradient exists for one of two competitors for the carrier (101). It is difficult to imagine how counterflow could occur if arabinose influx were by phagotrophy. Some caution should be exercised in interpreting Cirillo's results, however. Some of the increases in intracellular concentrations observed may have been due to shrinkage of the cells following addition to the medium of large concentrations of sugar.

It was mentioned above (Section VII) that *Tetrahymena* is able to recover changes in cell volume resulting from changes in extracellular osmolarity partially by adjusting the intracellular concentration of free amino acids. No correlation was found between the composition of the intracellular pool and protein-bound amino acids (99). The size of the intracellular pool of free amino acids may be increased not only by breakdown of intracellular proteins but by increased influx of amino acids as well.

When 0.24 mM of L-phenylalanine (equal to its internal concentration) was added to the external solution, the internal phenylalanine increased to a maximum value within 20 min, representing concentration gradient of intracellular/extracellular phenylalanine of 15:1 (46). Schaeffer (81) reported that strain W was able to accumulate leucine and alanine against concentration gradients as well. The induction of food vacuoles by the addition of SP-Sephadex beads (1 μm in diameter) was shown to have no effect on the influx of L-methionine (46). It is suggested that *Tetrahymena* is able to accumulate amino acids across its cell surface by a mediated process which is independent of food vacuole formation.

The relationship between rate of influx and the external concentration of leucine, alanine, or phenylalanine (independently determined) demonstrated saturation kinetics, usually considered to be a characteristic of carrier-mediated influx. The maximum rates of influx for leucine and alanine were 0.67 and 0.25 mmole/h $\times$ kg of cells, respectively, while the external concentrations of these amino acids at which rates of influx were half-maximum were 0.33 and 0.22 mM, respectively (81). A value for the maximum rates of phenylalanine influx (calculated from ref. 93 of 4.2 mmoles/h $\times$ kg cells was substnatially higher than those for alanine or leucine influx.

The Q_{10} of influx for phenylalanine between 12 and 20°C was 2.19 (93), similar to that for arabinose (11). A metabolic inhibitor, DNP, at a concentration of 5 $\times$ 10^{-5} M reduced phenylalanine influx by 86% (93; it was not specified whether the rate or extent of influx was inhibited). The rate of leucine influx was reduced by 86–90%, and the rate of alanine influx was reduced by 75% by 5 $\times$ 10^{-5} M DNP (81). The rates of influx of leucine and alanine were also reduced by 1 $\times$ 10^{-5} M N-ethylmaleimide (81). These observations suggest that a significant portion of the influx of each of these amino acids is dependent on metabolic energy.

Evidence for the stereospecificity of amino acid influx has also been presented. The rates of influx of the D isomers of alanine and leucine were significantly less than the rates for the L isomers (81). Addition of the D isomer had no competitive effect on phenylalanine influx (93).

In view of these characteristics of neutral amino acid influx in *Tetrahymena,* that is, the demonstration of saturation kinetics and of stereospecificity, the dependency on metabolic energy, and the influx against concentration gradients, a carrier-mediated process is suggested. This process has been shown to be somewhat dependent on the presence of extracellular Na, either as part of a system of coupled transport of Na and amino acids, or as an activator of the amino acid transport system.

Hoffmann and Kramhøft (48) synchronized cell division in suspensions of *Tetrahymena* by subjecting them to a series of heat shock treatments. Certain amino acids, including L-lysine and DL-*p*-fluorophenylalanine, inhibited cell division in these cells only when extracellular Na was present. The influx of phenylalanine was stimulated by increasing the extracellular concentration of Na.

If Na were required only as a cofactor, the intracellular concentration of Na would not be affected by the presence or absence of extracellular amino acids. However, the intracellular concentration of Na in cells incubated with 0.125 mM phenylalanine was 50% higher than in those in which the external amino acid was absent, although it was not determined whether Na influx was increased or Na efflux was decreased by the presence of the amino acid in the external solution (48).

The influx of certain amino acids was also shown to be affected by the simultaneous presence of other amino acids in the external medium. Alanine influx was inhibited by glycine, while the highly hydrophobic neutral amino acids [i.e., those with long hydrocarbon side chains, including phenylalanine, leucine, isoleucine, valine, methionine, proline, D-leucine and 1-aminocyclopentane-1-carboxylic acid (ACPC)] most effectively inhibited the influx of leucine, ACPC, alanine, and α-aminoisobutyric acid (AIB) (81).

Preincubation conditions were shown to have a significant effect on the kinetics of neutral amino acid influx (81). Generally, cells were harvested from growth medium into a chemically defined inorganic medium in which they were preincubated for 3 h, enabling them to reach a condition of steady state with this medium. When ^{14}C-labeled alanine or leucine was added to the medium and the rate of appearance of radioactivity in the cells was measured, an initial lag phase of 1 h was observed in which little or no radioactivity was detectable in the cells. However, if the desired amino acid was included in the preincubation medium and a trace amount of the ^{14}C-labeled amino acid added after the cells had reached steady state, radioactivity began to appear in the cells immediately. Although the extracellular concentration of amino acid at which the rate of influx was half-maximum was the same under both conditions, the maximum rate of influx of leucine was about three times greater when 1.0 mM leucine was present in the preincubation medium (2.0 mmoles/h $\times$ kg cells) than when leucine was absent (0.67 mmoles/h $\times$ kg cells). It was suggested that when leucine was present during preincubation there were a higher number of transport sites and/or a larger rate-limiting kinetic constant (which might be a function of the turnover time per transport site). Under both conditions, however, the affinity of the transport site for leucine was about the same.

Preincubation with 1.0 mM leucine also maintained (i.e., no lag phase) the influx of alanine, valine, AIB, and ACPC. Uptake of leucine was maintained by preincubation with 1.0 mM phenylalanine, leucine, isoleucine, valine, methionine, D-leucine, proline, or ACPC. It appears that the amino acids that compete most effectively with leucine influx were also those that most effectively maintained the influx of leucine. Schaeffer (81,82) concluded that these neutral amino acids were transported by a common carrier-mediated mechanism.

Tetrahymena pyriformis GL incorporates phenylalanine into protein at a rate of about 10^{-9} mole/min per 10^6 cells (calculated from ref. 24), the same order of magnitude as the maximum rate of influx of phenylalanine reported by Stephens and Kerr (93). Hoffman et al. (50) observed that the rate of influx of ^{14}C-labeled amino acids was nearly equal to the rate of incorporation of labeled amino acids into TCA-precipitable material. Extracellular methionine was shown to inhibit not only the influx of phenylalanine but the incorporation of [^{14}C]phenylalanine into TCA-precipitable material as well. However, the cellular TCA-precipitable protein fraction was found to be unaffected by the presence or absence of 10 mM methionine in the external solution. That is, the equilibrium between synthesis and degradation of proteins was not altered by extracellular methionine. Furthermore, [^{14}C]methionine previously incorporated into protein was eliminated even after addition of 10 mM methionine to the external solution, showing that degradation could continue under these conditions. From these results it was concluded that the inhibitory action of methionine on [^{14}C]phenylalanine incorporation (into protein) resided at the level of transport into the cell rather than at the level of macromolecular turnover.

Hoffmann and Rasmussen (49) showed that preincubation with phenylalanine resulted in an increased influx of phenylalanine which was not a result of a higher intracellular concentration of phenylalanine and therefore was not a result of an accelerated exchange diffusion. Addition of 10 mM phenylalanine to the external medium resulted in a loss of leucine from the cells, however, indicating that the amino acid carrier mechanism does have exchange properties (81).

It was suggested by Schaeffer and supported by Hoffman and Rasmussen that a neutral amino acid transport system could be induced in *Tetrahymena* under appropriate conditions. In the absence of any amino acid in the preincubation medium, influx of both leucine and alanine were inhibited by cycloheximide and actinomycin D, inhibitors of cytoplasmic protein synthesis and of RNA synthesis, respectively. Cycloheximide had no inhibitory effect on either alanine or leucine influx in cells preincubated with 1.0 mM leucine. From these observations Schaeffer concluded that a short-lived protein component, coded for by a short-lived messenger RNA and induced by the presence of neutral amino acids, is required for an active neutral amino acid transport system to operate. When the external concentration of leucine is high enough, and therefore the need for the transport system exists (e.g., cells preincubated with the amino acid), the short-lived components are continuously produced and the neutral amino acid transport system is maintained.

Conner and Ungar (20) demonstrated accumulation of [C^{14}]cholesterol in *T. pyriformis*. At low temperatures the time course of uptake had several phases, each with a different temperature coefficient. Again, it was not clear whether entry into the cell or metabolic conversion was limiting.

ACKNOWLEDGMENTS

We gratefully acknowledge the collaboration of Drs. Edna S. Kaneshiro, Robert D. Prusch, Larry C. Stoner, and John F. Schaeffer, and Mrs. Judith B. Pederson. Previously unpublished work from this laboratory was supported by USPHS grant NB-08089.

REFERENCES

1. Allen, R. D. 1967. J. Protozool. 14:553–565.
2. Andrus, W. D. and Giese, A. C. 1963. J. Cell. Comp. Physiol. 61:17–30.
3. Brandt, P. W. and Freeman, A. R. 1967. Science 155:582–585.
4. Browning, I. 1949. J. Exp. Zool. 110:441–460.
5. Cameron, I. L. and Burton, A. L. 1969. Trans. Amer. Microsc. Soc. 88:386–393.
6. Chapman-Andresen, C. 1962. C. R. Trav. Lab. Carlsberg 33:73–264.
7. Chapman-Andresen, C. and Dick, D. A. T. 1962. C. R. Trav. Lab. Carlsberg 32:445–469.
8. Chapman-Andresen, C. and Nilsson, J. R. 1968. C. R. T. Lab. Carlsberg 36:405–432.
9. Chatton, E. and Tellier, L. 1927. C. R. Soc. Biol. 97:780–784.
11. Cirillo, V. P. 1962. J. Bacteriol. 84:754–758.
12. Cline, S. G. 1966. J. Cell. Physiol. 68:157–164.
13. Cline, S. G. and Conner, R. L. 1966. J. Cell. Physiol. 68:149–156.
14. Conner, R. L. 1967. In M. Florkin and B. J. Scheer, eds., Chemical zoology, vol. 1, Protozoa (G. W. Kidder, ed.), Academic Press, New York, pp. 309–350.
15.ᵉ Conner, R. L., Chook, S. A., and Ray, E. 1963. J. Cell Biol. 19:15–16A,
16. Conner, R. L., and Cline, S. G. 1964. J. Protozool. 11:486–491.
17. Conner, R. L., Goldberg, R., and Kornacker, M. S. 1961. J. Gen. Microbiol. 24:239–246.
18. Conner, R. L., Kornacker, M. S., and Goldberg, R. 1961. J. Gen. Microbiol. 26:437–422.
19. Conner, R. L., and Pruett, P. O. 1963. In Ludvik, J., Lom, J., and Vavra, J., eds. Progress in protozoology, 1st International Congress on Protozoology, Publishing house of the Czechoslovak Academy of Sciences, Prague, pp. 143–147.
20. Conner, R. L. and Ungar, F. 1964. Exp. Cell Res. 36:134–144.
21. Corliss, J. O. 1953. Parasitology 43:49–87.
22. Corliss, J. O. 1953. Proc. Soc. Protozool. 4:3–4.
23. Crane, R. K. 1965. Fed. Proc. 24:1000–1006.
24. Crockett, R. L., Dunham, P. B., and Rasmussen, L. 1965. C. R. Trav. Lab. Carlsberg. 34:451–486.
25. Degen, A. 1905. Bot. Ztg. 63:163–226.
26. de Terra, N. 1960. Exp. Cell Res. 21:34–40.
27. Diamond, J. M. and Bossert, W. H. 1967. J. Gen. Physiol. 50:2061–2083.
28. Dick, D. A. T. 1966. Cell water. Butterworths, Washington, D.C.
29. Dryl, S. and Grebecki, A. 1966. Protoplasma 62:255–284.
30. Dunham, P. B. 1964. Biol. Bull. 126:373–390.
31. Dunham, P. B. 1965. Progress in protozoology, 2nd ed., International Congress on Protozoology. Ser. 91, 234.
33. Dunham, P. B. and Child, F. M. 1961. Biol. Bull. 121:129–140.
34. Dunham, P. B. and Stoner, L. C. 1967. J. Protozool. 14(suppl.);34.
35. Dunham, P. B. and Stoner, L. C. 1968. Proc. Int. Union Physiol. Sci. 7:119.
36. Dunham, P. B. and Stoner, L. C. 1969. J. Cell Biol. 43:184–188.
37. Eisenberg, E. 1926. Arch. Biol. 35:441–464.
38. Elliott, A. M. and Bak, I. J. 1964. J. Protozool. 11:250–261.
39. Elliott, A. M. and Clemmons, G. L. 1966. J. Protozool. 13:311–323.
40. Erickson, E. B. and Conner, R. L. 1967. J. Protozool. 14(suppl.):8.

41. Frisch, J. A. 1937. Arch. Protistenk. 90:123–161.
42. Grebecki, A. 1963. Protoplasma 56:89–98.
43. Hall, R. P. 1964. Protozoan nutrition. Blaisdell, New York.
44. Hamburger, K. and Zeuthen, E. 1960. C. R. Trav. Lab. Carlsberg 32:1–18.
45. Hartog, M. 1888. Rep. Brit. Ass. Adv. Sci. 58:714–716.
46. Hoffmann, E. K. 1971. J. Protozool. 18(suppl.):34.
48. Hoffmann, E. K. and Kramhøft, B. 1969. Exp. Cell Res, 56:265–268.
49. Hoffmann, E. K. and Rasmussen, L. 1972. Biochim. Biophys. Acta. 266:206–216.
50. Hoffmann, E. K., Rasmussen, L., and Zeuthen, E. 1970. C. R. Trav. Lab. Carlsberg 38:133–143.
51. Holz, G. G. 1964. In Hutner, S. H., ed., Biochemistry and physiology of protozoa, vol. III, Academic Press, New York, pp. 199–242.
52. Holz, G. G., Erwin, J. A., and Davis, R. J. 1959. J. Protozool. 6:149–156.
53. Johnson, D. F. 1936. Arch. Protistenk. 86:359–378.
54. Jurand, A. 1961. J. Protozool. 8:125–130.
55. Kaneshiro, E. S., Dunham, P. B., and Holz, G. G. 1969. Biol. Bull. 136:63–75.
56. Kitching, J. A. 1956. Protoplasmatologia III, D, 3, a:1–45.
57. Kitching, J. A. 1967. In Chen, T. T., ed., Research in protozoology, vol. I, Pergamon Press, New York, pp. 307–336.
59. Kropp, D. L. 1971. Ph.D. Thesis, Syracuse University.
60. Kropp, D. L. and Dunham, P. B. 1971. J. Protozool. 18(suppl.):25.
61. LeBoy, P. S., Cline, S. G., and Conner, R. L. 1964. J. Protozool. 11:217–222.
62. Ling, G. N. 1962. A physical theory of the living state: The association-induction hypothesis. Blaisdell, New York.
63. Loefer, J. B. 1939. Physiol. Zool. 12:161–172.
64. Mast, S. O. 1947. Biol. Bull. 92:31–72.
65. Miller, O. L. and Stone, G. E. 1963. J. Protozool. 10:280–288.
66. Müller, M. and Röhlich, P. 1961. Acta Morphol. Acad. Sci. Hung. 10:297–305.
67. Müller, M., Röhlich, P., and Törö, I. 1965. J. Protozool. 12:27–34.
68. Munk, N. and Rosenberg, H. 1969. Biochim. Biophys. Acta. 177:629–640.
69. Nilsson, J. R. and Williams, N. E. 1966. C. R. Trav. Lab. Carlsberg. 35:119–141.
70. Organ, A. E., Bovee, E. C., Jahn, T. L., Wigg, D., and Fonseca, J. R. 1968. J. Cell Biol. 37:139–145.
71. Organ, A. E., Bovee, E. C., and Jahn, T. L. 1968. J. Protozool. 15(Suppl.):23.
72. Pitelka, D. R. 1961. J. Protozool. 8:75–89.
73. Potts, W. T. W. 1968. Ann. Rev. Physiol. 30:73–104.
74. Prosser, C. L. 1950. In Prosser, C. L., ed., Comparative animal physiology. Saunders, Philadelphia, pp. 6–74.
75. Pruett, P. O., Conner, R. L. and Pruett, J. R. 1967. J. Cell. Physiol. 70:217–224.
76. Prusch, R. D. and Dunham, P. B. 1970. J. Cell Biol. 46:431–434.
77. Rasmussen, L. and Kludt, T. A. 1970. Exp. Cell. Res. 59:457–463.
78. Riddick, D. H. 1968. Amer. J. Physiol. 215:736–740.
79. Rosenberg, H. 1966. Exp. Cell Res. 41:397–410.
80. Rosenberg, H. and Munk, N. 1969. Biochim. Biophys. Acta. 184:191–197.
81. Schaeffer, J. F. 1970. PhD. Thesis, Syracuse University.
82. Schaeffer, J. F. and Dunham, P. B. 1970. J. Protozool. 17(suppl.):20.
84. Schmidt-Nielsen, B. and Schrauger, C. R. 1963. Science 139:606–607.
85. Schneider, L. 1960. J. Protozool. 7:75–90.
86. Seaman, G. R. 1955. In Hutner, S. H., and Lwoff, A., eds., Biochemistry and physiology of protozoa, vol. II, Academic Press, New York, pp. 91–158.
87. Seaman, G. R. 1961. J. Protozool. 8:204–212.
89. Slater, J. V. 1957. Biol. Bull. 112:390–399.

90. Slater, J. V. and Bonitati, J. A. 1957. J. Protozool. 4(suppl.):11.
91. Slater, J. V. and Tremor, J. W. 1962. Biol. Bull. 122:298–309.
92. Steinbach, H. B. 1962. Perspect. Biol. Med. 5:338–355.
93. Stephens, G. C. and Kerr, N. S. 1962. Nature 194:1094–1095.
94. Stoner, L. C. 1970. Ph.D. Thesis, Syracuse University.
95. Stoner, L. C. and Dunham, P. B. 1967. Biol. Bull. 133, 454.
96. Stoner, L. C. and Dunham, P. B. 1970. J. Exp. Biol. 53:391–399.
97. Tokuyasu, K. and Scherbaum, O. H. 1965. J. Cell. Biol. 27:67–81.
98. Whittam, R. and Wheeler, K. P. 1970. Ann. Rev. Physiol. 32:21–60.
99. Wigg, D., Bovee, E. C., and Jahn, T. L. 1967. J. Protozool. 14:104–108.
100. Wu, C. and Hogg, J. F. 1956. Arch. Biochem. Biophys. 62:70–77.
101. Wilbrandt, W. and Rosenberg, T. 1961. Pharmacol. Rev. 13:109–183.
102. Yamaguchi, T. 1960. J. Fac. Sci. Univ. Tokyo. 8:573–591.
103. Yamaguchi, T. 1963. Annot. Zool. Jap. 36:55–65.
104. Zeuthen, E. 1964. In Zeuthen, E., Synchrony in cell division, Interscience, New York, pp. 99–158.

Growth Characteristics of *Tetrahymena*

Ivan L. Cameron

Department of Anatomy
The University of Texas
Medical School at San Antonio
San Antonio, Texas

INTRODUCTION

In nature *Tetrahymena* are primarily phagotrophic feeders, but experimental work is difficult to control and to interpret when mixed populations of microorganisms are used, and for this reason only pure (axenic) culture studies have been included in this chapter.

The cell division cycle consists of one interphase and one cell division. Intimately linked to this reproductive process is the duplication and equal partition of complete sets of genetic information (DNA) to the two daughter cells. Loosely but vitally coupled to the cell replication process is the cell growth process (increase in cell mass and size). In this chapter the relation between, or degree of coupling of, cell growth and cell reproduction is stressed, and the influence that different environments (i.e., temperature, pH, media composition and concentration, cell number, starvation, and so on) have on the cell reproduction rate is emphasized.

The cell population growth cycle of *Tetrahymena* is also described and analyzed in terms of factors that influence the phases of the growth curve and the final growth yield.

The chapter ends with a discussion of the control of growth and macromolecular synthesis in *Tetrahymena,* especially as compared to the control of growth and macromolecular synthesis in bacteria.

II. THE CELL DIVISION CYCLE

A. Dissection of the Cell Division Cycle

Introduction of tritiated thymidine (a radioactive precursor which specifically labels newly synthesized DNA), in combination with autoradiography, is frequently used to dissect the cell division cycle of *Tetrahymena*. Figure 1 illustrates one method of dissecting the cell cycle. Groups of dividing *Tetrahymena* were selected from a logarithmically growing cell population with the aid of a braking pipet and a dissecting microscope. At intervals thereafter individual groups of these cells were briefly exposed (5-min pulse) to tritiated thymidine, harvested, and dried down on slides. The cells were then fixed, washed and autoradiographed. The developed autoradiographs were analyzed to see when in the cell cycle the isotope was incorporated into the nuclear DNA of the cells. After measuring the average cell generation time of the cells, a temporal dissection of the entire cell cycle is possible. By this means one can demonstrate that macronuclear DNA duplication occurs during a portion of interphase and that micronuclear DNA duplication occurs during cytokinesis. More accurate and elaborate methods of cell cycle dissection have now been introduced (11,47).

The cell cycle terminology introduced by Howard and Pelc (31) has been adopted for *Tetrahymena:* D, the division period; S, the macronuclear DNA synthetic period; G_1, the period from the end of cell division to the start of the S period; G_2, the period from the end of the S period to the start of division; GT, the mean cell generation time. The terminology for the micronuclear S period is given in Fig. 2. However, many strains of *T. pyriformis* do not have micronuclei, and the micronuclear terminology is not needed in these cases.

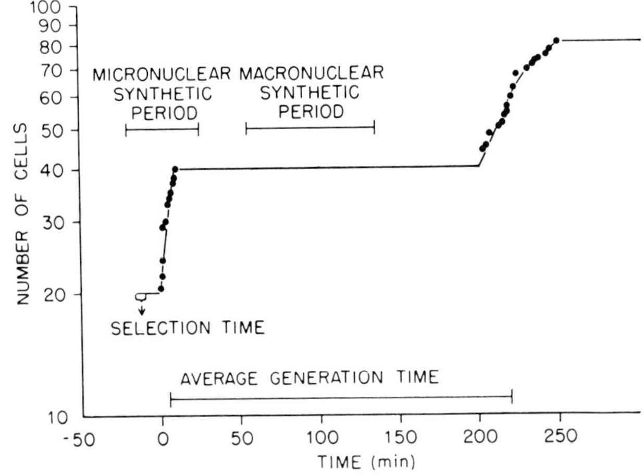

Fig. 1.

The division kinetics of 20 *Tetrahymena* which were selected and isolated during the division process. The approximate macronuclear and micronuclear DNA synthetic periods are indicated. [From Stone and Cameron (81); courtesy of Academic Press.]

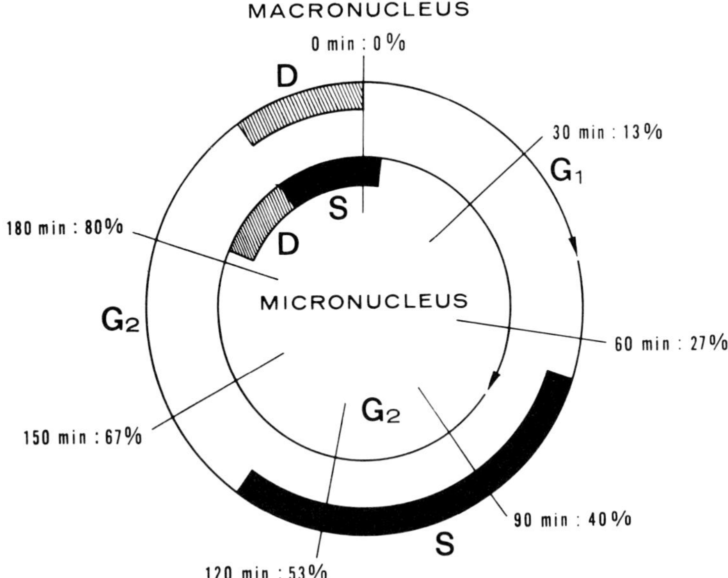

Fig. 2.

The cell division cycle of *T. pyriformis* HSM. Clockwise the outer circle indicates the macronuclear events, and the inner circle the micronuclear events. The total cycle time in this experimental situation was 225 min. [From Flickinger (23); courtesy of the Rockefeller Press.]

The temporal relation between the macronuclear S period and cell division is not a fixed temporal relationship because studies have shown that a decreased availability of nutrients causes the S period to be initiated proportionately later in the cell division cycle and that the proportion of the cell cycle spent in S period increases (Fig. 3) (11).

Three reports taken together suggest that the initiation and maintenance of the S period is temperature-dependent when *Tetrahymena* are grown in nutritionally poor media, whereas increasing the richness of the medium counteracts this temperature-dependent effect (11,17,46). Perhaps the richer growth medium supplies the cell with essential

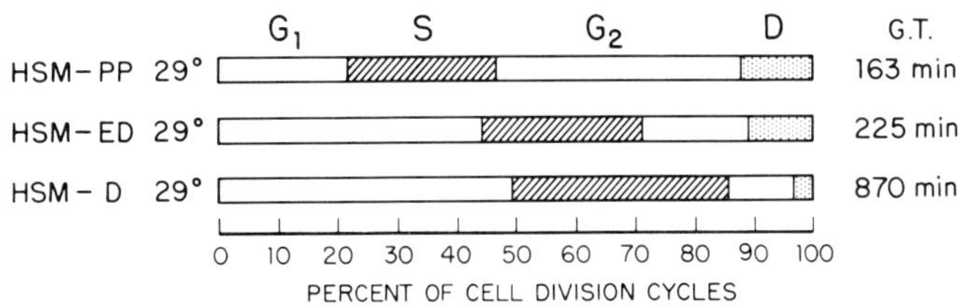

Fig. 3.

The proportionate time that *T. pyriformis* HSM spends in the phases of the cell division cycle when grown at 29°C on three different media: PP, liver extract-enriched proteose–peptone medium; ED, enriched defined medium; and D, chemically defined medium. The generation time (GT) is listed for each medium. [From Cameron and Nachtwey (11); courtesy of Academic Press.]

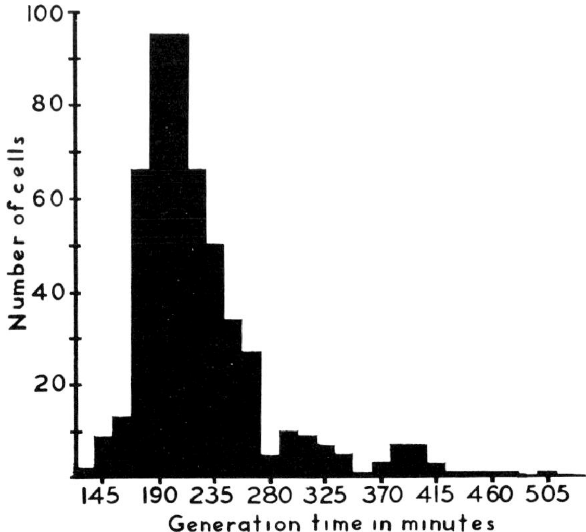

Fig. 4.

Frequency distribution of generation times of 515 clones of *T. pyriformis*. [From McDonald (82); courtesy of *The Biological Bulletin.*]

substrates and therefore the cell does not have to manufacture these substrates de novo. It appears as if the regulatory controls underlying macronuclear DNA synthesis are such that the S period can be shifted to any part of interphase by altering the growth conditions of the cell.

B. The Generation Time Distribution

Generation time is defined as the interval between two successive cell divisions. The way that generation time varies between cells in a population gives rise to a frequency distribution. Observation of individual *Tetrahymena* that share the same culture conditions demonstrates that the generation times of individual cells have a distinct frequency distribution (see Fig. 4). Examination of Fig. 4 suggests that the frequency distribution curve of the generation times is not the familiar symmetrical, bell-shaped relationship known as the normal distribution curve. In fact, the frequency distribution is somewhat skewed to the right. It turns out that, when the generation time frequency distribution of *Tetrahymena* and those of many other cell types are plotted on a logarithmic abscissa rather than on a linear abscissa, the distribution appears as a normal (symmetrical) frequency distribution (see ref. 47 for details).

A frequency distribution of this type, which appears as a normal distribution after being plotted on a logarithmic axis, is termed a log normal frequency distribution. Further evidence that the distribution is truly log normal is obtained by converting the original frequency distribution to a cumulative frequency distribution and then plotting the data values on probit paper using first a linear and then a logarithmic abscissa. Figure 5 illustrated a log normal cumulative plot, and there can be little doubt that the generation

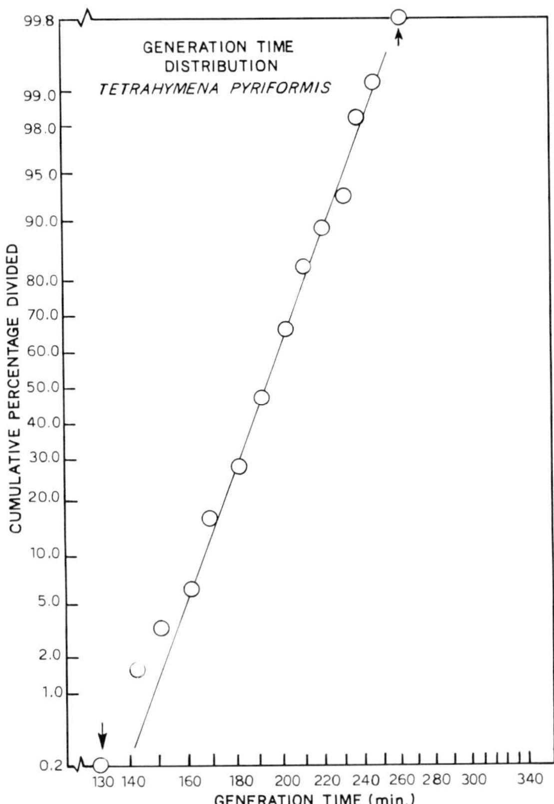

Fig. 5.

Illustration of a method of handling generation time data, as described in text. This cumulative frequency distribution of generation time data, similar to that in Fig. 4, shows a good fit to a log normal-type frequency distribution. [Redrawn from Nachtwey and Cameron (47); courtesy of Academic Press.]

time fits the log normal distribution. The point at which the line crosses the 50% value in Fig. 5 is the median value, and the mean value for this type of distribution is always somewhere to the right of the median value. In the normal distribution the mean and median values are always the same. The slope of the line in Fig. 5 is a measure of the variability around the median value. The steeper the slope the lower the coefficient of variation or the standard deviation. Other types of frequency distributions are conceivable, but in general they do not fit the data for *Tetrahymena* as well as the log normal distribution (47).

In populations of cells with a log normal distribution of generation times, cells with short generation times contribute more to the geometric increase in total cell number than cells with longer generation times. The net effect is that the culture doubling time is not the same as the mean cell generation time in mass cultures (for corrections and discussion see ref. 47). Of perhaps more interest are the theoretical implications underlying

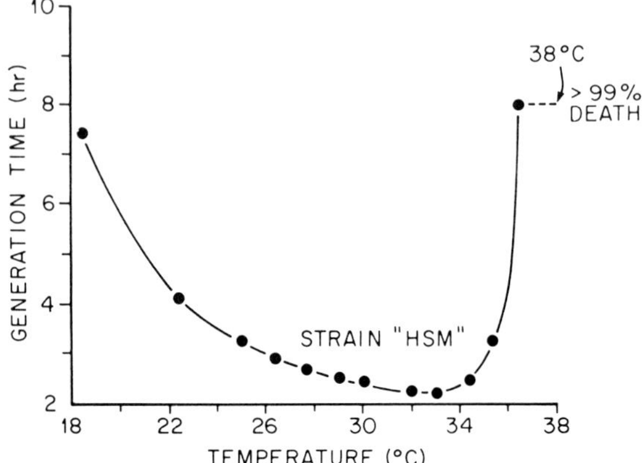

Fig. 6.

A plot of incubation temperature against generation time for *T. pyriformis* strain HSM. [Redrawn from Prescott (55).]

a log normal frequency distribution of generation times. A log normal distribution results from *proportionate or multiplicative* effects in a series of independent variables. However, a normal distribution would be generated by the *addition* of a large number of small independent variables around the cell cycle (40).

C. The Rate of Cell Reproduction

1. Temperature and Generation Time

Phelps (53) demonstrated that the generation time was increased in *T. pyriformis* T-P as the temperature was raised from 7 to 28°C. Figure 6 illustrates the relation between temperature and generation time for strain HSM when grown on enriched proteose–peptone. Cell reproduction rate is optimal at 32.5°C and ranges from 36.6°C to below 18.4°C. Temperatures above 36.6°C are lethal. James and Read (33) report substantial cell reproduction at 10°C for this same strain, and it can survive for weeks at 4°C.

The temperature optimum for cell replication is strain-dependent; for example, strain GL shows an optimum at 29°C and death at 34.6°C (55). Holz (29,30) tested 45 mating type strains of *T. pyriformis* for ability to grow at higher temperatures. His data suggest that temperature lethality has a genetic basis in the different strains.

Adaptability to higher temperatures over several cell cycles, as measured by generation times changes, was found by Thormar (74) and Holz (30). The upper lethal growth temperature of *T. pyriformis* can be increased by enriching a basic chemically defined medium with nucleic acid derivatives and with phospholipids (59).

2. pH and Generation Time

The relation between pH and generation time for *T. pyriformis* HSM is illustrated in Fig. 7. The pH optimum is 7.25–7.30. The absolute pH growth range for *Tetrahymena*

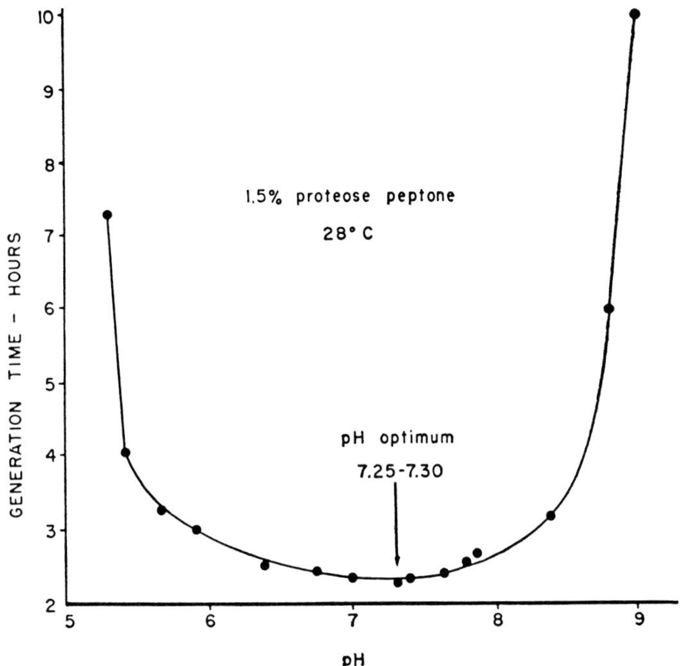

Fig. 7.

A plot of pH against generation time for *T. pyriformis* HSM. [From Prescott (56); courtesy of *Physiological Zoology.*]

is between pH 5.0 and 8.6 as reported by Elliott (20) and by Prescott (56). It is very difficult to be sure how a change in pH or any other environmental factor will affect other culture factors such as substrate solubility, osmotic conditions, gas tension, and so on. If the intracellular pH is actually changed by the environment, the pH change would be expected to affect intracellular enzyme reactions.

3. Media and Generation Time*

Medium concentration and medium composition affect generation time of *T. pyriformis* as illustrated in Fig. 8. The lower curve in Fig. 8 indicates cell generation times for different dilutions of proteose–peptone No. 3 enriched with liver extract in a 10:1 ratio. An optimum concentration of 1.0–2.5% is indicated. Above 5% or below 0.1%, there is a sharp drop in generation rate. Ten percent proteose–peptone was lethal to the cells, and 0.05% produced growth with a generation time of 28 h. The sharp rate change at either end of the medium concentration range may be caused by several factors such as toxic products in the extract, osmotic difficulties, starvation, or by any combination of these.

*For an excellent discussion and review of *Tetrahymena* nutritional growth requirements, see Kidder and Dewey (38) or Kidder (36).

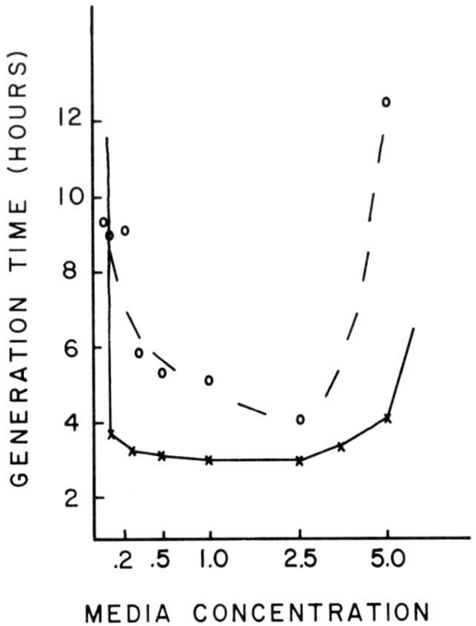

Fig. 8.

A plot of medium concentration against generation time for *T. pyriformis* HSM. The circles represent dilutions of a stock solution of proteose–peptone, and the crosses represent dilutions of a stock solution of proteose–peptone with liver extract (10:1 w/w). (Cameron, unpublished.)

The upper curve in Fig. 8 demonstrates the effect of omitting the liver extract. Clearly, liver extract promotes more rapid cell multiplication (35,37).

The finding of a medium concentration optimum in the range of 1–2.5% is contrary to an earlier report by Phelps (52), who indicated a constant generation time over an even wider range of medium concentrations. The difference in the two studies is probably due to the more refined methods of generation time determination now available. Various medium concentrations cause changes in the osmotic properties of the environment, which have not been compensated for in either of these studies.

Prescott (56) tested several different media preparations in an attempt to determine *the* optimum generation time for *Tetrahymena*. Interestingly, he found an optimum using a medium made from an extract of rapidly growing *Tetrahymena*. Apparently, the extract supplies the *Tetrahymena* with compounds that the cell otherwise has to synthesize de novo.

4. Osmolarity and Generation Time

To determine the effect of osmolarity on the reproductive rate of *Tetrahymena*, a series of culture vessels was prepared with a broth of 2% proteose–peptone enriched with 0.1% liver extract. To each vessel a different amount of sodium chloride was added, ranging from 0 to 1%. After inoculation and several transfers, the cells were considered to be adapted to the various salinities. Drop cultures (Nachtwey, personal communication) were then prepared and the generation times were measured by periodic

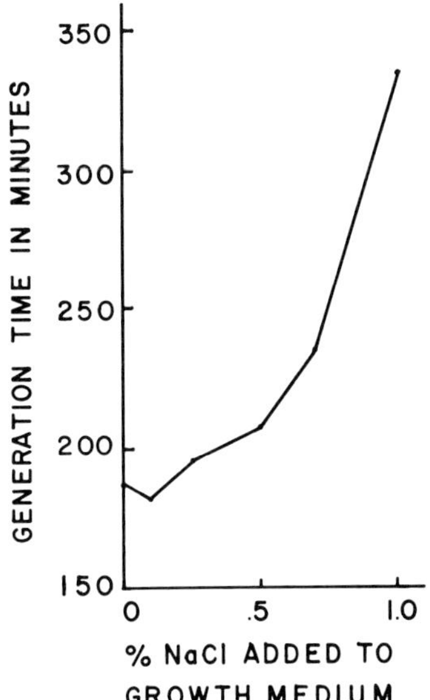

Fig. 9.

A plot of sodium chloride concentration and generation time for *T. pyriformis* HSM. (Cameron, unpublished).

counting of the number of cells in each drop culture. The results are shown in Fig. 9. In general, the generation times became progressively longer as the sodium chloride concentration was linearly increased. Figure 10 shows what happened to the generation time of *T. pyriformis* when increasing concentrations of sucrose were added to the growth medium. Once again the generation time was determined after several transfers of the cells through the various concentrations of sucrose. Sucrose was added to the growth medium as a means of increasing the osmotic pressure of the cells' environment. Sucrose was chosen because it is a nonelectrolyte, is not used as a food source for *Tetrahymena*, is not permeable to the plasma membranes of this cell, does not react with the medium ingredients, and is both soluble in solution and has a reasonable molecular weight. All these characteristics make it acceptable for the experimental design. We can therefore study the osmotic effects of sucrose without involving specific ion effects. The generation time becomes progressively longer as the sucrose osmolarity increases. Figure 10 also shows that addition of 0.003 M calcium chloride to the growth medium plus sucrose has an antagonistic effect on the generation time increase. Thus it appears that calcium chloride antagonizes the osmotic effect of sucrose. For additional discussion of the effects of osmolarity on *Tetrahymena*, see Chapter 6, or Holz et al. (30).

5. The Optimal Generation Time

By using the type of information given above, one can predict the growth conditions that should be used in order to achieve the fastest generation time for a given strain of *T. pyriformis*. In the case of strain HSM, the conditions are: temperature 32.5°C; pH 7.27 adjusted with sodium hydroxide; a medium relatively free of extra ionized

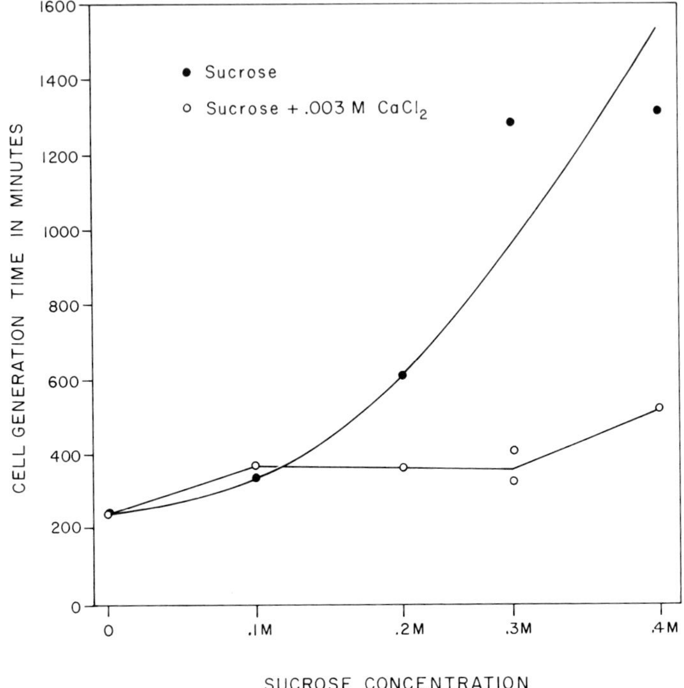

Fig. 10.

A plot of sucrose concentration and generation time (closed circles); the open circles show the results when 0.003 *M* calcium chloride is also added to such cultures. (Cameron, unpublished.)

substances such as sodium chloride; a nutrient consisting of 1–2% proteose–peptone enriched with 0.1–0.25% liver extract or, if possible, a medium made from the autolysate of log-phase *Tetrahymena*. Prescott (56) found that the enriched proteose–peptone medium prepared under the conditions listed above gave an optimum generation time of 1.83 h, whereas a medium made from *Tetrahymena* autolysate gave a generation time of 1.65 h.

Different strains of *T. pyriformis* are know to have different generation times when grown under identical culture conditions. Of six strains that we have tested, the GL-C strain (from the Carlsberg Laboratories) has the fastest generation time.

D. The size of *Tetrahymena*

Figure 11 illustrates the cell size frequency distribution of a logarithmically multiplying population of *T. pyriformis* HMS in a state of balanced growth (a condition in which the cell volume exactly doubles during the cell division cycle). One can see from this figure that the cell volume range falls almost exactly between a doubling of the cell volume, as is expected of a homogeneous population of cells in a state of balanced growth. Note also that the frequency distribution is skewed to the right. A replotting

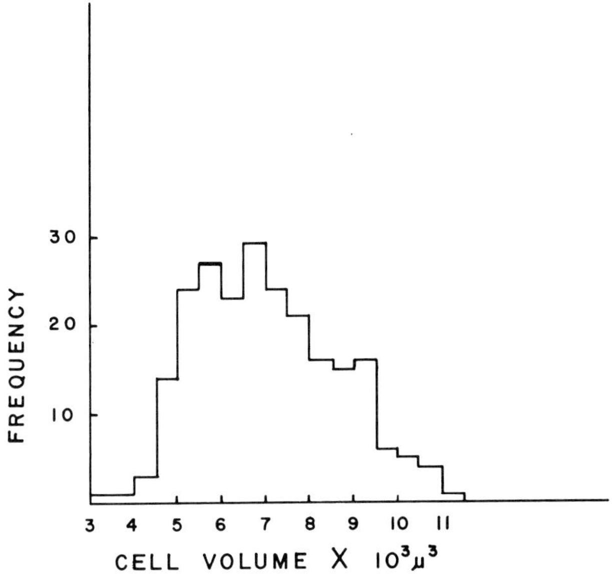

Fig. 11.

Frequency distribution of cell volume of 230 logarithically growing *T. pyriformis* HSM in balanced growth. (See legend for Fig. 12 for details of volume determination.)

of this data as a cumulative frequency distribution on probit paper using both a linear and a logarithmic axis confirms that cell volumes are log normally distributed. One should keep in mind that in log phase cell populations there is an age frequency distribution in which there are about twice as many newborn daughter cells as there are dividing mother cells. Thus one would expect more smaller cells than larger cells in the size frequency distribution plot. Figure 11 confirms this expectation.

Several other studies have been made of the cell volume distributions of logarithmically growing *Tetrahymena* (33,42,67,69,70,72), all of which indicate that the frequency distribution of cell volumes is log normally distributed.

Attempts have been made to use knowledge of the cell volume distribution and the mean generation time value in order to calculate cell volume growth rate between cell divisions (67,72). Careful analysis of these attempts by Schmid (70) indicates that this amount of information (taken by itself) is not sufficient for this purpose.

Loefer (43) reports that different strains of *T. pyriformis* grown under similar culture conditions are of different sizes. Apparently, size is related to the genetic composition of the strain. It is known, for instance, that different strains have different amounts of DNA in their macronucleis (14), and it seems likely that size is related rather directly to the amount of DNA in the cell nucleus (3).

Starvation studies, described later in this chapter, indicate an average cell volume difference of 13-fold between starved and well-fed *Tetrahymena*. Some individuals from these two populations of cells vary over more than a 20-fold size difference range.

Much attention has been given to the relationship between cell size and temperature.

James and Read (33) found three different volume classes in *T. pyriformis* HSM when the organisms were grown at temperatures of 10, 20, and 30°C. Cell volume averaged 16,250, 12,350, and 9375 μm^3, respectivey. Examination of the frequency distribution of the cell volumes indicated that there was a shift in the entire range of volumes in each case, and therefore that the change was not just a weighting of a single distribution (33). Thormar (75) extended the temperature studies to a larger range of temperatures and found that the size of *Tetrahymena* increased at low growth temperatures but also greatly increased at higher temperatures which are known to be above the optimum for the rate of cell multiplication. Therefore the cell volume increases as the generation time decreases either at sub- or supraoptimal temperatures. Schmid (69) also reports that the optimum temperature for generation time and that for cell growth in volume are appreciably different. Browning (5) noted that cell size is increased as generation time decreased in *Tetrahymena* subjected to different osmotic conditions.

E. Cell Growth over the Cell Division Cycle

Several different attempts have been made to measure the size or growth rate of *Tetrahymena* over the cell division cycle. None of the methods used is completely satisfactory. Individual cells were found to increase in volume at a linear rate throughout interphase, but then to increase more rapidly during cell division (13). A rapid imbibing of water may account for the rapid increase in cell volume at cell division. Zeuthen's studies on the respiratory rate of single cells also show a linear increase during interphase, but the rate plateaus or decreases during division. Respiratory rate may reflect a direct relation to respiratory machinery, and this may in turn reflect a direct measure of cell growth. The repression or activation of existing respiratory machinery over the cell cycle must also be considered as an explanation for this data, however. During interphase and cell division, Prescott (57) found that the incorporation of $[^{14}C]$methionine into protein goes on at a linear rate. Although all three of these studies suggest a linear increase during interphase, they differ from one another as to what happens at the time of cell division. As Prescott (57) points out, this seeming discrepancy may simply mean that the quantities used to measure growth may vary independently of one another. In the ciliated protozoan *Paramecium aurelia,* Kimball et al. (39) showed that mass increase, as measured by interference microscopy and x-ray absorption methods, is clearly exponential. Their data are probably subject to less criticism than are other methods. Løvlie (44), who used a Cartesian diver balance, also suggests that growth curves for single cells may be exponential when mass per cell doubles from one cell division to the next (in other words, a condition of balanced growth occurs). Løvlie indicates that exponential growth may indeed be concealed if cell growth is unbalanced.

Schmid (70) has recently subjected all the data mentioned above to statistical analysis to determine if the kinetics of growth show a better fit to a linear or to an exponential growth process. Taken together the data support the contention that division-to-division growth is probably exponential. In addition, it has recently been shown that the growth rate of *Tetrahymena* (as measured by volume increase) is exponential rather than linear in the presence of constant high temperature (69).

The exponential growth curves recorded under conditions of balanced growth may be explained if one assumes that the nucleus is producing ribosomes at a rather constant rate throughout interphase. According to this proposed explanation, a number of ribosomes would accumulate throughout interphase (assuming that ribosome turnover is minimal in rapid growth states), and this would lead to a greater overall cellular protein synthetic and growth potential at the end of interphase than at the beginning. This theory also assumes that ribosomes are the limiting factor to growth of the cell. That ribosome concentration is the limiting factor that controls growth rate of bacteria is postulated by Maaløe and Kjeldgaard (45). Halving of the nucleus at cell division would in turn halve the production rate of ribosomes, as well as the total number of ribosomes in a new daughter cell. The nucleoli do not disappear during the nuclear division of *Tetrahymena,* and RNA synthesis continues uninterrupted. The accumulation of ribosomes would then continue in the newborn cell. This explanation of an increasing cell cycle growth rate pattern may not be the same as for the single-cell growth patterns of mitotically dividing cells which are known to cease nuclear RNA synthesis during the condensation of chromosomes, a process known to take place at the time of prophase when the nucleoli disappear.

At present, there is no compelling reason for selecting any particular pattern of single-cell growth as characteristic of all cells or even of the same cell type grown under different growth conditions. Indeed, according to Williams (78), the maximum difference between an exponential and a linear growth curve pattern is only 6%. This fact and others leave little wonder that the true growth curve is difficult to distinguish (78).

III. THE CELL POPULATION GROWTH CYCLE, INCLUDING FACTORS THAT INFLUENCE THE PHASES OF THE GROWTH CURVE AND THE FINAL GROWTH YIELD

Introduction of a few *Tetrahymena* into a sterile flask containing a nutritionally adequate medium results in growth as measured either by an increase in the number of cells or by an increase in protoplasmic material (weight or volume increase). A great deal of information about the control and regulation of cell growth and reproduction can be obtained by studying the growth curves of a cell population. As one follows a cell population growth cycle, it is of considerable advantage to measure both the increase in cell number and the increase in protoplasmic material. As will be illustrated in the case of *T. pyriformis* HSM, the two growth parameters seldom increase at the same rate or in parallel.

A typical cell population growth curve for *T. pyriformis* is shown in Fig. 12A. The curve can be divided into phases: lag, acceleration, exponential (log), maximum stationary, decline, and logarithmic decline. The lag phase is characterized by an initial period of no increase in cell number or sometimes even a slight drop in cell number. This is followed by a phase of accelerated increase in cell number. During the exponential (log) phase, cell number increases geometrically at a constant maximal rate characteristic of the strain and of the culture conditions. If cell number is plotted on a logarithmic axis versus time on a linear axis, this phase is represented by a straight line. The log phase is followed by a phase of decelerating increase in cell number, and then by a

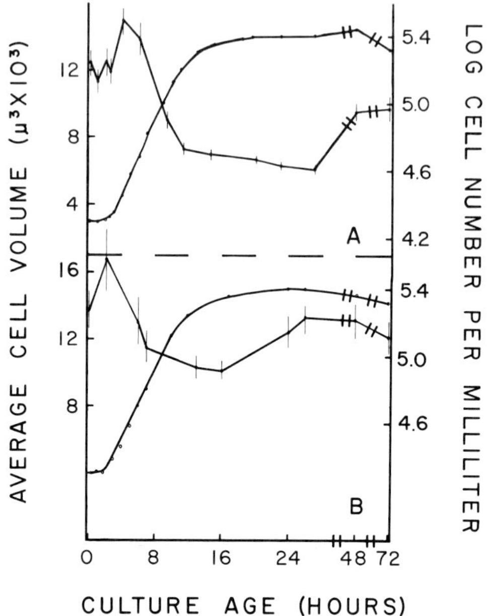

Fig. 12.

Illustration of changes in the average cell volume of *T. pyriformis* HSM during the cell population growth cycle. Perpendicular bars indicate standard error of each value. A, Stirred and aerated culture; B, nonstirred and nonaerated culture. Note that culture-B cells are twice the size of culture-A cells, although the cell number is the same in both cultures. Volume was determined by taking phase-contrast photomicrographs of a living cell flattened to a known thickness under a cover slip; the photographs were then enlarged; the long and short cell axes were measured, and the cell surface area calculated by using the formula for a prolate spheroid. After corrections were made for photographic enlargement, the cell thickness times the cell area gave the absolute cell volume (Cameron, unpublished).

stationary phase in which the cell number is maximal and rather constant for a time; with time the culture enters a decline phase which can be divided into a phase of accelerating cell death and one of logarithmic cell death.

The entire curve is influenced by the relation between the rate of cell birth, the rate of cell death, and cell lysis or the inability of a cell to reproduce. In *Tetrahymena* little or no cell death takes place during early periods following culture inoculation if the inoculum is not too old (i.e., if the inoculum is taken at a point before the decline growth phase). Few if any cells die or are nondividers during the log growth phase (54). Observations made in our laboratory indicate that the stationary phase is a phase of dynamic equilibrium in which cell birth continues but is balanced by cell death and lysis. Culture factors that differentially affect cell birth, death, or lysis can influence the exact shape of the curve. Cell death followed by cell lysis is obviously the major contributor to the decline phase.

The data shown in Fig. 12 are from unpublished experiments carried out in our laboratory. They illustrate some important points about the growth characteristics of

Table 1.
Analysis of Cell Composition under Different Culture Conditions[a]

	A		B	
	At beginning of lag phase	At maximum stationary phase (24 h)	At beginning of lag phase	At maximum stationary phase (24 h)
Log of cell number	4.3	5.4	4.3	5.4
Dry weight (mg/10^6 cells)	4.19	2.40	4.22	4.25
Average cell volume (μm^3)	12,400	6,100	13,700	13,300
Protein (mg/10^6 cells)	2.28	0.83	2.64	2.23
Carbohydrate (mg/10^6 cells)	0.523	0.191	0.495	0.562
RNA (mg/10^6 cells)	0.196	0.207	0.210	0.200

[a]A, Stirred and aerated; B, nonstirred and nonaerated.

Tetrahymena. The culture conditions of Fig. 12A and B were similar except that the culture represented in Fig. 12A was constantly aerated and stirred throughout the experiment, whereas the culture shown in Fig. 12B was neither aerated nor stirred. The curves indicating an increase in cell number are almost identical in all respects (Fig. 12A and B). The curves indicating the mean cell volume in these figures are quite different. In both cultures initial inocula were taken from a 3-day-old nonstirred and nonaerated culture. In both experiments cell volume increased during the lag and accelerating growth phases and then decreased during the log phase. During the deceleration growth phase and during the early stationary phase, cell volume continued to decrease in the stirred and aerated culture but increased significantly in the nonstirred and nonaerated culture. A similar observation was also made by Scherbaum (62). Figure 12A indicates that average cell volume decreased during log phase [this was also observed by Scherbaum (62)]. A state of unbalanced growth is therefore evident.

The product of the cell number and the average cell volume is a rough measure of the total cell mass per milliliter at any one time. Comparison of the cell masses at stationary phase (inferred from Fig. 12A and B) shows that there is twice the volume of protoplasmic material in the nonstirred and nonaerated culture as in the aerated and stirred culture. Table 1 reveals some important information about the control of cell growth and macromolecular synthesis. There is approximately twice as much cell mass and cell volume in an average nonstirred and nonaerated cell. There is about three times as much carbohydrate and protein in these cells. The cellular content of RNA, however, is almost the same in the cells grown under the two different culture conditions. Thus the amount of RNA per cell and the number of cells per milliliter are the only common parameters found in the two different culture conditions. We return to these comparisons later in the chapter.

Much of the work reported below was influenced by the earlier reports of Phelps (51,52). In his studies on the effect of different concentrations of nutrient material upon growth of *Tetrahymena*, Phelps (52) found that, within relatively wide limits, the growth

Table 2.
Effect of Nutrient on Cell Growth Yield and on the Efficiency of Conversion of Nutrient to Protoplasm[a]

Concentration	Maximum cell yield per milliliter[b]	Dry weight per 10^9 cells (mg)	Efficiency (%)[c]
Proteose–peptone with liver extract[d]:			
0.05	9,000	—	—
0.10	90,000	0.82	0.67
0.25	212,000	1.16	8.90
0.50	346,000	1.33	8.40
1.00	432,000	2.63	10.30
2.50	608,000	3.85	8.50
3.50	486,000	4.25	5.30
5.00	278,000	4.26	2.20
10.00	Cell death	—	—
Proteose–peptone without liver extract:			
0.05	394	—	—
0.10	909	—	—
0.20	2,264	—	—
0.30	3,098	—	—
0.50	6,436	—	—
1.00	13,482	—	—
2.50	28,300	1.67	1.9
5.00	24,360		
One percent proteose–peptone with different concentrations of liver extract:			
0.001	85,730	2.30	1.9
0.005	168,660	2.25	3.8
0.01	220,500	2.20	4.8
0.05	354,500	2.52	8.4
0.10	420,000	2.80	10.6
0.50	394,500	2.90	7.5
Liver extract without proteose–peptone:			
0.060	35,600	1.12	6.6
0.125	83,200	1.32	8.9
0.25	138,100	1.44	7.9
0.50	258,000	1.76	8.7
1.00	424,000	2.65	11.2

[a]All cultures were grown at 25°C ± 1°C in 250-ml Ehrlenmeyer flasks containing 50 ml of medium. In all cases at least three culture transfers were made before readings were taken. A Coulter particle counter Model F was used for cell counts (counting error was less than 3%).

[b]Cell counts were made at various times throughout stationary phase, and only the maximum cell count value is listed here.

[c]Efficiency was determined by calculating the conversion of mass of nutrient (expressed as dry weight per milliliter in uninoculated culture medium) to mass of protoplasm (expressed as dry weight per milliliter of washed cells at maximum stationary phase). The initial dry weights of the cell inocula were so small that they could be ignored as a correction factor in the calculations.

[d]A stock solution of 10% proteose–peptone no. 3 plus 1% liver extract concentrate (Nutritional Biochemicals Corporation) was made and sterilized; it was then diluted with sterile distilled water to the various concentrations listed in the table.

yield, expressed in number of cells per milliliter in maximum stationary phase, is directly proportional to the concentration of nutrient material. Comparison of cells grown in proteose–peptone plus liver extract to cells grown in proteose–peptone alone (Table 2) indicate that excretory waste products of *Tetrahymena* are probably not the limiting factor on the final cell yield attained in stationary phase. Indeed, we have found that the addition of vitamins to stationary-phase cells grown in proteose–peptone alone causes another cycle of cell replication which presumably would not occur if waste products were limited to the number of cells in the culture. Cell concentrations even higher than those listed in Table 2 can be produced by growing *Tetrahymena* in a medium of 2% proteose–peptone, 0.2% glucose, 0.1% yeast extract, and 0.003% sequestrine (13% iron EDTA) with aeration. Cell counts of more than a million cells per milliliter have been attained by this procedure (Rosenbaum, personal communication). Parenthetically, Conner and Cline (18) report that iron deficiency is often limiting to growth of *Tetrahymena*. Because of the total number of cells present under different growth conditions is strikingly different, the physical crowding of cells or concentration of excretory waste products is probably seldom limiting to growth yield. Phelps reached similar conclusions (51,52).

What factors influence the duration of the various growth curve phases? Phelps (51) showed that within wide limits the size of the inoculum (ranging from 10 to 1200 cells/ml) had no detectible effect on the duration of the lag phase. In comparable studies we found no effect of the size of the inoculum on lag-phase duration (in medium inoculated at concentrations of 5000 to 42,000 cells/ml). Thus unlike the situation in bacteria, fungi, and chick cells (25,50,60,73), the size of initial inoculum has no effect on the duration of the lag period in *Tetrahymena*. That the size of the inoculum does not alter the duration of this adjustment phase indicates that the cells are adjusting internally rather than causing a marked "conditioning" of the medium which must precede cell multiplication. Conditioning of the growth medium is known to occur with bacteria, fungi, and chick cells, and this indicates a fundamental difference between *Tetrahymena* and these other cell types.

Both Phelps (51) and Prescott (56) demonstrated that the age of the inoculum greatly influences the duration of the lag phase. Prescott grew a mass culture of *Tetrahymena* and took samples at different times throughout the growth phases. He found that the duration of the lag phase was negligible during early log phase and increased almost linearly during the remainder of the log, growth deceleration, and stationary phases (see Fig. 13). There is a clear change in the physiological state of *Tetrahymena* as a function of culture age even during log phase.

As might be anticipated, inoculation with a small inoculum permits a longer logarithmic phase than when large inocula are employed (51). However, cells inoculated in low concentrations and cells inoculated in high concentrations eventually reach the same final cell number in maximum stationary phase. Phelps reasons that, because the cells live in the culture medium longer in the case of the low-concentration inoculum, the nutrient concentration or value of the medium is more of a limiting factor in determining the number of cells attained in a growth population than are other factors such as waste products or oxygen deficiency.

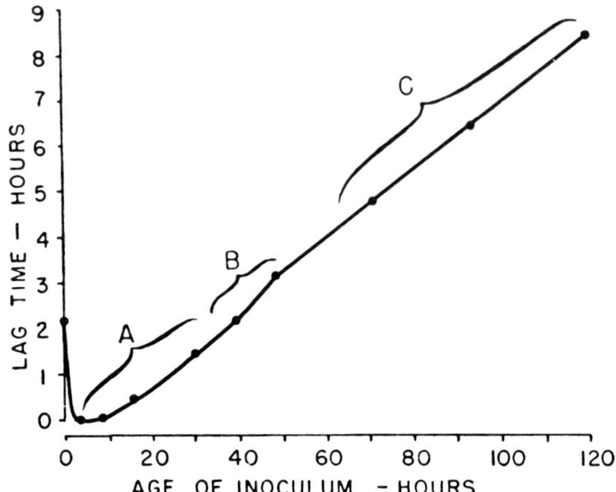

Fig. 13.

Relation between duration of lag time and age of inocula in *T. pyriformis* HSM. Inocula were taken from cultures in log phase (A), phase of decelerating growth (B), and stationary phase (C). [From Prescott (54); courtesy of Academic Press.]

The data listed in Table 2 indicate that the maximum yield of protoplasm as measured in milligrams of dry weight per milliliter of culture media, or the maximum yield as measured in total cell number per milliliter, occurs in a 2.5% proteose–peptone medium enriched with liver extract. This growth condition, of all others tested, gave the largest yield of protoplasm. When the quantity of nutrient used (expressed in dry weight per milliliter of uninoculated growth medium) is compared to the maximum yield or harvest (expressed in dry weight of protoplasm per milliliter of growth medium), the efficiency of the conversion process is 8.5%. In this regard, however, a 1% enriched proteose–peptone system is even more efficient (about 10%). Thus from the standpoint of efficiency, one can obtain the most protoplasm for the least nutrient in a 1% enriched proteose–peptone system. The maximum yield of protoplasm is less than half that of a 2.5% enriched proteose–peptone system, however. The lowest and highest concentrations of enriched proteose–peptone nutrient have low efficiencies. Cells grown in liver extract alone have a good yield efficiency in several concentrations.

Comparisons of all of the media systems listed in Table 2 show that proteose–peptone medium, without enrichment with liver extract, does not give a good yield of cells. Enrichment of proteose–peptone with even a small amount of liver extract markedly increases the yield. Stationary-phase cells grown in 3.5 and 5.0% enriched proteose–peptone have a higher dry weight of protoplasm than cells grown under any other growth condition listed in the table.

The maximum cell number yield drops off in enriched proteose–peptone medium after the 2.5% concentration. Total cell death occurs at a 10% concentration. This drop in yield most likely represents an osmotic problem. In this regard Browning (5) showed that the total number of *Tetrahymena* that can be supported by the medium decreases as the osmotic pressure increases (addition of sucrose), but that some of the osmotic effect can be antagonized by increasing the calcium ion concentration.

IV. GROWTH IN THE ABSENCE OF CELL DIVISION

As pointed out earlier in this chapter, it is a generalized phenomenon that average cell volume increases when cell generation time is less than optimal. This occurs, for example, at temperatures that arc sub- or supraoptimal for cell division. Thus it follows that lowering the temperature or raising the temperature away from that for the optimum generation time tends to increase the size of *Tetrahymena*. The uncoupling of the cell growth process from cell division was reported by Scherbaum and Zeuthen (68). They believed that a series of six to eight supraoptimal heat shocks might suppress events leading to cell division but permit events leading to cell growth. (For explanation see refs. 66 and 80.)

Schmid (69) demonstrated that a single shift in temperature from 28°C to a supraoptimal temperature is followed by a brief period of continued cell division, and then by a prolonged period of blocked cell division. Cell growth in size continued at an exponential rate during this period of cell division blockage. Schmid also demonstrated that a series of six to seven high-temperature shocks (30 min each) also cause exponential cell growth in the absence of cell multiplication.

In a similar manner Scherbaum and Zeuthen (68) and Zeuthen (79,80) demonstrated that a shift of cells to a low temperature blocks cell division but permits cell growth. The blockage of cell division at the lower temperatures is not permanent but permits cell growth to continue for some time and oversized cells are produced.

Browning (5) describes osmotic conditions that cause cell growth in the absence of cell division. Osmotic pressure was raised by increasing the sucrose concentration. During this treatment *Tetrahymena* often die or are stalled during the process of cell cleavage. Many of the cells that do survive become large or "monster" cells, that is, growth continues in the absence of division.

Weis (77) showed that the volume of starved *Tetrahymena* increased from 1900 to 20,000 μm^3 upon refeeding and prior to the first cell division.

Perhaps the most commonly observed example of cell growth in the absence of cell division in *Tetrahymena* occurs during the lag period following inoculation of aging cells into fresh medium (see Fig. 12A and B as an illustration of this phenomenon). This may simply represent another example of the starvation-refeeding phenomenon.

V. CELL DIVISION IN THE ABSENCE OF GROWTH, INCLUDING A THEORETICAL APPROACH TO INDUCED CELL SYNCHRONY

The most extreme case of cell division in the absence of growth is that found in cleaving eggs (1). Let us not forget, however, that the extreme degree of unbalanced growth observed in cleaving eggs is preceded by an extreme case of cell growth in the absence of cell division, which occurs during oogenesis (22). Fully aware of this phenomenon in eggs, Zeuthen (80) and Scherbaum and Zeuthen (68) set out to dissociate experimentally growth and cell division in *Tetrahymena*. That they were successful in this pursuit is now well documented and is also a tribute to their insight into the problem.

A series of six to seven heat shocks to *Tetrahymena* causes cell growth in the absence

Table 3.
Changes in Starved *Tetrahymena*[a]

Duration of starvation (h)	Average cell volume ± S.E. (μm^3)	RNA per 10^6 cells (μg)	Increase in cell number (%)	Average number of cell divisions of initial cell
0	15,726 ± 954	394	0	0
4	7,360 ± 396	—	135	1.12
8	4,741 ± 227	—	215	1.36
24	—	—	1140	3.42
96	1,233 ± 75	55	1140	3.42

[a]The cells were starved in a phosphate buffer (15). Cells were in early log phase at the beginning of the starvation period; they were washed three times by gentle hand centrifugation and resuspension; care was taken to avoid cell breakage.

of division. After the last heat shock, the cells are considerably larger than log-phase cells (80). Most of these large cells divide synchronously in about 60–90 min either in the presence or in the absence of nutrient medium. Apparently, the heat shocks allow growth events to continue but interfere with division-directed event(s), so that at the end of the last heat shock all the cells begin the same sequence of steps leading to cell division. Even in nonnutrient, sterile, inorganic buffer, these large cells divide in synchrony at least two or three times (26,27). Because *T. pyriformis* is known to require a complex organic medium for saprophytic growth and reproduction, the ability of the cells to divide in the absence of required nutrient was therefore attributed to the utilization of a surfeit of essential endogenous intracellular reserves which accumulated during the heat shock treatment. In confirmation of this idea, Zeuthen (80) cites biochemical data on log-phase and heat-shocked *Tetrahymena,* which show that the cell mass and the quantities of major macromolecules in the cells more than double during heat shocks, that is, protein, RNA, and DNA content; dry weight; and cell volume (63,80).

We carried out a series of experiments that demonstrated the ability of nonsynchronized *Tetrahymena* to carry out cell division in the absence of growth. Our approach was to remove late lag-phase, exponentially multiplying and stationary-phase cells from their rich nutrient medium and resuspend them in nonnutrient buffer. Small drop cultures of these cells were followed to determine cell proliferation under starvation conditions. Examination of the data from Table 3 shows that suspension of *T. pyriformis* HSM in nonnutrient phosphate buffer permits cell division. Table 3 indicates that cells that are just finishing lag phase continue for an average of 3.4 divisions before cell division ceases. Analysis of 22 cell clones from this optimal culture indicates that an individual cell isolate tends to give rise to approximately 4, 8, or 16 daughter cells representing two, three, and four cell divisions, respectively. These results suggest that the number of cells produced by a single cell isolate is not random but is determined by some physiological state of the parent cell isolate. Is there some cellular parameter that can give us an indication of the physiological state of the parent cell and therefore enable us to predict how many cells can be produced? One such parameter might be cell size as measured by cell volume. Indeed, as indicated in Table 3, cell volume seems to

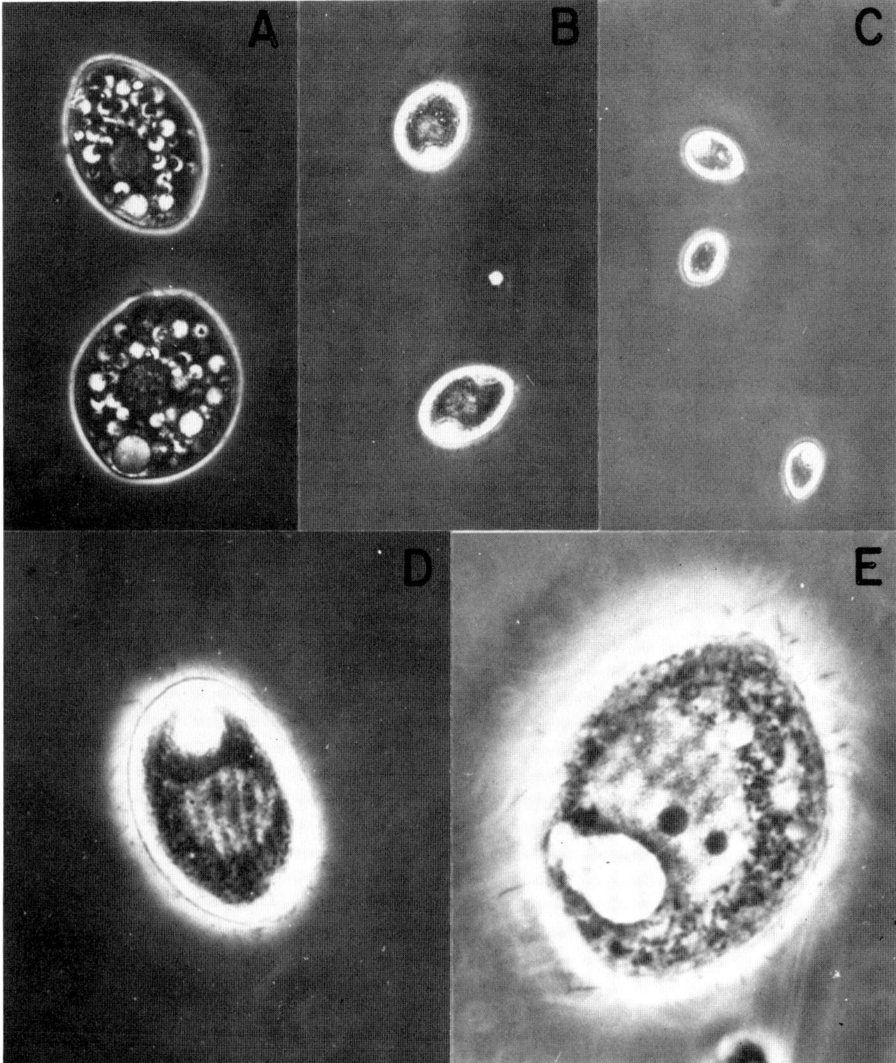

Fig. 14.

Phase-contrast photomicrographs of starved *Tetrahymena*. Duration of starvation: 0 h (A); 11 h (B); 24 h (C). (A), (B), and (C) are all at the same magnification; (D) and (E) are enlarged photographs of 24-h starved cells and show a proportionate increase in the nuclear cytoplasmic volume ratio as compared to the cells in (A). (Cameron, unpublished.)

halve after each cell doubling, so that the final cell volume of the starved cells is directly related to the initial cell volume of the nonstarved parent cells when one considers the average number of cell divisions that have occurred during the starvation period. This finding indicates that the volume of protoplasm in the culture did not change during starvation, but only the number of cells. Figure 14 illustrates the striking change in cell size that takes place during the starvation period. Closer cytological observations

Table 4.
Effect of Culture Age on Proliferative Response When Suspended in Phosphate Buffer

Age of culture (h)	Phase of culture growth cycle	Time to reach 50% increase in cell number (h)	Final increase in cell number (%)
6	Late log–early log	1.3	1140
37	Early stationary	3	190
51	Maximum stationary	4	170
97	Decline	5	155
102	Later decline	7	140

(Fig. 14A, D, and E) of the starved and nonstarved cells indicate that most of the decrease in cell volume is due primarily to a decrease in cytoplasmic volume and not in nuclear volume. For example, measurements indicate that 3–7% of the total cell volume of the nonstarved cells is occupied by the nucleus, whereas the nucleus occupies 20–33% of the total cell volume in the much smaller starved cell (for examples, compare Fig. 14A with Fig. 14D and E). In other words, nuclear materials are being replicated, conserved, and maintained during starvation at the expense of cytoplasmic reserves. Based on counts of cell number and on average cell volume measurements (see Table 3), it appears that the nucleoplasm has increased severalfold in the culture at the expense of cytoplasm, while the net amount of protoplasm stays almost constant.

From the foregoing description it becomes evident that cell division in *Tetrahymena* is not determined by cell size but by biosynthetic events which are related more directly to cell and nuclear division.

The data in Table 4 show that in progressively older cultures the initial adaptation of the cells to the nonnutrient buffer takes longer before the first cell divisions are initiated. There is also a marked decrease in the number of cell divisions that these cells can undergo; however, even late stationary and decline cells are able to divide. That late stationary-phase cells can divide in the nonnutrient buffer but did not divide as rapidly in the nutrient culture medium suggests that some of the division-restraining mechanisms imposed in the stationary-phase cultures have been reversed by placing the cells in the nonnutrient buffer.

The lag period is believed to be a measure of the time the cells need to retool themselves for cell division events. The early log-phase cells showed no lag phase. The ultimate number of divisions that the cell can undergo is a measure of both the amount of retooling and the net endogenous reserves of the cell at the beginning of starvation. The cell size data illustrated in Fig. 12A and B (and see refs. 19,28,49) show that cell size decreases progressively in aging (declining) cultures. The data taken together allow us to draw certain conclusions. The age of the cell culture determines the duration of the lag phase in buffer, and the duration of the lag phase is an indicator of the rate and amount of retooling that must occur before cell division can occur. This lag period is therefore a good indicator of the physiological state of the average cell in an aging culture. The initial size of the starved cells (endogenous reserves that a cell contains) probably determines in a rather direct manner the number of cell divisions a cell can undergo. It also appears, however, that the limiting factor that determines the final cessation of cell division in a starved cell is the inability of the cell's nuclear material to undergo reductive divisions which would decrease the nuclear volume and DNA content of the cell below an absolute minimal value (equivalent to the G_1 condition). McDonald (personal communication) found that starved cells do not lower the DNA content below the G_1 value. Perhaps there are not enough cytoplasmic reserves to bring about another round of nuclear and cortical replication.

In general terms, it is as if the starved cells have shifted their objective to the maximum production of cells, each of which contains complete sets of the nuclear genetic material. Barring encystment or a dormant state, which are not known to occur in *T. pyriformis*, it appears that the production of numerous cells endowed with complete sets of genetic material gives them the greatest survival advantage.

It becomes clear that cell growth does not of itself cause cell division; however, the opposite may very well be true. That is, nuclear replication followed by cell division may indeed double the growth potential. It seems that starvation experiments with *Tetrahymena* offer a system in which the biosynthetic pathways leading to cell growth are turned off or limited in favor of those biosynthetic pathways that lead to cell reproduction. Thus this system should prove favorable for the study of cell reproduction without the added complications of cell growth.

VI. IS *TETRAHYMENA* EVER IN A STATE OF BALANCED GROWTH?

In the present context, *balanced growth* means that every component of an individual cell doubles in quantity between successive cell divisions (16). If during one or more cell division cycles the cells grow larger or smaller, and the average cellular composition changes, then an *unbalanced growth* state occurs.

Enteric bacteria have been maintained in balanced growth states. Maaløe and Kjeldgaard (45) analyzed bacteria grown in several different balanced growth states. These analyses have enabled them to establish definite correlations between chemical composition and cell reproduction rate. This type of correlation indicates that the size of the protein-synthesizing machinery of the bacteria (ribosomes and polysomes) seems to be economically and effectively controlled under balanced growth conditions. From a practical point of view, balanced growth is achieved in mass culture by keeping the population density reasonably low and constant by dilution. Under the conditions described for bacteria, one can harvest large enough quantities of cells to verify and measure a doubling of quantities of cellular components as a function of the cell reproduction rate in the culture. However, we have not been able to obtain balanced growth of *Tetrahymena* in population densities high enough to yield sufficient amounts of cells for most biochemical analyses (Fig. 12A and B). It is likely that balanced growth can be attained by keeping cell density low, below 50,000 cells/ml (in enriched proteose–peptone media), while maintaining a constant and maximum cell reproduction rate. The technical aspects of such an approach, however, make it a difficult task to accomplish in mass cultures.

As an alternative approach, one may be able to attain balanced growth states by use of a chemostat, although the cost of maintaining mass cultures of *T. pyriformis* in a minimal defined culture medium is prohibitive.

From the above it should be clear that a balanced growth state for *Tetrahymena* can be effectively accomplished by frequent subcultures of logarithmically growing cells. One can, with relative ease, produce cells in a balanced growth state in quantities sufficient for cytological and cytochemical studies, but much more effort is required to scale up the growth conditions so as to attain sufficient quantities for chemical analysis. Making correlations between chemical composition and cell reproduction rate under a variety of balanced growth conditions will be much more difficult for *Tetrahymena* than it is for enteric bacteria. Hamburger and Zeuthen (27) reported balanced growth under one growth condition experiment, and Leick (42) studied balanced cultures of *Tetrahymena* to determine the dependency of cellular protein and nucleic acid composition on the growth rate of the culture. However, Leick's report does not make a convincing case for his assumption that the cells were indeed in a state of balanced growth in each of his three different growth rate experiments. Johnson (34), working with fission yeast and reviewing the literature on other microorganisms, concludes that growth of cells in culture is usually unbalanced, and that it should be assumed that growth under any circumstance is unbalanced until it is demonstrated not to be.

We can nevertheless study the patterns of macromolecular composition and synthesis during unbalanced but carefully controlled growth situations in mass cultures of *Tetrahymena*. An unbalanced growth situation is generally produced by increasing ("shifting up") or decreasing ("shifting down") the richness of the nutrient medium. The next section deals with this type of approach.

VII. THE CONTROL OF GROWTH AND MACROMOLECULAR SYNTHESIS

The macromolecular composition of *Tetrahymena* is in general similar to that of other eucaryotic animal cells. As is indicated in Table 1, about 50% of the dry weight of *Tetrahymena* is protein. In general, RNA content is about 1/10 that of protein or about 5% of the dry weight of the cell. It has been found that DNA content is likewise about 1/10 the cellular content of RNA (8,80); DNA is therefore a minor component of the total cell mass. As Table 1 indicates, the carbohydrate (primarily glycogen) content of *Tetrahymena* is quite variable and depends on the growth condition of the cells. In general, it ranges from 4% or less to as much as 50% of the cells' dry weight (64,65). Extraction of the lipids from *Tetrahymena* indicate that lipids generally constitute about 15% of the dry weight of the cell (4,21). Summing up the quantities of these major groups of compounds indicates that we can account for almost all of the dry weight of *Tetrahymena* in these major molecular classes.

Review of the literature (42,63,66,80) along with the data listed in Table 1, gives a clear indication that the macromolecular composition of *Tetrahymena* is very dependent on the growth conditions.

How then can we obtain a unifying idea of how a cell controls its macromolecular synthesis and growth rate in response to the environment? Two approaches to this problem can be used. First, balanced growth can be obtained under several growth conditions, then each steady-state condition can be defined by several cellular parameters such as mass, protein content, and RNA and DNA content. Listing the parameters and the ratios

Table 5.

Temporal Comparisons of Increase in Cellular Parameters between Bacteria and *Tetrahymena* Following a Shift-up in the Growth Medium

	Time before increase is noticed (min)			
	RNA	Protein	DNA	Cell number
Bacteria[a]	Immediate	5–10	15	75
Tetrahymena[b]	Less than 15 min	75	150	360

[a]From Maaløe and Kjeldgaard (45).
[b]From the pyrimidine replacement experiment of Cameron (8).

of these cell components under different steady states of growth has already proved to be a rewarding approach in several bacterial species, however, the difficulty of attaining balanced or steady-state growth situations in *Tetrahymena* makes this approach difficult. The studies on balanced growth cultures of bacteria at different growth rates can be summarized as follows. Variations in mass per cell are due to changes in the number of nucleoids per cell and in the mass per nucleoid. An increase in the ribosomes (RNA) at higher rates of growth can account for the increase in mass per nucleoid. The findings indicate that the protein synthetic rate per individual ribosome is constant at all growth rates.

Another approach to the problem is to investigate the temporal pattern of change in a cell's macromolecular composition as the cell undergoes a transition between growth states. This approach has also been used in bacteria, and the results are summarized as follows. Synthesis of cellular RNA is the first obvious macromolecular component to respond to a change in richness of the growth medium. The rates of protein and DNA synthesis change later. The last to change is the cell division rate (45,48,61). It seems that the size of the protein-synthesizing machinery of the cell is economically and effectively controlled.

How does *T. pyriformis* respond to a shift in growth rate and how does its response compare to that of bacteria?

Tetrahymena has an absolute requirement for pyrimidines in its growth medium. Several reports (8,9,41,76) indicate that removal and replacement of pyrimidines have a profound effect on the macromolecular events in *Tetrahymena*. These reports taken together are summarized in Table 5. Clearly, the pattern of macromolecular events is similar to that for bacteria. Cameron et al. (9) showed that during the early pyrimidine replacement period there is an increase in the cytoplasmic protein-synthesizing machinery (polysomes) prior to increased protein synthesis. That *Tetrahymena* polysomes are actually sites of protein synthesis was also demonstrated in this study.

For technical reasons a shift-up experiment is easier to perform than a shift-down experiment (45), and we therefore concentrate on the shift-up situation. A temporal comparison between bacteria and *Tetrahymena* during a shift-up is listed in Table 5. Although the macromolecular pattern is similar in procaryotic and eucaryotic cell types, the absolute time scales are very different (see the Addendum for further comment). The bacterium responds much faster than *Tetrahymena*.

It seems important to ask, What type of factors account for these temporal differences? One factor may be related to the absence of a nuclear envelope in bacteria and its presence in *Tetrahymena*. Prescott (58) showed that 10–12 min elapse between the time of RNA synthesis in the nucleus of *Tetrahymena* and its appearance in the cytoplasm. Thus at least 10–12 min are required between genome transcription in the nucleus and the appearance of RNA in the cytoplasm where protein synthesis occurs. No such structural barrier exists in bacteria and any newly transcribed RNA is probably immediately available to the cytoplasm. Another factor may be the accessibility of the chromatin for gene expression. In *Tetrahymena* the nuclear DNA is complexed with basic proteins (2), whereas it is generally known that bacterial DNA is not complexed with basic protein. This may mean that the genome is readily accessible for transcription in the case of bacteria but not as readily accessible in the case of *Tetrahymena*. There is also some indication that the rate of translation of a genetic message (cistron) in the case of procaryotes (called the "transit time") is in the order of 5 s in *E. coli* (24), whereas in eucaryotic (mammalian) cells it is about 25 s (71). Even the rate of duplication of molecular DNA is known to be faster in bacteria than in eucaryotic (mammalian) cells (30 μm/min versus 2.5 μm/min) (6,7,32). The temporal differences listed in Table 5 may therefore be accounted for by any or all of the above factors. Obviously, these factors may include the rate of gene transcription and/or the migration of genetic information from the genome to the site of translation, as well as the rate of translation of the transcribed information itself.

It should be apparent from this review that *Tetrahymena* is a eucaryotic cell type which is both easy to grow and easy to manipulate experimentally. It is also apparent that much is already known about its growth characteristics. Furthermore, *Tetrahymena* promises to be of even greater usefulness in our efforts to understand the control of cellular growth and macromolecular synthesis in higher cell types.

In this chapter the growth characteristics of *T. pyriformis* have been emphasized. The main concern has been the dynamic characteristics of the living cell. We feel strongly that to obtain a proper understanding of or perspective on the living cell one should first examine and experiment with the intact cell in its environment before turning to a reductionist approach involving a study of the lower and smaller units of its organization.

ADDENDUM

A starvation-refeeding regime which leads to cell synchrony has been reported (Cameron, I. L. and Jeter, J. R., Jr., 1970, J. Protozool. 17:429). The use of this synchrony system revealed that increased protein synthesis occurs prior to increased RNA synthesis, which is a sequence of macromolecular events different from those observed in studies reported in Section VII (Cameron, I. L., Griffin, E. E. and Rudick, M. J., 1971, Exp. Cell Res. 65:265). Thus it appears that, unlike the situation in bacteria, the macromolecular response of *T. pyriformis* to a new environment is not fixed but is dependent upon the cell's previous environmental history.

ACKNOWLEDGMENTS

The technical assistance of M. Agi and D. Hicks is gratefully acknowledged.

Thanks are extended to Nicholas Terebey who carried out some of the unpublished starvation experiments, and to Guy Johnson, III, and David Golden.

Thanks are also due to Drs. G. G. Holz, Jr., B. B. McDonald and D. S. Nachtwey for helpful suggestions and criticisms concerning this chapter.

The unpublished studies described in this article were supported in part by USPHS grant no. GM 14442 and in part by NSF Grant no. GB3158.

REFERENCES

1. Agrell, I. 1964. In E. Zeuthen, ed., Synchrony in cell division and growth, John Wiley, New York, p. 39.
2. Alfert, M. and Goldstein, N. O. 1955. J. Exp. Zool. 130:403.
3. Allfrey, V. G., Mirsky, A. C., and Stern, H. 1955. Adv. Enzymol. 16:411.
4. Allison, B. M. and Ronkin, R. R. 1967. J. Protozool. 14:313.
5. Browning, I. 1949. J. Exp. Zool. 110:441.
6. Cairns, J. 1963. J. Mol. Biol. 6:208.
7. Cairns, J. 1966. J. Mol. Biol. 15:372.
8. Cameron, I. L. 1965. J. Cell Biol. 25:9.
9. Cameron, I. L., Cline, G. B., Padilla, G. M., and Miller, O. L., Jr. 1966. National Cancer Inst. Monogr. 21:361.
10. Cameron, I. L. and Guile, E. E., Jr. 1965. J. Cell Biol. 26:845.
11. Cameron, I. L. and Nachtwey, D. S. 1967. Exp. Cell Res. 46:385.
12. Cameron, I. L., Padilla, G. M., and Miller, O. L., Jr. 1966. J. Protozool. 13:336.
13. Cameron, I. L. and Prescott, D. M. 1961. Exp. Cell Res. 23:354.
14. Cameron, I. L. and Stone, G. E. 1964. Exp. Cell Res. 36:510.
15. Cameron, I. L. and Terebey, N. 1967. J. Protozool. 14(suppl.):70. (Abstra.)
16. Campbell, A. 1957. Bacteriol. Rev. 21:263.
17. Cleffmann, G. 1967. Z. Zellforsch. Mikrosk. Anat. 79:599.
18. Conner, R. L. and Cline 1964. J. Protozool. 11:486.
19. Corbett, J. J. 1958. Exp. Cell Res. 15:512.
20. Elliott, A. M. 1933. Biol. Bull. 65:45.
21. Everhart, L. P., Jr., and Ronkin, R. R. 1966. J. Protozool. 13:646.
22. Fautrez-Firlefyn, N. and Fautrez, J. 1967. Int. Rev. Cytol. 22:171.
23. Flickinger, C. J. 1965. J. Cell Biol. 27:519.
24. Goldstein, A., Goldstein, D. B., and Lowney, L. I. 1964. J. Mol. Biol. 9:213.
25. Graham-Smith, G. S. 1920. J. Hyg. 19:133.
26. Hamburger, K. and Zeuthen, E. 1957. Exp. Cell. Res. 13:443.
27. Hamburger, K. and Zeuthen, E. 1960. C. R. Trav. Lab. Carlsberg, 32:1.
28. Harding, J. P. 1937. J. Exp. Biol. 14:431.
29. Holz, G. G., Jr. 1957. Anat. Rec. 128:566.
30. Holz, G. G., Jr., Erwin, J. A., and Davis, R. J. 1959. J. Protozool. 6:149.
31. Howard, A. and Pelc, S. R. 1953. Heredity, suppl. 6:261.
32. Huberman, J. A. and Riggs, A. D. 1968. J. Mol. Biol. 32:327.
33. James, T. W. and Read, C. P. 1957. Exp. Cell Res. 13:510.
34. Johnson, B. F. 1968. Exp. Cell Res. 49'59.
35. Kidder, G. W. 1941. Biol. Bull. 80:50.
36. Kidder, G. W. 1967. In M. Florkin and B. J. Scheer, eds., In G. W. Kidder, ed., Chemical zoology, vol. I, Academic Press, New York, pp. 93–159.
37. Kidder, G. W. and Dewey, V. C. 1944. Biol. Bull. 87:121.
38. Kidder, G. W. and Dewey, V. C. 1951. In A. Lwoff, ed., Protozoa, vol. I, Academic Press, New York, pp. 323–400.

39. Kimball, R. F., Casperson, T. O., Svensson, G., and Carlson, L. 1959. Exp. Cell Res. 17:160.
40. Koch, A. L. 1966. J. Tehoret. Biol. 12:276.
41. Lederberg, S. and Mazia, D. 1960. Exp. Cell Res. 21:590.
42. Leick, V. 1967. Co. R. Trav. Lab. Carlsberg 36:113.
43. Loefer, J. B. 1952. J. Morphol. 90(3):407.
44. Lovlie, A. 1963. C. R. Trav. Lab. Carlsberg 33:337.
45. Maaloe, O. and Kjeldgaard, N. O. 1966. Control of macromolecular synthesis. A study of DNA, RNA, and protein synthesis in bacteria, W. A. Benjamin, New York.
46. Mackenzie, T. B., Stone, G. E., and Prescott, D. M. 1966. J. Cell Biol. 31:633.
47. Nachtwey, D. S. and Cameron, I. L. 1968. In D. M. Prescott, ed., Methods in Cell physiology vol. III, Academic Press, New York, pp. 213–259.
48. Neidhardt, F. C. and Fraenkel, D. G. 1961. Cold Spring Harbor Symp. Quant. Biology 26:63.
49. Ormsbee, R. A. 1942. Biol. Bull. 82:423.
50. Penfold, W. J. 1914. J. Hyg. 14:215.
51. Phelps, A. 1935. J. Exp. Zool., 70:109.
52. Phelps, A. 1936. Exp. Zool. 72:479.
53. Phelps, A. 1946. J. Exp. Zool. 102:277.
54. Prescott, D. M. 1957. Exp. Cell Res. 12:126.
55. Prescott, D. M. 1957. J. Protozool. 4:252.
56. Prescott, D. M. 1958. Physiol. Zool. 31:111.
57. Prescott, D. M. 1959. Exp. Cell Res. 16:279.
58. Prescott, D. M. 1961. In T. W. Goodwin and O. Lindberg, eds., biological structure and function, vol. II, Academic Press, New York, pp. 257–536.
59. Rosenbaum, J. N., Erwin, J., Beach, D., and Holz, G. G., Jr. 1966. J. Protozool. 13(4):535.
60. Rubins, H. 1967. In B. D. Davis and L. Warren, eds., The specificity of cell surfaces, Prentice-Hall, Englewood Cliffs, New Jersey. pp. 181–194.
61. Schaechter, M. 1961. Cold Spring Harbor Symp. Quant. Biol. 26:53.
62. Scherbaum, O. 1956. Exp. Cell Res. 11:464.
63. Scherbaum, O. H. 1963. In L. Levine, ed., The cell in mitosis, Academic Press, New York, pp. 125–157.
64. Scherbaum, O. H. and Levy, M. 1961. Pathol. Biol. Sem. Hop. 9:514.
65. Scherbaum, O. H. 1964. In E. Zeuthern, ed., Synchrony in cell division and growth, John Wiley, New York, p. 185.
66. Scherbaum, O. H. and Loefer, J. B. 1964. In S. Hutner, ed., Biochemistry and physiology of protozoa, vol. III, Academic Press, New York, pp. 9–59.
67. Scherbaum, O. and Rasch, G. 1957. Acta Pathol. Microbiol. Scand. 41:161.
68. Scherbaum, O. and Zeuthen, E. 1954. Exp. Cell Res. 6:221.
69. Schmid, P. 1967. Exp. Cell Res. 45:460.
70. Schmid, P. 1967. Exp. Cell Res. 45:471.
71. Stanners, C. P. 1968. Biophysical J. 8:231.
72. Summers, L. G. 1963. J. Protozool. 10:288.
73. Sussman, A. S. and Halvorson, H. O. 1966. Spores – Their dormancy and germination. Harper and Row, New York, p. 147.
74. Thormar, H. 1962. Exp. Cell. Res. 27:585.
75. Thormar, H. 1962. Exp. Cell. Res. 28:269.
76. Watanabe, Y. 1963. Jap. J. M. ed., Sc. Biol. 16:107.
77. Weis, D. 1954. J. Protozool. 1(suppl.) (Abstr. 50).
78. Williams, F. M. 1967. J. Theoret. Biol. 15:190.
79. Zeuthen, E. 1958. Adv. in Biol. Med. Phys. 6:37.
80. Zeuthen, E. 1964. In E. Zeuthen, ed., Synchrony in cell division and growth, John Wiley, New York, p. 99.
81. Stone, G. E., and Cameron, I. L. 1964. In D. M. Prescott, ed., Methods in cell physiology, vol. 1, Academic Press, New York, pp. 127–140.
82. McDonald, B. B. 1958. Biol. Bull. 114:71.

CHAPTER 8

Effects of Some Environmental Factors on the Biochemistry, Physiology, and Metabolism of *Tetrahymena**

Michael R. Levy

Department of Biological Sciences
Southern Illinois University
Edwardsville, Illinois

I. INTRODUCTION

Tetrahymena, in its natural habitat, is subjected to extreme fluctuations, both seasonal and diurnal, in temperature, pH, amounts and types of nutrients, and concentrations of dissolved gases. These changes can have profound effects on the activities and the makeup of the cell, and the organism must be able to adjust to these changes if it is to be successful. Many investigators have, intentionally or otherwise, studied the response of the cell to environmental stresses and changes. These responses can be seen at the physiological, biochemical, and morphological levels. *Tetrahymena* has been grown in a variety of media, at many temperatures, and under conditions that favor a rapid or a slow exchange of gases. Consequently, great differences have been reported in the levels of enzymes or reserve materials, the activities of metabolic pathways, and the appearance and abundance of cell organelles. One purpose of this chapter is to present and discuss some of the cellular changes that occur in response to environmental variations. Inherent in this presentation is the hope that those who use this organism will be aware that even seemingly insignificant changes in culture conditions can greatly

*The following abbreviations are used in this chapter. AP, Acid phosphatase; cAMP, cyclic 3′,5′-adenylic acid; FDPase, fructuose-1,6-diphosphatase; GT, generation time; IDH, isocitrate dehydrogenese (TPN-dependent); IL, isocitrate lyase; LAO, lactate oxidase; LDH, lactate dehydrogenase; MDH, malate dehydrogenase; MS, malate synthase; PEP, phosphoenolpyruvate; PK, pyruvate kinase; RER, rough endoplasmic reticulum; SDH, succinate dehydrogenase; UDPG, uridine diphosphoglucose.

affect the activities of this ciliate. This point has too often been overlooked, and reports stating, for example, that *Tetrahymena* has a given level of an enzyme or of another chemical constituent can be quite misleading unless the physiological state of the cell at the time of sampling is also known. For example, 200-fold variations occur in glycogen content, while the specific activities of some enzymes undergo variations of 20-fold or more depending upon growth conditions and culture age. Thus the fact that a given enzyme is present at very low levels may indicate only that the metabolic route in which this enzyme participates is nonoperative at the time of sampling. In the following discussion it can be seen that even slight variations in culture depth or in concentration of a nutritional supplement can alter the biochemical makeup and the activities of the cell.

The importance of stating the exact growth conditions employed when presenting data on microorganisms has often been stressed, but it is perhaps worth mentioning again. The following information should be included: (1) exact nature and concentration of the nutrients; (2) growth temperature; (3) size and age of the inoculum; (4) size and shape of the culture vessel; (5) volume of the medium; (6) whether or not the cultures were shaken, and if so, the rate and the nature of shaking (gyratory or reciprocal); (7) *exact* method of harvesting, especially when metabolic studies are to be performed on the cells; (8) culture age at the time of harvesting; and (9) *correct* growth stage, and perhaps the number of cells or the amount of cell protein per volume of culture. The erroneous use of turbidometric means of determining culture growth has often led to incorrect statements concerning the growth rate or the growth stage of a culture. The inclusion of the above Information will provide a basis for judging whether or not processes such as glycolysis, respiration, glyconeogenesis, lipogenesis, or cellular degradation were more or less likely to be occurring at the time of harvest.

Another purpose of this chapter is to stress the great potential of the organism as a research tool. *Tetrahymena* grows almost as rapidly as bacteria, yet its metabolism greatly resembles that of a mammalian kidney or liver, and it has typical eucaryotic organelles. Since its metabolism can be readily changed by manipulation of the growth conditions, it can be used for studying the ultrastructural and biochemical changes that accompany or that are responsible for changes in cellular metabolism. It is also useful for studies on the effects of drugs, hyperthermia, and other injurious agents at the ultrastructural, metabolic, and biochemical levels.

Certain aspects of the data discussed below are reviewed elsewhere (27,57, 65,68,74,134).

II. EFFECTS OF TEMPERATURE ON BIOCHEMISTRY, PHYSIOLOGY, AND MORPHOLOGY OF *TETRAHYMENA*

There has been considerable interest in the effects of temperature on *Tetrahymena*. Undoubtedly, much of this had its origin in attempts to determine the factors responsible for the induction of division synchrony. In this section some of the effects of temperature changes and of temperature extremes on *Tetrahymena* are discussed.

A. Temperature and Growth

Holz and his co-workers (63,128) have presented data on the upper temperature limits for *Tetrahymena*. For most strains optimal GTs occur between 27 and 35°C, with growth ceasing a few degrees above the optimum. The temperature tolerance seems to be quite sensitive to the nutritional environment, however.

The upper limit for *Tetrahymena* is about 40°C. Variety 1, mating type II grows at this temperature provided cells are first grown at 35°C, a crude medium (casein, yeast autolysate, glucose) is used, and cells are frequently transferred to fresh medium (66). The HS strain grows at 39°C, presumably by selection of heat-resistant cells, since this temperature kills more than 99% of the cells (119).

Rosenbaum et al. (128) studied the effects of hyperthermia on growth and found that addition of nucleic acid derivatives or phospholipids to a chemically defined medium increased the maximal tolerable temperature by several degrees. The phospholipid requirement was specific for lecithins or cephalins, although a triglyceride-rich fraction isolated from *Tetrahymena* grown at 35°C was also active. Asolectin, a crude preparation of plant phospholipids, also permits growth at 40°C, although the yield is low (44). Fatty acids purified from cells grown at 35°C also permitted growth if added with bovine serum albumin and α-tocopherol.

A preliminary report suggests that thermophily in *Tetrahymena* is cytoplasmically inherited (118). Clones from all exconjugants of crosses between thermophilic and normal cells tolerated elevated temperatures, and the trait was maintained through six generations of backcrossing. Apparently, no further work has been done on this interesting observation.

The lower limit for growth and division seems to be about 7°C, although a strain isolated from icy waters can grow at lower temperatures (38). The fission rate is irregular below 7°C, and growth ceases between 5 and 6°C (117). Cultures have been maintained for "prolonged periods" at 4°C (120). Survival at 0–6°C can be prolonged by addition of lecithin, calcium succinate, or $MgSO_4$ to the medium (125). Lecithins plus antioxidants prolonged survival to 6 wk.

Growth at temperature extremes for one generation results in a partial adaptation to that temperature (149). Thus the GTs for the second and subsequent divisions in cells placed at 10 or 32°C were much shorter than the first GT at the new temperature. Also, the "cold-adapted" cells were more heat-sensitive, and the "heat-adapted" cells more cold-sensitive. While it was suggested that the adjustment results from a "stabilization" of macromolecular structures, it could result from the change in the ratio of saturated to unsaturated fatty acids of membrane phospholipids that accompanies adaptation to extreme temperatures (45; cf. 47,48). Changes in temperature did not seem to alter the relative time spent in each stage of the cell cycle (98). Thus G_1 accounted for about 44%, S for about 28%, and G_2 for about 28%, despite a great variation in GT.

B. Physiology and Biochemistry

There have been studies both on the effects of temperature on the respiratory rate and on the respiratory capacity of cells grown at various temperatures. Cells (HS strain) grown at 10, 20, or 30°C were washed and tested at 20°C with glucose as substrate (70). Cells grown at 10°C respired three times as fast as those grown at 30°C. However, the rates would be about equal if expressed on a protein basis, indicating that the growth temperature has little effect on respiratory capacity. Unfortunately, glucose is a poor substrate for measurements of this type. Its fate is dependent upon the growth phase and growth conditions (59,91,94), the nature of the suspending medium (24), and the glucose concentration (78), and it often fails to stimulate the respiratory rate above the endogenous level (20,131).

A comparison of O_2 consumption and anaerobic CO_2 production in washed GL cells showed that both processes proceeded about twice as fast at 37°C as at 25°C when measured over at 5-h period. The ability of the cells to respire this long at 37°C is surprising, as 34°C is considered lethal for the GL strain. It was noted that the cells grew well at 37°C. In the W strain, the respiration rate of cells from both "young" (3–4 day) and "old" (7–8 day) cultures increased sharply between 10 and 25°C and then declined (116).

The heat sensitivity of several respiratory enzymes has been tested by subjecting homogenates to elevated temperatures. *Tetrahymena* enzymes are notoriously unstable in vitro, even without the benefit of temperature shock (147), so the significance of the results is questionable. Several of these enzymes do seem to be quite labile, however, and the following were largely inactivated by a brief exposure to elevated temperatures: succinoxidase and DPNH oxidase $\left[5 \text{ min at } 37°C (30,31)\right]$; SDH $\left[5 \text{ min at } 40°C \text{ destroyed}\right.$ activity toward methylene blue and 50% of activity toward phenazine mesosulfate $\left.(32)\right]$; DPNH-cytochrome C reductase from a postmitochondrial fraction $\left[10 \text{ min at } 40°C (71)\right]$; LAO $\left[20 \text{ min at } 50°C (36)\right]$. It is possible that the functioning of the cytochrome chain is important in determining the upper temperature limit for growth.

The temperature sensitivity of oxidative phosphorylation in homogenates could be due to the production of lysophosphatides or fatty acids which, by their detergent action, could disrupt various cellular membranes (32,33). Thus an enzymically produced material from *Tetrahymena* particulates inhibits electron transport and phosphorylation, and incubation of boiled particulates with lecithinase A yielded a similar product. The inactivation of several of the above enzymes was prevented by bovine serum albumin, presumably because of a binding of fatty acids. *Tetrahymena* has a potent system of lipolytic enzymes which become activated upon cell disruption (147). They are primarily in the soluble fraction and have pH optima in the alkaline range, indicating that they are not lysosomal. It is not known if phospholipid breakdown is stimulated in vivo at elevated temperatures.

DPNH oxidase activity, the ratios of oxidized to reduced forms of nicotinamide adenine dinucleotides, and the ability to carry out oxidative phosphorylation have been tested in homogenates from cells subjected to repetitive temperature shocks (111,112). The specific activity of DPNH oxidase was slightly lower (on a protein, but not on a cellular basis) at the end of the series of temperature shifts. TPNH and DPNH levels (protein basis) were also reported to be slightly lower. It is not possible to determine if this is the case when the cells are in culture, however, as the ratios of reduced to oxidized dinucleotides change as the cells are harvested and washed.

A decrease in phosphorylation, but not in oxidation, also reportedly occurred as a result of heat shocks (112). Again, it is not possible to determine if these effects occur

Table 1.
Effects of Prolonged Temperature Shock on Enzyme Levels[a]

	Specific activity (percent of 0-h control)	
	12 h 34°C	12-h control
Lactate oxidase	49	100
Isocitrate lyase	38	96
D-Amino acid oxidase	65	100
Catalase	74	102
Isocitrate dehydrogenase	87	108
Succinate dehydrogenase	113	118
Pyruvate kinase	87	108
Acid phosphatase[b]	125	86
Acid proteinase[b]	167	—

[a] Early stationary-phase cultures, grown without shaking at 28°C, were either transferred to 34°C or maintained at 28°C. Cells were harvested when indicated.
[b] Tested at 24 h.

in vivo and at the elevated temperature, as the assays were done at room temperature on cells that had been harvested and fractionated. Also, the cells were fractionated in sucrose concentrations that permit the swelling of lysosomal membranes (105). The assay conditions were such that leakage of enzymes or of breakdown products from these particles would be expected. Since hyperthermia leads to an increased number of autophagic vacuoles (90), it is quite possible that products (lysolecithins?) released from these vacuoles could interfere with electron transfer and phosphorylation. It might also be noted that the ''mitochondrial'' fraction tested showed no enrichment over the total homogenate in regard to the rate of substrate oxidation.

The specific activities of several enzymes have been tested in cells incubated for several hours at supraoptimal temperatures (Table 1) (92). Interestingly, the levels of several peroxisomal enzymes decreased as a result of this treatment, including LAO, IL, catalase, and D-amino acid oxidase. SDH and PK, which are localized in the mitochondria and cytosol, respectively, were not affected. It is not known if the decrease is due to a direct effect on the enzymes or on their synthesis, or to a loss of the structural integrity of peroxisomes. It has been noted that peroxisomes undergo morphological changes after a brief exposure of cells to high temperature (90).

Enzymes involved in intra- and extracellular digestive processes seem to be relatively heat-stable, both in vivo and in vitro. Thus an intracellular RNase is stable for 30 min at 50°C and at 100°C in the presence of substrate (81). An extracellular RNase also seems to be relatively heat-stable (34). The specific activities of acid RNase and phosphodiesterase may actually increase for a time when *Tetrahymena* is incubated at supraoptimal temperatures (82). Two other hydrolases, AP and cathepsin, did not decrease when the E strain was incubated for 24 h at 33.5°C (Table 1), at which time the cells were nearly dead (90). The effects of hyperthermia on the localization of the two hydrolases are discussed in Section II,C.

The effects of hyperthermia on RNA and protein synthesis in *Tetrahymena* have been studied. In the GL strain there was little increase in culture protein at 33.7°C (87), while in the E strain protein in cultures incubated at 33°C increased 22% in 4 h as compared with an increase of 81% in the controls (90). Cells shaken with $[^{14}C]$serine for 20 min after various periods at 34°C showed a gradual decrease in the ability to incorporate label into protein (69% of the control value during the first 20 min, 40% at 4 h). The rate of protein turnover at 34°C was similar to that of the controls, both in cultures and in washed cells (except after prolonged periods of temperature shock).

Incorporation of $[^{14}C]$amino acids into proteins at supraoptimal temperatures has also been studied by Byfield and Scherbaum (16). Transfer of GL cells from 29 to 34°C led to a slight decrease in the rate of incorporation into cell protein over a 30-min period. The results were similar when a strain with a higher temperature optimum (WH-14) was subjected to a shift from 34 to 43°C. GL cells incubated for 2 h at 34°C incorporated about 80% as much isotope into protein as the controls. Since growth was occurring in the latter culture, the uptakes, if expressed on a cellular basis, might be equal. The rate of uptake of $[^{14}C]$uracil into RNA was also decreased by this treatment. An increased rate of decay of RNA was observed at 34°C (15,17). It was suggested that at least part of this was mRNA and that the reduced ability to form proteins at 34°C resulted from an increased rate of destruction of mRNA. Since the uptake of tracer levels of $[^{14}C]$glycerol into lipids was not inhibited, it was concluded that the energy source was not depleted (16). However, cells exposed to 34°C for 60 min have a greatly reduced ability to incorporate tracer levels of acetate into glycogen, and net glyconeogenesis from acetate is greatly reduced at this temperature (92).

The decrease in protein synthesis could result from an impairment of ribosome function (14), as in two strains both ribosome melting and a reduced efficiency of mRNA translation occurred at temperatures several degrees above the optimal.

Lazarus (81) also studied the effects of hyperthermia on RNA synthesis and presented a somewhat different picture. Incorporation of ^{32}P into RNA was severely restricted even after 10 min at 34°C. Since the specific activity of RNase and phosphodiesterase increased during the first few hours of temperature shock, despite the lack of RNA synthesis, it was suggested that mRNA might be stable at 34°C.

Recently, the capacity for mRNA synthesis at 34°C has been studied by testing for enzyme induction at this temperature. Isocitrate lyase is readily induced by transferring shaken cultures to static conditions, and the process is dependent upon RNA and protein synthesis (86). When this transfer was made at 34°C, an actinomycin-D-sensitive increase of IL occurred (92). Although the increase was less than in the controls, the results indicate that both the synthesis and translation of a specific mRNA can occur at 34°C.

Several groups have studied the effects of growth temperature on the ratio of saturated to unsaturated fatty acids. Cells grown at 35°C had relatively more saturated fatty acids than cells grown at 25°C (45). The increase seemed to occur at the expense of monounsaturated fatty acids. Increases in unsaturated fatty acids and in the degree of unsaturation in membrane fractions occurred when the growth temperature was lowered from 25 to 15°C (25). The ratio of tetrahymenol to phospholipids also decreased. It was suggested that both changes would lead to a decreased packing of the membrane at lower temperatures and help to retain the mobility believed necessary for normal membrane function. Such a view is consistent with current models of cellular membranes, which suggest that a semiliquid state is necessary for normal function (139). It seems to be a general phenomenon that cold-adapted organisms have relatively more unsaturated fatty acids than they do normally, and that psychrophilic species have more than their mesophilic counterparts (47,101). The partial adaptation of *Tetrahymena* to temperature extremes (149) could result from a change in the composition of the fatty acids of membrane phospholipids (see Section II, A).

The sensitivity to high temperatures could result from an inhibition of synthesis of polyunsaturated fatty acids, with a subsequent inability to form phospholipids (44). Thus variety 1, mating type II when maintained at 40°C, is unable to grow, is morphologically abnormal, and has a much lower phospholipid content than when grown at 35°C. Addition to the medium of asolectin (a mixture of plant phospholipids containing primarily unsaturated fatty acids) or fatty acids from cells grown at 35°C permitted growth at 40°C. A mixture of synthetic phospholipids containing primarily palmitate gave poor growth, while free palmitate or its di- and triglyceride derivatives failed to support growth. The asolectin seemed to be serving as a source of unsaturated fatty acids, while the synthetic phospholipids were apparently incorporated directly into the membranes.

Chou noted an accumulation of several phosphate-containing compounds in GL cells incubated at 34°C (21). One of these, 2-aminoethyl phosphonic acid (22), accumulates within 30 min after a temperature shift. This compound is a membrane component (129), and the finding that its accumulation results from cellular breakdown rather than synthesis (21) supports the suggestion that elevated temperatures may interfere with normal maintenance of cellular membranes (128).

C. Morphology and Ultrastructure

Incubation of *Tetrahymena* at sub- or supraoptimal temperatures results in many structural lesions. (The effects on the oral apparatus are covered elsewhere in this volume.)

Rosenbaum et al. (128) noted many structural abnormalities in cells incubated at supraoptimal temperatures. Both anucleate and binucleate cells were observed, as were cells with an abnormally large nucleus. There were extreme variations in cell size and shape, and the presence of fractured and displaced kineties was noted. Phospholipids restored the ability for growth but offered relatively little protection against the morphological damage. Organisms maintained at 0–6°C are often misshapen, suggesting that membrane damage occurs at low temperatures as well (125).

Gross (55) also observed nuclear enlargement, but no multinucleation, in cells maintained at high temperatures. Nuclear disruption occurred if the treatment was continued.

The effects of repetitive and prolonged temperature shock on cell ultrastructure have been studied. Elliott et al. (40) noted an alternate appearance and disappearance of "RNA bodies" in nuclei of cells subjected to repetitive heat shock. Stationary-phase cells were used in this study, and it has been suggested that the RNA bodies were nucleolar fusion bodies that had been formed when the cultures entered the stationary phase (19). However, they can also be formed in cells subjected to repetitive temperature shock (18,19). These bodies are characterized by an outer layer of particles, which resemble ribosomes, surrounding an inner, somewhat amorphous layer. They are also present in starved cells (89,109), suggesting that nucleolar fusion is associated with a cessation of cell growth. Cameron et al. (19) did not detect rhythmic patterns of disaggregation, but it did occur when cells were returned to conditions that favored growth.

Most types of organelles were damaged in cells incubated for prolonged periods at supraoptimal temperatures (89). Mitochondria gradually became devoid of tubules, and the latter were sometimes seen within autophagic vacuoles. The RER became greatly reduced, and smooth-walled vesicles became prominent in the cytoplasm. Nuclear damage was readily apparent, as very large bodies were found within the nucleoplasm (and sometimes in the cytoplasm). They may have arisen from a continuation of the process of nucleolar fusion described above. Peroxisomes became swollen at a very early stage, and lipidlike bodies with vacuolated spaces became prominent. Large autophagic vacuoles, filled with cellular debris, were common. Apparently, the cell retained its ability to sequester intracellular debris and damaged organelles.

The widespread effects of hyperthermia on membrane-bound organelles (90), as well as the ability of phospholipids to protect cells kept at high temperatures (128), suggested that some of the damage might stem from disruption of lysosomes and the subsequent release of acid hydrolases into the cytoplasm. A study to test this possibility revealed, however, that no solubilization of AP or cathepsin occurred (90). Over 80% of each enzyme was readily sedimentable in homogenates prepared either from normal cells or from cells incubated at 34°C for periods of up to 24 h. These results are consistent with those obtained in other systems, in that lysosomal disruption, if it occurs at all, seems to be a result rather than a cause of cellular damage (29,140). The possibility that toxic breakdown products may leak out of the lysosomes is not excluded, however.

D. Discussion

The effects of temperature extremes on microorganisms has been the subject of many symposia and review articles (10,47,48,82,122,143). Many of these have dealt with factors that might determine the maximal and minimal temperatures at which growth can take place. Possibilities commonly offered include: (1) denaturation of specific enzymes, (2) interference with protein synthesis, (3) disorganization of cellular membranes, and (4) lack of nutrient uptake, possibly as a result of interference with permease activity. Interference with the regulation of enzyme synthesis (67,110), and infidelity of mRNA translation (51), have also been suggested. In considering the effects of temperature extremes on the cell, it is difficult to distinguish between primary and secondary lesions. For example, a lack of ATP, resulting from an interference with oxidative phosphorylation, could lead to a variety of other effects, both structural and biochemical. Similarly, a direct effect on membrane organization could result in the abnormal functioning of any or all cell organelles. Caution must also be shown when extrapolating, to the living system, results obtained from in vitro experiments. While certain enzymes or enzyme systems are very temperature-labile in cell homogenates (30,31,32,71), this may result from a dependency upon the structural integrity of the organelle upon which they are localized or from activation of hydrolytic enzymes.

Brock (10) has suggested that tolerance of temperature extremes was related more to the stability of cellular membranes than to that of any individual enzyme. Many of the biochemical and morphological lesions observed in *Tetrahymena* are consistent with such a view. Rosenbaum et al. (128), in discussing the effects of hyperthermia on this ciliate, suggest that a derangement of membranes could account for many of the experimental observations such as swelling and disruption of mitochondria, uncoupling of oxidative phosphorylation (if this does indeed occur), and various structural abnormalities. The ability of various phospholipids and fatty acids to restore normal morphology and to permit growth at otherwise lethal temperatures suggests that membrane damage could be a primary lesion in heat-shocked cells. The release of 2-aminoethylphosphonic acid shortly after a temperature increase (21) is also consistent with such a view. Interestingly, the only membrane-bound organelles that seem to be undamaged after prolonged heat shock are the lysosomes (90).

An abundance of information is now available concerning the effects of temperature extremes on bacteria and on the cellular basis of temperature tolerance in these organisms. *Tetrahymena* provides an excellent system for the carrying-out of such studies on eucaryotic cells. A comparison of strains with high- and low-temperature optima might readily permit an evaluation of the roles of enzyme structure, on the one hand, and of organizational integrity, on the other, in determining the temperature tolerance of the cell.

III. THE GASEOUS ENVIRONMENT

Perhaps the most overlooked factor in studies on *Tetrahymena* is the gaseous environment. The amounts of dissolved gases in the medium can vary considerably, depending upon the surface-to-volume ratio of the culture, manner and rate of shaking, cell number, age of the culture, and nature of the nutrient medium. Changes in the gaseous environment can have profound effects on the biochemistry and physiology of the cell.

A. Growth

1. Effects of Oxygen on Growth

Tetrahymena is an aerobic organism but grows well at low O_2 concentrations. The cells deplete the medium of O_2, and the rate of consumption is linear at an O_2 concentration as low as $4 \times 10^{-6} M$ (5). Under certain conditions cells can be maintained anaerobically for relatively long periods (131). There are conflicting reports concerning the ability of *Tetrahymena* to grow at various O_2 concentrations, and some of these differences are probably due to the manner in which the cultures were aerated.

Pace and Ireland (115) studied the effects of O_2 and CO_2 on growth in cultures with very low surface-to-volume ratios. Cells died when the pO_2 above the medium was 0.5 mm, and the cell yield was three times greater in cultures grown under O_2 than those grown under air. Cell size and lipid reserves were reduced in cells grown under O_2. Browning et al. (11), however, by bubbling gaseous mixtures into the medium,

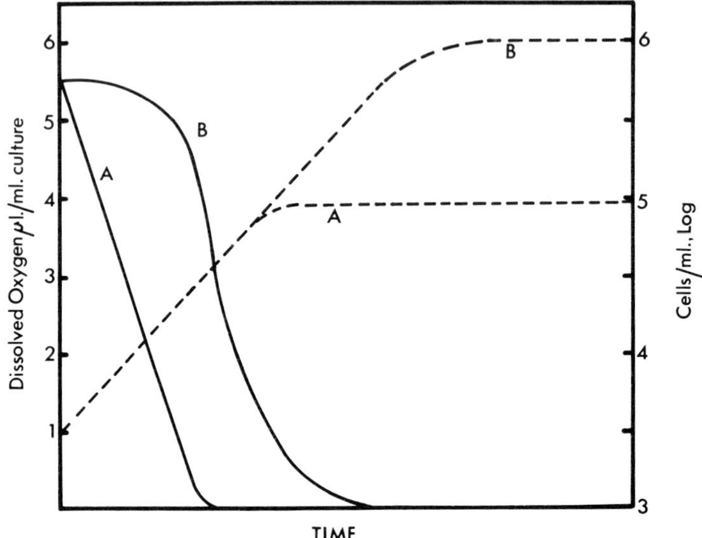

Fig. 1.

Culture growth and dissolved O_2 content of the medium in shaken and static cultures. Solid lines, μl O_2 per milliliter of medium; dashed lines, log of number of cells per milliliter. A, Static cultures; B, shaken cultures. Cells were grown on 2% (w/v) peptone with 0.1% (w/v) liver fraction. Initial volume was 1400 ml/2.5-liter low-form culture flask. Diagramatic, after ref. 94.

obtained a maximal yield when the air phase contained 5–20% O_2. Cells grew well on 1% O_2, and the yield in pure O_2 was only 20% of the maximal. Gaseous mixtures giving maximal yield also tended to give the shortest GTs. The discrepancies between the two studies were attributed to the slow rate of diffusion of O_2 into the medium in the unbubbled cultures. The failure of Pace and Ireland to observe O_2 toxicity was probably due to the low concentration of this gas in the medium. Elliott et al. (41) also noted a high mortality of cells exposed to pure O_2.

Levy and Scherbaum (94) followed growth and O_2 concentration of the medium in vigorously and gently shaken cultures (Fig. 1). In the latter the rate of O_2 utilization began to exceed the rate of entry into the medium at a population of less than 10,000 cells/ml, in a medium that supports up to 10^6 cells/ml. Oxygen was depleted when the population reached about 50,000 cells/ml, and the final yield was about 90,000 cells/ml. The corresponding values in vigorously shaken cultures were about 10 times higher. The GTs were similar in the two cultures. If larger surface-to-volume ratios were used, cells remained in log phase to populations of nearly 10^6 cells/ml. The maximal yield seems to be 3 mg cell protein per milliliter of culture (probably about 1.5 to 2×10^6 cells/ml), obtained by bubbling air into the medium (57). The total culture protein in cultures shaken under 1% O_2 is less than 20% of that of the controls (91).

The above data indicate that in statically grown cultures the O_2 concentration is already falling at a time when growth is hardly perceptible, and that even the use of large surface-to-volume ratios may not provide cells with sufficient O_2 in unshaken cultures. In fact, growth can be brought to an abrupt halt by transferring shaken cultures to static conditions at surface-to-volume ratios under which the rate of diffusion of O_2 into the medium becomes limiting (94). The following figures are convenient for determining the approximate respiration rates of cells in static cultures. *Tetrahymena* grown in peptone–liver fraction have a respiratory *capacity* of about 5 μl $O_2/10^6$ cells/min (about 3 μl/mg cell protein per minute) when in log phase. The *maximal* rate of diffusion of O_2 from air into nonagitated distilled water (assuming the water lacks O_2) is about 0.05 μl O_2/cm^2 per minute (113). By using these figures it was predicted that for 150 ml of culture in a 500-ml flask the dissolved O_2 of the medium would already be dropping at a cell population of 10,000/ml, and this was indeed the case (94). Many studies purportedly show exponential growth occurring for 2 or 3 days in cultures with much lower surface-to-volume ratios than in the example just given. This can usually be traced to the improper use of turbidometric means in determining growth curves. The concentration of microorganisms, including *Tetrahymena* (141), *is directly proportional* to the optical density. Hence when this means is used to follow growth, the *log* of the optical density should be plotted against time. Failure to do this has led to the publication of curves in which a period of linear, postlog-phase growth has been termed the exponential growth phase, while the latter has been overlooked. A shortcoming in the use of turbidometric procedures is that the glycogen content of *Tetrahymena* can range from 1% to more than 250% of cell protein (95). Much of this accumulates after growth, as measured by cell number or protein content, has ceased. It is suggested that, when the effects on growth of nutrients that can be readily converted to glycogen are to be tested, this not be done turbidometrically.

Koyama and Tsukada (79) also noted an immediate cessation of cell division when cultures were transferred from shaken to static conditions. Protein continued to increase, as the amount per cell quadrupled 48 h after the shift. The final population in the shaken cultures was 10 times that of the nonshaken, but the increase in cellular protein did not occur.

Just to confuse matters, Malecki et al. (99) found that cessation of shaking of stationary-phase cultures (grown under 3% O_2) led to a ninefold *increase* in cell number. Furthermore, cultures grown to stationary phase under static conditions (with extremely high surface-to-volume ratios) under 3% O_2 had final populations of 570,000 cells/ml, versus 84,000 cells/ml for shaken cultures. These results conflict with those discussed above, in which much greater yields were obtained in shaken cultures. These investigators suggest an O_2-dependent ''mutual inhibition factor'' which serves to limit cell growth in shaken cultures. This mutual inhibition is compared with contact inhibition and may result from the collision of cells during shaking. It was also found that yields were about 10^6, 10^5, and 20,000 cells/ml in cultures shaken under air, 3% O_2, and 0.7% O_2, respectively.

It is of interest that cells from nonshaken stationary-phase cultures have at least twice as much DNA (103) and protein (79), and several times as much glycogen (131), as cells from log-phase cultures, so that biosynthetic processes *can* occur during stationary phase. The fact that division does not occur suggests that a prerequisite for division is not met when O_2 is restricted. The cells divide if aerated, even in tap water (13,87). Swann has suggested that when energy is limiting cell division is inhibited before growth, and this seems to be the situation here. The sensitivity of cells to anaerobiosis, respiratory inhibitors, and uncoupling agents is maximal at the onset of nuclear division (63), suggesting that this process may be O_2-dependent. It seems likely that cellular diffusion gradients of O_2 exist even within bacterial cells when the concentration of this gas in the medium is low (50). This implies that O_2-dependent processes occur primarily at the cell periphery, especially since large numbers of mitochondria are generally found in this area (39,42,132). Thus aerobic processes in the cell interior could be selectively prevented because O_2 would not reach this part of the cell.

2. *Effects of CO_2 and Ammonia on Growth*

Jahn (69) has noted that *Tetrahymena* apparently does not require CO_2 for growth. Growth rates and final cell populations were equal in cultures aerated in the presence or absence of CO_2. Travis et al. (150) found that growth (measured after 24 h) decreased progressively when the pCO_2 was increased from 40 to 350 mm Hg. Inhibition was 50% at 90 mm, and 90% at 220 mm. Cells survived for 72 h under the latter conditions. Pace and Ireland (115) found that *Tetrahymena* grew best in CO_2-free air (although cells were grown in deep layers.) Good growth was obtained at up to 122 mm CO_2, and cells died at 244 mm CO_2. The pH dropped to 6.2 under the latter conditions, but death was attributed to excess CO_2, as cells could withstand the low pH.

Hill and van Eys (58) followed ammonia levels in cultures throughout the growth cycle. The level was less than 0.5 μmoles/ml for the first 30 h, then began to increase sharply about one generation before culture growth ceased. At 120 h the level was 12 μmoles/ml and increasing. Nardone and Wilbur (108) reported similar results. Ammonia is a potent inhibitor of the Krebs cycle (75), and some of the effects on metabolism that have been attributed to a lack of O_2 may be due to excessive ammonia levels, as conditions favoring one also tend to favor the other.

B. Biochemistry and Physiology

The effects of the gaseous environment on the biochemistry and physiology of *Tetrahymena* are difficult to evaluate. The concentrations of dissolved O_2, CO_2, and ammonia change continuously during culture growth, as well as with growth conditions, and may have little resemblance to the levels in the air phase above the culture. Changes in the amounts and nature of storage reserves, enzyme levels, respiratory capacity, and RQ occur as cultures age, and it is difficult to determine if they are due to cessation of growth, depletion of certain nutrients, or changes in the levels of certain gases, or to a combination of these factors. Also, it is difficult to relate the metabolism of cells as measured, for example, in a washed suspension, to that occurring when the cells are in culture.

No attempt is made here to cover specifically respiration and fermentation in *Tetrahymena,* as this work has been extensively reviewed (27,65,134).

1. Variations in Glycolytic and Related Enzymes

The specific activities of many enzymes in *Tetrahymena* are influenced by variations in the rate of gaseous exchange. These changes can be accentuated by variations in the culture medium. The differences in specific activities are often an order of magnitude or more.

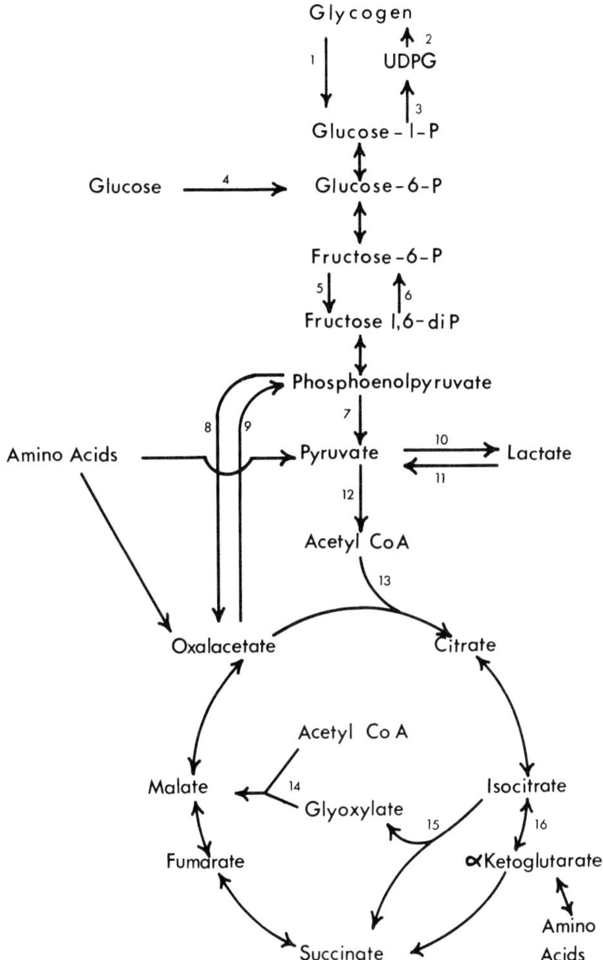

Fig. 2.

Pathways of carbohydrate metabolism. The main enzymic steps in the breakdown and formation of carbohydrates are shown. The common or proper name of each enzyme is listed below, together with its EC number. 1, Phosphorylase (2.4.1.1); 2, glycogen synthetase (2.4.1.11); 3. glucose-1-phosphate uridyl transferase (2.7.7.9); 4, hexokinase (2.7.1.1); 5, phosphofructokinase (2.7.1.11); 6, fructose diphosphatase (3.1.3.11); 7, pyruvate kinase (2.7.1.40); 8, PEP carboxylase (4.1.1.31); 9, PEP carboxykinase (4.1.1.32); 10, lactate dehydrogenase (1.1.1.27); 11, lactate oxidase (1.1.3.2); 12, pyruvate dehydrogenase complex; 13, citrate synthase (4.1.3.7); 14, malate synthase; 15. isocitrate lyase (4.1.3.7); 16, isocitrate dehydrogenase (1.1.1.42).

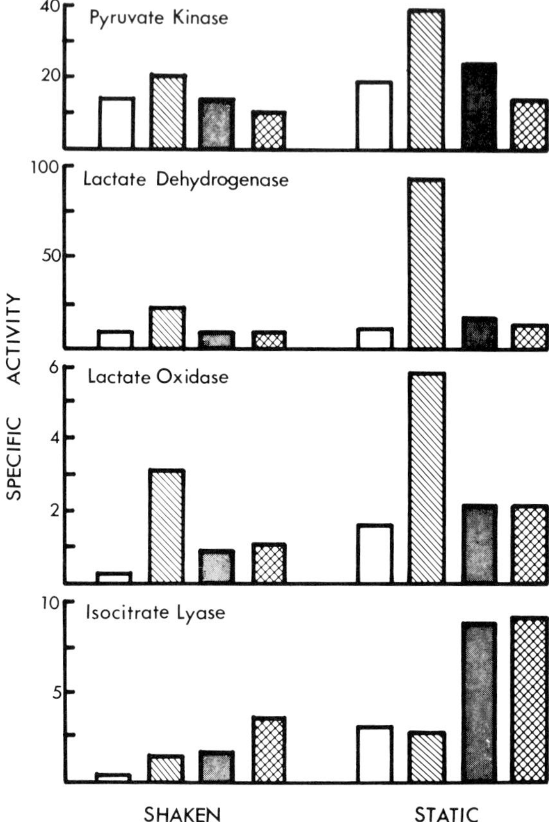

Fig. 3.

Effects of culture conditions on some enzymes related to carbohydrate metabolism. Cultures were grown with or without agitation for 40–44 h. Culture medium contained 1% (w/v) peptone with no additions (open bars) or supplemented with 0.5% (w/v) glucose (hatched bars), 0.5% Sodium lactate (stipled bars), or 0.5% (w/v) Sodium pyruvate (crosshatched bars). (From ref. 95.)

Warnock and van Eys (154), using cells from glucose-supplemented cultures, showed the presence of all the glycolytic enzymes, although the levels of phosphofructokinase and LDH were low (cf. Fig. 2). The specific activity of both LDH and PK are influenced by changes in the growth conditions (95). Both enzymes are low in well-shaken peptone cultures, higher in either shaken cultures supplemented with 0.5% (w/v) glucose or in nonshaken cultures, and very high in nonshaken, glucose-supplemented cultures (Fig. 3). Thus both enzymes increase under conditions that favor breakdown of carbohydrates. The value for LDH in static, glucose-supplemented cultures was about 20 times that of the shaken cultures and about 200 times the value reported by Warnock and van Eys. The extreme lability of this enzyme in cell homogenates may account for some of this difference (87). In vertebrate tissues the M form of LDH can be induced by a lowering of the O_2 tension (53), and it would be of interest to determine which forms are present in *Tetrahymena* under various growth conditions. Eichel et al. (35) suggested that LDH is in the mitochondria in *Tetrahymena*, and this has been confirmed by Risse and Blum (126) who also found hexokinase to be associated with the mitochondria when cells were disrupted at pH 7.9.

The existence of an L-(+)-lactate oxidase in *Tetrahymena* was shown by Eichel and Rem (36). The enzyme (or family of enzymes that uses O_2 to oxidize several α-hydroxy acids to the corresponding keto acid plus H_2O_2) was found primarily in the particulate fraction (35) and specifically in peroxisomes (6,107).

The specific activity of LAO varies greatly with culture conditions (Figs. 3 and 4), a finding that may shed some light on the metabolic role of peroxisomes. Thus high levels were found under culture conditions that favor rapid rates of glycolysis. These

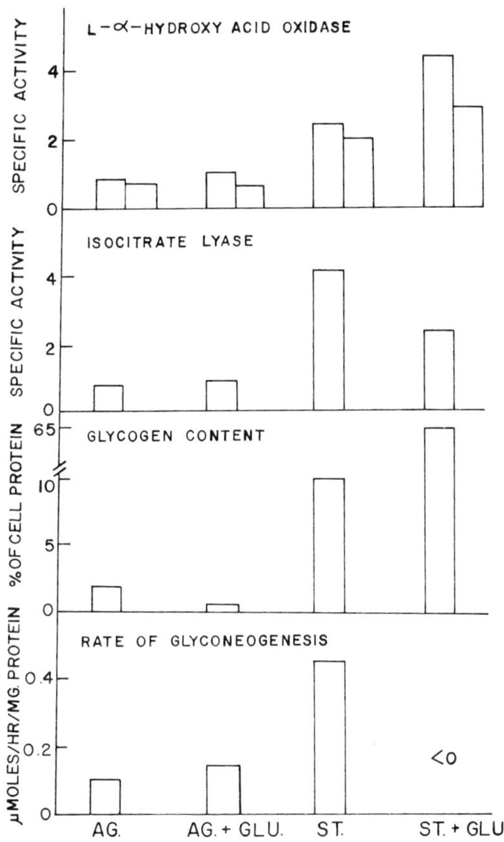

Fig. 4.

Effects of culture conditions on the levels of certain peroxisomal enzymes, the cellular capacity for glyconeogenesis, and cellular glycogen content. Cultures were grown for 24–28 h in the presence or absence of 0.1% (w/v) glucose (GLU). Flasks were either agitated (AG) or maintained without shaking (ST). The values shown for α-hydroxy acid oxidase represent those obtained when D.L-α-hydroxybutyrate and L-(+)-lactate were used as substrates. Enzyme activity is expressed as μmoles of product formed per hour per milligram of protein. (Data from ref. 91.)

include growth under 0.5 or 1% O_2 (91), as well as the conditions mentioned above that led to high LDH and PK levels. In fact, changes in the levels of the three enzymes paralleled one another, suggesting a metabolic relationship (95). Also, the three increase at about the same rate when shaken cultures are made O_2-deficient and supplemented with glucose (87), indicating that their synthesis is coordinately controlled. Addition of either lactate or pyruvate to the medium also led to increased levels of LAO (85), and it has been suggested that the peroxisomes may be involved in the metabolism of excess lactate. Synthesis of peroxisomal enzymes is not under coordinate control, as conditions that favor the highest levels of LAO are not those that lead to maximal levels of the enzymes of the glyoxylate bypass (Figs. 3 and 4). Also, peroxisomal enzymes such as catalase, TPN-dependent IDH, and acetyl-CoA synthetase are relatively unaffected by changes in growth conditions (93,95). It should be noted that maximal levels of LAO occur in cells incapable of performing glyconeogenesis. While the existence of the glyoxylate bypass on *Tetrahymena* peroxisomes clearly implicates this organelle in this process, it seems that LAO participates in some other process carried out in peroxisomes. Since LAO has a low affinity for O_2 (36), it is of interest that it may be functional in cultures that are nearly anaerobic.

2. Glycogen Concentration

The effects of environmental changes on the glycogen content of *Tetrahymena,* the cellular capacity for glyconeogenesis, and the specific activities and the intracellular localization of glyconeogenic enzymes have been studied extensively. These factors are influenced by interactions between the gaseous and the nutritional environment, so the effects of these two variables are discussed here.

The glycogen content of *Tetrahymena* ranges from 1% to more than 250% of cell protein. The amount is generally low in cells from log-phase peptone cultures and increases as cells enter stationary phase (94,131). Manners and Ryley (100) noted that *Tetrahymena* could store vast quantities of glycogen which served as an energy source in anaerobic cells (131).

The glycogen content has been followed cytochemically over a 27-day period in cells grown on tryptone, glucose, and acetate under a very low surface-to-volume ratio (76). Glycogen increased between days 3 and 9, began to fall after day 12, and disappeared by day 21.

Cells from glucose-supplemented cultures may have glycogen contents of more than 250% of cell protein (95) (Fig. 5). The high value occurs in nonshaken cultures. Electron micrographs show cells with a low glycogen content to have only a few isolated granules. In cells with a high content, this material may occupy most of the cytoplasm, with organelles being present as small islands within vast fields of glycogen (89).

Cells grown under various conditions also vary greatly in their glyconeogenic capacity. Glycogen increases rapidly when statically grown stationary-phase cultures, or washed cells from such cultures, are shaken (62,85,93,94,154). In either case tracer or substrate levels of $[2\text{-}^{14}C]$acetate are rapidly incorporated into glycogen (62,85,94), sometimes quantitatively. The glyconeogenic capacity is much lower in cells from shaken cultures, while cells from glucose-containing cultures cannot convert acetate to glycogen (61,85,91).

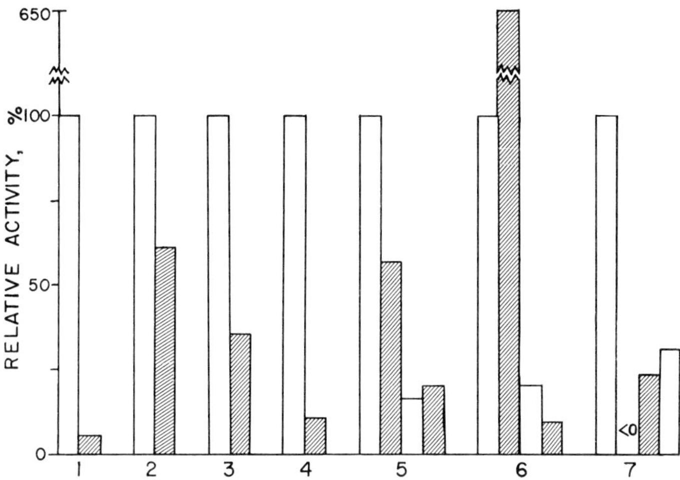

Fig. 5.

Glycogen content, rates of glyconeogenesis, and levels of glyconeogenic enzymes in cells grown under various culture conditions. In each case the specific activity in cells grown under static conditions in medium lacking glucose is set at 100%. Hatched bars represent values obtained from cells grown on glucose-supplemented medium. The last two bars in groups 5 through 7 represent values from shaken cultures. Cells were grown for 48 h (numbers 1 through 4) or 24 h (5 through 7). 1, PEP carboxykinase in cytosol; 2, PEP carboxykinase in "mitochondrial" fraction; 3, fructose diphosphatase in cytosol; 4, malate dehydrogenase in cytosol; 5, IL; 6, glycogen content; 7, cellular capacity for glyconeogenesis. [From ref. 135 (numbers 1 through 4) and ref. 91 (numbers 5 through 7).]

Wagner (152) found the increase in glycogen in washed stationary-phase cells could be accounted for by a decrease in phospholipids. The data of Warnock and van Eys (154) are in agreement with this insofar as the increase was not accompanied by a loss of cell protein. However, we have noted a sharp decrease in protein in cells transferred to inorganic medium (89). Ultrastructural studies show that considerable mitochondrial degeneration occurs in the early stages of starvation (89), so that either mitochondrial phospholipids or protein or both could provide precursors for glyconeogenesis. Ribonucleic acid is also broken down in the early stages of starvation (23).

The cellular glycogen content and the concentration of dissolved O_2 in the medium have been followed in gently shaken cultures (94). As the medium became depleted of O_2, there was a cessation of culture growth and the glycogen content began to increase. Thus when the availability of O_2 decreases, *Tetrahymena* responds by storing carbohydrate reserves. The accumulation of glycogen is accompanied by an increased capacity to form glycogen (Table 2), as cells from O_2-deficient cultures exhibit high rates of glyconeogenesis, especially when returned to conditions of adequate aeration. A period of O_2 deficiency of 1.5 h was sufficient to activate the glyconeogenic pathway (85,86,94). This treatment was effective both in cells in culture and cells in inorganic medium, and results, at least in part, from an increase in the relative amount of acetate metabolized via the glyoxylate bypass. In either case the effect was abolished by glucose (85). Interestingly, while the increased capacity for glyconeogenesis is *acquired* when cells are under nearly anaerobic conditions, and while considerable glycogen does accumulate under

Table 2.
Effects of Glucose, Acetate, and Culture Condition on Isocitrate Lyase Activity and the Cellular Capacity for Glyconeogenesis[a]

	Isocitrate lyase	Glycogen content	Rate of glycogen formation
Shaken cultures			
0 h	0.56	1.7	0.078
3.5 h	0.81	1.5	0.13
3.5 h + acetate	1.35	1.7	0.17
3.5 h + glucose	0.76	21.2	0.022
Static cultures			
3.5 h	6.2	4.5	0.46
3.5 h + acetate	5.0	6.8	0.31
3.5 h + glucose	1.85	36.1	0.037
3.5 h + glucose and acetate	1.85	34.7	<0

[a]Shaken, log-phase, peptone cultures were either shaken for 3.5 h or transferred to static conditions and maintained for 3.5 h. Glucose and acetate, when added, were present at concentrations of 1 mg/ml. At 0 or 3.5 h cells were tested for IL (expressed as μmoles of glyoxylate formed per hour per milligram of protein), glycogen content (as percent of cell protein), and the rate of glyconeogenesis (μmoles of glucose formed per hour per milligram of protein). The last was tested by shaking washed cells for 1 h with sodium acetate (1 mg/ml).

Table 3.
Isocitrate Lyase Activity, Glyconeogesis, and Incorporation of $[2\text{-}C^{14}]$ Acetate into Glycogen in Cells Incubated under Static Conditions or under 0.5% O_2.[a]

	Isocitrate lyase (μmoles glyoxylate formed/h/mg protein)	Net glyconeogenesis (μmoles glucose formed/h/mg protein)	Incorporation of $[2\text{-}C^{14}]$ acetate into glycogen (counts/min/mg protein)
0 h, shaken	0.72	0.095	755
3 h, shaken under 0.5% O_2–1% CO_2	1.1	0.27	3500
3 h, static	3.6	0.26	4050

[a]Glyconeogenesis was tested as described for Table 2. Varying the pCO_2 mixture from 0 to 5% did not influence the results. (Levy, unpublished.)

these conditions, *maximal* rates of synthesis occur when the cells are subsequently returned to aerobic conditions. Other reports also show that glycogen accumulates when the availability of O_2 is low. Thus Shrago et al. (135) noted an increased content with increasing depth of the culture, and Levy (87) found an accumulation when atmospheres containing 0.5–2.0% O_2 were passed over cultures. The latter treatment also activated the glyoxylate bypass (Table 3).

Shug et al. (137) found the glycogen content of cells from shaken peptone cultures was influenced by the iron content of the medium. Maximal growth and glycogen content occurred in cultures supplemented with 0.4 μg iron/ml. Glycogen accounted for 37 and 15% of the dry weight, respectively, of cells from supplemented and unsupplemented cultures. There was also a great increase (about 175-fold) in the uptake of $[2\text{-}C^{14}]$acetate into glycogen in shaken, 2-day cultures. Acetate incorporation into lipids was also stimulated. Addition of iron to poorly aerated (nonshaken) cultures had little effect on growth or glycogen content. It is not clear if the increases in glycogen synthesis are due to a direct effect of iron, or if they result from a greater degree of anaerobiosis stemming from the greater cell yield. Cells from young cultures incorporated acetate into glycogen at negligible rates, despite the presence of iron.

Another report from the same laboratory (138) showed that antimycin A also led to an increased glycogen content in cells from shaken cultures. Thus 10 μg/ml of the inhibitor caused the glycogen to increase from 1.9 to 9.1% of cell protein. Two glyconeogenic enzymes, PEP carboxykinase and MDH, increased severalfold in the cytosol, although there was little change in total enzyme. Cells grown with antimycin A also incorporated labeled acetate and amino acids into glycogen to a much greater extent that did control cells. These effects did not seem to be due to an inhibition of respiration and did not occur in nonshaken cultures. It is of interest that both iron and antimycin A cause well-aerated cells to assume some of the metabolic characteristics found in poorly aerated cells. Perhaps these findings will help to clarify the mechanism through which glyconeogenesis is activated in oxygen-deficient cells.

3. Regulation of Glyconeogenic Enzymes

The synthesis of key glyconeogenic enzymes is apparently under coordinate control in *Tetrahymena*. Figure 5 compares the relative levels of these enzymes and the glyconeogenic capacity of cells grown under various culture conditions.

Acetate is rapidly converted to glycogen in *Tetrahymena* via the glyoxylate bypass. The key enzymes of this pathway (Fig. 2), IL and MS, are localized on the peroxisome (107). This pathway permits the net synthesis of succinate from two molecules of acetate. This may then be converted to sugars via PEP carboxykinase and a net reversal of glycolysis. The presence of the two enzymes was first shown by Hogg (60) and by Reeves et al. (124). It was reported that both were repressed by glucose and that acetate could induce IL and largely relieve the glucose repression (60). Only cells from acetate-supplemented cultures were reported to be capable of converting acetate to glycogen. It was also reported that IL and MS were localized in a particulate fraction in cells capable of glyconeogenesis, but not in cells lacking this ability (61). However, this claim has been withdrawn (106). Later, it was found that the great glyconeogenic capacity of stationary-phase cells (94) could be accounted for by increased levels of IL and MS (85,93).

In order to study the enzymic changes that lead to an increased capacity for glyconeogenesis, cultures were transferred from shaken to static conditions (93). As described above, cells rapidly develop the ability to convert acetate to glycogen at high rates after such a transfer. The key enzymes of the glyoxylate bypass were induced by such a treatment (Table 2). The level of both MS and IL doubled within 3 h in the GL strain (93), while IL increased 5 to 10-fold in the E strain (85). This treatment had little effect on the level of MDH or acetyl-CoA synthetase. The increased importance of the glyoxylate bypass was reflected in an increase in the relative amount of $[2\text{-}^{14}C]$acetate entering this pathway (93). Nutrients were not required for either the increase in IL or in glyconeogenic capacity (85,86). Both occurred when cultures, freshly harvested cells, or cells that had been starved for several hours were transferred to static conditions. In nonnutrient medium, however, the increase in IL was reduced.

The increased capacity for glyconeogenesis was not dependent upon enzyme synthesis and occurred even in the presence of concentrations of inhibitors of RNA or protein synthesis that prevented the increase of IL (86). An increase in glyconeogenesis but not in the amount of IL also occurs when cells are shaken under atmosphere of 0.5–2% O_2 (Table 3). Apparently, glyconeogenesis can be regulated by activation, as well as induction of key enzymes. This also seems to be the case in mammalian systems (56,123).

Tetrahymena IDH, which competes for isocitrate in the peroxisomes, is subject to a concerted feedback inhibition by oxalacetate and glyoxylate (88). Each compound is also capable of inhibiting the enzyme. Inhibition of IDH could perhaps activate glyconeogenesis by shunting metabolites through the glyoxylate bypass.

Glucose is a potent inhibitor of glyconeogenesis in *Tetrahymena* and completely blocks the ability to convert acetate to glycogen in cells transferred to static conditions (Table 2). Synthesis of IL is only partially blocked, however. Similarly, cells grown for 1 day or more on glucose cannot convert acetate to glycogen (61,91) (Fig. 4), although they have high levels of IL if grown without shaking (91, 95).

The metabolites responsible for IL induction (derepression) in *Tetrahymena* remain to be determined. While it was reported that acetate induces this enzyme (61), the studies just described indicate that both the synthesis of IL and MS and the increased capacity for glyconeogenesis do not require acetate or, for that matter, any nutrients (Table 2). Similarly, acetate did not relieve the glucose repression of IL synthesis as had been reported (61), nor did it have much of an effect on IL if included in the medium from the time of inoculation (95). Phosphoenolpyruvate seems to inhibit both the synthesis and the activity of IL in microorganisms (cf. 77) and could account for the lowered level of this enzyme and the inability to convert acetate to glycogen in glucose-grown cells. The inclusion of either lactate or pyruvate in the medium leads to very high levels of IL, however, especially in O_2-poor cultures (95), suggesting that one of these compounds, or a related metabolite, could be an inducer. Lactate could also be responsible for the induction of IL in static cultures (85,93), as an accumulation could occur when O_2 is limiting.

It is of interest that two apparently functionally unrelated systems on the peroxisomes are induced by O_2 deficiency. Lactate oxidase, which seems to be metabolically related to glycolysis and IL and MS, which are glyconeogenic, all reach their highest levels in O_2-deficient cells. The peroxisome has been described as a primitive respiratory organelle (28). Perhaps it was designed to function at low O_2 concentration.

Phosphoenolpyruvate carboxykinase in *Tetrahymena* has been studied in relation to glyconeogenesis (135,136). This enzyme was found both on cell particulates and in the cytosol. The latter form may be involved in glyconeogenesis, as it was repressed by glucose (Fig. 5). The presence of acetate in the medium had no effect on either form. The specific activity was greater in deep than in shallow cultures (Fig. 6), in

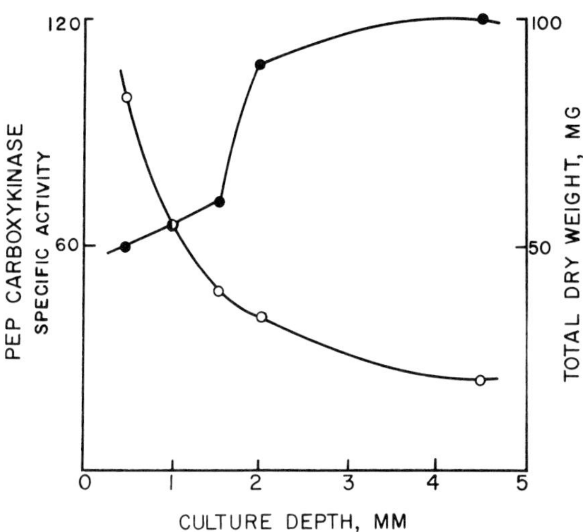

Fig. 6.

Effects of culture depth on cell growth and PEP carboxykinase activity. Cells were grown for 48 h in cultures of various depths. Enzyme activity is expressed as μmoles of PEP formed per hour per milligram of protein. (From ref. 135.) Closed circles, PEP carboxykinase. Open circles, cellular dry weight.

agreement with the finding that O_2-deficient cells have high glyconeogenic capacities. The glyconeogenic enzyme fructose diphosphatase was also present and subject to glucose repression (Fig. 5), as was MDH. Another group found the former enzyme did not change with culture conditions (126). The effects on MDH are surprising, since extremely high levels of MDH are present in cells incapable of glyconeogenesis, and conditions that induce IL and activate the glyoxylate bypass had little effect on the enzyme (93). Interestingly, conditions reported to give low levels of MDH (low O_2, high glucose) also lead to high levels of SDH (95).

Recent data suggest that aldolase A is the predominant form of this enzyme in cells from shaken cultures, and aldolase B in cells from static cultures (79). The latter form is believed to participate in glyconeogenesis (130), and an increased level in statically grown cells would be consistent with such a view. It was also noted that several new isozymes of catalase appeared following the transfer of cultures from shaken to static conditions (79).

The regulation of tyrosine aminotransferase, another glyconeogenic enzyme, has been studied in *Tetrahymena* (102). The specific activity increases as cells grown on basal medium enter stationary phase. Omission of acetate from the medium led to increased levels in late log-phase cells. Peptone-grown cultures had higher specific activities than those grown on synthetic medium, and glucose repression occurred in either case. The cultures were grown under conditions of O_2 deficiency. Possibly, well-aerated cultures would have lower levels of the enzyme.

Blum studied the regulation of enzymes directly concerned with glycogen synthesis and breakdown. When cells were grown without glucose, both glycogen synthetase and phosphorlyase increased as the surface-to-volume ratio decreased (8). However, a great change in the ratio led to only a twofold increase in enzyme. In cells grown with glucose, the synthetase decreased as the relative surface area decreased, which is surprising, as such cells have very high levels of glycogen. Phosphorylase activity was not affected. Acetate had little effect on either enzyme. Theophylline (which inhibits cAMP phosphodiesterase) led to an increase in synthetase and a decrease in phosphorylase. An increase in glycogen content in cells exposed to theophylline was attributed to these changes (implying that cAMP would *increase* glycogen in *Tetrahymena*).

Cook et al. (26) studied several enzymes involved in glycogen metabolism. Phosphoglucomutase and UDPG pyrophosphorylase were somewhat higher in stationary-phase than in log-phase cells. Glycogen synthetase activity was variable, and there was little variation in the relative amounts in the D and I forms.

IV. CULTURE AGE

Many physiological, biochemical, and structural changes occur as cultures of *Tetrahymena* pass through the growth cycle. Some of these undoubtedly arise from changes in the concentrations of various gases or a depletion of nutrients, while others may be more directly related to changes in growth phase. The changes in levels of glycogen and glyconeogenic enzymes discussed in Section III seem more related to environmental changes, as they can be induced at will by manipulation of the gaseous environment (86,93,94). In most cases, however, the factors leading to the various changes are not known. In this section some of the differences among cells in various growth stages are discussed.

A. Biochemistry and Physiology

1. Respiration

Several workers have measured the respiratory capacity of cells from cultures of various ages. Baker and Baumberger (5) found the respiratory capacity was greatest in cells from 2-day cultures and decreased steadily with culture age. Cytochrome content was similar in cells from 5- and 12-day cultures, despite a 75% decrease in respiratory capacity.

Pace and Lyman (116) also noted that cells from young cultures had a greater respiratory capacity than cells from older (7–8 day) cultures. The difference was observed when cells were tested in inorganic medium or in peptone. The RQ of old cells in peptone was 2.8. Wingo and Cameron (156) obtained similar results. Cells from 115-h cultures respired at only 25% of the rate of cells from 40-h cultures.

Ormsbee (114), comparing cells from log-phase (20–30 h) and early stationary-phase (50-h) cultures, found the respiration rates to be similar on a dry weight basis, although the rate was greater in the latter if expressed on a cell basis.

Waithe (153) measured glucose uptake in washed cells and found the rate in cells from 108-h cultures was about 10% that of cells from 74-h cultures. Glucose disappearance from the medium was measured, so the results probably do not reflect the rate of glucose utilization since glucose can often be converted primarily to glycogen (59,89).

2. Enzyme Levels

Several studies. indicate that the levels of some respiratory enzymes decrease with culture age. These findings are consistent with the decreased respiratory capacity of cells from older cultures. There is also evidence for mitochondrial degeneration in such cells (Section IV, C). Thus SDH, as measured by a modified cytochemical method, was highest on day 3 and gradually declined, reaching a low level on day 9 (76). Cells were grown under a low surface-to-volume ratio. Succinate dehydrogenase decreases much sooner in well-aerated cultures, and the presence of glucose delays the decrease (95). Diaphorase and cytochrome c reductase were higher in log- than in stationary-phase cells, although no difference was found in DPNH oxidase (157). Glutamate-oxalacetate transaminase was much higher in stationary-phase cells. The specific activity of glutamate dehydrogenase was the same in the two growth phases on a dry weight basis, but was twice as high in stationary-phase cells on a cell basis (148). Isocitrate dehydrogenase decreased gradually during the first 3 days of culture growth (95). Changes in LAO were discussed earlier. In shaken cultures there is little activity after 2 days unless the medium contains glucose (91,95). Activity is much higher and is lost less rapidly in static cultures.

Recently, the activities of several respiratory enzymes were studied in cells grown on peptone under forced aeration (96). Succinoxidase and α-ketoglutarate oxidase dropped to almost undetectable levels by day 4. By day 6, DPNH- and TPNH-cytochrome c reductases dropped to 19 and 39%, respectively, of their day-2 values. There was little change in catalase, except in very old cultures. The appearance of new isozymes of catalase in stationary-phase cells has been reported (79).

Changes occur in the isozyme pattern of esterases (1,75), alkaline phosphatase (49), and AP (2,3) during the growth cycle. The significance of these findings is not known, however.

There are conflicting reports concerning changes in AP during the growth cycle. Lazarus and Scherbaum (82) found a decrease in specific activity as cells entered log phase, and an increase as they entered stationary phase. Ribonuclease increased as cells entered stationary phase. (The "log"-phase samples were taken at the end of the period of exponential growth.) Changes in latency were noted, 85 and 20% of the AP activity being in the "free" form in cells from "log" and maximal stationary phase, respectively. It was suggested that rapidly dividing cells have a large fraction of their acid hydrolases outside of lysosomes, and that these enzymes may have a role in cell division. However, the *free* activity, was measured, not the *soluble* activity. In several other studies employing cultures of various ages and treatment, little AP was found in the postmitochondrial fraction (90,104,105).

Acid phosphatase-positive bodies, presumably lysosomes, are abundant in cells from old, but not from young, cultures (49). (See Section IV, C for ultrastructural studies.) A role of AP in catabolic processes was suggested, as increased levels were found in cells subjected to adverse conditions. Thus the specific activity began to increase after 12 days of growth, at about the time the cell number began to decline. The timing of the two events coincided under several growth conditions. Similarly, AP was lower in cells grown at an optimal pH than in cells grown at a pH that permitted only slow growth.

These findings led to the suggestion that AP synthesis was increased in cells growing under adverse conditions. While this is not necessarily the case (under some conditions an increased specific activity of AP results from a general loss of other cell proteins, not from an increased synthesis; see ref. 89), these data are consistent with the view that lysosomes are actively involved in cellular degradation and increase in numbers during environmental stress (89,90). Quite possibly, the increase in lysosomes and/or lysosomal enzymes in cells from old cultures is a means of supplying the cell with nutrients, by breakdown either of cell parts or of materials taken up from lysing cells. (See Sections IV, C and D.)

Thompson (146), in an attempt to determine the basis for the instability of enzymes in homogenates of *Tetrahymena*, found the cell to have a potent system of enzymes capable of degrading phospholipids to fatty acids. The rate of breakdown was much faster in homogenates from stationary-phase cells than in those from log-phase cells. A mixture of phospholipases was probably responsible for the breakdown. The bulk of these enzymes appeared in a postlysosomal fraction. An enzyme capable of cleaving glyceryl ethers did not change during the growth cycle (72).

3. Lipid Reserves

Several groups have reported changes in the levels of triglycerides as cultures pass through the growth cycle. Fennell (49) noted that lipids were localized anteriorly in cells from 3–10 day cultures, and that they were distributed throughout the cytoplasm in cells from older cultures. Engemann (43) reported similar results. In cells grown in upright test tubes, lipids (primarily neutral fat) increased as the culture aged and then decreased after 18 days. The decrease began sooner when the surface area was doubled. Lipid reserves were greatly reduced in cells grown under O_2 or under a CO_2-rich atmosphere (115). These changes were attributed to a loss of synthetic abilities in the presence of CO_2, and to a high metabolic rate in the presence of O_2. Fatty acids accounted for about 5% of the dry weight of log-phase cells, and 10% of that of stationary-phase cells (45). The increase was due to the storage of triglycerides. The ratio of monounsaturated to saturated fatty acids decreased as cultures aged.

Everhart and Ronkin (46), using cells grown on peptone and with a very low surface-to-volume ratio, found that total cell lipid increased throughout the log and early stationary growth phases but then fell sharply. Most lipid components varied little with culture age, but the triglycerides increased about eightfold from the first to the third day of culture growth.

Allison and Ronkin (4) also studied changes in lipid reserves as a function of culture age. Cultures were grown in upright test tubes and lipid droplets were detected cytochemically. In each of four strains tested, the percentage of cells that contained lipid droplets increased in early stationary phase. There was then a decline, followed by a secondary peak during the third week, just after the onset of a rapid decrease in cell number. Elliott and Bak (39) also found that the number of lipid droplets increases with age of the culture, as evidenced both by staining with oil red-0 and by electron microscopy. (See Section IV, C.)

It would be of interest to determine whether the accumulation of triglycerides results from an environmental stimulus or from a cessation of growth at a time when nutrients are still abundant. In all these studies triglycerides accumulate under conditions of severe O_2 deficiency and in the presence of high concentrations of ammonia and presumably CO_2, conditions that tend to inhibit the operation of the Krebs cycle. The specific activities of lipogenic enzymes in the various growth stages are not known. However, palmitate is incorporated only into phospholipids in log-phase cells and primarily into glycerides and other nonphosphate-containing lipids in stationary-phase cells (145). Incorporation of palmitate or chimyl alcohol into the membrane component, 2-aminoethylphosphonic acid was also lower in stationary-phase cells (146).

B. pH and Culture Age

Changes in the pH of the medium during growth are dependent upon the relative amounts of glucose and peptone in the medium and can also be influenced by the rate of gas exchange. Depletion of O_2 can lead to acid production, but this is usually more than offset by ammonia formation (58,108) and the pH generally increases. In cultures grown in the presence of high glucose concentrations, however, the pH can drop until cell death occurs (87). Large amounts of acid were formed in cells grown in test tubes on glucose (127). This did not occur in cells grown in flasks. Similarly, high levels of certain amino acids led to a sharp decrease in the pH of test tube cultures, whereas flask cultures became alkaline. These differences can probably be accounted for by the greater availability of O_2 in flask cultures. The presence of O_2 would not only lower the rate of glucose breakdown (131,154), but keto acids formed by deamination of amino acids would presumably be oxidized (or perhaps be converted to glycogen) rather than fermented to various acids.

The effects of O_2 on the pH of the medium have been studied (12). Cultures were grown on phosphate-buffered peptone, and various gaseous mixtures were bubbled through the flasks. The pH of the medium increased from an initial value of 6.9 to 7.9 on day 4 and 7.7 on day 8 when air was used. For a given number of cells, the pH was higher when pure O_2 was the gas phase than when 50% O_2, and highest in 20% O_2, which was the order of increasing cell number.

Prescott (121) followed the pH of cultures grown in very shallow layers in a similar medium and buffered to pH 7.4. The pH remained constant until the end of log growth and then increased to 8.3 at 120 h. Lazarus (81) noted that the pH of cultures was initially 6.45 and increased to 7.6 after 32 h.

C. Changes in Ultrastructure

Elliott and Bak (39) studied the ultrastructure of *Tetrahymena* as cultures passed into early and late stationary-growth phases. Cultures were grown without shaking in glucose-supplemented peptone. In cells from 24- to 72-hr cultures, the mitochondria seemed to be elongated and were found primarily along the cell membrane. In cells from 5-day cultures, they were more spherical, tended to be distributed throughout the cell, and often contained large inclusion bodies. Cells from older cultures had numerous degenerating mitochondria, some of which were present within large autophagic vacuoles. The breakdown of mitochondria probably provides the cell with a source of nutrients. Lipid droplets also increased in number as the cultures aged. Sato (132) also noted a peripheral localization of mitochondria, but we did not find such a distribution in cells grown under several conditions (89). Cells from 2- or 3-day cultures grown as described above have extremely high levels of glycogen, and this material occupies much of the cell cytoplasm (89, Fig. 17). Conceivably, the peripheral occurrence of the mitochondria could result from physical dislocation as the cell becomes filled with glycogen.

In another study (41) cell ultrastructure was followed during the growth cycle in cells from shaken and static cultures. Cells from shaken cultures seemed to have fewer mitochondria, and these were peripherally located. These cells also seemed to have fewer autophagic vacuoles and lipid droplets.

The ultrastructure of the nucleolus of cells passing from log to stationary growth phase has been studied (18). In late log phase nucleoli began to aggregate and fuse, until in stationary phase only a few large vacuolated fusion bodies remained. These disaggregated when cells were transferred to fresh medium. Again, since nucleolar fusion also occurs in starved and in heat-shocked cells, this may be a general response to a cessation of cell growth (19,40,90). (See Section II, C.)

D. Starvation and the Aging Process

Many of the ultrastructural and physiological changes observed in cells from older cultures undoubtedly result from a depletion of nutrients, and indeed, many of the ultrastructural features of such cells appear shortly after the transfer of cells to inorganic medium (89). The protozoan cell, growing in medium such as liver-supplemented peptone, is geared for high rates of growth and division. Consequently, large numbers of organelles such as mitochondria and ribosomes are present and are potentially available as an energy source when nutrients are depleted. Within 30 min after the transfer of *Tetrahymena* to an inorganic medium, numerous degenerating mitochondria are seen. In the initial stages of degeneration, they are often enclosed within a double-layered membrane (89), and AP may be localized in the space between the surrounding membranes and the mitochondrial membranes. The gradual decrease in respiratory capacity (5,114,116,156) and in respiratory enzymes (76,95,96) that occurs as cells are incubated in inorganic medium may result from the destruction of mitochondria. Respiration in starved cells seems to occur at the expense of phospholipids (62,152,154), and these compounds could perhaps arise from mitochondrial membranes.

Mitochondrial breakdown products can apparently be stored as triglycerides or as glycogen, depending upon the history of the culture. Cells taken from *shaken* cultures show a transient increase in triglyceride droplets during the first several hours of starvation (89). These droplets seemed to arise within the cisternae of the endoplasmic reticulum, and possible connections were seen between degenerating mitochondria and forming lipid droplets. These lipids are used in preference to glycogen, as the droplets rapidly disappear whereas glycogen levels remain relatively constant for some time. Addition of acetate to cell suspensions led to a great increase in the abundance of triglyceride droplets, but they disappeared within 25 h while glycogen was still increasing. In cells taken from *static* cultures, the mitochondrial breakdown products are presumably converted primarily to carbohydrate rather than to lipids, as these cells have high endogenous rates of glyconeogenesis (61,85,91,94).

With continued starvation autophagic vacuoles, which are often AP-positive and which contain a variety of degenerating organelles, become quite prominent (89). There is no increase in the total activity of either AP or cathepsin. The former decreases with cell protein by about 70% after 24 h. Cathepsin triples in specific activity but not in amount, as the increase can be accounted for by the loss of cell protein.

Müller (104) has recently suggested that a special population of lysosomes is released from the cell during starvation. Thus a dense lysosomal fraction was absent from cells starved for 5 h, and a large percentage of many of the cellular acid hydrolases was present in the medium at this time. The released hydrolases may serve to break down the RNA released from the cell during the first few hours of starvation (23). We observed numerous bodies which appear to be releasing AP-positive material into the medium in electron micrographs of starved cells (87). An increase in the relative amount of acid hydrolases in denser lysosomes seems to occur in both starved and old cells (96).

Large nucleolar fusion bodies become prominent during the first 3 h of starvation (89). Many nucleoli seem to pass into the cytoplasm by a blebbing process. In the cytoplasm they seem to be present in structures that resemble autophagic vacuoles, where they apparently are degraded. This process may in part be responsible for the large decrease in cellular RNA that occurs in the early stages of starvation (23).

It would be of interest to determine the factors that trigger the destruction of organelles in starved cells. It has been suggested that each organelle may contain hydrolases which can become activated and destroy the organelle when it becomes nonfunctional (151). There is considerable evidence now for the existence of phospholipases and proteases in mitochondria and other organelles (52,54; cf. 7,9,142), and the tentative presence of a mitochondrial AP in *Tetrahymena* has been shown (97). Possibly, the lack of oxidizable substrates could serve to activate these hydrolases, or to render their substrates susceptible to attack, and thereby account for the large number of degenerating mitochondria seen in the early stages of starvation.

The resemblances between starved cells and cells from old cultures are remarkable. In both cases the following are observed: large numbers of degenerating mitochondria, large, debris-filled autophagic vacuoles, a transient increase in lipid droplets, formation of large nucleolar fusion bodies, and decreases in respiratory capacity and respiratory enzymes. It appears that in both cases the metabolism of the cell has shifted so that unneeded cell parts are now being used to provide the organism with nutrients.

REFERENCES

1. Allen, S. L. 1964. J. Exp. Zool. 155:349.
2. Allen S. L. 1968. Ann. N. Y. Acad. Sci. 151:190.
3. Allen, S. L., Misch, M. S., and Morrison, B. M., 1963. J. Histochem. Cytochem. 11:706.
4. Allison, B. M. and Ronkin, R. R. 1967. J. Protozool. 14:313.
5. Baker, E. G. S. and Baumberger, J. P. 1941. J. Cell Comp. Physiol. 17:285.

6. Baudhuin, P. M., Müller, M. and de Duve, C. 1965. Biochem. Biophys. Res. Commun. 20:53.
7. Bartley, W. and Birt, L. M. 1970. In W. Bartley, H. L., Kornberg, and J. R. Quayle, eds., Essays in cell metabolism, John Wiley, New York, p. 1.
8. Blum, J. J. 1970. Arch. Biochem. Biophys. 137:65.
9. Bohley, O., Kirscke, H., Langer, J., Ansorge, S., Wiederanders, B., and Hanson, H. 1971. In A. J. Barrett and J. T. Dingle, eds., Tissue proteinases North Holland, Amsterdam, p. 187.
10. Brock, T. D. 1967. Science 158:1012.
11. Browning, I., Bergendahl, J. C., and Brittain, M. S. 1952. Texas Rep. Biol. Med. 10:790.
12. Browning, I. and Brittain, M. S. 1952. Texas Rep. Biol. Med. 10:778.
13. Browning, I., Brittain, M. S. and Bergendahl, J. C. 1952. Texas Rep. Biol. Med. 10:790.
14. Byfield, J. E., Lee, Y. C., and Bennett, R. L. 1969. Biochem. Biophys. Res. Commun. 37:806.
15. Byfield, J. E. and Scherbaum, J. E. 1967. J. Cell Physiol. 68:203.
16. Byfield, J. E. and Scherbaum, O. H. 1966. Life Sci. 5:2263.
17. Byfield, J. E. and Scherbaum, O. H. 1967. Science 156:1504.
18. Cameron, I. L. and Guile, E. E., Jr. 1965. J. Cell Biol. 26:845.
19. Cameron, I. L., Padilla, G. M., and Miller, O. L., Jr. 1966. J. Protozool. 13:336.
20. Chaix, P., Chauvet, J., and Fromageot, C. 1947. Antoine van Leewoenhoek, J. Microbiol. Serol. 12:145.
21. Chou, S. C. and Scherbaum, O. H. 1965. Exp. Cell Res. 40:217.
22. Chou, S. C. and Scherbaum, O. H. 1966. Exp. Cell Res. 45:31.
23. Cline, S. and Conner, R. L. 1966. J. Cell Physiol. 13:336.
24. Conner, R. L. and Cline, S. G. 1967. J. Protozool. 14:22.
25. Conner, R. L. and Murnane, S. 1971. J. Protozool. 18(suppl.):9.
26. Cook, D. E., Rangaraj, D. E., Best, N., and Wilkin, D. R. 1968. Arch. Biochem. Biophys. 127:72.
27. Danforth, W. F. 1967. In T. T. Chen, ed., Research in protozoology, vol. I, Pergamon Press, New York, p. 201.
28. DeDuve, C. 1969. Ann. N. Y. Acad. Sci. 168:369.
29. Dianzani, M. U. 1963. In A. V. S. de Rueck and M. P. Cameron, eds., Ciba Foundation Symposium on lysosomes, Little, Brown, and Co., Boston, p. 335.
30. Eichel, H. J. 1956. J. Biol. Chem. 222:137.
31. Eichel, H. J. 1959. Biochem. J. 71:106.
32. Eichel, H. J. 1959. Biochim, Biophys. Acta. 34:589.
33. Eichel, H. J. 1960. Biochim. Biophys. Acta. 43:364.
34. Eichel, H. J. and Figueroa, E. 1963. J. Protozool. 10(suppl.):6.
35. Eichel, H. J., Goldenberg, E. K., and L. T. Rem. 1964. Biochim. Biophys. Acta. 81:172.
36. Eichel, H. J. and Rem, L. T. 1962. J. Biol. Chem. 237:940.
37. Elson, C., Hartman, H. A., Shug, A. L., and Shrago, E. 1970. Science 168:385.
38. Elliott, A. M. Personal communication.
39. Elliott, A. M. and Bak, I. J. 1964. J. Cell Biol. 20:113.
40. Elliott, A. M., Kennedy, J. R., and Bak, I. J. 1962. J. Cell Biol. 12:515.
41. Elliott, A. M., Travis, D. M., and Bak, I. J. 1962. Biol. Bull. 123:495.
42. Elliott, A. M., Travis, D. M., and Work, J. A. 1966. J. Exp. Zool. 161:177.
43. Engemann, J. G. 1958. J. Protozool. 5(suppl.):13.
44. Erwin, J. A. 1970. Biochim. Biophys. Acta. 202:21.
45. Erwin, J. and Bloch, K. 1963. J. Biol. Chem. 238:1618.
46. Everhart, L. P. and Ronkin, R. R. 1966. J. Protozool. 13:646.
47. Farrell, J. and Rose, A. H. 1965. Adv. Appl. Microbiol. 7:335.
48. Farrell, J. and Rose, A. H. 1967. Ann. Rev. Microbiol. 21:101.
49. Fennell, R. A. 1951. J. Elisha Mitchell Sci. Soc. 67:219.
50. Fisher, R. B. 1964. In F. Dickens and E. Neil, eds., Oxygen and the animal organism, Pergamon Press, New York, p. 339.
51. Friedman, S. M. and Weinstein, I. B. 1964. Proc. Natl. Acad. Sci. U.S. 52:988.
52. Fowler, S. D. 1967. J. Cell Biol. 35(suppl.):41A.
53. Goodfriend, T. L., Sokol, D. M., and Kaplan, N. O. 1966. J. Mol. Biol. 15:18.

54. Gray, R. W., Arsenis, C., and Jeffay, H. 1970. Biochim. Biophys. Acta. 222:627.
55. Gross, J. A. 1962. J. Protozool. 9:415.
56. Haynes, R. C. 1965. Adv. Enzyme Regul. 3:111.
57. Hill, D. H. 1972. The biochemistry and physiology of *Tetrahymena.* Academic Press, New York.
58. Hill, D. L. and J. van Eys. 1965. J. Protozool. 12:259.
59. Hogg, J. F. 1957. 132nd Meet. Abstr. Amer. Chem. Soc. 10c.
60. Hogg, J. F. 1959. Fed. Proc. 18:247.
61. Hogg, J. F. and Kornberg, H. L. 1963. Biochem. J. 86:462.
62. Hogg, J. F. and C. Wagner. 1956. Fed. Proc. 15:275.
63. Holz, G. G., Jr. 1957. Anat. Rec. 128:566.
64. Holz, G. G., Jr. 1960. Biol. Bull. 118:84.
65. Holz, G. G., Jr. 1964. In S. H. Hutner, ed., Biochemistry and physiology of protozoa, vol. 3, Academic Press, New York, p. 199.
66. Holz, G. G., Jr., Erwin, J. A., and Davis, R. J. 1959. J. Protozool. 6:149.
67. Horiguchi, T. and Novick, A. 1961. Cold Spring Harbor Symp. Quant. Biol. 26:247.
68. Jahn, T. L. 1934. Cold Spring Harbor Symp. Quant. Biol. 2:167.
69. Jahn, T. L. 1936. Proc. Soc. Exp. Biol. Med. 33:494.
70. James, T. W., and Read, C. P. 1957. Exp. Cell Res. 13:510.
71. Kamiya, T. and Takahashi, T. 1961. J. Biochem. 50:277.
72. Kapoulas, V. M., Thompson, G. A., and Hanahan, D. J. 1969. Biochim. Biophys. Acta. 176:250.
73. Katunuma, N. and Okada, M. 1963. Biochem. Biophys. Res. Commun. 12:252.
74. Kidder, G. W. and Dewey, V. C. 1951. In A. Lwoff, ed., Biochemistry and physiology of protozoa, vol. I, Academic Press, New York, p. 324.
75. Klamer, B. and Fennell, R. A. 1963. Exp. Cell Res. 29:166.
76. Koehler, L. D. and Fennell, R. A. 1964. J. Morphol. 114:209.
77. Kornberg, H. L. 1966. Biochem. J. 99:1.
78. Koroly, M. J. and Conner, R. L. 1968. J. Protozool. 15(suppl.):14.
79. Koyama, S. and Tsukada, H. 1970. Tumor Res. 5:29.
80. Langridge, J. 1963. Ann. Rev. Plant Physiol. 14:441.
81. Lazarus, L. H. 1966. PhD. Thesis, University of California, Los Angeles.
82. Lazarus, L. H. and Scherbaum, L. H. 1966. J. Cell Physiol. 68:95.
83. Lee, D. 1970. J. Cell Physiol. 76:17.
84. Lennarz, W. J. 1970. Ann. Rev. Biochem. 39:359.
85. Levy, M. R. 1967. Comp. Physiol. Biochem. 21:291.
86. Levy, M. R. 1967. J. Cell Physiol. 69:247.
87. Levy, M. R. Unpublished.
88. Levy, M. R. 1972. Arch. Biochem. Biophys. 152:463.
89. Levy, M. R. and Elliott, A. M. 1968. J. Protozool. 15:208.
90. Levy, M. R., Gollon, C. E., and Elliott, A. M. 1969. Exp. Cell Res. 55:295.
91. Levy, M. R. and Hunt, A. E. 1967. J. Cell. Biol. 34:911.
92. Levy, M. R., Naeve, J. A. and Irish, L. E. Unpublished.
93. Levy, M. R. and Scherbaum, O. H. 1965. Arch. Biochem. Biophys. 109:116.
94. Levy, M. R. and Scherbaum, O. H. 1965. J. Gen. Microbiol. 38:221.
95. Levy, M. R. and Wasmuth, J. J. 1970. Biochim. Biophys. Acta. 201:205.
96. Lloyd, D., Brightwell, R., Venables, S. E., Roach, G. I., and Turner, G. 1971. J. Gen. Microbiol. 65:209.
97. Loeh, A. L. and Levy, M. R. 1969. J. Cell Biol. 43:82a.
98. Mackenzie, T. B., Stone, G. E., and Prescott, D. M. 1966. J. Cell Biol. 31:633.
99. Malecki, M. T., Licko, V., and Eiler, J. J. 1971. Curr. Mod. Biol. 3:291.
100. Manners, D. J., and Ryley, J. F. 1952. Biochem. J. 52:480.
101. Marr, A. G. and Ingraham, J. L. 1962. J. Bacteriology. 84:1260.
102. Mavrides, C. and D'Iorio, A. 1969. Biochem. Biophys. Res. Commun. 35:467.
103. McDonald, B. M. 1958. Biol. Bull. 114:71.
104. Müller, M. 1972. J. Cell Biol. 52:478.

105. Müller, M., Baudhuin, P., and de Duve, C. 1966. J. Cell Physiol. 68:165.
106. Müller, M. and Hogg, J. F. 1967. Fed. Proc. 26:284.
107. Müller, M., Hogg, J. F., and de Duve, C. 1968. J. Biol. Chem. 243:5385.
108. Nardone, R. M. and Wilbur, C. G. 1950. Proc. Soc. Exp. Bio. Med. 75:559.
109. Nilsson, J. R. and Leick, V. 1970. Exp. Cell Res. 60:361.
110. Ng, H., Ingraham, T. L., and Marr, A. G. 1962. J. Bacteriol. 84:331.
111. Nishi, A. and Scherbaum, O. H. 1962. Biochim. Biophys. Acta. 65:411.
112. Nishi, A. and Scherbaum, O. H. 1962. Biochim. Biophys. Acta. 65:419.
113. Odum, H. T. 1956. Limnol. Oceanogr. 1:102.
114. Ormsbcc, R. A. 1942. Biol. Bull. 82:423.
115. Pace, D. M. and Ireland, R. L. 1945. J. Gen. Physiol. 28:547.
116. Pace, D. M. and Lyman, E. D. 1947. Biol. Bull. 92:210.
117. Phelps, A. 1946. J. Exp. Zool. 102:277.
118. Phelps, A. 1955. J. Protozool. 2(suppl.):9.
119. Prescott, D. M. 1957. Exp. Cell Res. 12:126.
120. Prescott, D. M. 1957. J. Protozool. 4:252.
121. Prescott, D. M. 1958. Physiol. Zool. 31:111.
122. Prosser, C. L., ed. 1967. Molecular mechanisms of temperature adaptation.
     Am. Assn. Adv. Sci., Washington, D.C.
123. Ray, P. D., Foster, D. O., and Lardy, H. A. 1964. Fed. Proc. 23:482.
124. Reeves, H., Papa, M., Seaman, G.. and Ail.. S. 1961. J. Bacteriol. 81:154.
125. Reid, R. 1967. J. Protozool. 14(suppl.):8.
126. Risse, H. S., and Blum, J. J. 1972. Arch. Biochem. Biophys. 149:329.
127. Rockland, L. B. and Dunn, M. S. 1949. J. Biol. Chem. 179:511.
128. Rosenbaum, N., Erwin, J., Beach, D., and Holz, G. G., Jr. 1966. J. Protozool. 13:535.
129. Rosenberg, H. 1964. Nature 203:299.
130. Rutter, W. J., Rajkumar, T., Penhoet, E., Kochman, M., and Valentine, R. 1968.
     Ann. N. Y. Acad. Sci. 151:102.
131. Ryley, J. F. 1952. Biochem. J. 52:483.
132. Sato, H. 1960. Anat. Rec. 138:381.
133. Scherbaum, O. H. and Levy, M. R. 1961. Biol.-Pathol. 9:514.
134. Seaman, G. R. 1955. In S. H. Hutner and A. Lwoff, eds., Biochemistry and physiology of
     protozoa, Academic Press, New York, p. 91.
135. Shrago, E., Brech, W., and Templeton, K. 1967. J. Biol. Chem. 242:4060.
136. Shrago, E. and Shug, A. L., 1966. Biochem. Biophys. Acta. 122:376.
137. Shug, A. L., Elson, C., and Shrago, E. 1969. J. Nutr. 99:379.
138. Shug, A. L., Ferguson, S., and Shrago, E. 1968. Biochem. Biophys. Res. Commun. 32:81.
139. Singer, S. J., and Nicolson, G. L. 1972. Science. 175:720.
140. Slater, T. F., Greenbaum, A. L. and Wang, D. Y. 1963. In A. V. S. de Rueck and M. P. Cameron, ed
     Ciba Foundation symposium on lysosomes. Little, Brown, and Co., Boston, p. 311.
141. Slater, J. V. Ph.D. Thesis. University of Michigan.
142. Smith, A. D. and Winkler, H. 1969. In J. T. Dingle and H. B. Fell, eds., Lysosomes in biology
     and pathology, part 1, North-Holland, Amsterdam, p. 155.
143. Stokes, J. L. In C. L. Prosser, ed., Molecular mechanisms of temperature adaptation.
     Am. A. Adv. Sci., Washington, D. C., 1967, p. 311.
144. Swann, M. M. 1954. Colston Pap. 7:185.
145. Thompson, G. A. 1967. Biochem. 6:2015.
146. Thompson, G. A. 1969. Biochim. Biophys. Acta. 176:330.
147. Thompson, G. A. 1969. J. Protozool. 16:397.
148. Thompson, W. D. and Wingo, W. J. 1959. J. Cell. Comp. Physiol. 53:413.
149. Thormar, H. 1962. Exp. Cell Res. 28:269.
150. Travis, D. M., Elliott, A. M., and Bak, I. J. 1962. Biol. Bull. 123:487.
151. Van Lanker, J. L. 1963. Fed. Proc. 23:1050.
152. Wagner, C. 1956. Ph.D. Thesis. The University of Michigan.

153. Waithe, W. I. 1964. Exp. Cell Res. 35:100.
154. Warnock, L. G. and van Eys, J. 1962. J. Cell Comp. Physiol. 60:53.
155. Warnock, L. G. and van Eys, J. 1963. J. Cell Comp. Physiol. 61:309.
156. Wingo, J. J. and Cameron, L. E. 1952. Texas Rep. Biol. Med. 10:1075.
157. Wingo, W. J. and Thompson, W. D. 1960. Fed. Proc. 19:243.

CHAPTER 9

Life Cycle and Distribution of *Tetrahymena*

Alfred M. Elliott

Department of Zoology
The University of Michigan
Ann Arbor, Michigan

I. LIFE CYCLE

There is some variation in the life cycle of the various species of *Tetrahymena*. This has been discussed in Chapter 1, consequently this section deals with the life cycle of *T. pyriformis* only.

Tetrahymena pyriformis, similar to most ciliates, maintains its numbers by asexual reproduction which is dependent on various environmental conditions. Under certain conditions, usually when food is scarce, the ciliate undergoes a sexual phase involving conjugation during which meiosis and fertilization occur. The progeny then divide mitotically for a varying period of time after which conjugation once more is initiated. Following successful conjugation the progeny are sexually immature, that is, they do not conjugate until they have divided a certain number of times which varies in the different varieties. Once the ciliates are sexually mature, they remain so for varying periods of time. Sooner or later, however, they become senile and finally die. The usual signs of senility are loss of the micronucleus, and selfing (mating within the clone). Amicronucleate ciliates are sexually dead and usually fail to mate. They occur in surprisingly high numbers (30–50%) under natural conditions (13,21,25). How amicronucleate strains come about has been determined by Allen et al. (2). Many of our strains have become amicronucleate in laboratory culture over a period of 17 yr.

A small number of selfers appear in our collections, as well as in the progeny of certain crosses. As suggested by Sonneborn (65) for *Paramecium aurelia,* selfing in *T. pyriformis* may be age-induced. Normal nonselfing strains in syngen 6 later became selfers as shown by Wells (67). Selfing usually results in death of the conjugants, hence is one way of eliminating these unstable cells from the population.

The above observations have all been made in the laboratory; whether or not this life cycle occurs under natural conditions is impossible to say, although it is likely that it does.

A. Asexual Reproduction

Very little is known about the morphological details of mitosis in *T. pyriformis.* The events can be clearly observed in Feulgen preparations with the light microscope, or in living material under the phase-contrast microscope. Just prior to division the cells are large and the micronucleus swells and moves out of the indentation of the macronucleus. Thin, thready chromosomes are visible at metaphase but not before. They move to the poles in conventional fashion, and the nucleus divides. The nuclear envelope remains intact throughout the division cycle.

Electron microscopy has added little to our knowledge of mitosis. Several stages of the dividing micronucleus have been observed (13). The interphase chromosomes become condensed but show little detail (Fig. 1). At metaphase and anaphase microtubules

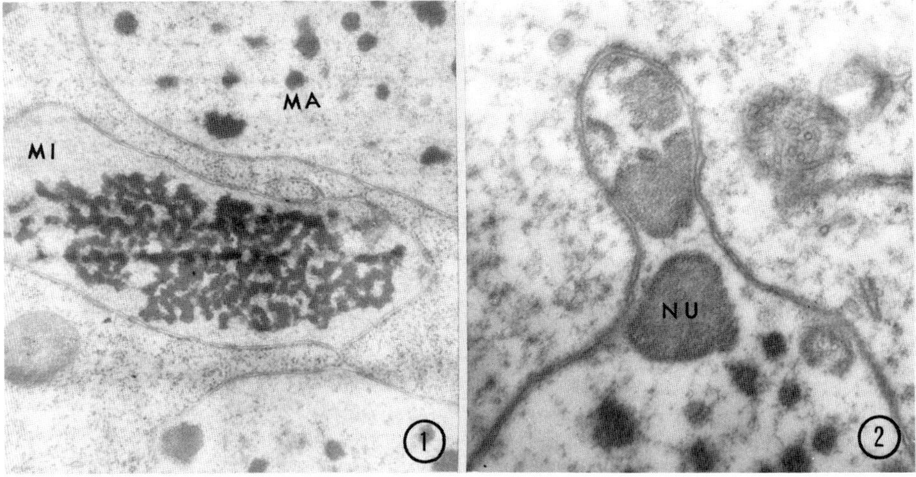

Fig. 1.

An interphase micronucleus (MI) lying in the indentation of the macronucleus (MA). Note the continuity of the macro- and micronuclear envelopes. The chromosomes are dense and highly coiled. ×22,000.

Fig. 2.

A bleb containing nucleoli (NU) extends from the macronucleus into the cytoplasm of a normally growing ciliate. Such blebs probably pinch off, depositing their nucleolar ribosomes and other macronuclear material in the cytoplasm. ×28,000.

(spindle fibers) appear but they can be seen better in the crescent during conjugation (Figs. 13 and 14).

Under the light microscope the macronucleus appears to pinch into two equal parts distributing Feulgen-positive chromatin bodies and nucleoli to the daughter macronuclei. There is no morphological evidence that mitosis is occurring although cytochemical tests show that the DNA doubles during interphase (see Chapter 10). At the electron microscope level, the chromatin bodies and nucleoli can be seen to be more numerous prior to division. This is particularly obvious in cells that have undergone several heat shocks to induce synchrony in division (see Chapter 7). Ribonucleic acid is rapidly synthesized during this treatment. It seems to concentrate in centrally located bodies, called RNA bodies (23), which eventually fuse with nucleoli. This was observed only in heat-shocked cells (strain WH-6) but was not confirmed by Cameron and Guile (7) in an amicronucleate strain (GL). The discrepancy may be due to a strain difference.

The macronuclear envelope remains intact throughout fission. The question how RNA leaves the macronucleus may be answered from ultrastructural studies. During a brief period between the third and fourth heat shock, conspicuous blebs containing nucleolar material extend into the cytoplasm (23). These also occur during the first division following the heat shock treatment. The blebs apparently pinch off as fragments, subsequently releasing their contents (ribosomes and some chromatin material) to the cytoplasm. Similar blebbing occurs in normal fission (Fig. 2), hence is not induced, but may be accentuated, by heat shock.

B. Sexual Reproduction

Nuclear reorganization normally occurs during conjugation. Autogamy, common in some ciliates, has not yet been observed in *T. pyriformis*. Aberrations in conjugation perhaps occur more frequently than was once thought (see Chapter 11).

1. Conjugation

The first account of conjugation in *T. pyriformis* was that of Maupas in 1889 (43) in *Leucophrys patula* which is now thought to be *T. pyriformis* (9,10). The process was reported again over 60 yr later by Horn (34) and Elliott and Nanney (24). All these observations were made on selfers which left no progeny. Fruitful conjugation with surviving offspring was found in several strains isolated from a Cape Cod, Massachusetts, freshwater pond in 1952 by Elliott and Gruchy (19). Subsequently, mating types were defined for the first variety, and during the next few years 12 varieties were identified (see below).

The term "variety" was originally used to distinguish these interbreeding populations, but the term "syngen" is currently receiving more acceptance by many workers in the field and is used in this chapter.

In the laboratory several factors have been found to influence the mating process. Starving the cells by washing them in distilled water induces conjugation within a few hours (20). Temperature and light variations have very little effect on the process, but anoxia and agitation prevent it. Well-fed ciliates do not mate. A lack of nutrilites is probably the most likely factor in nature that induces conjugation. An effort to discover

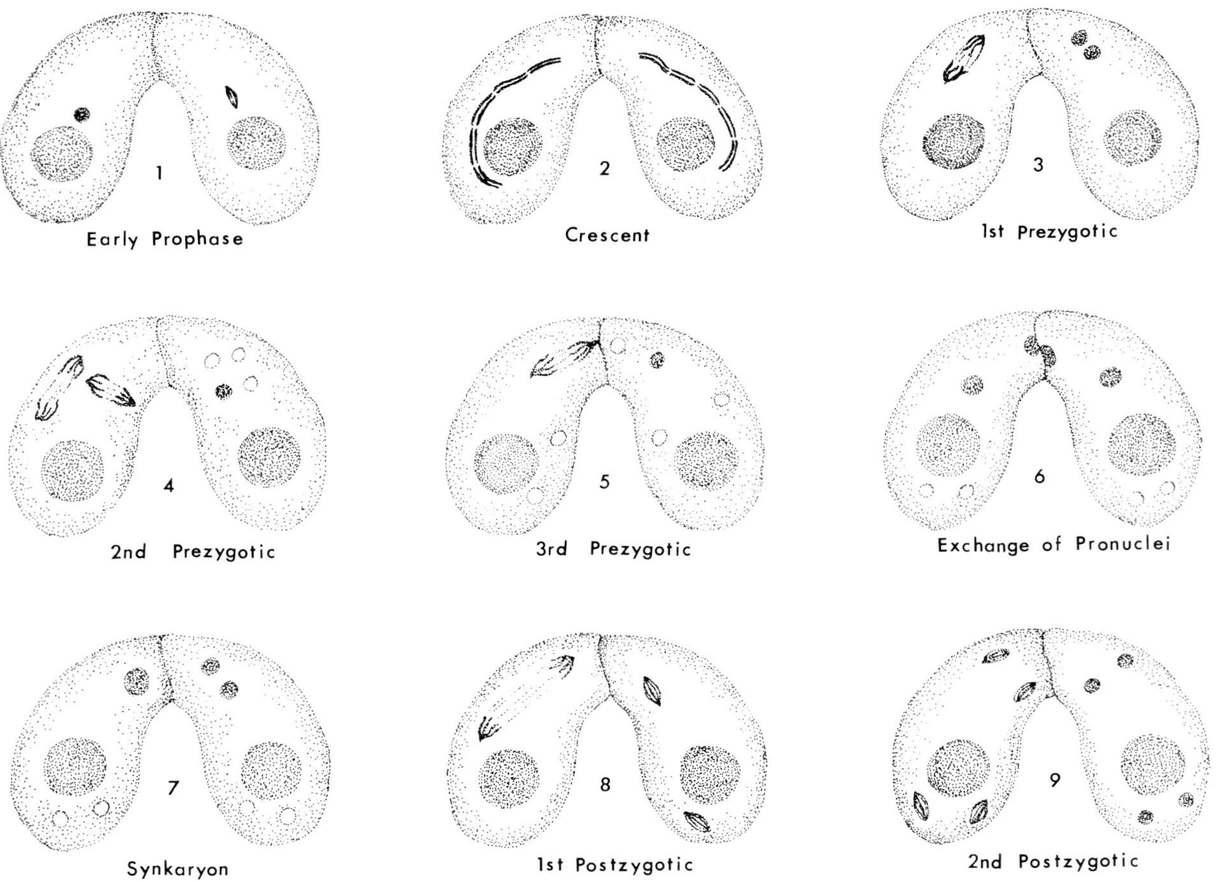
1
Early Prophase
2
Crescent
3
1st Prezygotic
4
2nd Prezygotic
5
3rd Prezygotic
6
Exchange of Pronuclei
7
Synkaryon
8
1st Postzygotic
9
2nd Postzygotic

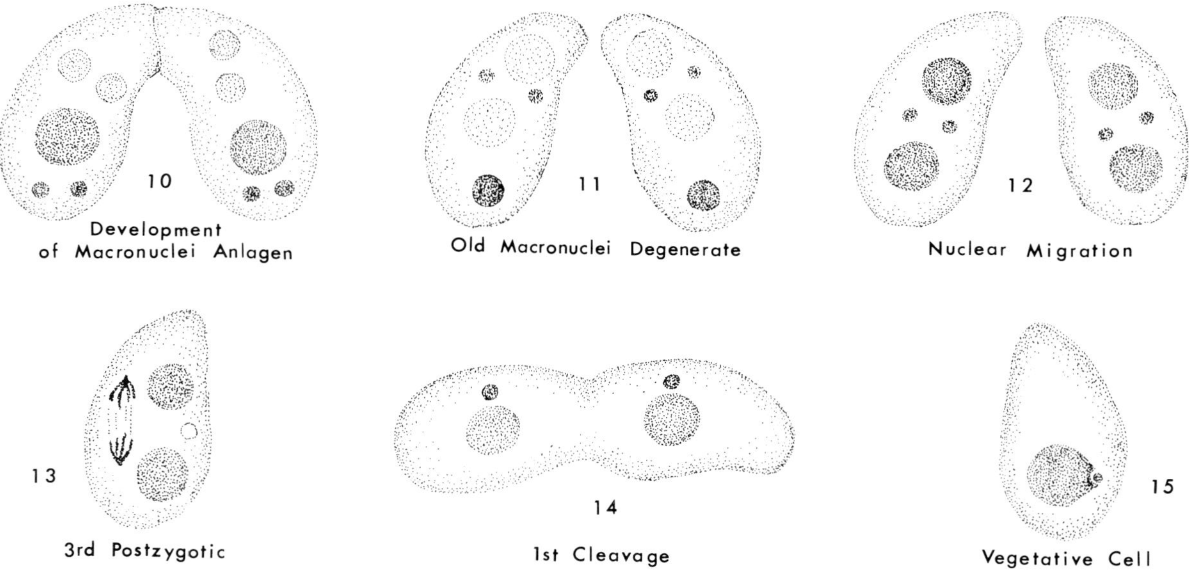

Fig. 3.

A schematic representation of the various stages of conjugation. See text for explanation.

which nutrilites, singly or in combination, of the 18 required for growth of strain E induced or inhibited mating proved inconclusive (12). Insofar as we have been able to determine, no sex hormones are released into the medium in which starved cells are maintained. Such hormones may be present but adhere to the external surfaces of the mates and function only upon contact. It is true that conjugation takes place more readily when the cells are concentrated than when dispersed, suggesting that the ciliates make contact fortuitously rather than being attracted to one another by some exogenous attractant (hormone).

The gross cytological events of conjugation at the light microscope level have been worked out in both selfing strains (43,47) and in mating types (20,60). A brief account of the events follows. The stages are sketched in Fig. 3.

The mates become attached to a specific region just anterior to the mouth. The micronuclei move away from the macronuclei and enlarge, forming a sac with unfolded chromosomes which are at first single but later become paired. The sac elongates and the double chromosomes attach end to end, forming the *crescent stage* which extends the length of the cell. The crescent then shortens, and at this time five pairs of chromosomes' are clearly visible. There are three of similar dimensions, one larger, and one smaller, all with submedian or median centromeres. Tetrads result when the chromosomes duplicate. They then separate into dyads at the first meiotic division. This is followed by the second meiotic division, resulting in four haploid nuclei three of which disintegrate leaving one functional nucleus. This one undergoes a single division, giving rise to two gametic pronuclei. One of these in each conjugant migrates through the membranes where the cells are attached and fuses with the pronucleus in the reciprocal mate. The resulting synkaryon undergoes two mitotic divisions, producing four nuclei two of which move to the anterior and two to the posterior regions of each mate. The anterior nuclei differentiate into macronuclei, and the posterior nuclei become micronuclei one of which degenerates. The conjugants separate at about this time. During these latter events the old macronucleus in each mate becomes very dense and gradually decreases in size until shortly after the conjugants separate, when they are no longer visible. After separating, the conjugants undergo one fission in which the micronuclei divide but the macronuclei do not. The resulting daughter cells then contain a single macronucleus and micronucleus. Vegetative fissions follow until again interrupted by conjugation. The cytological details of these events have been described exceptionally well by Ray (60,61) in strains WH-6 and WH-14. (See the legends for Fig. 4.)

Nothing new regarding the events of conjugation has been learned from ultrastructural studies. Some morphological details have thrown light on certain aspects of the process, however.

Starved cells of opposite mating type, when mixed, possess pronounced fingerlike processes anterior to the undulating membrane which is the area of future contact between the mates (Fig. 5). Each process contains pellicular membranes with regularly spaced granules lying between the outer and the next inner membranes, and a cytoplasmic core. The latter is bounded by the inner pellicular membrane, and an empty space occupies the tip of the process. Several microtubules (five to six), shown in cross section, lie at the tip of the cytoplasmic extension. A kinetosome and cilium lie between the processes. All these structures may be involved in mating.

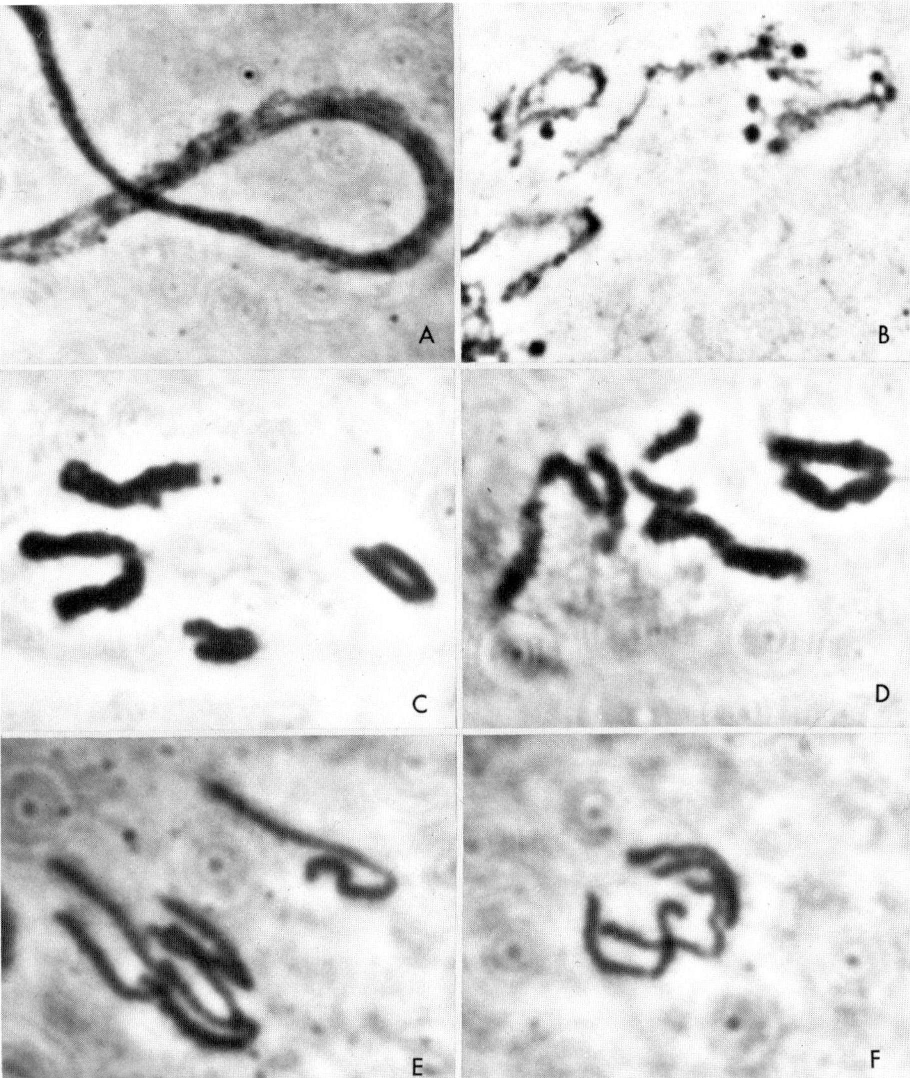

Fig. 4.

Light micrographs of chromosomes during various stages of conjugation. (A) Part of a micronucleus at late crescent stage during first meiotic division at conjugation. Note discontinuities along the micronucleus. (B) Chromosomes at a slightly later stage than shown in (A) Chromosomes at late pachytene of first meiotic division during conjugation. Note heterochromatic knobs and segments. (C) Four of the five pairs of chromosomes at premetaphase of the first meiotic division; the smallest fifth pair of chromosomes is outside the field of view. (D) Five pairs of chromosomes at premetaphase of the first meiotic division. (E) Five chromosomes at one pole of late anaphase of the second meiotic division. (F) Five chromosomes at one pole of late anaphase of the third prezygotic division. The third prezygotic division is a mitotic division which produces the two gametic pronuclei. (Micrographs kindly supplied by C. Ray, Jr.)

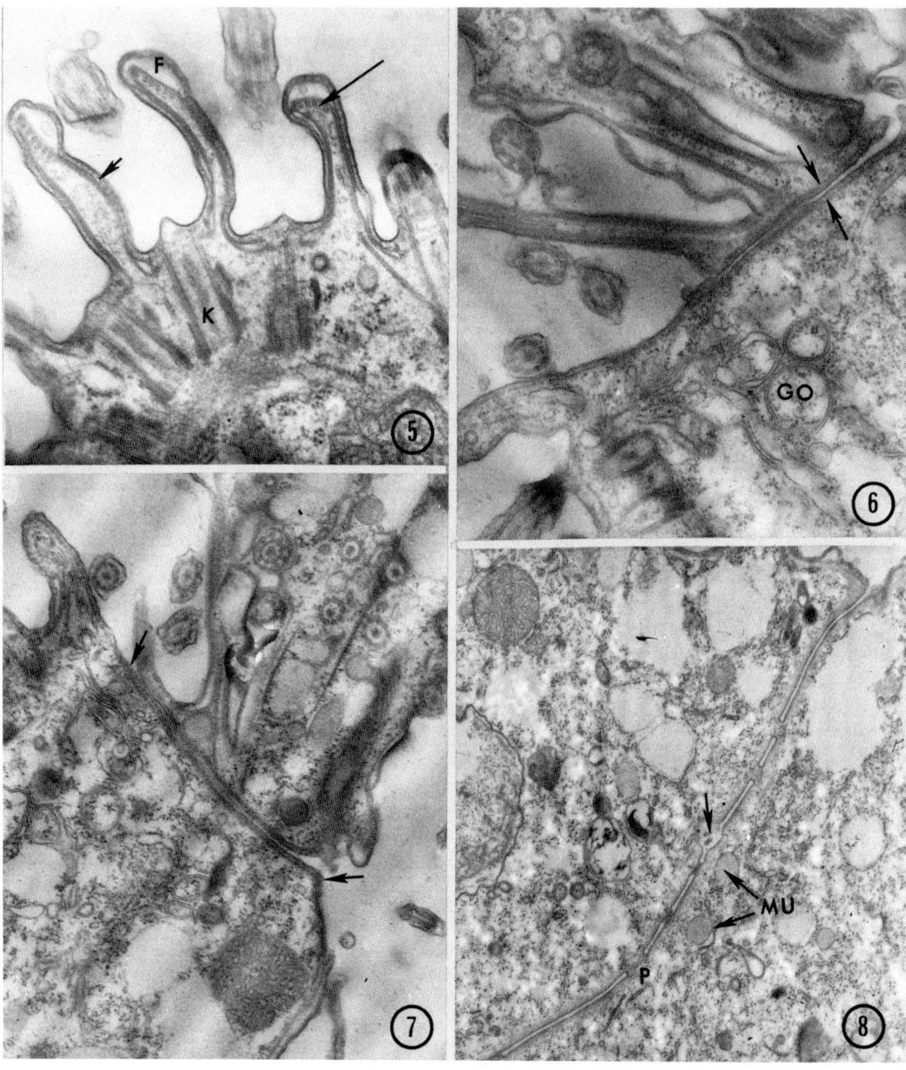

Fig. 5.

Fingerlike processes (F) appear in the region anterior to the mouth where future contact will be made by conjugating cells. Note the kinetosomes (K) lying between the processes. The plane of section is such that the cilia show only in cross section. Regularly spaced granules (short arrow) lie between the pellicular membranes, and several microtubules (long arrow) appear in cross section. ×44,000.

Fig. 6.

The initial stage of contact between mates shows the fingerlike processes making contact with a modified part of the pellicle lying anterior to the oral apparatus. The processes fuse at their tips, forming a continuous flattened structure which lies next to the modified pellicle of the opposite mate. Internal to the region of contact of both mates, the pellicular membranes are fused, thick, and dense (arrows). Golgi saccules (GO) lie near the contact membranes. ×40,000.

During the initial stages of conjugation, the processes come in contact with a modified region of the pellicle of the mate (Fig. 6). This area lies just anterior to the oral apparatus and is bounded by a ridge forming a cuplike structure (Figs. 6 and 7). There are no cilia in this structure, and the pellicular membranes are quite dense, probably as a result of fusion. The processes of one mate fuse near the point of contact with the other mate to form a flat surface which lies very close yet does not touch the mate, suggesting that some substance lies between. The contact membranes are dense and thick at this stage. Flattened saccules with swollen ends and small vesicles nearby lie beneath the contact membranes.

Later in conjugation the area of contact between the mates enlarges until the entire anterior ends of the organisms are securely attached (Fig. 8). Most of the oral structures have been resorbed, although some remnants of the undulating membrane seem to persist.

The inner part of the contact membranes is less well defined than in earlier stages and appears fuzzy (Fig. 9). Pores of irregular dimensions are present in the membranes. Tiny vesicles, similar to those in the membranes at earlier stages, aggregate around the edges of the pores. Occasionally, unidentified material becomes trapped between the contact membranes (Fig. 8). Numerous mucocysts and saccules are seen in this region of the mates (Figs. 6, 8, and 9).

The region of contact between mates of other ciliates has been studied at the ultrastructural level by several investigators. The series of events in several species of *Paramecium* is similar to that in *T. pyriformis* (36,62,66), while in *Euplotes* it is markedly different (51).

A classic Golgi apparatus is induced prior to mating. The single saccules observed in vegetative cells (29,30) (see Chapter 2) appear to become stacked into two or more lamellae. In as little time as 5 min following mixing, these stacked saccules, located in the oral region, dilate and become elongated (Fig. 10). Small vesicles lying nearby are thought to originate from the tips of the saccules by a pinching-off process. The fact that the Golgi apparatus is concentrated in the oral region and that it seems to be packaging some product prior to conjugation suggests that its appearance is induced by the mating process (26). Once the mates are firmly attached, the Golgi apparatus becomes inactive, that is, it no longer pinches off vesicles (Fig. 9). Some time before the mates separate (10–15 h after mixing), however, the Golgi apparatus is occasionally observed producing vesicles.

It is tempting to speculate on the function of the Golgi apparatus in *T. pyriformis*,

---

Fig. 7.

A later stage in the fusion of mates. The processes are consolidated into a single structure, and the thickened membranes in the contact region lie close to one another. The extent of the fusion area of the lower mate is indicated by arrows. ×38,000.

Fig. 8.

The fusion region between the mates is large after several hours of contact. The pellicular membranes are less dense, and well-defined pores (P) provide free flow of cytoplasmic particulates between the mates. Mucocysts (MU) and vacuoles of various kinds are seen in the cytoplasm of both mates. Unidentified material often becomes trapped between the contact membranes (arrow). ×17,000.

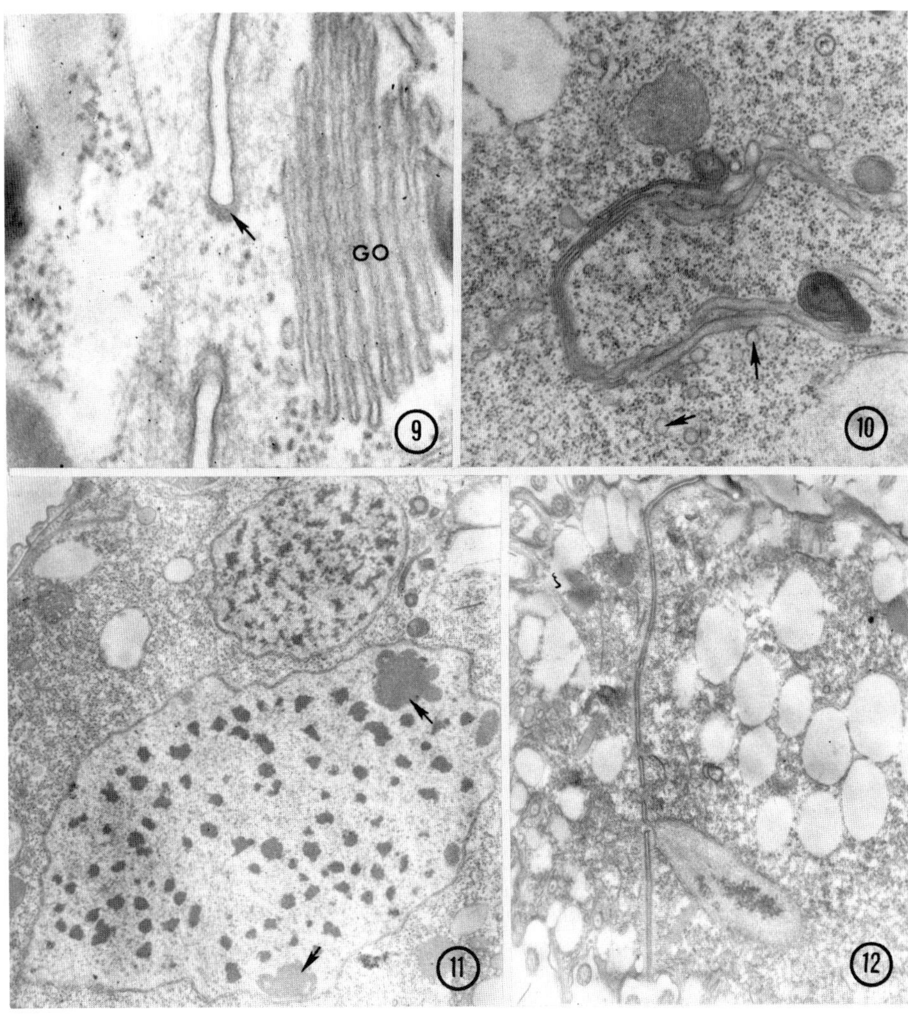

Fig. 9.

At higher magnification the details of the pore can be seen. Tiny vesicles form the inner limits of the pore (arrows). The Golgi apparatus (GO) consists of tightly stacked saccules without nearby vesicles. ×70,000.

Fig. 10.

A conspicuous Golgi apparatus appears a few minutes after opposite mating types are mixed. The saccules are elongated at their tips, and small vesicles seem to be pinching off from them (arrows). ×32,000.

Fig. 11.

Once the mates become attached, the micronucleus moves away from the macronucleus and swells as shown in this micrograph. The chromosomes are fine and thready. The large nucleoli (arrows) result from fusion of the smaller interphase nucleoli. ×9000.

Fig. 12.

The anterior end of the crescent is shown here lying close to the contact membrane of one mate. ×12,000.

and possibly ciliates in general. Similar dense-membraned saccules have been seen in the oral regions of other ciliates (27, 39, 59). These are usually isolated single saccules with different shapes, but the membranes are similar in the different ciliates. In *T. pyriformis*, when in close association with the opposite mating type, the isolated saccules form stacks and begin packaging (perhaps even synthesizing) some product that is essential for successful mating. This could be a substance that functions as a sex attractant, or as a cementing agent which makes the fusion of the mates possible. The fact that the Golgi apparatus seems to be active before actual contact between the mates is made, and only when they are mixed, suggests that either of these two functions, or perhaps both, are possible. If this is so, it would be one of the few cases in which the induction of a Golgi apparatus can be observed.

The ultrastructural details of the nuclear events during meiosis and fertilization have not been completed, although some stages have been observed (unpublished).

Once the mates are attached, the micronucleus moves away from the macronucleus and enlarges, and the chromosomes appear as fine threads (Fig. 11). The nucleoli in the macronucleus consolidate into several large structures, while the chromatin bodies show no change in structure, although there is a steady increase in their numbers. The crescent forms and is clearly seen as an elongated sac containing numerous microtubules (spindle fibers) and poorly defined chromosomes (Figs. 12–14). The double-stranded chromosomes can best be seen in cross section (Fig. 14). They are composed of dense granules. The microtubules are concentrated in the central region of the crescent in close association with the chromosomes and also near the nuclear membrane on one side in Fig. 14, which may not be typical. The external nuclear membrane is studded with ribosomes, and the envelope seems to contain no pores.

Following the third prezygotic division, the migratory pronuclei move toward (Fig. 15), and finally lie flattened against, the contact membranes (Fig. 16) which disintegrate later as the pronuclei pass through (Fig. 17). How the membranes reestablish themselves has not been observed.

During the later stages of conjugation, the two anterior nuclei develop into macronuclei by replication of their contents at a very rapid pace. All one sees in electron micrographs is a striking increase in size of the entire structure and a proliferation of the chromatin bodies which are at first tiny bodies of low density (Fig. 18). By the time the macronuclei are fully developed, the chromatin bodies are numerous and dense. How and when the nucleoli are formed has not been observed. The old, degenerating macronuclei decrease in size and become extremely dense (Fig. 18) before the cells separate. Some internal structure remains as the degrading process occurs (Fig. 19). The oral structures, most of which have been resorbed during the fusion process, begin to reappear before the mates separate (Fig. 18).

2. Mating Types and Syngens

The concept of mating types and syngens in ciliates was established in 1938 by Sonneborn (64) and Jennings (37), who worked with two species of *Paramecium*. This concept has been followed by investigators working with other ciliates including *T. pyriformis*. Briefly, it states that conjugation occurs only between complementary

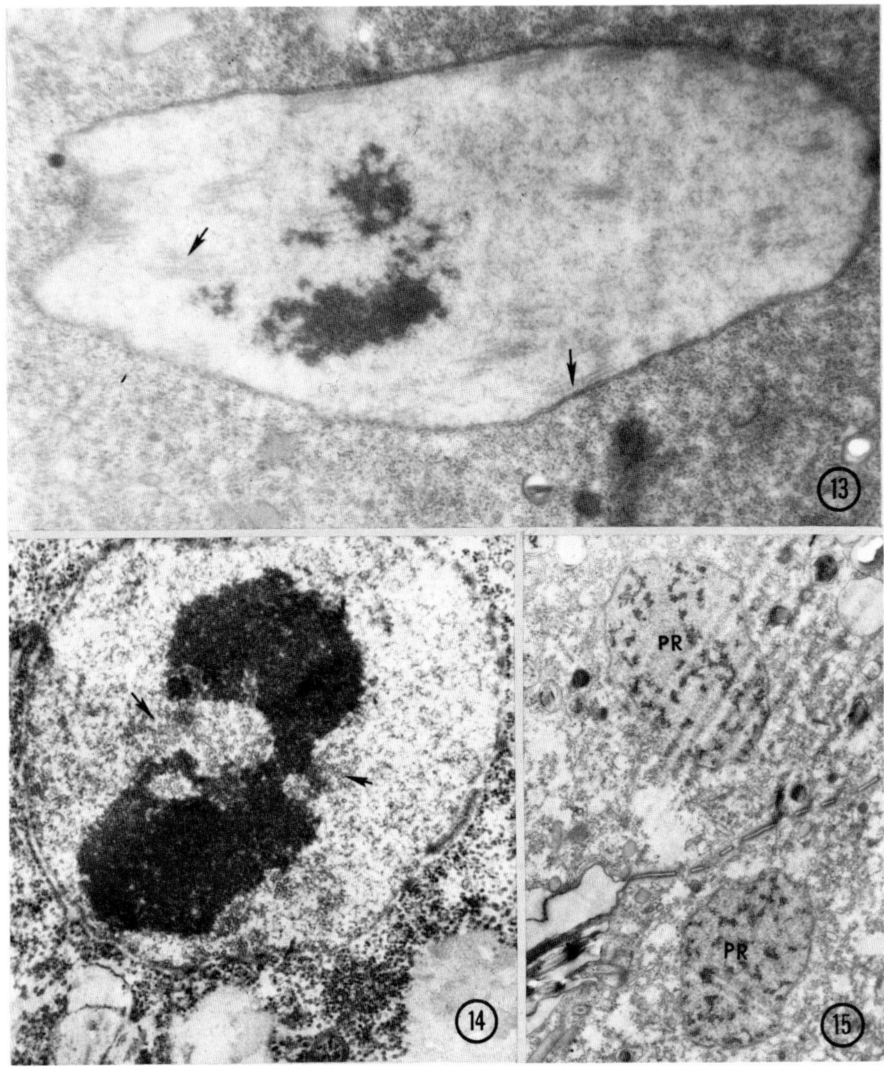

Fig. 13.

An enlarged view of a portion of the crescent showing scattered microtubules (arrows) and fragments of the chromosomes. There seem to be no pores in the envelope. ×29,000.

Fig. 14.

In cross section the double-stranded chromosomes and microtubules (arrows) of the crescent can be seen. ×32,000.

Fig. 15.

The third-prezygotic migratory pronuclei (PR) approach the contact membranes. ×10,000.

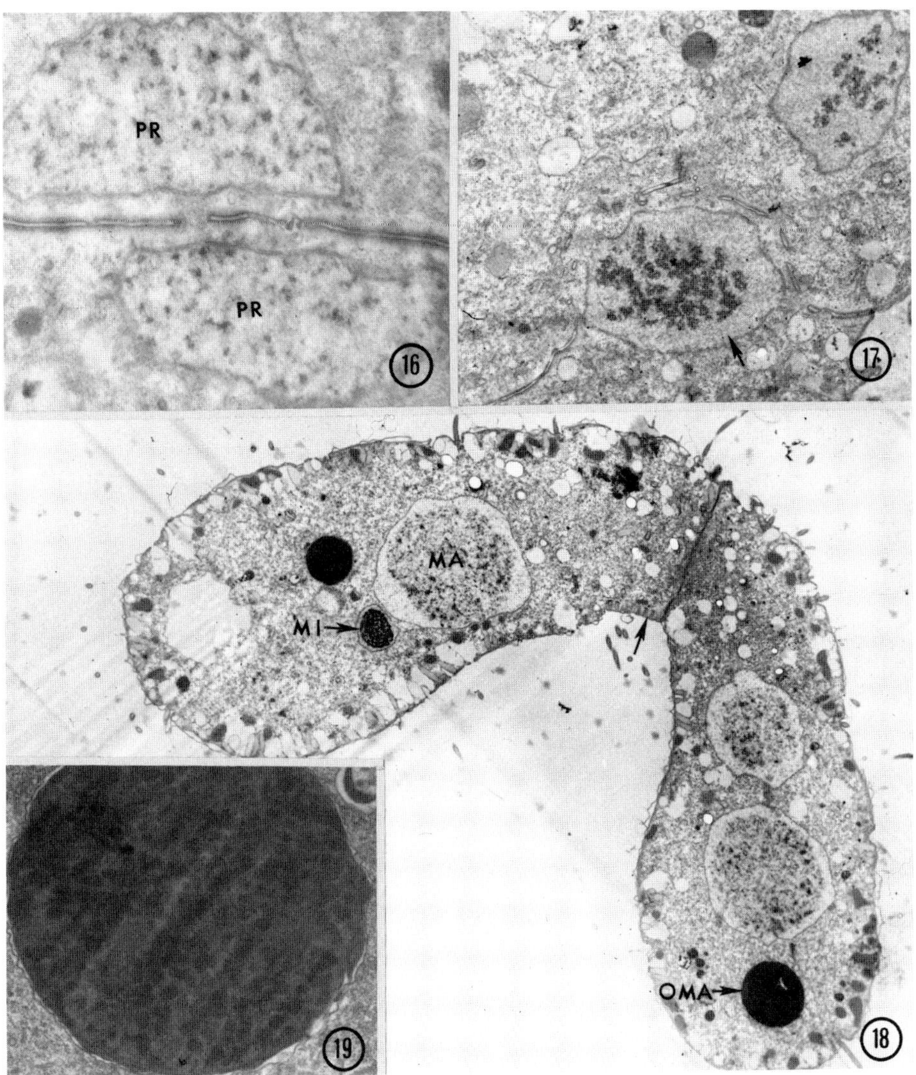

Fig. 16.

The third-prezygotic pronuclei (PR) come to lie flattened against the contact membranes. ×17,000.

Fig. 17.

A third-prezygotic pronucleus of one of the mates is shown passing through the contact membranes (arrow), while one of the pronuclei of the other mate lies near the contact membranes. This could be either a stationary or a migratory pronucleus. Note that at the point of penetration of the pronucleus the contact membranes seem to be disintegrating. ×10,000.

Fig. 18.

Toward the end of conjugation the macronuclear anlagen enlarge (MA), and the chromatin bodies increase in number and density. The old, degenerating macronuclei are spherical and extremely dense (OMA). One micronucleus (MI) is caught in the section. The oral apparatus (arrow) is apparently regenerating. ×3500.

Fig. 19.

An enlarged view of a degenerating macronucleus. The spherical dense bodies are presumably degenerating chromatin. ×17,000.

Fig. 20.

This map shows the world distribution of *T. pyriformis* as described in this chapter. Each point represents several collections yielding mating types. The syngens are indicated by their corresponding numbers.

physiological classes of individuals which are called mating types. They usually do not mate within the mating type. The character is inherited. When many morphologically similar clones are isolated from different habitats, some are found to be complementary and will mate, whereas others, although morphologically similar, will not mate with the first group but will mate with complementary members of their own group. These are the varieties, or syngens, that are populations sexually isolated from one another. There may be many syngens within a species, which is true of *T. pyriformis*.

Tetrahymena pyriformis, similar to the several species of *Paramecium* and other ciliates that have been carefully examined, includes many reproductively isolated populations (syngens), each with a sharply defined gene pool. If the concept of a biological species is adhered to, each would qualify as a species. The relative merits of such designations has been thoroughly discussed by Sonneborn (65) and Hairston (32).

Whereas conjugation occurs in other species of *Tetrahymena* (Chapter 1), and perhaps mating types exist in them, the following discussion includes only *T. pyriformis*.

The first interbreeding population of *T. pyriformis,* taken from freshwater samples collected on Cape Cod, Massachusetts, was discovered in 1951 (19). Since that time nearly 1300 freshwater habitats from the major continents, as well as Pacific Islands, have been examined for this ciliate (Fig. 20, Table 1). Thirty-nine percent of the habitats contained the ciliate, but this varied widely (10–70%) in different localities. Among the 7115 clones isolated, 50 mating types were categorized into 12 syngens. Some minor morphological and physiological variations exist among them. Whereas an attempt is made to distinguish them from one another based on the few characteristics that have been revealed, the one single character that clearly separates them genetically is

Table 1.

Location	Approx. sq. miles examined[a]	No. of habitats examined	No. positive	No. of clones isolated	No. of syngens	Refs.
U.S. and Canada	10,000	783	154(20%)	3,400	8(1–8)	19, 31
Mexico	500	39	11(28%)	62	1(2)	21
Panama	200	72	22(30%)	104	2(2,9)	21
Colombia	100	24	7(29%)	101	1(9)	21
Continental Europe	1,000	162	17(10%)	340	2(3,6)	15
England	100	26	10(39%)	200	3(4,6,10)	15
Asia	—	—	—	25	1(4)	B. R. Seshashar (unpublished)
Africa	—	6	1(16%)	44	1(6)	R. E. Kuntz (unpublished)
Australia	1,000	133	94(70%)	2,000	2(11,12)	25
New Zealand	500	44	21(48%)	400	0	25
Fiji	50	4	(50%)	40	1(9)	25
Hawaii	50	3	(33%)	26	1(9)	25
Japan	100	12	7(58%)	94	0	25
Hong Kong	50	14	6(43%)	95	0	25
Philippines	100	13	11(85%)	184	0	25
Totals	3750[b]	1290[c]	364(39%)[c]	7115	12	

[a]These very rough figures are based on the approximate miles travelled assuming that the water samples are representative of the area.
[b]Exclusive of Africa and Asia.
[c]Exclusive of Asia.

breeding pattern. There is little if any gene flow among the syngens because crossbreeding rarely occurs.

In an effort to discover biochemical differences that may exist between and among the 12 syngens, representatives of each were screened for their nutritional requirements. With a very few exceptions, all require 11 amino acids and 7 B vitamins (12,17,18,20,21,25), indicating their remarkable homogeneity in regard to their metabolic processes. No nutritional tests have been devised that could be used to separate the syngens. However, eight of the syngens have been distinguished from one another serologically by using immobilization as a criterion for separation (16). It is likely that the other four syngens would also fall into distinct groups. As pointed out by Loefer and Scherbaum (40), however, the serotype of a strain can be altered by temperature and other cultural conditions, which makes its usefulness in taxonomy unreliable.

The present method employed is the mating reaction. As long as the type testers are sexually active and thrive under laboratory conditions this works well. However, inroads made by persistent aging require periodic cross-mating within syngens in order to maintain viability as well as active mating. Unless conjugation interrupts the aging process, these characters are gradually lost, resulting in death of the lines.

Table 2.

Syngen	Location	No. of mating types	Length (u)	Immaturity period[a]	Breeding system	Refractory period (hours)	Temperature tolerance (8C)
1	North America	7	41–44	80	out and inbreeder	2–4	10–40
2	North and Central America	9	42–49	12–192	outbreeder	4–6	2–37[b]
3	North America, Europe	7	42–50	158	outbreeder	4–6	2–37[b]
4	North America Europe Asia	3	43–48	0	out & inbreeder	4–6	2–37
6	North America, Europe, Africa	3	56–64	24–150?	—	4–6	2–37
7	North America	3	50–51	0	inbreeder	24	6–37
8	North America	3	57–61	120–150	outbreeder	46	2–37
9	Central and South America, Fiji, Hawaii	5	57–63	0	out and inbreeder	36	10–37
10	England	2	36–37	0	inbreeder	4–6	6–37
11	Australia	3	49–52	0	inbreeder	4–6	2–37
12	Australia	4	50–53	0	inbreeder	2–4	10–35

[a]Number of fissions before sexual maturity is reached.
[b]Only a few mating types can barely survive at 2°C.

a. General Characters of the Syngens. All syngens have certain characteristics in common. *Tetrahymena pyriformis* feeds on bacteria in nature, and it flourishes axenically in test tubes when offered different bacteriological media. It can also be maintained on a chemically defined medium consisting of 20 nutrilites (17,18). Nearly all strains have these nutritional requirements, hence have been designated "wild type." A very small number show slight variations from this pattern and have been called "mutants" (17,18). Such mutants are not correlated with syngens, consequently nutritional requirements cannot be used as a syngenic character. Other biochemical variations exist among the syngens (5).

Morphologically, all syngens are very similar. They are usually pyriform in shape, with variations possibly sufficient in some cases to be used as a syngenic character (Fig. 21). In a study of their dimensions under controlled conditions (unpublished), it was found that, whereas there is a great deal of overlap, some syngens can be identified by size. A specified number of ciliates (100/ml) were inocculated into a 2% proteose–peptone medium and incubated at 25°C. When they were in logarithmic growth (48 hr), representative samples (100 cells) were measured (length and width). It turned out that some syngens were longer (varieties 6,8, and 9) than others (variety 10), but most fell between the extremes (Table 2). The ciliary meridians vary from 17 to 23, but this amount of variation may occur within one clone, hence is of no value as a

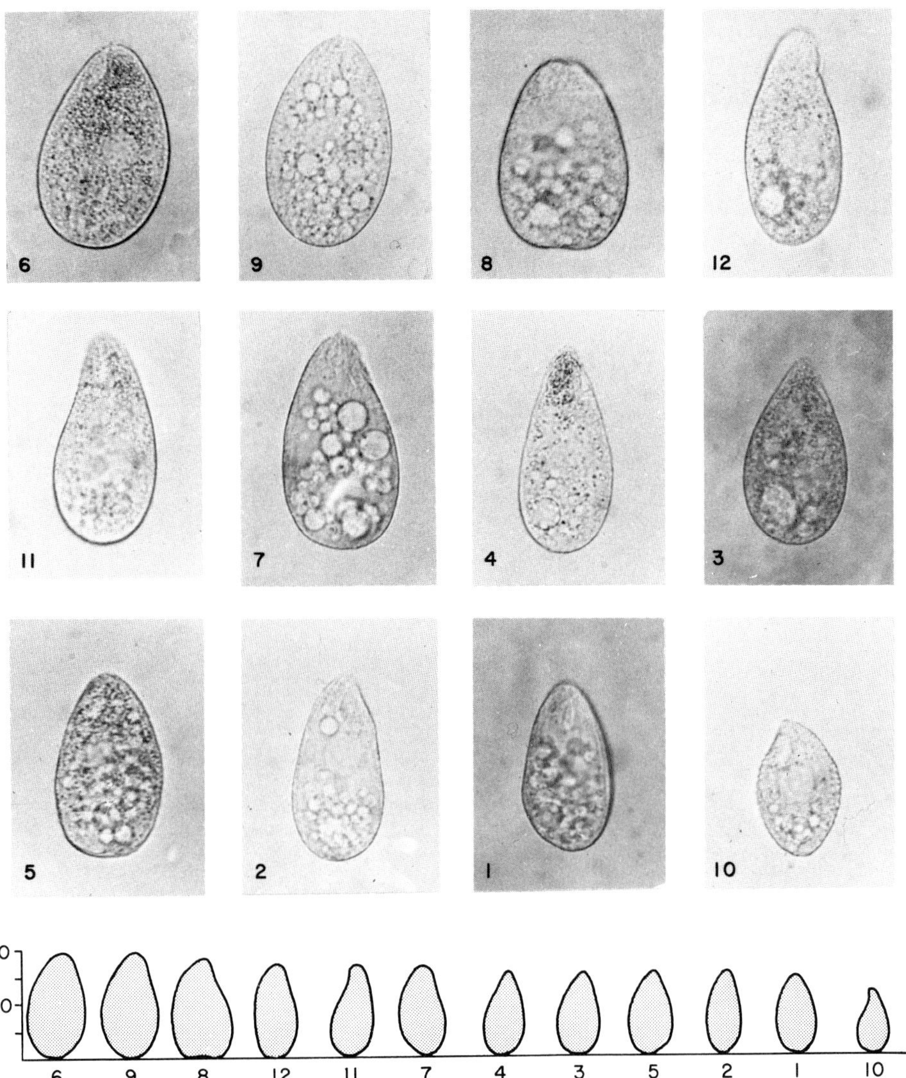

Fig. 21.

Light micrographs of representatives of the various syngens. Each cell was taken from a 48-h culture in which the cells were undergoing logarithmic growth. The relative sizes indicated schematically (bottom) are the averages of measurements of 100 cells.

diagnostic character. The buccal apparatus is similar in all syngens (9,10). The irregularly spherical macronucleus and the smaller micronucleus lying in an indentation in the macronucleus are very much alike in all syngens. Larger ciliates possess a somewhat larger macronucleus than smaller ones, whereas the micronucleus seems to be the same size in all.

During conjugation the old, degenerating macronucleus usually lies in the posterior region of the conjugants. In syngen 9, however, it lies in the anterior third of both mates (22). This is a striking character that occurs only in this syngen.

The time required for conjugation to begin, once the mating types are mixed, is called the refractory period (20); it is relatively consistent for all mating types in a syngen when clones are grown axenically in broth and washed twice in distilled water before mixing (unpublished). This character can be used to distinguish some syngens.

The length of time of sexual inactivity following successful conjugation is defined as the immaturity period (65). This sometimes differs among syngens and can be used as a distinguishing character in some. In some syngens no immaturity period has been found. Exconjugants, when permitted to undergo vegetative fission until a sufficient number are available for a test, can mate immediately. Others require varying periods of time before they are sexually mature, some as long as 30 days when frequently subcultured. Those with no immaturity period are thought to be inbreeders because they have little or no opportunity to disperse before they can mate again (65). Those with a long immaturity period, however, may spread into new environments before conjugation can again occur; thus they are thought to possess outbreeding tendencies. These breeding characteristics may influence the distribution of many syngens.

The immaturity period is not completely reliable owing to certain cytogenetic errors that frequently occur. Sometimes the cells mate but fertilization does not occur because of defective micronuclei, faulty meiosis, or for some reason there is no exchange of pronuclei. This phenomenon is called genomic exclusion (1–5, 49). Strains suffering from this aberration possess no immaturity period. This could account for some of the syngens that seem to have no immaturity period. Such errors in conjugation must be considered when using the immaturity period as a syngenic character. At the time the data were collected on the syngens described below, this phenomenon was unknown, hence some adjustment may be warranted.

The system of mating type inheritance has been studied in some syngens following the tenets set down by Sonneborn (64) for *P. aurelia*. Since the data are incomplete for all the syngens except syngen 1, it is inadvisable to include the available information as a syngenic character.

Tolerance to temperature can be used to distinguish some syngens, as pointed out by Holz et al. (33). They studied the high-temperature tolerance of representative strains belonging to all the mating types of syngens 1 through 9. Their work was verified and extended in our laboratory (unpublished). Representative strains in the 12 syngens were subjected to low (2–6°C) and high (37°C) temperatures for a period of 10 days. All the strains grow well at 25°C. Some tolerate and even grow at 2°C, and these have been designated "cold strains." Others, identified as "hot strains," survive at temperatures as high as 40°C. Low-temperature tolerance may be significant in the distribution of some syngens.

The geographical distribution of some syngens is sufficiently precise to be used for identification. This is particularly true of syngens 9 through 12. Syngens 1, 2, 7, and 8 are confined to the American continents, and syngens 3, 4, and 6 are distributed over four continents. In other words, one would expect to find syngen 10 only in England, syngens 11 and 12 only in Australia, syngens 1, 2, 7, and 8 only in the Americas, but 3, 4, and 6 can be found in Eurasia and Africa as well as in the Americas.

The characters described above are the only ones that have come to our attention that are useful in distinguishing one syngen from another. They are summarized in Table 2.

b. The Syngens. The characteristics of each are given below.

Syngen 1: Sexually active strains of *T. pyriformis* were first isolated from freshwater ponds near Woods Hole, Massachusetts in 1951 (19). They consisted of three strains (WH-6, WH-14, and WH-52) which were categorized as mating types I, II, and III, respectively, and the interbreeding group was assigned to syngen 1. Shortly thereafter two other strains, PR-1 and PR-2, were established from a lake in northern Minnesota. These belonged to mating types I and II. Seven mating types were subsequently derived in the laboratory from WH-6 and WH-14 by Nanney and Caughey (50). Gruchy (31) reported mating type IV (strains UM221 and UM241) from a Vermont habitat. Nanney (personal communication) found ALP-4 and IL-12 in Michigan. More recently, Phillips (56) reported mating types III (LW-A) and VI (LW-B) from Illinois. The distribution of this syngen seems to be sparse, although widespread in the United States.

As this syngen has been investigated more than any other, a great deal is known about its breeding system as well as its genetics (see Chapter 11). It possesses a short refractory period (2–4 h) and has an immaturity period lasting through at least 80 fissions (48). The maturity period lasts for several years as indicated by the fact that WH-6 and WH-14 still conjugate after 17 yr in culture, although offspring viability has dropped from 17% to less than 1% (unpublished). Since Nanney (48) was able to improve progeny viability with inbreeding, syngen 1 strains may inbreed in nature. This is substantiated by its relatively sparse distribution and by its mating type inheritance (47). However, multiple mating types with a relatively long immaturity period, which permits dispersal when they are sexually inert, are outbreeders. Therefore syngen 1 seems to show both outbreeding and inbreeding characteristics.

No nutritional mutants have been identified in this syngen as yet. None of the strains tolerate temperatures of 6°C, but all grow at 37°C and some survive at 40°C providing the cells are exposed to 35°C before being placed at the higher temperature (33).

Syngen 2: Representatives of this syngen have been found in widely separated localities in the Americas; it appears more often in collections than any other syngen. It was first found in water samples taken from Cape Cod, Massachusetts, and later from Minnesota, Michigan, Indiana, New York, Montana, Utah, South Dakota, Oklahoma, Florida, and Ontario, Canada (31). It has also been found in Mexico and Central America (21). It may occur in South America as well, although collections near Bogota, Colombia, were negative. It has not been taken from habitats outside the Americas.

Ten mating types were originally identified by Gruchy (31), and an eleventh by Elliott and Hayes (21). When these were reexamined by Hurst (35), however, only nine were considered valid, V and XI being relegated to obsolescence. Even so, it contains the largest number of mating types of any of the syngens, emphasizing its outbreeding potentialities. It has a variable refractory period (4–12 h) and an immaturity period that is also highly variable. This extreme variation in the immaturity period may have significance in explaining the widespread distribution of this syngen in the Americas.

A few tropical strains belonging to this syngen grow without pyridoxine, which is an absolute requirement for the wild type (17). There is some variation among the strains regarding their tolerance to temperature. Some grow poorly at 6°C, while others survive 2°C. None can withstand 37°C, but Holz et al. (33) report that all mating types survive and grow well at 35°C.

Syngen 3: Members of this syngen were isolated from habitats in Mississippi, Louisiana, North Carolina, Alabama, Vermont, and Michigan (31). Later it was found in Austria (15). In all, eight mating types were identified by Gruchy (31). An intensive study of the breeding system by Byrd (6), however, revealed only seven and he was unable to add any new mating types. The refractory period is relatively short (4–6 h). This syngen possesses strong outbreeding tendencies documented by its long immaturity period (158 fissions for one-half to become mature). It remains mature and sexually active for a long time (at least 13 yr). Moreover, its progeny survival following conjugation is enhanced more when outbred than inbred. These facts, together with its numerous mating types and wide distribution, emphasize its outbreeding economy.

No nutritional mutants have been found in this syngen. A few strains tolerate low temperatures, surviving at 6°C. Holz et al. (33) report that all strains grow moderately well at 35°C. In our laboratory none survived at 37°C.

Syngen 4: Strains belonging to this syngen were taken from Minnesota, New York, Nebraska, and Michigan (31). It was later found in England (15) and more recently in India, hence it also has a wide distribution. Three mating types were isolated in 1954 (31), all of which showed symptoms of senility after several years of culture. A rigorous breeding program has resulted in reactive mating types which yield viable progeny. In 1964 B. R. Seshashar very generously sent us clones of *T. pyriformis* taken near Delhi, India. These all turned out to belong to syngen 4, mating type II, and they are the only sexually active strains taken from Asia.

Members of this syngen have a short refractory period (4–6 h) and an immaturity period of approximately 3 days of laboratory culture, which means that they may have none at all [but see Allen et al. (2)]. It is possible that these strains are ridden with defective micronuclei, which would mean that fruitful conjugation did not occur in our tests, thus explaining the lack of an immaturity period. Its long maturity period (13 yr) and wide distribution suggest that it may outbreed in nature.

Strains of this syngen grow well at 6°C and survive at 2°C, but none tolerate 37°C. Mating type III, though not I and II, survives 35°C (33).

Syngen 5: When first isolated from natural habitats in Massachusetts, mating types I and II failed to produce viable offspring (31). Numerous breeding experiments since then have also been negative. It has never appeared in other collections, hence it may not be a true syngen. This may be a rare case of intersyngenic mating. Within recent years mating type I was lost in culture, consequently this syngen is not included in Table 2.

Syngen 6: Strains belonging to this syngen have a rather interesting history. Mating type I was isolated from Michigan waters and mating type II from a Florida habitat (31). A subsequent effort was made to derive additional mating types from these two which was successful, yielding mating type III. Other strains belonging to mating types I and II were found in England, France, Holland, and Italy (15). More recently (1964), through the courtesy of R. E. Kuntz, water samples taken within 10 miles of Cairo, Egypt, contained *T. pyriformis*. These all belong to syngen 6, mating type II (unpublished). This syngen, similar to syngens 3 and 4, is widely distributed over several continents.

The refractory period is relatively short (4–6 h), and the immaturity period lasts from 4 days to 4 wk (31,67). The maturity period is apparently very long, because our original strains still mate after 15 yr in culture. They have steadily declined in sexual activity over the years, however. Efforts to revitalize the various strains of this syngen by an intensive breeding program were successful in 1964. What its status is at present is not known.

Nutritional mutants occur in this syngen, some from Europe requiring 5 amino acids and 2 B vitamins in addition to the 18 required by the wild type (15). All from Egypt and Tanganyika, however, grew abundantly without thiamine or thioctic acid or both (unpublished). This syngen, similar to syngen 4, apparently possesses a wide spectrum of genomes controlling its synthetic capabilities.

All strains tested grow well at 6°C, and some survive at 2°C. None survive at 37°C.

Syngen 7: Representatives of this syngen were first isolated from a single habitat in North Carolina (31). Two mating types were established. Outka (55), in a careful analysis of the breeding system, derived mating types III and IV from the original parental types. He also demonstrated mating type instability which is characteristic of this syngen. Physiological conditioning by means of the adjustment of the osmolarity with sucrose was required for successful conjugation. The refractory period is unusually long (24 hr), and no immaturity period was observed. More recently, however, Phillips (57) found an immaturity period in strains of this syngen. She also induced competence for mating through the use of cell-free fluids (58).

No nutritional mutants were found in this syngen. Our present strains, belonging to three mating types, barely survive at 6°C but die at 37°C.

Syngen 8: Strains belonging to this syngen were isolated from several habitats in Minnesota, from which two mating types were established (31). Later another mating type, III, was added by Orias from Michigan (52–54). He also found members of this syngen in Wyoming but did not report the mating type. The refractory period was shown by Gruchy (31) to be 4–6 h at 25°C, but Orias says it is very long (46 hr). He also found that the immaturity period lasts for 120 to 150 fissions in routine culture. It probably outbreeds in nature. Orias (54) reports that it crossbreeds with syngen 6.

No nutritional mutants were found among the strains tested in our laboratory. Strains of this syngen tolerate 2°C and grow at 6°C. They cannot survive at 37°C.

Syngen 9: Five mating types belonging to this syngen were first identified from 19 clones isolated from habitats in Panama and the Canal Zone in 1954 (21). Among several clones taken near Bogota, Colombia, four belonged to mating type IV. Mating type III was found on Oahu Island, Hawaii, and Viti Levu, Fiji (25). Since the syngen has been found only near or within the tropics, it may be restricted to these latitudes.

It possesses one morphological character that separates it from all others investigated. During the later stages of conjugation, the degenerating macronuclei lie in the anterior region of each conjugant, whereas in all other syngens these structures are located in the posterior region (22).

Members of this syngen have a long refractory period (36 h), and no immaturity period could be demonstrated (22). Since progeny survival is about the same whether they are inbred or outbred, they seem to possess both inbreeding and outbreeding characteristics.

Strains of this syngen do not survive at 6°C nor do they tolerate high temperatures (37°C). A few serine mutants were found in this syngen (18).

Syngen 10: Water samples taken from the tributaries of the Thames River in the vicinity of London, England, contained members of this syngen (15). From the small number of habitats examined which contained four reactive clones, two mating types were isolated. It was not found in any other collections on the continent of Europe, hence it may be confined to the British Isles. Syngens 4 (mating type III) and 6 (mating type II) were collected in the same localities in England as well as on the Continent. It is interesting that these two syngens have successfully spread to England, but syngen 10 apparently did not succeed in reaching the Continent.

Strains of this syngen are generally smaller than representatives of other syngens. Their mean length is 36 μm, which is the lower limit for all syngens. There is a slight difference in the growth rate between the mating types, I being slower than II. The refractory period is 4–5 h at 25°C, and no immaturity period could be demonstrated. When first isolated in 1958, mating was vigorous, but at that time no effort was made to determine progeny survival. In 1968, however, mating could be induced in only one cross of the several strains in our collection, and none of the progeny survived. Apparently, aging had progressed rapidly during the 6 yr of laboratory culture. With only two mating types, no immaturity period, and a very limited distribution, one may conclude that syngen 10 is an inbreeder.

No nutritional mutants were found in this syngen. The strains grow poorly at 6°C and die at 37°C.

Syngen 11: A rather extensive collecting schedule in eastern Australia in 1960 revealed that this syngen is rather generally distributed in the province of New South Wales (25). It was not found in Queensland nor in the Northern Territory (near Darwin). It may be confined to the southern latitudes because the sampling was sufficiently extensive to reveal its presence if it occurred in any abundance in the northern provinces. This syngen, as well as syngen 12, has been found only on the continent of Australia.

Three mating types were established from the many clones isolated. The refractory period is short (4–6 h), and no immaturity period could be demonstrated. Conjugation is strikingly out of synchrony, and the conjugants remain attached unusually long (24–48 h). No information was obtained on progeny survival when strains of this syngen were first isolated. However, 4 yr later approximately 5% of the offspring survived (unpublished). The strict limitation of this syngen to the continent of Australia is adequate for identification purposes.

Several nutritional mutants were found in this syngen requiring several nutrilites in addition to the 18 essential for the wild type. Aspartic acid was required by most strains; however, some required biotin and proline in addition to aspartic acid (25). All the strains grow at 6°C and some survive at 2°C; all die at 37°C (unpublished).

Syngen 12: Representatives of this syngen were discovered in Australia, being taken in both New South Wales and Queensland (25). They seem to be more widely distributed than syngen 11, although the actual numbers isolated were fewer (38%) than for syngen

11 (62%). Conjugation is sharply synchronized, the refractory period is short (2–4 h), and no immaturity could be demonstrated. The conjugants separate within 24 hr. These characters alone separate syngens 11 and 12.

A few nutritional mutants requiring biotin and aspartic acid were found. They cannot survive at 6 or 37°C, hence are the least tolerant of temperature variation of all of the syngens.

c. Unidentified T. pyriformis. *Tetrahymena pyriformis* clones were isolated from all landmasses, but a great many of these (30–50%) turned out to be amicronucleate and were discarded. Even though numerous clones were micronucleate, it was not possible to assign all of them to syngens owing to their failure to mate either among themselves or with the known testers. This was probably due to the fact that they belong to a new syngen and only one mating type was isolated, or they were senile when isolated, or for reasons that are not clear. Nonreactive micronucleate clones were taken from habitats in Kwangtang Province, Hong Kong, the Philippine Islands near Manila, New Zealand (both North and South Islands), Japan, Switzerland, Sweden, and Central Africa. R. E. Kuntz very kindly supplied water samples from Africa, in addition to those from Egypt mentioned earlier, which contained micronucleate *T. pyriformis.* Clones isolated from habitats in Tanzania unfortunately did not mate with our 12 syngenic testers nor among themselves. Those isolated from water samples taken near Nairobi, Kenya, however, did mate among themselves but not with our testers. They may well be a new syngen.

II. DISTRIBUTION*

From the foregoing account it is clear that *T. pyriformis* is ubiquitous in its distribution. It occurs in freshwater ponds, small streams, lakes, and rapidly moving water. It may live as a facultative parasite. It has been found in natural waters at temperatures ranging from 4°C (Kiruna, Sweden) to 30°C. It exists at sea level and at 10,000 ft in mountain streams. It is probably safe to assume that it has invaded nearly all freshwater habitats. How it, as well as other species of protozoa, has been able to spread over the surface of the earth is a fascinating problem about which we know next to nothing.

Based on morphology it may be assumed that all of the syngens probably had a common origin. If this is so, and most taxonomists agree, then through time geographical isolation must have been the primary mechanism in bringing about the numerous syngens we see today. *Tetrahymena pyriformis,* being a highly prolific organism with strong powers of movement, has the capacity to become widespread. The constant shifts in the earth's surface with concomitant alterations in patterns of flow of water through geological time have also facilitated its ubiquitous distribution over the present landmasses. It is not so easy to explain how it reached isolated landmasses such as the Pacific Islands.

Any attempt to trace the evolution of the numerous syngens from a common ancestor, if one exists today, is fraught with difficulties owing primarily to a lack of sufficient data both on syngenic characteristics and on their present location. As pointed out earlier,

*Much of this section has been previously published (14).

we do not know enough about the genetics of each syngen and collections have been too sparse in many regions of the world to be representative, although they may be in some, the United States, for example. Even with the current fragmentary data it is possible to show some relationships between the known syngens.

A. The Americas

It was pointed out earlier that syngen 2 has the widest distribution of all in the American continents. It has been found in collections from Massachusetts to Utah and from Canada to Panama. All strains interbreed freely, suggesting that dispersal of mating types has been relatively recent. No members of the syngen have been found elsewhere, hence it seems to be confined to the Americas.

Syngens 1, 7, and 8 have been found only in the United States, and they seem to exist as isolated populations. Syngen 1 is the most widespread of the three. It has appeared in collections from Massachusetts, Vermont, Minnesota, Michigan, and Illinois. Syngen 8 has been taken only in the Midwest, Michigan, Minnesota, and Wyoming. Syngen 7 seems to be the most isolated one of all, having been found only in North Carolina and more recently in Michigan (57).

Syngens 3, 4, and 6 are widely distributed over the United States (31), as well as in other parts of the world. These three syngens tolerate low temperatures. This character may have made it possible for them to migrate to other continents, as is discussed later.

Syngen 9 was found only in Central and South America and on two Pacific Islands. It has not appeared in collections taken from northern latitudes, hence it may be confined to the tropics.

B. Eurasia and Africa

Of all the syngens that occur in the Americas today, only 3, 4, and 6 have been found in Eurasia and Africa. The collections from Asia and Africa were very few, but in Europe they may be sufficiently numerous to include the commonly occurring syngens (15). An interesting question is, How did the syngens reach distant lands? Some speculation may be warranted even though the data are fragmentary. Eurasia and North America were continuous via the Bering Strait throughout most of the Tertiary period, and there was much mammalian migration during Eocene times (42). This land connection existed again during the Pliocene and Pleistocene epochs (28,63). It is reasonable to assume that syngens 3, 4, and 6 could have arrived in Eurasia via this land bridge, paralleling the path taken by the horse and other animals. Subsequently, syngen 6 could have reached Africa following its spread throughout Asia. Syngens 3 and 4 may have reached this southern continent also, but they were not found in the very limited collections made. Since all the Eurasian and African syngens mate with North American corresponding syngens, it could be assumed that they evolved in the Americas before they crossed the Bering Strait. They then could have spread throughout Eurasia and Africa, two of them (4 and 6) crossing into England which had a continental connection in the past (28,38).

It is interesting to note that syngens 3, 4, and 6 are cold strains. They can survive, and 4 and 6 even grow, at 6°C or less. Only a few strains in syngen 2 can barely tolerate these low temperatures. This may explain why this syngen was not found in Europe; it could not withstand the low temperatures of northern latitudes, a necessity if it were to cross the Bering Strait land bridge. The only other American syngen that tolerates low temperatures is syngen 8 which also was not found in Europe. Perhaps factors other than low-temperature tolerance, such as sparse distribution, accounted for this failure. It is impossible to say that low-temperature tolerance is the one character that was responsible for 3, 4, and 6 spreading into Eurasia, but it could have been a contributing factor.

C. The South Pacific Islands and Australia

The pathway taken by *T. pyriformis* in reaching Australia was most likely from southeastern Asia, island hopping as did the monotremes and marsupials. There are records of monotremes from the late Triassic or early Jurassic periods and for marsupials from the late Cretaceous or early Tertiary periods (63). *Tetrahymena pyriformis* may have crossed the Bering Strait during the early Tertiary period when there was much faunal interchange. Syngens 3, 4, and 6, any one or all three, could have reached Australia during this period and subsequently evolved into syngens 11 and 12 which are the only ones that were found on this continent.

Certain similarities exist between European and Australian syngens. Strains from all the syngens except syngen 12 are cold strains, and they all have similar refractory periods. Both syngens 11 and 12 have no immaturity period, however, which suggests that they are more closely related to syngen 4 than to 3 or 6.

It is likely that *T. pyriformis* has been in Australia for a long time, probably antedating man. Had it been carried to this continent by man it does not seem reasonable that such distinct syngens as 11 and 12 could have evolved in so short a time. Moreover, representatives of one or the other syngen, or both, should have appeared in some of the Eurasian collections had they evolved on the continent.

How *T. pyriformis* traversed thousands of miles of ocean and became established on tiny Pacific islands is an interesting question. If it is assumed that this ciliate cannot adapt to seawater and does not form resistant cysts, then vast distances must have been spanned by mechanical means. From laboratory experience one may conclude that this ciliate does not produce cysts, nor can it survive in seawater. In nature, however, there is always the possibility that it may do both. In only one case an estuarine *T. pyriformis* clone survived in both sea- and fresh water (25). Cysts have been reported in some cultures in syngen 2 (35). These are isolated cases, and judging from our experience in handling thousands of cultures in the laboratory, it is unlikely that they constitute a part of the normal life cycle. The following discussion ignores either possibility as important in the dispersal of *T. pyriformis*. It is assumed that dispersal to these remote islands was accomplished by mechanical means such as massive air disturbances (hurricanes), and when water was transported by man during his migrations.

There is no question about the possibility of ciliates being carried along with hurricanes, but no records show that the pathways of violent wind disturbances follow those that must have been taken by *T. pyriformis* as it moved from the Americas westward to Hawaii and Fiji. There is some evidence, however, that this ciliate could have been carried by man as he migrated to the Pacific Islands from South America.

It has already been pointed out that syngen 9 inhabits South and Central America, Hawaii, and Fiji. Representatives from these localities all interbreed, indicating that they have close affinities and probably have not been separated for any great period of time. Moreover, both Hawaii and Fiji are "young" islands by geological time standards. The Hawaiian Islands may possibly date back as far as Cretaceous times (11). Most of the rocks making up the Fiji Islands are from the Tertiary period, laid on older bases. Fossils from Fiji also date back to Tertiary times (41). Therefore animals reaching these islands must have come at some later time, perhaps many came much later. Since one principal way protozoa could be transported such great distances was in fresh water carried in containers, and since man is the only animal capable of making long voyages while carrying his drinking water with him, one might conclude that human migrations westward may have played a part in distributing this protozoan to the Pacific Islands. That such voyages did occur is supported by botanical evidence such as the early introduction of the American potato into Polynesia (44–46). There is strong evidence that the Pacific Islands were not inhabited until approximately 1500 years ago (8), which means that *T. pyriformis* probably arrived since that time.

The available data on syngen 9 seem to fit the hypothesis that the Pacific Islands were populated with this ciliate when it was carried by man in his migrations westward from South and perhaps Central America. If these protozoa had migrated eastward from Asia, they should still exist in Eurasia. They were not found in the European collections which were sufficiently numerous to include them if they were present. Here again the lack of data rules out the possibility of drawing any conclusions because the Asian collections were too small to be representative. The free interbreeding of the Pacific Island syngen-9 strains with those from the mainlands of Central and South America is very good evidence that they invaded the islands only recently. Moreover, only mating type III was found in both Fijian and Hawaiian waters, which means that perhaps only one mating type made the trip. One cannot rule out the possibility of overlooking other mating types, however, since the collections were very limited on both islands.

Based on very limited evidence, it seems that geographical isolation was primarily responsible for the numerous syngens that we see today. Having been isolated for long periods of time, the present syngens have evolved genomes which are compatible within each syngen. In other words, the same isolating mechanisms that have given rise to new species in other organisms have operated in similar fashion in this protozoan.

ACKNOWLEDGMENT

Much of the work reported was supported for many years by grant AI 01416 from the National Institute of Allergy and Infectious Diseases, NIH, USPHS.

REFERENCES

1. Allen, S. L. 1963. J. Protozool. 10(4):413–420.
2. Allen, S. L., File, S. K., and Koch, S. L. 1966. Genetics 55(4):823–837.

3. Allen, S. L., and Lee, P. H. T. 1971. J. Protozool. 18:214–218.
4. Allen, S. L., and Weremiuk, S. L. 1971. J. Protozool. 18:509–515.
5. Allen, S. L., and Weremiuk, S. L. 1971. Biochem. Genet. 5:119–135.
6. Byrd, J. R. 1959. Ph.D. Thesis, University of Michigan.
7. Cameron, I. L., and Guile, E. E. 1965. J. Cell Biol. 26:845–855.
8. Coon, C. S. 1954. The story of man, Knopf, New York.
9. Corliss, J. O. 1952. Trans. Am. Microsc. Soc. 71:159–184.
10. Corliss, J. O: 1953. Parasitology 43:49–87.
11. Degener, O. 1945. Plants of Hawaii National Park. Illustrative of plants and customs of the South Seas. Edwards Brothers, Ann Arbor, Michigan.
12. Elliott, A. M. 1959. Ann. Rev. Microbiol. 13:79–96.
13. Elliott, A. M. 1963. In L. Levine, ed., The cell in mitosis, Academic Press, New York, pp. 107–124.
14. Elliott, A. M. 1970, J. Protozool. 17:162–168.
15. Elliott, A. M., Addison, M. A., and Carey, S. E. 1962. J. Protozool. 9(2):135–141.
16. Elliott, A. M. and Byrd, J. R. 1959. J. Protozool. 6(suppl.):18.
17. Elliott, A. M. and Clark, G. M. 1958. J. Protozool. 5(4):235–240.
18. Elliott, A. M. and Clark, G. M. 1958. J. Protozool. 5(4):240–246.
19. Elliott, A. M. and Gruchy, D. F. 1952. Biol. Bull. 103:301.
20. Elliott, A. M. and Hayes, R. E. 1953. Biol. Bull. 105:269–284.
21. Elliott, A. M. and Hayes, R. E. 1955. J. Protozool. 2:75–80.
22. Elliott, A. M. and Kennedy, J. R. 1962. Trans. Amer. Microsc. Soc. 81(3):300–308.
23. Elliott, A. M., Kennedy, J. R. and Bak, I. J. 1962. J. Cell Biol. 12:515–531.
24. Elliott, A. M. and Nanney, D. L. 1952. Science 116(3002):33–34.
25. Elliott, A. M., Studier, M. A., and Work, J. A. 1964. J. Protozool. 11(2):250–261.
26. Elliott, A. M. and Zieg, R. G. 1968. J. Cell Biol. 36(2):391–398.
27. Faure-Fremiet, E., Favard, P., and Carasso, N. 1962. J. Microsc. 1:287.
28. Flint, R. F. 1957. Glacial and Pleistocene geology, John Wiley, New York.
29. Frank, W. W., Eckert, W. A., and Krein, S. 1971. Z. Zellforsch. 119:577–604.
30. Frank, W. W. and Edkert, W. A. 1971. Z. Zellforsch. 122:244–253.
31. Gruchy, D. F. 1955. J. Protozool. 2:178–185.
32. Hairston, N. G. 1958. Evolution 12(4):440–450.
33. Holz, G. G., Erwin, J. A., and Davis, R. J. 1959. J. Protozool. 2(2):149–156.
34. Horn, G. F. 1951. M. S. Thesis, University of Washington.
35. Hurst, D. D., 1958. Ph.D. Thesis, University of Michigan.
36. Inaba, F., Imamoto, K., and Suganuma, Y. 1966. Proc. Jap. Acad. 42(4):394–398.
37. Jennings, H. S. 1938. Proc. Natl. Acad. Sci. U.S. 24:112.
38. Jukes-Browne, A. J. 1922. The building of the British Isles, Edward Stanford, London.
39. Kennedy, J. R. 1965. J. Protozool. 12:542–61.
40. Loefer, J. B. and Scherbaum, O. H. 1963. Syst. Zool. 12(4):174–177.
41. Mansfield, W. C. 1926. Pap. Dept. Mar. Biol. Carnegie Ins. Wash. D.C., 23:87–107.
42. Matthew, W. D. 1939. Climate and evolution, 2nd ed., New York Academy of Sciences, New York.
43. Maupas, E. 1889. Arch. Zool. Exp. Genet. 7:149, 517.
44. Merril, E. D. 1946. Merrileana. A selection from the general writings of Elmu Drew Merrill, F. Verdorn, ed., Chronica Botanical, Waltham, Massachusetts, 10(3–4).
45. Merrill, E. D., 1950. Ceiba, 1(1):2–26.
46. Merrill, E. D. 1954. The botany of Cook's voyages and its unexpected significance in relation to anthropology, biography and history, Chronica Botanica, Waltham, Massachusetts.
47. Nanney, D. L. 1953. Biol. Bull. 105(1):133–148.
48. Nanney, D. L. 1957. Genetics 42:137–146.
49. Nanney, D. L. 1963. Genetics 48:737–744.
50. Nanney, D. L. and Caughey, P. A. 1953. Proc. Natl. Acad. Sci. U.S. 39:1057–1063.
51. Nobili, R. 1967. Monitore Zool. Ital. NS. 1:73–89.
52. Orias, E. 1959. J. Protozool. 6(suppl.):19.
53. Orias, E. 1960. PhD. Thesis, The University of Michigan.

54. Orias, E. 1963. Genetics, 48:1509–1518.
55. Outka, E. E. 1961. J. Protozool. 8:179–183.
56. Phillips, R. B. 1968. Genet. Res. Camb. 11:211–214.
57. Phillips, R. B. 1969. Genetics 63:349–359.
58. Phillips, R. B., 1971. J. Protozool. 18:163–165.
59. Pitelka, D. R. 1965. J. Microsc. 4:373–394.
60. Ray, C. 1956. J. Protozool. 3:88–96.
61. Ray, C. and Elliott, A. M. 1954. Anat. Rec. 120:812.
62. Schneider, L. 1963. Photoplasma 56:109–140.
63. Simpson, G. G. 1953. Evolution and geography, State System of Education, Eugene, Oregon.
64. Sonneborn, T. M. 1938. Proc. Amer. Phil. Soc. 79:411–434.
65. Sonneborn, T. M., 1957. In E. Mayr, ed., The species problem, Publ. 50, Am. As. Adv. Sci., Washington, D. C., 155–324.
66. Vivier, E. 1962. C. R. Soc. Biol. 156:1115–1116.
67. Wells, C. 1958. Asso. Southeast. Biol. Bull. 5:17.

Nucleic Acids in *Tetrahymena* during Vegetative Growth and Conjugation

Barbara B. McDonald

Department of Biology
Dickinson College
Carlisle, Pennsylvania

I. VEGETATIVE GROWTH

A. Distribution of the Nucleic Acids

Standard techniques for staining nucleic acids—the Feulgen procedure (162), azure B (58), methyl green–pyronin (84)—demonstrate the abundance of these compounds in *Tetrahymena*. Vegetatively growing cells have RNA/DNA ratios of roughly 20:1 (28,87,145). In addition to the macronucleus and micronucleus, DNA is present in mitochondria (36,118,158) which are located close to the cortex during growth (50,142), and possibly also in the kinetosomes of the cortex itself (57,127,128). Most of the RNA is in the cytoplasm, but about 2% is in the macronucleus (88), and numerous small nucleoli lie just beneath the macronuclear membrane in growing cells (28,52,116,137,140).

Prior to fission the kinetosomes multiply and some participate in the formation of an additional mouth (61, 71); the micronucleus (when one is present) moves away from its usual location, in a depression at the side of the macronucleus, and undergoes mitosis [but its individual chromosomes can be distinguished only during meiosis (44,46,132,134)]. Only then does the polyploid macronucleus elongate and pull apart as the cell itself constricts to form two daughters. Progressive stages of macronuclear amitosis in *T. pyriformis* strain H, which lacks a micronucleus, are shown in Figs. 1–4 (92). (In a micronucleate strain the micronucleus divides at about the stage shown in Fig. 1.) Although both daughter cells usually receive a fair share of cellular constituents

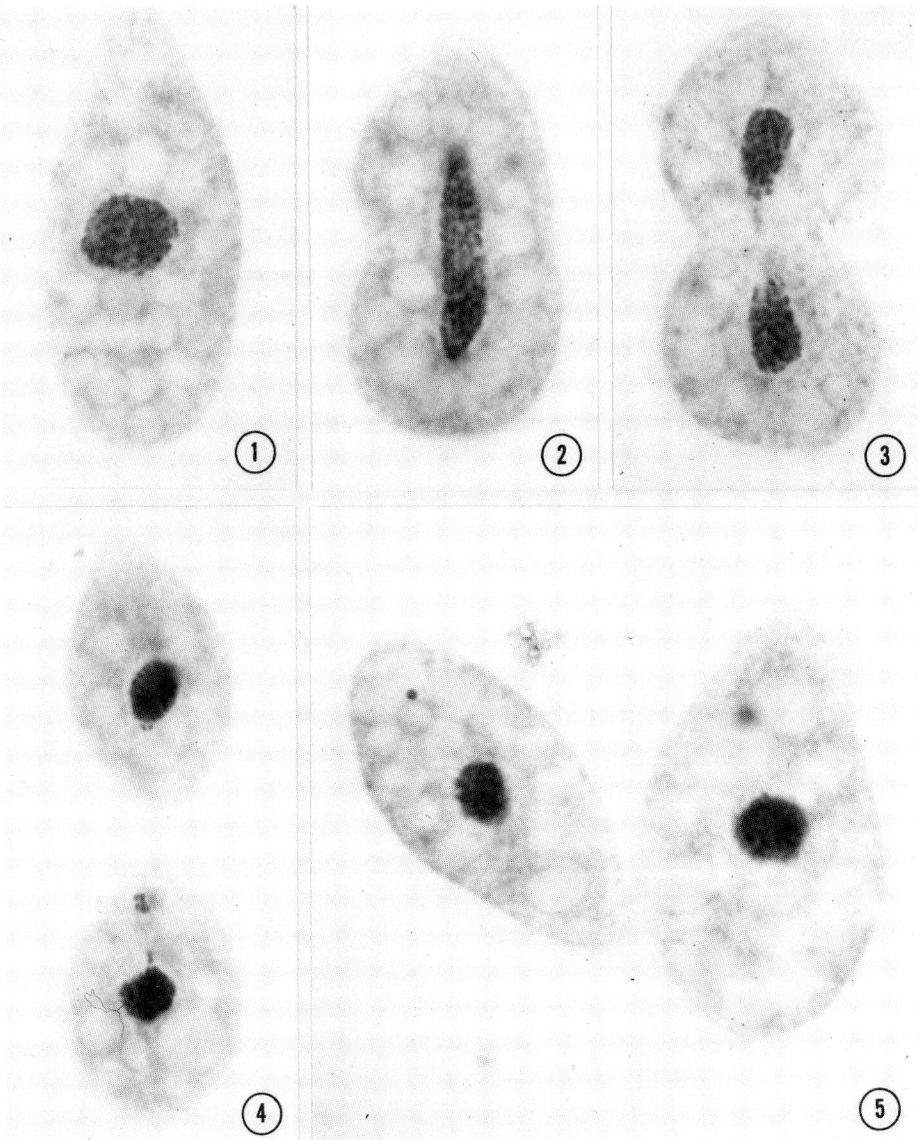

Figs. 1–5.

Stages in division and interphase of *T. pyriformis* amicronucleate strain H (92). Magnification 1200 ×. (Reproduced by permission of *The Biological Bulletin.*)

(45,92,110,120,142,143), bits of macronuclear chromatin (called "extrusion bodies" or "subnuclear aggregates") may be left behind in the cytoplasm, where they eventually disappear (49,62,82,92,147,152). Examples of extrusion bodies are shown in Figs. 4 and 5 (left of the two interphase cells).

Condensed macronuclear chromatin is visible in these dividing cells. As viewed by electron microscopy, chromatin from burst macronuclei consists of granules interconnected by slender threads (182); at division discrete groups of the granules (subnuclei?) are evident in amicronucleate strains (137,115) but not in those that contain micronuclei (59,137). Longitudinally oriented microtubules have been observed in dividing macronuclei (55,75,115,137).

Tetrahymena pyriformis, of various syngens and strains, was used for most of the work reported in this chapter, but some other species of *Tetrahymena* are also represented.

Investigations of nucleic acid metabolism during growth have utilized mass cultures in log phase (30,92,98,169), cells "naturally synchronized" by selection at division from log-phase cultures (39,122,157) or by prior starvation (29), and paired sisters and their progeny (42,92,120,121,158). [Because this chapter is primarily concerned with "normal" growth and division, no attempt has been made to include all the work inspired by Scherbaum and Zeuthen's classic article on populations induced to divide synchronously by temperature cycling (148).]

Food sources have included axenic medium [plain and supplemented proteose–peptone (30,92,128,169) or defined medium (51,122,157,160)], and bacteria (24). Other factors known to affect the growth rate, such as age of the strain or culture and cell concentration (108,139,178), temperature (30,41,), and availability of air (53), have varied.

The techniques of cytophotometry and autoradiography form the basis of many of the reports mentioned in this chapter. By cytophotometry of Feulgen-stained nuclei, relative amounts of DNA have been measured (25,32,42,49,92,160,169,183); however, precautions must be observed in such work, since hydrolysis curves of *Tetrahymena* macronuclei have two absorption peaks instead of one (1) and Feulgen stainability is reduced by ferric ions (33).

Autoradiographs of cells briefly or continuously exposed to radioactive nucleosides (usually tritiated thymidine for DNA, and tritiated uridine or cytidine for RNA) provide evidence for times of synthesis and for the location of nucleic acids. To insure that the low-energy beta particles emitted by tritium reach the overlying autoradiographic film, the cells have been flattened (94,157), sectioned (6,68,158), or disrupted (118,128). During storage tritiated compounds undergo self-radiolysis, and this results in nonspecific cellular labeling (170,171). Even with newly purified $[^3H]$ thymidine, small amounts of activity have been found in RNA, lipids, and proteins of *Tetrahymena* (2). Both uridine and cytidine (probably through conversion to uracil) can satisfy pyrimidine requirements for growth (184). The importance of confirming autoradiographic results by nuclease digestion is obvious, but a lengthy treatment is required to extract DNA from mitochondria (118) and even this procedure is not always successful (77,94,97).

Several useful experimental techniques have been summarized by Stone and Cameron (157).

B. The Nuclei

1. Deoxyribonucleic Acid

a. Synthesis. Analyses of macronuclear DNA from different species of *Tetrahymena* have indicated a rather low content of guanine plus cytosine, ranging from about 22

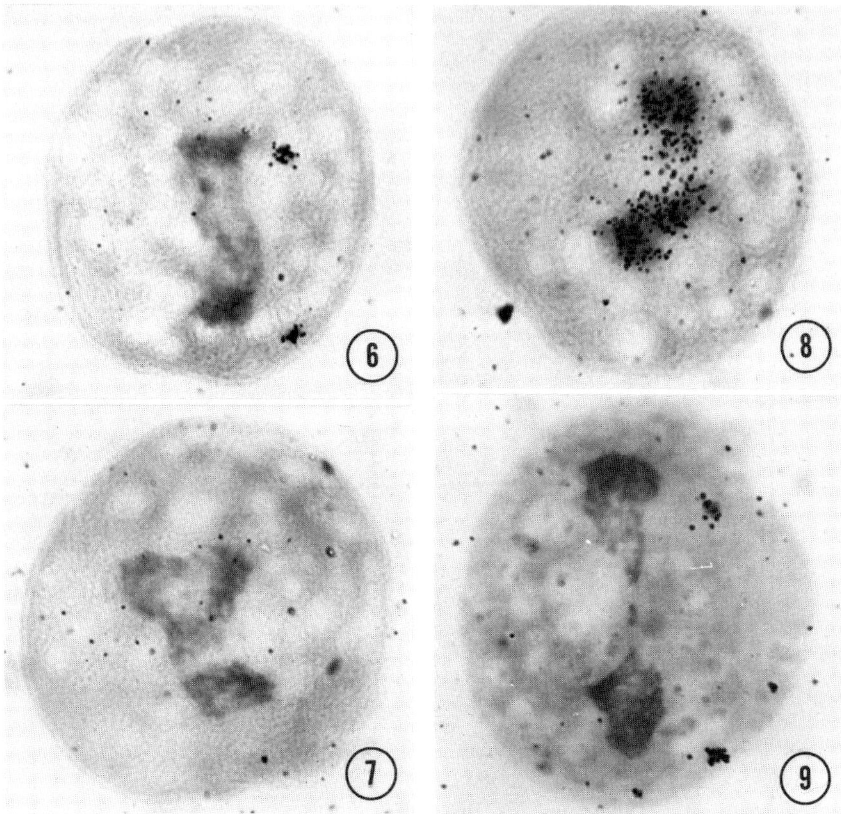

Figs. 6–9.

Autoradiographs of *T. pyriformis*, mating type II of syngen 1, pulse-labeled with $[^3H]$ thymidine and killed at division (88). Magnification 1000 ×. (Reproduced by permission of *The Journal of Cell Biology*.)

to 31% (12,20,65,79,145,163,177). A replication rate of more than 3 μm/min has been estimated from electron microscope autoradiographs of DNA strands isolated from cells briefly pulsed with $[^3H]$ thymidine (186).

The nuclei of *Tetrahymena* synthesize their DNA at different times. Macronuclear S occurs during an intermediate part of the life cycle (92,94,122), and micronuclear S (as indicated by $[^3H]$ thymidine incorporation) begins about 1/2 h before cell fission, immediately after its own division, and continues for a short time thereafter (69,94,126). During most of the life cycle, therefore, the micronucleus is in G_2; the length of G_1 is negligible, and S lasts about 30–60 min. In cells with a generation time of 4 h or less, an approximate distribution of macronuclear interphase—G_1 20–60 min, S 60–90 min, G_2 40–80 min—is consistent with data from cytophotometry (92) and $[^3H]$ thymidine autoradiography (24,30,32,40,41,77,94,98,122,160). The DNA of macronuclear nucleoli is replicated late in the S period (35,100).

The autoradiographs in Figs. 6–9, focused at the level of the silver grains, represent cells of *T. pyriformis* syngen 1 pulsed with $[^3H]$ thymidine at the ages of 5, 30, and

60 min, and at the start of division (94). The first three examples were returned to nonactive medium until fission began; hence all have two micronuclei, visible at the right. The cells have distorted shapes because they were rinsed in water and air-dried. Radioactive micronuclei can be seen in Figs. 6 and 9, and a radioactive macronucleus in Fig. 8, but there is no nuclear activity in Fig. 7.

Cytophotometric evidence that S may span most of interphase has been reported for mass cultures of two amicronucleate strains (GL and W) (136,169), but this may in part reflect variability of "basic" DNA content and generation time within random populations (42,56,92). Autoradiographic studies of these strains have demonstrated that they do have G_1 and G_2 periods (30,77,135).

In general, G_1 tends to be short when generation time is short (24,94,122), and under some conditions (growth in defined medium or in defined medium lacking essential amino acids), G_2 may be negligible (160). Although the generation time and therefore the duration of interphase stages may be affected by environmental conditions, such as medium (30) and temperature (41,98), there usually is no correlation between quantity of macronuclear DNA and the length of S or generation time (32,92).

Autoradiography can pinpoint the start and end of S more precisely than can cytophotometry, but the results of pulse-labeling experiments would be misleading if cells retain pools of $[^3H]$ thymidine (or derivatives) for appreciable periods of time. Apparently, they do not (3,4,76,94,159) [even when labeled bacteria provide the source of radioactivity (24)] unless pulsing occurs late in S; then cells of strain HSM have been found to retain a water-soluble derivative of $[^3H]$ thymidine within the macronucleus, through G_2 and fission, which they incorporate into DNA during the following S period (159). In defined medium that lacks amino acids but includes $[^3H]$ thymidine, G_1 cells that enter S replicate only 20% of their DNA and eschew the labeled nucleoside (160,161). Taken together, these results imply that macronuclear DNA synthesis actually begins somewhat earlier than is indicated by direct incorporation of the labeled nucleoside from the medium.

Cells maintained in pyrimidineless medium can use their intracellular RNA for DNA synthesis and subsequently divide several times (25,85). When divisions have essentially ceased, during starvation or in stationary phase, some populations of *Tetrahymena* have been found to include both G_1 and G_2 cells (97,138). Incorporation of $[^3H]$ thymidine from fresh medium indicates that the cells in G_2 may nevertheless synthesize nucleolar DNA before dividing (97).

b. Regulation of the Amount of Macronuclear DNA. In *Tetrahymena*, sister cells usually receive similar quantities of macronuclear DNA at fission (42,45,92,115). In some cell populations the basic (G_1) amount varies widely (42,92) [less so in others (25,32,160)] but is fully replicated between divisions (25,42,92). Extra synthesis can result, however, from certain chemical or physical treatments such as actinomycin D (39,40), colchicine (69), vinblastine (156), temperature cycling (45,145,147,188) [unless amino acids are lacking (172)], visible light under certain conditions (181), ultraviolet irradiation (152) [the irradiation is followed by increased production of DNA polymerase (81), and excision repair of DNA (19)]. After a few divisions under favorable growth conditions, the original range is usually restored. Before that happens, the DNA may increase further (34,69,146), although it may not double (91,152), and consecutive divisions can occur without additional synthesis (77,91,156). Chromatin extrusion is not uncommon in dividing *Tetrahymena* (42,49,62,92) but is pronounced in cells containing excessive macronuclear DNA (49,147,152). Prior to extrusion this chromatin, as well as that extruded by normal cells (49), was metabolically active; when the macronuclear DNA had been labeled with $[^3H]$ thymidine, the extrusion bodies were labeled (152). In the cytoplasm, however, extruded chromatin fails to incorporate $[^3H]$ thymidine (49) and is generally resorbed before the next cell division. The ways in which abnormal cells with excessively high macronuclear DNA revert to lower amounts thus include reduced synthesis (or no synthesis) between divisions, and extrusion of large amounts of chromatin.

In normal dividing cells unequal macronuclear division and chromatin extrusion would seem to result in depletion and eventual loss of the macronucleus itself. The reason this does not happen has been discovered by Cleffmann (42); when DNA in the macronucleus approaches a low level, it replicates twice instead of once before the next division (to accommodate this double S period, the generation time is extended). The frequency of this phenomenon has been estimated at 2%.

Within certain limits, therefore, the amount of macronuclear DNA in *Tetrahymena* can be regulated—decreased in abnormal cells when it is too high, and doubled in normal cells when it is too low.

c. Relative Amounts of DNA in the Nuclei. For estimates of macronuclear polyploidy in *Tetrahymena*, measurements of DNA must be made for both micro- and macronuclei in micronucleate strains and must take into account the fact that their S periods are out of phase. Additional care must be exercised because diploid micronuclei have been replaced in some cultures with aneuploid (14,178), tetraploid (133), or polytene (5) micronuclei.

On the basis of genetic evidence submitted to mathematical analysis (144), Allen and Nanney (13) proposed that in syngen 1 of *T. pyriformis* a G_1 macronucleus contains 45 subnuclei, each equivalent to a diploid micronucleus; a G_2 macronucleus therefore would have 90 diploid subnuclei segregating as units at division. A more recent proposal by Allen (10) (see Chapter 11) concerning macronuclear constitution involves fragmentation of chromosomes during development, a process that has been demonstrated in some hypotrichs (15,129).

In a strain of mating type I, syngen 1, of *T. pyriformis,* Woodward et al. (183) have reported DNA quantities of about 10.16 pg in G_2 micronuclei, 21.29 pg in G_2 macronuclei, and 0.86 pg in G_2 micronuclei. The ratios obtained therefore are: G_1 macronucleus/G_2 micronucleus, 11.85:1; and G_2 macronucleus/G_2 micronucleus, 23.7:1. The latter figure is close to half of 45:1 (the ratio expected if a G_2 macronucleus contains 90 diploid subnuclei), and so they suggested that subnuclei might be haploid rather than diploid. An analysis of morphological evidence from electron micrographs favors the latter interpretation (115).

Such a straightforward relationship between amounts of nuclear DNA is not found in all populations of *Tetrahymena* and indeed could hardly be expected when unequal macronuclear division and chromatin extrusion occur. For strain HSM of *T. pyriformis* $[$which contains a micronucleus but has not been assigned a syngen number (175)$]$, Cleffmann (42) obtained G_2 macronucleus/G_2 micronucleus ratios ranging from about 25:1 to 50:1 and averaging about 40:1.

For populations of three other *Tetrahymena* species, Dysart (49) has reported G_1 mac-

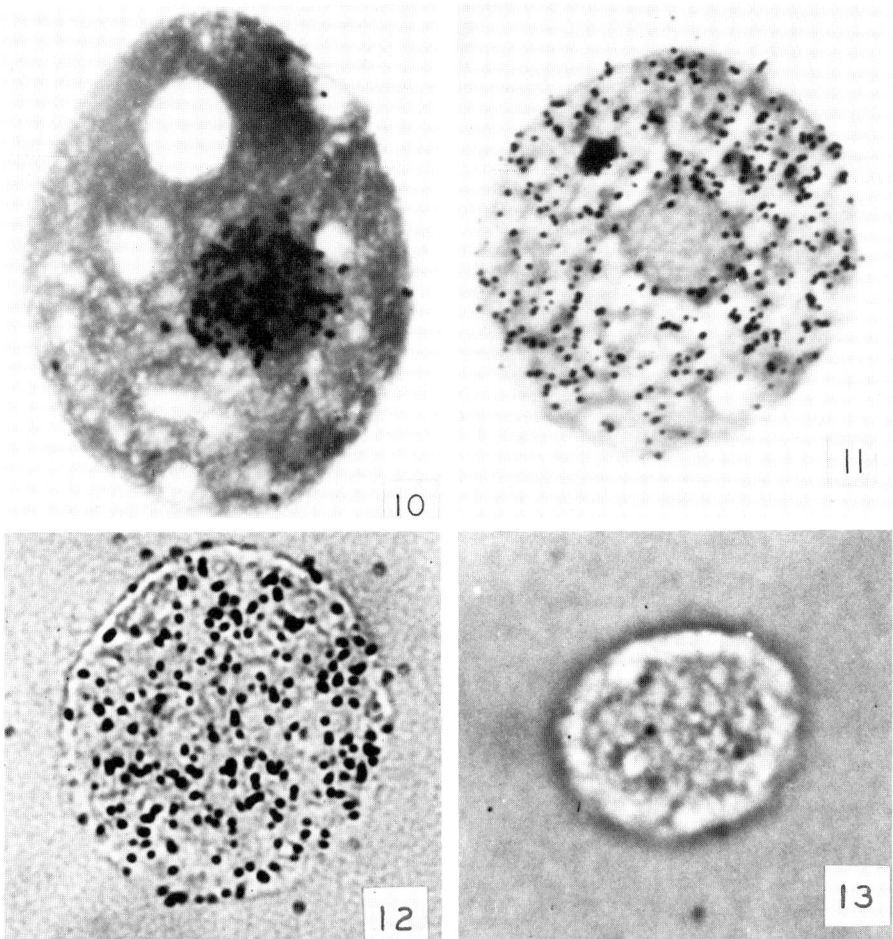

Figs. 10 and 11.

Incorporation of [³H] cytidine by the nucleus and transfer of activity to the cytoplasm in cells of strain HSM (123). (Reproduced by permission of D. M. Prescott and *The Journal of Histochemistry and Cytochemistry.*)

Figs. 12 and 13.

Nucleated and anucleated fragments of strain HSM exposed to [³H] cytidine (123). (Reproduced by permission of D. M. Prescott and *The Journal of Histochemistry and Cytochemistry.*)

ronuclear polyploid levels ranging from three- to sixfold (basing her calculations on a haploid micronuclear value: 30 to 90 in *T. rostrata;* 15 to 90 in *T. limacis*; 150 to 900 in *T. patula*). Extrusion bodies contained DNA in multiples of the diploid micronuclear amount, providing some support for the diploid subnuclear hypothesis.

2. Ribonucleic Acid

a. Macronuclear Synthesis. The cytoplasm of *Tetrahymena* contains a large amount of RNA, which is supplied to it by the macronucleus, as demonstrated by Figs. 10–13

(123). These autoradiographs represent cells (or parts of cells) that were provided with $[^3H]$ cytidine. After a 15-min exposure in radioactive medium, the label was localized in the nucleus (Fig. 10); after a 12-min exposure followed by more than 1 h in nonactive medium, essentially all the label had migrated to the cytoplasm (Fig. 11). Figures 12 and 13 represent living fragments of cells that were exposed to $[^3H]$ cytidine for 1 h; the labeled fragment had a nucleus and the other did not. [Such anucleate fragments, however, are able to continue protein synthesis for several hours (123).] Isolation from live *Tetrahymena* of macronuclei (66,125,136) retaining RNA polymerase (86,101) has been described by several investigators.

More precise information about the kinetics of RNA synthesis has been obtained through density gradient ultracentrifugation of nuclei or whole cells from log-phase cultures, following brief exposure to an appropriate $[^3H]$- or $[^{14}C]$nucleoside (for incorporation into RNA) or methyl-labeled methionine (for methyl labeling of RNA) (83,88,124). The bulk of cellular RNA is ribosomal, and its formation has been followed from a 35 S precursor molecule in the nucleus to appearance in the cytoplasm, within a few minutes, of ribosomal RNA units (17 and 26 S) which are found in cytoplasmic ribosomes. After a chase with nonactive medium, input from labeled nucleoside precursor pools continues for at least 45 min, whereas methyl-labeled methionine is rapidly depleted; within 10 min (5-min exposure, 5-min chase), the radioactivity pattern matched that of whole-cell RNA (predominantly 26, 17 and 4 S) (124).

In the course of culture growth, a close correlation has been observed between the quantity of cellular RNA (139) and the nucleolar organization revealed by electron microscopy (28,140). In early log-phase cultures, RNA production is very high, and the abundant small nucleoli [estimated at 300 to 1000 (28,116)] are distributed just beneath the nuclear membrane; as the culture approaches the stationary phase, RNA production decreases (139) and nucleoli begin to fuse into larger bodies (28,140). Enlarged nucleoli are typically found in the stationary phase and in cells whose growth or metabolism has been disturbed [for example, by treatment with actinomycin D (140), starvation (89,114), heat cycling, but not cold cycling (31)]. The change is reversed upon restoration of favorable growth conditions. Nucleoli persist through macronuclear division (59,137), although some are reported to be cast into the cytoplasm (143). During heat cycling extrusion from macronuclei of RNA blebs has been observed by some investigators (52) but not by others (31).

Stages of cortical morphogenesis in *Tetrahymena* are convenient markers in determining progress toward cell division (60,61,71,100) and can be arrested or reversed by physical or chemical treatments (60,61,64,179) which, directly or indirectly, must affect RNA synthesis. So, too, is transformation from microstome to macrostome forms, in *T. vorax*, controlled by time-related synthesis of RNA (113). (Conceivably, macronuclear RNA synthesis might not be implicated here; the production site of kinetosome-associated RNA remains unsettled; see Section I,D.)

Certain aspects of RNA synthesis related to a particular end point (usually synchronous division) have been studied in cells naturally synchronized by divider selection (39) and in mass cultures brought to a similar physiological state by such treatments as hot or cold temperature cycling (18,21–23,38,102,103,105,106,180), high hydrostatic pressure (90,185), or starvation (27,138), often in combination with inhibitors of RNA or protein synthesis. Reduced RNA synthesis and degradation or disruption of existent

RNA have been observed under conditions of stress [temperatures both high (21–23,38) and low (180); high pressure (90,185)]. Return to a more favorable environment results in a substantial increase in RNA (22,38,102,185), and when synchronized cultures are allowed to progress beyond a critical time (39,130,187) fission occurs [even after emacronucleation or treatment with chemicals that interfere with macronuclear division (64,168)]. Evidence has been presented that in heat-cycled cells a specific "division protein" is formed at about the critical time (173,174) [as proposed by Zeuthen (187)]. In cells starved and refed, however, RNA synthesis was found to be necessary up to the beginning of cytokinesis (138).

Only a small percent of *Tetrahymena* DNA hybridizes with the RNA; the amount increases near the critical time (in heat-cycled cells) and decreases at division (38). Following division (of cold-cycled cells), polysomes break down (180). In naturally synchronized cells (by selection of dividers), the rate of RNA synthesis is low immediately following division (39,122). A dearth of polysomes has been observed in cells starved for 24 h, but some mRNA and sRNA apparently outlast this condition; upon refeeding, polysomes re-form and resume protein synthesis, even when new RNA synthesis is prevented by actinomycin D (27).

b. Micronuclear Synthesis. Both micronucleate and amicronucleate strains of *Tetrahymena* abound in nature, and both can be readily cultivated. When micronucleate cells lose their micronuclei in the laboratory, however, their growth is impaired and most die (11,14,108,176). For normally amicronucleate strains the possibility of micronuclear RNA synthesis obviously does not exist. If the micronucleus in micronucleate strains could synthesize RNA, however, this might help to account for its apparent importance during vegetative growth.

Negative reports concerning micronuclear RNA synthesis in strains of *T. pyriformis* syngen 1 (6,67,68) have now been followed by a positive report for strain HSM (104). Micronuclear incorporation of RNase-sensitive label from [³H] uridine and [³H] cytidine, detected by electron microscope autoradiography, was found to occur at about the time of micronuclear DNA synthesis. The investigators recognized the possibility that RNA located in the micronucleus might have migrated from the macronucleus.

C. Mitochondria

Cytoplasmic labeling by [³H] thymidine has been commonly observed in *Tetrahymena* (and other cells) and generally can be attributed to incorporation of the nucleoside into DNA of mitochondria (36,118,120,158). Examples of labeled mitochondria are seen in the electron microscope–autoradiograph in Fig. 14 (158). The relative ineffectiveness of DNase toward mitochondrial DNA (20,63,112,165) may help to explain contradictory reports about whether (118,120,146,158) or not (77,94) treatment with the enzyme removes cytoplasmic label in autoradiographed *Tetrahymena*. Activity in the cytoplasm, however, is not always restricted to mitochondria; other components can incorporate label (2) and undoubtedly do so if [³H] thymidine has undergone self-radiolysis (170,171). In addition, the presence of mitochondria within digestive vacuoles of stationary (50) and starving (89) cells suggests another possible source of nonspecific labeling.

Mitochondrial DNA from several strains of *Tetrahymena* has base ratios that are similar

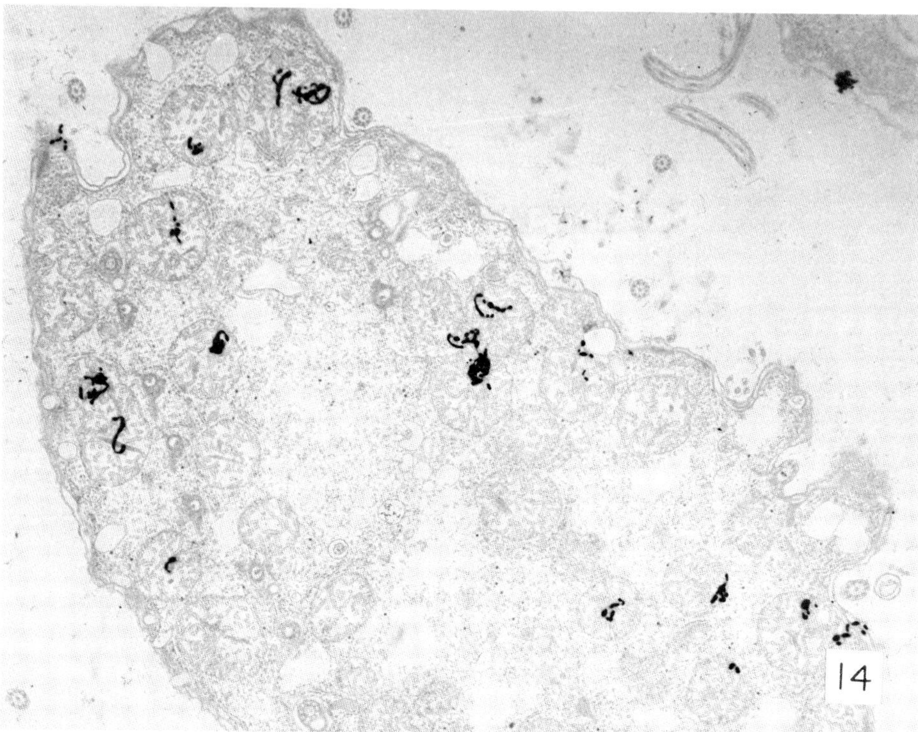

Fig. 14.

Electron micrograph autoradiograph of strain HSM showing mitochondria labeled with [³H] thymidine (158). Magnification 6150 ×. (Reproduced by permission of G. E. Stone and O. L. Miller, Jr., and *The Journal of Experimental Zoology.)*

(57,119,165) or identical (20) to those of the nuclear DNA (determined by buoyant density in cesium chloride density gradients). It is double-stranded (119,164,165), and each molecule weighs about 30–40 daltons (20,164).

High-power electron micrographs have revealed DNA filaments within mitochondria of *Tetrahymena* (166), and upon lysis the release of seven or more linear strands (averaging about 17 μm in length) from individual organelles (164). Evidence for at least eight double-stranded DNA molecules per mitochondrion has been obtained by following the distribution of [³H] thymidine-labeled cytoplasm through four cell generations (120).

Mitochondrial DNA synthesis occurs throughout the life cycle of normally growing *Tetrahymena* (26,36,118,120), an increased rate having been reported at macronuclear S (120) and micronuclear S periods (26). In the presence of [³H] thymidine, all mitochondria become labeled within one cell generation (120), and the label is conserved (118,120,158). The number of mitochondria in a new daughter cell (approximately 600 to 800) doubles before the next fission (142).

Ribosomes have been identified in electron micrographs of *Tetrahymena* mitochondria (167) and are distinguishable from those of the cytoplasm by unique physical characteristics (37).

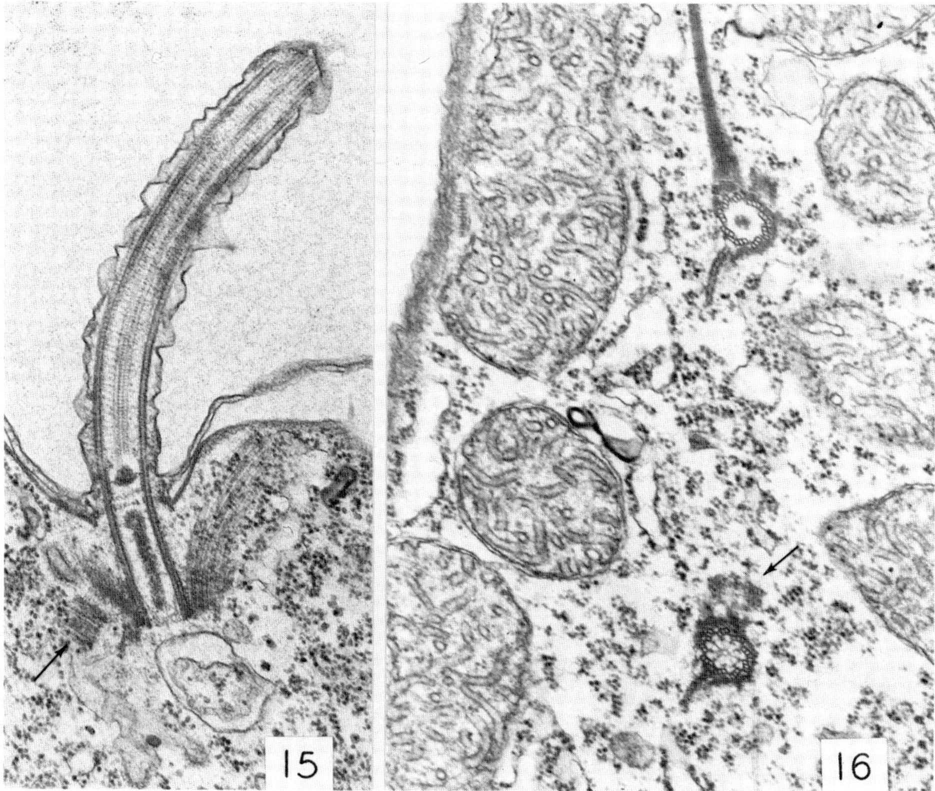

Figs. 15 and 16.

Electron micrographs of strain W showing basal bodies in longitudinal and cross section with adjacent probasal bodies (arrows) in longitudinal section (7). Magnification 37,500 ×. (Reproduced by permission of R. D. Allen and *The Journal of Protozoology*.)

D. Kinetosomes

Despite extensive research the presence of nucleic acids in the basal bodies, or kinetosomes, has been neither firmly established nor disproved. Ciliated protozoans have been choice subjects for these investigations, since they have so many kinetosomes at the cortex (one at the base of every cilium). Maintenance of cortical patterns independent of nuclear genes, as in *Paramecium* (17) and *Tetrahymena* (109), has been attributed to the kinetosomes.

New basal bodies normally develop in association with mature ones, and the process has been traced through electron micrographs of *Paramecium* (47) and *Tetrahymena* (8). Basal bodies with immature probasal bodies are numerous in *Tetrahymena* at the start of division (8) and are shown in two views in Figs. 15 and 16 (7). An analysis of specimens silver-stained for light microscopy indicates that the cortical units develop at a fairly constant rate during the *Tetrahymena* life cycle and at fission are divided approximately evenly (110).

An early report that basal bodies contain DNA and RNA (3 and 2% of dry weight) was based on isolated *Tetrahymena* kinetosomes whose purity was not checked by electron microscopy (151). Subsequent preparations examined by electron microscopy have shown varying degrees of contamination (16,70,141), and on analysis some were found to contain similar amounts of RNA but less DNA (16,70). Negative evidence for nucleic acids in kinetosomes has been obtained from electron micrographs of sectioned cells after nuclease treatments and specific staining (117,167), and from electron microscope autoradiographs of cells grown in $[^3H]$ thymidine (127,155).

Methods suitable for light microscopy, however, have provided evidence for nucleic acids, in pellicular fragments or ghosts from *Tetrahymena* (128) and *Paramecium* (153,154), at the sites of the kinetosomes. Following cell growth in $[^3H]$ thymidine, autoradiographs have revealed some labeling at the kineties when concentration of the nucleoside had been very high (128), or when the film had been exposed for a very long time (153). Kinetosome sites could be stained with fluorescent dyes usually considered specific for nucleic acids [(but see Kasten (80)]; however, a positive reaction for DNA was generally restricted to the latter part of interphase in *Paramecium,* and shortly before fission in temperature-cycled *Tetrahymena* [approximately the times that probasal body formation has been observed in electron micrographs (8,47)]. Attempts to induce contamination of the pellicles with whole-cell DNA were unsuccessful (128,153).

Nevertheless, the possibility of pellicular contamination has been demonstrated in other laboratories (57,72–74). When DNA was extracted from pellicles of *Tetrahymena* or *Paramecium* and centrifuged in a cesium chloride density gradient, the banding patterns corresponded to those of DNA from nuclei, mitochondria, or bacteria in the culture medium. Different investigators have reported evidence for (74) and against (154) the binding of radioactive DNA to nonactive *Paramecium* pellicles. The DNA associated with basal bodies in isolated regenerating oral bands from another ciliated protozoan, *Stentor,* has been attributed to mitochondrial contamination (99).

II. CONJUGATION

Conjugation in *Tetrahymena,* which occurs between complementary mating types of a syngen, normally results in genetic recombination (13,111) accompanied by complete reorganization of the nuclei (51,107,132) (see Chapter 9). Investigations of nucleic acids during conjugation have been conducted primarily with *T. pyriformis* syngen 1.

Before they will conjugate the organisms must be starved. This can be accomplished by centrifuging them from growth medium and resuspending them in water (51) or in an inorganic salt solution [such as Dryl solution (48,177)], or by feeding them a limited number of bacteria which they soon exhaust [usually *Aerobacter aerogenes* in Cerophyl extract (Cerophyl Laboratories, Inc., Kansas City, Mo.) (111)]. When the starved mating types begin to conjugate [usually within a few hours after they are mixed together (9,11,51,95)], food can be added to prevent more pairs from joining, so that the stages of development will be "synchronized" (9).

Autoradiographic studies have utilized cells supplied with appropriate radioactive compounds before or during conjugation. (When the oral regions are attached during conjugation, the labeled compound presumably enters via the cortex.) Sometimes only one

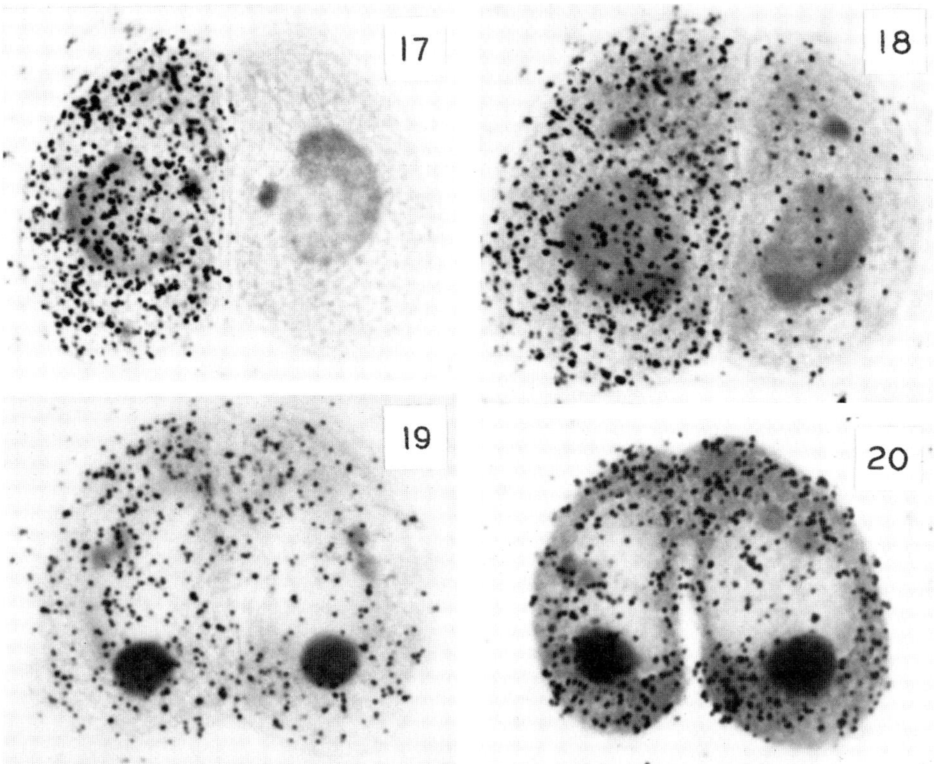

Figs. 17–20.

Exchange of cytoplasm between labeled and unlabeled mates, syngen 1 (95). Magnification 1200 ×. (Reproduced by permission of *The Journal of Protozoology.*).

of the mating types was labeled prior to mixing. In this case the possibility exists that radioactivity in the environment of the washed labeled cells [resulting from dead cells or excretions of live ones (43)] could be incorporated by cells of the unlabeled mating type after they are mixed. Apparently, the problem is not serious when an 18–20 h starvation period precedes mixing (95,97).

A. The Cytoplasm: Exchange of RNA and Protein

That cytoplasm is exchanged during normal conjugation in *T. pyriformis* syngen 1 was demonstrated by mixing an unlabeled mating type with one whose protein had been labeled with a radioactive amino acid (95,150), or whose RNA had been labeled with [³H] uridine (95), and preparing autoradiographs at different stages of conjugation.

Electron micrographs of conjugants belonging to this syngen provide evidence of the way in which exchange can be effected; small intercommunicating pores are present in the attachment region of the mates, and many ribosomes lie free in the cytoplasm (54).

The accompanying autoradiographs represent experiments in which one mating type had been labeled with [³H] histidine (Figs. 17–19) or with [³H] uridine (Fig. 20) (95).

When the mates first joined, only one was radioactive (Fig. 17), but soon exchange of some cytoplasm could be detected (Fig. 18, precrescent). By the macronuclear anlagen stage (Figs. 19 and 20), or earlier, radioactive equilibrium had been attained. During the course of conjugation, the cytoplasmic activity was conserved (95). Labeled RNA became quite evenly distributed between the mates, but a cell whose protein had been labeled always retained more activity, indicating that the radioactive amino acid had been incorporated into cytoarchitectural elements, such as those of the cortex, and cytoplasmic structures too large to penetrate the pores.

B. The Nuclei

1. Synthesis of DNA

At the beginning of conjugation in *Tetrahymena,* the micronucleus is assumed to be in G_2, as usual, and thus to contain 4C DNA. Dividing cells are frequently seen in a mixture of starved mating types at about the time that conjugation begins (95,131), but when [³H] thymidine is available micronuclei in the dividing cells incorporate the nucleoside, and early conjugants (at the stage shown in Fig. 17) include mates both with and without radioactive micronuclei (97). Apparently, cell division is not a prerequisite for conjugation, but cells that do divide replicate their micronuclear DNA.

In order for the final products of the second postzygotic division to be 2N, containing 2C DNA, the micronucleus therefore must synthesize DNA three times during the course of conjugation. Determination of these times has been attempted by supplying [³H] thymidine to conjugants for periods ranging from 30 to 90 min (96,97). Several problems complicate interpretation of the autoradiographic results: (1) the conjugants were not synchronized, and some stages of conjugation resemble each other; (2) considerable cytoplasmic activity which accumulated at some stages (particularly early and late ones) could have obscured nuclear activity; (3) treatment times were quite long, making it difficult to be sure whether [³H] thymidine was incorporated before or after a particular nuclear division. Because of these problems, the following periods are presented only as strong possibilities for times of DNA synthesis: before the first prezygotic division (during the crescent stage); before the third prezygotic division; and before the second postzygotic division.

Evidence for DNA synthesis after the second postzygotic division is more clear-cut (96), as illustrated in Figs. 21–23. At first the young macronuclear anlagen synthesize some DNA and the micronuclei do not (Fig. 21), but the situation is soon reversed; both micronuclei (even the one fated to degenerate) incorporate [³H] thymidine, and the anlagen do not—or at least do so at a very slow rate (Fig. 22). After the mates have separated, however, macronuclear DNA synthesis resumes in the exconjugants (Fig. 23, top cell; considerable cytoplasmic labeling is visible in this exconjugant and in one of the two nonconjugants included in the autoradiograph). Evidence for a steady increase in DNA in new macronuclei of *Tetrahymena* (78) has been cited in another report (15).

Breakdown products from the old, degenerating macronucleus obviously would be useful to the new macronuclei when they synthesize DNA. Utilization of this source of DNA has been studied by prelabeling cells with [³H] thymidine before conjugation

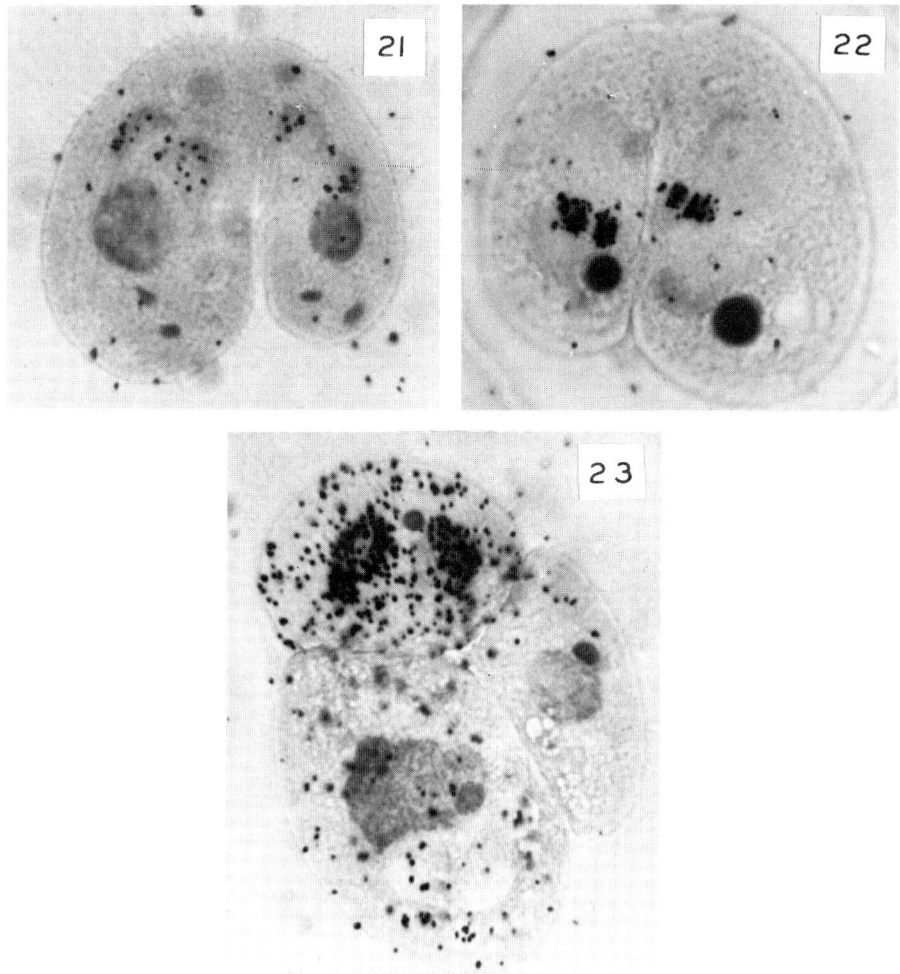

Figs. 21–23.

Incorporation of $[^3H]$ thymidine, following the second postzygotic division, by new macro- and micronuclei. Magnification 1200 ×.

(93,97). [Usually, little cytoplasmic activity was detected after the cells had starved, perhaps for a combination of reasons; the amount of labeling was relatively light because the cells had been grown in a low concentration of $[^3H]$ thymidine, and it probably was diluted during starvation by digestion of mitochondria (89) and division of some cells.] The autoradiographic results indicate that DNA from the old macronucleus is indeed salvaged; about 30–40% of the original macronuclear activity was found in new macronuclei of exconjugants that remained in water, and about half as much in those that were transferred, while still conjugating, to a nutrient medium (proteose–peptone). When only one of the mating types had been labeled, the new macronuclei of the originally unlabeled mate sometimes contained a small amount of radioactivity, indicating that breakdown products from the old macronucleus had been included in cytoplasmic exchange. (Some nuclear activity of course would have been introduced by the migratory pronucleus from the labeled mate.) The opportunity for exchange of macronuclear breakdown products is probably limited, however, because the mates often separate while the old macronucleus (shrunken but highly condensed) is still present. In addition, as described above and illustrated in Fig. 23, intensive DNA synthesis by new macronuclei has been observed only after the mates have separated.

2. Synthesis of RNA

The potential mates in *Tetrahymena* starve before conjugation, and after the oral regions are attached the conjugants continue to starve, even if particulate food becomes available. Any RNA synthesis that occurs during the course of conjugation therefore normally requires recycling of cellular resources. Some RNA is retained during starvation [apparently including messenger RNA and soluble RNA (27)] and conjugation, as demonstrated by azure-B staining and autoradiography (95) (see Section II, A).

The ability of conjugants at different stages of development to synthesize RNA has been studied by exposing them to [^{3}H] uridine (67,96). When the time was long enough (about 1 h) to allow accumulation of RNA in the cytoplasm, considerable activity was observed in autoradiographs of very early conjugants (indicating that the old macronucleus was still functional), and very late conjugants and exconjugants (indicating that the new macronuclei had become functional) (96). In fact, after short exposure to [^{3}H] uridine, radioactivity in very young macronuclear anlagen has been reported (67). At the beginning and end of conjugation, presumably, specific mRNA is needed to direct remodeling of the oral regions.

ACKNOWLEDGMENTS

Thanks are due to the following individuals for providing original photographs and/or unpublished information: Drs. R. D. Allen, I. L. Cameron, M. A. Gorovsky, L. A. Hufnagel, O. L. Miller, Jr., D. M. Prescott, M. V. N. Rao, G. E. Stone, H. Swift, and J. Woodward. Appreciation is also expressed to the editors of *The Biological Bulletin, The Journal of Cell Biology, The Journal of Experimental Zoology, The Journal of Histochemistry and Cytochemistry,* and *The Journal of Protozoology* for authorizing the use of illustrations that have appeared in these publications.

Library research for this chapter was aided by a Dickinson College Faculty Research grant. Previously unpublished experimental work by the author was made possible by grants G-22076 and GB-3289 from the National Science Foundation.

REFERENCES

1.   Agrell, I. and Bergqvist, H. 1962. J. Cell Biol. 15:604.
2.   Albach, R. A. 1967. J. Protozool. 14:271.
3.   Albach, R. A. and Hull, R. W. 1962. J. Protozool. 9(suppl.):22.
4.   Albach, R. A. and Hull, R. W. 1963. J. Protozool. 10(suppl.):20.
5.   Alfert, M. and Balamuth, W. 1957. Chromosoma 8:371.

6.   Alfert, M. and Das, N. K. 1959. Anat. Rec. 134:523.
7.   Allen, R. D. 1967. J. Protzool. 14:553.
8.   Allen, R. D. 1969. J. Cell Biol. 40:176.
9.   Allen, S. L. 1967. Genetics 55:797.
10.   Allen, S. L. 1971. Genetics 68:415.
11.   Allen, S. L., File, S. K., and Koch, S. L. 1967. Genetics 55:823.
12.   Allen, S. L. and Gibson, I. 1971. J. Protozool. 18:518.
13.   Allen, S. L. and Nanney, D. L. 1958. Amer. Nat. 92:139.
14.   Allen, S. L. and Weremiuk, S. L. 1971. J. Protozool. 18:509.
15.   Ammerman, D. 1971. Chromosoma 33:209.
16.   Argetsinger, J. 1965. J. Cell Biol. 24:154.
17.   Beisson, J. and Sonneborn, T. M. 1965. Proc. Natl. Acad. Sci. U.S. 53:275.
18.   Bernstein, E. and Zeuthen, E. 1966. Compt. Rend. Trav. Lab. Carlsberg 35:501.
19.   Brunk, C. F. and Hanawalt, P. C. 1967. Science 158:663.
20.   Brunk, C. F. and Hanawalt, P. C. 1969. Exp. Cell Res. 54:143.
21.   Byfield, J. E. and Lee, Y. C. 1970. Exp. Cell Res. 61:42.
22.   Byfield, J. E. and Lee, Y. C. 1970. J. Protozool. 17:445.
23.   Byfield, J. E. and Scherbaum, O. H. 1967. Proc. Natl. Acad. Sci. U.S. 57:602.
24.   Calkins, J. and Gunn, G. 1967. J. Protozool. 14:210.
25.   Cameron, I. L. 1965. J. Cell Biol. 25(2,pt.2):9.
26.   Cameron, I. L. 1966. Nature 209, 630.
27.   Cameron, I. L., Griffin, E. E., and Rudick, M. J. 1971. Exp. Cell Res. 65:265.
28.   Cameron, I. L. and Guile, E. E., Jr. 1965. J. Cell Biol. 26:845.
29.   Cameron, I. L. and Jeter, J. R., Jr. 1970. J. Protozool. 17:429.
30.   Cameron, I. L. and Nachtwey, D. S. 1967. Exp. Cell Res. 46:385.
31.   Cameron, I. L., Padilla, G. M., and Miller, O. L., Jr. 1966. J. Protozool. 13:336.
32.   Cameron, I. L., and Stone, G. E. 1964. Exp. Cell Res. 36:510.
33.   Cerroni, R. E. and Goldstein, N. O. 1960. Exp. Cell Res. 20:258.
34.   Cerroni, R. E. and Zeuthen, E. 1962. C. R. Trav. Lab. Carlsberg 32:499.
35.   Charret, R. 1969. Exp. Cell Res. 54:353.
36.   Charret, R. and André, J. 1968. Cell Biol. 39:369.
37.   Chi, J. C. H. and Suyama, Y. 1970. J. Mol. Biol. 53:531.
38.   Christensson, E. C. 1970. J. Protozool. 17:496.
39.   Cleffmann, G. 1965. Z. Zellforsch. 67:343.
40.   Cleffmann, G. 1966. Z. Zellforsch. 70:290.
41.   Cleffmann, G. 1967. Z. Zellforsch. 79:599.
42.   Cleffmann, G. 1968. Exp. Cell Res. 50:193.
43.   Cline, S. G. and Conner, R. L. 1966. J. Cell Physiol. 68:149.
44.   Corliss, J. O. 1965. Ann. Biol. T. IV, Fasc. 1–2:49.
45.   DeBault, L. E. and Ringertz, N. R. 1967. Exp. Cell Res. 45:509.
46.   Devidé, Z. and Geitler, L. 1947. Chromosoma 3:110.
47.   Dippell, R. V. 1968. Proc. Natl. Acad. Sci. U.S. 61:461.
48.   Dryl, S. 1959. J. Protozool. 6(suppl.):8.
49.   Dysart, M. P. 1963. J. Protozool. 10(suppl.):8.
50.   Elliott, A. M. and Bak, I. J. 1964. J. Cell Biol. 20:113.
51.   Elliott, A. M. and Hayes, R. E. 1953. Biol. Bull. 105:269.
52.   Elliott, A. M., Kennedy, J. R., Jr., and Bak, I. J. 1962. J. Cell Biol. 12:515.
53.   Elliott, A. M., Travis, D. M., and Work, J. A. 1966. J. Exp. Zool. 161:177.
54.   Elliott, A. M. and Tremor, J. W. 1958. J. Biophys. Biochem. Cytol. 4:839.
55.   Falk, H., Wunderlich, F., and Franke, W. W. 1968. J. Protozool. 15:776.
56.   Jauker, F. and Cleffmann, G. 1970. Exp. Cell Res. 62:477.
57.   Flavell, R. A. and Jones, I. G. 1971. J. Cell Sci. 9:719.
58.   Flax, M. and Himes, M. 1952. Physiol. Zool. 26:297.
59.   Flickinger, C. J. 1965. J. Cell Biol. 27:519.

60. Frankel, J. 1962. C. R. Trav. Lab. Carlsberg 33:1.
61. Frankel, J. 1965. J. Exp. Zool. 159:113.
62. Furgason, W. 1940. Arch. Protistenk. 94:224.
63. Fussel, C. P. 1968. J. Cell Biol. 39:264.
64. Gavin, R. H. and Frankel, J. 1966. J. Exp. Zool. 161:63.
65. Gibson, I. R. 1966. J. Protozool. 13:650.
66. Gorovsky, M. A. 1970. J. Cell Biol. 47:619.
67. Gorovsky, M. A. and Woodard, J. 1968. J. Cell Biol. 39:54a.
68. Gorovsky, M. A. and Woodard, J. 1969. J. Cell Biol. 42:673.
69. Harrington, J. D. 1960. Ph.D. Thesis, Biological Studies no. 57, Catholic University, Washington, D.C.
70. Hoffman, E. J. 1965. J. Cell Biol. 25:217.
71. Holz, G. G., Jr. 1960. Biol. Bull. 118:84.
72. Hufnagel, L. A. 1966. 6th Intl. Cong. Electron Microsc., 239.
73. **Hufnagel, L. A. 1968. J. Cell Biol. 39:63a.**
74. Hufnagel, L. A. 1969. J. Cell Sci. 5:561.
75. Ito, J., Lee, Y. C., and Scherbaum, O. H. 1968. Exp. Cell Res. 53:85.
76. Jacobson, K. B. and Prescott, D. M. 1964. Exp. Cell Res. 36:561.
77. Jeffery. W. R., Stuart, K. D., and Frankel, J. 1970. Exp. Cell Res. 46:533.
78. Johansson and Zech. Cited in ref. 15.
79. Jones, A. S. and Thomspon, T. W. 1963. J. Protozool. 10:91.
80. Kasten, F. H. 1967. Int. Rev. Cytol. 21, 141.
81. Keidig, J. and Westergaard, O. 1971. Exp. Cell Res. 64:317.
82. Kidder, G. W. and Diller, W. F. 1934. Biol. Bull. 67:201.
83. Kumar, A. 1969. J. Cell Biol. 40:716.
84. Kurnick, N. B. 1955. Int. Rev. Cytol. 4:221.
85. Lederberg, S. and Mazia, D. 1960. Exp. Cell Res. 21:590.
86. Lee, Y. C. and Byfield, J. E. 1970. Biochemistry 9:3947.
87. Leick, V. 1967. C. R. Trav. Lab. Carlsberg 36:113.
88. Leick, V. 1969. Eur. J. Biochem. 8:221.
89. Levy, M. R. and Elliott, A. M. 1968. J. Protozool. 15:208.
90. Lowe-Jinde, L. and Zimmerman, A. M. 1971. J. Protozool. 18:20.
91. Lowy, B. A. and Leick, V. 1969. Exp. Cell Res. 57:277.
92. McDonald, B. B. 1958. Biol. Bull. 114:71.
93. McDonald, B. B. 1959. J. Protozool. 6(suppl.):17.
94. McDonald, B. B. 1962. J. Cell Biol. 13:193.
95. McDonald, B. B. 1966. J. Protozool. 13:277.
96. McDonald, B. B. 1968. Proc. XII Int. Congr. Genet. 1:140.
97. McDonald, B. B. Unpublished.
98. Mackenzie, T. B., Stone, G. E., and Prescott, D. M. 1966. J. Cell Biol. 31:633.
99. Margulis, L., Banerjee, S., and Younger, K. B. 1971. Abstr. Amer. Soc. Cell Biol., 177.
100. Miller, O. L., Jr., and Stone, G. E. Cited in ref. 31.
101. Mita, T., Shiomi, H., and Iwai, K. 1966. Exp. Cell. Res. 43:696.
102. Moner, J. G. 1965. J. Protozool. 12:505.
103. Moner, J. B. and Berger, R. O. 1966. J. Cell Physiol. 67:217.
104. Murti, K. G. and Prescott, D. M. 1970. J. Cell Biol. 47:460.
105. Nachtwey, D. S. 1965. C. R. Trav. Lab. Carlsberg 35:25.
106. Nachtwey, D. S. and Dickinson, W. J. 1964. J. Protozool. 11(suppl.):24.
107. Nanney, D. L. 1953. Biol. Bull. 105:133.
108. Nanney, D. L. 1959. J. Protozool. 6:171.
109. Nanney, D. L. 1968. Science 160:496.
110. Nanney, D. L. 1971. Develop. Biol. 26:296.
111. Nanney, D. L. and Caughey, P. A. 1955. Genetics 40:388.
112. Nass, S., Nass, M. M. K., and Hennix, U. 1965. Biochim. Biophys. Acta 95:199.

113. Nicolette, J. A., Buhse, H. E., Jr., and Robin, M. S. 1971. J. Protozool. 18:87.
114. Nilsson, J. R. 1970. C. R. Trav. Lab. Carlsberg 38:107.
115. Nilsson, J. R. 1970. J. Protozool. 17:539.
116. Nilsson, J. R. and Leick, V. 1970. Exp. Cell Res. 60:361.
117. Outka, D. E. 1971. J. Protozool. 18(suppl.):18.
118. Parsons, J. A. 1965. J. Cell Biol. 25:641.
119. Parsons, J. A. and Dickson, R. C. 1965. J. Cell Biol. 27:77a.
120. Parsons, J. A. and Rustad, R. C. 1968. J. Cell Biol. 37:683.
121. Prescott, D. M. 1959. Exp. Cell Res. 16:279.
122. Prescott, D. M. 1960. Exp. Cell Res. 19:228.
123. Prescott, D. M. 1962. J. Histochem. Cytochem. 10:145.
124. Prescott, D. M., Bostock, C., Gamow, E., and Lauth, M. 1971. Exp. Cell Res. 67:124.
125. Prescott, D. M., Rao, M. V. N., Evenson, D. P., Stone, G. E., and Thrasher, J. D. 1966. In
      D. M. Prescott, ed., Methods in cell physiology, vol. 2, Academic Press, New York, p. 131.
126. Prescott, D. M. and Stone, G. E. 1967. In T. T. Chen, ed., Research in protozoology, vol. 2,
      Pergamon Press, New York, p. 117.
127. Pyne, M. C. 1968. C. R. Acad. Sci. Paris, Ser. D 267:755.
128. Randall, J. and Disbrey, C. 1965. J. Roy. Soc., Ser. B 162:473.
129. Rao, M. V. N. and Ammerman, D. 1970. Chromosoma 29:246.
130. Rasmussen, L. and Zeuthen, E. 1962. C. R. Trav. Lab. Carlsberg 32:333.
131. Ray, C., Jr. 1955. Ass. Southeast. Biol. Bull. 2(1):9.
132. Ray, C., Jr. 1956. J. Protozool. 3:88.
133. Ray, C., Jr. 1958. Proc. X Int. Cong. Genet. 2:229.
134. Ray, C., Jr., and Elliott, A. M. 1954. Anat. Rec. 120(3):228.
135. Read, S. A. and Buetow, D. E. 1971. Exp. Cell Res. 64:239.
136. Ringertz, N. R., Bolund, L., and DeBault, L. E. 1967. Exp. Cell Res. 45:519.
137. Roth, L. E. and Minick, O. T. 1961. J. Protozool. 8:12.
138. Rudick, M. J. and Cameron, I. L. 1972. Exp. Cell Res. 70:411.
139. Satir, B. 1967. Exp. Cell Res. 48:253.
140. Satir, B. and Dirksen, E. R. 1971. J. Cell Biol. 48:143.
141. Satir, B. and Rosenbaum, J. L. 1965. J. Protozool. 12:397.
142. Sato, H., 1960. Anat. Rec. 138(3):381.
143. Sato, H., 1963. Proc. XVI Int. Congr. Zool., 2:294.
144. Schensted, I. V. 1958. Amer. Nat. 91:161.
145. Scherbaum, O. H. 1957. Exp. Cell Res. 13:24.
146. Scherbaum, O. H. 1960. Ann. N. Y. Acad. Sci. 90:565.
147. Scherbaum, O. H., Louderback, A. L., and Jahn, T. L. 1958. Biol. Bull. 115:269.
148. Scherbaum, O. and Zeuthen, E. 1954. Exp. Cell Res. 6:222.
149. Schildkraut, J. L., Mandel, M., Levisohn, S., Smith-Sonneborn, J. L., and Marmur, J. 1962.
      Nature 196:795.
150. Schooley, C. N. 1958. J. Protozool. 5(suppl.):24.
151. Seaman, G. R. 1960. Exp. Cell Res. 21:292.
152. Shepard, D. C. 1965. Exp. Cell Res. 38:570.
153. Smith-Sonneborn, J. and Plaut, W. 1967. J. Cell Sci. 2:225.
154. Smith-Sonneborn, J. and Plaut, W. 1969. J. Cell Sci. 5:365.
155. Stevens, A. R. 1966. In D. M. Prescott, ed., Methods in cell physiology, vol. 2,
      Academic Press, New York, p. 255.
156. Stone, G. E. 1968. J. Cell Biol. 30:556.
157. Stone, G. E. and Cameron, I. L. 1964. In D. M. Prescott, ed., Methods in cell physiology,
      vol. 1, Academic Press, New York, p. 127.
158. Stone, G. E. and Miller, O. L., Jr. 1965. J. Exp. Zool. 159:33.
159. Stone, G. E., Miller, O. L., Jr., and Prescott, D. M. 1965. J. Cell Biol. 25:171.
160. Stone, G. E. and Prescott, D. M. 1964. J. Cell Biol. 21:275.
161. Stone, G. E. and Prescott, D. M. 1965. In C. P. LeBlond and K. B. Warren, eds., The use of

radioautography in investigating protein synthesis, Symposia of the International Society for Cell Biology, vol. 4, Academic Press, New York, p. 95.

162. Stowell, R. E. 1946. Stain Technol. 21:137.
163. Sueoka, N. 1961. Cold Spring Harbor Symp. Quant. Biol. 26:35.
164. Suyama, Y. and Miura, K. 1968. Proc. Natl. Acad. Sci. U.S. 60:235.
165. Suyama, Y. and Preer, J. R., Jr. 1965. Genetics 52:1051.
166. **Swift, H. 1965. Amer. Nat. 99:201.**
167. Swift, H., Adams, B. J., and Larsen, K. 1964. J. Roy. Microsc. Soc. 83:161.
168. Tamura, S., Tsuruhara, T., and Watanabe, Y. 1969. Exp. Cell Res. 55:351.
169. Walker, P. M. B., and Mitchison, J. M. 1957. Exp. Cell Res. 23:258.
170. **Wand, M. 1968. J. Protozool. 15(suppl.):36.**
171. Wand, M., Zeuthen, E., and Evans, E. A. 1967. Science 157:436.
172. Watanabe, Y. 1971. Exp. Cell Res. 68:437.
173. Watanabe, Y. and Ikeda, M. 1965. Exp. Cell Res. 39:443.
174. Watanabe, Y. and Ikeda, M. 1965. Exp. Cell Res. 39:464;
175. Wells, C. 1960. J. Protozool. 7:313.
176. Wells, C. 1961. J. Protozool. 8:284.
177. Wells, C. 1962. Amer. Zool. 2:457.
178. Wells, C. 1965. J. Protozool. 12:561.
179. Whitson, G. L. and Padilla, G. M. 1964. Exp. Cell Res. 36:667.
180. Whitson, G. L., Padilla, G. M., and Fisher, W. D. 1966. Exp. Cell Res. 42:438.
181. Wille, J. J., Jr., and Ehret, C. F. 1968. J. Protozool. 15:785.
182. Wolfe, J. A. 1967. Chromosoma 23:59.
183. Woodard, J., Gorovsky, M., and Kaneshiro, E. 1968. J. Cell Biol. 39:182a.
184. Wykes, J. R. and Prescott, D. M. 1968. J. Cell. Physiol. 72:173.
185. Yuyama, S. and Zimmerman, A. M. 1972. Exp. Cell. Res. 71:193.
186. Zadylak, A. H., Ehret, C. F., and Wille, J. J. 1970. Atomlight 71:1.
187. **Zeuthen, E. 1961. In T. W. Goodwin and O. Lindberg, eds., Biological structure and function,** vol. 2, Academic Press, New York, p. 537.
188. Zeuthen, E. and Scherbaum, O. 1954. In J. A. Kitching, ed., Recent developments in cell physiology, Proceedings of the 7th Symposium of the Colston Research Society, London, p. 141.

CHAPTER 11

Genetics of *Tetrahymena*

Sally Allen

Departments of Botany and Zoology
University of Michigan
Ann Arbor Michigan

and

Ian Gibson

School of Biological Sciences
University of East Anglia
Norwich, Norfolk, England

I. INTRODUCTION

Tetrahymena pyriformis presents a paradox. A unicellular eucaryote, it is morphologically complex, yet it is amazingly uniform in its ground plan. From this ground plan, however, great diversity of phenotype has been recorded, and it has evolved a number of distinct gene pools. It has successfully invaded many different parts of this planet. It is found in temperate zones, in the torrid tropics, and in the freezing Arctic. How has it managed to survive these extremes of climate and adapted to the vicissitudes of its diverse little worlds? In short, how does it achieve such overwhelming diversity in the face of its common morphology?

In particular, we are concerned with the two types of nuclei and the parts they play in the genetics and evolution of this organism. We would like to understand how the complex organization of *Tetrahymena* is transmitted from generation to generation, how certain characters are inherited, and how the genetic systems are maintained or changed during the course of evolution.

First, however, we must ask if there is simple Mendelian inheritance in *Tetrahymena*. Is there a system of inheritance involving chromosomes? What are the relative roles of the two nuclei in the hereditary process and in the control of phenotypes?

We show that *Tetrahymena* possesses a system for Mendelian inheritance for most

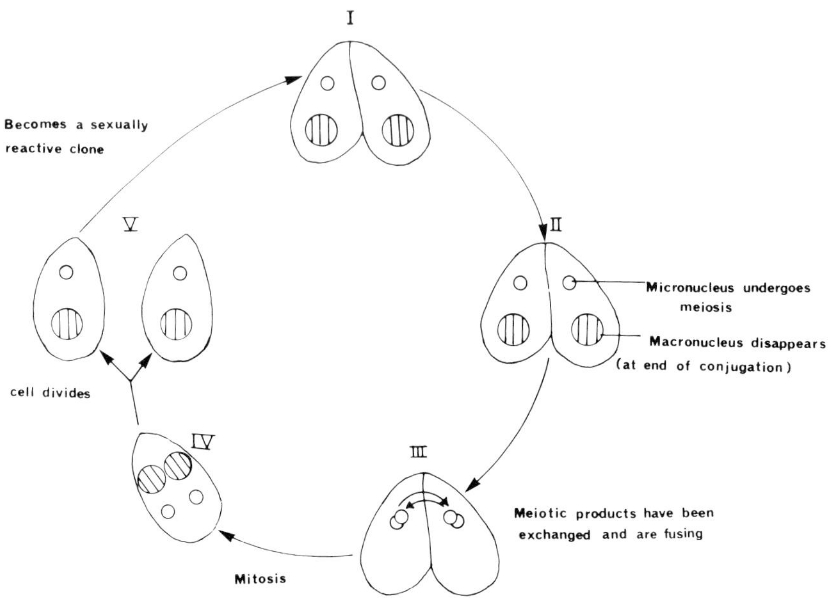

Fig. 1.

An outline of the life cycle of the nuclei following conjugation. I, The cells are paired; II, the micronucleus undergoes meiosis; the old macronucleus is resorbed, although not until stage IV (to simplify the diagrams it is not shown in subsequent stages); III, a fusion nucleus forms in each conjugant following the exchange of meiotic products between them; IV, the fusion nucleus divides twice to produce four nuclei, two becoming macronuclei and two becoming micronuclei; V, the normal nuclear complement is restored; this is accomplished by the following steps: (1) before division one micronucleus is resorbed; (2) at division the remaining micronucleus divides; (3) the two macronuclei do not divide but are distributed, one to each daughter cell.

of its characters. In addition, variant forms of some characters appear to depend upon "epigenetic" systems. These systems have been shown to be located in one, or both, or neither, of the nuclei. Our primary interest is in the nuclear systems. Thus we start by summarizing what is known of the origin, structure, replication, and function of these two nuclei.

Both nuclei have a common ancestry. The events that give rise to the two types of nuclei during the life cycle are shown in Fig. 1. During conjugation the "old" macronucleus is discarded, and a new one is derived from the old micronucleus. A zygotic nucleus forms from fusion of the meiotic products of the old micronucleus. This zygotic nucleus then divides twice, and two of the products synthesize large amounts of DNA relative to the other two nuclei. After cell division each daughter cell has one new micronucleus and one new macronucleus. This is the normal nuclear complement of *Tetrahymena*.

Spectrophotometric measurements of the relative amounts of DNA in the two nuclei show there is 20 times as much DNA in the macronucleus, taking into account the differences in their time of replication during the cell cycle (40,64,90,171,172).

Five pairs of chromosomes have been observed in the micronucleus during meiosis, but no conventional mitotic configurations have been observed during cell division. There are no visible chromosomes in the macronucleus of *Tetrahymena,* and the macronucleus divides by amitosis. Kloetzel (84) has suggested that the polytene chromosomes seen in the macronuclei of other protozoa (150) are intermediate stages in the fragmentation of chromosomes. Various reports have appeared which describe the organization of macronuclear DNA into small bodies (164). In a recent study Wolfe (170) and Nilsson (127) observed that these small bodies can be seen under phase-contrast microscopy and they appear to be bound together by fibers. Whether these are whole chromosomes or fragments of chromosomes remains yet to be determined.

During cell division replication of DNA is not in phase in the two nuclei. Micronuclear DNA synthesis occurs much earlier in the cell cycle than does macronuclear DNA synthesis. This suggests that the controls over replication are different for the two nuclei. There are also controls, the molecular basis of which is unknown, over the total amount of DNA synthesized in the macronucleus. If the amount of DNA falls below a certain threshold, then cell division is delayed and extra rounds of DNA replication occur (40). If the total amount exceeds the normal level, then extrusion of chromatin occurs.

It appears that during the vegetative cycle the DNA within the micronucleus is inactive, while at least some of the DNA in the macronucleus is active. Experiments with heterocaryons, in which the macronucleus possesses one gene and the micronucleus possesses an alternative form, have shown that the macronuclear genotype, and not the micronuclear genotype, exerts its influence over a particular phenotype (8). This suggests that no RNA is produced in the micronucleus. In support of this, large bodies containing RNA are found in the macronucleus (164), but no RNA seems to be found in the micronucleus (65). In autoradiographic experiments involving the labeling of the two nuclei in *Tetrahymena,* one group of investigators showed that while the macronucleus synthesizes RNA the micronucleus does not (67,69), although another group came to a different conclusion (93). Furthermore, differences in the histones associated with each of these nuclei have been found, specific fractions being confined to each type of nucleus. (66,68)

The micronucleus is not always essential to the survival of *Tetrahymena.* In natural populations 30–50% of the cells collected do not have micronuclei and are perfectly functional although they are incapable of mating. Yet, in laboratory strains of *Tetrahymena* the cells usually die if the micronucleus is lost. Thus some function normally essential for survival must be ascribed to the micronucleus, but it can occasionally be abrogated in some unknown way.

The behavior of the two nuclei needs to be taken into account when we discuss the inheritance of various characters. We emphasize at this point, however, that in sexual forms of *Tetrahymena* genes are transmitted through the micronucleus from one sexual generation to the next, while the macronucleus is derived from a micronuclear product and appears to have the role of controlling phenotype during vegetative divisions.

The genetic systems of *Tetrahymena,* as in other organisms, have evolved. The "purpose" of any genetic system must surely be to create the diversity upon which natural selection can operate during evoluton. How is it that the genetic material of *Tetrahymena* responds to changes in the environment. How is the ability to change built into the genotype of the organism? At the same time the genetic system which creates diversity must be restrained, otherwise the organism will suffer from too high a genetic load.

Tetrahymena pyriformis has evolved into several isolated breeding groups, designated syngens. Different strains within a syngen are able to exchange genes and create progeny with recombinant genomes. Gene flow does not occur between syngens, however, despite the fact that they all have the same micronuclear chromosome number ($2N = 10$). The existence of syngens shows that genetic diversification has occurred and that different systems have been preserved. What kinds of changes occur in these systems to bring about the differences between syngens? How important are mutations, recombination of genes, and epigenetic events in bringing about change? Has the possession of two types of nuclei played a role in creating genetic diversification? Most of the genetic work has been done with syngen 1. It is possible that different syngens may have evolved different mechanisms for creating diversity upon which natural selection acts.

This chapter is an attempt to reinterpret the genetics of *T. pyriformis*. We hope to point up the problems and to suggest new ways of thinking about these problems in the hope of stimulating wider interest in the genetics of this organism. Other reviews with different points of view have been published on the genetics of *T. pyriformis, Paramecium aurelia,* and other ciliated protozoa (10,29,63,83,114,124,145, 156, 157, 159).

In this review we discuss the methodology used in studying the genetics of *Tetrahymena* (Section II), the genetics of this organism (Section III), clonal variations in syngens 1 and 7 (Section IV), variation between syngens (Section V), and evolution in *Tetrahymena* (Section VI), in which we combine these observations and examine possible mechanisms which may be important in creating diversity.

Our focus in this chapter is on the two types of nuclei found in *T. pyriformis,* as in other ciliated protozoa, and their roles in the life cycle and evolution of this organism. We hope to show that the dimorphism of the nuclei and the evolution of this species of ciliate are closely related.

II. METHODOLOGY

A. Derivation of Cell Lines

1. Wild Strains

Strains of *T. pyriformis* have been collected from a variety of locales from many parts of the world (46). A clone is initiated from a single isolate and tested against two or more mating types from all 12 syngens. A high frequency of such clones do not mate either because they are amicronucleate or immature. Clones (strains) that mate can be tentatively assigned to a particular syngen by their sexual reactivity with testers of that syngen. Confirmation of this assignment is obtained if viable progeny are produced. This has not always been carried out; therefore it is possible that the existing collections contain more than 12 syngens.

Table 1.
Collection of Syngen-1 Wild Strains

Strain	Location	Collector and date	Phenotype	Breeding behavior	Ref.
WH-6	Woods Hole, Massachusetts	Elliott and Gruchy, 1952	I, Ha, Ta, E-1b, E-2b, P-1b	Infertile	(50)
WH-14	Woods Hole, Massachusetts	Elliott and Gruchy, 1952	II, Hd, Tc, E-1b, E-2b, P-1b	Infertile	(50)
WH-52	Woods Hole, Massachusetts	Elliott and Gruchy, 1952	III, Hb	Infertile	(50)
UM-226	Bennington, Vermont	Gruchy, 1955	VI, He, Ta, E-1c, E-2c, P-1b	Infertile	(73)
ALP-4	Alpena, Michigan	Nanney, 1956	VI, Hd, Ta, E-1b, E-2b, P-1a	Infertile	(98)
IL-12	Indian Lake, Michigan	Nanney, 1956	VI, Hd, Ta, E-1b, E-2b, P-1a	Infertile	(98)
LW-A	Lake-of-the-Woods, Illinois (A)	Phillips, 1966	III, Hd, Ta, E-1b, E-2c, P-1b	Fertile	(139)
LW-B	Lake-of-the-Woods, Illinois (B)	Phillips, 1966	VI, Ha, Ta, E-1bc, E-2b, P-1a	Partially fertile	(139)

Table 2.
Derivation and Genotypes of Inbred Strains in Syngen 1

Inbred strain	Origin	Genotype[a]					
		mt	*H*	*T*	*E-1*	*E-2*	*P-1*
A	WH-6 × WH-14	A	A	A	B	B	A
A1	A_4 × B_3	A	D	A	B	B	A
A3[b]	A_{11} × C_{11}	A	E	B	B	C	B
B	WH-6 × WH-14	B	D	A	B	B	B
B2	B_{14} × $D1_{12}$	B	C	C	B	B	A
C[b]	UM-226 × B_2	C	E	B	C	C	B
C1	A_{11} × C_{11}	C	A	A	C	B	B
C2[c]	A/C × C*	C	E	B	C	C	B
D	ALP-4 × B_2	D	D	C	B	B	A
D1	B_2 × D_7	D	C	C	B	B	A
E	IL-12 × B_9	E	D	A	B	B	B
F	ALP-4 × B_9	F	D	A	B	B	B

[a]Homozygous for allele specified.
[b]Extinct.
[c]Replaces C; derived by genomic exclusion.

The collection of wild strains of syngen 1 is shown in Table 1. Many of these strains are now infertile. This is also the case for the wild strains of the other syngens, largely collected by A. M. Elliott and his students.

2. Inbred Strains and "Homozygous Strains"

Strains were derived from crosses within the syngen-1 collection, and these have been inbred (1,98,136). These have been referred to as "families" or "inbred strains," and they are designated by the letter (A, B, and so on) corresponding to the mating type allele that is present (mt^A, mt^B, and so on). Their derivation is shown in Table 2. They are highly homozygous because of the practice of close inbreeding (selfing) for at least 12 and up to 24 generations, and to rigorous selection for the genes shown in the table. Inbred strains have now been initiated from the recent syngen-7 collections (140).

Genomic exclusion (see Section II,D,4) was used to generate "instant" homozygous cell lines in syngen 1, as well as congenic lines of strain D (17). Most of these have different gene combinations from the standard inbred series shown in Table 2.

B. Maintenance of Fertile Cell Lines

Stocks may be maintained asexually in the laboratory for years; however, their breeding performance deteriorates with time. This point is well illustrated by the current performance of the wild strains of syngen 1 (Table 1). When such strains are crossed, the mating pairs either die or give rise to progeny by genomic exclusion (4,104). The most likely cause for this deterioration is the behavior of the micronucleus in laboratory strains.

The micronucleus becomes defective, and some lineages produce amicronucleate cells. These die in the laboratory (23,96). When lines with defective micronuclei are used in crosses, normal progeny are not produced (20). Therefore considerable effort must be exerted to prevent deterioration by selecting lines that have a micronucleus and give rise to normal progeny. This can be accomplished by yearly rounds of inbreeding of all strains. The recent introduction of freezing (155) permits more convenient maintenance of lines.

C. Growth of Cells for Genetic Work

Genetic studies have utilized cells grown either in proteose–peptone or in bacterized medium (*Aerobacter aerogenes* grown in diluted peptone, nutrient agar, or Cerophyl rye grass). The progeny are then subcultured in the medium appropriate for testing a particular phenotype, including a complex synthetic medium which contains about 28 different components (56) and a recently developed nearly minimal medium (132). In bacterized medium large numbers of organisms can be manually manipulated without sterile technique. However, crosses can be made using axenically grown cells, and the entire operation carried out with an aseptic procedure.

D. Techniques for Crossing Cells

1. Sexual Processes

Conjugation appears to be the only means by which gene recombination occurs in *T. pyriformis*. Autogamy, a form of self-fertilization which results in homozygosity for all genes, has not been found in *T. pyriformis*. It does, however, occur in *T. rostrata* during encystment (41). Cysts were observed in certain strains of syngen 2 in *T. pyriformis* (77,79), but excystment could not be induced reliably so it was not possible to determine if autogramy had occurred. Hurst believed that there might be genetic factors that control cyst formation (79). Cytogamy, in which autogamy is induced by two cells in contact, has not been reported (130).

Conjugation may be normal—in which case reciprocal exchange of gametic nuclei occurs between mates. It may also be abnormal, as in genomic exclusion (4,104). When genomic exclusion occurs, one of the two genetically identical gametic nuclei of one mate is transferred to its partner and each gametic nucleus becomes diploid in some unknown way (9). As a result, exconjugants are genetically identical and homozygous for all genes. Genomic exclusion therefore mimics autogamy in its genetic consequences. Actually, it involved two consecutive matings and a complicated series of nuclear maneuvers. This feature permits its separation from normal conjugation which involves only one mating. Depending upon the type of cross desired, either normal conjugation or genomic exclusion can be employed.

2. Induction of Mating

Conjugation is induced if mixtures of different mating types are made. Mating pairs normally form only in mixed cultures and do not form in pure cultures, with the exception of cultures of selfing clones. Cell contact appears to be necessary for invoking the mating reaction both in syngen 1 (120) and in syngen 7 (141). In the latter syngen a heat-stable, low-molecular-weight substance, present in cultures activated for mating, can induce mating in unreactive mixtures of cells.

The requirements for the induction of mating have been worked out for most of the syngens (37,38,51,53,55,57,73,78,79,131,134,141). Mating does not occur in proteose–peptone nor in well-fed bacterized cultures. Mild starvation appears to be a requirement for mating. Bacterized cultures mate when the bacteria are exhausted. Peptone cultures must be washed with several changes of distilled water or Dryl's solution (43) before mixing the cultures. Apparently, amino acids inhibit mating (44). A low nitrogen requirement is a prerequisite for mating in *Chlamydomonas* (152). This may be the case for *Tetrahymena,* although it has not been proved. The addition of nutrient prevents new matings in a culture that contains conjugating pairs (9).

3. Normal Conjugation

The sequence of stages in normal conjugation has been described by Nanney (94), Elliott and Hagen (51), and Ray (151). In crosses it is important to monitor whether or not normal conjugation has in fact occurred. Examining stained preparations excludes pairs that have not completed conjugation, but it does not eliminate pairs that have undergone genomic exclusion. A more reliable method involves the use of timed matings in which pairs are isolated from a mating mixture within 10 h, or before a second mating can occur, and tests of the sexual maturity of the progeny. In syngens with a known immature period (syngens 1, 2, 3, 6, 7, and 8), cells that have successfully completed normal conjugation become "immature," or sexually unreactive when tested against known reactive cells. In these syngens it is a simple matter to conduct a maturity test, hence to identify progeny that have undergone normal conjugation. In syngens without an immature period, it is presently not possible to determine if normal conjugation has occurred, since genetic markers are not available. With genetic markers it would be possible to detect the exchange of genes, hence that normal conjugation had occurred.

4. Genomic Exclusion

Genomic exclusion is associated with clones that have defective micronuclei (20,23,26). It differs from normal conjugation in several ways (9). There are two consecutive matings, the first being abnormal and the second normal. In the first mating (round 1) the exconjugants that arise are heterocaryons and are sexually mature, since (1) one of the mates contributes replicates of a single meiotic product to both mates; (2) each nucleus undergoes diploidization either by an extra division and fusion or by endoreduplication; (3) the old macronucleus is retained and the new macronuclei abort. The second mating (round 2) follows immediately and is a normal conjugation, thereby giving rise to exconjugants which are sexually immature. The progeny are homozygous for all genes if exconjugants from the same pair remate. If exconjugants from different pairs remate, some may be heterozygous, depending upon the genotype of the normal cell line used in the first mating (4,104). By the use of timed matings and the isolation of round-2 pairs, genomic exclusion can be exploited to generate homozygous cell lines (8,17).

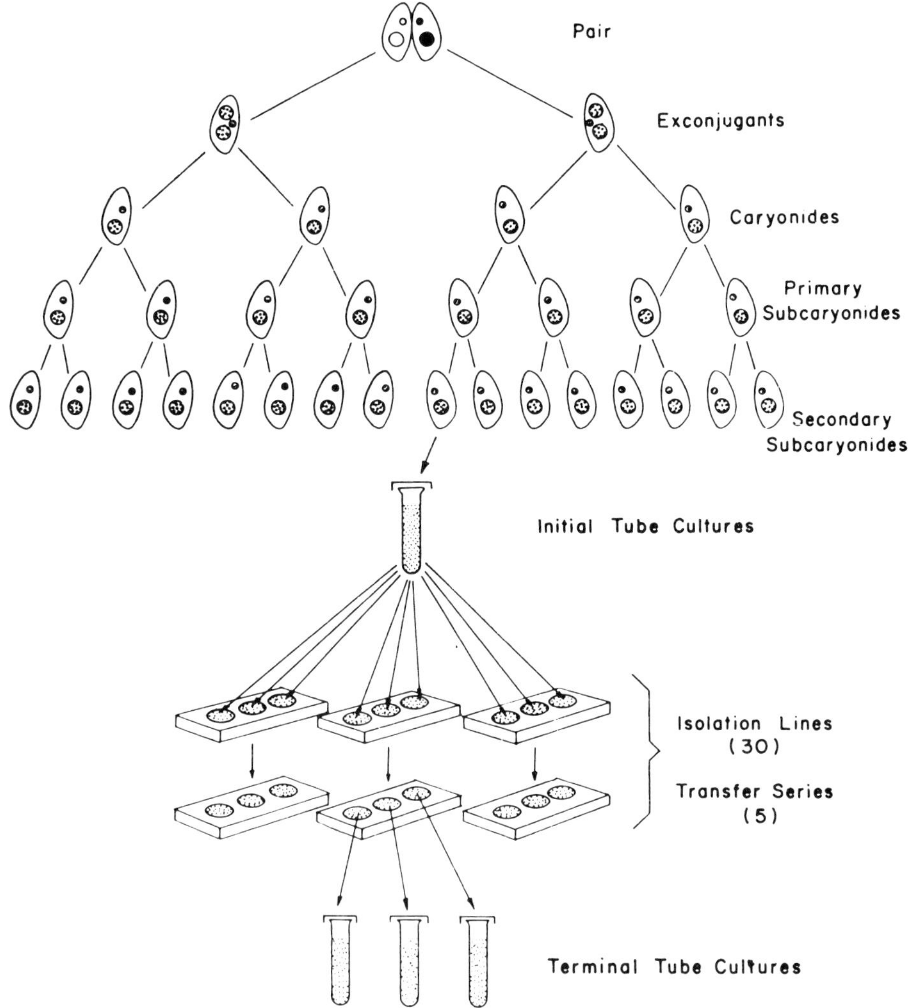

Fig. 2.

Establishment of vegetative pedigrees. [From Bleyman et al. (32).]

E. Scoring Phenotypes

After pairs are isolated into separate containers and the exconjugants come apart, several routes can be followed, depending upon the type of analysis needed for a specific problem. The two exconjugants may be allowed to develop populations in the same container. In this case the lineages are referred to as "synclones." If the two exconjugants are isolated into separate containers, the lineages are referred to as "exconjugant clones." Separate cultivation of the products of the first fission results in lineages referred to as "caryonides." With patience and stamina subsequent separation of second, third,

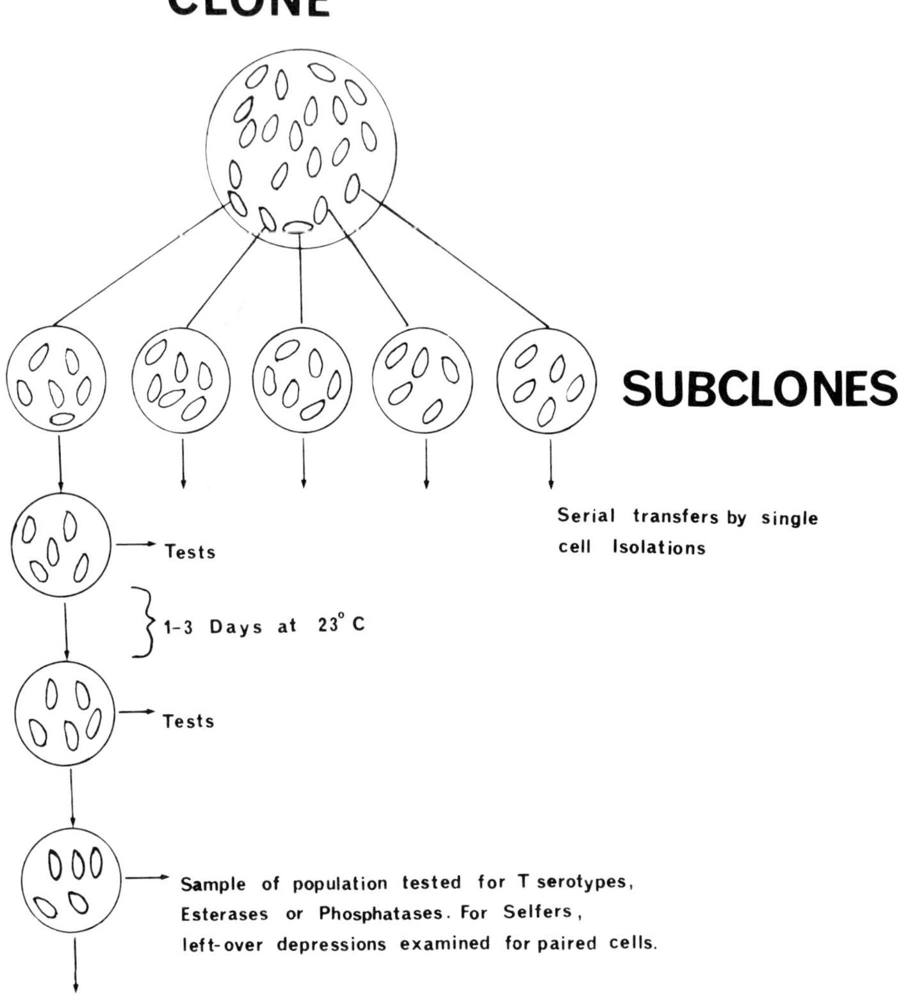

Fig. 3.

Clonal analysis. Subclones could be initiated from a caryonide, subcaryonide, or from a population of cells after several hundred fissions subsequent to conjugation. Many modifications of clonal analysis are possible other than the one illustrated.

and so on, fission products can also be made and leads to lineages referred to as 1° subcaryonides, 2° subcaryonides, and so on. A diagram of these manipulations summarizes the relationship of these lineages (Fig. 2). Analysis of synclones establishes the genetic basis for many traits in *Tetrahymena* (Section III, B). The role of cytoplasmic factors is assessed by analysis of exconjugant clones. Analysis of caryonides has shown that macronuclei developing in the same exconjugant can in some manner control different phenotypes. A more detailed cell pedigree analysis gives further information on this relationship. Such analyses have been carried out by Nanney and his students. In addition to these types of manipulations, a "clonal analysis" can be initiated at any fission following conjugation (Fig. 3). In this case several single cells are isolated from a population of cells and each is placed in a separate container to initiate subclones. These subclones may then be serially propagated at particular fission intervals by single cell transfer. This is a grosser type of analysis, but it has been particularly useful in studies of selfing clones (18) and certain heterozygotes (7,12,137).

Phenotypes of cell lineages are scored under the conditions optimal for their expression. This may mean transferring a sample of the population to a different growth medium or to a different temperature. This point is illustrated by the particular test systems used for assaying certain traits in syngen 1 (Fig. 4). For nutritional studies samples are transferred to synthetic media of varying composition, and the type of growth scored (48,49). Further discussion of phenotypes can be found in Section III,C.

III. GENETICS

A. Mutation

Mutations occur both in the micronucleus and macronucleus, but if a mutant is observed, it is necessary to breed it to determine its nuclear basis. Only a dominant micronuclear mutation will be observed in the F_1; recessive micronuclear mutations will appear in the F_2. If a macronuclear mutation has occurred, it will not be observed in either generation. Spontaneous mutations have been recorded for the micronucleus (Section III, C). Indeed, one case occurred in the laboratory in genetic tests of two subclones of a phosphatase heterozygote (24). Both subclones had a heterozygous phenotype. One subclone maintained at 16°C bred as if it were a heterozygote. The other subclone maintained at 23°C bred as if it were a homozygote. The latter thus behaved as if a micronuclear mutation had occurred. Another case involved the *mt*[A] allele in syngen 8 (131). Mutations interpreted as restricted to the macronucleus have also been observed, as exemplified by the morphological variants tested by Nanney (97). No solid information has been reported on rates of mutation for either nucleus, although Orias (131) has reported "mutation rates" as high as 10^{-3} for the micronucleus in syngen 8. However, these may not be mutations but changes in the function of regulatory genes.

Mutations can be induced by several agents. Micronuclear mutations have been sought, but their detection has been hampered until recently by suitable selective techniques. Early efforts to induce nutritional mutations by x rays, ultraviolet light, cobalt-60, and certain chemical mutagens failed (52). Recently, Carlson (39) induced micronuclear mutations with 0.25% ethyl methanosulfonate and took advantage of the clonal changes that occur in heterozygotes (Section IV) to select for mutants resistant to caffeine and allyl alcohol. Nitrosoguanidine has also been used to induce early maturity mutants (30) and temperature-sensitive mutants (Phillips, personal communication).

B. Mendelian Inheritance

How does one demonstrate that Mendelian inheritance occurs for a particular trait? Since exchange of genetic material is accomplished by conjugation, the way in which

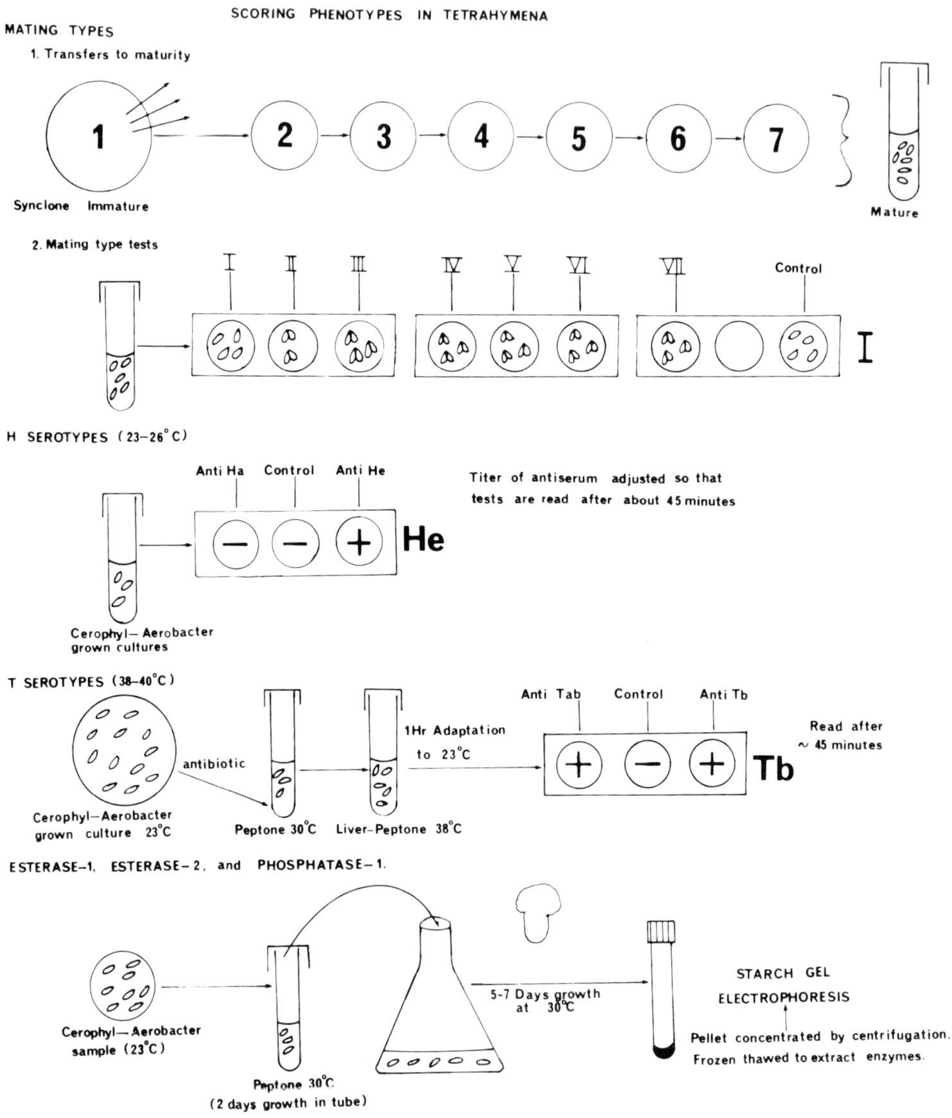

Fig. 4.

Scoring phenotypes in *T. pyriformis* syngen 1.

the nuclei behave during conjugation leads to certain genetic expectations. In *normal* conjugation duplicate copies of the gametic nuclei are made in each conjugant, and one copy is exchanged between them. After fusion the diploid syncarya are therefore identical in both exconjugants. Thus in a mating of *AA* × *aa* both exconjugants are *Aa* (Fig. 5, line 1). In such a cross different mating *pairs* would also be genotypically identical for their *A* genes.

1. AA x aa

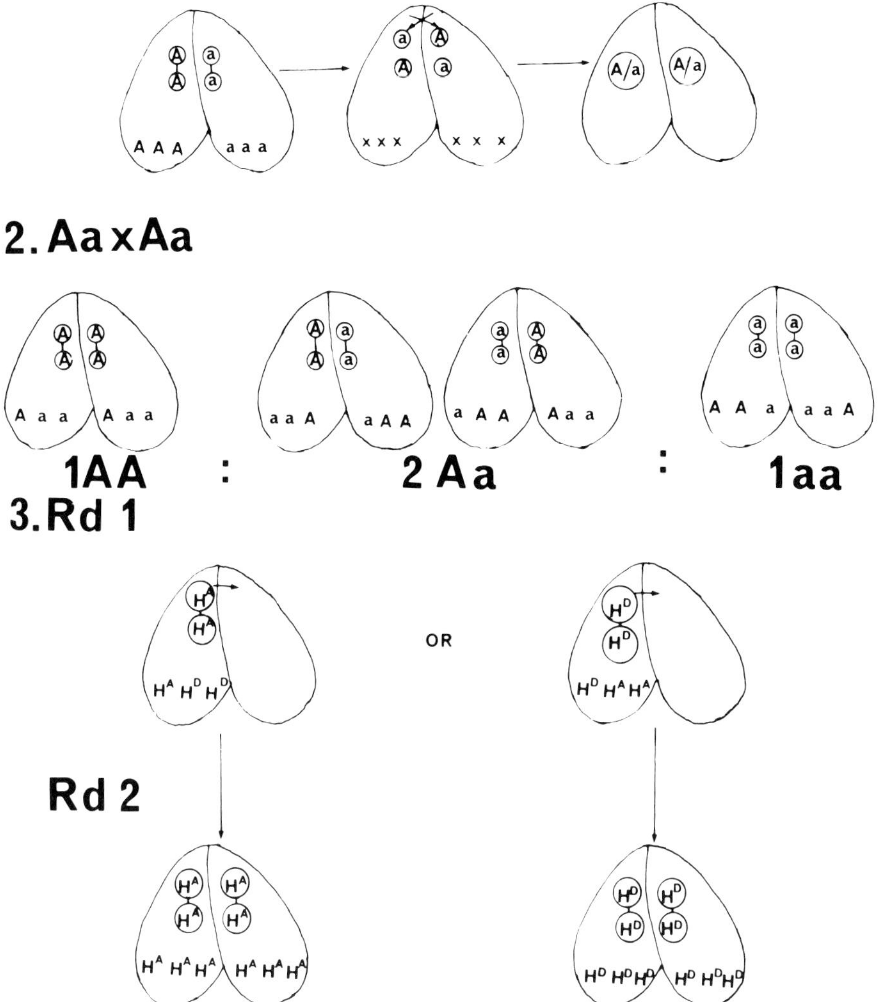

Fig. 5.

Nuclear behavior and transmission of genes during conjugation.

In a cross of *Aa* × *Aa* gametic nuclei are either *A* or *a* in genotype. Only one of the four meiotic products is retained, however. Whether an *A* or *a* nucleus is retained is independently determined in the two conjugants. As a result, three classes of pairs are formed (Fig. 5, line 2), and in the 1:2:1 ratio expected for a cross of two heterozygotes.

As an example, Mendelian inheritance can be demonstrated for the H immobilization antigens in syngen 1. Some inbred strains have one type of H antigen; other strains have a different type (123). These are distinguished by specific antisera. If a cross is made between a strain with the Ha antigen and another strain with the Hd antigen, the progeny respond to *both* antisera and are therefore Had. The progeny from all pairs are Had and both exconjugants from the same pair are similar in phenotype. If the F_1 progeny are intercrossed, three classes of *pairs* appear in the F_2 generation: Ha, Had, and Hd, in a 1:2:1 ratio. Those that are Ha, when intercrossed, give rise to only Ha progeny; those that are Hd, when intercrossed, give rise to only Hd progeny; while three classes of pairs appear in intercrosses of Had cell lines. Backcrosses of the F_1 progeny to the inbred parents give rise to two classes of pairs, half of which are Had and the other half Ha or Hd, depending upon the type of cross. These observations are typical for a trait controlled by an allelic pair of genes.

Genomic exclusion was first detected as a deviation from the above pattern of transmission expected for simple Mendelian inheritance. When it was realized that two consecutive matings take place, a cross was set up using a normal parent which was $H^A/H^D \, P\text{-}1^A/P\text{-}1^B$ mt^A/mt^B, and round-1 exconjugants from the same pair were remated (9). Half the crosses of remated exconjugants gave rise to only Ha progeny; the other half to Hd progeny. The progeny bred true when intercrossed or tested against inbred strains of the same presumed genotype. They were therefore homozygous for *H*. Similar observations were made for *P-1* and for *mt;* only homozygous progeny appeared, and in the different crosses in 1:1 ratios. Therefore round-2 progeny were also homozygous for *P-1* and *mt*. These genetic results can be accounted for on the basis of the abnormal nuclear behavior that occurs during genomic exclusion. Key stages are illustrated diagrammatically for *H* (Fig. 5, line 3). Here the genotypes of the gametic nuclei in the exconjugants of the two matings (rounds 1 and 2) are given.

C. Genotypes and Phenotypes

What kinds of traits have been subjected to genetic analysis in *T. pyriformis?* Some are cell surface traits, such as mating type, serotype (immobilization antigens), or corticotype (kinetal pattern). Some are structural variations in enzymes observed by differences in their electrophoretic mobility in starch gels. Others are nutritional variations which result in differences in their ability to grow in synthetic medium of varying composition. Still others are lethals, some of which also have effects in the heterozygote on the fission following conjugation at which sexual reactivity first appears. Genes have been found for all traits mentioned except for corticotypes (Table 3).

For the above traits certain extrinsic and other intrinsic factors affect the expression of phenotype. Extrinsic factors determine whether or not a given phenotype is expressed at all. For example, temperature is important for the expression of a particular serotype, or for determining whether certain lethals are expressed (temperature-sensitive mutants).

Nutritional state is important for mating and therefore for the expression of mating types. It is also important for observing nutritional mutants and for the expression of certain enzymes.

Intrinsic factors also play a role. Thus cells must be mature before mating can occur. Genes are involved in an unknown way with maturation (30,31). They may also play a role at other stages of the life cycle—with the maintenance of the micronucleus (23,96), or perhaps effect changes within the macronucleus.

Table 3.
Genes of *T. pyriformis*

			Syngen 1[a]		
Gene	Alleles	Phenotype	Number of stable types in a single homozygote	Number of stable types in a single heterozygote	Ref.
mt	A(D,E,F) C B	Mating types	I,II,III,V,VI I,II,III,(V),VI II,III,IV,V,VI,VII	5 in A/C, C/D, and so on 7 in A/B, B/C, and so on	(98,122)
H	A,C,D,E	High-temperature serotypes (expressed at 20–30°C in Cerophyl–*Aerobacter* or in peptone)	1	2	(123)
St	A,C	Salt serotypes (expressed in 100 m*M* NaCl in peptone	1	2	(71)
T	A,B,C	Torrid temperature serotypes (expressed at 38–40°C in liver–peptone)	1	2	(136)
E-1	B,C	Variants of a *group* of five to six propionyl esterase isozymes	1	2	(2)
E-2	B,C	Variants of a butyryl esterase	1	2	(2)
P-1	AB	Variants of an acid phosphatase (interaction in heterozygote, with a total of five isozymes, but variation in different subclones)	1	3	(12,24)
Em-1	*Em-1,em-1*	Early maturer: *Em-1/em-1*; *Em-1/Em-1* is lethal	—	Diversity among heterozygous clones in time of maturation	(31)

Table 3 *(cont.)*

		Syngen 1[a]			
Gene	Alleles	Phenotype	Number of stable types in a single homozygote	Number of stable types in a single heterozygote	Ref.
Em-2	*Em-2,em-2*	Early maturer; $\overline{Em\text{-}2}/\text{—}$; homozygous dominant not lethal	—	—	(30)
Em-3	*Em-3,em-3*	Early maturer: *Em-3/em-3*; homozygous dominant is lethal	—	—	(30)
F	*F,f*	"Fat"; *f/f* lethal; heterozygote normal	—	None reported	(130)
T	*T,t*	"Tiny"; *t/t* lethal; heterozygote normal	—	None reported	(130)
caf₁	*cat₁⁺,caf₁ʳ*	*caf₁ʳ* grows in 10 m*M* caffeine	—	Potentially 2	(39)
caf₂	*caf₂⁺,caf₂ʳ*	*caf₂ʳ* grows in 10 m*M* caffeine	—	Potentially 2	(39)
aa	*aa⁺,aaʳ*	*aaʳ* survives 250 m*M* allyl alcohol for 12 h	—	Potentially 2	(39)

		Syngen 2		
Gene	Alleles	Phenotype	Remarks	Ref.
P	P,p	Nutritional	P/— dominant requirement for pyridoxine; *p/p* grows without pyridoxine	(48)

			Syngen 7		
Gene	Alleles	Phenotype	Number of stable types in a single homozygote	Number of stable types in a single heterozygote	Ref.
mt	A B	Mating type	II,IV III,V	4	(140)
P-1	A,B,C	Variants of an acid phosphatase	1	2	(139,143)
R	A,B	Room-temperature serotypes, R^A dominant	1	1?	(142)

			Syngen 8		
Gene	Alleles	Phenotype	Genotype	Mating type	Ref.
mt	A,B,C	Mating types	A/— B/B, B/C C/C	I III II	(128,131)

			Syngen 9	
Gene	Alleles	Phenotype	Remarks	Ref.
S	S,s	Nutritional	S/— dominant requirement for serine; *s/s* grows without serine	(49)

[a]Additional genes have been identified in syngen 1 (e.g., ref. 33) but have not been published in detail.

In syngens 1 and 7 heterozygotes may express only one of the two alleles in some lineages and therefore have the same phenotype as one of the homozygous genotypes. This is true for the genes identified in syngen 1 and for *P* in syngen 7. This means that if we were presented with a clone from syngen 1 of unknown ancestry, we could not assign its genotype without testcrosses. For example, all the wild strains except for LW-B have phenotypes expected for homozygous genotypes. Yet it can be shown from either breeding analysis or examination of the genotypes of the derived inbred lines that several of these wild strains are in fact heterozygotes.

D. Genes and Specific Phenotypes

1. Mating Types

The collections of *T. pyriformis* are classified into syngens on the basis of their mating behavior. Each syngen has two or more mating types which mate with each other but not with mating types from other syngens. An exception to this rule are the cross-reactions observed between the mating types of syngens 6 and 8 (129). The mating type character is stable and heritable in asexual reproduction, since with few exceptions the progeny of a cell express the same mating type. Different types may, however, arise as a result of conjugation. How then is this character transmitted during sexual reproduction? And how do multiple mating types arise?

Many syngens (1, 2, 3, 6, 7, and 8) have known immaturity periods of different lengths. For example, syngen 1 has an immature period lasting 50 to 80 fissions, syngen 7 has one that lasts 70 to 100 fissions (140), and syngen 8 has one that lasts 150 fissions (128,131). Mating types can be examined only in mature cells; therefore it is necessary to bring the progeny of crosses to maturity by serial transfers of cells in order to test for mating types. The techniques are illustrated in Fig. 4 for syngen 1.

Two major types of mating type inheritance have been described in the ciliated protozoa (75). In some species mating types are inherited as if controlled by simple Mendelian differences; in other words, there is a direct correlation with differences in mating type and differences in genotype. In other species this correspondence between genotype and phenotype does not hold. In this case genes may limit the expression to one type, or restrict the array of types that appears. The actual type expressed by a cell, however, is correlated with macronuclear differences which arise with the separation of the two new macronuclei at the first cell division. Each caryonide *may* express a different mating type; therefore this type of mating type inheritance is referred to as caryonidal inheritance.

In *Tetrahymena* both types of inheritance have been reported in different syngens. Direct genetic control occurs in syngen 8 (128,131) and perhaps in syngen 2 (78,79), whereas caryonidal inheritance has been established for syngens 1 and 7 (95,140) and is suspected for syngen 3 (37,38).

There are three mating types in syngen 8. When crossed in various combinations, all four caryonides manifest the same mating type; however, the types that appear in different synclones and their ratios depend upon the genotype of the parents. The results of several crosses suggested that there was a mating type locus with three alleles (128,131). If a cell has the mt^A allele, it is mating type I (mt^A/mt^A, mt^A/mt^B, mt^A/mt^C). A cell

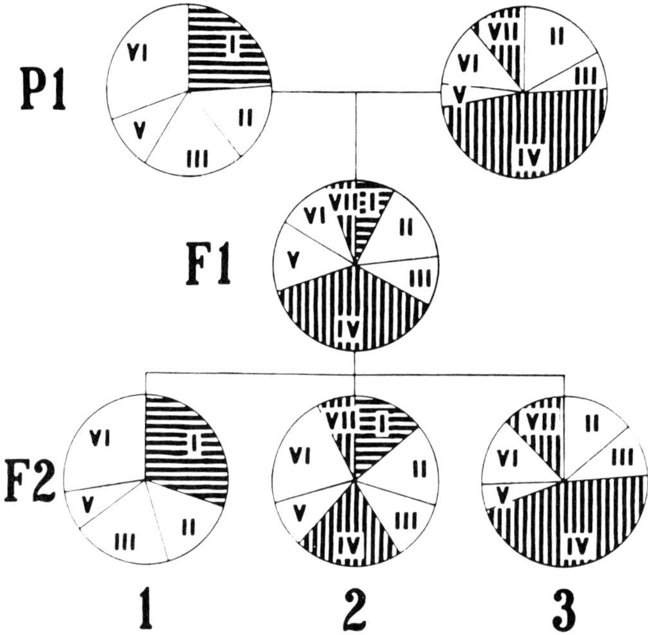

Fig. 6.

Diagrammatic representation of crosses; mating type frequencies are indicated as sectors in a circle. The P_1 frequencies are derived from the accumulated data for families A and B; the F_1 frequencies are derived directly from the F_1; the F_2 frequencies for the three classes are based on progeny tests in which sister caryonides from single F_2 pairs were crossed. The horizontal hatching designates the type found uniquely in family A; the vertical hatching designates the types found uniquely in family B. [From Nanney et al. (122).]

is III, if it is mt^B/mt^B or mt^B/mt^C; and it is II, if it is homozygous for mt^C. Thus there are three alleles that can be placed in a linear series of increasing dominance: $mt^C < mt^B < mt^A$. This fact, coupled with the rare appearance of selfers (see below) and unexpected phenotypes among the progeny, was interpreted by Orias as evidence for the regulatory nature of the mt locus. That is, it is concerned with the selection of which mating type specificity is expressed. Whether it also has information for the specificities themselves is not known. Orias (128) notes that by shifting the time of the action of a regulatory gene from post- to mid-, or even preconjugation that the pattern of mating type inheritance shifts from caryonidal inheritance to one exhibiting direct genetic control.

There are seven mating types in syngen 1 (120). However, not all genotypes are capable of producing all seven types. Inbred strain A cannot produce types IV and VII, and inbred strain B is unable to produce I. Moreover, the types produced appear in certain characteristic frequencies. Genetic analysis was carried out by crossing the two strains (122). All seven types appeared in the F_1 in certain frequencies. When the F_1 were intercrossed, three classes of progeny appeared in the F_2: (1) those that produced all seven mating types (2) those lacking the ability to produce IV and VII;

Table 4.
Frequency of Mating Types in Homozygous Genotypes[a]

Inbred strain	Allele	Frequency of mating types							Total number of progeny
		I	II	III	IV	V	VI	VII	
A	mt^A	0.23	0.17	0.17	0	0.08	0.35	0	1147
C, C1	mt^C	0.47	0.14	0.07	0	0.01	0.31	0	1575
B	mt^B	0	0.15	0.09	0.51	0.04	0.11	0.09	1711

[a]Combined data Nanney, Allen (published and unpublished).

and (3) those lacking the ability to produce I. In addition, the frequencies with which the mating types appeared in class 1 were similar to the frequencies observed in the F_1 generation; in class 2 to the frequencies observed in inbred strain A; and in class 3 to the frequencies observed in strain B. The results are illustrated in Fig. 6. They were interpreted as evidence for a *mt* locus with two alleles. One allele (mt^A) found in the A strain restricts the development of mating types to I, II, III, V, and VI. The other allele (mt^B) restricts them to II, III, IV, VI, VI, and VII. The heterozygote (mt^A/mt^B) can express any one of the seven types. Only a fraction of the possible mating types actually appears in the progeny of a single mating pair, therefore it is necessary to examine many pairs and to testcross each pair in order to assign its genotype. The *mt* gene thus determines which of the array of seven mating types a cell can produce. It may also determine the probability, or frequency, with which different types appear, although there appear to be other genetic factors that alter the frequencies (98). Therefore outcrosses often result in shifts in the frequencies of the array of types. Different inbred strains manifesting I, II, III, V, and VI produce them in slightly different frequencies. Hence there is some question as to whether mt^D, mt^E, and mt^F are really different from mt^A. Differences in the frequencies of types II through VII also occur in strains carrying the mt^B allele. For example, marked differences in the frequency of IV are observed in inbred strains B and B2. Table 4 summarizes data on the frequencies of mating types observed in the A strain (mt^A/mt^A) and in the B strain (mt^B/mt^B). It also gives the frequencies of I, II, III, V, and VI in strains C and C1 (pooled). These frequencies depart from the "other" mt^A alleles in that a very high frequency of mating type I and a very low frequency of V are observed. In fact, no V appeared in over 600 clones from the C1 strain. This allele may be distinct enough from mt^A to warrant calling it mt^C.

Caryonidal inheritance has been shown to occur for all these genotypes. Thus (1) there is no correlation between the mating types of the parents and progeny (95); (2) there is a lack of correlation between the mating types of different caryonides (95); (3) there is a marked effect of temperature during the period of macronuclear development on mating type frequencies (99); (4) genetic control is exerted on the array of types that can be produced (122).

Five mating types have been established in syngen 7, but mating type I occurred only in a defective micronucleate line. Crosses with recent collections yielded mating types II through V. The system is somewhat similar to syngen 1. Mating type inheritance is caryonidal, and there is a mating type locus with two alleles that restrict the array of types that appear (140). One allele permits expression of mating type II or IV, the other allele allows expression of III or V, and the heterozygote can express any one of the four types. Temperature effects were again shown to influence the determination of mating types.

Selfing clones are an exception to the rule that a clone is "pure" for a single mating type. Intraclonal mating is observed in selfers. Selfers have been reported to occur in syngens 1, 2, 3, 6, 7, and 8 (37,38,78,121,128,134,165), as well as in collections not classified into syngens (73). None were observed in the collections of syngens 10, 11, and 12 (55,57). Selfers appear to be a heterogenous group. Some, such as the Ann Arbor strain and an exceptional selfing clone in syngen 1, persisted as selfers (54,119). Others found in syngen 1 and also in syngen 8 give rise to subclones which are "pure" for mating type (18,128). Finally, one case was reported in syngen 6 of clones, "pure" for a mating type, that selfed after many asexual generations (165). The appearance of selfers must be accommodated in any scheme of mating type inheritance, and often this is not easy to do. For example, the selfers of syngen 8 are not at all easily explained on the hypothesis of direct genetic control of mating types. This led Orias (128) to propose a modified form of caryonidal inheritance to account for them. The selfers of syngen 1 are further discussed in Section IV.

2. Serotypes

A soluble protein associated with the surface of *Tetrahymena* is antigenic in rabbits. The antigen can be detected in two ways. Living cells that contain it are immobilized by homologous antibodies in an immobilization test, or a precipitate is formed by the interaction of the antigen and antibody in double-diffusion tests. Cells that can be shown by these tests to contain the antigen are said to belong to a particular serotype. Different antigens can, however, be detected in the same cells if they are grown under different conditions. Thus a cell exhibits different serotypes under different conditions but only *one* serotype at any one time. Similar observations have been made in other ciliated protozoa, such as *P. aurelia*. Here it is known that the immobilization antigen is a large fibrous protein with a molecular weight of 310,000 or more (144,160). The H and T antigens are much smaller proteins with molecular weights of 29,000 and 23,000, respectively (36).

Antigens have been observed in different syngens by the immobilization test (47), but in order to make comparisons and to determine relationships, more information about conditions for their expression is essential. Syngen 1 forms several antigens under known conditions, as well as some that are less well characterized. The serotypes of syngen 1 have been detected by preparing specific antisera to cells grown under different conditions (86,87,91). At 15–20°C the L antigen is formed. This serotype varies, however, depending on the temperature and whether the cells are grown in axenic or bacterized medium. At 20–30°C several antigens may be formed which vary with the medium, which are induced by antiserum, or which appear sporadically. The principal antigen formed is called H, and it appears in both axenic and bacterized media. There are

Table 5.

Inheritance of H Serotypes[a]

	Serotypes of progeny			
	Hc	Hcd	Hd	Total
1. $H^C/H^C \times H^C/H^C$	72	0	0	72
2. $H^D/H^D \times H^D/H^D$	0	0	98	98
3. $H^C/H^C \times H^D/H^D$	0	76	0	76
4. $H^C/H^D \times H^C/H^D$	95	215	125	435
5. $H^C/H^D \times H^C/H^C$	44	45	0	89
6. $H^C/H^D \times H^D/H^D$	0	49	47	96

[a]Data from Nanney and Dubert (123).

several allelic forms of H, as well as minor forms (86,101,123) or other antigens such as S (3), which may be related in some way to H. The I serotype is induced by growing H cells in the presence of anti-H sera. There is probably an array of different I types, which appear to be unstable, since cells of one I type quickly diversify into others (81,82). When adapted to 100 mM sodium chloride in peptone medium, the St serotype appears (70). At 38–40°C the T antigen is produced but only in axenic medium. Transformation occurs from one serotype to another when the conditions of growth are changed. Phillips (136) followed the course of transformation from T to H and from H to T and observed that it was completed within two to three fissions for both. Transformation from H to St takes 3–4 days under certain salt conditions, while the reverse transformation, St to H, passes through a stage in which the cells are refractory to both anti-St and anti-H (70).

Five distinct H serotypes were identified among inbred and wild strains of syngen 1: Ha, Hb, Hc, Hd, and He, and other forms may also occur (86). Hb is found only in WH-52 which is infertile. The other four types were shown to be specified by alleles at a single locus designated the *H* locus (123). Three distinct T serotypes have been detected: Ta, Tb, and Tc, and these are specified by alleles at the *T* locus (136). Two distinct St serotypes have also been detected, Sta and Stc, and these are specified by alleles at the *St* locus (71). There are strain differences in the I types, and there may be minor strain differences in the response to temperature of the L types; however, genetic studies have been complicated by the instability of these types and by temperature effects and have not been definitive (81,82). In syngen 7 strain differences in serotypes are determined by a single locus, *R*, which unlike syngen 1 shows dominance. There is no cross-reaction between the antigens of syngens 1 and 7 (142).

As an illustration of the pattern of genetic transmission for the H or T serotypes, a cross of two strains having different *H* alleles is discussed. When grown at 20–30°C cells of inbred strain D1 are immobilized by a specific antiserum (anti-Hc) and unaffected by an antiserum prepared against cells of strain B (anti-Hd). Strain B, however, is immobilized by anti-Hd and not by anti-Hc (Table 5, lines 1 and 2). If crosses are made between strains B and D1, the F$_1$ are immobilized by both anti-Hc and anti-Hd

Table 6.
Crosses of H^C/H^B Heterozygotes Differing in Phenotype[a]

Parental serotypes	Serotypes of progeny			Total
	Hc	Hcd	Hd	
Hc × Hc	35	70	51	156
Hc × Hd	29	65	34	128
Hcd × Hd	9	27	17	53
Hd × Hd	24	53	24	101

[a]From Nanney and Dubert (123).

and are therefore classed as Hcd (Table 5, line 3). Intercrosses between F_1 hybrids give rise to Hc, Hcd, and Hd progeny in a 1:2:1 ratio (Table 5, line 4). Backcrosses of the F_1 to strain D1 give rise to Hc and Hcd in a 1:1 ratio, and backcrosses to strain B yield Hcd and Hd in a 1:1 ratio (Table 5, lines 5 and 6). These results are typical for single-gene control with codominant alleles. Similar inheritance was found for Ha and He in crosses between inbred strains with these phenotypes, and in crosses to strains that were Hc or Hd. Thus four alleles were identified at the *H* locus in these crosses. In the same way—by crosses between inbred strains—the three alleles at the *T* locus and two alleles at the *St* locus were discovered.

Heterozygotes at these loci are not always immobilized by the antisera to both parental specificities. This difference may be quantitative or it may even be qualitative, if heterozygous subclones are tested. For example, in the case of H^C/H^D heterozygotes, some subclones may be Hc, some Hd, and others Hcd. What effect does this have in crosses? The results of crosses made of heterozygotes having different phenotypes are shown in Table 6. For all combinations of phenotypes shown, including crosses of Hc × Hc and Hd × Hd, 1:2:1 ratios were observed in the progeny. Thus no effect of parental phenotype was observed, and the parents *bred* as if they were heterozygotes. Similar observations have been made on other allelic combinations in *H* heterozygotes and in *St* and *T* heterozygotes, and in all cases no carryover of the phenotypic differentiation to subsequent sexual generations has been observed in crosses.

Phenotypic lag is commonly observed in crosses between cells of different serotypes in *P. aurelia,* in which each exconjugant maintains the serotype of its cytoplasmic parent (29). This cytoplasmic effect presumably operates through a feedback system upon the genes. It can be disrupted if cytoplasmic exchange is induced. Then both exconjugants have the same serotype. Is there a cytoplasmic effect on the serotype of *T. pyriformis?* One report suggested that there was (80). There is some doubt, however, that the pairs they studied were true conjugants. Another report came to the opposite conclusion, that is, there was no effect of cytoplasm upon the serotype (102). Here crosses were carried out between cells grown at different temperatures and expressing the H or L serotypes. The H × L cross was carried out at an intermediate temperature, pairs were isolated and, after the exconjugants had come apart, each was isolated in a different container and the resulting clone tested for serotype. In the cases in which successful conjugation occurred, both exconjugants reacted to H and L antiserum, and no effect of cytoplasm could be demonstrated upon the exconjugant clones. This lack of cytoplasmic effect is not surprising since McDonald (89) has since shown that massive cytoplasmic exchange normally occurs during conjugation in syngen 1. Thus both exconjugants would be expected to show the same serotype, which is what is observed. The results on H × L crosses have been confirmed and extended by Juergensmeyer (81,82).

3. Enzymes

Multiple forms of several enzymes have been observed in extracts of *T. pyriformis* syngen 1. These include: esterase (eight substrates); acid phosphatase (six substrates); dihydronicotinamide adenine dinucleotide phosphate ($NADPH_2$) oxidase; dihydronicotinamide adenine dinucleotide ($NADH_2$) oxidase; nicotinamide adenine dinucleotide phosphate (NADP)-linked isocitrate dehydrogenase (IDH); nicotinamide adenine dinucleotide (NAD)-linked IDH; NADP-linked malate dehydrogenase (MDH); NAD-linked MDH; glucose-6-phosphate dehydrogenase; α-glycerophosphate dehydrogenase; NADP-linked glutamate dehydrogenase; choline oxidase; L-leucyl-β-naphthylamidase; α-hydroxy acid oxidase; NADP-MDH (decarboxylating); tyrosine aminotransaminase; tetrazolium oxidase; aldolase; hydroxybutyrate dehydrogenase; fumarase; lactate dehydrogenase; fructose-1,6-diphosphate; and glyceraldehyde-3-phosphate dehydrogenase (11,33). Multiple forms of esterases, phosphatases, and other enzymes have also been observed in the other syngens (19,138; D. Borden, personal communication).

The electrophoretic forms of these enzymes differ in their molecular relationship. Some molecules have overlapping specificities, as observed for the esterases. These differ from each other in their genetic control, biochemical properties, response to growth conditions, and intracellular distribution. Others are more closely related and have similar biochemical properties. Some are multiple products of a single gene, occur in homozygotes, and may arise by configurational isomerism or by epigenetic change. Some are the products of different alleles and some are observed only in heterozygotes.

Structural genes have been identified for seven enzymes in syngen 1: esterase-1, esterase-2, phosphatase-1, tyrosine aminotransaminase, tetrazolium oxidase, NADP-IDH, and NADP-MDH (decarboxylating) (2,24,33). A structural gene has also been identified for phosphatase-1 in syngen 7 (138,143). For each gene there are two codominant alleles which produce forms of an enzyme with different electrophoretic mobility. As a demonstration of their pattern of inheritance, the observations on phosphatase-1 in syngen 1 are discussed.

There are about 20 resolved electrophoretic forms of acid phosphatase in crude extracts of *T. pyriformis* syngen 1 (25); yet it was early observed that variations occurred in a series of phosphatases which migrated toward the anode—all within a region located about 2.5–4.5 cm from the origin under the electrophoretic conditions used. A slow form (isozyme 1) was present in some inbred strains, and in other strains this was absent and a faster form (isozyme 5) was present. Each of these bred true in crosses within a strain. If crosses were made between strains with isozymes 1 and 5, however, the F_1 had isozymes 1 and 5 as well as a new form, isozyme 3, and trace amounts of two others, isozymes 2 and 4. When the F_1 were intercrossed, three classes appeared

Table 7.
Inheritance of Phosphatase-1[a]

	Phosphatases of progeny			Total
	P-1a	P-1ab	P-1b	
$P\text{-}1^A/P\text{-}1^A \times P\text{-}1^A/P\text{-}1^A$	20	0	0	20
$P\text{-}1^B/P\text{-}1^B \times P\text{-}1^B/P\text{-}1^B$	0	0	38	38
$P\text{-}1^A/P\text{-}1^A \times P\text{-}1^B/P\text{-}1^B$	0	28	0	28
$P\text{-}1^A/P\text{-}1^B \times P\text{-}1^A/P\text{-}1^B$	41	74	39	154
$P\text{-}1^A/P\text{-}1^B \times P\text{-}1^A/P\text{-}1^A$	10	11	0	21
$P\text{-}1^A/P\text{-}1^B \times P\text{-}1^B/P\text{-}1^B$	0	41	52	93

[a]Data from Allen, et al. (24).

in the F_2; one with only isozyme 1, one with all five isozymes, and one with only isozyme 5, and these classes appeared in a 1:2:1 ratio. The recovered types with only isozyme 1 or only isozyme 5 bred true; and the class with all five isozymes gave rise to three classes of progeny. Backcrosses of the F_1 hybrid gave two of the three classes. This pattern of inheritance is typical for single-gene control with codominance of alleles. The gene was designated *P-1*. Strains with isozyme 5 were assigned the genotype $P\text{-}1^A/P\text{-}1^A$, and those with isozyme 1 the genotype $P\text{-}1^B/P\text{-}1^B$. The original data upon which this conclusion was reached are reproduced in Table 7. The data shown (and its breakdown; see ref. 24, Table 3) exclude two unlinked loci (*AA bb* and *aa BB*) but do not exclude the possibility that closely linked loci may be involved. Phosphatase-1 in syngen 7 also appears to be specified by a *P* gene (138,143).

Further comment should be made about heterozygotes. First, the electrophoretic pattern observed in heterozygotes may simply be the sum of the two parental phenotypes—as in the case of esterase-1. Or, it may have a unique pattern, as observed for phosphatase-1 in syngen 1. Second, heterozygous subclones varied in their enzyme phenotypes in the same way as recorded for the H and T serotypes; that is, some lines appeared having the same phenotypes as homozygotes. Thus for phosphatase-1 some lines were

Table 8.
Crosses of Heterozygotes Differing in Phenotype[a]

Type of cross[b]	Phenotypes of parents	Phosphatases of progeny			Total
		P-1a	P-1ab	P-1b	
$F_1 \times F_1$	P-1a × P-1a	8	14	10	32
	P-1b × P-1b	9	16	7	32
$P_1 \times F_1$	P-1b × P-1a	0	5	9	14
	P-1b × P-1b	0	12	18	30
	P-1b × P-1ab	0	12	17	29

[a]Data from Allen, et al. (24).
[b]$P_1 = P\text{-}1^B/P\text{-}1^B$

Table 9.

Chemical Characterization of Enzymes from Syngen 1[a]

Treatment	Esterase-1	Esterase-2	Phosphatase-1
Stability[b]			
Heat	63–70°C	47°C	42–48°C
pH	3	4.5	3, 7–8
Urea	–	4 *M*	2–4 *M*
Thiols	–	–	?
PCMB	–	+	–
IANH$_2$	–[c]	+	?
Activity[d]			
pH optimum	7.4	6.5	5.0
Taurocholate	A	I	–
Triton X-100	A	I	–
Eserine	I	–	–
d-Tartaric acid	?	?	I
Sodium fluoride	?	?	I
Thiols	I	I	?
PCMB	–	I	–

[a]Based on data presented in refs. 7,21,24,25.

[b]Crude extracts treated before electrophoresis. +, Effect; –, no effect, unless specified as to condition under which stability is affected.

[c]Additional isozymes form which migrate further toward cathode; native isozymes converted into new forms with increasing time or concentration.

[d]Assayed during incubation of starch gels. A, Activation; I, inhibition; –, no effect.

P-la in phenotype and others P-lb. Such lines have been crossed in all possible phenotypic combinations, some of which are shown in Table 8. The classes of progeny that appear are those expected of crosses involving a heterozygote. Intercrosses of F_1 gave rise to three classes, and backcrosses to two classes—regardless of the phenotypes of the F_1 clone used in the crosses. Similar observations have been made for phenotypic variants of *E-1* or *E-2* heterozygotes. These lines are therefore heterozygous as far as their micronuclear genes are concerned. Phillips (143) also reports phenotypic variation in the *P* heterozygotes of syngen 7, which she interprets as similar to the phenomenon observed in syngen 1. The significance of this phenomenon is detailed in Section IV.

Esterase-1, esterase-2, and phosphatase-1 of syngen 1 differ not only genetically but in several other properties. The effects of certain chemical treatments are compared in Table 9. Esterase-1 is the most stable of three enzymes. It is a propionyl esterase, although it also splits α-naphthyl acetate and β-naphthyl acetate. It is inhibited by eserine sulfate and certain thiols and activated by sodium taurocholate and Triton X-100. Esterase-2 is less stable than esterase-1 and is inactivated by urea, *p*-chloromercuribenzoic acid (PCMB), and iodoacetamide (IANH$_2$). It is a butyryl esterase, although it also splits α-naphthyl valerate. It is inhibited by certain concentrations of sodium taurocholate,

PCMB, and certain thiols. Phosphatase-1 is the least stable of the three enzymes, and isozyme 1 is less stable than isozyme 5, the intermediate isozymes having intermediate stabilities. It is inactivated by urea and pH values lower than 3 and higher than 7–8. It splits several synthetic phosphate compounds including α-naphthyl acid phosphate. It is inhibited by *d*-tartaric acid and sodium fluoride.

Esterase-2 appears to be unaffected by the growth conditions used so far and is localized to the microsomal fraction in the cell. Esterase-1 forms isozymes in *homozygotes*. The number resolved varies with the pH of the buffer used during electrophoresis. These exhibit complex changes as a function of the growth cycle and in different media, are distributed to different cell fractions (5), and are converted by $IANH_2$ into new forms. These then appear to be "conformers"—or may arise by secondary changes in structure. Phosphatase-1 forms isozymes in *heterozygotes* (and also in one of the homozygotes). These migrate in parallel as the pH of the buffer used during electrophoresis is changed. This enzyme is lysosomal in location; its presence is facilitated by media that provoke vacuole formation; and it appears to be a tetramer in structure. The intermediate isozymes appear to arise by subunit interaction, although these interactions do not appear to be random, but instead involve three types of dimers. Not all dimers seem to be synthesized by all cells as indicated by the different cell types observed in heterozygotes. These types give rise to qualitatively different kinds of subclones, three of which are stable types, including a stable type (P-lab) which has isozyme 3 exclusively. Further discussion and interpretation of these data in terms of the tetramer structure can be found elsewhere (7,10–12).

4. *Corticotypes*

The surface structures of the ciliated protozoa are revealed by the silver staining method (42). The most prominent organelles are the cilia. Each cilium is a complex of many different molecules, and it is organized into a cortical unit which includes the cilium, a cylinder of fibrous elements called a kinetosome at the base of the cilium, a vesicle which surrounds the base of the cilium, and a kinetodesmal fiber which joins the units (see Chapter 2). The unit has an anterior-posterior axis and is organized from the left side of the organism to the right. Rows of these units are organized in an anterior-posterior direction into kineties. The kineties, with other surface structures are arranged in various patterns on the surface of the cell, depending on the species of ciliated protozoa. *Tetrahymena* has a general surface pattern which occurs in all cells within a culture, in all strains, and in different species (Fig. 7). It has kineties running from the anterior to the posterior of the cell, which are interrupted at the anterior end by the mouth. The kineties are numbered starting with the ciliary row terminating on the right side of the cell's mouth. This is row 1. It is also designated the stomatogenic kinety, since on it is the site of formation of the new oral primordium. The other ciliary rows are numbered consecutively to the cell's right. The structures around the mouth consist of four oral membranes (which gives *Tetrahymena* its name). At the posterior end of kinety 1 is the cell anus or cytoproct, and halfway around the side of the cell are the openings to the two contractile vacuoles.

Although the total number of cortical units approaches a constant (116,117), other

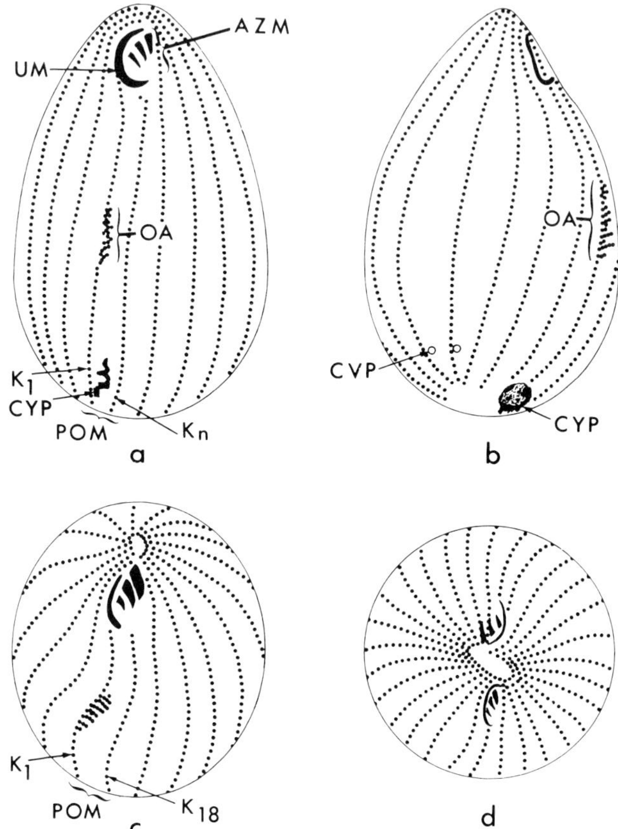

Fig. 7.

Cortical patterns in *T. pyriformis*. (a) Ventral view (b) Right lateral view. (c) apical view of 18 rowed singlets. (d) apical view of a homopolar doublet. UM, Undulating membrane; AZM, adoral zone of membranelles; OA, oral anlage; CYP, cytoproct; CVP, contractile vacuole pore; K_1, the first stomatogenic kinety (meridian or ciliary row); K_n, the last kinety; POM, postoral meridians. [From Nanney (106).]

features of the surface pattern vary, such as the number of ciliary rows (meridians, or kineties), the number and position of contractile vacuole pores, the number of postoral meridians, the number of stomatogenic meridians, and the number of mouths (106,107). However, the variations are correlated since these features are different manifestations of the geometry of the cell surface.

Nanney introduced the concept of corticotype, which defines a cell by the number of ciliary rows. Each strain can produce several corticotypes, for example, 17 or 18, but there is a limit to the observed variations. The rates of change are low especially when close to the favored corticotype, or the "stability center." Different strains can be defined by their range of corticotypes and their varying rates of change (108–110,112,113,115).

An analysis of the differences between strains of cells with different corticotypes was carried out (109). Meridian number was studied in three types of crosses. The

Table 10.

Corticotype Distributions in Exconjugant Clones in Cross 1 A × B[a]

Clones	Corticotype								
	19	20	21	22	23	24	25	26	27–32
A parent	—	1	49	—	—	—	—	—	—
B parent	—	—	—	—	—	2	6	9	34
Progeny									
1a	—	—	50	—	—	—	—	—	—
1b	—	—	—	3	19	25	3	—	—
2a	—	1	49	—	—	—	—	—	—
2b	—	—	—	3	19	26	2	—	—
3a	—	2	48	—	—	—	—	—	—
3b	—	—	—	—	5	17	22	6	—
4a	1	3	46	—	—	—	—	—	—
4b	—	—	—	1	8	30	11	—	—
5a	—	2	48	—	—	—	—	—	—
5b	—	—	—	2	23	25	—	—	—

[a]From Nanney (109).

first was between cells, one of which was from a population of corticotype 21, and the other from a population which ranged from 24 to 32. The second involved cells of types 18 and 19, and the third cells of types 16 and 22. The results of the first cross are shown in Table 10. One exconjugant clone had 20 to 21 meridians, while the other ranged from 22 to 26. There was no overlap between the two exconjugants in any of the pairs. The other crosses again showed differences in the phenotypes of the exconjugants in terms of their corticotypes.

Since differences between exconjugants persist through sexual reproduction, corticotype differences do not appear to be determined by nuclear gene differences between the two strains but by cytoplasmic differences. The cytoplasmic difference, however, must not be capable of being exchanged, since cytoplasmic exchange normally occurs during conjugation in *Tetrahymena* (89). There may be cortical genes. Or, the cortical units may be qualitatively the same but differ in their distribution in the two different corticotypes. The latter hypothesis appears to be more plausible from the observation that surface variations occur within the same clone in which similar corticogenes might be expected, and from the observation that in an abnormal clone organelles can be displaced and develop in new regions (111). The latter would be expected only if the cortical units were equivalent.

These experiments have been interpreted in terms of two hereditary mechanisms involved during the division of *Tetrahymena:* one in which the nuclear genes specify the macromolecules, and one in which the structural organization of the cell—the association between organelles—determines the organellar pattern in the next generation. The mechanism involved in these latter events has been termed cytotaxis (158).

Long-range control of the stable kinety pattern might be exerted through the action of nuclear genes, whereas cytotactic mechanisms might generate the short-term variations

(74). This hypothesis could be checked by isolating natural variants of *Tetrahymena* and determining what the role of nuclear genes is in this organism and how they interact with the cytotactic process.

E. Linkage Relationships

There are five pairs of chromosomes in the micronucleus. If random distribution of nonhomologous chromosomes occurs during meiosis, as it usually does in other eucaryotic organisms, then markers on different chromosomes should assort independently and markers on the same chromosome should show linkage. This seems to be the case for *T. pyriformis,* despite some of the peculiarities that occur during meiosis. Thus linkage maps can be constructed for this organism. Most of the tests have involved markers in syngen 1, and a few in syngen 7.

How is evidence on linkage relations between genes obtained? Crosses are made and recombination is examined in the F_2 or backcross hybrid generations. Deviations from free assortment (9:3:3:1 in the F_2 and 1:1:1:1 in the backcross generations) suggest linkage. Or, genomic exclusion can be used and recombination looked for in the progeny. Here a 1:1:1:1 ratio would be expected with free assortment. For most markers evidence for free assortment or linkage can be obtained directly from the observed ratios in the hybrid generations. For the mating type locus, testcrosses in subsequent generations are usually necessary to identify the alleles carried by the hybrid progeny.

What is the evidence on the linkage relationship of genes in syngen 1? There are five published studies. In the first, Nanney (100) crossed strains which were mt^D/mt^D H^C/H^C and mt^B/mt^B H^D/H^D and produced the F_2 generation. Recombination for *H* was examined in the genotypes mt^D/mt^D, mt^B/mt^D, and in the partial genotype $mt^B/?$. In none of these genotypes were departures from 1:2:1 ratios observed for *H*. Thus *mt* and *H* appeared to assort independently, although loose linkage cannot be excluded from these data.

In a second study the linkage relationships of *H*, *E-1*, *E-2*, and *P-1* were examined (6). Here a "triple cross" was employed. A heterozygote between the A and C strains was constructed and then crossed to the B strain, and recombination was examined in the progeny. The A and C strains differ in their allelic constitution at all four of the named loci. Strain B has a third allele at the *H* locus. Thus by examining the progeny for *H*, this cross served as a means of screening progeny produced by normal conjugation versus genomic exclusion. Sixteen genotypes are expected, and all 16 were observed in the 112 progeny examined. Independent assortment was observed for all four genes, although linkages with 40% crossing-over could not be eliminated. Evidence from additional two-factor crosses also supported independent assortment. In the same study the linkage relationships of *mt*, *E-1*, and *E-2* were examined in the backcross generations of F_1 and F_2 hybrids from a cross between the B and C inbred strains. Here only some of the backcross data could be used because of genomic exclusion. Of the 52 backcross progeny testcrossed for *mt*, independent assortment was observed for *mt* and *E-2* and for *E-1* and *E-2*, but linkage was observed between *mt* and *E-1* (21:7:6:18) with recombination being 25%.

In the third study, Phillips (136) examined linkage relationships of *T* with respect

Table 11.

Segregation Data from Genomic Exclusion

	Ha(Hd):He	Ta:Tb	E-1b:E-1c	E-2b:E-2c	P-1a:P-1b
A/C × C*	37:39	50:26[a]	44:32	35:41	34:42
B/C × C*	26:16	24:18	22:20	21:21	0:42[b]

[a] $p < 0.01$ for 1:1.

[b] Both strains B and C are homozygous for *P-1*[b]/*P-1*[b], therefore segregation not expected.

to *mt, H, E-1, E-2*, and *P-1*. She devised an ingenious test for *T* and *mt* in which she used key mating types to test for linkage, taking into account that these types are represented in half the frequency in heterozygotes. The distribution of types among *T* genotypes was calculated for free assortment, complete linkage, and 25% recombination in the F_2 progeny. The observed numbers ruled out the latter two hypotheses but not the first. This suggested that *mt* and *T* assort independently. F_2 data were also used for assessing the extent of recombination between *T* and each of the other loci. The statistical tests ruled out close linkages for all comparisons, but loose linkages could not be excluded for *T* and the enzyme loci. This study thus indicated that *T* appears to assort independently of all five loci tested.

In the fourth study Bleyman (30) found that the early maturer genes, *Em-1, Em-2*, and *Em-3*, were not linked to each other nor to *mt* or *H*, nor was *Em-1* linked to *P-1*. In the fifth study, Grass (71) found that *St* assorted independently of *mt, H, T*, and *E-1*.

Additional linkage data in syngen 1 have been obtained using genomic exclusion and are presented in Tables 11 and 12. Heterozygotes between the A and C strains

Table 12.

Linkage Data from Genomic Exclusion[a]

Loci	A/C × C*		B/C × C*		Total	
	p	r	p	r	p	r
H/T	33	43	24	18	57	61
H/E-1	37	39	20	22	57	61
H/E-2	40	36	21	21	61	57
H/P-1	33	43	—	—	—	—
T/E-1	40	36	20	22	60	58
T/E-2	39	37	25	17	64	54
T/P-1	36	40	—	—	—	—
E-1/E-2	41	35	19	23	60	58
E-1/P-1	36	40	—	—	—	—
E-2/P-1	39	37	—	—	—	—
E-1/mt	—	—	17	4[b]	—	—

[a] p, Parental; r, recombinant.

[b] $P < 0.01$ for 1:1.

or B and C strains were constructed and then crossed to C* (a clone with a defective micronucleus). Strains A and C differ in their alleles at *mt, H, T, E-1, E-2,* and *P-1,* and strains B and C at all loci except *P-1.* Seventy-six progeny from the A/C × C* cross and 42 progeny from the B/C × C* were examined for alleles at all loci except *mt.* Normal segregation of alleles at all loci was observed except for T in the cross A/C × C* (Table 11). In this case the discrepancy is probably technical, since it is not always easy to discriminate Tb from Ta. When the linkage relationships between these loci were examined, independent assortment was observed for all 10 combinations of two loci (Table 12). Linkages with 30% recombination are statistically eliminated for all 10 combinations, and linkages with 40% recombination are eliminated for all except *T* and *E-2.* From the B/C × C* cross 21 selected progeny were also examined for recombination between *mt* and *E-1.* Seventeen had parental phenotypes and four recombinant phenotypes, showing linkage between *mt* and *E-1,* recombination having been 19%. Taken together, these data support previous observations that all loci so far tested in syngen 1 assort independently except for *mt* and *E-1.*

Very recently additional linkages with new genetic markers have been found in syngen 1 (P. Doerder, personal communication). More important, crosses involving different combinations of strains have not behaved uniformly with regard to recombination of *mt* and *E-1* (P. Doerder, personal communication; J. W. McCoy, personal communication). The significance of these strain differences for linkage studies in general is not clear.

Only three markers are available in syngen 7: *mt, R,* and *P.* Phillips (142) found that they appeared to assort independently.

F. Population Genetics

We have as yet little information on population genetics in *T. pyriformis.* The collections of syngen 1 for which we have genetic information are limited to six sites in the United States (Table 1). Three strains (WH-6, WH-14, and WH-52) were collected from the same pond at Woods Hole, Massachusetts; one strain was collected from a lake near Bennington, Vermont (UM-226); one from a drainage ditch near Alpena, Michigan (ALP-4); another from Indian Lake, Michigan (IL-12); and two from different lakes at Lake-of-the-Woods, Mahomet, Illinois (LW-A and LW-B). With one exception they express only one of the alleles at each of the loci examined. However, there is evidence from breeding analysis and from the genotypes of derived strains that these wild strains are heterozygous for many of their loci. IL-12 was shown by Orias (130) to be heterozygous for the recessive lethals fat and tiny. Complete genotyping of LW-A and LW-B was performed by Phillips (139). She found that LW-A was heterozygous for *T* and *E-2,* and that LW-B was heterozygous for *mt* and *E-1.* WH-14 is heterozygous for *mt* (98). Additional heterozygosity of the Woods Hole strains may be inferred from the genotypes of strains derived from them in crosses (Table 2). Four alleles at the *H* locus were present in the same pond: H^B in WH-52, H^A in WH-6, H^D in WH-14, and one of the latter must have been heterozygous for H^C. All wild strains that have been tested are Ta in phenotype except for WH-14 which is Tc. The T^C allele is not presently found in strains A and B which were derived from a cross of WH-6 and WH-14. Either it was present in earlier generations of strain B, hence its reappearance in strain D, or ALP-4 may be heterozygous for T (T^A/T^C). The T^B allele is found in inbred C which was derived from a cross of UM-226 and strain B. UM-226 is Ta in phenotype. Either UM-226 is T^A/T^B, or one of the Woods Hole strains also carried T^B. Finally, heterozygosity for P-1 of one of the Woods Hole strains is inferred from the fact that strain A carries the P-1^A allele, although both WH-6 and WH-14 are P-1b in phenotype. Thus, even with this small sample of strains, it appears that a given wild strain is heterozygous for at least one of the eight loci mentioned, and some may be heterozygous for as many as four.

It appears then that there is genetic polymorphism in natural populations of *Tetrahymena*. However, little attention has been paid so far to the behavior of genes in populations—in an attempt to understand why such polymorphism exists. One piece of work stands in splendid isolation. This is an analysis by Orias and Rohlf (133) of the mating system in syngen 8 which has three alleles and three mating types. They developed a mathematical model which gives the equilibrium distribution of genotypes during many generations, and they attempted to explain the derivation of this three-allele system from a two-allele system. Certain assumptions were made about population size, differential mutation, rates of migration, and selective reproduction, since these parameters are unknown in *Tetrahymena*. They predicted that at equilibrium the three mating types should be present in equal numbers and that certain genotypes should be more frequent. The distribution of types and alleles in the samples so far collected, although small, were not incompatible with their expectations.

IV. PHENOTYPIC DIVERSITY IN CLONES

A. Introduction

Intraclonal variations have been observed for several different kinds of phenotypes. Some are relatively unstable, some are highly stable, but reversible, and some are not only stable but irreversible. This last class includes variations in heterozygotes, which involve differences in the expression of alleles, and differences in mating types which arise in homozygotes. These share certain properties that are derived from initial macronuclear heterogeneity and changes that occur in the macronucleus during fission. They are therefore of special interest and are considered together in this section.

Phenotypic changes have been observed within all heterozygous clones of syngen 1, and with one heterozygote in syngen 7. If we take as an example A/a, both alleles are expressed immediately after conjugation, but after several hundred fissions most of the cells express only one of the alleles (either A or a). Lines that express A or a are completely stable, while those few expressing both alleles are unstable and give rise to new lines which express either A or a. These phenotypic changes appear to be controlled by the macronucleus, since the "missing" allele reappears in the next generation after normal conjugation, when the old macronucleus is destroyed. The source of the missing allele is of course the micronucleus. During genomic exclusion the old macronucleus is not destroyed but is retained, and under these conditions the phenotypic "differentiation" persists in the progeny. The correlation between the disappearance or persistence of the phenotype with the behavior of the macronucleus, but not the micronucleus, localizes the source of the effect to the macronucleus.

Observations similar in kind have been made with the mating type locus in homozygotes in exceptional clones referred to as selfers. Selfers are "unstable" clones and give rise to stable lines "pure" for a mating type. As in the case of heterozygotes, no effect of parental phenotype is observed after normal conjugation; moreover, the basic pattern of distribution of mating types is caryonidal. That is to say, there is a correlation in the distribution of mating types with the disjunction of the two new macronuclei into separate cell products.

The basis for these examples of phenotypic diversity clearly resides in the macronucleus. A change takes place in the macronucleus which results in altered phenotypic expression. What is the nature of this change? When does it occur? How is the appearance of the altered phenotype geared to cell division? Observations on different genes give definition to this phenomenon. These are set forth as a series of eight properties. Some properties are common to all genes and to both heterozygotes and homozygotes. Other properties are gene-specific.

B. Properties

1. Selfers Give Evidence for Assortment of Macronuclear Units during Cell Division

Selfers are usually ditypic (31,119,121). They give rise to lines pure for two different mating types but usually in different proportions. One type is referred to as the majority type, since it is produced more frequently than the other, the minority type. In recently derived sublines of a selfer, only the majority type is found. With continued cloning, however, the minority type also appears. The rates of appearance may be initially diverse in different lineages but with continued fissions the rate of appearance of the majority type decreases and the rate of appearance of the minority type increases; eventually, after many fissions the two rates become equivalent. Much sooner, however, the total rate of stabilization of both types approaches a constant per fission. In an early study of selfers (18), equilibrium was reached by 25 fissions and the equilibrium value was 0.0113.

These observations were expressed in terms of a mathematical model by Schensted (153). In setting up this model, she made certain assumptions: (1) that there are several units within the macronucleus; (2) that each unit replicates once during the interfission cycle; (3) that the total number of units at the time of macronuclear division remains constant; (4) that equal numbers of units are assorted to daughter macronuclei; and (5) that assortment of the units occurs randomly. She found that the total rate of stabilization at equilibrium was related to the number of units by the following equation:

$R_f = 1/(2N - 1)$, where R_f = rate per fission and N = number of units. Since $R_f = 0.0113$, $N = 45$. With $N = 45$ she calculated the cumulative fractions of stable lines from 0 to 200 fissions for different initial (or "input") ratios of the two types of units A and a. Representative values for different input ratios of A and a are shown in Table 15.

Equilibrium rates similar to 0.0113 have been observed in the production of stable

Table 13.
Equilibrium Rates of Stabilization per Fission for Different Genes

Gene	Genotype	R_f	Ref.
mt	mt^A/mt^A	0.0113 ± 0.0004	(18)
H	H^C/H^D	0.0112 ± 0.0018	(123)
T	T^B/T^C	0.0118 ± 0.0004	(137)
	T^A/T^C	0.0114 ± 0.0001	(137)
P-1	$P\text{-}1^A/P\text{-}1^B$	0.0113 ± 0.0012	(12)
St	St^A/St^C	0.0153 ± 0.0078	(71)
aa	aa^+/aa^r	0.0126	(39)
caf_1	caf_1^+/caf_1^r	0.0097	(39)
caf_2	caf_2^+/caf_2^r	0.0109	(39)

lines from heterozygotes for seven different genes (Table 13). No detailed kinetic data are available for *E-1* and *E-2*, or for *P* in syngen 7, but the accumulated fractions of stable lines after a certain number of fissions are compatible with the fractions expected from the Schensted equations. Thus the same number of units ($N = 45$) is implied in studies of heterozygotes for all genes.

The equilibrium rate of 0.0113 applies to cells grown without starvation. If the cells are allowed to starve, an increase in the frequency of stable types seems to occur. This effect is well documented for selfers (121) but has been reported only for the *H* heterozygotes. Here Nanney and Dubert (123) observed an increase in the rate from 0.0112 to 0.013. Other effects on the rate have not been systematically explored, except for temperature in the case of selfers (18). Here no effect was observed.

A prediction of the Schensted model, as it was formulated, is that *individual* rates of the two stable types become equal. This is assumed in calculating what is referred to as the "output" ratio, or the proportion of cells expressing different stable phenotypes. The output ratio is also an estimate of the "input" ratio, or the relative numbers of the two types of units initially present in the macronucleus. In calculating output ratios there are three classes considered: X = number of stable A lines, Y = number of stable a lines, and XY = number of unstable lines. The frequency of A can be determined from the formula $100[(2X + XY)/2N]$. If A = 60, then a = $100 - 60$, or 40. These output ratios imply that there were 60% $\times$ 45, or 27 A units initially present in the macronucleus, and 40% $\times$ 45, or 15 a units. The assumption is made in calculating these ratios that XY will yield equal numbers of X and Y.

2. The Number of Assorting Stable Types Depends on the Genotype

Most heterozygotes with codominant alleles (*H, T, St, E-1,* and *E-2* in syngen 1, and *P* in syngen 7) give rise to two stable cell types (2,71,123,137,143). The phenotypes of these two types are similar to homozygotes. For example, stable lines derived from the H^A/H^D heterozygote are Ha and Hd in phenotype. Three stable types are observed for the phosphatase locus (*P-1*) in syngen 1 (24). Two of these have phenotypes similar to homozygotes (P-1a and P-1b), but the third type has a hybrid phenotype (it has isozyme 3). Seven stable types are observed for mt^A/mt^B heterozygotes in syngen 1, and four in syngen 7 (122,140).

Where one allele shows dominance, only the recessive (resistant) stable type could be detected for *aa*, caf_1, and caf_2 in syngen 1 by the screening techniques (39). In syngen 7 in which serotype R^A is dominant to R^B, Rb cell lines were not observed, but only 40 cell lines were examined (142).

Most homozygotes have a single stable type. The *mt* homozygotes are the exceptions (98,122,140).

3. The Assorted Type Is Stable and Cannot Be Reversed

Once a line becomes differentiated for a particular type, it is completely stable during asexual reproduction. No reversion has been observed, nor have conditions been found that bring about reversion. Heterozygous lines expressing Ha, for example, can be transformed to another antigenic type (L, I, St, or T) in which the H specificity is lost. They can be maintained under these new conditions for many cell generations but, when returned to conditions under which H is expressed, they express Ha (81). Phosphatase heterozygotes expressing isozyme 1 or 3 or 5 are stable even when the macronucleus is forced to go through the morphological changes that occur with genomic exclusion (7). Several chemical and physical treatments have also been applied, all with no effect. Therefore the acquired phenotype appears to be irreversible. The only known way of losing the phenotype is by destroying the macronucleus.

4. Assortment Can Occur in the Absence of Gene Expression

Assortment has been observed only while the gene is expressed for the enzyme loci, but assortment normally occurs in the absence of gene expression for *mt* during the period of immaturity. Assortment for the serotypes can occur in the presence or absence of gene expression (71,137).

5. Phenotypes Assort Independently for Different Genes, Even When Linked

When comparisons are made of lines stable for different phenotypes, no correlations are observed in which types assort together. Such observations have been made for certain gene combinations: for *mt* and *H* (100), for *mt*, *H*, and *T* (32), for *H* and *P-1* (12), and for *H*, *E-1*, *E-2*, and *mt* (7). For example, lines that are E-1b may be either E-2b or E-2c in phenotype, and lines that are E-1c may be either E-2b or E-2c. In other words, there is complete independence in the association of types. This is also true for *E-1* and *mt* which are linked genes. In this analysis heterozygotes that were $mt^B E\text{-}1^B/mt^C E\text{-}1^C$ were examined. If linked characters tend to be associated, then mating type IV or VII (produced by the mt^B allele) should be associated more frequently with the E-1b phenotype and mating type *I* (mt^C) should associate more frequently with the E-1c phenotype. Such preferential associations were, however, not observed.

Table 14.
Output Ratios and Fissions at Which Determination Occurs for Six Different Allelic Combinations at *H* Locus[a]

E *A*	*E* *D*	*E* *C*
81 19	90 10	98 2
1–2 fissions	0–2 fissions	0 fissions
A *D*	*A* *C*	*D* *C*
87 13	96 4	57 43
0 fissions	2 fissions	1 fissions

[a]Based on data from Refs. 32,125,126.

6. The Ratio of Assorting Stable Types Is Gene-Specific and Allele-Specific

Output ratios for different genes fall into two classes. For some genes (*T*, *E-1*, and *E-2*) these ratios are nearly 1:1. For example, Phillips (137) reports output ratios of 54:46 for the T^B/T^C heterozygote with variation between 36 and 53 in the percentage of Tc. Output ratios for the three phosphatase types can be computed by modifying the equation to include three unstable types and taking into account the fact that there are restrictions on which stable types can arise from particular unstable types (12). In this way output ratios of 44:16:40 are obtained. The majority types are those with homozygous phenotypes (P-1a and P-1b), and they are nearly equal. The minority type has the hybrid phenotype.

The output ratios for *mt*, *H*, and *St* are highly eccentric. The majority mating type is usually present in high frequency in ditypic caryonides (31,119). For example, in some I-VI caryonides ratios of 2:98 are observed. Similar eccentric ratios are found for the *H* locus, but here different allelic combinations vary in their eccentricity. There is a peck order in the dominance relationships of the alleles with $H^E > H^A > H^D > H^C$ (126). This is reflected in the output ratios (Table 14). Whenever H^E is present, it is always the majority type, and the output ratio reflects the degree of dominance, the most eccentric ratio being found in combination with H^C. The least eccentric ratios are found for the H^C/H^D combination, although there is considerable variation in different lineages. Output ratios appear only to be allele-specific for the *H* locus, although it has not been definitely ruled out for some combinations of mating types.

7. The Number of Divisions before Assortment Is Observed Is Gene-Specific and Allele-Specific

Table 15 shows the expected fractions of stable lines at successive fissions for different input ratios. Table 16 lists the observed fractions for eight different genes.

Assortment of mating types is observed only when maturity is reached, or at about 80 fissions. The observed fraction of lines pure in mating type was 0.98. If we examine

Table 15.
Cumulative Fraction of Lines Stabilized at Different Fissions after Determination for Different Input Ratios of A/a[a]

Number of fissions after determination	Input ratio A/a					
	1/44	5/40	10/35	15/30	20/25	22/23
1	0.25	0.00	.00	0.00	0.00	0.00
10	0.74	0.20	0.04	0.01	0.00	0.00
20	0.84	0.40	0.15	0.05	0.02	0.01
30	0.88	0.52	0.26	0.12	0.07	0.06
40	0.91	0.60	0.35	0.20	0.13	0.12
50	0.92	0.66	0.43	0.28	0.21	0.20
60	0.93	0.70	0.49	0.35	0.28	0.27
70	0.94	0.74	0.54	0.41	0.34	0.34
80	0.95	0.77	0.59	0.47	0.41	0.40
90	0.95	0.79	0.63	0.52	0.46	0.45
100	0.96	0.82	0.68	0.58	0.54	0.53
120	0.97	0.85	0.73	0.64	0.59	0.59
140	0.97	0.87	0.77	0.69	0.65	0.65
160	0.98	0.89	0.81	0.74	0.70	0.70
180	0.98	0.91	0.83	0.77	0.74	0.73
200	0.98	0.92	0.85	0.80	0.77	0.76

[a]From Phillips (137). Values were derived from the printout from an MIDAC digital computer (153).

Table 15, we see the value 0.95 at 80 fissions if the input ratio is 1:44. Similarly, in the example shown for *H* the observed fractions at different fissions fit best the expected fractions for an input ratio of 5:40. This also applies to the data shown for *St*. For all three loci assortment seems to have started near the first fission, since the observed fractions correspond to one of the sets of expected fractions.

By contrast, if we look at the observed fractions for the other genes (*T, E-2, E-1,* and *P-1* in syngen 1, and *P* in syngen 7), we note that no assortment is observed until 40 or more fissions. Moreover, if we look at the observed fractions at 70 to 80 fissions, we see that they range from 0.08 to 0.15. None of the expected fractions are this low. Even with an input ratio of 22:23, 0.34 to 0.40 of the lines should be stable at this time. This suggests that for these genes assortment does not begin immediately and that expression of an allele only follows after some other event occurs. In other words, two steps are required before a stable line appears: (1) *determination*, the event that results in differences in the units; and (2) *assortment* of the differentiated units to produce a homogeneous macronucleus. The genes *T, E-2, E-1, P-1,* and *P* therefore do not appear to undergo determination before the first fission but only after several fissions have transpired. This is also true for the three recessives caf_1^r, caf_2^r, and aa^r.

The onset of determination can be measured if the set of observed fractions is compared to the expected fractions for an input ratio of 22:23. To illustrate, 0.03, 0.13, 0.26, and 0.40 stable *T* lines were observed at 50, 70, 90, and 110 fissions, respectively; whereas 0.01, 0.12, 0.27, and 0.40 are the expected fractions at 20, 40, 60, and 80

Table 16.

Cumulative Fraction of Lines Stabilized at Successive Fission Intervals for Different Genes[a]

Fissions	Expected fraction 22A/23a	mt	St	H	T	E-2	E-1	P-1	Syngen 7, P	Time of determination
0–4	0.000	—	0.04	0.02	—	—	—	—	—	← mt, H, St
10	0.001	—	—	0.08	—	0	0	0	—	—
16	—	—	0.10	—	—	—	—	—	—	—
20	0.01	—	—	—	—	—	—	—	—	← caf_1
25	0.04	—	—	0.19	—	0	0	0	—	—
30	0.06	—	0.28	—	—	—	—	—	—	← T
40	0.12	—	0.60	0.42	—	0.04	0.02	0	—	← E-$1, E$-2
50	0.20	—	—	0.59	0.03	—	—	0.01	—	← P-$1, P, caf_2$
60	0.27	—	—	—	—	—	—	—	—	← aa
65	.32	—	—	0.69	—	—	—	0.05	—	—
70	0.34	—	—	—	0.13	—	—	—	—	—
80	0.40	0.98	—	0.78	—	0.15	0.14	0.10	0.08	—
90	0.45	—	—	0.80	0.26	—	—	0.15	—	—
100	0.53	—	—	0.83	—	—	—	0.21	—	—
110	0.56	—	—	—	0.40	—	—	—	—	—
120	0.59	—	—	0.85	—	—	—	0.32	0.27	—
140	0.65	—	—	—	—	—	—	—	—	—
160	0.70	—	—	—	—	—	—	0.56	0.44	—
182	0.73	—	—	—	—	—	—	—	0.56	—
208	0.77	—	—	—	—	—	—	—	0.69	—
234	0.80	—	—	—	—	—	—	—	0.79	—

[a] Data on mt, E-1, and E-2 from Allen (7); data on T from Phillips (137); data on H and P-1 from Allen (12); data on St from Grass (71); data on aa, caf_1, and caf_2 from Carlson (39); data on P in syngen 7 from Phillips (143).

fissions, respectively. The difference between these two sets of numbers is 30. This suggests that the determination for the T locus begins at about 30 fissions. Similar comparisons lead to the conclusion that E-1 and E-2 are determined at 40 fissions, and the two phosphatase genes in different syngens (P-1 and P) at 50 fissions. Note that all the genes that are determined "late" have output ratios near equality in contrast to genes that are determined early which have highly eccentric output ratios.

A more accurate measure of the time of determination of mt and H can be made by cell pedigree analysis (see Fig. 2). Here complete pedigrees may be obtained through secondary (2°) subcaryonides. Each 2° subcaryonide is then expanded to 30 lines so that quantitative variations in different types can be recorded and an output ratio assigned to each 2° subcaryonide. One can then examine the branches of the pedigree to see at which fission discontinuities occur. Table 17 gives an example of a cell pedigree. If we look at the mating types, we see that there arc four different groups, but that there is qualitative similarity of the four cells making up each group. Each group of four represents a single caryonide; therefore mating type determination occurs *before* the first fission and discontinuities are observed between caryonides.

Table 17.
Serotype Output Ratios and Mating Types[a]

2° Sub-caryonides	T	H	mt
1	43	10	V-VI
2	42	12	V-VI
3	50	20	V-VI
4	42	25	V-VI
5	42	18	I
6	43	13	I
7	43	5	I
8	47	7	I
9	52	5	I-VI
10	47	17	VI-I
11	38	17	I-VI
12	47	22	VI-I
13	50	13	II
14	53	5	II
15	55	5	II
16	45	5	II

[a]Data from Bleyman et al. (32). Output ratios for minority T or H type are shown above.

Differences between lineages in *H* are quantitative and are expressed by the different output ratios. To assess where discontinuities occurred in quantitative terms, an analysis of variance was applied (125). In the example shown most of the variance was attributable to differences in the output ratios of primary (1°) subcaryonides. Hence determination occurs for this *H* heterozygote after the first fission. The output ratios for the *T* serotypes also appear in Table 17. No discontinuities could be found, which is further evidence that determination has not occurred in the fission intervals examined.

All six combinations of alleles at the *H* locus have been examined by cell pedigree analysis (32,125). The results are shown in Table 14. They indicate that each *H* heterozygote has a characteristic time of determination (or distribution of times). For example, determination occurs before the first fission for H^C/H^E, after the first fission for H^C/H^D, and after the second fission for H^A/H^C, but it may occur at any of these fission intervals for H^D/H^E. Recent disclosures of nonrandomness of assortment in the early fissions (31) cast doubt on the significance of the differences in timing for the H serotypes. Determination may not be allele-specific, but only locus-specific, and may occur before the first division.

To summarize, these studies indicate that:

1. *Determination is a separate process from assortment.*
2. *The time of determination is gene-specific.* It begins very early for *mt*, *H*, and *St* (0 to 2 fissions) but later for other genes (20 fissions for caf_1, 30 fissions for *T*, 40 fissions for *E-1* and *E-2*, 50 fissions for *P-1*, *P*, and caf_2, and after 60 fissions

for *aa*). Early maturer genes which alter the time when mating types are expressed do not affect the timing of other genes (30).

3. *There is a relation between the output ratio and the time of determination.* Late genes have equal output ratios; early genes have eccentric output ratios.

8. Determination Occurs When the "Entire Complement" of Units Is Present

The number of units immediately before division of the macronucleus is 90, according to the Schensted model. During replication of these units, a geometric progression might be represented: 2, 4, 8, 16, 32, and so on. How many of these units are present—or are involved—during determination? This problem was tackled by examining the dispersion of output ratios for the H serotypes using the formula $(p + q)^N$, where $N =$ the number of units (125). The observed distributions fitted best to $(p + q)^{32}$. This indicated that a minimum of 32 units might be present when determination occurs. The distributions indicated independence of the units.

A similar number of units was implicated for the mating types by the observation that the minority output fractions fitted the series 1/32, 2/32, 3/32, and so on (31). However, different units do not seem to be independent when determination takes place (119). With this number of units, a high frequency of polytypic caryonides would be expected, containing three or more mating types. Most caryonides are monotypic, and many are ditypic. This indicates "coordination" between units when determination takes place.

2. Interpretation

1. Essential Features

To accommodate all the observations, the following five features are essential to any hypothesis:

1. There are units, the total number being the same for all genes. These are assorted at random with respect to kind, and each daughter cell receives an equal number of units ($N = 45$).
2. Assortment occurs during cell division.
3. For each gene there are three periods in relation to fission numbers: predetermination, determination, and postdetermination when the stable types are assorted. The length of the first period in terms of numbers of fissions is gene-specific. For late-determined genes the information for both alleles must persist for many fissions.
4. In terms of types and frequencies, what is determined is gene-specific.
5. Determination takes place regardless of whether or not a gene is expressed, and is irreversible.

Any hypothesis must take into account what is known of the relative amount of DNA in the macronucleus compared to the micronucleus, the replication of DNA, and the organization of the genetic material in the macronucleus. We know that there is sufficient DNA in the macronucleus for 45 copies of the *haploid* genome (64,171,172). We also know that the total amount of macronuclear DNA fluctuates in response to growth conditions (40), but that certain fractions of DNA respond differentially (167,168). We know there are DNA-containing elements in the macronucleus, and we suspect they may be organized in some way, but cytological studies are conflicting as to whether or not there are chromosomes in the mature macronucleus (27,72,84,127,148,170). Recent measurements of the DNA molecules present in the mature macronucleus of *Stylonychia mytilus* show that they are fragments (146).

Theoretically, the units revealed in the genetic studies could be conceived of as involving different levels of organization of the genetic material: (1) individual genes, (2) whole chromosomes, (3) haploid chromosome sets, and (4) diploid chromosome sets. As Nanney (105) pointed out, it is difficult to account for differences in the timing of determination and output ratios of different genes if the units are (2) or (3). He preferred a hypothesis in which the units were diploid chromosome sets. In view of the recent findings on the amount of macronuclear DNA, his ideas have been modified (D. L. Nanney, personal communication), and he conceives of the units as containing only a portion of the genome but what is included is diploid. In 1968, when this review was first written, we came to a more radical view that the units were smaller than chromosomes and essentially individual genes or clusters of genes. A similar view was independently formulated by Wille and Ehret in their "replicon soup" model (personal communication; 167).

At present these different views of the units may not be quite so far apart. The essential feature of Nanney's hypothesis is that the units are *diploid*. It is our view that they may be *haploid*. Depending upon whether the units are diploid or haploid, different hypotheses can be constructed. An outline of some of the possibilities is shown in Table 18. Two of these hypotheses are discussed in greater detail below.

2. Subnuclear Hypothesis

a. Nature of Units. The macronucleus is viewed as containing several subnuclei (156), each of which contains the diploid complement of certain genes (D. L. Nanney, personal communication). The number of diploid subnuclei in the macronucleus of *T. pyriformis* would be 90 before, and 45 after, division (105). As shown in Table 18, it is also possible that the units are diploid replicons. For each gene there would be 90 copies before, and 45 copies after, division of the macronucleus.

b. Determination. Determination is viewed as a regulatory event which occurs within each subnucleus and is localized to individual genes. "Allelic repression" is thought to result in differentiated subnuclei in heterozygotes (103,105). In each subnucleus one allele is repressed. Allelic repression thus occurs independently in each subnucleus. Each gene is also regulated independently at different times (fissions), and the probability that a given repression occurs would be allele-specific to account for differences in output ratios. The resulting macronucleus would contain subnuclei which are differentiated in different ways for each pair of alleles at each locus.

Coordination in mating type determination would be explained as a result of the *lack* of independence in the regulatory events that select which mating type is to be expressed in different subnuclei. Once an initial differentiation to a particular mating type occurs in a subnucleus, it is speculated that it "spreads" to others.

Table 18.

Hypotheses for Macronuclear Units

Diploid units:
Gene diminution in mature macronucleus
1. Units are diploid subnuclei
   All genes in a single unit
   Each subnucleus replicates; number of copies, 45 × 2
   Alleles "interact" functionally
   Assortment of differentiated subnuclei
2. Units are diploid replicons
   Each replicon duplicates; number of copies, 45 × 2
   Alleles "interact" functionally
   Assortment of differentiated replicons

Haploid units:
All genes present in mature macronucleus
1. Units are haploid replicons
   Alleles "interact" in terms of replication
   For each allele—one master
   Masters make slaves (nonreplicating)
   Total number of copies for each pair of alleles is 45
   Assortment is genetic
2. Units are haploid replicons
   Alleles "interact" in terms of replication
   For each allele—several masters
   Total number of copies for each pair of alleles is 45
   Assortment is genetic

Haploid units:
Gene diminution in mature macronucleus
Units are haploid replicons
Alleles "interact" in terms of replication
For each allele—several masters
Total number of copies for each pair of alleles is 90
Assortment is genetic

---

c. Assortment. Once determined, the differentiated subnuclei are viewed as being passively assorted to daughter macronuclei in subsequent fissions. After a certain number of fissions, a macronucleus is produced which contains subnuclei differentiated all alike. Such a lineage would thereafter be stable in phenotype.

d. Fit of Hypothesis to Data. This is a particularly attractive hypothesis, since all the observations can be accommodated. It is a two-stage hypothesis (see Fig. 8). Specificity is introduced in the first stage, while the second stage is a passive process. Since determination is viewed as a regulatory event, differences in gene action can be easily accounted for by building in new assumptions. No attempt has been made to define the nature of allelic repression or how mating types are regulated, but it is usually thought to involve transcriptional controls (but see ref. 114 and Section 4). The kind of control that is envisioned is of the type that occurs with differentiation in higher organisms, since it can work in the absence of gene expression.

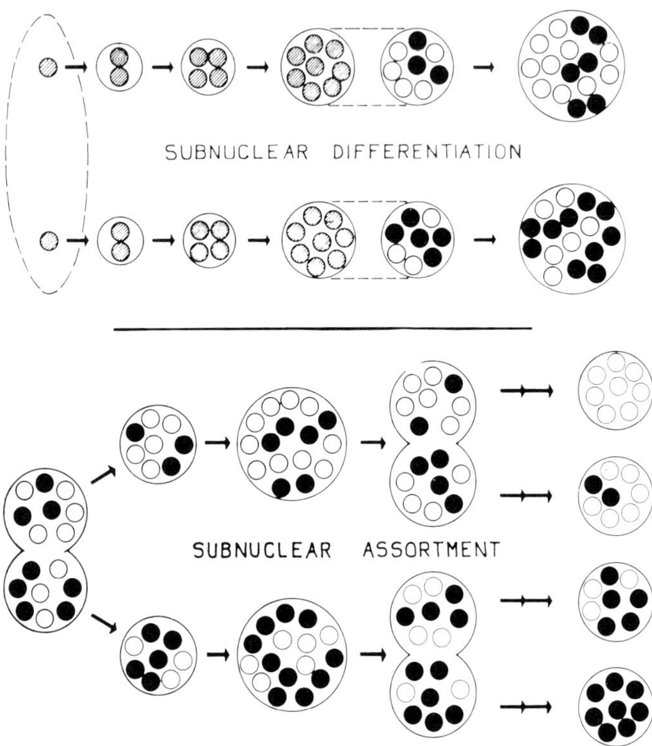

Fig. 8.

Subnuclear differentiation and subnuclear assortment in the macronucleus of *T. pyriformis* syngen 1. [From Nanney (103).]

3. Master-Slave Hypothesis

a. Nature of Units. The macronucleus is viewed as containing several haploid units which are smaller than chromosomes. These units are "replicons" and differ in their ability to replicate. Stable cell lines with a particular phenotype appear as a result of the loss of units that specify alternative phenotypes.

These units arise as a result of chromosome fragmentation during the development of the macronucleus (Fig. 9). The units of homologous chromosomes include homologous stretches of DNA. In Fig. 9 we show similar lengths of DNA for homologous units (A,a; B,b; and C,c). This implies reproducible breakage points in a particular chromosome, although this may not be a necessary assumption. These units now replicate and produce a certain number of copies.

Several versions of hypotheses with haploid units are shown in Table 18. They differ in whether gene diminution occurs, in the total number of copies produced by each pair

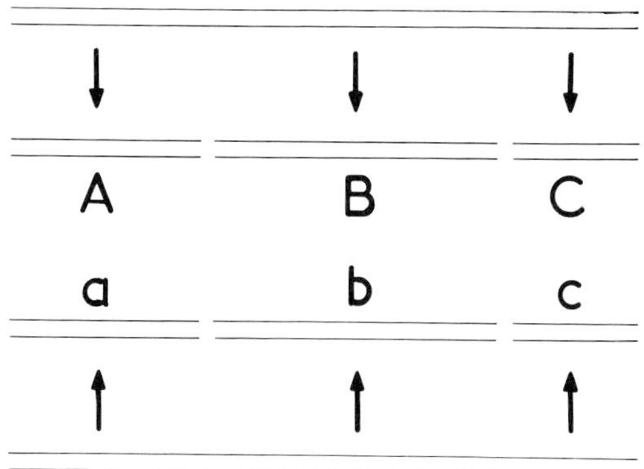

Fig. 9.

Fragmentation of a pair of homologous chromosomes to produce master units. Master units could be of different lengths (A versus B versus C). They could contain one or several genes. Homologous units (A,a) are shown as being the same length.

of alleles, and whether or not the copies can replicate. Only the first satisfactorily copes with the problem of the timing of determination and the maintenance of the heterozygous state for many fissions. We call this the "master-slave" hypothesis. Here all genes are present in the mature macronucleus, and the total number of copies is 45 for homologous units (A,a) after cell division. Separate control of replication occurs for each pair of alleles, but the total number of copies approaches a similar threshold. With this hypothesis only the original chromosome fragments can replicate and are referred to as "master units." The master units replicate once per fission and make copies which cannot replicate. These are referred to as "slaves." The total number of A and a slaves is 45.

b. Determination and Assortment. We assume that there are two kinds of units, master units and slave units. There are two homologous master units corresponding to A and a, which replicate once each fission. The slave units (A and a) are produced from the masters during each cell generation and are distributed at random during fission. The master units then produce extra slaves to make up the numbers present before fission. There are not necessarily equal numbers of slaves, but the total number of A and a slaves is 90 before cell division and 45 after division. The slaves determine the phenotype directly.

During a cell generation there is a probability (perhaps in about 1 in 100 cells) that one of the masters will misreplicate so that one of the daughter cells will be missing a master copy, the other daughter retaining both masters. This represents the process of determination, and since it involves the loss of a gene in one of the daughter cells, it is irreversible. Determination could, however, involve the failure of a master to make slaves. Misreplication (or failure to make slaves) could occur during early fissions for some genes, but it could be delayed for many fissions for other genes.

Following the loss of a master unit, the assortment out of a stable cell line would depend on the number of the two types of slaves. If there are only 45 slaves, this should occur rapidly, and within a few fissions stable lines of one phenotype would occur. If determination occurred at only one set fission in many cells simultaneously,

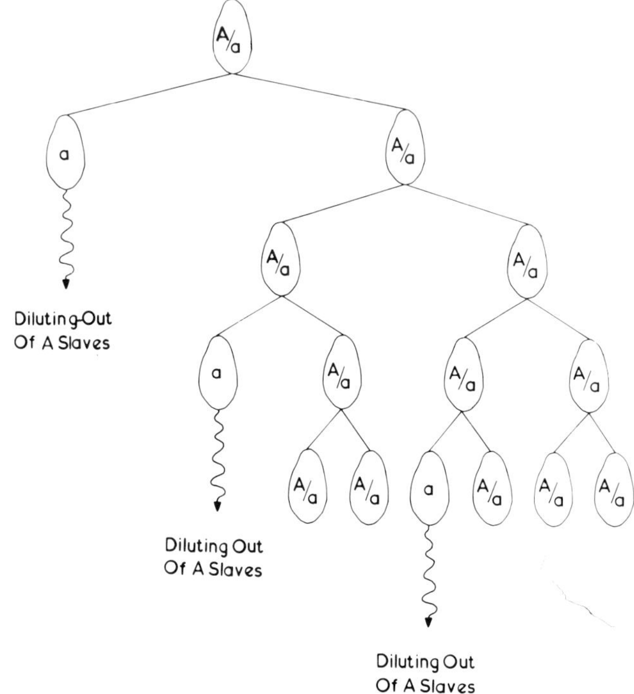

Fig. 10.

The process of determination according to the master-slave hypothesis. This occurs by the overlap of two events: misreplication of master unit A, followed by dilution-out of the A slaves at different fissions in a population of cells.

the long lag period that is observed would not be expected. To account for this lag, there would need to be 10^{22} slaves, if only dilution of slaves were involved. If, however, the "time" of determination were spread out over several fissions, and we consider the *population* of cells in time, a much longer lag period could result. At the time of the first misreplication, most of the cells in the population retain both master units. Further misreplication at later fissions will give rise to more stable lines, and after many more fissions only a few cells which contain both master units will be present in the population. The relative numbers of the two stable phenotypes will depend on the frequency of misreplication of each master unit and on whether misreplication of the masters occurs at random. To account for the long lag effects observed, we therefore visualize the process as involving overlap between the two events of misreplication of a master unit and dilution of the slaves of the lost master in a population of cells (Fig. 10). The "time" of determination would be a statistical average over a number of fissions, and the rate of appearance of a stable line would be a statistical summation of two events. The kinetics would be more complicated than those originally proposed by Schensted but would imitate them.

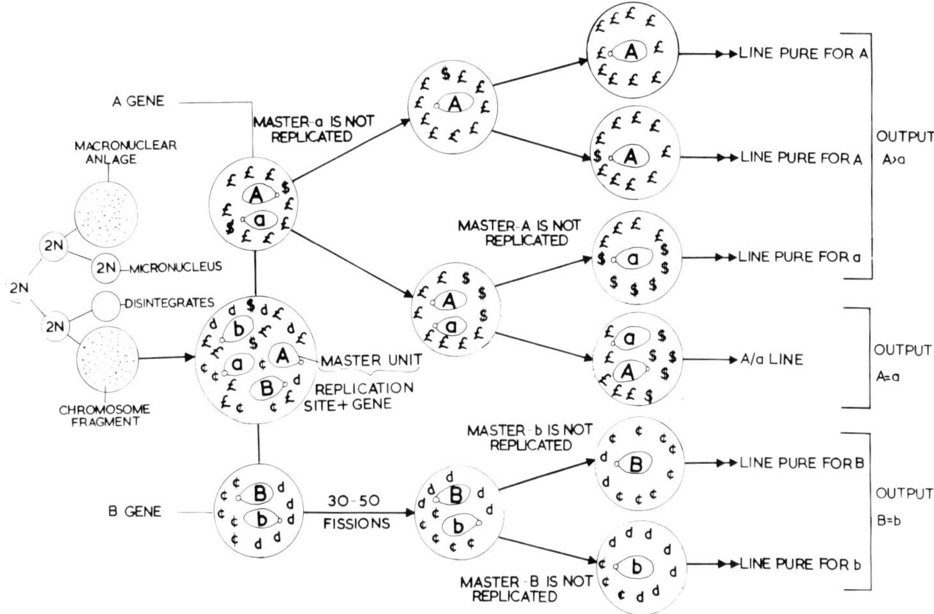

Fig. 11.

The formation of stable cell lines from a heterozygote by misreplication of a master unit and loss of slaves. A and a are a pair of homologous units; B and b are a different pair of homologous units. A, Master unit which produces slaves (£); a, master unit which produces slaves ($); B, master unit which produces slaves (¢); b, master unit which produces slaves (d).

To summarize, the major properties of the master and slave units are:

1. The master unit can replicate.
2. Determination involves loss of one of the master units.
3. Assortment occurs when all the slave copies of the lost master are diluted out. Since misreplication occurs at different times in a cell line, assortment is spread over a long period in terms of fissions.

*c. **Application to Heterozygotes.*** In heterozygotes genes appear to be "determined" at different fissions following conjugation. To explain the differences in the behavior of early and late genes, we assume that homologous masters make slaves at different rates for the early genes but at similar rates for the late genes. This difference in behavior might have another explanation, if nonrandomness in the early fissions were attributable to extra copies of DNA being made from perhaps one of the masters of early genes, or if there were nonrandom distribution of the copies made by early genes. In terms of differential replication rates, we illustrate what we mean in Fig. 11. In our illustration we have simplified the situation, and the total number of slaves equals 10. First, let us take a pair of alleles (A,a) which are determined "early." Master unit A makes eight slaves copies (represented by the monetary symbol £) for every two slaves ($)

made by master unit a. If master unit a is misreplicated, then one daughter cell may receive one, both, or neither, of the slaves of gene a. A cell that receives neither master unit a nor the a slaves will become phenotype A. In another cell within the clone, master unit A may not replicate, and cells without A slaves will eventually appear. These will become phenotype a. The loss of the master units can occur at different fissions within the clone, so that stable lines will appear at different fissions. To give the unequal proportions of the two phenotypic classes, we assume that the same unit misreplicates more often by chance than the other. We further assume that there is a relationship between the ability of the master unit to replicate and to produce slaves. Thus when the *second* master unit begins to misreplicate, then in cells that still possess both units, the rate of misreplication becomes the same for both master units, even though they were initially lost at different rates. This means that there is a selection for equal rates of replication in later fissions, and from these selected lines the two phenotypic classes will be derived in equal proportions.

Late-determined genes produce equal numbers of the two phenotypic classes. To explain this observation we assume that the two master units of late genes produce slaves at equal rates and also that there is an equal chance for either of the units to be misreplicated In Fig. 11, B and b represent a pair of alleles of a late-determined gene. Master B produces slaves (ϕ) equal in number to the slaves (d) produced by master b. Master B and master b replicate with equal efficiency for many fissions, but when misreplication does occur, there is an equal chance for either master unit to be misreplicated, and this could even occur at the same fission.

d. Application to Homozygotes. There are certain features of the mating type system that have led us to adopt additional assumptions in order to explain them. Differentiation occurs in homozygotes, and there are five or six possible mating types that a homozygous cell in syngen 1 can form. However, we restrict our attention to the mt^A/mt^A genotype with five mating types and examine ditypic caryonides.

We suggest that there are five linked genes which occur on the same chromosome fragment. There is a replication site which can vary in position, and replication is polarized, the gene next to the replication site being the only gene in the unit usually capable of making slave copies. A homologous fragment with the same five genes, but with the replication site next to a different gene, results in slaves of a different type being formed. If more slaves of one gene were initially produced on one of the master units than from the gene on the other master unit, this would lead to the observation of "coordination." If it happened that the next gene on one of the master units also made slaves, this would give rise to the possibility of two groups of slaves from one unit and three mating types in a clone. Misreplication of a master unit, or replication accompanied by the loss of the ability to make slaves, could result in previous slaves being diluted out (the latter is shown in Fig. 12) and a stable mating type could arise. In monotypic caryonides in which only one mating type occurs, the replication site is at the same position on both master units, and the same gene makes slaves.

In Fig. 12 we show that gene III in one master has a greater probability of making slaves (represented as hearts) than genes V, II, I, and VI. In the homologous unit, gene I makes slaves (represented as flowers). Following fission a master unit may be produced which does not synthesize any slaves, and with dilution of the old slaves in subsequent fissions, sublines of cells appear which are pure for either hearts or flowers.

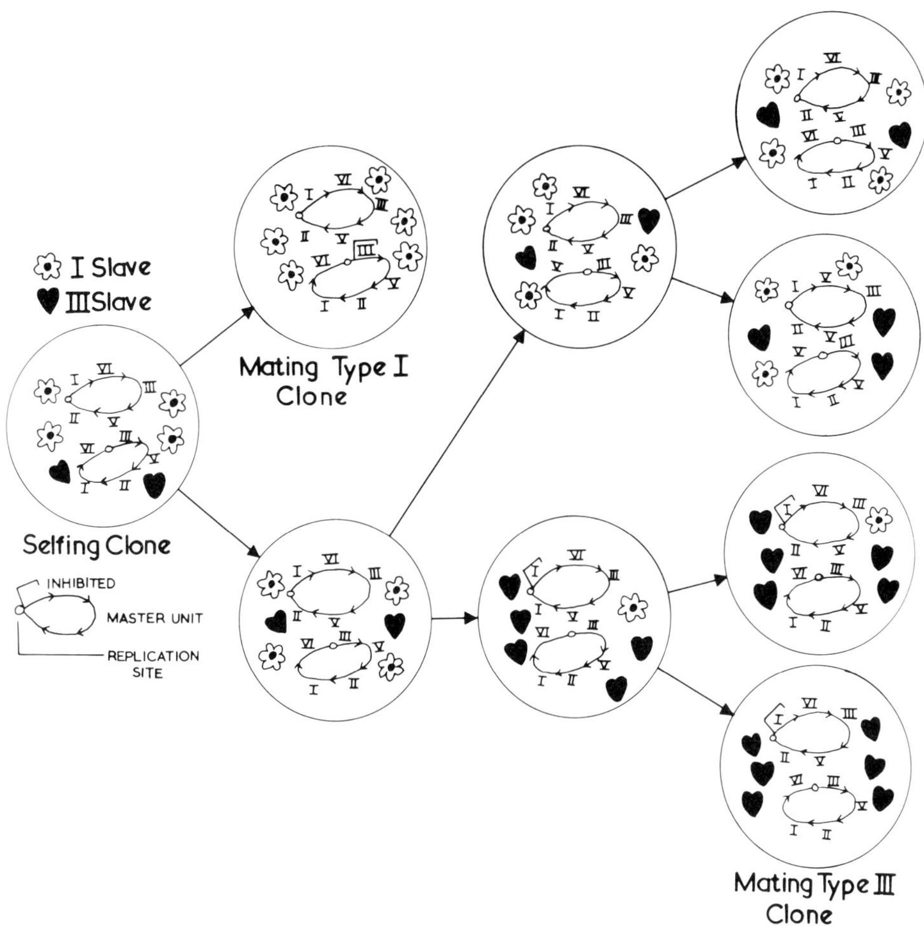

Fig. 12.

The formation of cell lines pure for one mating type in a homozygous clone. The replication site of a master unit is inhibited, and no slaves are formed. I, VI, III, V, II, Genes for these different mating types. III produces slaves, represented by hearts; I produces slaves, represented by flowers.

These lines would be mating type III or I. The relative proportions of these two types would depend on the timing of the inability to synthesize slaves. Figure 12 shows the appearance of a ditypic caryonide (referred to as a selfing clone) and clones pure for mating types I and III.

    e. Fit of Hypothesis to Data. The master-slave hypothesis intrigues us because most of the data can be accommodated, and it fits with observations made on the total amount of DNA in the macronucleus without having to invoke gene diminution. Determination involves differences in replication of homologous units, while assortment depends on the frequency of misreplication of master units and the dilution-out of slaves. The kinetics of assortment do not follow the Schensted formula but are more complicated since they include two events which are spread over many fissions. With more complex forms of this hypothesis (in which there are multiple masters and no slaves), the Schensted kinetics could be applied directly.

Within the context of the hypothesis presented above, certain aspects of phenotypic differentiation within clones and the macronucleus require special mention. First, nonrandomness of assortment has been reported for early fissions, and randomness for late fissions (31; L. K. Bleyman and E. M. Simon, personal communication). Possibly, extra copies of DNA are synthesized in the early fissions, and they may be nonrandomly distributed during cell division. This might then give rise to the observations on unequal ratios of cell types, which are characteristic of the early genes. Excessive amounts of DNA are found in the macronucleus of *P. aurelia* up to the third fission (173). Some of this DNA could of course be due to excessive synthesis of ribosomal genes, which in *Tetrahymena* is sensitive to growth conditions (168,169).

Second, temperature effects are known to influence the frequency of mating types actually produced by a genotype. This is known not to be due to effects on the assortment process but rather to occur during anlagen development. In such cases temperature might influence the position of the replication site on different master units, misreplication of master units, the number of slave copies made, or some combination of these effects. It is interesting that the period in which temperature exerts its effect is early, that nonrandom effects occur early, and that an early gene is affected. It might be predicted that temperature might have an effect on the ratio of types produced by *H* or *St*, also early genes.

Third, starvation effects have been recorded in the frequency of occurrence of pure mating types in selfing clones and stable *H* types in heterozygous clones. In starvation the frequency of selfing or of heterozygous types decreases. For the master-slave hypothesis, this might involve the destruction of slave copies present in the least numbers.

Fourth, there is the problem of explaining the occurrence of stable cell lines which form a hybrid phosphatase molecule. Such cells have a distinct phenotype unlike the heterozygote immediately following conjugation or like either of the parental phenotypes. They are as stable as cell lines with phenotypes similar to homozygotes but are much less frequent in occurrence. Such a cell line might arise as a result of the exchange of genetic material between master units, or the incorporation of both masters into a single replicon. These rare recombinant genomes might then produce the hybrid enzyme.

The master-slave units must be considered in terms of our knowledge of macronuclear structure. The master units are *separate* as far as their replication is concerned, but it is possible they are physically joined together at certain stages to ensure that each daughter nucleus receives a complete set of master units and an equal number of slaves at cell division. Such a mechanism may also be important in preventing the loss of both master units. Wolfe (170) has indeed seen structures that are joined together, but we do not know whether these structures are fragments of chromosomes, whole chromosomes, or more complicated structures. Our knowledge of the morphology of the macronucleus lags a long way behind genetic interpretations involving either subnuclei or master-slave units.

4. Distinction between Hypotheses

The assumptions of the subnuclear hypothesis and the master-slave hypothesis are compared in Table 19. The key difference between these hypotheses resides in the nature

Table 19.

Comparison of Subnuclear and Master-Slave Hypothesis

Diploid subnuclear hypothesis	Master-slave hypothesis
The unit is a diploid subnucleus.	There are two kinds of units derived from a chromosome fragment. One is a replicating unit (master), the other a nonreplicating unit (slave) derived from the master. For each kind of unit, there are two types representing each pair of alleles. Totally, there are four types: master A, master a, slave A, and slave a.
Each diploid subnucleus divides once each fission.	Each master divides once every fission; the slaves do not divide but are synthesized from the masters. New slaves are made every fission. Old slaves are passively diluted.
The total number of diploid subnuclei is 90 before division.	The total number of both types of slaves is 90 before division.
Assortment out of the total number of subnuclei involves equal numbers to both daughters. Each daughter receives 45 diploid subnuclei.	Assortment out of the total involves an equal number of slaves to both daughters, but the two types of slaves (A and a) may not necessarily be equal. Each daughter receives 45 slaves inclusive of the two types.
Determination involves the functioning of only one allele per diploid subnucleus. This gives rise to two types of differentiated subnuclei (A or a).	Determination involves the loss of one of the masters (A or a).
Each subnucleus is functionally active.	The slaves are functionally active.
There may or may not be equal numbers of the two types of differentiated subnuclei.	There are two masters per cell until determination occurs. The two types of slaves may or may not be in equal numbers because of their differing rates of synthesis.
By random assortment cells appear with differing numbers of the two types of differentiated subnuclei. Some cells appear that are pure for only one type and are stable (A or a).	By random assortment cells appear with differing numbers of slaves. Some cells appear which have only one type of master and one type of slave and are stable (A or a).
The rate of appearance of stable cell lines depends on the relative numbers of the two types of differentiated subnuclei.	The rate of appearance of a stable line A depends on the frequency of misreplication of master a plus the dilution out of the a slaves.
The ratio of the two stable cell lines (A versus a) depends on the relative number of the two types of differentiated subnuclei present at the time of determination.	The ratio of the two stable cell lines (A versus a) depends on the specificity of misreplication.
Gene-specific properties are explained by gene-specific differences in regulatory events.	Gene specific properties are explained by gene (and allele) specific differences in replication of master and numbers of slaves made by master.
DNA quantities are determined by the total number of diploid subnuclei.	DNA quantities are determined by the total number of masters and slaves.

of the units and determination. According to the subnuclear hypothesis, the macronuclear units contain the diploid complement of half the genes present in the micronucleus, and determination hinges upon the meaning of allelic repression and selection of mating types. This is usually thought to involve a transcriptional event, although E. Orias (personal communication) believes that somatic recombination has not been ruled out. This latter possibility would be less applicable, however, if the units were diploid replicons. In the case of the master-slave hypothesis, homologous units are separate replicons, derived from chromosome fragments, and determination involves misreplication of one of the master units. The Schensted formula is applicable to the subnuclear hypothesis, since the rate of appearance of stable lines depends upon the relative numbers of two types of differentiated subnuclei. This formula is not, however, directly applicable to the master-slave hypothesis, in which the rate of appearance of stable cell lines depends upon the frequency of misreplication of one of the masters over several fissions and the dilution-out of the slaves. The kinetics here would be more complicated, but we feel that the observations are not incompatible with a statistical summation of the two events, with the "time" of determination being spread over several fissions.

The subnuclear hypothesis differs from the master-slave hypothesis with regard to the organization, replication, and nucleotide sequences of the DNA present in the macronucleus, as discussed elsewhere (16). We would expect a genome size difference between the micronucleus and macronucleus according to the subnuclear hypothesis, and about 90 copies of the reduced genome in the G_1 macronucleus. We would expect a similar genome size in both types of nuclei according to the master-slave hypothesis, and 45 copies of the genome in the G_1 macronucleus. Renaturation experiments showed that in *normal* strains of *Tetrahymena* the size of the macronuclear genome was 50 times that of *E. coli* and similar to that of the micronucleus (16).

The mode of replication of the DNA molecules that make up the units should also differ for the two hypotheses. We would expect semiconservative replication of the DNA of the subnuclei. For the master-slave hypothesis, since the slaves make up a large percentage of the DNA and cannot replicate, we would expect that some of the DNA of the macronucleus would be conserved. Andersen and Zeuthen (28) report that all DNA previously labeled with $[H^3]$thymidine became hybrid in density after one full cycle of growth in the heavy-atom label 5-bromo-2-deoxyuridine in heat-shocked synchronized cells. However, Wille (166, and personal communication) found that a fairly large fraction of the DNA did not become hybrid in density by the end of one full cycle in heat-shocked synchronized cells. He also observed unusual patterns of DNA replication using other methods to induce synchrony.

Finally, the two hypotheses differ in what we might expect to find with regard to the DNA nucleotide sequences present in the differentiated macronucleus. For the master-slave hypothesis (and for other hypotheses with haploid units), we would expect qualitative differences in the DNA molecules among clones as a result of differences between macronuclei, and these differences would be correlated with the differentiated phenotypes. We would not expect to observe this correlation if the units are differentiated subnuclei unless somatic recombination were involved. For both diploid and haploid unit types of hypotheses, we would expect to observe differences in the nucleotide sequences present in the micronucleus and macronucleus.

V. PHENOTYPIC DIVERSITY BETWEEN SYNGENS

A. Introduction

The syngens of *T. pyriformis* share many properties and differ in others. They may vary in the details of their breeding systems, in temperature tolerance, in nutritional requirements, and in their geographic distribution (46). Other types of comparisons may give information on phenotypic diversity that is specifically intersyngenic and which may aid in assessing relationships between the syngens. Comparisons of macromolecules have been useful in probing relationships between other organisms. A modest beginning has been made for the syngens of *T. pyriformis*. Thus comparisons have been made of the nucleic acids and the electrophoretic patterns of certain enzymes. Another approach is morphological and makes capital of the basis on which *Tetrahymena* is classified: that is, on surface pattern. Here specific features of this pattern have been delineated and compared in different syngens. It is obvious that these different kinds of comparisons measure different parameters. They could provide different answers. Taken together, however, they may shed some light on the degree to which the syngens have diverged and on the mechanisms by which intersyngenic diversity has arisen.

B. Comparison of Nucleic Acids

A way of measuring genetic relationships among organisms is to look for similarities among them in overall base compositions or in nucleotide sequences. Many studies have been carried out in bacteria, viruses, and in higher animals and plants in terms of these properties in an attempt to understand how the genetic material has evolved. In theory, then, this approach should be particularly useful in exploring the syngens of *T. pyriformis* with regard to their genetic relationships.

When DNA is extracted from *Tetrahymena*, it can be examined in terms of its base composition by various techniques, for example, equilibrium density gradient centrifugation. In Table 20 we show the results from several laboratories on the base composition of nuclear and mitochondrial DNA from different syngens of *Tetrahymena*. These results indicate a variability in base composition of nuclear DNA which is more similar to that observed in bacteria than in higher animals. The percent of guanine plus cytosine of DNA of different syngens (1 through 12) varies between 24 and 33%. Syngens 1, 2, 3, 5, 6, 7, 8, and 12 have the lowest values (25–27%); syngen 9 varies between 28 and 30%; and syngens 4, 10, and 11 have the highest values (30–33%). In general, there is a smaller spread within a syngen than between syngens.

Mitochondrial DNA from different syngens may be similar in base composition, and in some syngens (4, 9) and strains (GL, ST, L-I, T, and W) may differ from nuclear DNA, according to Suyama (162) and Flavell and Jones (59,60). Differences in base composition between mitochondrial and nuclear DNA were not found, however, by Brunk and Hanawalt (35).

There is disagreement about the presence of DNA in kinetosomes (61,147,149). Also, there are conflicting reports on the presence of satellite DNAs. Satellite DNAs identified as mitochondrial in origin were observed by Suyama (162), Suyama and Preer (163),

Table 20.

Base Compositions of DNA from *T. pyriformis*

Source of DNA	Obtained by		Percent G + C	Ref.
	$T_m(^\circ C)$	CsCl		
Nuclear DNA:				
Syngen 1				
A strain	—	a	25	(161)
A strain	—	1.685	25	(59)
A-1464?	—	1.685	25	(162)
WH-52	—	a	25	(161)
IL-12	—	a	25	(161)
Strain 7	79.5[b]	—	25	(15)
Strain 8	80.0[b]	—	26	(15)
Strain 21	79.0[b]	—	25	[d]
D-20693	63.0[c]	1.683	24	(15,[d])
D/congenic-1	63.0[c]	1.683	24	(15,[d])
C*	63.0[c]	1.683	24	(15,[d])
Syngen 2				
2-1 (UM-3?)	—	a	25	(161)
Syngen 3				
3-1 (UM-700?)	—	a	27	(161)
Syngen 4				
UM-981 (4/III)	—	1.692	33	(162)
IN-3 (4/II)	65.5[c]	1.689	30	(15,[d])
Syngen 5				
5-1 (R 5/I?)	—	a	25	(161)
UM-30 (5/II)	—	1.687	27	(161)
Syngen 6				
UM-1060 (6/I)	—	1.685	25	(162)
HSM	—	1.685	25	(135)
6/III	—	1.685	25	(135)
Syngen 7				
7-1 (UM-1215?)	—	a	27	(161)
UM-1215 (7/I)	63.5[c]	—	25	(15)
UM-1216 (7/II)	63.5[c]	—	25	(15)
UC-651 (7/III)	80.0[b]	—	26	(15)
Syngen 8				
8-2	—	a	25	(161)
Syngen 9				
9-1 (TC-105?)	—	a	28	(161)
TC-?	—	1.690	30	(162)
TC-89 (9/V)	80.7[b]	—	28	[d]
Syngen 10				
EN-131 (10/II)	66.0[c]	—	31	(15)

Table 20 *(cont.)*

Source of DNA	Obtained by		Percent G + C	Ref.
	T_m(°C)	CsCl		
Syngen 11				
11/AU-50-1	65.5[c]	—	30	(15)
11/AU-94-10	65.5[c]	—	30	(15)
Syngen 12				
12/AU-F₁2	63.5[c]	—	25	(15)
12/AU-115-3	63.5[c]	—	25	(15)
Strain E	—	a	31	(161)
Strain GL	—	1.688	28	(162)
Strain GL	—	1.686	26	(59)
Strain GL-R	—	1.689	29	(59)
Strain GP	—	1.688	28	(59)
Strain L-I	—	1.687	27	(59)
Strain L-II	—	1.688	28	(59)
Strain ST	—	1.692	33	(162)
Strain T	—	1.688	28	(59)
Strain W	81.2[b]	1.690	30	(154)
	—	1.692	33	(59)
Unknown (Kitching)	81.5[d]	—	30	(62)
Mitochondrial DNA:				
Syngen 1				
A-1464?	—	1.685	25	(162)
A strain	—	1.685	25	(59)
Syngen 4				
UM-981	—	1.686	26	(162)
Syngen 5				
UM-30	—	1.686	26	(162)
Syngen 6				
UM-1060	—	1.685	25	(162)
HSM	—	1.671	—	(135)
6/III	—	1.671	—	(135)
Syngen 9				
TC-?	—	1.684	24	(162)
Strain GL	—	1.684	24	(162)
Strain GL	—	1.684	24	(59)
Strain ST	79.5[b]	1.686	26	(162)
Strain L-I	—	1.684	24	(59)
Strain T	—	1.685	25	(59)
Strain W	—	1.686	26	(59)

[a]Figure not quoted.　　　　　　　　　　[c]In 0.1 × SSC.

[b]In 1 × SSC.　　　　　　　　　　[d]S. L. Allen and I. Gibson, unpublished.

Table 21.

Hybridization between DNA Molecules of Different Syngens of *T. pyriformis*[a]

Source of DNA competitor (C) or DNA filter (F)	Percent related to syngen 1, strain 7	
	C	F
Syngen 1		
Strain 7	100	100
$\left[(7 \times 8) \times 7\right]_{10}$	92	88
Strain 8	59	—
Strain 17	48	—
Strain 21	48	—
Syngen 8		
UM-1286	—	45
Syngen 9		
TC-89	41	—
Syngen 10		
EN-131	36	38
Syngen 12		
AU-115-3	—	18
Paramecium aurelia (syngen 1)	21	20
Xenopus laevis	—	1

[a]In the competition experiments (C) agar gels containing DNA from strain 7, syngen 1, were incubated at 72°C with radioactive DNA fragments from the same source in the presence of competitor molecules from different sources. In the filter experiments (F) radioactive DNA fragments from strain 7, syngen 1, were incubated at 72°C with filters containing DNA from different sources.

and Flavell and Jones (59). Other satellite or "contaminant" molecules, nonmitochondrial in origin, were found in whole-cell DNA by these same researchers (59,163). In contrast, Brunk and Hanawalt (35) report the lack of satellites in *Tetrahymena* DNA and that mitochondrial DNA has the same buoyant density as nuclear DNA. At present, the studies on base composition conflict as to whether or not there is heterogeneity within the DNA molecules of a cell. They do tell us that there is variation among the DNAs of different syngens, but they cannot tell us how these changes in the DNA molecules come about.

A beginning has been made using another technique for probing relationships between nucleic acid molecules. This is the technique of nucleic acid hybridization, and it is interpreted as giving a quantitative measure of the genetic relatedness of different organisms. Details of the technique are given elsewhere (34,76,88). This technique has been applied to the nucleic acids from different syngens of *Tetrahymena* (13; S. L. Allen and I. Gibson, unpublished). Some results with DNA molecules are shown in Table 21, and with rRNA molecules in Table 22. In the latter the two components (large and small) that comprise the rRNA molecules of eucaryotes and procaryotes were separated out on sucrose gradients, and each hybridized in separate experiments. In addition to the experiments shown in Tables 21 and 22, other experiments were carried out on hydroxyapatite columns (16). In these experiments two strains of syngen 1 had 53%

Table 22.
Hybridization between rRNA Molecules of Different Syngens of *T. pyriformis*[a]

Source of DNA on filter (or RNA competitor[b])	Percent related to syngen 1, strain 7	
	Large component	Small component
Syngen 1		
Strain 7	100	100
Strain 17	98[b]	96[b]
Strain 21	99[b]	98[b]
Syngen 3		
UM-700	74	74
Syngen 7		
R-II	83	85
Syngen 8		
UM-1286	92	90
Syngen 9		
TC-89	84	84
Syngen 10		
EN-131	81	87

[a]Radioactive large and small ribosomal components from syngen 1, strain 7, were isolated on sucrose gradients. Each was then separately incubated at 72°C with filters containing DNA from different sources.

[b]Each of these was incubated at 69°C with agar gels containing DNA from strain 7 in the presence of unlabeled ribosomal components from other strains of syngen 1.

of their sequences in common, and 48% of these were hybrids between molecules that had a low thermal stability. 4% homology was shown by strains from different syngens (1 and 7).

There are two interesting features of these results. The first is that the magnitude of the differences among DNA molecules is greater than that among rRNA molecules. This observation has also been made in other organisms and has been interpreted as conservation in the portion of the genome that specifies rRNA (92). Thus differences were observed in the DNA molecules of different strains within syngen 1, but no significant differences could be detected between rRNA molecules. The second feature is the magnitude of the differences among DNA molecules. Large differences were found among the syngens in terms of their DNA molecules. Smaller differences were found among strains within a syngen; yet they could be differentiated. Moreover, we even detected small, but repeatable, differences in the DNA molecules of different clones within a strain. At present, these data must be interpreted with caution in assessing syngenic differences, since we also observed differences within a syngen. However, if they are meaningful, then the syngens could be placed in the following order relative to syngen 1: syngen 8, 9, 7, 10, 3, and 12.

The results on the formation of hybrid nucleic acid molecules must be taken as tentative in view of two considerations: (1) the genome of most eucaryotic organisms seems to contain several classes of DNA in terms of their renaturation rates, and (2) the hybrid reaction depends on the conditions under which it is carried out.

Studies on the DNA of several strains of *Tetrahymena* have been made in terms of their renaturation rates. It has been shown in fact that 15–30% of the DNA behaves as repeated sequences, and the other 70–85% as unique sequences, under the conditions used for growth of the cells (16). The differences among the DNA molecules described above may involve both or either fraction of DNA. Preliminary observations already reveal little homology between the unique sequences of different syngens (S. L. Allen and C.-I. Li, unpublished). Further fractionation of DNA and characterization of hybrids in clones, strains, and syngens will allow us to assess the degree of molecular divergence between different syngens. One curious observation, however, should be mentioned with regard to comparing the degree of diversity observed in the syngens of *T. pyriformis* with that in *P. aurelia*. In *Paramecium* the similarities in DNA molecules are higher when comparisons are made among syngens than they are in *Tetrahymena* (16).

C. Comparison of Electrophoretic Patterns of Enzymes

The electrophoretic techniques discussed in Section III, D, 3 have been applied to the analysis of the propionyl esterases, butyryl esterases, and acid phosphatases obtained from all 12 syngens. Our reference group was syngen 1, and our approach was to compare the electrophoretically separated forms of an enzyme of the other syngens to those we had some knowledge of in our reference group—in electrophoretic mobility, enzymic properties, and stability. In all these studies we used the same electrophoretic conditions that were found optimal for syngen 1.

Certain homologous esterases seem to be present in several syngens (19). Intrasyngenic variation is observed in some syngens. There is acid phosphatase activity in all the syngens, and considerable polymorphism in the observed patterns. Some phosphatases seem to have mobility similar to ones observed in syngen 1. When we tested for PCMB sensitivity, however, we found no phosphatases that behaved similarly to phosphatase-1 in syngen 1. Thus all we can point to at present is the existence of considerable variation, some of which may be intersyngenic and some of which is certainly intrasyngenic.

These investigations have been extended (D. Borden, unpublished), and several other enzymes have been shown to exhibit intersyngenic variation (e.g., glutamic dehydrogenase, and so on).

The outstanding feature of these studies has been diversity in electrophoretic pattern within and among the syngens of *T. pyriformis*. This observation stands in contrast to observations made on the esterases (14,22), acid phosphatases (S. L. Allen, P. H. T. Lee, and F. J. Malinoff, unpublished), and other enzymes (164a) in the 14 syngens of *P. aurelia*. In *Paramecium* the enzyme patterns vary, but they are considerably less diverse both among the syngens as well as within syngens (with some exceptions) than in *Tetrahymena*.

D. Comparisons of Corticotypes

The concept of the corticotype was introduced in Section III, D, 4. In a survey of strains in different syngens, Nanney (108) found that the number of ciliary rows, or

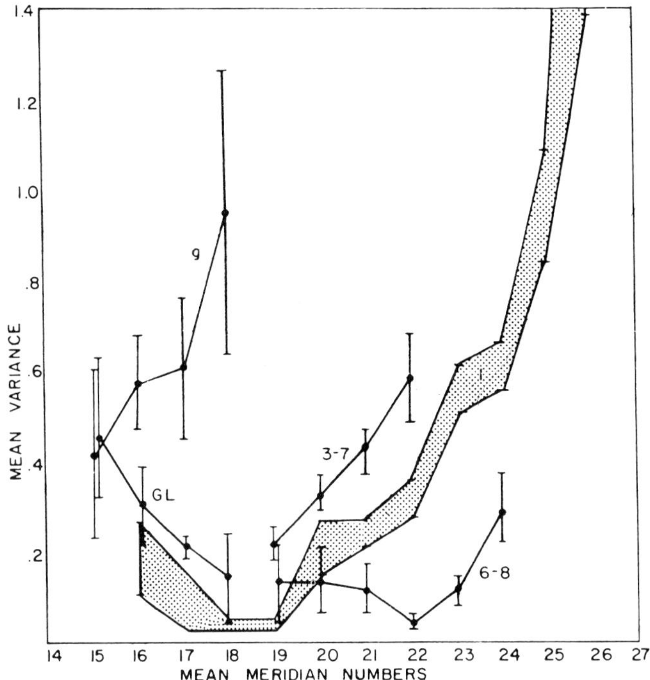

Fig. 13.

Mean variances ±SE of 20-fission clones of various strains of *T. pyriformis* plotted against mean meridian numbers of the clones. The syngen-1 pattern is shaded. [From Nanney (113).]

meridians, was correlated with other surface features, including the position and number of contractile vacuole pores, the number of meridians with contractile vacuole pores, the number of postoral meridians, and the stability of the cell type.

Of all these features only two were found to be reliable criteria of intersyngenic differences. The first was the position of the contractile vacuole pores. Within a syngen the position of these pores moves relative to the ciliary rows as a function of the total number of ciliary rows, but in different syngens the mean position of the two contractile vacuole pores shifts when different numbers of meridians are present. Nanney (110,118) found that the syngens could be arranged in a continuous series in the order 1, 3, 7, 6, 8, 11, 12, 9, 2, 5, GL, 4, and 10.

The second feature by which syngens could be distinguished was their stability pattern (113). By taking a large number of clones, the stability pattern of a strain can be delineated in terms of the rate at which cells modulate to a stable corticotype. Mathematically, it is expressed in terms of the variance of meridian number compared to the mean number of meridians. When plotted, different curves are obtained for different groups of syngens (Fig. 13). By this method syngens 3 and 7 cannot be separated, nor can 6 and 8, but the two groups can be separated from each other as well as from syngen 1 and from syngen 9 and GL.

E. Conclusions

What can be concluded from these different types of comparisons? Certainly, all comparisons indicate that phenotypic diversity occurs at the syngenic level, but we really do not know the extent of this diversity. Different degrees of diversity and different relationships between syngens apparently are being assessed by these different methods. Thus an ordering of syngens with respect to one another was obtained for the corticotypes (1, 3, 7, 6, 8, 11, 12, 9, 2, 5, 4, and 10) different from that based on nucleotide sequences (1, 8, 9, 7, 10, 3, and 12). It is possible that mechanisms that operate to regulate overall surface architecture may not work in the same evolutionary direction as those that control the supply of macromolecules.

Comparisons of base composition and nucleotide sequences reveal wide differences among syngens. This is also true for the electrophoretic patterns of certain enzymes. It has been concluded by computer analysis that all the characters so far studied show that the extent of the differences in syngens are similar to those of genera (D. Borden, personal communication). Moreover, the degree of difference seems to be greater for the syngens of *T. pyriformis* than for *P. aurelia*. If mutations, chromosomal aberrations, and so on, are involved in creating these differences, it would suggest that the syngens of *T. pyriformis* have an ancient evolutionary history and one that is more ancient than that of *P. aurelia*. It is also clear, however, that intrasyngenic variation is extensive and that clonal differences may also contribute to the observed variations. Therefore it is possible that the extent of the syngenic differences in *T. pyriformis* may be masked by effects of "macronuclear noise," by the organization of the DNA (repeated sequences, and so on), or perhaps it may also be technical. Until these contributions are sorted out, and the basis for phenotypic diversity known, the extent of phenotypic diversity between syngens must remain an open question. It is possible that they may not be as diverse as they seem.

VI. EVOLUTION IN *TETRAHYMENA*

All strains of *T. pyriformis* are morphologically similar, yet it has been possible to group them into syngens according to their mating relationships (157). The syngen could be comparable to a genus or even to some higher taxonomic level. This problem has recently been discussed by Laird and McCarthy (85), who point out the difficulties of equating taxon rank when comparing different groups of organisms. In general, animal groups more closely related to man are more finely divided than remote groups. We do not really know how diverse the different groups are under the taxonomic umbrella *T. pyriformis*. Comparisons of macromolecules suggest wide differences among syngens, but these differences may be overinflated by macronuclear effects which result in phenotypic differences within clones. At a minimum the taxon *T. pyriformis* is probably equivalent to a genus in vertebrates, but it could be comparable to an even higher order of taxonomic grouping, if the differences between syngens are greater.

How has phenotypic diversity been brought about during the evolution of *Tetrahymena*? In other organisms the genetic mechanisms that create diversity are (1) gene mutation,

(2) recombination of the genetic material, (3) chromosomal changes, and (4) regulation of gene action bringing about the appearance of different phenotypes from the same genetic information. Do all these mechanisms operate to create intersyngenic differences in *Tetrahymena?* From genetic analyses carried out between strains, we know that single gene effects bring about phenotypic differences and that recombination takes place between unlinked and linked genes. As for chromosomal changes, all syngens have the same number of chromosomes. Nevertheless, there may be duplications, inversions, or other chromosomal rearrangements hidden within this overall similarity in number. We also know that genes appear to be regulated in response to environmental stimuli. This is best illustrated by the serotypes, in which each locus is expressed under a different set of conditions. It does seem clear then that some of, if not all, the genetic mechanisms found in other organisms apply to *Tetrahymena*. We do not know their relative importance in generating the diversity between syngens, however.

A complication that has emerged from studies on the genetics of *Tetrahymena* is that within a syngen not only are there strain differences but even within a clone there is phenotypic diversity. We do not know the relative importance of the mechanisms that create clonal diversity compared to the phenotypic diversification produced by well-known genetic processes such as mutation. We do know, however, that they must be taken into consideration and understood if we are to appreciate their effects in the evolution of this group.

Clonal diversity in phenotype is well documented for various characters. Some involve nuclear effects, and others do not. Some of the changes are unstable and can easily be reversed, while others are stable. Some of the stable changes are irreversible, notably the phenotypic changes that occur within heterozygous and homozygous clones of syngens 1 and 7. The phenomenon of phenotypic differentiation affects all the heterozygotes that have been examined in any detail, and mating types in homozygous clones. The detailed properties of this phenomenon have been described, and hypotheses offered to explain these observations. These hypotheses involve units within the macronucleus. One assumes diploid subnuclei which become differentiated. These assort out, and lines of cells appear with different phenotypes. Another hypothesis assumes haploid units smaller than chromosomes and differences in replication of these units to produce lines that differ in their content of macronuclear genes. Regardless of its molecular basis, we know that the macronucleus is clearly involved in these differences. We know that it appears to be an important source of phenotypic variation in at least two of the syngens of *Tetrahymena,* but we do not know if it is a general feature of the species complex. For the known genes in a heterozygote of syngen 1, it has been calculated that about 5376 different combinations of stable phenotypes may occur in different lineages. Where it does occur, it could be a novel way of generating cell lines having phenotypes similar to homozygotes in an organism that does not indulge in autogamy.

Clonal diversity may also be brought about by genetic systems that do not involve the nucleus, as illustrated by the corticotypes. Although our knowledge of them is scanty, these genetic systems may also play an important role in the evolution of syngens.

With the exception of syngens 6 and 8, the different syngens do not interbreed and therefore constitute distinct gene pools. Because they are separated from one another, different regulatory mechanisms may have also evolved. Does phenotypic differentiation occur in clones of syngens other than 1 and 7? Are the properties of the phenomenon described in these two syngens applicable to the other syngens? We can also ask if amicronucleate cells occur in all syngens. This is implied by the fact that isolates containing a high proportion of amicronucleate cells have been obtained from almost all geographic areas. How and why do amicronucleate cells arise? Does genomic exclusion occur and are its genetic consequences similar to those in syngen 1? Differences in these features could occur in various syngens. Certainly, differences are indicated between various groups of ciliates. For example, heterozygotes in *P. aurelia* do not seem to behave in the same way as those in *T. pyriformis*. Instead, this ciliate relies on the process of autogamy to generate homozygous genotypes.

In the subsequent discussion we assume, however, that syngen 1 is not atypical of the species complex. A common feature shared by all the syngens, which must have a bearing on the genetic systems and their evolution, is that all sexually competent cells have two nuclei. The micronucleus transmits mutations and recombinant genomes through sexual reproduction. The macronucleus, which is derived from a micronuclear product and is discarded during the next sexual cycle, is the active nucleus during asexual reproduction and determines the phenotypic characters. Events within the macronucleus cause phenotypic diversity within clones during vegetative divisions. This system thus involves two nuclei, one that serves as a germ line and another that serves as a somatic line and is free to change since it is ultimately destroyed. How does this system operate in the evolution of the syngens?

We suggest that mutations occur in the micronucleus and that these are transmitted from one sexual generation to the next. The micronucleus is largely inert and does not affect the phenotype. However, the macronucleus is active in this regard. Micronuclear mutations could be amplified through the macronucleus which permits expression of the mutations in at least some cell lines. The mechanism in the macronucleus involved in the creation of clonal diversity might therefore have been selected in order to bring to expression at a greater rate mutations that occur in the diploid micronucleus. It would also have a buffering effect, since some lines would have nonmutant phenotypes although mutant in their micronuclear genotype.

It is expected then that the micronucleus would carry a large number of mutations and that a large degree of heterozygosity would occur in the micronuclear genomes of these ciliates. A large degree of heterozygosity is not unusual in other diploid organisms. However, they usually have means for regulating the expression of their genetic reservoir to dampen the deleterious effects of their genetic load. In *Tetrahymena* the macronucleus also serves to buffer the effects of mutation. There may also be other ways in which this problem is met, however.

The production of amicronucleate cells might be one way of resolving this problem. Such cells commonly occur in nature and probably in all syngens. These cells are frozen, and any further phenotypic changes come through the macronucleus. Heterozygotes would be resolved into stable lines. Any further changes would then depend on macronuclear mutations. Such mutations may be rare, although how rare, we do not know. Another way of resolving heterozygosity is by means of genomic exclusion. During crosses of laboratory strains that are losing their micronucleus with strains with normal micronuclei, the defective strains receive a normal micronucleus which gives rise to a homozygous genotype. Thus genomic exclusion may be another way of generating stable phenotypes.

We suggest that it is no accident that *Tetrahymena* has two nuclei, and that this system has been selected during the course of evolution as a means for amplifying and diversifying gene expression. Mutations can be accrued in the micronucleus, but they are not expressed. The macronucleus brings the mutations to expression, and lines that are well adapted are selected. Within a single macronuclear lineage many different phenotypes can appear. Depending upon the environment, different phenotypes can be selected out. The resulting polymorphism must surely have been highly advantageous to an organism faced with adapting to changing environments and would have facilitated its rapid evolution.

In the light of these considerations, it becomes crucial to understand the mechanism that generates phenotypic diversity within the macronucleus and the process by which the micronuclear DNA is rendered inactive. By studies that will lead to an understanding of the steps involved, we can then hope to find out if the scheme we have proposed has any reality for the evolution of *T. pyriformis*.

ACKNOWLEDGMENTS

We would like to thank Alan Cavill, Chris Cullis, Norman Martin, and Mary Watson of the University of East Anglia, and Sally Farrow, Patricia Lee, Frances Malinoff, and Wanda Newcomb of the University of Michigan, for their assistance in preparing this manuscript. Our researches are supported by grants from the U.S. Public Health Service (GM-15879) and the British Medical Research Council.

REFERENCES

1. Allen, S. L. 1960. Genetics 45:1051.
2. Allen, S. L. 1961. Ann. N. Y. Acad. Sci. 94:753.
3. Allen, S. L. 1962. Amer. Zool. 2:502.
4. Allen, S. L. 1963. J. Protozool. 10:413.
5. Allen, S. L. 1964. J. Exp. Zool. 155:349.
6. Allen, S. L. 1964. Genetics 49:617.
7. Allen, S. L. 1965. Brookhaven Symp. Quant. Biol. 18:27.
8. Allen, S. L. 1967. Science 155:575.
9. Allen, S. L. 1967. Genetics 55:797.
10. Allen, S. L. 1967. In M. Florkin and B. Scheer, eds., Chemical zoology, vol. I, Protozoa (G. W. Kidder, ed.), Academic Press, New York, p. 617.
11. Allen, S. L. 1968. Ann. N. Y. Acad. Sci. 151:190.
12. Allen, S. L. 1971. Genetics 68:415.
13. Allen, S. L. and Gibson, I. 1967. Science 158:523. (Abstr.)
14. Allen, S. L. and Gibson, I. 1971. Biochem. Genet. 5:161.
15. Allen, S. L. and Gibson, I. 1971. J. Protozool. 18:518.
16. Allen, S. L. and Gibson, I. 1972. Biochem. Genet. 6:293.
17. Allen, S. L. and Lee, P. H. T. 1971. J. Protozool. 18:214.
18. Allen, S. L. and Nanney, D. L. 1958. Amer. Nat. 92:139.
19. Allen, S. L. and Weremiuk, S. L. 1971. Biochem. Genet. 5:119.
20. Allen, S. L. and Weremiuk, S. L. 1971. J. Protozool. 18:509.
21. Allen, S. L., Allen, J. M., and Licht, B. M. 1965. J. Histochem. Cytochem. 13:434.
22. Allen, S. L., Byrne, B. C., and Cronkite, D. L. 1971. Biochem. Genet. 5:135.
23. Allen, S. L., File, S. K., and Koch, S. L. 1967. Genetics 55:823.
24. Allen, S. L., Misch, M. S., and Morrison, B. M. 1963. Genetics 48:1635.

25. Allen, S. L., Misch, M. S., and Morrison, B. M. 1963. J. Histochem. Cytochem. 11:706.
26. Allen, S. L., Weremiuk, S. L., and Patrick, C. A. 1971. J. Protozool. 18:515.
27. Ammermann, D. 1971. Chromosoma 33:209.
28. Andersen, H. A. and Zeuthen, E., 1971. Exp. Cell Res. 68:309.
29. Beale, G. H. 1954. The Genetics of *Paramecium aurelia,* Cambridge, University Press, London, 179 pp.
30. Bleyman, L. K. 1971. In I. L. Cameron, G. M. Padilla, and A. Zimmerman, eds., Developmental aspects of the cell cycle, Academic Press, New York, p. 67.
31. Bleyman, L. K. and Simon, E. M. 1968. Develop. Biol. 18:217.
32. Bleyman, L. K., Simon, E. M., and Brosi, R. 1966. Genetics 54:277;
33. Borden, D., Miller, E. T., Nanney, D. L., and Whitt, G. S. 1971. Isozyme Bull. 5:29.
34. Britten, R. J. and Kohne, D. E. 1968. Science 161:529.
35. Brunk, C. F. and Hanawalt, P. C. 1969. Exp. Cell Res. 54:143.
36. Bruns, P. J. 1971. Exp. Cell Res. 65:445.
37. Byrd, J. R. 1959. J. Protozool. 6(suppl.): 14. (Abstr.)
38. Byrd, J. R. 1959. Ph.D. Thesis, University of Michigan.
39. Carlson, P. S. 1971. Genetics 69:261.
40. Cleffmann, G. 1967. Exp. Cell Res. 50:193.
41. Corliss, J. O. 1952. C. Re. Acad. Sci., Paris, 235:339.
42. Corliss, J. O. 1953. Stain Technol. 28:97.
43. Dryl, S. 1959. J. Protozool. 6(suppl.):25.(Abstr.)
44. Ducoff, H. S. 1956. J. Protozool. 3(suppl.):3. (Abstr.)
45. Elliott, A. M. 1963. In L. Levine, ed., The cell in mitosis, Academic Press, New York, p. 107.
46. Elliott, A. M. 1972. In A. M. Elliott, ed., Biology of *Tetrahymena,* Dowden, Hutchinson and Ross, Stroudsburg, Pennsylvania, p. 0000.
47. Elliott, A. M. and Byrd, J. R. 1959. J. Protozool. 6(suppl.):18. (Abstr.)
48. Elliott, A. M. and Clark, G. M. 1958. J. Protozool. 5:235.
49. Elliott, A. M. and Clark, G. M. 1958. J. Protozool. 5:240.
50. Elliott, A. M. and Gruchy, D. F. 1952. Biol. Bull. 103:301.
51. Elliott, A. M. and Hayes, R. E. 1953. Biol. Bull. 105:269.
52. Elliott, A. M. and Hayes, R. E. 1954. J. Protozool. 1(suppl.):2. (Abstr.)
53. Elliott, A. M. and Kennedy, J. R. 1962. Trans. Amer. Microsc. Soc. 81:300.
54. Elliott, A. M. and Nanney, D. L. 1952. Science 116:33.
55. Elliott, A. M. Addison, M. A., and Carey, S. E. 1962. J. Protozool. 9:135.
56. Elliott, A. M., Hogg, J. F., Slater, J. V., and Wu, C. 1952. Biol. Bull. 103:301.
57. Elliott, A. M., Studier, M. A., and Work, J. A. 1964. J. Protozool. 11:370.
58. Flavell, R. A. and Jones, I. G. 1970. Biochem. J. 116:155.
59. Flavell, R. A. and Jones, I. G. 1970. Biochem. J. 116:811.
60. Flavell, R. A. and Jones, I. G. 1971. FEBS Lett. 14:354.
61. Flavell, R. A. and Jones, I. G. 1971. J. Cell Sci. 9:719.
62. Gibson, I. 1966. J. Protozool. 13:650.
63. Gibson, I. 1970. Adv. Morphog. 8:159.
64. Gibson, I. and Martin, N. 1972. Chromosoma 35:374.
65. Gorovsky, M. A. 1965. J. Cell Biol. 27:37A.
66. Gorovsky, M. A. 1970. J. Cell Biol. 47:631.
67. Gorovsky, M. A. and Woodard, J. 1968. J. Cell Biol. 39:54a.
68. Gorovsky, M. A. and Woodard, J. 1968. J. Cell Biol. 39:54a.
69. Gorovsky, M. A. and Woodard, J. 1969. J. Cell Biol. 42:673.
70. Grass, F. S. 1972. J. Protozool. 19:505.
71. Grass, F. S. 1972. Genetics. 70:521.
72. Grell, K. G. 1964. In J. Brachet and A. E. Mirsky, eds., The cell, vol. 6, Academic Press, New York, p. 1.
73. Gruchy, D. F. 1955. J. Protozool. 2:178.
74. Heckmann, K. and Frankel, J. 1968. J. Exp. Zool. 168:11.

75. Hiwatashi, K. 1968. Genetics 58:373.
76. Hoyer, B. H. and Roberts, R. B. 1967. In H. J. Taylor, ed., Molecular genetics vol. II, Academic Press, New York, p. 425.
77. Hurst, D. D. 1957. J. Protozool. 4(suppl.):18. (Abstr.)
78. Hurst, D. D. 1957. J. Protozool. 4(suppl.):17. (Abstr.)
79. Hurst, D. D. 1958. Ph.D. Thesis, University of Michigan.
80. Inoki, S. and Matsushiro, A. 1958. Med. J. Osaka University 8:763.
81. Juergensmeyer, E. B. 1967. Ph.D. Thesis, University of Illinois.
82. Juergensmeyer, E. B. 1969. J. Protozool. 16:344.
83. Kimball, R. F., 1964. In S. H. Hutner, ed., Biochemistry and physiology of protozoa, vol. III, Academic Press, New York, p. 243.
84. Kloetzel, J. A. 1970. J. Cell Biol. 47:395.
85. Laird, C. D. and McCarthy, B. J. 1968. Genetics 60:303.
86. Loefer, J. B. and Owen, R. D. 1961. J. Protozool. 8:387.
87. Loefer, J. B., Owen, R. D., and Christensen, E. 1958. J. Protozool. 5:209.
88. McCarthy, B. J. 1967. Bacteriol. Rev. 31:215.
89. McDonald, B. B. 1966. J. Protozool. 13, 277.
90. McDonald, B. B. 1972. In A. M. Elliott, ed., Biology of *Tetrahymena*, Dowden, Hutchinson and Ross, Stroudsburg, Pennsylvania, p. 00.
91. Margolin, P., Loefer, J. B., and Owen, R. D. 1959. J. Protozool. 6:207.
92. Moore, R. L. and McCarthy, B. J. 1968. Biochem. Genet. 2:75.
93. Murti, K. G. and Prescott, D. M. 1970. J. Cell Biol. 47:460.
94. Nanney, D. L. 1953. Biol. Bull. 105, 133.
95. Nanney, D. L. 1956. Amer. Nat. 90:291.
96. Nanney, D. L. 1957. Genetics 42:137.
97. Nanney, D. L. 1959. J. Protozool. 6:171.
98. Nanney, D. L. 1959. Genetics 44:1173.
99. Nanney, D L. 1960. Physiol. Zool. 33:146.
100. Nanney, D. L. 1960. Genetics 45:1351.
101. Nanney, D. L. 1962. J. Protozool. 9:485.
102. Nanney, D. L. 1963. J. Protozool. 10:152.
103. Nanney, D. L. 1963. In R. J. C. Harris, ed., Biological organization at cellular and supercellular levels, Academic Press, New York, p. 91.
104. Nanney, D. L. 1963. Genetics 48:737.
105. Nanney, D. L. 1964. In M. Locke, ed., The role of chromosomes in development, Academic Press, New York, p. 253.
106. Nanney, D. L. 1966. Amer. Nat. 100:303.
107. Nanney, D. L. 1966. J. Exp. Zool. 161:307.
108. Nanney, D. L. 1966. J. Protozool. 13:483.
109. Nanney, D. L. 1966. Genetics 54:955.
110. Nanney, D. L. 1967. J. Protozool. 14:553.
111. Nanney, D. L. 1967. J. Exp. Zool. 166:163.
112. Nanney, D. L. 1968. Science 160:496.
113. Nanney, D. L. 1968. J. Protozool. 15:109.
114. Nanney, D. L. 1968. Ann. Rev. Genet. 2:121.
115. Nanney, D. L. 1970. J. Exp. Zool. 175:383.
116. Nanney, D. L. 1971. J. Exp. Zool. 178:177.
117. Nanney, D. L. 1971. Develop. Biol. 26:296.
118. Nanney, D. L. 1971. J. Protozool. 18:33.
119. Nanney, D. L. and Allen, S. L. 1959. Physiol. Zool. 32:221.
120. Nanney, D. L. and Caughey, P. A. 1953. Proc. Natl. Acad. Sci. U.S. 39:1057.
121. Nanney, D. L. and Caughey, P. A. 1955. Genetics 40:388.
122. Nanney, D. L., Caughey, P. A., and Tefankjian, A. 1955. Genetics 40:668.
123. Nanney, D. L. and Dubert, J. M. 1960. Genetics 45:1335.

124. Nanney, D. L. and Rudzinska, M. A. 1961. In J. Brachet and A. E. Mirsky, eds., The cell, vol. 4, Academic Press, New York, p. 109.
125. Nanney, D. L., Nagel, M. J., and Touchberry, R. W. 1964. J. Exp. Zool. 155:25.
126. Nanney, D. L., Reeve, S. J., Nagel, J., and De Pinto, S. 1963. Genetics 48:803.
127. Nilsson, J. R. 1970. J. Protozool. 17:539.
128. Orias, E. 1959. Ph.D. Thesis, University of Michigan.
129. Orias, E. 1959. J. Protozool. 6(suppl.):19. (Abstr.)
130. Orias, E. 1960. J. Protozool. 7:64.
131. Orias, E. 1963. Genetics 48:1509.
132. Orias, E. and Frank, M. 1972. J. Protozool. In press.
133. Orias, E. and Rohlf, F. J. 1964. Evolution 78:620.
134. Outka, D. E. 1961. J. Protozool. 8:179.
135. Parsons, J. A. and Dickson, R. C. 1965. J. Cell Biol. 27:77A.
136. Phillips, R. B. 1967. Genetics 56:667.
137. Phillips, R. B. 1967. Genetics 56:683.
138, Phillips, R. B. 1968. Genetics 60:211.
139. Phillips, R. B. 1968. Genet. Res. Camb. 11:211.
140. Phillips, R. B. 1969. Genetics 63:349.
141. Phillips, R. B. 1971. J. Protozool. 18:163.
142. Phillips, R. B. 1971. Genetics 67:391.
143. Phillips, R. B. 1972. Develop. Biol. 29:65.
144. Preer, J. R. 1959. J. Immunol. 83:385.
145. Preer, J. R. 1968. In T. T. Chen, ed., Research in protozoology, vol. 3, Pergamon Press, New York, p. 129.
146. Prescott, D. M., Bostock, C. J., Murti, K. G., Lauth, M. R., and Gamow, E. 1971. Chromosoma 34:355.
147. Pyne, M. C. 1968. C. R. Acad. Sci., Paris 267:755.
148. Raikov, I. B. 1969. In T. T. Chen, ed., Research in protozoology, vol. 3, Pergamon Press, New York, p. 1.
149. Randall, J. and Disbrey, C. 1965. Proc. Roy. Soc. Ser. B. 162:437.
150. Rao, M. V. N. and Ammermann, D. 1970. Chromosoma 29:246.
151. Ray, C., Jr. 1956. J. Protozool. 3:88.
152. Sager, R. and Granick, S. 1954. J. Gen. Physiol. 37:729.
153. Schensted, I. 1958. Amer. Nat. 91:161.
154. Schildkraut, C. L., Mandel, M., Levisohn, S., Smith-Sonneborn, J., and Marmur, J. 1962. Nature 196:795.
155. Simon, E. M. and Hwang, S. 1967. Science 155:694.
156. Sooneborn, T. M. 1947. Adv. Genet. 1:263.
157. Sonneborn, T. M. 1957. In E. Mayr, ed., The species problem, AAAS Symposium, Washington, D.C., p. 155.
158. Sonneborn, T. M. 1963. In J. M. Allen, ed., The nature of biological diversity, McGraw Hill, New York, p.165.
159. Sonneborn, T. M. 1967. In R. A. Brink, ed., Heritage from Mendel, University of Wisconsin Press, Madison, Wisconsin, p.375.
160. Steers, E. 1965. Biochemistry, 4:1896.
161. Sueoka, N. 1961. Cold Spring Harbor Symp. Quant. Biol. 26:35.
162. Suyama, Y. 1966. Biochemistry, 5:2214.
163. Suyama, Y. and Preer, J. R. Jr., 1965. Genetics 52:1051.
164. Swift, H., Adams, B., and Larsen, K. 1964. J. Roy. Microsc. Soc. 83:161.
164a. Tait, A. 1970. Biochem. Genet. 4:461.
165. Wells, C. 1958. Ass. Southeast. Biol. Bull. 5:17.
166. Wille, J. J., Jr. 1966. Argonne Natl. Lab. 7278:203.
167. Wille, J. J., Jr. 1972. Biochem. Biophys. Res. Commun. 46:677.
168. Wille, J. J., Jr. 1972. Biochem. Biophys. Res. Commun. 46:692.

169. Wille, J. J., Jr., Barnett, A., and Ehret, C. F. 1972. Biochem. Biophys. Res. Commun. 46:685.
170. Wolfe, J. 1967. Chromosoma 23:59.
171. Woodard, J., Gorovsky, M., and Kaneshiro, E. 1968. J. Cell Biol. 39:182a.
172. Woodard, J., Kaneshiro, E., and Gorovsky, M. A. 1972. Genetics 70:251.
173. Woodard, J., Woodard, M., Gelber, B., and Swift, H. 1966. Exp. Cell Res. 41:55.

CHAPTER 12

Cortical Development in *Tetrahymena*

Joseph Frankel and Norman E. Williams

Department of Zoology
University of Iowa
Iowa City, Iowa

I. INTRODUCTION

The ciliate cortex presents opportunities for the study of significant developmental problems at the cellular and subcellular levels. As an expansive region studded with organelles, it poses questions regarding the mechanisms of organelle formation and regulation. Questions may be asked experimentally about the regulated synthesis and assembly of the molecular building blocks that make up the cortical organelles. Moreover, unknown integrative mechanisms are at work which determine developmental events (both neo-formations and regressions) in specific regions of the cell cortex, and which maintain the species-specific pattern of the cortex as a whole. The problem of organelle integration can probably be approached more easily in ciliates than in other types of eucaryotic cells. Recent attention has focused on hereditary properties now demonstrated to reside in the cortex of ciliates (5). Studies of cortical inheritance in ciliates have clearly established that preformed structures in the cortex play an essential role in determining the organization of new structures, a process for which Sonneborn has suggested the term "cytotaxis" (96). In addition, it has been shown in two instances that the pattern of organelle development in the cortex is under the long-term control of nuclear genes (40,41,51). Both cytotaxis and gene-controlled syntheses are likely to be involved in the control and regulation of organelle formation and overall cortical pattern. The fundamental problem of potential significance for the study of cell development as a whole is to understand the interaction of gene-dependent processes and cytotaxis in the control of cortical morphogenesis.

Tetrahymena provides excellent opportunities for analysis of cortical development, in part owing to the possibilities for bringing nutritional, biochemical, and genetic methods to bear on the resolution of developmental problems. Analytical studies of cortical development in *Tetrahymena* have focused on three broad problem areas.

(1) Pathway selection: Each intensively studied species has been found to be capable of selecting alternative developmental pathways. These pathways involve stomatogenesis, alternatives being possible in the site of stomatogenesis (equatorial versus anterior) and in the dimensions of the structure formed (microstome versus macrostome). Equatorial stomatogenesis is associated with cell division, and anterior stomatogenesis with oral replacement. Some species of *Tetrahymena* can form macrostome oral structures; in these the selection of alternative pathways brings about polymorphism. The selection process appears in all cases (whether or not polymorphism is involved) to depend on nutritional inputs. Recent studies suggest that the different pathways may be variations of a common cycle, oral development sharing control mechanisms in common with cell division.

(2) *Stomatogenesis—synthesis, assembly,* and *maintenance:* The analysis of stomatogenesis (both anterior and equatorial) is facilitated by its synchronizability and by the fact that oral structures can be isolated for chemical analysis or autoradiography. This type of study, combined with inferences derived from studies utilizing inhibitors, has provided evidence for the involvement of different classes of protein subject to different forms of regulation in relation to stomatogenic events. Additional information is available which shows that "pattern factors" that are sensitive to colchicine and high temperatures control spatial orientation of organelles, and a preformed but normally latent degradative system exists which can, when appropriately triggered, undo developmental progress by resorbing newly forming oral structures.

(3) *Cytogeometry:* The analysis of spatial pattern in the cortex is facilitated by the great potential for quantitative variability of certain cortical structures (such as the ciliary meridians), and by the possibility of making crosses, which allows one to test the influence of genetic exchange on the inheritance of cortical phenotypes. By taking advantage of these possibilities, it has been demonstrated that preexisting cortical configurations have a tendency to perpetuate themselves through many cell generations, and also that the placement of new organelles (such as oral areas and contractile vacuole pores) is determined by relational properties of the cortex rather than by unique local determinants.

In the following discussion the alternative sequences of cortical development are described in some detail. Following this, analytical studies are reviewed.

II. THE COURSE OF CORTICAL DEVELOPMENT

A. Oral Development

1. Formation of New Oral Structures during Division

In *Tetrahymena,* as in most other ciliates, a new feeding organelle system which is inherited by the posterior division product (opisthe) is formed during the latter part

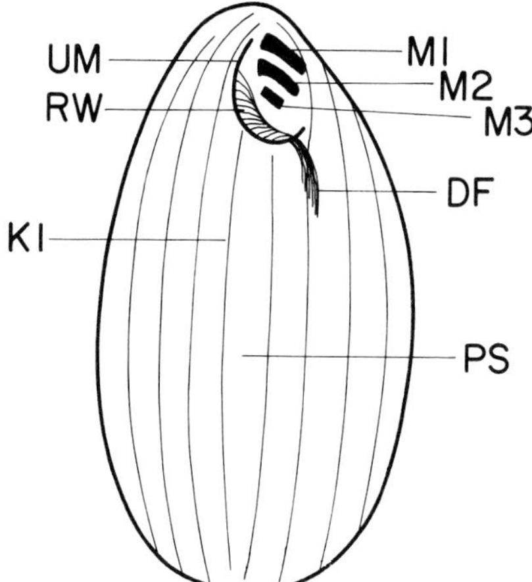

Fig. 1.

A schematic diagram of the cortical structure of *Tetrahymena*. DF, Deep fiber; KI, kinety 1 (or meridian 1); MI, membranelle 1; M2, membranelle 2; M3, membranelle 3; PS, oral primordium site; RW, ribbed wall; UM, undulating membrane. For further explanation see the text.

of the cell cycle. The old oral system is inherited by the anterior division product (proter). The usual site of formation of the new oral apparatus during division in *Tetrahymena* is an equatorial region of ectoplasm posterior to the old oral apparatus (Fig. 1). The first sign of stomatogenesis (and also of the coming division) is formation of a field of kinetosomes at this site (Figs. 1 and 2). This primordium develops into a mature oral apparatus, essentially by means of a gradual ordering of the stomatogenic field kinetosomes into the species-specific oral pattern, and the concomitant formation of additional structural elements in association with this patterned array of kinetosomes (Fig. 2). The fission furrow forms just anterior to the new oral apparatus, and the latter assumes its final position near the anterior end of the opisthe after cleavage is complete. A detailed consideration of these processes is given below.

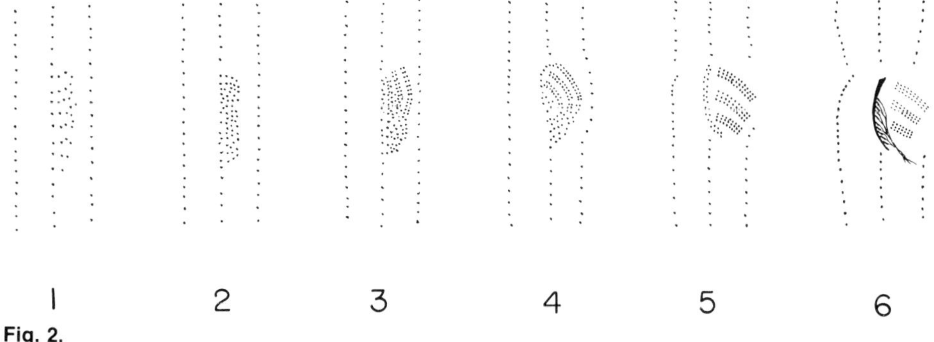

Fig. 2.

The sequence of events occurring during predivision oral primordium development in *T. pyriformis*. Each diagram represents the midregion of meridian 1 (kinety 1) plus two adjacent kineties. The oral primordium is shown as it appears at six arbitrarily designated stages of development. For explanation see the text.

In most species of *Tetrahymena,* the overall result of stomatogenesis and cell division is the formation of two daughter cells with oral structures identical to those previously found in the parent cell. There is, however, an interesting exception. Two different cell types, called "macrostomes" and "microstomes," are found in the polymorphic species *T. patula* and *T. vorax.* The two cell types in each species differ primarily, although not exclusively, in the oral structures they possess. Although the organizational pattern is similar in both, the oral apparatus of the macrostome is very much larger than that of the microstome and differs from it in significant details. The macrostome will divide to produce daughter macrostomes only if there is an abundance of ciliate prey to feed on (usually *T. pyriformis*). If ciliate prey is not present, a macrostome will either divide to produce two microstomes (8,98), or will encyst and divide several times in succession to produce cells called "tomites" (45,109,110). In either case the new oral apparatus formed initially by the macrostome in its midregion is of the microstome variety and therefore fundamentally different from the one present in the anterior end of the cell (8, Figs. 8 and 9). These divisions do not, however, produce two daughter cells of different phenotype because the large anterior oral apparatus undergoes a "remodeling" (discussed below) which transforms it into a microstome also. The ability of the cell to form either of two types of oral apparatus during division, and to make the "correct" choice between them, is impressive.

It should be mentioned that division is not always accompanied by stomatogenesis. Although instances of this are extremely rare, they demonstrate that stomatogenesis is not an absolute prerequisite for division (28,71).

The oral primordium in hymenostomes was first observed by Klein (57), and its formation was described in detail by Chatton et al. (19). From the beginning it was noted that the oral primordium arises in association with a particular row of somatic cilia, the right postoral row, which is now the starting point for the numbering system of ciliary rows. This row is designated meridian 1 (Fig. 1), and the others are numbered successively proceeding to the right as one views the ciliate from the anterior end. Because of its association with the early primordium, meridian 1 was called the *Hauptmeridian* by Klein, the *strie stomatogène* by the French workers, the *Richtungsmeridian* by von Gelei (48), and the "stomatogenous meridian" by Furgason (43). The first kinetosomes formed in the stomatogenic field normally arise either within a subequatorial segment of meridian 1, or immediately to its left. In certain strains, however, the oral primordium frequently arises adjacent to ciliary rows other than meridian 1 (71), demonstrating that this meridian is not uniquely specialized to play a role in primordium formation.

The French workers proposed that new kinetosomes are formed by the division of preexisting kinetosomes in the region of meridian 1 (61). Ultrastructural studies (113) have shown that the new kinetosomes of the oral field do indeed arise near old ones, but not by division. Instead, the nascent kinetosome initially appears as a short cylinder of microtubules oriented at a right angle to the basal portion of a preexisting kinetosome (see p. 390). In the stomatogenic field, unlike the somatic kineties (3), new kinetosomes can form adjacent to other immature ones (113).

Studies employing protargol staining (118, and unpublished) and scanning electron microscopy (13,91) have shown that the kinetosomes of the stomatogenic field are initially unciliated, but ciliary outgrowth begins prior to the formation of membranelles and proceeds gradually throughout the remainder of the process of stomatogenesis. At any given time there is an anterior-posterior gradient of ciliary length (13,91). Formation of new basal bodies also continues until the organization of the membranelles and undulating membrane is complete (stage 5, Fig. 2) (113). Prior to the formation of membranelles, the primordium is not organized into any obvious pattern, although kinetosome distribution is not always entirely random. The term "stomatogenic field" is probably preferable to the frequently used "anarchic field" as a descriptive term for this stage.

The organization of membranelles from the kinetosomes of the stomatogenic field begins with the aggregation of kinetosomes into a more compact field (29,30). Very soon after this the kinetosomes align into short double files which then orient end-to-end to form the three membranelles (30). This organization begins at the upper left part of the stomatogenic field and proceeds toward the lower right (Fig. 2); the gradient of organization is similar to that of ciliary length.

The electron microscope allows further analysis of details of membranelle formation. In advanced stomatogenic fields each kinetosome is associated with a row of accessory radial ribbons oriented in a unique "convergent" pattern with respect to the kinetosome (113). These kinetosomes and accessory ribbons are embedded within a meshwork of randomly oriented thin-walled microtubules. Membranelle formation is accomplished through movement of kinetosomes and accessory ribbons, such that the kinetosomes become oriented in double files and the ribbons all become situated to the animal's right of the right row within each incipient membranelle; these ribbons are the rudiments of the membranellar connectives (81,115).

The undulating membrane begins to form soon after membranelle organization is under way (Fig. 2). A double row of kinetosomes organizes itself along the right margin of the stomatogenic field, proceeding from the anterior end. In the incipient undulating membrane, the accessory radial ribbons come to lie on the left side of the inner (left) row; these ribbons later grow extensively and form the ribbed-wall microtubular ribbons (113).

About the time that the undulating membrane becomes organized, a third row of kinetosomes is added on the left side of each membranelle. These new kinetosomes are formed in the conventional manner adjacent to the kinetosomes of the second row. This third row becomes ciliated later than the other two; the outer row of the undulating membrane also becomes ciliated relatively late (stages 5 and 6), while the inner row remains unciliated (81,115).

Up to this point (stage 5), oral development has occurred on the surface of the cell. Subsequently, the buccal cavity forms. Associated with this, the deep fiber (81) and the cross-connectives (81,115) appear (113). These are microtubular structures associated with the bases of the kinetosomes which form the underpinnings of the oral apparatus. The ribbed wall (81,115) also develops at this late stage, and the filamentous reticulum (115) is first seen at this time.

The process of stomatogenesis may be summarized by presenting the staging series followed extensively in experimental work (29,30).

Stage 1: A loose field of stomatogenic kinetosomes.
Stage 2: The kinetosomes form a compact field.

Stage 3: Anterior parts of one or two membranelles are visible as double files of kinetosomes.

Stage 4: All three membranelles are visible; the undulating membrane is differentiated only in its anterior portion.

Stage 5: Membranelles are now complete and consist of triple rows of kinetosomes. The undulating membrane is complete or nearly so. No buccal cavity or deep fiber is as yet present.

Stage 6: The membranelles have sunk beneath the surface and the deep fiber is in the process of formation.

The relative position and duration of each developmental stage in the cell cycle has been determined in exponentially growing cells of strain GL-C at 28°C (Fig. 3, top); the same time relations have been observed in this strain at 31°C, and in inbred lines of syngen 1 at 24°C (77). Comparison with data on times of DNA synthesis (14,54) indicates that macronuclear DNA synthesis normally begins prior to primordium initiation and is completed before membranelle organization begins. In strain WH-6, which has a micronucleus, micronuclear division begins at stomatogenic stage 2 and is completed during stage 5. Macronuclear division begins in early stage 5 and is completed in early stage 6. The fission line (equatorial breaks in kineties) develops in late stage 4, and division furrowing begins in middle or late stage 5.

All cells in populations synchronized by repetitive heat shocks possess stage-1 primordia at the end of the synchronizing treatment (EST) (Fig. 3, bottom) (53,118). The primordia remain in this condition for approximately half of the interval from EST to division, presumably arrested in this condition as a consequence of the heat shocks (29,53,118). Thereafter "reactivation" occurs with the transition to stage 2, and the primordium

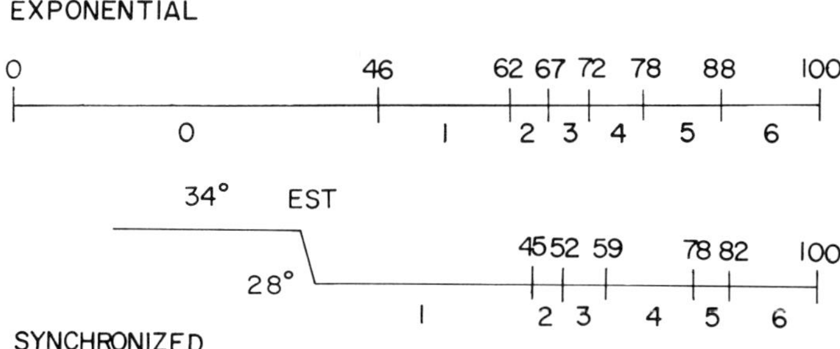

Fig. 3.

The time schedules of oral primordium development in exponentially growing and synchronized cells of *T. pyriformis* GL-C. The upper horizontal bar represents the time from one cell division to the next, at 28°C, while the lower bar represents the time between EST and the completion of the first synchronous division. Both bars are drawn on the same time scale. The amount of time spent in each developmental stage is represented by the length of the corresponding segment of the bar. The *relative* time at the beginning of each stage is expressed in terms of percent of time from one division to the next (top bar), or from EST to division (bottom bar). The estimates for exponential cells are based on averaged results of 14 experiments, while those for synchronized cells are based on 6 experiments. The stages are the same as those shown in Fig. 2. For further explanation see the text.

continues to develop with timing not greatly different from that observed in exponentially growing cells. Stage 4, however, is somewhat prolonged in synchronized relative to exponential cells, and formation of the cleavage furrow begins at late stage 4 in the former, as opposed to stage 5 in the latter. Thus there is no rigid coupling of stomatogenesis to cleavage. DNA synthesis is even more radically uncoupled from oral development, since in the synchronized cells less than half of the cells engage in DNA synthesis in the interval between EST and cell division (15,54).

The time relations of stomatogenesis and nuclear divisions in the interval between the first and second synchronous divisions (52) are similar to those of the normal cell cycle, except that the onset of stomatogenesis is probably somewhat earlier relative to the cycle as a whole.

2. Changes in the Old Oral Apparatus during Division

Although the old oral apparatus, destined for the anterior fission product, remains intact throughout the developmental process, it does undergo some reversible changes. These changes involve the regression of certain of the anterior oral structures *in situ* just prior to the onset of cleavage, and their gradual re-formation during cleavage. These include: (1) disappearance of the deep fiber bundle during stage 5, and re-formation during division (44,119); (2) regression or loss of the cilia of the undulating membrane with regrowth during division (13,91); (3) a probable reversible shortening of the cilia of the membranelles (91) which, however, are never completely lost; and (4) reversible loss of the buccal cavity, so that the membranellar bases become temporarily situated on the same plane as the surrounding somatic cilia (29). As a result of the regressive changes, at the onset of cleavage the anterior oral apparatus is very similar to the stage-5 oral primordium. The completion of development of the primordium and the restructuring of the lost parts of the anterior oral system proceed simultaneously during division (91,119). The period of regression and restructuring corresponds rather closely to the time at which dividing cells are known to be incapable of forming food vacuoles (16,66). However, close comparison of the timing of the morphological (119) and physiological (16,66) changes suggests that cessation of feeding precedes the obvious signs of structural regression by about 10 min.

As mentioned earlier, when *T. patula* or *T. vorax* macrostomes transform into microstomes (a process that involves division), marked changes occur in the preexisting anterior oral apparatuses of these cells. The changes have been studied during microstome formation by Buhse (8), who found that during primordium stages 5 and 6 the anterior oral apparatus gradually shrinks to become a microstome oral apparatus without the disappearance of the membranelles and undulating membrane. During this conversion the buccal cavity is lost and the remaining structures come to lie on the cell surface. These changes, although more extensive, are thus similar to those that occur in the anterior oral apparatus during normal division. They probably represent two variations of the same fundamental process.

In one newly described species, *T. bergeri*, the anterior oral apparatus does not persist through division but is replaced by a new oral apparatus formed from an oral replacement primordium which appears just prior to fission (88). This situation is interesting in that it involves simultaneous yet asynchronous development of two oral primordia within a single cell. A comparable situation of asynchronous development has also been observed in *T. patula* dividing within reproductive cysts (45) but is not observed in normal vegetatively dividing cells of this species.

3. Oral Replacement

Under certain circumstances oral primordia develop not at the usual equatorial site but rather in the anterior half of the cell, generally a short distance posterior to the old oral apparatus (or posterior to the site the oral apparatus had previously occupied). These oral primordia differentiate into feeding structures which replace the preexisting oral apparatus, which is resorbed either prior to or during the replacement process (Fig.

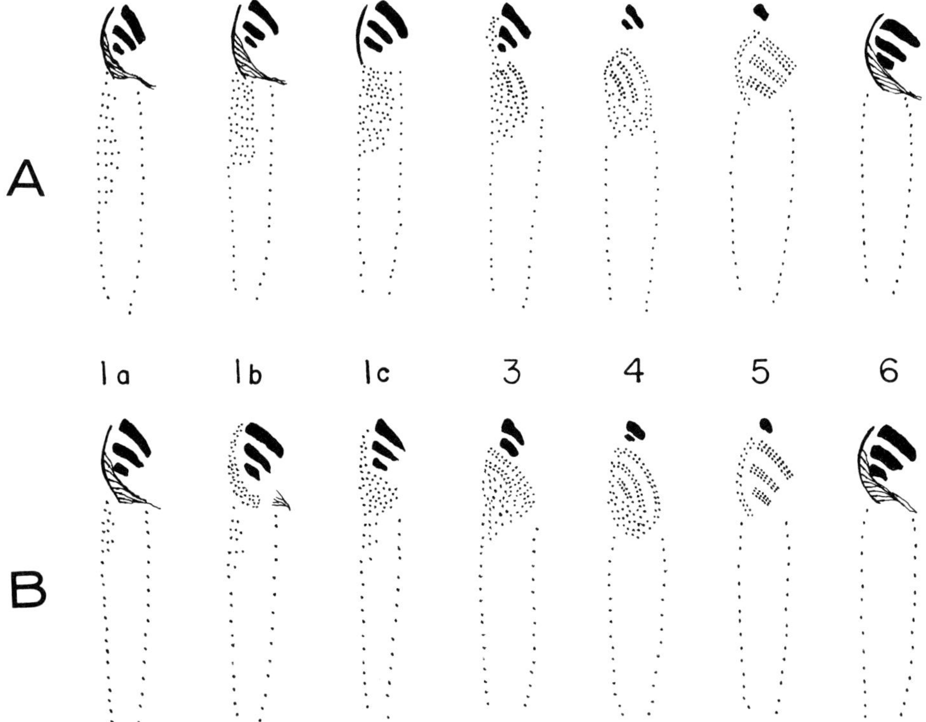

Fig. 4.

Developmental sequences of oral replacement. Two sequences are shown, the typical one (A) and the one observed in *T. pyriformis* GL-C, in which there is considerable involvement of the old undulating membrane (B). The diagrams (except for the one of the extreme right) include the old oral apparatus or portions thereof, the developing oral replacement primordium, and kineties 1 and *n*. The numbers between the two sequences of diagrams are arbitrarily designated stages. The diagrams, from left to right, show progressively later stages in the development of the new oral primordium and the resorption of the old oral membranelles; at the extreme right of both sequences (stage 6), the old structures are no longer visible, having completely regressed, while the new oral apparatus is completing its development. For further description see the text.

4). While this differentiation is taking place, the cells may temporarily round up; however, they do not divide except in the unusual cases noted above.

Oral replacement may or may not result in a change in the size and form of the oral apparatus. It does not do so in *T. pyriformis,* in which preexisting oral structures are replaced by new ones indistinguishable from the old. In the polymorphic forms *T. patula* (98,109) and *T. vorax* (8,12), however, oral replacement is the means by which cells undergo transformation from the microstome to the macrostome form. The developing oral primordium is large and differentiates into a macrostome oral area which replaces the old microstome oral apparatus.

Oral replacement in ciliates is characteristically observed in one of four circumstances: (1) following certain environmental alterations, including some nutritional changes; (2) following damage or removal of feeding structures; (3) during sexual reorganization in the form of conjugation or autogamy; (4) in some cases during cell division. We briefly describe each of these as they are observed in *Tetrahymena.*

(1) Oral replacement in undamaged vegetative cells has been observed in *T. pyriformis* in a variety of situations: following heat shocks applied at late developmental stages (32,46), during maintenance in an amino acid-free medium (38), and in the stationary phase of culture growth (Nanney, personal communication; 36). In *T. vorax* V-2 oral replacement leading to macrostome formation takes place following exposure of micro-stomes to potential prey such as *T. pyriformis* (9), or to a substance released by the prey, called "stomatin" (10). In some sublines of *T. vorax* V-2 (110), however, as well as in *T. patula* L-FF (98,109), oral replacement bringing about macrostome formation can occur in the absence of prey, this transformation occurring preferentially at the onset of the stationary phase, or when small excysted organisms (tomites) transform to macrostomes (45).

(2) Development of new oral primordia following excision of preexisting feeding structures is well documented in large ciliates such as *Stentor* (101, Chapter VII). In *T. pyriformis,* reconstitution of normal form and renewed cell division was observed by Albach and Corliss (1) following excision of the anterior third of the cell, including the old feeding structures. Stomatogenesis was not specifically investigated in their study; however, it is likely that the reconstitution that preceded the resumption of cell division involved the formation and differentiation of an oral primordium in the region just posterior to the cut. Such a process has been demonstrated in other hymenostome ciliates following excision of the anterior end of the cell (65), or after localized ultraviolet irradiation of the oral apparatus (27).

(3) The oral apparatus remains intact during the early phase of conjugation in *T. pyriformis,* with the site of union of the conjugating partners located anterior to the oral apparatus (86). No information as to what happens later is available in this species. In closely related forms (*Colpidium colpoda* and *Glaucoma chattoni*), however, exconju-gants are known to be astomatous (62; Frankel, unpublished). This indicates that the oral apparatus must be resorbed at some time during conjugation. New oral structures appear several hours following conjugation, shortly prior to the first postconjugation fission. Details of how old oral structures are resorbed and new ones later appear are not available.

(4) Replacement of the anterior oral apparatus during cell division occurs in *T. bergeri* (88) and in reproductive cysts of *T. patula* (45). In the latter case there is no definite time relationship between the onset of division and oral replacement.

In addition to the four situations described above, in which oral replacement is more or less predictable, sporadic occurrence of "spontaneous" oral replacement has been observed even under apparently normal growth conditions (109; Frankel, unpublished).

The sequence of developmental events during oral replacement has been described in *T. patula* (98,109) in *T. vorax* (8,12), and in *T. pyriformis* (36; Nanney, personal communication). In each of these species the sequence of developmental events observed *within* the developing primordium is the same as that observed in the formation of new oral structures during cell division, sufficiently so that we have felt justified in including ultrastructural information derived from study of *T. pyriformis* undergoing oral replacement in the general account of stomatogenesis during division (Section II,A,1).

Oral replacement differs from predivision stomatogenesis in regard to the site of appearance of the primordium and in regard to the fate of the preexisting oral structures. The site of appearance is always close to the oral structures, but there is some variation in the degree of involvement of the old undulating membrane. Two alternate sequences

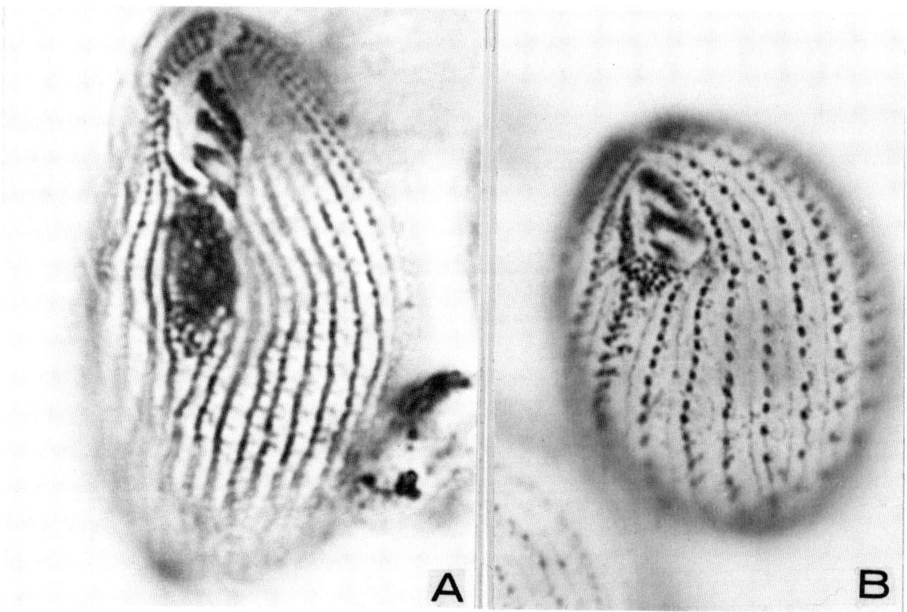

Fig. 5.

Photographs of silver-impregnated cells fixed at an early stage of oral replacement in two strains of *Tetrahymena*. (A) An oral replacement primordium in a cell of *T. patula* undergoing oral replacement to form a macrostome mouth (109). The old (microstome) oral area is still intact and has a well-defined undulating membrane. An extensive stomatogenic field is situated posterior to the oral area, adjacent to kinety 1. The old oral area is clearly separate from the stomatogenic field. ×1400. (B) An oral replacement primordium of *T. pyriformis* undergoing oral replacement in an amino acid-free medium (38). The main stomatogenic field is in the position formerly occupied by the undulating membrane of the old oral area. ×2000.

are shown in Fig. 4. What may be considered the typical sequence (observed in *T. vorax, T. patula,* and *T. pyriformis* syngen 1) is illustrated in Fig. 4A; the stomatogenic field first appears adjacent to the anterior portion of kinety 1 (Fig. 5A), although in somewhat later stages (stage 1c, Fig. 4A) the posterior portion of the undulating membrane may become incorporated. In starved *T. pyriformis* GL-C, the participation of the old undulating membrane is more extensive; the entire undulating membrane loses its ciliature and becomes converted into a field of kinetosomes which makes up the bulk of the stomatogenic field (Figs. 4B and 5B). Whether the individual kinetosomes of the old undulating membrane themselves enter the field, or regress after formation of new ones, will require detailed ultrastructural study for resolution.

Nanney (personal communication) has observed an unusual process of oral replacement in long-term, stationary-phase cultures of *T. pyriformis* syngen 1, in which the anterior portions of the postoral and other neighboring kineties are first retracted or resorbed, and later a stomatogenic field forms adjacent to the new anterior end of kinety 1. The primordium subsequently achieves a more anterior position, while the old oral structures are concurrently resorbed.

In the sequences of oral replacement shown in Fig. 4, resorption of membranelles (and also of the anterior portion of the undulating membrane in sequence 4A) takes place while membranelles are being formed within the new oral structures. This resorption has been investigated ultrastructurally (64,113) and has been shown to involve either the stepwise *in situ* disassembly of cilia, or retraction of cilia followed by disassembly within the cytoplasm. Envelopment of regressing structures within membrane-bound vacuoles (autophagic vacuoles or cytolysomes) is *never* observed, suggesting that the regressing structures are probably disassembled into their protein subunits rather than digested to amino acids (64,113).

Although regression of old structures typically proceeds concurrently with the development of new ones, this type of temporal relationship is not invariant. During conjugation and within *T. patula* cysts (45), regression of old structures is completed before development of the new ones begins. The converse situation, development of an anterior primordium *without* regression of preexisting structures, probably does not occur. Such a situation would result in the appearance of two adjacent or tandem oral apparatuses, which are not observed. In oral replacement the processes of development and regression are in some way geared together. While cases of completion of regression when development is aborted have been observed (102) or inferred (28) in certain other ciliates, in *T. pyriformis* artificial arrest of continued development of stage-1 oral replacement primordia results in considerable retardation of the resorption of the remaining old structure (Williams, unpublished).

As the cell is carrying out oral replacement, it progressively rounds up and becomes virtually spherical during stages 5 and 6, thus appearing very similar to division products.

The time required for oral replacement can be estimated in systems in which this process has been synchronized. Such estimation has been accomplished, by various means, for macrostome oral replacement in *T. patula* (98) and *T. vorax* (10), and for oral replacement in *T. pyriformis* GL-C (36,38). In the last-named species, predivision stomatogenesis and oral replacement stomatogenesis have been synchronized by identical methods employing single (36) or multiple (38) heat shocks. The time required for the two processes was very similar except that, following synchronization by multiple heat shocks, the organization of membranelles (stages 2 to 4) proceeded more rapidly in cells undergoing oral replacement than in cells preparing for division, probably because of the considerably smaller size of the oral replacement primordia.

B. Development of the Somatic Cortex

In *T. pyriformis* and closely related ciliates, increase in the number of kinetosomes within the kineties occurs throughout the cell cycle. A detailed study by Nanney (77) of the temporal pattern of increase in cells of syngen 1 suggests that the overall rate of addition of kinetosomes within kineties is approximately constant, although there may be some depression during early phases of stomatogenesis (possibly reflecting the active addition of kinetosomes within the oral primordium at that time). Nanney's study also showed that the temporal pattern of kinetosome formation differed somewhat in different kineties, with selective retardation during stomatogenesis in a postoral kinety (kinety 2), while in other kineties increase continued throughout stomatogenesis. Williams and Scherbaum (118) had reported earlier that in synchronized cells of strain GL-I there was very little increase in kinetosome number in a nonpostoral kinety, $n - 2$, between the end of a series of synchronizing heat shocks and the onset of the synchronized division. In this system the increase had occurred during the period of synchronizing shocks, and a "saturating" level had presumably been reached. There appears to be no invariant coordination between stages of stomatogenesis and patterns of increase in kinetosome number within kineties.

Observations employing the protargol technique, which stains both basal bodies and cilia, revealed that kinetosomes devoid of cilia as well as others with short cilia were found at all levels of the cell and at all stages of the cell cycle (111,118). Nonciliated kinetosomes are characteristically seen very close to ciliated ones, with the nonciliated granule situated anterior to the ciliated structure (Fig. 6). This pattern of association suggests that the nonciliated kinetosome is a "daughter" which is formed in close association with the ciliated "parent," as in the formation of the early oral primordium. Recent electron microscope studies have confirmed this inference; nascent kinetosomes are seen adjacent to preexisting kinetosomes within kineties, with the daughter kinetosome invariably situated anterior to the parent (3) (Fig. 7). The other subpellicular elements of the kinetide, for example, the kinetodesmal fiber and the posterior and transverse tubules (2,23,84) develop in association with the new kinetosome after it has fully elongated (3); the new cilium forms last.

In nondividing *T. pyriformis* the kineties are continuous. During the period of membranelle differentiation (stages 3 and 4, Fig. 2) the continuity of the two postoral kineties is interrupted as a result of the growth of the oral primordium; the posterior segments of these kineties become the postoral kineties of the posterior fission product. Somewhat later, during stage 5, gaps appear equatorially in the other kineties. These gaps collectively define the fission line which marks the site of the division furrow. The ultrastructural basis of this equatorial zone of discontinuities is unknown. The division furrow, when it forms, is first evident as a "notch" immediately anterior to the oral primordium; it then spreads in both directions around the cell until a complete ring is formed. In the course of constriction, the kineties on both sides of the oral primordium become bent in the direction of the primordium; when constriction is completed, the anterior ends of these kineties converge just in front of the new oral area, at the anterior end of the opisthe. The preoral suture is hardly evident in the early spherical division products; it becomes somewhat more prominent after the postdivision elongation of the cells, probably because of an increased crowding of kineties at the narrowing cell apex. This suture always remains short, however, and does not reflect extensive preoral growth such as occurs during fission in *Paramecium* (95). In *Tetrahymena*, unlike certain more highly specialized ciliates such as *Paramecium* and *Euplotes*, extensive growth or displacement of parts does not occur during cell division.

Another cortical landmark that remains to be discussed is the contractile vacuole pore (CVP). The CVPs, as seen by electron microscopy, are depressions of the cell surface adjacent to the contractile vacuoles (CV) (22); the black rings appearing near the CV site in silver-impregnated specimens are presumably the walls of the CVP. Just prior to the beginning of division furrowing, a second CV can be seen in the midregion of the living cell, not far from the developing oral area; in silver preparations of this same stage, one, two, or three (generally two) CVPs appear at a site just anterior to

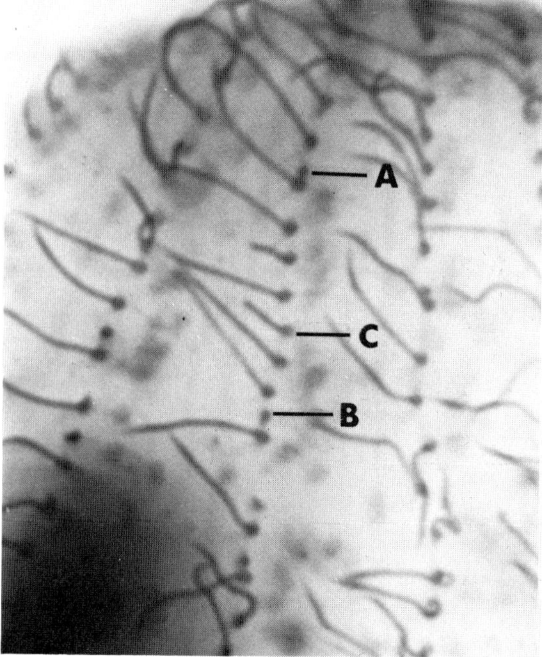

Fig. 6.

A protargol preparation showing what appear to be stages in the development of somatic kinetosomes and cilia in *T. pyriformis* GL-I (111). The anterior direction on the animal is upward in this micrograph. New kinetosomes first appear in close proximity and anterior to the older, ciliated ones (A). Next they move away from the parent (B) and sprout cilia (C). ×3000.

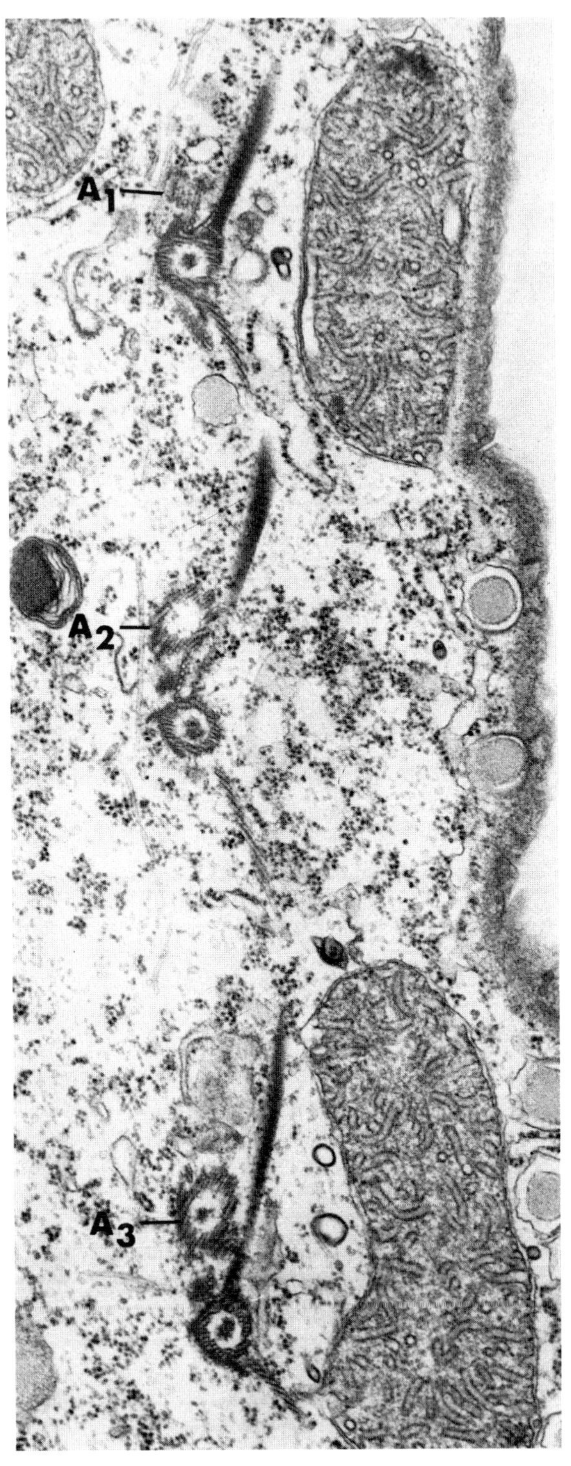

the fission line, about a quarter of the circumference of the cell to the animal's right of the oral primordium. (See Section III,D for a more detailed discussion of CVP cytogeometry.) The pores appear between kineties, generally closer to the kinety on the animal's right. Nothing is known about the stages in the formation of the CVP, however, or of the CV underlying it; there is no indication of participation of kinetosomes in the formation of these organelles. Furthermore, it is unclear why, with only one CV, there should often be two or three CVPs visible in silver preparations. After cell division the new CV and CVPs come to be located near the posterior end of the anterior division product. The original CV and CVPs, situated near the posterior end of the dividing cell, apparently remain intact and are passed on to the posterior division product. However, there is good evidence, in *Tetrahymena* (71) and also in *Paramecium* (56) and *Nassula* (105), that CVPs need not remain indefinitely at the site at which they were originally established; both longitudinal (56,105) and lateral (71) shifts are possible; presumably, the pores can either migrate over the surface (105), or else be resorbed and then re-formed at a new site.

III. ANALYSIS OF CORTICAL DEVELOPMENT

A. The Formation of Unit Organelles: Kinetosome and Cilium

The characteristic organelle of the cell surface of ciliates is the kinetosome, or ciliary basal body. This organelle serves as a topographic focus and possibly a developmental organizing center for other surface structures such as cilia, kinetodesmal fibers, and certain microtubular systems (3,17,18,23,50,84). In addition, proliferation and alignment of kinetosomes appear to play an important part in the development of complex organelle systems such as the oral apparatus. The apparent central significance of the kinetosome in cortical development in ciliates such as *Tetrahymena* makes the problem of its synthesis and assembly one of overriding importance.

Chatton and Lwoff (see ref. 61) considered kinetosomes to be self-reproducing bodies, on the basis of light microscope observations demonstrating the mutual proximity of kinetosomes during phases of proliferation. Observations of a similar kind made in *Tetrahymena*, employing the protargol technique which specifically stains kinetosomes and cilia, were described earlier (Section II,B, Fig. 6). Observations of this type, however, are equally consistent with a "generative" model of basal body formation, in which new kinetosomes arise from a "seed" adjacent to an old one (42,50,63,111). Recent

---

Fig. 7.

An electron micrograph showing developing kinetosomes in a portion of a kinety of *T. pyriformis* W (3). As in Fig. 6, the anterior direction on the animal is upward. Cross sections through three mature kinetosomes are shown, with associated fibrillar structures (kinetodesmal fiber and transverse tubules anterior to the kinetosomes, posterior microtubules behind). Anterior to the mature kinetosomes are developing kinetosomes in different stages of maturation. (A_1) An early prokinetosome oriented at right angles to the mature parent kinetosome. (A_2) A more advanced nascent kinetosome at an angle of less than 90° to the parent but still lacking the dense central material characteristic of mature basal bodies. (A_3) A still more advanced nascent kinetosome which has acquired the central dense material. However, the accessory fibrillar systems have not yet formed, and the daughter kinetosome has not yet moved away from the parent. All three stages in this figure correspond to Stage (A) in Fig. 6 (Micrograph courtesy of R. D. Allen.)

detailed electron microscope studies on proliferation of kinetosomes in *T. pyriformis* and *Paramecium aurelia* have supported this generative model (3,21,113). A nascent basal body initially appears as a short cylinder of microtubules oriented at a right angle to the basal portion of a preexisting kinetosome (Fig. 7). It is known, for *Paramecium*, that the probasal body differentiates as a result of the sequential formation of microtubules within a plaque of dense material, which is close to but not directly contiguous with the old basal body (21). These earliest stages have not yet been unequivocally identified in *Tetrahymena* (but see ref. 113, Fig. 10). In both *Paramecium* and *Tetrahymena,* elongation of the probasal body and torsion so as to achieve an orientation perpendicular to the surface of the cell both occur after a complete ring of triplet microtubules has formed (3). At all stages of development of the new basal body, there are no visible connections to the parent structure.

The morphological picture of the site and mode of formation of new kinetosomes leaves unanswered questions concerning the mode of synthesis and assembly of the molecular components of these structures, but raises the possibility that old kinetosomes may play some role in these processes. In this connection it is important to know whether kinetosomes contain DNA, as was originally claimed by Seaman (92). This question has been subjected to an excellent critical review by Fulton (42), and we will not repeat Fulton's considerations here. It is sufficient to reiterate his conclusion that "it seems fair to view the evidence for DNA in basal bodies with some caution and skepticism" (42, p. 186). Since Fulton's review appeared, Flavell and Jones (26) have reported that they could not find DNA in isolated oral apparatuses of *T. pyriformis*. Since each isolated oral apparatus contains about 170 densely packed kinetosomes (81), one should expect that if DNA were present in kinetosomes it would be detectable in this preparation.

The question whether RNA is present in kinetosomes is equally undecided and less investigated. Electron micrographs show no sign of polysomes within kinetosomes, however, although some may be present nearby. It is highly doubtful that the kinetosome is an active site of protein synthesis.

There is little doubt that the kinetosome is made up primarily of protein. Furthermore, a large proportion of the kinetosomal protein is identical or at least very similar to the microtubule proteins of cilia (85). Rannestad and Williams (85) demonstrated that isolated oral apparatuses, which lack cilia, contain as major components two proteins which migrate identically to ciliary axonemal proteins under exacting electrophoretic conditions. This is not surprising in view of the fact that two of the three triplet microtubules of the wall of the kinetosome are structurally continuous with the outer doublet microtubules of cilia, which make up the bulk of the axonemal preparation from which the microtubule proteins are isolated (87).

The problem of control of the synthesis and assembly of kinetosomes and cilia is a difficult one, since there is evidence that overall regulation of the formation of the organelles involves more than simply control of synthesis of the major identified protein of these organelles, the microtubule protein. There is now good evidence, both for the oral apparatus (see Section III,B) and for cilia, that preformed microtubule proteins may be very extensively utilized to make new basal bodies and cilia (117). As might be expected, microtubule proteins interact with one another to form an assembled structure; the drug colchicine, which is known to bind specifically to this protein (6), prevents the interaction and also rapidly and reversibly inhibits ciliary outgrowth (89) and oral development (79) in *Tetrahymena*. If such assembly of preformed proteins were all that is involved, we would expect that inhibitors of protein synthesis would have only a limited and delayed effect on the formation of basal bodies and cilia. Yet concentrations of cycloheximide sufficient to inhibit protein synthesis prevent the appearance of oral primordia in *T. pyriformis* WH-6 (47) and GL-C (38), and prevent recovery of motility by artificially deciliated cells of *T. pyriformis* W (20,89) and GL-C (Nelsen, personal communication). The blockage of ciliary outgrowth is, however, probably not absolute, since cilia of the developing oral primordium of *T. pyriformis* WH-6 completed their normal outgrowth following inhibition of protein synthesis at a stage when the cilia were still very short (47). Oral cilia, but not somatic cilia, can grow out following deciliation in the presence of cycloheximide (Nelsen, personal communication). The very limited developmental performance in the presence of cycloheximide, when combined with the evidence alluded to above for extensive pools of preformed microtubule protein, suggests that assembly of basal bodies and initiation of ciliary outgrowth require other proteins which must be newly synthesized. The nature of such proteins is unknown. Proteins other than microtubular proteins have been obtained from isolated oral apparatuses (85), but their relation to specific structures is unknown (see Section III,B). Cilia also contain other proteins, including a well-characterized ciliary ATPase, dynein, which makes up the "arms" attached to the outer ciliary doublet microtubules (49).

Actinomycin D prevents the appearance of oral primordia in partially synchronized cells of *T. pyriformis* HSM (108), and in exponentially growing cells of strain GL-G (33). This suggests that perhaps new transcription as well as translation is required for kinetosome formation, at least in some situations.

B. Oral Morphogenesis: Formation of an Organelle System

1. Synthesis

One major aspect of the problem of oral apparatus formation is the synthesis of the molecular building blocks of the structure. This problem is particularly interesting because of the possibility that development of the system as a whole may be controlled by regulation of the supply of structural materials. It is known, however, that molecular species that are not incorporated into differentiated structures play critical roles in morphogenesis in at least some cases (120). The general objective of the study of synthesis in relation to differentiation, therefore, should be to discover the patterns of synthesis for all molecular species that can be demonstrated to play any significant role in structural differentiation.

Some information is available about the structural molecules present in the oral apparatus of *Tetrahymena*. The major components of oral cilia are undoubtedly those found in somatic cilia, primarily microtubule proteins (87) and membrane components (82). A method for isolating an oral apparatus fraction has been published (85,121), and chemical analysis of this fraction gives the components of the intracytoplasmic elements; the morphological elements present include basal bodies, microtubule connectives, filamentous reticulum, and pellicle. The major component is protein (2.3% of the total cell protein, or about 36 pg per oral apparatus, ref. 121) and the major proteins are microtubule proteins. Two microtubule proteins have been identified, and these are identical to those found in cilia (85).

The first studies of synthesis in relation to oral primordium development have examined the correlations between the occurrence of stomatogenesis and the synthesis of cellular DNA, RNA, and protein. The correlations are sometimes negative for DNA (see Section II,A,1) but always positive for RNA and protein (see ref. 11 for an analysis of these correlations in *T. vorax*). This suggests no immediate relationship to DNA synthesis but raises the possibility that RNA and protein synthesis may be required for stomatogenesis.

The effects of inhibitors of RNA and protein synthesis on oral development in two strains of *T. pyriformis* are summarized in Fig. 8. As indicated, the consequence for oral development of blocking synthesis is dependent upon the stage of development shown by the oral primordium at the time the inhibitor takes effect; early primordia are arrested (crosshatching), organizing primordia are often resorbed (stipple), and late primordia are relatively insensitive and continue to develop (29,33,34). The general conclusion is that primordium development is dependent upon synthesis during earlier stages and largely independent in later stages. The transition from sensitivity to insensitivity of the primordium to inhibition of synthesis is called the "stabilization point" (29).

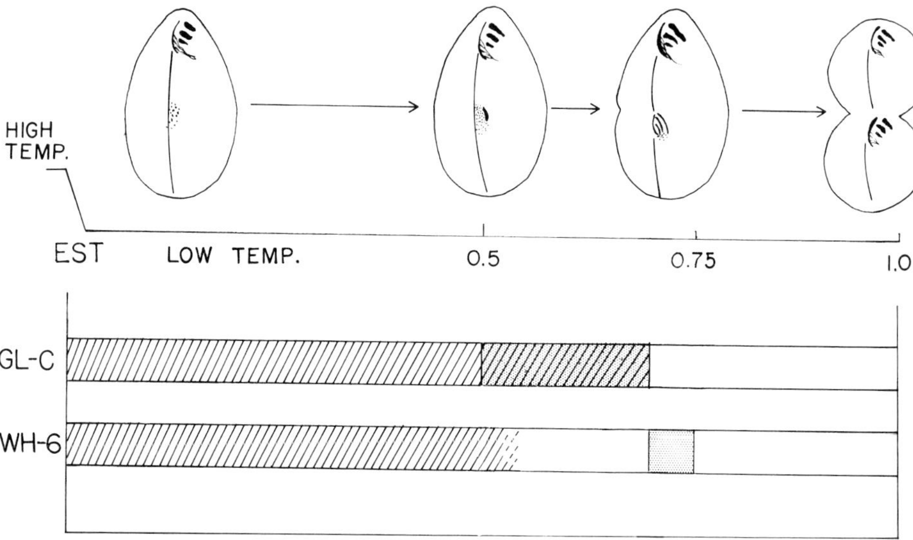

Fig. 8.

Summary diagram of morphogenetic properties in synchronously dividing *T. pyriformis*. Development of the oral primordium and of the fission furrow are indicated above the time axis. The decimals on the axis refer to the proportion of time elapsed between EST and completion of the first synchronous division. The horizontal bars below the axis present information regarding the relation of protein synthesis to oral development and cell division in two strains of *T. pyriformis*. The crosshatched zones represent the periods during which protein synthesis is essential for the completion of oral development and cell division; the stippled areas represent the periods during which resorption can take place.

Experiments of this type were first made with synchronized cells of strain GL-C (29). Logarithmically growing cells of the same strain were investigated subsequently, and the general conclusions were found to apply here also (31,33). However, stabilization occurred during stage 5 rather than near the end of stage 4 as in synchronized cells (31). This difference is probably related to differences in the relative duration of these stages (Fig. 3) and to the somewhat variable relationship of stomatogenesis to division furrow formation (see p. 381). In strain GL stabilization always occurs just prior to division furrowing, regardless of the precise morphogenetic stage attained at that time (112). Thus the onset of the period of insensitivity is not at a uniquely defined stomatogenic stage but rather at a point of cell development that is most reliably correlated with the appearance of the division furrow.

The studies of strain GL-C involving cycloheximide (37) show that a partial inhibition of protein synthesis blocks primordium development if effected prior to the stabilization point. This might mean that over these intervals either (1) oral development requires synthesis of relevant proteins in normal amounts, or (2) the synthetic machinery for producing specific proteins required for oral development is particularly sensitive to inhibition by puromycin and cycloheximide.

Cells that are in stage 1 of primordium development when protein synthesis is inhibited respond by cessation of development. Careful examination of affected stage-1 primordia suggests that there might be some attenuation of structure, although this remains to be characterized at the ultrastructural level. It is, however, a very slow process and very different from the rapid, all-or-none resorption response obtained by treating primordia in the process of membranelle organization with inhibitors of synthesis. The latter response is discussed in Section III, B, 3. It has recently been recognized that the insensitivity to inhibitors of synthesis that is found after the stabilization point is relative rather than absolute. Concentrations of puromycin and cycloheximide that significantly affect protein synthesis slow the progress of poststabilization cells toward division. Complete arrest of cleavage sometimes occurs at concentrations that reduce protein synthesis by more than 90%; however, many cells eventually divide at these levels of inhibition (35,37).

Study of the relation of protein synthesis to oral development in strain WH-6 of *T. pyriformis* has shown some significant differences between this strain and strain GL-C (Fig. 8). Working with strain WH-6, Gavin and Frankel (47) found that cycloheximide (30 μg/ml) completely inhibited protein synthesis within about 2 min after addition. This prevented further development of the oral primordium if administered during stages 1 and 2. However, complete inhibition of protein synthesis with cycloheximide, initiated *immediately* after the beginning of membranelle formation, resulted in continued development to stage 5, then arrest and subsequent resorption of the primordium as indicated by the stippled area in Fig. 8. When cycloheximide was added after this time, oral development generally continued to completion and the cells divided, albeit more slowly than control cells, with some cases of blockage during fission. Therefore *each* of the steps in oral development after stage 2 in strain WH-6 can occur in the complete absence of protein synthesis, but the entire program of steps cannot be run in the absence of synthesis. It appears that advancing past stage 5 requires proteins that are synthesized in stage 3.

The general conclusion from the study of the relation of protein synthesis to oral development in dividing cells in both strains of *T. pyriformis* studied is that concurrent translation is required for the initiation of membranelle formation to occur; it is required for different lengths of time in the two strains during the process of membranelle formation; and is not essential following the completion of this process.

The effects of cycloheximide on oral replacement in *T. pyriformis* have also been studied (38). Response characteristics identical to those found in dividing cells of the same strain (GL-C) were observed. For *T. vorax,* however, there is a brief report that oral replacement in microstomes undergoing transformation to macrostomes is sensitive to *p*-fluorophenylalanine at all times until the process is completed (9). Possibly, this extended period of sensitivity is related to a requirement for new protein for formation of a large pharyngeal pouch at the conclusion of macrostome oral replacement.

The relation of RNA synthesis to oral development has been studied in strain GL-C by Frankel (33), in strain WH-6 by Gavin and Frankel (47), and in strain HSM by Whitson and Padilla (108). In a study involving actinomycin D, the last-mentioned investigators came to the conclusion that inhibition of RNA synthesis in cells that are in any part of the process of forming new oral apparatuses does not block completion of stomatogenesis. This conclusion has not been confirmed by the more extensive studies of Frankel and Gavin. In strain GL-C actinomycin D and 5-fluorouridine (FUR) stopped primordium development at stages up to morphogenetic stage (late) 4 (33). The response was the same as that obtained in studies of inhibition of protein synthesis; prior to initiation of membranelle formation development was arrested, whereas during membranelle formation primordium resorption generally occurred. As might be expected, however, the time lag between addition of the inhibitor and the appearance of a developmental effect was longer with agents affecting RNA synthesis than with inhibitors of protein synthesis. The same stabilization point was found with these inhibitors as with inhibitors of protein synthesis, after which no effect on development was observed (Fig. 8).

The effects of actinomycin D on stomatogenesis, RNA synthesis, and protein synthesis have also been studied in strain WH-6 (47), with somewhat different results. Prior to the initiation of membranelle formation, actinomycin D arrested stomatogenesis but had no effect on oral development after that time. It was concluded from the effects of this inhibitor on RNA and protein synthesis that RNA synthesis was effectively stopped in these experiments; therefore RNA synthesis is essential for initiation of membranelle formation in this strain but is not required for subsequent development of the oral primordium.

In summary, then, it may be said that RNA synthesis occurring a short time prior to the event is probably required in dividing cells of both strains for the initiation of membranelle formation, is required for completion of membranelle formation in strain GL-C but not in strain WH-6, and is not essential subsequently. It is also known that oral replacement in *T. pyriformis* GL-C (38) and *T. vorax* microstomes (9,80) requires RNA synthesis, although less detailed information is available in these systems.

All this information strongly suggests that *de novo* synthesis of some type (or types) of protein made from newly synthesized messenger RNA is required for at least the early phases of oral primordium development. What is the nature of this synthetic requirement? Because the oral apparatus is largely made up of microtubules, one suggestion might be that the requirement is for a supply of microtubule subunit proteins. This possibility is consistent with the finding of Rannestad and Williams (85) that logarithmically growing cells apparently do not have large free cytoplasmic pools of microtubule proteins and appear to make microtubule proteins for the new oral apparatus in each cell cycle. A recent study by Williams and Nelsen (117), however, shows that the synthesis requirement is probably not for microtubule proteins. The most critical experiment involved measuring the synthesis of microtubule proteins of the oral apparatus during synchronous oral replacement in *T. pyriformis* GL-C. There was no indication of induced synthesis of these proteins in this experiment. During the synchronous oral replacement, which took about 2 h, all cells in the population developed new oral apparatuses and it was demonstrated that 94% of the microtubule proteins used to make these structures preexisted within the cells and only 6% was synthesized *de novo* during development of the oral apparatuses. It was concluded that the requirement is not for the major supply of microtubule subunit proteins but is probably for synthesis of one or more morphogenetic regulators, that is, a protein of one type which is required for assembly of a structure composed largely of other types of proteins. Early primordium development, then, may be controlled by gene products that serve as morphogenetic regulators.

Earlier studies provided evidence that some oral apparatus proteins are apparently pooled, even in logarithmically growing cells, for use over periods that may span several cell generations. This was first inferred by Williams et al. (116) from autoradiographic studies of isolated oral apparatuses in which synthesis of total oral apparatus protein was investigated. This pattern of regulation was found later by Rannestad and Williams (85) in several proteins separated from the oral apparatus by acrylamide gel electrophoresis. Although it is not known from which oral apparatus structures these proteins come, the data point up the fact that the synthesis of various proteins within a common organelle system may be regulated in different ways.

Finally, the synthesis studies have indicated that even morphostatic oral apparatuses, that is, those not undergoing formation or regression, constantly gain and lose protein subunits. It was first shown autoradiographically (116) that labeled amino acids are incorporated in proteins of morphostatic oral structures, and that such structures lose label in pulse-chase experiments. Some experiments of this type have recently been made with oral apparatus microtubule proteins (117), and they have shown that these individual proteins enter and leave morphostatic oral structures. Thus even a morphostatic oral apparatus is probably dynamic in a molecular sense. As concluded by Williams et al. (116), the oral apparatus may be viewed as a morphogenetic field which imparts order to molecular materials which are in a constant state of flux within the system.

2. Assembly

Some of the central problems presented by the assembly of cell organelles can be viewed in terms of (1) the timing of assembly steps, (2) the form of the assembled structure, and (3) the positioning of the structure within the cell. The timing of assembly steps, or proper sequencing, in early oral primordium development might in theory be controlled by sequential synthesis of subunits or regulators, because it is known that protein synthesis is an absolute requirement for early oral primordium development. After the stabilization point, however, the sequencing of assembly processes must be controlled in some other way because morphogenesis will occur in the absence of protein synthesis. It is probable that the sequencing is at the level of protein–protein interactions. A precedent for this idea is found in bacteriophage T4 morphogenesis, in which it is known that all proteins for virus structure are synthesized simultaneously and yet a specific sequence of assembly steps occurs subsequently (55).

The form of the assembled structures may ensue from the properties of the subunits, as in the case of tobacco mosaic virus (58), or be codetermined by subunits and form regulators of the type demonstrated to operate in bacterial flagella formation (4). Nothing is known as yet about this in *Tetrahymena*. Preliminary results of ultrastructural analysis of oral apparatuses formed in the presence of low concentrations of colchicine, a drug believed to be a specific inhibitor of microtubule protein assembly, show that abnormal forms of cilia and basal bodies are sometimes assembled (Nelsen, unpublished). The same thing is seen in oral apparatuses formed during prolonged heat shocks (Jaeckel, unpublished), but neither study has been completed.

More is known about positioning because the normal process has been described ultrastructurally (114) and experimental treatments affecting it have been studied (30,79). Positioning can be accomplished in two ways, and both occur in *Tetrahymena* oral primordium development. First, structures may form initially in their final positions. The outer row of ciliated basal bodies of each of the three membranelles forms in this way, appearing after the membranellar pattern has been established (113). The second method of positioning involves formation of a structural element followed by orientation as a separate event. All ciliary units coming from the stomatogenic field follow this pattern. The accessory (radial) ribbons adjacent to field basal bodies impart polarity and permit the observation, also made in *Stentor* (83), that both lateral and rotational movements occur after basal bodies are assembled. These movements give the final pattern of structures within the oral apparatus (pattern formation). The tilting movements of probasal bodies during their maturation provide another example of positioning movements in oral apparatus development.

Two agents that affect positioning, or orientation, movements in the oral primordium of *Tetrahymena* have been studied. Frankel (30) first observed that prolonged exposure of cells of strain GL-C to temperatures just below that which permanently blocks development results in abnormality in the spatial arrangement of membranelles and of the kineties. For example, at 33°C development is blocked for 4 h and then resumes. The membranelles formed at this temperature consist of the usual parallel files of kinetosomes, indicating that the initial lining up of kinetosomes to form membranelle rudiments is taking place, but the initial membranelle segments fail to orient properly with respect to each other and the geometry of the cell as a whole. As a result, the pattern of the completed membranelles is haphazard (Fig. 9). The high-temperature-induced disorganization is reversed after cells are returned to optimal growth temperatures; new oral apparatuses are generally normal in pattern. Abnormal patterns have also been generated by high-temperature treatment of strain GL-C undergoing oral replacement (Jaeckel, unpublished) and *T. vorax* macrostomes undergoing division (Williams, unpublished).

A concerted attempt was made to mimic the high-temperature effect on oral pattern

Fig. 9.

A silver-impregnated cell of *T. pyriformis* GL-C after 5 h of exposure to 33°C. The oral primordium is in stage 5, however, the membranelles are abnormal in size, number, and orientation. The anterior oral structures, which were formed prior to the beginning of the heat shock, are situated at the surface of the cell (as is typical at stage 5) and display normally oriented membranelles. ×1700.

by employing agents affecting RNA and protein synthesis (33). None of these was successful, therefore the disorganization of pattern by high temperature is not a result of an effect on metabolism but most likely upon assembly mechanisms specifically. This suggestion is confirmed by the study of Nelsen (79), who found that the pattern effects could be duplicated by colchicine, a drug presumably specific for assembly of microtubules. The colchicine results also implicate microtubule formation as a causal factor in achieving pattern. The question of which microtubules may play such a role raises a general consideration. One might imagine that the positioning of assembled elements such as basal bodies and radial ribbons is due to some intrinsic property of these structures themselves, a kind of self-assembly at the level of basal body–basal body interaction. If so, the pattern effects of colchicine might be due to effects upon the structural microtubules within these elements, which in some way alter basal body "bonding" properties. However, positioning of basal bodies might be brought about by extrinsic factors. The colchicine data suggest that microtubule formation might be part of such an external mechanism. In this connection it may be significant that electron microscope studies of oral apparatus development have disclosed the presence of great quantities of thin-walled microtubules randomly oriented around all basal bodies of the stomatogenic field. They greatly resemble the astral tubules surrounding centrioles, and it has been proposed that they may play some role in the orientation of basal bodies within the developing oral apparatus (113). Ultrastructural studies now in progress may aid in determining the correct interpretation.

It has been proposed that the period in oral primordium development following stabilization (late stage 4 in strain GL-C) may involve assembly processes exclusively (94). This is because the only two agents known to block oral development after the stabilization point are colchicine (79), which is probably specific for microtubule assembly, and high hydrostatic pressure (94), which is known to affect protein–protein interactions. The latter also affects protein synthesis, but the fact that this particular pressure effect is not duplicated by inhibitors of RNA and protein synthesis is evidence that the blockage of development is through some other mechanism. Cell division, however, is not prevented by these agents when they are applied late in oral development, and the consequence is daughter cells with incomplete oral structures. This incompleteness is temporary, however, and eventually corrected by oral replacement. The effects of colchicine and pressure administered prior to the stabilization point are the same as those produced by inhibitors of synthesis, that is, primordium arrest in early primordia and resorption of primordia in the process of membranelle formation. Thus there is no stabilization point for agents that block assembly.

It has also been found that the anterior oral apparatus in dividing cells, which is never susceptible to inhibitors of synthesis and which usually is not affected by assembly inhibitors, becomes sensitive to pressure and colchicine for a brief period when the oral primordium is in late stages of development (79,94). This is the time (see Section II,A,2) when the anterior oral apparatus undergoes its cycle-dependent remodeling, and the colchicine and pressure effects support the notion that this is a time of intensive morphogenetic activity. Anterior oral apparatuses blocked in assembly eventually are resorbed and replaced by normal structures in the daughter cells that receive them.

3. Maintenance

A major condition, which must be met in order for oral primordium development to progress, is continued inactivity of the resorption mechanism. This concept of the maintenance of development is a consequence of a study by Frankel (34) showing that resorption is brought about by an active cellular mechanism which is triggered by inhibitors of metabolism and assembly. The most important findings on which this conclusion is based are (1) the resorption process is a stereotyped all-or-none sequence of events, which proceeds to completion even if the agent that triggers it is no longer present; (2) the resorption process is energy-dependent, being arrested or delayed by low temperature or dinitrophenol; (3) the triggering of resorption can be separated from the later carrying-out of the process; a low-temperature or dinitrophenol treatment which blocks both development and resorption predetermines a later resorption, which takes place after return to normal conditions (34).

The resorption of oral primordia in *Tetrahymena* therefore contrasts sharply with the regression of labile microtubule systems such as the mitotic apparatus (90) or axopodia of heliozoans (104). In these last-mentioned systems disassembly of microtubules is a direct physical response to conditions such as low temperature, and reassembly occurs shortly after return to normal conditions. In the *Tetrahymena* oral apparatus, however, all microtubular systems are cold-stable (114). Regression of these structures must therefore be viewed as an active degradative process, presumably enzymic, for which low temperature or other environmental insults serve as a trigger.

In strain GL-C resorption in response to most chemical agents, heat, and cold can occur only if triggered prior to the stabilization point. Resorption is accompanied by blockage of cell division. However, high pressure and colchicine can trigger resorption even after exposure during stage 5 (79,94). In these cases cell division continues and the completion of resorption is often delayed and accompanied by oral replacement.

In strain WH-6 heat shocks elicit prompt resorption throughout most of the period of membranelle formation (46). Inhibition of protein synthesis is effective only during the beginning of this process (stage 3); inhibition at this time does not elicit immediate resorption but rather predestines subsequent resorption during stage 5, with blockage of cell division (47,Fig.8). Present information suggests that the basic characteristics of the resorption response are the same in both strains, while the timing both of the synthetic events related to resorption and the expression of resorption differ in the two strains. Such differences are not surprising if it is recalled that resorption is not a passive response of microtubular systems to environmental insult, but rather an active response of the cell. Some of the ultrastructural aspects of active resorption have recently been published (64,113) (see Section II,A,3).

C. Pathway Determination

We have already seen (Section IIA) that there are two forms of stomatogenesis in *Tetrahymena:* predivision stomatogenesis, with equatorial development of the oral primordium, and oral replacement stomatogenesis, with development of the oral primordium in the anterior region of the cell. These two forms of stomatogenesis are affected in a similar way by heat shocks and by inhibitors of RNA and protein synthesis (Section III,B), indicating that they share common requirements and have comparable phases. In the section that follows, we develop the view that the two processes share much in common, and that the cell can be switched from one to another of these parallel developmental pathways by changes in nutritional conditions. In the polymorphic forms, *T. vorax* and *T. patula,* a switch from one pathway to another can result in a substantial alteration in cell size, form, and living habits. This is not the case in *T. pyriformis.* We discuss pathway determination first for *T. pyriformis,* then for the polymorphic species.

1. Determination of Oral Replacement in T. pyriformis

When cells of *T. pyriformis* (strain GL-C) are transferred, during the exponential phase of growth, from a standard nutrient medium to a medium lacking amino acids, purines, and pyrimidines, culture growth continues slowly for about 6 h, and then ceases. Prior to the cessation of cell division there is, as might be expected, a decline in the frequency of cells observed in predivision stomatogenesis. All development does not stop, however; concurrent with the above-mentioned decline there is a complementary increase in the proportion of cells undergoing oral replacement. This proportion attains a level of 10–15%, which remains steady for 24–28 h while cell number remains approximately constant. A similar phenomenon is noted when cell division ceases in cultures entering stationary phase. Thus cells kept in the absence of inhibitors at temperatures allowing development appear to undergo oral replacement when nutrient conditions do not favor cell division.

The oral replacement process can be synchronized, either by a multiple heat shock treatment identical to that used to synchronize cell division (38), or by single, long heat shocks (36). The multiple heat shocks accumulate cells in stage 1 of oral replacement stomatogenesis, analogous to what is observed in the synchronized dividing system (see Section II,A,1). Furthermore, both the timing of development following release of the heat shocks and the response to inhibitors are similar in the synchronized division and oral replacement systems (38). Oral replacement is completed, and cells become maximally rounded (38,100,106), at about 80 min after EST. A second synchronous oral replacement, accompanied by a second synchronous rounding, is often observed at a time comparable to that of the second synchronous division of cells synchronized in a nutrient medium (38). These parallel features suggest that in spite of the absence of growth, DNA synthesis (38,107), and division, the process of oral replacement shares many control features in common with the stomatogenesis that precedes and accompanies cell division. We therefore may, following Tartar (102), view oral replacement as a "pseudodivision" which takes place when some but not all of the conditions for cell division are present. In the absence of some critical growth requirement, cells are shunted from the division pathway to a parallel oral replacement pathway, stomatogenesis being a "common denominator" (103) in both pathways.

2. Pathway Determination in T. vorax and T. patula

These species are polymorphic; three stages are represented: the microstome, the macrostome, and the encysted tomites. Microstomes and macrostomes may persist indefinitely by fission, whereas cells in any of the three states may transform to any other under appropriate conditions (109,110). The conditions that determine the phenotype are nutritional. Both species respond to nutritional changes in the same general way, although *T. patula* is a more highly adapted carnivore with the result that the macrostome phenotype is found under a wider range of conditions that in *T. vorax*. In general, however, microstomes divide in broth cultures and transform into macrostomes when this rich food supply is curtailed. Macrostomes thus appear in stationary-phase broth cultures (98,110). They can also be induced by potential prey (*T. pyriformis*) in a nonnutrient medium (7), and by an aqueous secretion of prey organisms: the stomatin preparation of Buhse (10). Recent studies on stomatin, and on the metabolism of microstomes undergoing transformation in stomatin, suggest that it is probably not a specific inducing agent but rather a constellation of nutritional factors "excreted" by the prey organism (Buhse, personal communication). All the present data therefore suggest that a specific reduction (and/or change) in available exogenous nutrients induces macrostome formation. Although less well studied, the other transformations in *T. vorax* and *T. patula* also seem to be controlled by nutritional inputs. Macrostomes and excysted forms (tomites) transform into microstomes in broth cultures; macrostomes and microstomes encyst in distilled water; and tomites transform into macrostomes in the presence of potential prey.

The question then arises, Can polymorphic transformations, like oral replacement in *T. pyriformis*, be viewed as nutritionally controlled processes closely related to division, that is, as pseudodivisions? First, some of the transformations regularly involve division, as mentioned earlier. Macrostomes and microstomes divide during encystment, and macrostomes always divide when forming microstomes. Oral morphogenesis in these transformations is temporally correlated with cell division, as it is in ordinary binary fission. The interest here lies in the fact that the new oral structures are usually not the same as those found in ordinary binary fission. The polymorphic species therefore have more than one morphogenetic sequence which may accompany division, and nutritional factors apparently determine the choice between them.

The remaining question is whether any kinship with division can be found for those transformations that do not actually involve division. One such transformation, from microstome to macrostome, has been studied extensively. The primary event is the resorption of the old oral apparatus and its replacement by a new, larger structure at the same site. The methods developed by Buhse (7,10) for synchronization of the microstome to macrostome transformation in mass cultures of *T. vorax* involve washing and suspending stationary-phase microstomes in an inorganic medium and then adding either washed *T. pyriformis* or the stomatin preparation extracted from *T. pyriformis* cells. Under these conditions 70–90% of the microstomes transform into macrostomes within 7 or 12 h after mixing, the precise timing depending on whether live prey or stomatin is used. If, however, the washed microstomes are left in inorganic medium without the addition of prey or stomatin, they undergo a synchronous division instead. Without entering into a discussion concerning the reason for the synchrony, it appears that either the microstomes divide *or* they undergo transformation into macrostomes. These facts suggest that oral replacement associated with cell-type transformations in the polymorphic species *T. vorax* and *T. patula* may indeed be a kind of pseudodivision, in the sense that this term was applied to oral replacement in the monomorphic species *T. pyriformis*.

In summary, we can say that *Tetrahymena* cells can: (1) divide (rarely) without the formation of oral structures; (2) duplicate preexisting types of oral structures along with division; (3) form oral structures during division which are different from the old ones possessed by the cell; (4) resorb oral structures and replace them with new ones of the same type with or without simultaneous division; and (5) resorb oral structures and replace them with a different type, without division. The choice is predicated on the type of species and its nutritional environment. We can speculate that the mechanism for forming oral structures evolved as an extension of the division machinery in the cell. Later evolutionary developments probably included the ability to form more than a single type of oral structure during division, and the ability to disengage stomatogenesis from division. These adaptations would in general be advantageous, except that it is difficult to imagine the adaptive significance of oral replacement in *T. pyriformis*, in which the new oral apparatus is apparently identical to the one that is resorbed.

D. Cortical Architecture

A considerable amount of evidence has accumulated indicating that cortical patterns of ciliates possess a high degree of autonomy (5,24,74,95,97,99,101). The classic demonstrations of this autonomy were based largely on microsurgical analysis of large heterotrich ciliates (99,101,103). Smaller ciliates such as *Tetrahymena, Paramecium,* and *Euplotes* are not as amenable to microsurgical approaches; however, they have compensating advantages, such as high multiplication rates, cortical patterns that are readily visualizable by silver techniques and, above all, well-analyzed breeding systems which make possible genetic analysis of cortical differences. The studies on *Tetrahymena* that are relevant to the problem of pattern control are reviewed below.

1. Determination of Kinety Number

Kinety number, which is readily quantifiable in *Tetrahymena,* is one of the most obvious of cortical phenotypes since in this genus the kineties are generally parallel, unbranched, and extend from one end of the cell to the other (except for the "postoral kineties" which terminate at the posterior margin of the oral area). The number of kineties in a strain is not rigidly constant, but may vary depending on growth conditions and on the previous history of the strain. The dependence of kinety number on growth conditions is most striking in the facultative parasites *T. limacis* (59) and *T. rostrata* (60), in which kinety number is considerably higher in cells maintained in their normal gastropod hosts than in clones derived from these hosts and cultivated in axenic or bacterized culture media. These changes are reversible (59).

Under conditions of continual axenic culture with daily transfers, average kinety number in strains GL-C and WH-6 tends to fluctuate within clearly defined limits over a period of many years (39). The total range of kinety numbers observed under these conditions remains virtually constant. No striking *stable* polymorphism appears to have arisen within these strains. Such polymorphism is in fact present in *T. vorax* V-2 (93,110). In this strain three stable sublines have appeared within a single clone; these sublines are distinct in cell size, nuclear diameter, kinety number, competence for macrostome and cyst formation (110), and immobilization antigens (93). These sublines have been called D (dimorphic), M (monomorphic), and S (small). The range of kinety number, under conditions of reasonably frequent transfer, is different in the three sublines. These differences persist, with only occasional spontaneous transformations from one type to another, even though the sublines are maintained under identical conditions. Cultivation at moderately high temperature (25°C) somewhat increases the frequency of transformation from the D to the M type, and the latter type is stable at 20°C, the standard temperature used for maintenance of *T. vorax* cultures. The only procedure discovered that reliably induces transformations is rapid serial subculturing, at 24-h intervals, which induces transformation to the S type from either D or M. The change to the S type is sudden, occurring within the interval of a single transfer. Such a very rapid change in phenotype indicates that the change is induced directly by the treatment rather than as a result of selection of preexisting cells of the S type. Unfortunately, strain V-2 is amicronucleate, precluding direct genetic analysis. The mode of alteration of type indicates that it is mediated by an epigenetic rather than a genetic mechanism, however, which has far-reaching consequences with respect to the morphology of the cell.

Transient polymorphism in kinety number has been induced and analyzed in *T. patula* by Fauré-Fremiet (25) and in *T. pyriformis* by Nanney (67,68,70,73,74,78). In *T. patula* (known earlier as *Leucophrys patula*) the normal kinety number is nearly doubled when formation of regular homopolar doublets is induced by treatment of dividing cells with dilute formalin. This treatment blocks division and results in doublet formation as a result of translocation forward of the opisthe (25). The duplex kinety number then gradually modulates downward, in the course of about 100 cell generations, to a number only slightly higher than that characteristic of normal singlets; in the course of this downward modulation, the doublets revert to singlets.

A considerably more extensive analysis of a comparable phenomenon was carried out by Nanney, employing inbred lines of *T. pyriformis* syngen 1. These lines were routinely maintained by monthly transfers, so that cells spent most of the time under conditions of anoxia and starvation in late stationary phase. Under such rigorous conditions considerable variability in kinety number and other cortical characteristics appears, variability far greater than that found in cultures maintained by frequent transfers. Variability was found within and between stocks. Interestingly, however, while kinety number varied greatly, the number of kinetosomes per kinety was negatively correlated with the kinety number in such a way that the total number of kinetosomes ("ciliary units") in the cell remained approximately constant regardless of the kinety number (76).

The most significant aspect of Nanney's analysis concerns what happens to the variability in kinety number after single cells from late stationary-phase cultures are inoculated into fresh medium and allowed to grow exponentially for about 20 fissions. Three critical observations were made on the "20-fission" clones derived from these single cells:

(1) Clones derived from stationary-phase cultures tended to have distributions of kinety numbers different from those in the original cultures. The modulation was downward in clones derived from cultures with high kinety numbers, and upward in clones derived from cultures with low kinety numbers (67).

(2) The apparent rate of modulation is correlated with the kinety number (called "corticotype" by Nanney) of the initial clone; the more extreme the initial distribution of corticotypes, the more rapid the modulation. At certain corticotypes (18 to 19 in syngen 1) this rate is near zero; this point (or narrow range) is known as the "stability center." Clones derived from cultures with extreme corticotypes modulate toward the stability center at a rate proportional to their initial distance from this center (67,68). After 20 fissions of exponential growth clones derived from such cultures have not yet reached this center, however, demonstrating that differences persist for at least 20 cell generations, probably many more.

(3) The rate of modulation toward the stability center is not influenced by conjugation between cells from cultures with different initial corticotypes (68); this rate remains the same as it would have been if no conjugation had taken place, demonstrating that the persisting differences between these clones are not a result of differences in either nuclear genes or free-flowing endoplasm, which are exchanged at conjugation in *Tetrahymena* (68).

From these three basic observations, two complementary deductions were made: (a) Any given strain (possessing a common genotype) is characterized by a maximally stable kinety number, or narrow range of maximally stable kinety numbers, known as the stability center. (b) Deviations from the stability center, acquired during a period of maintenance under suboptimal conditions, tend to persist even after return to identical optimal conditions of growth. The duration of persistence and, more importantly, the time course of modulation indicate that the basis of this persistence is *not* the presence of a population of macromolecules undergoing passive dilution; if this were so, the rate of modulation should increase as cells approach the stability center, whereas the opposite is observed. Therefore there must be some mechanism that actively maintains these differences. Since the modulation rate is not affected by exchange of genes or internal cytoplasm, the mechanism must be situated in the cortex. Since the differences in question are differences in *number* of elements, all of which are basically similar in kind (see Section II,B), this mechanism probably involves an active perpetuation of preexisting cortical *patterns* rather than differences in presumed cortical genes (68,74). Evidence for the active self-perpetuation of cortical patterns has been intensively investigated. The involvement of cytotaxis implies that there are mechanisms at the supramolecular level for perpetuation of preexisting patterned configurations. The specific nature of these mechanisms has not yet been elucidated. However, the highly specific arrangement of the fibrillar systems within the kinety may provide the necessary directional guidance (3).

It has been demonstrated, in *Euplotes minuta,* that the stability center is under genic control (40,51). If this were so in *Tetrahymena* also, then the cortical pattern at any given time might be envisioned as an outcome of a dual mechanism: (1) a cytotactic mechanism which tends to insure propagation of the preexisting state in successive cell generations (for example, by formation of new kinetides within a preexisting kinety, oriented in the same fashion as the old ones); and (2) a superimposed genic mechanism that controls which of many possible cortical states is most stable under a given set of conditions. By generating corticotypes distant from the stability center, Nanney demonstrated the cytotactic mechanism. Demonstration of genic control of the stability center in *Tetrahymena* would be possible if strains within a syngen were found that had clearly different stability centers. While differences have been demonstrated for strains of different syngens (70,73,75), intrasyngenic differences suitable for a breeding analysis have not yet been found, but probably can be. The existence of alternative stability centers in *T. vorax* V-2 suggests the additional existence of a third level of control, possibly involving differential gene activation or some other epigenetic mechanism.

2. Relational Geometry of Cortical Organelle Formation

The perpetuation of cortical features other than the kineties has been investigated extensively, also primarily by Nanney (67,69,71,72,78). Studies of the variability in a wide range of cortical properties have led to the conclusion that ''the individual cortical variations are not independent of each other, but are highly correlated and reflect a basic polymorphism involving the total geometric pattern of the cell suface'' (67). Presumably, all cortical morphogenesis, therefore, and not only the perpetuation of the kineties, appears to be determined to some extent by the preexisting cortical pattern.

The data on the number and positions of oral apparatuses and contractile vacuole pores are particularly instructive and therefore are discussed here.

The usual position of the oral primordium site has been described in Section I and is indicated in Figs. 1 and 10A. Nanney (71) studied two strains of *T. pyriformis* in which the primordium frequently arises adjacent to the ciliary rows immediately to the left and (less frequently) to the right of meridian 1 (Fig. 10B). He points out that, in the course of successive fissions and repeated stomatogenic "slippage," the primordium site completely circumnavigates the cells. This effectively rules out any notion of special morphogenetic capacities in specific kineties. Furthermore, although the oral primordium usually forms in the equatorial region of a kinety, it is known that primordium formation

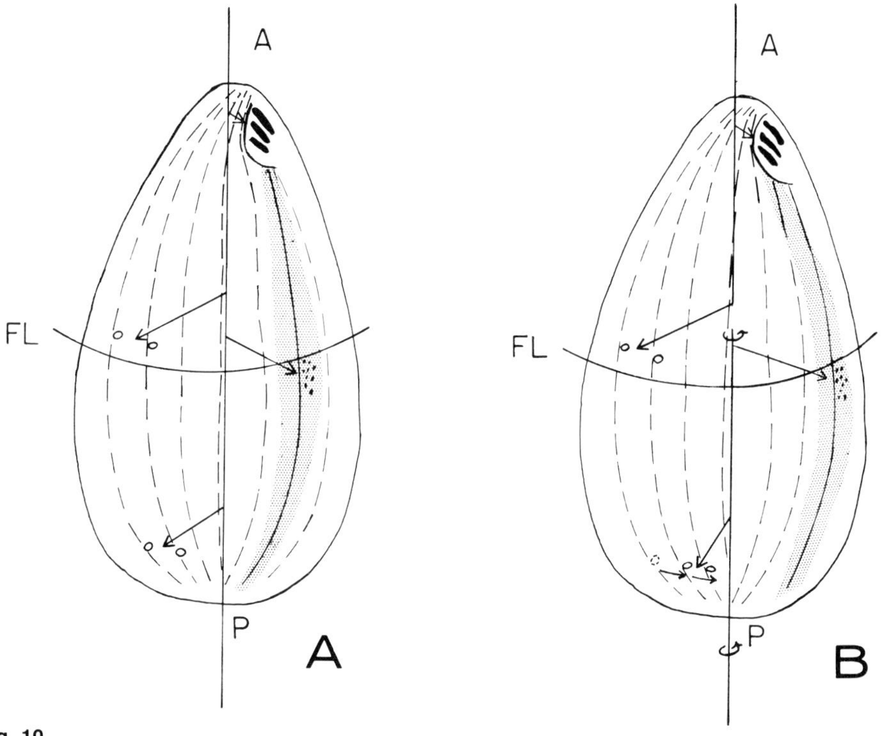

Fig. 10.

A schematic representation of normal and "displaced" geometrical patterns of morphogenesis in *T. pyriformis*. In both diagrams A–P is the longitudinal axis and FL is the site of the future fission line. The shaded area is the inductive field for stomatogenesis; in predivision stomatogenesis the oral primordium forms in the region of this inductive field just posterior to the fission line. The meridians along which the anterior CVPs form are determined by an inductive angle (see text) defined in relation to the site of the anterior oral apparatus; the location of the posterior CVPs are specified in relation to the site of the oral primordium. (A) Normal morphogenesis. The stomatogenic meridian is kinety 1 (the right postoral meridian), and the CVPs of the proter form adjacent to kineties 5 and 6, which are the same as the CVP meridians of the opisthe. (B) Displaced morphogenesis. The inductive field for stomatogenesis has shifted to the left, so that the stomatogenic kinety is meridian *n*, (left postoral meridian). The CVPs of the proter still form adjacent to kineties 5 and 6, while the CVPs of the opisthe shift one kinety to the left (4 and 5) in coordination with the shift of the oral primordium. (Modified from Fig. 2 of ref. 71 with permission of the author.)

can be induced at various levels along the anterior-posterior axis of a kinety. In oral replacement, for example, the primordium forms near the anterior end of kinety 1, as described above. In addition, Simpson and Williams (94) showed that, as a result of exposing cells in certain developmental stages to high hydrostatic pressure, primordia can be induced to form posterior to the original primordium site and halfway between it and the anterior oral area. These studies show that there is no localization of morphogenetic capacities within any particular kinety or in any particular region of a kinety. The entire cortex seems to be potent in oral primordium formation, and we are left with the notion of a gradient of potentialities. The place where the oral primordium *actually* forms is apparently determined by the presence and disposition of preexisting cortical structures, because it can be defined and predicted with reference to them.

Nanney (69) analyzed CVP sites and has presented a particularly attractive and instructive geometric model for the induction of CVP formation. A systematic analysis of the variations in cortical morphology has shown that the number of CVPs and their positions vary with kinety number (69). In general, the midpoint of the CVP group moves progressively to the right as the total meridian number increases. The correlation is not linear, however, but is a step function. In syngen 1, for example, cells with 18 to 20 meridians have CVP's associated with meridians 5 and 6, whereas cells with 22 to 24 meridians typically have them associated with meridians 6 and 7. Cells with 21 meridians represent the transitional condition, both patterns of association being abundant. These observations and others can be rationalized with the aid of Nanney's geometric model. One imagines a dorsoventral plane, defined by the stomatogenic meridian and the anterior-posterior axis of the cell (Fig. 10). If one then projects a line at the level of the CVP group from the central axis to the midpoint of the CVP group, it will define a specific angle from the dorsoventral plane. If this geometrical projection defines the locus of inductive action for CVP formation in the cortex, the CVP position will move to the right as the kinety number increases. If the "inductive point" is not a point but an inductive field corresponding to an arc, then the covariation of meridian number and CVP position as a step function can also be explained (see ref. 69 for further details).

The importance of relational geometry in the determination of CVP sites is particularly well demonstrated in strains that undergo "cortical slippage" (71). In these strains the primordium may form adjacent to the meridian to the left of meridian 1 (discussed above). In cells of this type, the new CVPs that develop in the equatorial region form in a position that represents a specific inductive angle with respect to the *old* oral apparatus, whereas the old CVPs in the posterior end shift their position to the left so as to maintain the same angle with respect to the *new* oral apparatus (Fig. 10B). It appears as though there has been a relative displacement of the inductive apparatus and the cortical response system in the posterior half of the cell.

The inductive angle shows only rather small variations among different strains within a syngen (72), or during prolonged growth of a given strain (39). It varies, however, *between* syngens, the range falling between 78 and 96° (72,75). The inductive angle for determination of the sites of CVPs, similar to the stability center for the corticotype, may be thought of as a genically determined characteristic of a biological species (syngen) of *T. pyriformis*. Proof of this inference is not yet available, however.

Observations on CVP position on doublets show that the concept of an inductive angle must not be taken literally as implying "that the cell uses a protractor and compass in organizing its gross architecture" (69). These observations have shown that the cell instead apparently measures the distance between one stomatogenic meridian and another and regulates field size and position in relation to this distance (69).

The classic view of the kinetosome as a self-duplicating structure (61) and the recent claims of DNA in the cortex of ciliates (reviewed in ref. 42) might suggest cortical pattern control by means of direct continuity of cortical reservoirs of nucleic acids. The information reviewed above, however, consistently supports the conclusion that morphogenetic processes in localized regions of the cortex are not determined by regionalized genic properties but are instead determined by inductive forces directed to specific regions which are defined in relation to the preexisting cortical differentiations. One can hardly improve on the lucid summarizing comments of Nanney, who states that "The concepts of fields and gradients so widely employed in a description of organismic integration in multicellular forms are equally as applicable (and equally as sterile perhaps) in a consideration of organization at the cellular level" (69). The cortex of *Tetrahymena* emerges as a "multidimensional information storage and transmission system whereby the pattern, in a sense, maintains itself" (74).

ACKNOWLEDGMENTS

The authors thank Drs. Howard E. Buhse and David L. Nanney for making available unpublished information which is utilized in this review. The drawings were executed by Miss E. Ann Kirkpatrick. Research by the authors described in this review was supported by NIH grant GM-11012 and NSF grant GB 32408 to J. F., and by NSF grant GB-24901 to N. E. W.

REFERENCES

1. Albach, R. A. and Corliss, J. D. 1959. Trans. Amer. Microsc. Soc. 78:276–284.
2. Allen, R. D. 1967. J. Protozool. 14:553–565.
3. Allen, R. D. 1969. J. Cell Biol. 40:716–733.
4. Asakura, S., Eguchi, G., and Iino, T. 1966. J. Mol. Biol. 16:302–316.
5. Beisson, J. and Sonneborn, T. M. 1965. Proc. Natl. Acad. Sci. U.S. 53:275–282.
6. Borisy, G. G. and Taylor, E. W. 1967. J. Cell Biol. 34:525–534.
7. Buhse, H. E., Jr. 1966. J. Protozool. 13:429–435.
8. Buhse, H. E., Jr. 1966. Trans. Amer. Microsc. Soc. 85:305–313.
9. Buhse, H. E., Jr. 1967. Trans. Amer. Microsc. Soc. 86:65–66.
10. Buhse, H. E., Jr. 1967. J. Protozool. 14:608–613.
11. Buhse, H. E. Jr., and Cameron, I. L. 1968. J. Exp. Zool. 169:229–236.
12. Buhse, H. E., Jr., Corliss, J. O., and Holsen, R. C. 1970. Trans. Amer. Microsc. Soc. 89:328–336.
13. Buhse, H. E., Jr., Stamler, S. J., and Corliss, J. O. 1973. Trans. Amer. Microsc. Soc. 92:95–105.
14. Cameron, I. L. and Nachtwey, D. S. 1967. Exp. Cell. Res. 46:385–395.
15. Cerroni, R. E. and Zeuthen, E. 1962. C. R. Trav. Lab. Carlsberg 32:499–512.
16. Chapman-Andresen, C. and Nilsson, J. R. 1968. C. R. Trav. Lab. Carlsberg 36:405–432.
17. Chatton, E., Lwoff, A., and Lwoff, M. 1929. C. R. Acad. Sci., Paris, 188:1190.
18. Chatton, E., Lwoff, A., and Lwoff, M. 1931. C. R. Acad. Sci., Paris, 193:670.
19. Chatton, E., Lwoff, A., Lwoff, M., and Monod, J. 1931. C. R. Soc. Biol., Paris 107, 540–544.

20. Child, F. M. 1965. J. Cell Biol. 27:18A.
21. Dippell, R. W. 1968. Proc. Natl. Acad. Sci. 61:461–468.
22. Elliott, A. M. and Bak, I. J. 1964. J. Protozool. 11:250–260.
23. Elliott, A. M. and Kennedy, J. R. 1973. In A. M. Elliott, ed., The biology of *Tetrahymena*, Dowden, Hutchinson, and Ross, Stroudsburg, Pennsylvania.
24. Fauré-Fremiet, E. 1945. Bull. Biol. France Belg. 79:106–50.
25. Fauré-Fremiet, E. 1948. Arch. Anat. Microsc. Morphol. Exp. 37:183–203.
26. Flavell, R. A. and Jones, I. G. 1971. J. Cell Sci. 9:719–726.
27. Frankel, J. 1960. J. Exp. Zool. 143:175–194.
28. Frankel, J. 1961. J. Protozool. 8:250–256.
29. Frankel, J. 1962. C. R. Trav. Lab. Carlsberg 33:1–54.
30. Frankel, J. 1964. J. Exp. Zool. 155:403–436.
31. Frankel, J. 1964. Exp. Cell Res. 35:349–360.
32. Frankel, J. 1964. J. Protozool. 11:514–526.
33. Frankel, J. 1965. J. Exp. Zool. 159:113–148.
34. Frankel, J. 1967. J. Exp. Zool. 164:435–460.
35. Frankel, J. 1967. J. Cell Biol. 34:841–858.
36. Frankel, J. 1969. J. Protozool. 16:26–35.
37. Frankel, J. 1969. J. Cell Physiol. 74:135–148.
38. Frankel, J. 1970. J. Exp. Zool. 173:79–96.
39. Frankel, J. 1972. J. Protozool. 19:648–654.
40. Frankel, J. 1973. J. Exp. Zool. 183:71–94.
41. Frankel, J. 1973. Develop. Biol. In press.
42. Fulton, C. 1971. In J. Reinert and H. Ursprung, eds., Origin and continuity of cell organelles, Springer Verlag, New York, pp. 170–221.
43. Furgason, W. H. 1940. Arch. Protistenk. 94:224–266.
44. Furgason, W. H., Small, E. B., and Loefer, J. 1960. Abstr. X Congr. Intl. Biol. Cell., 230.
45. Gabe, P. R. 1972. Ph.D. Thesis, University of Iowa.
46. Gavin, R. H. 1965. J. Protozool. 12:307–318.
47. Gavin, R. H. and Frankel, J. 1969. J. Cell Physiol. 74:123–134.
48. Gelei, J. 1935. Biol. Zentralbl. 55:436–445.
49. Gibbons, I. R. 1965. Proc. Natl. Acad. Sci. U.S. 50:1002–1010.
50. Hanson, E. D. 1967. In M. Florkin and B. J. Scheer, eds., Chemical zoology, vol. 1, Protozoa (G. W. Kidder, ed.), Academic Press, New York, pp. 395–539.
51. Heckmann, K. and Frankel, J. 1968. J. Exp. Zool. 168:11–37.
52. Holz, G. G. 1960. Biol. Bull. 118:84–95.
53. Holz, G. G., Sherbaum, O. H., and Williams, N. E. 1957. Exp. Cell Res. 13:618–621.
54. Jeffery, W. R., Stuart, K. D., and Frankel, J. 1970. J. Cell Biol., 46:533–543.
55. King, J. 1971. J. Mol. Biol. 58:13–34.
56. King, R. L. 1954. J. Protozool. 1:121–130.
57. Klein, B. M. 1927. Arch. Protistenk. 58:55.
58. Klug, A. 1972. Fed. Proc. 31:30–42.
59. Kozloff, E. N. 1956. J. Protozool. 3:20–28.
60. Kozloff, E. H. 1957. J. Protozool. 4:75–79.
61. Lwoff, M. 1950. Problems of morphogenesis in ciliates, John Wiley, New York.
62. Maupas, E. 1889. Arch. Zool. Exp. Gen. 7:149–517.
63. Mazia, D., Harris, P. J., and Bibring, T. 1960. J. Biophys. Biochem. Cytol. 7:1–20.
64. Moore, K. C. 1972. J. Ultrastruct. Res. 41:499–518.
65. Mugard, H. and Lorsignol, L. 1956. Bull. Biol. France Belg. 40:116–154.
66. Nachtwey, D. S. and Dickinson, W. J. 1967. Exp. Cell. Res. 47:581–595.
67. Nanney, D. L. 1966. Amer. Nat. 100:303–18.
68. Nanney, D. L. 1966. Genetics 54:955–968.
69. Nanney, D. L. 1966. J. Exp. Zool. 161:307–317.
70. Nanney, D. L. 1966. J. Protozool. 13:483–490.
71. Nanney, D. L. 1967. J. Exp. Zool. 166:163–169.

72. Nanney, D. L. 1967. J. Protozool. 14:690–697.
73. Nanney, D. L. 1968. J. Protozool. 15:109–113.
74. Nanney, D. L. 1968. Science 160:496–502.
75. Nanney, D. L. 1971. J. Protozool. 18:33–37.
76. Nanney, D. L. 1971. J. Exp. Zool. 178:177–181.
77. Nanney, D. L. 1971. Develop. Biol. 26:296–305.
78. Nanney, D. L. 1972. Ann. N.Y. Acad. Sci. 193:14–28.
79. Nelsen, E. M. 1970. J. Exp. Zool. 175:69–84.
80. Nicolette, J. A., Buhse, H. E., Jr., and Robin, M. S. 1971. J. Protozool. 18:87–90.
81. Nilsson, J. R. and Williams, N. E. 1966. C. R. Trav. Lab. Carlsberg 35:119–139.
82. Nozawa, Y. and Thompson, G. A. 1971. J. Cell Biol. 49:712–721.
83. Paulin, J. J. and Bussey, J. 1971. J. Protozool. 18:201–204.
84. Pitelka, D. R. 1968. In T. T. Chen, ed., Research in protozoology, vol. III, Pergamon Press, New York, pp. 280–368.
85. Rannestad, J. R. and Williams, N. E. 1971. J. Cell Biol. 50:709–720.
86. Ray, C., Jr. 1956. J. Protozool. 3:88–96.
87. Renaud, F. L., Rowe, A. J., and Gibbons, I. R. 1968. J. Cell Biol. 36:79–90.
88. Roque, M., dePuytorac, P., and Savoie, A. 1970. Protistologia 6:343–351.
89. Rosenbaum, J. L. and Carlson, K. 1969. J. Cell Biol. 40:415–425.
90. Roth, L. E. 1967. J. Cell Biol. 34:47–59.
91. Ruffolo, J. J. and Frankel, J. 1972. J. Protozool. (Abstr.) 19:20.
92. Seaman, G. R. 1959. Exp. Cell Res. 21:292–302.
93. Shaw, R. F. and Williams, N. E. 1963. J. Protozool. 10:486–491.
94. Simpson, R. E. and Williams, N. E. 1970. J. Exp. Zool. 175:85–97.
95. Sonneborn, T. M. 1963. In J. M. Allen, ed., The nature of biological diversity, McGraw-Hill, New York, pp. 165–221.
96. Sonneborn, T. M. 1964. Proc. Natl. Acad. Sci. 51:915–29.
97. Sonneborn, T. M. 1970. Proc. Roy. Soc. Lond., Ser. B. 176:347–66.
98. Stone, G. E. 1963. J. Protozool. 10:74–80.
99. Suzuki, S. 1957. Bull. Yamagata Univ., Nat. Sci. 4:85–192.
100. Tamura, S., Toyoshima, Y., and Watanabe, Y. 1966. Jap. J. Med. Sci. Biol. 19:85–96.
101. Tartar, V. 1961. The biology of *Stentor,* Pergamon Press, New York.
102. Tartar, V. 1966. Exp. Cell Res. 42:357–370.
103. Tartar, V. 1967. In T. T. Chen, ed., Research in protozoology, Pergamon Press, New York, pp. 1–116.
104. Tilney, L. G. and Porter, K. R. 1967. J. Cell Biol. 34:327–343.
105. Tucker, J. B. 1971. J. Cell Sci. 9:539–567.
106. Watanabe, Y. 1963. Jap. J. Med. Sci. Biol. 16:107–124.
107. Watanabe, Y. 1971. Exp. Cell Res. 68:437–443.
108. Whitson, G. L. and Padilla, G. M. 1964. Exp. Cell Res. 36:667–671.
109. Williams, N. E. 1960. J. Protozool. 7:10–17.
110. Williams, N. E. 1961. J. Protozool. 8:403–410.
111. Williams, N. E. 1964. In E. Zeuthen, ed., Synchrony in cell division and growth, Interscience, New York, pp. 159–175.
112. Williams, N. E. 1964. J. Protozool. 11:566–572.
113. Williams, N. E. and Frankel, J. 1973. J. Cell Biol. 56:441–457.
114. Williams, N. E., Frankel, J., and Jaeckel, R. F. 1972. J. Protozool. (Abstr.) 19:20–21.
115. Williams, N. E. and Luft, J. H. 1968. J. Ultrastruct. Res. 25:271–292.
116. Williams, N. E., Michelsen, E., and Zeuthen, E. 1969. J. Cell Sci. 5:143–162.
117. Williams, N. E. and Nelsen, E. M. 1973. J. Cell Biol. 56:458–465.
118. Williams, N. E. and Scherbaum, O. H. 1959. J. Embryol. Exp. Morphol. 7:241–256.
119. Williams, N. E. and Zeuthen, E. 1966. C. R. Trav. Lab. Carlsberg 35:101–118.
120. Wood, W. B. and Henninger, M. 1969. J. Mol. Biol. 39:603–618.
121. Zelnis, T. 1967. M.S. Thesis, University of Iowa.

CHAPTER 13

Tetrahymena as a Nutritional Pharmacological Tool

S. H. Hutner

Haskins Laboratories at Pace College
New York, New York

Herman Baker

Departments of Medicine and Preventive Medicine
College of Medicine and Dentistry of New Jersey
East Orange, New Jersey

Oscar Frank

Department of Medicine
College of Medicine and Dentistry of New Jersey
East Orange, New Jersey

and

Dudley Cox

Haskins Laboratories at Pace College
and Department of Biology
Pace College
New York, New York

I. INTRODUCTION

Respect for human dignity restricts the use of patients for pharmacological trials, thus enhancing the need to know the pharmacological activities *Tetrahymena* detects. The steps in clinical drug testing are tortuous indeed (7). Government regulation favors errors of omission over errors of commission (161). If organic chemists and microbiologists are not to confront a huge backlog of compounds awaiting preliminary biological tests, the in vitro screening has to be a simple yet good imitation of in vivo conditions, notably in detecting cytotoxic substances. *Tetrahymena* is used to detect cytotoxic substances but, probably not so as to exploit fully its sensitivity.

It is only a matter of time before the entire pharmacopoeia is ransacked, especially those drugs employed as biochemical tools, to pry into synchronous cell division in *Tetrahymena;* hydroxyurea (23), *p*-fluorophenylalanine, L-canavanine, and azaserine were so employed (124), to select some examples.

Structures for nearly all pharmaceuticals mentioned here and a précis of each can be found in comprehensive pharmacology texts (45,70).

II. MAINTENANCE

A. General Notes

The more often cultures are transferred, the greater the risk of contamination. Our conservation rituals follow.

For any important strain three lines are maintained, A, B, and C, each on a different medium. The A line is used least; the B line is used to start new C lines. C lines are for inocula for experiments; therefore every few months the C line is discarded and a new one started from B. Thus C is a high-risk line, B the inoculum reserve line, and A the lowest-risk line, hence the last reserve. The A line should be kept so as to require the longest possible interval between transfers. Also, the fewer the cell divisions, the greater the genetic stability. Our experience is almost entirely with syngen 1, mating types I and II. The A series is kept on a biphasic medium distributed in 125 × 20 mm screwcap tubes. The solid phase had (grams per 100 ml final medium): dextrin, 0.8; sodium acetate·3H$_2$O, 0.06; yeast hydrolysate, 0.5; Liver L, 0.06; Hycase SF, 0.5; and agar, 1.6; pH 7.1–7.4. The agar base is melted and distributed 5 ml per tube. The tubes, caps loosened, are autoclaved slanted for 15 min at 118–121°C and allowed to cool slowly in the autoclave to room temperature. The caps are tightened immediately on removing the tubes. Just before inoculation the agar is overlaid with ~3 ml of sterile distilled water dispensed with a 10- or 11-ml pipet, and then inoculated with 1 or 2 drops of a previous culture. One A line is incubated at 37°C. A duplicate line is incubated at 30°C. The 37°C line is kept at 37°C; it requires transfer about every 2 wk. After a 24-h incubation the 30°C line is placed at 15–18°C and is transferred every 4–5 wk.

The B line has been kept on various defined biphasic media at pH 6.3–6.8. Later media (36) have been designed to be stored frozen as dry mixes compounded from premix blocks. The C line is kept in broth (proteose–peptone 2.0 g/100 ml plus dextrin 0.05 g/100 ml, pH ~7.2), distributed 5 ml per tube at 30°C, and transferred weekly. Inoculation into defined media may then be made, the vessel depending on the nature of the experiment, for example, for vitamin assays 50- or 125-ml screwcap Bellco micro-Fernbach flasks conveniently contain 10–20 ml of "depletion" media; these larger volumes minimize carryover. For details of vitamin assays, see ref. 4.

Syngen-1 strains have a high temperature optimum (~37°C) and an upper temperature limit for growth—~40°C (133). Ionagar no. 2 (Colab Laboratories, North Chicago, Illinois) yields a very clear, firm medium. Yeast hydrolysate (Nutritional Biochemicals Corporation) is as good as the more expensive soluble yeast "extracts" and "autolysates."

Liver (fraction) L (Wilson's; obtainable from laboratory supply houses), the cheapest soluble liver digest, serves well. Hycase SF, an acid hydrolysate of casein (Sheffield Laboratories), is probably replaceable by any of the commercial pancreatic or peptic digests of casein or lactalbumin, for example, N-Z Amine, N-Z-Case, Tryptone, Trypticase, or Edamin. Agar media must not fall below about pH 6.0, or else the agar may hydrolyze on autoclaving. Liquid media, as for the C series, can be more dilute, and the dextrin reduced or omitted, prolonging viability. There are almost as many recipes for stock culture broths as workers, for example 2% proteose–peptone plus 1% glucose, with viability at least for 100 days for strain H (130); 1% proteose–peptone plus 0.1% yeast extract (125); 2% proteose–peptone plus 0.1% liver L (124). This profusion of recipes may stem in part from some batches of proteose–peptone–yeast extract media being iron-deficient—easily corrected by adding iron chelate, for example, iron EDTA (32). A substrate that may be important in peptone is serine (36) which is suppressed by glucose (42).

Rosenberg (134) kept strain WH-14 up to 6 mo on sterile infusions of rat brain under paraffin as described by Plesner et al. (122).

A simple procedure might entail the use of dehydrated organ or organ extract powders. Lyophilized whole glands and tissues are commercially available. Suspensions of such materials as sole nutrients might also serve in comparing phagotrophic abilities as well as fostering longevity. If phagotrophic or histotrophic strains are of advantage, it might be desirable to maintain selection pressure for phagotrophy by providing stock cultures with particulate foods corresponding to the kind of materials to be assayed. This problem may have arisen for *Ochromonas danica* and *O. malhamensis* maintained in rich, soluble media (98).

B. Chilled (0–5°C) and Heated Cultures

Freeze-storing was achieved at the American Type Culture Collection, but the freezing rate in liquid nitrogen N was critical (85,140), as emphasized by Wang and Marquardt (164). Another difficulty was that cultures stored at 0–5°C dried out in a few days. Crude soy lecithin protected. Activity resided in the sterols, for example, β-sitosterol, not in the phospholipids. Cultures stayed alive at 0–5°C for 4.5 mo in a rich, crude medium (grams per 100 ml final medium): trypticase, 0.5; liver L, 0.04; yeast hydrolysate, 0.5; sodium acetate (anhydrous), 0.01; dextrin, 0.05; pH 7.2 (adjusted with tris); supplemented with 0.004% soy lecithin (126). Adding soy lecithin had the advantage over adding sterols as such in that addition of antioxidant, for example, Ionox 330 (Ethyl Corporation, Baton Rouge, Louisiana), became unnecessary.

Airmailed cultures may alternate between near-freezing and tropical temperatures. Air-conditioning is subject to equipment and power failure. Therefore the medium might contain crude plant lecithin to provide phytosterol, phospholipid, and antioxidant, thus extending both temperature tolerances. Perhaps starch should be included to absorb toxic long-chain fatty acids as well as an autoclavable, utilizable carbohydrate. Lipidic media are demanded by *T. setifera* and *T. corlissi;* as noted later, certain lipid-requiring, voracious *Tetrahymena* may prove best for assessing proteins.

C. Agar Cultures; Antibiotics

Tetrahymena forms flat grayish colonies on moist agar. Contaminants on slants are easier to detect; if the *Tetrahymena* stay along a single line, most contaminants will fall outside it (and may form distinguishable colonies if they grow at all). Since *Tetrahymena* are readily axenized by a mixture of 250 μg/ml each of penicillin (presumably the sodium or potassium salt) and streptomycin (sulfate?) (50), it might be worthwhile to ascertain the solubility of this mixture in 70% ethanol—a nearly self-sterilizing solvent —thus avoiding having to filter-sterilize the antibiotics. Streptomycin even at 1.0 mg/ml did not inhibit *Tetrahymena* (96). This tolerance for antibiotics may simplify rescue of bacterially contaminated cultures.

If an assortment of antibiotics could be assembled that eliminates not only all bacteria and quick-growing yeasts and molds, even the subtle mycoplasmas, yet is harmless to *Tetrahymena,* experimentation with thermolabile or particulate materials would be aided. We have not found antifungal antibiotics and fungicides with relatively low toxicity for *Tetrahymena.* In agreement with older results noted below, broad-spectrum antibiotics had an uncomfortably narrow zone between antibacterial and *Tetrahymena*-inhibiting concentrations.

Cycloheximide (actidione) is quite toxic to *Tetrahymena* (104). Nothing was found that reversed its toxicity for this ciliate (34). Inhibition by low levels of puromycin was overcome by guanylic acid (17). Tetracyclines and chloramphenicol were highly toxic to *Tetrahymena* (72) in defined media, confirming earlier work on chloramphenicol (33,91,92,102,103). Neomycin had little toxicity for *Tetrahymena,* as was to be expected from its membership in the streptomycin family.

Several experimental antibiotics have been tested on *Tetrahymena* and other protozoa:

Demetric acid (39). Related to the antimycins and phyllomycins; unstable unsaturated acid. Minimal inhibitory concentrations (micrograms per milliliter): *Tetrahymena,* 12.5; *O. malhamenis,* 25; *Crithidia fasciculata,* 25; yeasts: *Kloeckera brevis,* 6.3; *Candida utilis,* 6.3; most bacteria required 100 μg/ml or more.

Leucinamycin (113). Weakly basic polypeptide, isolated from a *Streptomyces.* Antibacterial, antifungal, and antiprotozoal. Contains sulfur; foams strongly; LD_{58} for mice on intraperitoneal injection, 23 mg/kg. Minimal inhibitory concentrations (micrograms per milliliter): *Tetrahymena* (grown in half-strength nutrient broth), 50; *Trichomonas vaginalis,* 6.25–12.5; *Cryptococcus neoformans,* 0.78; *Saccharomyces cerevisiae,* 100; *Bacillus anthracis,* 0.78.

Hadamycin (20). Composition unknown. Detected in a fermentation broth by inhibition of mouse sarcoma 180. Inhibits gram-positive bacteria and HeLa cells and induced lysogeny. Exceedingly toxic to HeLa cells; minimal inhibitory concentration, 0.13 μg/ml, which is in the range of such antibiotics as antinoleukin, actinomycin D, and streptonigrin. Solutions decompose even at 6°C. Minimal inhibitory concentrations (micrograms per milliliter): *Tetrahymena,* 0.8; *Crithidia,* 12.5; *Ochromonas,* 1.6.

Pyrrolnitrin (71). This pseudomonad product, 3-chloro-4-(2-nitro-3-chloro-phenyl)pyrrole, is effective against dermatophytic fungi. Minimal inhibitory values (micrograms per disk): *Tetrahymena,* 2.5; *O. malhamensis,* 2.5; *T. vaginalis, Euglena gracilis, Chlorella,* and *Scenedesmus,* 10 each. *Trichophyton mentagrophytes,* a ring-worm fungus, was inhibited by 0.19 μg/disk.

III. BIOLOGICAL VALUE OF PROTEINS

Cereals are generally deficient in lysine; legumes in methionine plus cystine. Success in infant feeding by blending these plant proteins to make up for each other's deficiencies is spurring efforts to eliminate the chronic protein deficiency (kwashiorkor) that cripples millions of children in poor countries. The *Tetrahymena* assay for the biological value of proteins could be a powerful weapon against demographic misery. Sometimes fishmeal is assayed with *Tetrahymena* (25). Evidence is rapidly mounting that poor maternal and infant diets may impair mind as well as body (138). A recent United Nations' survey (157) is outstanding; it included an evaluation of proteins with allusions to *Tetrahymena.*

The value of *Tetrahymena* as a protein-quality reagent rests on its requiring the same amino acids needed by domestic animals and man, above all those generally deficient in plant foodstuffs. In rough order of decreasing importance are: lysine and methionine plus cystine, then tryptophan, threonine, and isoleucine and leucine (139). *Tetrahymena* amino acid requirements need not, then, coincide exactly with those of higher animals so long as it requires the limiting amino acids. *Tetrahymena* was early used to assay tryptophan (129,137) which tends to be destroyed on acid hydrolysis.

Tetrahymena should ideally also parallel higher animals in respect to:

1. Sensitivity to amino acid imbalances.
2. Utilization of nonessential amino acids (nonessential amino acids spare essential amino acids).
3. Demands on amino acids other than for protein synthesis.
4. Sparing use of protein as fuel by nonprotein fuels, for example, fat and carbohydrate.

Tetrahymena's capacity to catabolize amino acids—essential and nonessential—is uncharted. It brings up the question whether *Tetrahymena* has steroid hormones which stimulate gluconeogenesis (progesterone and the corticoids) or anabolism (androgens). There is as yet no evidence for *Tetrahymena* having steroid hormones, as discussed later.

Sheffner (139) cautions that in using *Tetrahymena* to gauge protein quality, ". . . culturing and measuring growth of the protozoan is complex and subject to many variables." Possible ways to improve the assay follow.

A. Choice of Strain

The W strain has been used almost exclusively; Kidder and Dewey (97) used it for their classic development of defined media. But its usual incubation temperature—25–27°C presupposes temperate zone norms or air-conditioned laboratories. Syngen-1 strains may be better; they prefer higher temperatures, while sharing the main vertebrate amino acid requirement pattern except that at least one strain requires serine. A carbohydrate-free medium has been devised; it appears suitable for clarifying relationships among substrates and other nutrients. This flexibility may surpass that now afforded by higher animals even where cost and convenience are not a consideration. Moreover, while the W strain is amicronucleate, hence asexual, various mating types of syngen 1 are available. Nutrition as affecting mating thus opens to analysis.

B. Basal Media

The old recipes of Kidder and Dewey (97) seem curiously sacrosanct to protein analysts. The media had hygroscopic constituents, for example, iron, calcium, copper, zinc, and choline chlorides, in addition to other variably hydrated, precipitate-prone compounds, for example, K_2HPO_4. *Tetrahymena* media have been formulated as dry mixes having fewer hygroscopic ingredients (49,145). Reynolds (127,128) recommended thermostable dextrin over glucose for faster growth and better responses to amino acids; Teunisson (149) suggested soluble starch. Dextrin is recommended for conservation media. Dry mixes are easily reconstituted in hot water at pH values below 6.0; CO_2 is then completely expelled and phosphate precipitates are virtually eliminated; media can then be adjusted to higher pH values with stable tris. A recent medium is suitable for pH values as low as 5.4, thanks to raising inorganic constituents (36).

Much is made of *Tetrahymena* phagotrophy in contributing to its efficiency in evaluating proteins. Nonetheless it is still necessary to predigest proteins with cyanide-activated papain (19,146) or pepsin (139). Some proteins, for example, soybean protein and egg albumin, are better utilized by *Tetrahymena* when heat-treated (162). A double digestion, imitating the animal gut—for example, peptic digestion followed by tryptic digestion, with the *Tetrahymena* assays carried out on a clear final filtrate—might be better. Elimination of particulates would obviate the need for indirect methods of measuring growth, for example, reduction in tetrazolium salt in final populations, then spectrophotometry of the acetone-extracted formazan reduction product (2), as is done for protein in oil (62) and legume (63) seeds. The tetrazolium method was deemed unsatisfactory (64); direct counts have therefore been used for large-scale comparative studies (19,146). Separation of cells from particulate matter was achieved by fine grinding, sedimentation, and screening (150).

Tetrahymena corlissi is avidly histophagous (78). When fed chunks of earthworm, only setae remained. This lipid-requiring strain therefore seems worth a trial for protein assay.

The need for simplification is emphasized by recent estimations of the availability of lysine and methionine in heat-processed proteins; chemical estimations of lysine were much higher than rat assay values (16).

Antiinsect fumigation of bulk foodstuffs is routine, for example, methyl bromide is used in grain elevators. 1,2-Dichloroethane has also come into wide use. It causes marked destruction of histidine and cystine, the latter attributable to formation of S,S'-ethylenebiscysteine; the synthetic compound did not inhibit *Leuconostoc mesenteroides*, *Tetrahymena*, or the rat (24).

IV. VITAMIN ASSAYS

Hindrances to the *Tetrahymena* assay of proteins also apply to vitamin assays. Nevertheless, *Tetrahymena* was early recommended for folic (48,95) and lipoic (thioctic) (145) acids. Its first clinical application seems to have been for estimating nicotinic acid, and then vitamin B_6, in tissues and body fluids (5). The *Tetrahymena* B_6 requirement has not yet been spared or bypassed by other metabolites. D-Alanine bypasses the B_6 requirement of *Streptococcus faecalis* (123). Furthermore, *Tetrahymena* responds more to pyridoxal and pyridoxamine—the forms of B_6 in tissues—than to pyridoxine; the widely used *Saccharomyces carlsbergensis* responds to these three forms of B_6 equally. B_6-Catalyzed reactions are of prime neurological interest since B_6 availability may govern levels of psychotropic (biogenic) amines in the brain. Comparison of the *Tetrahymena* B_6 requirement with that of *Crithidia* may nicely test the validity of both assays, and the *Tetrahymena* folic acid assay may likewise test the validity of the *Lactobacillus casei* assay for folic acid "activity" in blood serum (4).

Riboflavin assay for clinical use followed (5). Conjoint *Tetrahymena* assays for these three vitamins in blood, along with the *L. casei* assay for serum folates (an index of folate status), protozoan assays for thiamine, B_{12}, and biotin, and chemical determination of ascorbic acid, vitamin A and carotene, vitamin E, and total blood and serum cholesterol and triglycerides, were applied to 642 school children in New York City (5). Inadequate protein uptake led 74–78% of this population to fall below the mean of the total population in levels of thiamine, biotin, and ascorbate. Thiamine and biotin were determined with *O. danica*. One goal then is to make *Tetrahymena* assays match *O. danica* assays in simplicity. The basal media for *Tetrahymena* for these clinical studies were overcomplicated, but practical urgencies had dictated standardization of media for the sake of direct comparisons of clinical data stretching over a decade. Better media, for example, those described in ref. 36, were developed too recently for these surveys.

Tetrahymena assays for riboflavin, B_6, pantothenate, and nicotinic acid, and the rationale underlying their use in clinical work, are detailed in ref. 4.

The response of *Tetrahymena* to folic acid, together with *Crithidia* responses to unconjugated pteridines, was used to identify pteridines extracted from *Tetrahymena* (43), among them cileapterin, regarded by Brown (21) as the L-*threo* form of biopterin; biopterin has the *erythro* configuration.

An odd, still unconfirmed, finding is Ebringer's (47) observation that bacitracin (0.1–0.2 ppm) can satisfy the thiamine requirement in the presence of the pyrimidine moiety. Bacitracin has a thiazoline ring. Another oddity is the claim that sulfonamides spare the thiamine requirement of *Tetrahymena* (153); this work, ignored in recent reviews, seems not to be explicitly refuted.

V. PHOTODYNAMIC TEST FOR CARCINOGENIC HYDROCARBONS; ANTIOXIDANTS

Benzpyrene in air is an accepted index of air quality along with SO_2 content. As Hull (82) pointed out, wherever organic materials are heated above 750°C hydrocarbons form, carcinogens among them. A remarkably sensitive and reasonably specific test for polycyclic carcinogens is the photodynamic test; it detects carcinogenic hydrocarbons at 0.01 μg/ml in aqueous suspension. Earlier work was with *Paramecium* (for references see refs. 82 and 83); much later work was with *Tetrahymena*. The photodynamic test, essentially a photosensitized O_2-dependent oxidation (143), has been reviewed in relation to other in vitro tests of carcinogenicity (3). Presumably, epoxide formation in hydrocarbons, above all in carcinogenic ones, leads to hydroperoxidation of lipids in the cell membrane.

Experimental details are given by Hull (82) and Epstein et al. (54,57). Ciliates in depression slides are exposed to suspensions of carcinogen for ~2 h in the dark and then irradiated. Hull recommends a mercury vapor lamp emitting between 400 and 1500 μW/m^2 at a narrow peak at 368 nm. Small et al. (142) used ultraviolet light at 300–400 nm, peak 366 nm. All visible light is filtered out. Carcinogenic activity is measured as the reciprocal of the concentration (micrograms per milliliter) killing 90% in 30 min. The total time of observation seldom exceeds 1.5 h. Protective compounds, for example, antioxidants, are rated as the reciprocals of the concentrations producing a 2-fold prolongation of the time required for immobilization of 90% of the motile ciliates (LT_{90}), as determined from plots of concentration of test compound against median LT_{90} values. Benzpyrene (more precisely, benzo[a]pyrene) serves as the standard carcinogen and is dispensed from a freshly prepared aqueous suspension 0.1 μg/ml freshly prepared from stock acetone solutions, 100 μg/ml (57). Peptone media are suitable with the reservation that, there being evidence that carcinogenic hydrocarbons enter mainly by deceiving the tryptophan transport system (171), for greatest sensitivity media perhaps should not be as high in tryptophan as those based on casein digests. Epstein et al. (57) used 1.5% proteose–peptone plus 0.15% yeast extract at pH 6.8. Tween 60 interfered with benzpyrene uptake by *Tetrahymena* (56). Tests on 157 polycylics with *Paramecium* (60) revealed that high photodynamic potency was limited to compounds with four or five fused rings. Larger compounds were inactive, perhaps because of poor solubility or poor uptake. Compounds with high photodynamic activity had four times the probability of being carcinogenic as the others.

False negative results were obtained, especially with certain compounds lacking co-planarity. Conversely, some noncarcinogens were positive, for example, some linear polycycles. But this should not detract from the utility of the photodynamic test as a presumptive test in monitoring carcinogenic crudes. Chloroform extracts of treated drinking water in the northeastern United States had considerable activity (61), likewise a pooled sample of atmospheric particulates (55). Analysis of 217 subfractions from this sample showed that high activity was largely restricted to fractions high in benzpyrene, benz[a]anthracene, and 7-*H*-benz[d,e]anthracen-7-one. The validity of the photodynamic test was substantiated by comparing photodynamic activity with *Paramecium* of benzene extracts of atmospheric particulates from six American cities with (1) chemical identification of carcinogens and (2) carcinogenicity for neonatal mice injected subcutaneously (53). To illustrate the correlation: Cincinnati air, yielding the most photodynamically active extracts, yielded by far the most tumorous mice. At another extreme, extracts from Los Angeles air, low in photodynamic potency and in benzpyrene, yielded the fewest tumors.

If peroxidation of membrane lipids underlies photodynamic injury, lipid-soluble antioxidants should protect. In tests with benzpyrene this was so (58). *Tetrahymena* treated with 25 antioxidants and 4 photosensitizers (52) (benzpyrene and 3 other carcinogenic polycyclic hydrocarbons) showed linear relationships between antioxidant and sensitizer. The photodynamic activity of the sensitizers was similar whether in fine suspension or solubilized with the aid of serum proteins. Protection of liver mitochondria paralleled protection by whole cells.

A survey (59) then covered 97 compounds. They included vitamin E and some of its water-soluble derivatives, and most of the lipid-soluble hindered phenols used commercially as rancidity inhibitors in foods. Comparisons were made with toxicity for *Tetrahymena* (peptone medium) and with the literature on toxicity of these antioxidants fed to rats. The common rancidity inhibitor, 2,6-di-*tert* -butyl-*p*-cresol (BHT), had low toxicity to *Tetrahymena* but low activity by the photodynamic test. Another rancidity inhibitor, nordihydroguaiaretic acid, having low toxicity to rat and high toxicity to *Tetrahymena,* scored high.

Apart from public health importance, the photodynamic test with nonpolar compounds appears to be a way to study the ciliate cortex. Electron microscopy of photodynamic damage is needed to identify the site of damage. The blebbing that precedes photodynamic lethality (83) denotes membrane damage. Blebbing also marks injury at near-lethal elevated (133) or low temperatures (126).

As for water-soluble sensitizers, with methylene blue the sensitizer and energized by 700-nm light, photodynamic killing of *Tetrahymena* was inhibited by quinacrine and chloroquine; protection was not inhibited by quinacrine and chloroquine; protection was not attributable to their light absorption (22).

In *Tetrahymena* under 2.5 atm of 100% O_2, 90% died in 12 h after swelling to spheres (51). Again this suggests damage to the cell membrane or to the "water expulsion vesicle" which empties it (119).

VI. DIRECT EFFECTS OF CARCINOGENS

This is an accessible subject with the advent of water-soluble compounds approaching the structure of "proximate" carcinogens, that is, the active forms of carcinogens. Polycyclic benzenoid carcinogens administered to animals stimulate formulation of enzymes of liver microsomes, which in turn hydroxylate the hydrocarbons (18) and perhaps then oxidize them to compounds having free-radical properties; these may be the proximate carcinogens. The picture is clearer for azo dye carcinogens, for example, butter yellow. They are *N*-hydroxylated (108), then perhaps converted to free radicals via nitroso compounds. Accordingly, the *N*-hydroxy derivative of acetamidofluorene quickly killed *Tetrahymena* in a peptone medium, whereas the parent compound did not (147).

This hindsight explains why carcinogens have, as Tartar (148) puts it, been merely noxious, not teratogenic, as in the case of *Tetrahymena* treated with (154) 2-aminofluorene and *β*-naphthylamine (135).

At this time the carcinogen that may best combine availability, intense, immediate toxicity, and general carcinogenicity—signs that it is nearly the proximate carcinogen?—is

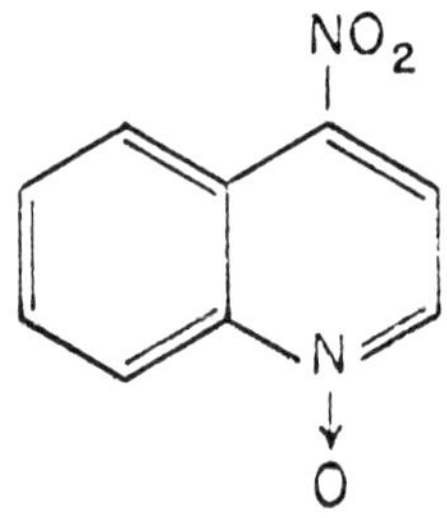

Fig. 1.

4-Nitroquinoline *N*-oxide. (From ref. 171.)

4-nitroquinoline *N*-oxide (NQO) (Fig. 1). Its target, as tested mainly on *Euglena* chloroplasts, appears to be the cytochrome site in the chloroplast corresponding to the cytochrome b site in mitochondria (80). The potency of NQO may reflect the easy reduction of the nitro group to a hydroxyamino group which is then oxidized to a nitroso group (96a).

Mita et al. reported (110,111) that NQO induces pleomorphic clones in the nonmating strain GL. Crosses of abnormal clones with normals by means of variety 1, mating types III and IV, led to progeny having uneven macromolecular divisions (112). Every subline from 50 hybrids divided normally. It was concluded that the abnormality was a recessive character, perhaps localized in the macronucleus and lost in the macronuclear reorganization after conjugation. Carcinogenic activity of 4-nitroquinoline 1-oxide derivatives and induction of unequal nuclear division in synchronized *Tetrahymena* ran parallel (109).

Aflatoxins B and G are exceedingly toxic, potent carcinogens produced by *Aspergillus flavus*. The W strain inactivated crystalline aflatoxin B, as shown by the appearance of a bright-blue fluorescence; G was not degraded (149). At 2 μg/ml neither affected mitosis but in 72 h many of the cells in G were rounded and nonmotile. *Tetrahymena* was less sensitive to aflatoxin than crustaceans, fish fry, or embryonated hen's eggs (40,41).

Interest in studies of carcinogenesis centers on the DNA repair process. The prerequisite for carcinogenicity may be a dual action: binding to DNA leading to strand breaking and induction of DNA repair (unprogrammed) syntheses but inhibition of repair. This does not contradict the idea expressed above that the site of action of NQO is a cytochrome, for mitochondrial DNA may be attached to one cytochrome. Much interest therefore attends the preliminary reports by Mouton (115) and Mouton and Fromageot (116) that the oncogenic mycotoxin luteoskyrin from *Penicillium islandicum*, even at 2×10^{-7} *M*, which had no effect on the multiplication and morphology of nonirradiated *Tetrahymena*, when added to the growth medium after ultraviolet irradiation showed an antirepair capacity equivalent to 10^{-3} *M* caffeine. Ueno and co-workers (155,156) had documented intense toxicity of luteoskyrin and related toxins to *Tetrahymena*, emphasizing the induction of nuclear abnormalities.

Hydrocarbon carcinogens may strongly attach to cell membranes. This attachment may interfere with the uptake of various cyclic metabolites, especially of tryptophan.

The effect of hydrocarbon carcinogens on invertebrates has been reviewed (79); regenerative cells are stimulated, but only unorganized cell masses form.

VII. *TETRAHYMENA* AS A PHARMACOLOGICAL TOOL

The use of *Tetrahymena* as a pharmacological tool had been foreshadowed by several articles, but it seems to have remained for Jírovec to use *Tetrahymena* explicitly to survey the toxicity of antibiotics. Tyrothricin and patulin were quite toxic; chloramphenicol, hydroxytetracycline, chlortetracycline, and subtilin less so (92). Jírovec noted that *Tetrahymena* tolerated 2% acetone. The year before he had reviewed use of *Tetrahymena* ". . . as a test object in pharmacology and physiology" (91). The advent of penicillin spotlighted the gap between procaryotes and eucaryotes.

A. Antitumor-Agent Screening

Tetrahymena, in a defined Tween 80-containing medium, was 1 of 15 organisms tested by Foley et al. (65) on 198 cytotoxic compounds; the aim was to evaluate screening procedures. Cultured KB human liver cells served for comparison. This immense survey should be consulted by those interested in the in vitro toxicity of most cytoxic compounds studied before 1958; only some highlights are given here. *Tetrahymena* proved sensitive in the sense that, of 89 compounds with reported antitumor activity, 82% inhibited it; "negative" compounds did not inhibit below 1 mg/ml. In contrast, only 39% of the compounds were positive, that is, inhibitory, for *Candida albicans.* The cell culture detected 79% of the compounds, but the superiority over bacteria was slight; *L. casei* detected 76%. Several of the test compounds were antimetabolites of purines and pyrimidines and therefore effective only in media with minimal purines and pyrimidines. In an even more extensive survey (152) (3837 compounds), *Tetrahymena* was compared with *Escherichia coli, L. casei,* KB cell cultures, and the Cancer Chemotherapy National Service Center three-tumor screen. The *Tetrahymena* results correlated best with those with the KB cells but were less selective.

The use of *Tetrahymena* for examining putative antitumor (oncolytic) agents may have had its first detailed treatment in an article in 1955 on 8-azaguanine (120), an antimetabolite then promising. An advantage of *Tetrahymena* in elucidating the cellular mode of action has not yet been demonstrated for any antitumor agent, but its use for screening for cytotoxic substances is growing. Foley et al. (64) state: ". . . microbiological assay systems can be devised readily by choice of proper strain and medium to detect compounds exhibiting antimetabolite activity against any specific essential metabolite." In general, their 12 main microbiological systems detected the same false positive compounds, that is, toxic compounds lacking antitumor activity. All missed these tumor-active compounds: formamide, *N*-methylformamide, *N*-methylacetamide, strophanthin K, and colchicine; the last mentioned, however, was extremely poisonous to the cell cultures. A combination of four microbiological systems responded to 95% of the 89 compounds in the series reported to have experimental antitumor activity. Based on the total 180 compounds tested, 30–35% were positives. The surprising correlation between the results with *Tetrahymena* and *S. carlsbergensis* was brought out in

Table 1.

Toxicity of Cytotoxic Compounds (66)

Active at 10 μg/ml	Active at 1.0 μg/ml
Actinomycin D	Hydrocortisone
8-Azaguanine	Netropsin HCl
Sodium azide	Protopine HCl
Cycloheximide	Pyrimethamine
	Amethopterin
	8-Hydroxyquinoline
	Puromycin

another study (65) in which the previous compounds were also tested on four different transplantable tumors. Mammalian cell cultures seemed to be superior to microbiological systems because they had fewer false positives—21%—compared with 56% for *Tetrahymena* and 54% for *S. carlsbergensis*.

Table 1 lists compounds of outstanding toxicity for *Tetrahymena*.

The shortcomings of *Tetrahymena* in antitumor screening emerged again in a study by Johnson et al. (93), who used Kidder and Dewey's medium A. They confirmed the superiority of mammalian cancer cells (HeLa and NCTC 1742 murine cells in medium 199 plus 10% horse serum) over *Tetrahymena, O. malhamensis, E. gracilis,* and *Scenedesmus basilensis. Tetrahymena* was sensitive to cycloheximide (toxic to all the organisms) and, in descending order, to amphotericin B, azaserine, and neomycin. *Tetrahymena* was rather insensitive to chlorotetracycline, chloramphenicol, streptomycin, polymyxin, 6-diazo-5-oxo-L-norleucine (DON), and the alkylating agents mitomycin, triethylene melamine, busulfan (Myleran, 1,4-dimethanesulfonoxybutane), and nitrogen mustard. White (167) found *Tetrahymena* also insensitive to mannitol Myleran; it was not inhibited by 0.1 mg/ml. Johnson et al. (94) also found *Tetrahymena* rather insensitive to amethopterin, cortisone and, in agreement with Foley et al. (65,166), to colchicine.

Essentially similar results had been obtained by Johnson et al. (92) with a 2% peptone medium. *Tetrahymena,* similar to *E. gracilis* and *O. malhamensis,* was insensitive to alkylating agents and cortisone.

Outweighing all these shortcomings may be its sensitivity, unlike the flagellates, to vinblastine (formerly vincaleukoblastine). Another *Vinca* alkaloid, vincristine, studied by Slotnick et al. (141), gave almost complete growth arrest of *Tetrahymena* at 25 μg/ml in a peptone–yeast extract or defined medium. Riboflavin or riboflavin mononucleotide at 10 μg/ml completely reversed the inhibition; the dinucleotide was ineffective. From incorporation studies with DNA precursors (leucine, uridine, thymidine), Slotnick concluded that vincristine selectively inhibits DNA synthesis in log-phase cells.

The importance of *Vinca* alkaloids as oncolytic agents, especially in acute leukemias (69), urges follow-up of this *Tetrahymena* work, especially to shed light on the neurotoxicity that limits their usefulness (169).

Work by the Bristol Laboratories group on responses of *Tetrahymena, O. malhamensis,* and *C. fasciculata* to 35 tumor-active and 26 tumor-inactive antibiotics substantiated the aforementioned conclusions as to the superiority of HeLa cells over the protozoa

(20). The three protozoa were grown in peptone media, and the HeLa cells in a synthetic medium fortified with 10% calf serum. At the cutoff level of 1 μg/ml, the HeLa cells were inhibited by 63% of the antibiotics, *Tetrahymena* by 20%, *Ochromonas* by 9%, and *Crithidia* by 6%. The converse was also observed; out of 100 antibiotic beers tested, 1 of the 50 tumor-active beers was active against all three. Despite the false positives, for example, among the tumor-inactive antibiotics 9 out of the 50 had antiprotozoal but not HeLa activity, it was concluded that protozoa were valuable in screening antibiotic beers for, in addition, the only in vitro activity displayed by several fermentation beers with antitumor activity was against protozoa.

Shortcomings in the knowledge of intracellular coordination in *Tetrahymena* are brought out by consideration of cancer chemotherapy. Stock (144) wonders whether the future of cancer chemotherapy lies not with the rather unselective "killer" drugs, for example, alkylating agents, but with subtler agents interfering with regulatory mechanisms. If *Tetrahymena* is to detect such agents, effects must be detected subtler than death through interference with the familiar metabolic systems, as evidenced by *Tetrahymena* nutritional requirements. Failure has accompanied all attempts to push protozoan hormones into the open by applying compounds that act in multicellular animals as antihormones by interfering with their synthesis or function. Thus growth inhibitions induced by hypocholesteremic agents, for example, triparanol, are annulled by long-chain fatty acids and not by sterol.

B. Water-Soluble Animal Hormones

The legend of the motorium as coordinator of ciliary metachrony (121) tempted many to equate it with a central nervous system. The availability of drugs blocking epinephrine, norepinephrine, serotonin, histamine, and acetylcholine provided attractive opportunities; sensitivity to antihormones, with reversal of toxicity or behavioral effects by the hormones, would substantiate the motorium idea. *Tetrahymena* might then serve for screening antihormones and eventually help elucidate the cellular mode of action of neurohormones. However, hydrodynamic viscous coupling, not a motorium, almost certainly underlies ciliary metachrony (88). Hope quickened with isolation of serotonin (89) and epinephrine and norepinephrine (90) from *Crithidia* and *Tetrahymena*, and of 3,4-dihydroxyphenylalanine (dopa) from *Crithidia* (90). These isolations do not prove that these hormones function in *Tetrahymena* in the protozoan version of a nervous system. *Tetrahymena* presumably synthesizes biopterin (43) whose tetrahydro coenzyme form catalyzes hydroxylations of aromatic metabolites leading to serotonin from tryptophan and epinephrine from tyrosine. The capacity of *Tetrahymena* to form these hormones is therefore not clearly controllable except perhaps by the use of biopterin inhibitors, among which may be numbered the 2,4-diaminopteridine and pyrimidine antifolics, for example, pyrimethamine.

Aaronson and Bensky (1) proposed the pharmacological approach to decide whether or not there was a ". . . protozoan analog of the metazoan nervous system." The anticholinesterase, physostigmine, at 0.03% did not inhibit growth of *Tetrahymena*. Exogenous choline was esterified; acetylcholine esterase could not be identified. Jahn and Bovee (87), after reviewing other work on protozoa with inhibitors of acetylcholine, including effects on *Paramecium* motility of neostigmine, atropine, and work on *Tetrahymena* on physostigmine and hexamethionium (160,165), concluded that acetylcholine has ". . . yet to be shown to have any role in the function of protozoa cilia or flagella . . . and hexamethonium . . . cannot be considered a specific acetylcholine esterase blocking agent for cilia, as it seems to be for vertebrate neuromuscular functions."

So far, polypeptide or protein toxins have been devoid of pharmacological properties for protozoa, for example, tetanus toxin for *Paramecium* (170) and staphylococcal enterotoxin (40,41). Insulin falls in this category; in a peptone medium it did not affect the rate of utilization of glucose (163).

Using an unspecified medium, Duerre et al. (46) claimed that 2.5×10^{-10} to 2.5×10^{-9} M L-thyroxine in well-aerated cultures accelerated growth but did not affect final growth; it increased the respiratory rate of growing cells. Higher concentrations inhibited both rate and extent of growth. Acceleration of growth with triiodothyronine was greater than with thyroxine.

Blum (10) sought information about a hormonal control mechanism by applying adrenergic blockers. He used either a crude medium (1% proteose–peptone plus 0.05% liver extract) or a defined medium fortified with 0.04–0.07% peptone (10). In accordance with earlier results (13), 2.1×10^{-5} M reserpine in the synthetic medium fortified with 0.05% peptone inhibited growth and reduced the cell content of epinephrine and norepinephrine (10); this inhibition was partly annulled by glucose. Dibenzylene, an α-adrenergic blocker, slightly inhibited growth at 0.1 mM and considerably potentiated growth inhibition by reserpine. Triiodothyronine inhibited above 0.1 mM. Propanolol, a β-adrenergic blocker, initially inhibited growth at 0.1 mM. Dichloroisoproterenol, another very potent β-adrenergic blocker, interfered with glycogen synthesis but not growth at 0.1 mM. Caffeine at 1.3 mM gave some potentiation of growth inhibition by reserpine. At high concentrations of reserpine, both 5'-AMP and 3',5'-cyclic AMP, each at 1 mM, partly protected against growth inhibition. Various monoamine oxidase inhibitors also inhibited growth; tranylcypromine at 1.0 mM inhibited; Catron (α-methyl-phenethylhydrazine) inhibited growth above 10^{-5} M, with noticeable inhibition at 7.2×10^{-6} M; similar to tranylcypromine it potentiated reserpine inhibition. Desipramine, a dibenazepine antidepressant, likewise strongly potentiated growth inhibition by reserpine.

Theophylline markedly increased glycogen synthetase activity, while diminishing phosphorylase activity in cells grown in a peptone–liver extract medium (12); both caffeine and theophylline increased 3',5'-adenosine monophosphate phosphodiesterase activity.

These results seem largely uninterpretable so far as denoting the presence of a conventional nervous system. In a preliminary note, Blum (11) concluded that the adrenergically reactive drugs in *Tetrahymena* controlled activity of isocitrate lyase.

Iwata et al. (86), using a peptone medium, noted growth stimulated by low concentrations of ergotamine, dibenamine, pyrogallol, dichloroisoproterenol, and phenylethylhydrazine SO$_4$, and growth inhibition by 5×10^{-5} propanolol or reserpine. These investigators later (87) added chlorpromazine and dibenzamine to the list of stimulants.

Clofibrate (ethyl p-chlorophenoxyisobutyrate; Atromid-S), a hypercholesteremic with unknown mode of action, appeared to interfere with glyconeogenesis as inferred from increased isocitric dehydrogenase and decreased isocitric lyase, and depletion of glycogen in cells grown in the peptone medium with as little as 0.12 mM of the compound (14).

VIII. MISCELLANEOUS DRUGS

A. Miscellaneous Antivitamins

McLaughlan et al. (105), using a defined medium, screened 44 compounds for growth inhibition resulting from antivitamin effects. Antipantothenate activity was shown by several oral antidiabetics (chlorpropamide, metahexamide, and tolbutamide). The following were somewhat more inhibitory in B_6-deficient media: chloroquine and chlorpromazine. Other threshold inhibitory concentrations (micrograms per milliliter) were: antipyretic: aminopyrine, 50.0; antihistamines (see also Section VIII, B: chlorpheneramine maleate, 1; diphenhydramine, 10; hydroxyzine, 10; methapyrilene HCl, 25; tranquilizers: chlorpromazine, 1; promazine, 1; thiopropazate·2 HCl, 10; antihypertensive: hydralazine HCl, 5; psychic energizer; isoniazid, 25.

Krezanoski (101) found the related iproniazid completely inhibitory for *Tetrahymena* between 3×10^{-3} and 3×10^{-4} M, but pyridoxine depressed respiration and, presumably, growth almost to the same extent.

An agar overlay assay method was described by West et al. (166) for protozoa including *Tetrahymena*. Each dish received an agar base layer containing nutrients (peptone, yeast extract, glucose) plus 1.5% agar, and then an agar overlay consisting of 3% gelatin plus 0.4% New Zealand agar (Ionagar) to which was added 0.1 ml of penicillin–streptomycin solution plus 2 ml of inoculum. The agar concentration chosen represents a compromise between growth inhibition and restriction of movement so as to obtain sharp colonies. Compounds being tested were applied to paper disks placed on the overlay agar. The threshold inhibitory values for the standard cytotoxic compounds were (micrograms per disk): actinomycin D, 25; amethopterin, 100; mitomycin C, 100; and urethan, 100.

B. Antihistamines; Tranquilizers

The antihistamines diphenhydramine, tripelennamine, and pheniramine inhibited growth of *Tetrahymena* in a synthetic medium, and motility of suspensions, in the order listed. Histidine counteracted these inhibitions (136).

Phenothiazine and related tranquilizers are used on a vast scale as antipsychotics. Their ability to uncouple oxidative phosphorylation obviously does not predict their psychic effects. They have been reviewed many times, with the consensus (for example, ref. 159) that they have a very complex pharmacology; their effects on mammalian cell cultures seem directed at least partly at lysosomal and cell membranes (37).

In the presence of histidine, which potentiated immobilization by phenothiazines (118), Guttman and Friedman (75) compared 13 antipsychotic phenothiazines. When less phenothiazine was required to immobilize in the presence of histidine, this was taken as an index of increased permeability. Trifluoperazine had an index of 8, and chlorpromazine 7; their sulfoxides were both rated 1. The toxicity of chlorpromazine was annulled by lecithin or oleate-containing Tween 80 and 85; Tween 60 (palmitate) was inactive. It was suggested that because of this permeability-enhancing activity phenothiazines be used in conjunction with other drugs.

Rogers (131) found promazine to be the most effective inhibitor of growth in a defined medium, and trifluoperazine the most effective inhibitor of glucose uptake. Inhibition was associated in the exponential phase with a 60% increase in lipid phosphorus (132). Chlorpromazine in a peptone medium completely inhibited uptake of thymidine and uridine at 8 μg/ml, and to some extent uptake of amino acids; it inhibited the incorporation of acetate into lipid hardly at all, hinting that nucleic acid synthesis may be involved in chlorpromazine action (27). Chlorpromazine differed from 2,4-dinitrophenol in type of phosphorus distribution, showing that uncoupling of oxidative phosphorylation did not account for all the toxicity of chlorpromazine.

C. Hypocholesteremics

These drugs, aimed at lowering the cholesterol level of the blood, are treated elsewhere in this volume in relation to their value in investigating lipid metabolism. Inhibition of growth of *Tetrahymena* in a defined medium by 3-dialkylaminoethoxy steroids paralleled their hypocholestemic activity for the rat (81). These inhibitions were at best only partly overcome by cholesterol. The following hormonal steroids used as controls at 10 μM gave less than 48% growth inhibition: androst-4-ene-3,17-dione; corticosterone; dehydroepiandrosterone; estradiol; estrone; progesterone; and testosterone. As emphasized by Holmlund (79,80), the mode of action of growth inhibitory steroids is unclear; dehydroandrosterone (50 μM) and progesterone (10 μM) inhibited respiration 10–15% and blocked RNA synthesis within 30 min by 95 and 40% respectively. 3β-(β-Di-methylaminoethoxy)androst-5-en-17-one, which has potent hypocholesteremic properties had no effect on respiration or on incorporation of acetate into lipids, but sharply curtailed synthesis of tetrahymanol. Similar growth inhibition by deoxycortisone, progesterone, and diethylstilbestrol were noted in a peptone medium (177); progesterone preferentially localized at the oral region. Indoleacetate, gibberellin, insulin, thyroxine, and triiodothyronine did not stimulate growth. Hypercholesteremic agents promise to remain useful for investigating lipid metabolism in protozoa; for a succinct review of their pharmacology, see ref. 158.

D. Antimalarials; Antipteridines; Sulfonamides

In a series of 2,6-diaminopyridines, particularly the 4-phenoxy derivatives, growth inhibition of *Tetrahymena* was reversed by Tween 80, and of *Crithidia* by biopterin (107). Animal lecithin also had high reversing activity (44), as did cholesterol, sitosterol, and stigmasterol, but not oleate or the fatty acids obtained by saponification of the lipids of *Tetrahymena*. Activity resided in a lecithin fraction yielding an abnormal Lieber-mann–Burchard test and precipitable with digitonin; it was not tetrahymanol.

Inhibition by 2,4-diamino-6,7-isoquinoline of *Tetrahymena* and *L. casei* was reversed by DNA; analysis stopped there (9).

Strain GL synchronized through seven cycles in a peptone medium was 43% inhibited by quinacrine (atebrine) HCl (8.0 μg/ml at the first division and 36% at the second division—an effect ascribed to interaction with DNA (26). Quinacrine was long ago observed to inhibit growth of *Tetrahymena* (73). By using defined media interference with riboflavin or purines might well be demonstrable. Clancy (28), using syngen 1, mating type II, also with a peptone medium, likewise found that quinacrine was quite toxic, the minimal lethal dose (lowest concentration stopping motility in all cells) was 17 μg/ml. Primaquine, proguanil, and quinine were much less toxic; the minimal lethal dose for chloroquine was 1.5 mg/ml. Sensitivity was much greater at 37°C than at room temperature, the lethal dose of chloroquine being reduced to 79 μm/ml. Metronidazole was inactive at 0.3 mg/ml, sulfadiazine at 416 μg/ml, and nitrofurantoin at 0.16 μg/ ml.

The mystery enveloping the mode of action of the di- and polycyclic antimalarials has not been dispelled by the finding that quinine inhibited synchronized cell division in a peptone medium, DNA synthesis being more sensitive than RNA and protein synthesis (30); primaquine and chloroquine acted as uncouplers of oxidative phosphorylation and inhibited oxidation of some Krebs cycle intermediates, β-hydroxybutyrate and glutamate (31); quinacrine inhibited the incorporation of thymidine of uridine by intact cells but had little effect on uptake, hence may inhibit the synthesis of nucleic acids; it also inhibited the uptake of amino acids, while quinine, chloroquine, and primaquine inhibited the uptake of all the precursors used (thymidine, uridine, and uniformly labeled amino acids) (29). The antibacterial agent nalidixic acid inhibited the incorporation of thymidine more than uptake, showing an interference with nucleic acid metabolism (29).

Strontium accumulation in *Tetrahymena* occurred only in rapidly multiplying cells (6). Inhibition of accumulation in a peptone medium was generally more sensitive to drugs than growth; sulfadiazine, at a concentration reducing growth 75%, reduced strontium accumulation 25%. This held for dinitrophenol, dicumarol, and physostigmine.

E. Antibalantidial Chemotherapy

Moving Balantidiidae from the heterotrichs to the trichostomes (35) puts them in the subclass Holotrichia, where *Tetrahymena* resides, and sharpens the issue of the value of *Tetrahymena* in screening for anti-*Balantidium* agents. Balantidiasis is worldwide; the infection can be severe, even fatal (168). The effective drugs are arsenicals, 7-iodo-8-quinolinol-5-sulfonate, pyrimethamine, and chloroquine; emetine is very active, but tetracyclines offer the best chance of successful treatment (168), presumably by suppressing food bacteria.

F. Radioopaque Agents

These compounds are used as x-ray contrast media in clinical diagnosis. The acetrizoate (3-acetamido-2,4,6-triodobenzoate) series and 3,5-diiodo-4-oxo-1(4*H*)-pyridineacetic acid neutralized with various highly polar bases, for example, diethanolamine, or used as the sodium salts, are popular. Their toxicity has been a problem at times. Seven of these agents were tested on *Tetrahymena* grown in peptone media, reduction in motility being the index of toxicity (106). Toxicities were in fair agreement with local toxicity findings in higher animals. The work was extended by Bertoni (8), who found that sodium salts inhibited more than the methylglucamine salts.

G. Miscellaneous

In concentrations of 1/800–1/400 *M*, morphine inhibited *Tetrahymena* (38). A dermatitis-producing principle from the blue-green alga *Lyngbya majuscula* lysed *Tetrahymena* even at 0.02 mg/ml (114).

H. Virus Multiplication in *Tetrahymena*

Remarkable experiments have been described by Kóvacs and co-workers on the multiplication of mammalian arthropod-borne viruses in *Tetrahymena*. Partly purified encephalomyocarditis virus from L-cell cultures, fed to *Tetrahymena,* reduced the ciliate count, and with an increase in virus titer (99). Also, intact poliovirus was recovered from *Tetrahymena* fed virus. In further experiments (100) death of *Tetrahymena* was paralleled by hemagglutinin assay of recovered virus. Infected *Tetrahymena* tended to be sluggish, clumped, and poorly motile; this clumping was thought perhaps to be caused by a process simulating hemagglutination and hemadsorption.

It had long ago been demonstrated that *Tetrahymena* destroys influenza virus (74,75). Presumably, viruses differ in digestibility and ciliates in ability to digest viruses. *Tetrahymena* may perhaps serve as source of enzymes for virus dissection.

IX. EPILOGUE

Media for *Tetrahymena* are still a crude abstraction from natural conditions; in nature *Tetrahymena* must be exposed to many more metabolites than are represented in media. Some of these constituents may act as minor substrates. Synthesis of other constituents may depend on reactions catalyzed by *Tetrahymena* growth factors. If these metabolites permeate cells, they would act as sparing factors for major metabolites. If these minor substrates and sparing factors yielded the same mass of *Tetrahymena* from a smaller weight of medium, it might mean that such more complicated but less concentrated media would be more responsive to agents whose antimetabolite activity accounted for their main mode of action or toxic side actions. By this logic, media for *Tetrahymena* mode-of-action studies should epitomize all knowledge of biosynthetic pathways, consonant with ease of standardization. Perhaps at this stage of knowledge, for greatest sensitivity, defined media should be employed along with as-dilute-as-possible crude media.

Growth techniques no matter how refined may turn out to be crude compared with behavioral methods, especially in dealing with agents influencing hormone-sensitive states in higher animals. This may hold above all with psychotropic agents. Effects in *Tetrahymena* on growth or motility may go no deeper than blocking the uptake or utilization of essential nutrients. Catecholamine antagonists perhaps interfere only superficially with phenylalanine and tyrosine metabolism rather than influencing the actual effectors.

A cautionary instance of the ease with which antimetabolite effects can mislead was afforded by thalidomide. The toxicity of rather high concentrations for *Tetrahymena* was counteracted by nicotinic acid (67). But teratogenicity depends on thalidomide hydrolyzing to an alkylating agent, perhaps similar to the way the antitumor antibiotic mitomycin is metabolically transformed to an alkylating agent (144). Such results tell something about how a drug enters a cell even if they hardly explain its subsequent action.

The mass of pharmacological results to date with *Tetrahymena* may be quite superficial: hardly more than hints of the transport mechanisms available to permit entry of foreign compounds. The impression from work with psychotropic agents is as though one is practicing neurosurgery with crowbars and sledgehammers—but it is a necessary step in laying bare the deep-hidden metabolic homologies between neuronal activity in metazoa and intracellular communication in protozoa.

REFERENCES

1. Aaronson, S. and Bensky, B. 1963. J. Protozool. 10(suppl.):8.
2. Anderson, M. E. and Williams, H. H. 1951. J. Nutr. 44:335–343.
3. Arcos, J. C., Argus, M. F., and Wolf, G. 1968. Chemical induction of cancer, vol. I, Academic Press, New York, pp.447–449.
4. Baker, H. and Frank, O. 1968. Clinical vitaminology: Methods and interpretation. John Wiley, New York.
5. Baker, H. and Frank, O., Feingold, S., Christakis, G., and Ziffer, H. 1967. Amer. J. Clin. Nutr. 20:850–857.
6. Ballantine, R. and Burford, D. D. 1964. Life Sci. 3:1455–1458.
7. Bechtol, L. D. 1968. In J. L. Rabinowitz and R. M. Myerson, eds., Topics in medicinal chemistry, vol. 2, John Wiley, New York, pp. 277–305.
8. Bertoni, L. 1967. Experientia 23:59–60.
9. Bird, O. D., Oakes, V., Undheim, K., and Rydon, H. N. 1964. In W. Pfleiderer and E. C. Taylor, eds., Pteridine chemistry, (Proceedings of the 3rd International Symposium, Stuttgart, 1962), Pergamon Press, New York, pp. 417–425.
10. Blum, J. J. 1967. Proc. Natl. Acad. Sci. U.S. 58:81–88.
11. Blum, J. J. 1968. Fed. Proc. 27:807.
12. Blum, J. J. 1970. Arch. Biochem. Biophys. 137:65–74.
13. Blum, J. J., Kirshner, N., and Utley, J. 1966. Mol. Pharmacol. 2:606–608.
14. Blum, J. J. and Wexler, J. P. 1968. Mol. Pharmacol. 4:155–161.
15. Blumberg, A. J. and Loefer, J. 1952. Physiol. Zool. 25:276–282.
16. Boctor, A. M. and Harper, A. E. 1968. J. Nutr. 94:289–296.
17. Bortle, L. and Oleson, J. J. 1955. Antibiot. Ann., 1954/1955, 766–770.
18. Boyland, E. 1970. In D. Bergmann and B. Pullman, eds., Physico-chemical mechanisms of carcinogenesis. Israel Academy of Science, Jerusalem: Academic Press, New York, pp. 25–44.
19. Boyne, A. W., Price, S. A., Rosen, G. D., and Stott, J. A. 1967. Brit. J. Nutr. 21:181–206.
20. Braduer, W. T., Heinemann, B., and Gourevitch, A. 1967. Antimicrob. Agents Chemother. 1966, 613–618.
21. Brown, G. M. 1971. Adv. Enzymol. 35:35–77.
22. Buchnicek, J. and Turek, B. 1968. Arzneim.-Forsch. 17:760–761.
23. Byfield, J. E. and Scherbaum, O. H. 1967. Nature 216:1017–1018.
24. Campbell, J. A. and Morrison, A. B. 1966. Fed. Proc. 25:130–136.
25. Celliers, P. B. 1961. S. Afr. J. Agr. Sci. 4:191–204.
26. Chou, S. C., Ramanathan, S., and Cutting, W. C. 1968. Pharmacology 1:60–64.
27. Chou, S. C., Ramanathan, S., Heu, P., and Conklin, K. A. 1971. Pharmacology 6:1–8.
28. Clancy, C. F. 1968. Am. J. Trop. Med. Hyg. 17:359–363.
29. Conklin, K. A., and Chou, S. C. 1972. J. Pharmacol Exp. Therap. 180:158–166.
30. Conklin, K. A., Chou, S. C., and Ramanathan, S. 1969. Pharmacology 2:247–256.
31. Conklin, K. A., Chou, S. C. and Heu, P. 1971. Biochem. Pharmacol. 20:1877–1882.
32. Conner, R. L. and Cline, S. G. 1964. J. Protozool. 11:486–491.
33. Cooney, N. R. 1954. J. Protozool. 1(suppl.):7.
34. Cooney, W. J. and Bradley, S. G. 1963. Antimicrob. Agents Chemother., 1961, 237–244.

35. Corliss, J. O. 1967. In M. Florkin and B. J. Scheer, eds., Chemical zoology, vol. 1, Protozoa (G. W. Kidder, ed.), Academic Press, New York, pp. 1–20.
36. Cox, D., Frank, O., Hutner, S. H., and Baker, H. 1968. J. Protozool. 15:713–716.
37. Dawson, M. 1972. Cellular pharmacology. C. C. Thomas, Springfield, Illinois.
38. Dessi, P. 1965. Arch. Ital. Sci. Farmacol. 15:172–176.
39. DeVault, R. L., Schmitz, H., and Hooper, I. 1966. Antimicrob. Agents Chemother., 1965. 796–800.
40. De Waart, J. 1967. Antonie van Leeuwenhoek J. Microbiol. Serol. 33:230.
41. De Waart, J. 1969. Progress in protozoology, Proceedings of the 3rd International Congress on Protozoology, Leningrad, 1969. Swets and Zeitlinger, Amsterdam, 191.
42. Dewcy, V. C. and Kidder, G. W. 1966. J. Gen. Microbiol. 22:79–92.
43. Dewey, V. C., and Kidder, G. W. 1967. J. Chromatogr. 31:326–36.
44. Dewey, V. C., Kidder, G. W., and Markees, D. G. 1967. 7th Int. Congr. Biochem., Tokyo, 1967., abstr. J-188 (p. 993), Science Council Japan, Ueno Park, Japan.
45. DiPalma, J. R., ed. 1971. Drill's pharmacology in medicine, 4th ed., McGraw-Hill, New York.
46. Duerre, J. A., McDougall, J. C., Bowden, P. M., and Miller, C. H. 1967. Bacteriol. Proc. Amer. Soc. Microbiol. 67th Meeting, New York, 35.
47. Ebringer, L. 1963. In J. Ludvík, J. Lom, and J. Vávra, eds., Progress in protozoology, Proceedings of the International Congress on Protozoology, Prague, 1961. Czechoslovak Academy of Sicience, Prague; Academic Press, New York, pp. 198–199.
48. Eigen, E. and Shockman, G. D. 1963. In F. Kavanagh, ed., Analytical microbiology, vol. 2, Academic Press, New York, pp. 431–488.
49. Elliott, A. M., Brownell, L. E., and Gross, J. A. 1954. J. Protozool. 1:193–199.
50. Elliott, A. M. and Hayes, R. E. 1955. J. Protozool. 2:75–80.
51. Elliott, A. M., Travis, D. M., and Bak, I. J. 1962. Biol. Bull. 123:495.
52. Epstein, S. S., Forsyth, J., Saporoschetz, I. B., and Mantel, N. 1966. Radiat. Res. 28:322–335.
53. Epstein, S. S., Joshi, S., Andrea, J., Mantell, N., Sawicki, E., Stanely, T., and Tabor, E. C. 1966. Nature 212:1305–1307.
54. Epstein, S. S., Koplan, J., Jones, H., Mantel, N., and Hutner, S. H. 1963. J. Natl. Cancer Inst. 31:163–168.
55. Epstein, S. S., Mantel, N., and Stanely, T. W. 1968. Environ. Sci. Technol. 2:132–138.
56. Epstein, S. S. and Niskanan, E. E. 1967. Exp. Cell. Res. 46:211–234.
57. Epstein, S. S., Saporoschetz, I. B., Small, M., Park, W., and Mantel, N. 1965. Nature 208:655–658.
58. Epstein, S. S., Saporoschetz, I. B., and Mantel, N. 1966. Life Sci. 5:783–793.
59. Epstein, S. S., Saporoschetz, I. B., and Hutner, S. H. 1967. J. Protozool. 14:238–244.
60. Epstein, S. S., Small, M., Falk, H. L., and Mantel, N. 1964. Cancer Res. 24:855–862.
61. Epstein, S. S. and Taylor, F. B. 1966. Science 154:261–263.
62. Evans, R. J. and Bandemer, S. L. 1967. Cereal Chem. 44:417–426.
63. Fernell, W. R. and Rosen, G. D. 1956. Brit. J. Nutr. 10:143–145.
64. Foley, G. E., Eagle, H., Snell, E. E., Kidder, G. W., and Thayer, P. S. 1958. Ann. N. Y. Acad. Sci. 76:952–960.
65. Foley, G. E., McCarthy, R. E. M., Binns, V. M., Snell, E. E., Guirard, B. H., Kidder, G. W., Dewey, V. C., and Thayer, P. S. 1958. Ann. N. Y. Acad. Sci. 76:413–438.
66. Frank, O., Baker, H., Ziffer, H., Aaronson, S., and Hutner, S. H. 1963. Science 139:110–111.
67. Frank, O., Ning, M., Gellene, R. A., Hutner, S. H., and Leevy, C. M. 1960. Amer. J. Clin. Nutr. 18:123–133.
68. Goldin, A., Sandberg, J. S., Henderson, E. S., Newman, J. W. Frei, E., III, and Holland, J. F. 1971. Cancer Chemother. Rep. 55:309–507.
69. Goodman, L. S. and Gilman, A., eds. 1970. The pharmacological basis of therapeutics, 4th ed., Macmillan, New York.
70. Gordee, R. S. and Matthews, T. R. 1968. Antimicrob. Agents Chemother., 1967. 378–387.
71. Gross, J. A. 1955. J. Protozool. 2:42–47.
72. Groupe, V. 1945. Proc. Soc. Exp. Biol. Med. 60:321–323.

73. Groupe, V., Herrmenn, E. C., and Rauscher, F. J. 1955. Proc. Soc. Exp. Biol. Med. 60:321–323.
74. Groupe, V. and Pugh, L. 1952. Science 115:307–308.
75. Guttman, H. N. and Friedman, W. 1963. Trans. N. Y. Acad. Sci., Ser. II 26:75–89.
76. Hamana, K. and Iwai, K. 1971. J. Biochem. (Tokyo) 69:463–469.
77. Harman, W. J. and Corliss, J. O. 1956. Trans. Amer. Microsc. Soc. 75:332–333.
78. Harshbarger, J. C. 1967. Fed. Proc. 26:1693–1697.
79. Holmlund, C. E. 1971. Biochim. Biophys. Acta 248:363–378.
80. Holmlund, C. E. and Bohonos, N. 1966. Life Sci. 5:2133–2139.
81. **Hull, R. W. 1963. In J. Ludvík, J. Lom, and J. Vávra, eds., Progress in protozoology,** Proceedings of the 1st International Congress on Protozoology, Prague, 1961, Czechoslovak Academy Science, Prague; Academic Press, New York, pp. 182–186.
82. Hull, R. W. 1964. Develop. Ind. Microbiol. 6:35–43.
83. Hutner, S. H., Zahalsky, A. C., and Aaronson, S. 1967. In T. W. Goodwin, ed., Biochemistry of chloroplasts, vol. 2, Academic Press, New York, pp. 703–720.
84. Hwang, S. W., Davis, E. E., and Alexander, M. T. 1964. Science 144:64–65.
85. Iwata, H., Kariya, K., and Fujimoto, S. 1968. Jap. J. Pharmacol. 18:329–330.
86. Iwata, H., Kariya, K., and Fujimoto, S. 1969. Jap. J. Pharmacol. 19:275–281.
87. Jahn, T. and Bovee, E. C. 1967. In T. T. Chen, ed., Research in protozoology, vol. 1., Pergamon Press, New York, pp. 41–200.
88. Janakidevi, K., Dewey, V. C., and Kidder, G. W. 1966. Arch. Biochem. Biophys. 113:758–759.
89. Janakidevi, K., Dewey, V. C., and Kidder, G. W. 1966. J. Biol. Chem. 241:2576–2578.
90. Jírovec, O. 1950. Schweiz. Z. Pathol. Bakteriol. 13:129–138.
91. Jírovec, O. 1951. Pathol. Bakteriol. 14:653–666.
92. Johnson, I. S., Simpson, P. J., and Cline, J. C. 1962. Cancer Res. 22:617–626.
93. Johnson, I. S., Vlantis, J., Mattas, B., and Wright, H. F. 1961. 4th Canadian Cancer Conference, 1960, Academic Press, New York, pp. 339–353.
94. Jukes, T. H. 1955. Meth. Biochem. Anal. 2:121–151.
95. Kataoka, N., Imamura, A., Kawazoe, Y., Chihara, G., and Nagata, C. 1967. Bull. Chem. Soc. Japan 40:62–68.
96. Kerr, N. S. 1961. Amer. Zool. 1:364.
97. Kidder, G. W. and Dewey, V. C. 1951. In Lwoff, A., ed., Biochemistry and physiology of protozoa, vol. 1, Academic Press, New York, pp. 323–400.
98. Klein, S. 1972. J. Protozool. 19:140–143.
99. Kovacs, E. and Bucz, B. 1967. Life Sci. 6:347–358.
100. Kovacs, E., Kolompar, G., and Bucz, B. 1967. Life Sci. 6:2359–2371.
101. Krezanoski, J. Z. 1961. Pharm. Sci. 50:421–424.
102. Leofer, J. B. 1950. Proc. Am. Soc. Protozool. 1:8.
103. Loefer, J. B. 1951. Physiol. Zool. 24:155–163.
104. Loefer, J. B. and Matney, T. S. 1952. Physiol. Zool. 25:272–276.
105. McLaughlan, J. M., Shenoy, K. G., and Campbell, J. A. 1961. J. Pharm. Sci. 50:59–63.
106. Mark, M. F., Imparato, A. M., Hutner, S. H., and Baker, H. 1963. Angiology 14:383–389.
107. Markees, D. G., Dewey, V. C., and Kidder, G. W. 1968. J. Med. Chem. 11:126–129.
108. Miller, J. A. and Miller, E. C. 1970. In E. D. Bergman, and B. Pullman, eds., Physico-Chemical Mechanism of Carcinogenesis, Israel Academy of Science, Jerusalem; Academic Press, New York, pp. 237–261.
109. Mita, T., Kawazoe, Y., and Araki, M. 1969. Gann 60:155–160.
110. Mita, T., Tokuzen, R., Fukuoka, F., and Nakahara, W. 1965. Gann 56:293–299.
111. Mita, T., Tokuzen, R. Fukuoka, F., and Nakahara, W. 1966. Gann 57:273–278.
112. Mita, T., Tokuzen, R., Fukuoka, F., and Nakahara, W. 1967. Gann 58:479–480.
113. Mizuno, K., Ohkubo, Y., Yokoyama, S., Hamada, M., Maedo, K., and Umezawa, H. 1967. J. Antibiot., Ser. A 20:194–199.
114. Moikeha, S. N. and Chu, G. W. 1971. J. Phycol. 7:8–13.
115. Mouton, R. F. 1971. In E. Broda, ed., 1st European Biophysics Congress, vol. 2, Medical Akad., Vienna, pp. 241–250.
116. Mouton, R. F. and Fromageot, P. 1971. FEBS Lett. 15:45–48.

117. Nakano, N. 1968. Jap. J. Med. Sci. Biol. 21:351–357.
118. Nathan (Guttman), H. A. and Friedman, W. 1962. Science 135:793–794.
119. Organ, A. E., Bovee, E. C., and Jahn, T. L. 1938. J. Protozool. 15(suppl.):23.
120. Parks, R. E., Jr. 1955. In C. P. Rhoads, ed., Antimetabolites and cancer, AAAS, Washington, D.C., pp. 175–197.
121. Pitelka, D. and Child, F. M. 1964. In S. H. Hutner, ed., Biochemistry and physiology of protozoa, vol. 3, Academic Press, New York, pp. 131–198,
122. Plesner, P., Rasmussen, L., and Zeuthen, E. 1964. In E. Zeuthen, ed., Synchrony in cell division and growth, Interscience, New York. p. 543.
123. Raines, R. C. and Haskell, B. E. 1968. Anal. Biochem. 23:413–421.
124. Rasmussen, L. and Zeuthen, E. 1966. Trav. Lab. Carlsberg 35:81–100.
125. Reeves, H., Papa, M., Seaman, G., and Ajl, S. 1961. J. Bacteriol. 81:154–155.
126. Reid, R., Cox, D., Baker, H., and Frank, O. 1969. J. Protozool. 16:231–235.
127. Reynolds, H. 1964. J. Wash. Acad. Sci. 54:99–108.
128. Reynolds, H. and Wragg, J. B. 1962. J. Protozool. 9:214–222.
129. Rockland, L. B., and Dunn, M. S. 1946. Arch. Biochem. 11:541–543.
130. Rockland, L. B. and Dunn, M. S. 1949. J. Biol. Chem. 179:511–521.
131. Rogers, C. G. 1966. Can. J. Biochem. 44:1493–1503.
132. Rogers, C. G. 1968. Can. J. Biochem. 46:331–340.
133, Rosenbaum, N., Erwin, J., Beach, D., and Holz, G. G., Jr. 1966. J. Protozool. 13:535–546.
134. Rosenberg, H. 1966. Exp. Cell Res. 41:397–410.
135. Roth, J. S. 1954. Cancer Res. 14:346–351.
136. Sanders, M. and Nathan, H. A. 1959. J. Gen. Microbiol. 21:264–270.
137. Shockman, G. D. 1963. In F. Kavanagh, ed., Analytical microbiology, vol. 1, Academic Press, New York, pp. 566–677.
138. Scrimshaw, N. S. and Gordon, J. E., eds. 1968. Malnutrition, learning and behavior. MIT Press, Cambridge, Massachusetts.
139. Sheffner, A. L. 1967. In A. A. Albanese, ed., Newer methods of Nutritional Biochemistry, vol. 3, Academic Press, New York, pp. 125–195.
140. Simon, E. M. and Hwang, S. W. 1967. Science 155:694–696.
141. Slotnick, I. J., Dougherty, M., and James, D. H., Jr. 1966. Cancer Res. 26(pt. 1):673–675.
142. Small, M., Mantell, N., and Epstein, S. S. 1967. Exp. Cell Res. 45:206.
143. Spikes, J. D. 1968. In A. C. Giese, ed., Photophysiology, current topics, vol. 3, Acadmic Press, New York, pp. 33–64.
144. Stock, J. A. 1966, In R. J. Schnitzer, and Hawking, F., eds., Experimental chemotherapy, vol. 4, Academic Press, New York, pp. 451–459.
145. Stokstad, E. L. R., Seaman, G. R., Davis, R. J., and Hutner, S. H. 1956. Meth. Biochem. Anal. 3:23–47.
146. Stott, J. A. and Smith, H. 1966. Brit. J. Mutr. 20:663–673.
147. Suss, R., Kinzel, V., Volm, M., Wayss, K., and Scribner, J. 1970. Z. Krebsforsch. 74:338–343.
148. Tartar, V. 1967. In T. T. Chen, ed., Research in protozoology, vol. 2, Pergamon Press, New York, pp. 1–116.
149. Teunisson, D. J. 1961. Anal. Biochem. 2:405–420.
150. Teunisson, D. J. 1971. Appl. Microbiol. 21, 878–887.
151. Teunisson, D. J. and Robertson, J. A. 1967. Appl. Microbiol. 15:1099–1103.
152. Thayer, P. S., Gordon, H. L., and Macdonald, M. 1971. Cancer Chemother. Rep. 2:27–133.
153. Tittler, I. A. 1948. J. Exp. Zool. 108:309–325.
154. Tittler, I. A. and Bovell, C. 1954. Proc. Soc. Exp. Biol. Med. 85:495–496.
155. Ueno, Y. and Sahcki, M. 1968. Jap. J. Exp. Med. 38:157–164.
156. Ueno, Y. and Yamakawa, H. 1970. Jap. J. Exp. Med. 40:385–390.
157. United Nations. 1968. Feeding the expanding world population: International action to avert the impending protein crisis. Document E. 68 XIII. 2, (E/4343/Rev. 1). United Nations, New York, pp. 1–106.
158. Ursprung, J. J. 1967. Ann. Rep. Med. Chem., 1966, 187–198.

159. Valzelli, L. and Garattini, S. 1970. In W. G. Clark, ed., Principles of psychopharmacology. Academic Press, New York, pp. 255–267.
160. Van Eys, J. and Warnock, L. G. 1963. J. Protozool. 10:465–467.
161. Visscher, M. B. 1968. Antimicrob. Agents Chemother. 1967. 19–21.
162. Viswanatha, T. and Liener, I. E. 1955. Arch. Biochem. Biophys. 56:222–229.
163. Waithe, W. I. 1964. Exp. Cell Res. 35:100–107.
164. Wang, G. T. and Marquardt, W. C. 1966. J. Protozool. 13:123–128.
165. Warnock, L. G. and Van Eys, J. 1963. J. Cell. Comp. Physiol. 61:309–310.
166. West, R. A., Jr., Barbera, P. W., Kolar, J. R., and Murrell, C. B. 1962. J. Protozool. 9:65–73.
167. White, F. R. 1963. Cancer Chemother. Rep. No. 23:71–77.
168. Woolfe, G. 1963. In R. J. Schnitzer and F. Hawking, eds., Experimental chemotherapy, vol. 1. Academic Press, New York, pp. 657–659.
169. Young, C. W. and Karnofsky, D. A. 1957 and 1966. Ann. Rep. Med. Chem. 166–175.
170. Zacks, S. I. and Sheff, M. F. 1966. J. Gen. Microbiol. 44:89–94.
171. Zahalsky, A. C., Keane, M. M., Hutner, S. II., Lubart, K. J., Kittrell, M, and Amsterdam, D. 1963. J. Protozool. 10:421–428.

General Bibliography

This section includes complete citations of all major and most minor publications on *Tetrahymena* (over (1,700) that have appeared in the literature through 1972, and even a few in 1973.* All abstracts and Master's or Doctoral theses have been excluded, because usually the significant research included in these publications appeared later in research journals. Also excluded are the majority of articles describing studies in which *Tetrahymena* was used as a food organism for some other animal. All articles concerning organisms that are members of the real genera *Colpidium* and *Glaucoma* have been excluded. Some earlier workers mistakenly used these generic names when they were actually working with a species of *Tetrahymena*. This confusion is cleared up in Chapter 1.

The references were accumulated in my (A.M.E.) laboratory over the years and were assembled for this book. To insure accuracy they were submitted to John O. Corliss, an outstanding authority on ciliates and ciliate literature, for inspection. He agreed to remove errors that had crept into the original manuscript and to add references that had been missed. As a result of his efforts, this bibliography is as accurate and complete as it is possible to make it. We owe him a debt of gratitude for undertaking this arduous task.

*A guide to the literature on *Tetrahymena* which has used this Bibliography as its principal source of references is currently in preparation by Dr. John O. Corliss and is scheduled to appear in the July 1973 number of the *Transactions of the American Microscopical Society* (volume 92). Readers should find it a useful adjunct and a helpful kind of "companion piece" to our listing of the literature.

Aaronson, S. and Ardois, G. 1971. Selective inhibition of blue-green algal growth by ethionine and other amino acid analogs. J. Phycol. 7:18-20.

Aaronson, S. and Baker, H. 1961. Lipid and sterol content of some protozoa. J. Protozool. 8:274-277.

Aaronson, S., Baker, H., Bensky, B., Frank, O., and Zahalsky, A. C. 1964. Protozoan assays for vitamins and cytotoxicity of carcinogens and pharmacological agents. Develop. Ind. Microbiol. 6:48-58.

Aaronson, S. and Bensky, B. 1965. Study of the cellular action of drugs with protozoa. II. Comparative study of the inhibitions induced by hypocholesteremic agents on *Ochromonas danica* multiplication: Annulment by oleic acid. J. Protozool. 12:236-240.

Aaronson, S., Bensky, B., Shifrine, M., and Baker, H. 1962. Effect of hypocholesteremic agents on protozoa. Proc. Soc. Exp. Biol. Med. 109:130-132.

Agrell, I. P. S. 1968. DNAse activity during the division of animal cells. *In* N. van Thoai and J. Roche, eds., Homologous enzymes and biochemical evolution, Gordon and Breach, New York, pp. 313-329.

Agrell, I. P. S. and Bergqvist, H.-A. 1962. Cytochemical evidence for varied DNA complexes in the nuclei of undifferentiated cells of *Rana temporaria* and *Tetrahymena pyriformis*. J. Cell Biol. 15:604-606.

Agrell, I. P. S. and Bergqvist, H.-A. 1967. Cytochemical studies on DNA-complexes during cell multiplication and cell differentiation. Comp. Biochem. Physiol. 22:189-198.

Akaboshi, M., Maeda, T., and Waki, A. 1967. Determination of the nucleic acid content of protozoa by activation analysis. Biochim. Biophys. Acta 138:596-597.

Albach, R. A. 1967. Incorporation of tritium from thymidine-methyl-H^3 into DNA, RNA, lipids and protein in logarithmic and stationary phase *Tetrahymena pyriformis*, strain W. J. Protozool. 14:271-277.

Albach, R. A. and Corliss, J. O. 1959. Regeneration in *Tetrahymena pyriformis*. Trans. Amer. Microsc. Soc. 78:276-284.

Alexander, J. B. 1967. A new immobilization test for *Tetrahymena pyriformis*. Trans. Amer. Microsc. Soc. 86:421-427.

Alexander, J. B. 1968. The mucocysts of *Tetrahymena pyriformis*: Localization of the TCA-ethanol-soluble protein. Exp. Cell Res. 49:425-440.

Alexander, J. B., Silvester, N. R., and Watson, M. R. 1962. An ethanol-soluble protein from isolated cilia. Biochem. J. 85(3):27P.

Alfert, M. and Balamuth, W. 1957. Differential micronuclear polyteny in a population of the ciliate *Tetrahymena pyriformis*. Chromosoma 8:371-379.

Alfert, M. and Goldstein, N. O. 1955. Cytochemical properties of nucleoproteins in *Tetrahymena pyriformis*; a difference in protein composition between macro- and micronuclei. J. Exp. Zool. 130:403-421.

Allen, N. E. and Suyama, Y. 1972. Protein synthesis *in vitro* with *Tetrahymena* mitochondrial ribosomes. Biochim. Biophys. Acta 259:369-377.

Allen, R. D. 1967. Fine structure, reconstruction and possible functions of components of the cortex of *Tetrahymena pyriformis*. J. Protozool. 14:553-565.

Allen, R. D. 1968. A reinvestigation of cross-sections of cilia. J. Cell Biol. 37:825-831.

Allen, R. D. 1969. The morphogenesis of basal bodies and accessory structures of the cortex of the ciliated protozoan *Tetrahymena pyriformis*. J. Cell Biol. 40:716-733.

Allen, S. L. 1960. Inherited variations in the esterases of *Tetrahymena*. Genetics 48:1051-1070.

Allen, S. L. 1961. Genetic control of the esterases in the protozoan *Tetrahymena pyriformis*. Ann. N.Y. Acad. Sci. 94:753-773.

Allen, S. L. 1963. Genomic exclusion in *Tetrahymena*: Genetic basis. J. Protozool. 10:413-420.

Allen, S. L. 1964. Linkage studies in variety 1 of *Tetrahymena pyriformis*: A first case of linkage in ciliated protozoa. Genetics 49:617-627.

Allen, S. L. 1964. The esterase isozymes of *Tetrahymena*: Their distribution in isolated cellular components and their behavior during the growth cycle. J. Exp. Zool. 155:349-370.

Allen, S. L. 1965. Genetic control of enzymes in *Tetrahymena*. Brookhaven Symp. Biol. 18:27-54.

Allen, S. L. 1967. Genomic exclusion: A rapid means for inducing homozygous diploid lines in *Tetrahymena pyriformis*, syngen 1. Science 155:575-577.

Allen, S. L. 1967. Cytogenetics of genomic exclusion in *Tetrahymena*. Genetics 55:797-822.

Allen, S. L. 1967. Chemical genetics of protozoa. *In* M. Florkin and B. T. Scheer, eds., Chemical zoology, vol. 1, Protozoa (G. W. Kidder, ed.), Academic Press, New York, pp. 617-684.

Allen, S. L. 1968. Genetic and epigenetic control of several isozymic systems in *Tetrahymena*. Ann. N.Y. Acad. Sci. 151:190-207.

Allen, S. L. 1971. A late-determincd gene in *Tetrahymena* heterozygotes. Genetics 68:415-433.

Allen, S. L., Allen, J. M., and Licht, B. M. 1965. Effect of Triton X-100 upon the activity of some electrophoretically separated acid phosphatases and esterases. J. Histochem. Cytochem. 13:434-440.

Allen, S. L., File, S. K., and Koch, S. L. 1967. Genomic exclusion in *Tetrahymena*. Genetics 55:823-837.

Allen, S. L. and Gibson, I. 1971. The purification of DNA from the genomes of *Paramecium aurelia* and *Tetrahymena pyriformis*. J. Protozool. 18:518-525.

Allen, S. L. and Gibson, I. 1972. Genome amplification and gene expression in the ciliate macronucleus. Genetics 6:293-313.

Allen, S. L. and Lee, P. H. T. 1971. The preparation of congenic strains of *Tetrahymena*. J. Protozool. 18:214-218.

Allen, S. L., Misch, M. S., and Morrison, B. M. 1963. Genetic control of an acid phosphatase in *Tetrahymena*: Formation of a hybrid enzyme. Genetics 48:1635-1658.

Allen, S. L., Misch, M. S., and Morrison, B. M. 1963. Variations in electrophoretically separated acid phosphatases. J. Histochem. Cytochem. 11:706-719.

Allen, S. L. and Nanney, D. L. 1958. An analysis of nuclear differentiation in the selfers of *Tetrahymena*. Amer. Nat. 92:139-160.

Allen, S. L. and Weremiuk, S. L. 1971. Defective micronuclei and genomic exclusion in selected C* subclones of *Tetrahymena*. J. Protozool. 18:509-515.

Allen, S. L. and Weremiuk, S. L. 1971. Intersyngenic variations in the esterases and acid phosphatases of *Tetrahymena pyriformis*. Biochem. Genet. 5:119-133.

Allen, S. L., Weremiuk, S. L., and Patrick, C. A. 1971. Is there selective mating in *Tetrahymena* during genomic exclusion? J. Protozool. 18:515-517.

Allison, B. M. and Ronkin, R. R. 1967. Lipid cytochemistry and morphologic change in aging populations of *Tetrahymena pyriformis*. J. Protozool. 14:313-320.

Andersen, H. A. 1972. Induced elimination of DNA from macronucleus of *Tetrahymena pyriformis*. Exp. Cell Res. 74:610-613.

Andersen, H. A. 1972. Requirements for DNA replication preceding cell division in *Tetrahymena pyriformis*. Exp. Cell Res. 75:89-94.

Andersen, H. A., Brunk, C. F., and Zeuthen, E. 1970. Studies on the DNA replication in heat synchronized *Tetrahymena pyriformis*. C. R. Trav. Lab. Carlsberg 38:123-131.

Andersen, H. A. and Zeuthen, E. 1971. DNA replication sequence in *Tetrahymena* is not repeated from generation to generation. Exp. Cell Res. 68:309-314.

Anderson, M. E. and Williams, H. H. 1951. Microbiological evaluation of protein quality. A colorimetric method for the determination of the growth of *Tetrahymena geleii* W in protein suspensions. J. Nutr. 44:335-343.

Andrus, W. D. W. and Giese, A. C. 1963. Mechanisms of sodium and potassium regulation in *Tetrahymena pyriformis*, strain W. J. Cell. Comp. Physiol. 61:17-30.

Archibald, A. R. and Manners, D. J. 1959. Studies on carbohydrate-metabolizing enzymes. 2. Transglucosylation by extracts of *Tetrahymena pyriformis*. Biochem. J. 73:292-295.

Argetsinger, J. 1965. The isolation of ciliary basal bodies (kinetosomes) from *Tetrahymena pyriformis*. J. Cell Biol. 24:154-157.

Avins, L. R. 1968. Studies on alpha oxidation of stearic acid by *Tetrahymena pyriformis* and *Crithidia fasiculata*. Biochem. Biophys. Res. Commun. 32:138-142.

Baer, E. and Basu, H. 1972. Phosphonolipids. XXV. Synthesis of α-octadecyl-β-hexadecanoyl-L-α-glyceryl-(2-aminoethyl) phosphonate, a homologue of an α-monoether-phosphonocephalin occurring in *Tetrahymena pyriformis.* Can. J. Biochem. 50:988-992.

Baker, E. G. S. and Baumberger, J. P. 1941. The respiratory rate and cytochrome content of a ciliate protozoan *Tetrahymena geleii.* J. Cell. Comp. Physiol. 17:285-304.

Baker, J. A. and Ferguson, M. S. 1942. Growth of platyfish (*Platypoecilus maculatus*) free from bacteria and other microorganisms. Proc. Soc. Exp. Biol. Med. 51:116-119.

Baker, H. and Frank, O. 1968. Clinical vitaminology: Methods and interpretations. John Wiley, New York. 238 pp.

Baker, H., Frank, O., Feingold, S., Christakis, G., and Ziffer, H. 1967. Vitamins, total cholesterol, and triglycerides in 642 New York City school children. Amer. J. Clin. Nutr. 20:850-857.

Baker, H., Frank, O., Feingold, S., Gellene, R. A., Leevy, C. M., and Hutner, S. H. 1966. A riboflavin assay suitable for clinical use and nutritional surveys. Amer. J. Clin. Nutr. 19:17-26.

Baker, H., Frank, O., Ning, M., Gellene, R. A., Hutner, S. H., and Leevy, C. M. 1966. A protozoological method for detecting clinical vitamin B_6 deficiency. Amer. J. Clin. Nutr. 18:123-133.

Baker, H., Frank, O., Pasher, I., Hutner, S. H., and Sabotka, H. 1960. Nicotinic acid assay in blood and urine. Clin. Chem. 6:572-577.

Balbiani, E. G. 1887. Observations relatives à une note recente de M. Maupas sur la multiplication de la *Leucophrys patula.* C. R. Acad. Sci., Paris 104:80-83.

Ballentine, R. and Burford, D. D. 1960. Differential density separation of cellular suspensions. Anal. Biochem. 1:263-268.

Ballentine, R. and Burford, D. D. 1964. Preferential block of strontium accumulation in *Tetrahymena pyriformis* by sulfadiazine. Life Sci. 3:1455-1458.

Bamforth, S. S. 1963. Limnetic protozoa of southeastern Louisiana. Proc. La. Acad. Sci. 26:120-134.

Barber, A. A., Harris, W. W., and Padilla, G. M. 1965. Studies of native glycogen isolated from synchronized *Tetrahymena pyriformis* (HSM). J. Cell Biol. 27:281-292.

Barber, M. A. 1927. The food of anopheline larvae: Food organisms in pure culture. U.S. Pub. Health Rep. 42:1494-1510.

Barber, M. A. 1944. The rearing of sterile adult *Anopheles.* U.S. Pub. Health Rep. 59:1384-1387.

Barnett, A., Wille, J. J., Jr., and Ehret, C. F. 1971. Resolution of some component classes of complex RNA by molecular hybridization in the eukaryote *Tetrahymena pyriformis.* Biochim. Biophys. Acta 247:243-261.

Barthelmes, D. 1960. *Tetrahymena parasitica* (Penard 1922) Corliss 1952 als Parasit in Larven vom *Chironomus plumosus*-Typ. Z. Fisch. Hilfswiss. 9:273-280.

Baudhuin, P. 1969. Peroxisomes (microbodies, glyoxysomes). *In* A. Lima-de-Faria, ed., Handbook of molecular cytology, John Wiley, New York, pp. 1179-1195.

Baudhuin, P., Müller, M., Poole, B., and de Duve, C. 1965. Non-mitochondrial oxidizing particles (microbodies) in rat liver and kidney and in *Tetrahymena pyriformis.* Biochem. Biophys. Res. Commun. 20:53-59.

Beach, D., Erwin, J. A., and Holz, G. G., Jr. 1969. Tetrahymanol synthesis by *Tetrahymena* species. *In* A. A. Strelkov, K. M. Sukhanova, and I. B. Raikov, eds., Progress in protozoology, 3rd International Congress on Protozoology, Leningrad, July 1969, Nauka, Leningrad, pp. 133-134.

Berezina, I. G. 1970. Cytochemical study of lactate dehydrogenase (LDH) isoenzymes in some species of protozoa. Tsitologiya 12:1205-1208.

Berger, E. 1929. Unterschiedliche Wirkungen gleicher Ionen und Ionengemische auf verschiedene Teirarten. (Ein Beitrag zur Lehre vom Ionenantagonismus.) Pfluger's Arch. Ges. Physiol. 223:1-89.

Berger, H. and Hanahan, D. J. 1971. Isolation of phosphonolipids from *Tetrahymena pyriformis.* Biochim. Biophys. Acta 231:584-587.

Berger, H., Jones, P., and Hanahan, D. J. 1972. Structural studies on lipids of *Tetrahymena pyriformis.* Biochim. Biophys. Acta 260:617-629.

Berghe, L. van den 1934. Sur un ciliate parasite de pontes de mollusques d'eau douce *Glaucoma paedophthora,* n. sp. C. R. Soc. Biol. 115:1423-1426.

Bergstrom, S. 1970. Amount of induced avoidance behavior to light in the protozoa *Tetrahymena* as a function of time after training and cell fission. Scand. J. Psychol. 10:16-20.

Bergstrom, S. 1970. Avoidance behavior to light in the protozoa *Tetrahymena.* The effect of a gradual versus an abrupt boundary between dark and light. Scand. J. Psychol. 10:81-88.

Berner, D. L. and Hammond, E. G. 1969. Phylogeny of lipase specificity. Lipids 5:558-562.

Bernheimer, A. and Steele, J. M., Jr. 1955. Ribonuclease and ribonuclease-inhibitors among higher plants. Proc. Soc. Exp. Biol. Med. 89:123-126.

Bernstein, E. and Zeuthen, E. 1966. The relationship of RNA synthesis to temperature as an inducer of synchronous division. C. R. Trav. Lab. Carlsberg 35:501-517.

Bertoni, L. 1967. Tissue toxicity of radiologic contrast media evaluated with *Tetrahymena pyriformis*. Experientia 23:59-60.

Bhatia, B. L. 1936. The fauna of British India, including Ceylon and Burma. Protozoa, vol. 1, Ciliophora, Taylor and Francis, London. 493 pp.

Bick, H. 1958. Ökologische Untersuchungen an Ciliaten fallaubreicher Kleingewässer. Arch. Hydrobiol. 54:506-542.

Bick, H. 1972. Ciliated protozoa: An illustrated guide to the species used as biological indicators in freshwater biology. World Health Organization, Geneva. 198 pp.

Bimpson, T., Goad, L. J., and Goodwin, T. W. 1967. The stereochemistry of hydrogen elimination at C-7, C-22 and C-23 during the conversion of cholesterol (cholest-5-en-3β-ol) into cholesta-5,7,22-trien-3β-ol by *Tetrahymena pyriformis*. Biochem. J. 115:857-858.

Bird, O. D., Oakes, V., Undheim, K., and Rydon, H. N. 1964. Folic acid antagonists based on some deazapteridine ring systems. *In* W. Pfleiderer and E. C. Taylor, eds., Pteridine chemistry, Pergamon Press, New York, pp. 417-426.

Birns, M. 1960. The localization of acid phosphate activity in the ameba *Chaos chaos*. Exp. Cell Res. 20:202-205.

Bleyman, L. K. 1971. Temporal patterns in ciliated protozoa. *In* I. L. Cameron, G. M. Padilla, and A. Zimmerman, eds., Developmental aspects of the cell cycle, Academic Press, New York, pp. 67-91.

Bleyman, L. K. and Simon, E. M. 1967. Genetic control of maturity in *Tetrahymena pyriformis*. Genet. Res. Camb. 10:319-321.

Bleyman, L. K. and Simon, E. M. 1968. Clonal analysis of nuclear differentiation in *Tetrahymena*. Develop. Biol. 18:217-231.

Bleyman, L. K., Simon, E. M., Brosi, R. 1966. Sequential nuclear differentiation in *Tetrahymena*. Genetics 54:277-291.

Blum, J. J. 1967. An andrenergic control system in *Tetrahymena*. Proc. Natl. Acad. Sci. U.S. 58:81-88.

Blum, J. J. 1970. On the regulation of glycogen metabolism in *Tetrahymena*. Arch. Biochem. Biophys. 137:65-74.

Blum, J. J., Kirschner, N., and Utley, J. 1966. The effect of reserpine on growth and catecholamine content of *Tetrahymena*. Mol. Pharmacol. 2:606-608.

Blum, J. J. and Wexler, J. P. 1968. Effect of clofibrate on *Tetrahymena*. Mol. Pharmacol. 4:155-161.

Blumberg, A. J. and Loefer, J. B. 1952. Effect of neomycin on two species of free-living protozoa. Physiol. Zool. 25:276-282.

Bolund, L. and Ringertz, N. R. 1966. Intracellular distribution of histone-like proteins in *Tetrahymena pyriformis*. Exp. Cell Res. 44:606-613.

Bond, R. M. 1933. A contribution to the study of the natural food cycle in aquatic environments with particular consideration to microorganisms and dissolved organic matter. Bull. Bingham Oceanogr. Coll. 4(4):1-89.

Borisy, G. G. and Taylor, E. W. 1967. The mechanism of action of colchicine. Binding of colchicine-H[3] to cellular protein. J. Cell Biol. 34:525-533.

Bortle, L. and Oleson, J. J. 1955. Effect of puromycin on *Tetrahymena pyriformis* nutrition. Antibiot. Ann. 1954/55:766-770.

Bouman, F. L. 1959. Histochemical evidence for the presence of phosphatase activity in *Tetrahymena geleii*. Mich. Acad. Sci. Arts Lett. 37:129-132.

Bovee, E. C. 1957. Protozoa of Amazonian and Andean waters of Colombia, South America. J. Protozool. 4:63-66.

Bovee, E. C. 1960. Protozoa of the Mountain Lake region, Giles County, Virginia. J. Protozool. 7:352-361.

Boyne, A. W., Price, S. A., Rosen, G. D., and Stott, J. A. 1967. Protein quality of feeding stuffs. 4. Progress report on collaborative studies on the microbiological assay of available amino acids. Brit.

J. Nutr. 21:181-206.

Bradner, W. T., Heinemann, B., and Gourevitch, A. 1967. Hedamycin, a new antitumor antibiotic. II. Biological properties. Antimicrob. Agents Chemother. 1966:613-618.

Brandwein, P. F. 1935. The culturing of fresh-water protozoa and other small invertebrates. Amer. Nat. 69:628-632.

Braun, R., Dewey, V. C., and Kidder, G. W. 1963. On the biosynthesis of the quinone ring of ubiquinone. Biochemistry 2:1070-1072.

Brizzi, G. and Blum, J. J. 1970. Effect of growth conditions on serotonin content of *Tetrahymena pyriformis*. J. Protozool. 17:553-555.

Brooks, W. M. 1968. Tetrahymenid ciliates as parasites of the gray garden slug. Hilgardia 39:205-276.

Brown, M. G. 1940. Growth of protozoan cultures. II. *Leucophrys patula* and *Glaucoma pyriformis* in a bacteria-free medium. Physiol. Zool. 13:277-282.

Browning, I. 1949. Relation between ion action and osmotic pressure on a ciliated protozoan. J. Exp. Zool. 110:441-460.

Browning, I. 1951. Cytoplasmic inclusions of the protozoan, *Tetrahymena geleii*. Tex. Rep. Biol. Med. 9:3-7.

Browning, I., Bergendahl, J. C., and Brittain, M. S. 1952. Cellular reproduction efficiency in various oxygen concentrations. Tex. Rep. Biol. Med. 10:790-793.

Browning, I. and Brittain, M. S. 1952. pH variation during culture development of *Tetrahymena geleii*. Tex. Rep. Biol. Med. 10:778-781.

Browning, I., Brittain, M. S., and Bergendahl, J. C. 1952. Synchronous and rhythmic reproduction of protozoa following inoculation. Tex. Rep. Biol. Med. 10:794-802.

Browning, I., Lockingen, L. S., and Wingo, W. J. 1952. Theoretical formulae for reproduction of free-living protista. Tex. Rep. Biol. Med. 10:782-789.

Browning, I., Varnedoe, N. B., and Swinford, L. R. 1951. Fragility of *Tetrahymena geleii* in different salt concentrations. Tex. Rep. Biol. Med. 9:420-427.

Browning, I., Varnedoe, N. B., and Swinford, L. R. 1952. Time of nuclear, cytoplasmic and cortical division of the ciliated protozoan, *Tetrahymena geleii*. J. Cell. Comp. Physiol. 39:371-381.

Brunk, C. F. and Hanawalt, P. C. 1967. Repair of damaged DNA in a eucaryotic cell: *Tetrahymena pyriformis*. Science 158:663-664.

Brunk, C. F. and Hanawalt, P. C. 1969. The nature of the excision repair region of the DNA in the eucaryotic organism, *Tetrahymena pyriformis*. Radiat. Res. 38:285-295.

Brunk, C. F. and Hanawalt, P. C. 1969. Mitochondrial DNA in *Tetrahymena pyriformis*. Exp. Cell Res. 54:143-149.

Bruns, P. J. 1971. Immobilization antigens of *Tetrahymena pyriformis*. I. Assay and extraction. Exp. Cell Res. 65:445-453.

Brutkowska, M. 1969. Fluids intake by *Tetrahymena pyriformis*. In A. A. Strelkov, K. M. Sukhanova, and I. B. Raikov, eds., Progress in protozoology, 3rd International Congress on Protozoology, Leningrad, July 1969, Nauka, Leningrad, pp. 135-136.

Buchníček, J. and Turek, B. 1968. The inhibitory effect of mepacrine and chloroquine on the photodynamic action. Arzneim.-Forsch. 17:760-761.

Buetow, D. E. 1970. Preparation of mitochondria from protozoa and algae. In D. M. Prescott, ed., Methods in cell physiology, vol. 4, Academic Press, New York, pp. 84-113.

Buhse, H. E., Jr. 1966. Oral morphogenesis during transformation from microstome to macrostome and macrostome to microstome in *Tetrahymena vorax* strain V_2 type S. Trans. Amer. Microsc. Soc. 85:305-313.

Buhse, H. E., Jr. 1966. An analysis of macrostome production in *Tetrahymena vorax* strain V_2 type S. J. Protozool. 13:429-435.

Buhse, H. E., Jr. 1967. Microstome-macrostome transformation in *Tetrahymena vorax* strain V_2 type S induced by a transforming principle, stomatin. J. Protozool. 14:608-613.

Buhse, H. E., Jr., and Cameron, I. L. 1968. Temporal pattern of macronuclear events during the microstome-macrostome cell transformation of *Tetrahymena vorax* strain V_2S. J. Exp. Zool. 169:229-236.

Buhse, H. E., Jr., and Corliss, J. O. 1969. Application of the scanning electron microscope to the study of the morphogenetic events of macrostome formation in the protozoon *Tetrahymena vorax*. Proc. 2nd Ann. SEM Symp., Chicago, Ill., April 1969, pp. 169-176.

Buhse, H. E., Jr., Corliss, J. O., and Holsen, R. C. 1970. *Tetrahymena vorax*: Analysis of stomatogenesis by scanning electron and light microscopy. Trans. Amer. Microsc. Soc. 89:328-336.

Buhse, H. E., Jr., Stamler, S. J., and Corliss, J. O. 1970. Application of the scanning electron microscope to the study of cell division in *Tetrahymena pyriformis*. Proc. 3rd Ann. SEM Symp., Chicago, Ill., April 1970, pp. 171-176.

Buhse, H. E., Jr., Stamler, S. J., and Corliss, J. O. 1973. Analysis of stomatogenesis by scanning electron microscopy in *Tetrahymena pyriformis* strain W during synchronous cell division. Trans. Amer. Microsc. Soc. 92:95-105.

Burge, W. E. and Williams, M. 1927. The utilization of dextrose, levulose and galactose by animal and plant cells and the antagonistic action of insulin to thyroxin. Amer. J. Physiol. 81:307-314.

Burmeister, J. 1971. Metabolism of *Tetrahymena pyriformis* during the uptake of nutrients and non-nutritive particles. Z. Allg. Mikrobiol. 11:275-282.

Burnasheva, S. A. and Efremenko, M. V. 1961. Adenine nucleotides and adenosine triphosphatase activity of the holotrichous infusorian, *Tetrahymena pyriformis*. Dokl. Akad. Nauk SSR, Biochem. Sect. Engl. Transl. 137:203-205.

Burnasheva, S. A. and Efremenko, M. V. 1962. Role of adenosine triphosphate in the motility of *Tetrahymena pyriformis*. Biokhimiya 27:167-172. In Russian with English summary.

Burnasheva, S. A., Efremenko, M. V., Chumakora, L. P., and Zuyeva, L. V. 1965. Isolation of contractile proteins from the cilia of *Tetrahymena pyriformis* and a study of their properties. Biokhimiya 30:765-771. In Russian with English summary.

Burnasheva, S. A., Efremenko, M. V., and Lyubimova, M. N. 1963. Studies of adenosinetriphosphatase activity of isolated cilia and ATP-ase isolation on *Tetrahymena pyriformis*. Biokhimiya 28:547-551. In Russian with English summary.

Burnasheva, S. A. and Karausheva, T. P. 1967. Participation of ATP in the motive functions of *Tetrahymena pyriformis*. Biokhimiya 32:270-276. In Russian with English summary.

Burnasheva, S. A., Yurzina, G., and Beskrovnova, N. 1968. Ultrastructure of the fibers of cilia of *Tetrahymena pyriformis*. Tsitologiya 10:249-252.

Burnasheva, S. A., Yurzina, G., and Lyubinova, M. N. 1969. Ultrastructure of the flagella of *Strigomonas oncopelti* and of the cilia of *Tetrahymena pyriformis*, and the localization of ATP-ase in them. Tsitologiya 11:695-699.

Bütschli, O. 1887-1889. Infusoria und System der Radiolaria. Bronn's Klassen und Ordnungen des Thier-Reichs 1 (III Abt.), C. F. Winters, Leipzig, pp. 1098-2035.

Butterfield, C. T. 1929. Experimental studies of natural purification in polluted waters. III. A note on the relation between food concentration in liquid media and bacterial growth. U.S. Pub. Health Rep. 44:2865-2872.

Butterfield, C. T. 1933. A note on the food habits of *Colpidium*. U.S. Pub. Health Rep. 48:814-818.

Butterfield, C. T., Purdy, W. C., and Theirault, E. J. 1931. Experimental studies of natural purification in polluted waters. IV. The influence of the plankton on the biochemical oxidation of organic matter. U.S. Pub. Health Rep. 46:393-426.

Byfield, J. E., Chou, S.-C., and Scherbaum, O. H. 1962. Some observations on the isolation of the mitochondria from *Tetrahymena pyriformis* GL. Biochem. Biophys. Res. Commun. 9:226-230.

Byfield, J. E. and Lee, Y. C. 1970. The effect of synchronizing temperature shifts on the synthesis and translation of replication-supporting messengers in *Tetrahymena pyriformis*. Exp. Cell Res. 61:42-50.

Byfield, J. E. and Lee, Y. C. 1970. Do synchronizing temperature shifts inhibit RNA synthesis in *Tetrahymena pyriformis*? J. Protozool. 17:445-453.

Byfield, J. E. and Scherbaum, O. H. 1966. Temperature-dependent RNA decay in *Tetrahymena*. J. Cell. Physiol. 68:203-206.

Byfield, J. E. and Scherbaum, O. H. 1966. Suppression of RNA and protein accumulation by temperature shifts in a heat synchronized protozoan. Life Sci. 5:2263-2269.

Byfield, J. E. and Scherbaum, O. H. 1967. Temperature-dependent decay of RNA and of protein synthesis in a heat-synchronized protozoan. Proc. Natl. Acad. Sci. U.S. 57:602-606.

Byfield, J. E. and Scherbaum, O. H. 1967. Temperature effect on protein synthesis in a heat-synchronized protozoan treated with actinomycin D. Science 156:1504.

Byfield, J. E. and Scherbaum, O. H. 1967. Stability of division-related protein and nucleic acid fractions in synchronized *Tetrahymena*. J. Cell. Physiol. 70:265-274.

Byfield, J. E. and Scherbaum, O. H. 1967. Incomplete inhibition of synchronized cell division by hydroxyurea and its relevance to the normal cellular life cycle. Nature 216:1017-1018.

Byfield, J. E. and Scherbaum, O. H. 1968. Amino acid control of RNA synthesis in an animal cell (*Tetrahymena*) and its relation to some aspects of gene expression. Exp. Cell Res. 49:202-206.

Byfield, J. E., Young, C. L., and Bennett, L. R. 1969. Thermal instability of *Tetrahymena* ribosomes: Effects on protein synthesis. Biochem. Biophys. Res. Commun. 37:806-812.

Calkins, J. 1962. Photoreactivation after division. Nature 196:686-687.

Calkins, J. 1964. The lethal effects of radiation on six species of protozoans. Photochem. Photobiol. 3:143-151.

Calkins, J. 1966. The dose-response relation for the photoreactivation of ultraviolet inactivated stationary phase *Tetrahymena pyriformis*. Photochem. Photobiol. 5:787-795.

Calkins, J. 1967. An unusual form of response in X-irradiated protozoa and a hypothesis as to its origin. Int. J. Radiat. Biol. 12:297-301.

Calkins, J. and Gunn, G. 1967. The time of DNA synthesis during the interdivision growth cycle of *Tetrahymena pyriformis* fed on living bacteria. J. Protozool. 14:210-213.

Cameron, I. L. 1965. Macromolecular events leading to cell division in *Tetrahymena pyriformis* after removal and replacement of required pyrimidines. J. Cell Biol. 25:9-18.

Cameron, I. L. 1966. A periodicity of tritiated-thymidine incorporation into cytoplasmic deoxyribonucleic acid during the cell cycle of *Tetrahymena pyriformis*. Nature 209:630-631.

Cameron, I. L. and Burton, A. L. 1969. On the cycle of the water expulsion vesicle in the ciliate *Tetrahymena pyriformis*. Trans. Amer. Microsc. Soc. 88:386-393.

Cameron, I. L., Cline, G. B., Padilla, G. M., Miller, O. L., Jr., and van Dreal, P. A. 1966. Polysomes from *Tetrahymena* following pyrimidine deprivation and replacement. Natl. Cancer Inst. Monogr. 21:361-371.

Cameron, I. L., Griffin, E. E., and Rudick, M. J. 1971. Macromolecular events following refeeding of starved *Tetrahymena*. Exp. Cell Res. 65:265-272.

Cameron, I. L. and Guile, E. E. 1965. Nucleolar and biochemical changes during unbalanced growth of *Tetrahymena pyriformis*. J. Cell Biol. 26:845-855.

Cameron, I. L. and Jeter, J. R., Jr. 1970. Synchronization of the cell cycle of *Tetrahymena* by starvation and refeeding. J. Protozool. 17:429-433.

Cameron, I. L., Padilla, G. M., and Miller, O. L., Jr. 1966. Macronuclear cytology of synchronized *Tetrahymena pyriformis*. J. Protozool. 13:336-341.

Cameron, I. L. and Prescott, D. M. 1961. Relations between cell growth and cell division. V. Cell and macronuclear volumes of *Tetrahymena pyriformis* HSM during the cell life cycle. Exp. Cell Res. 23:354-360.

Cameron, I. L. and Stone, G. E. 1964. Relation between the amount of DNA per cell and the duration of DNA synthesis in three strains of *Tetrahymena pyriformis*. Exp. Cell Res. 36:510-514.

Campbell, J. A. and Morrison, A. B. 1966. Nutritional impact of modern food processing. Fed. Proc. 26:130-136.

Canella, M. F. 1954. Ricerche sulla microfauna delle acque interne ferraresi. Introduzione allo studio dei ciliati e dei rotiferi. Pub. Civ. Mus. St. Nat. Ferrara 4:1-154.

Canella, M. F. 1964. Strutture buccali, infraciliatura, filogenesi e sistematica dei ciliofori. Ann. Univ. Ferrara (N. S., Sect. III) 2:110-188.

Canella, M. F. 1971. Sur les organelles ciliaires de l'appareil buccal des hyménostomes et autres ciliés. Ann. Univ. Ferrara (N. S., Sect. III) 3 (suppl.):1-235.

Cann, J. R. 1963. A kinetic model of induced division synchrony in *Tetrahymena pyriformis*. C. R. Trav. Lab. Carlsberg 33:431-453.

Cann, J. R. 1968. On the leakage of ultraviolet absorbing materials and alamine from synchronized *Tetrahymena pyriformis*. C. R. Trav. Lab. Carlsberg 36:319-325.

Carlson, P. S. 1971. Mutant selection in *Tetrahymena pyriformis*. Genetics 69:261-265.

Carpenter, P. L. 1943. The effect of 2,4,6-trinitrophenol (picric acid) on the growth of the ciliate *Tetrahymena geleii*. Physiol. Zool. 16:313-321.

Carter, H. and Gaver, R. 1967. Branched-chain sphingosines from *Tetrahymena pyriformis*. Biochem. Biophys. Res. Commun. 29:886-891.

Caspi, E., Greig, J. B., and Zander, J. M. 1968. The biosynthesis of tetrahymanol *in vitro*. Biochem. J. 109:931-932.

Caspi, E., Greig, J. B., and Mandelbaum, A. 1969. Incorporation of deuterium from deuterium oxide into tetrahymanol biosynthesized from squalene. Chem. Commun. 1969:28-29.

Caspi, E. and Mulheirn, L. J. 1969. The biosynthetic origin of the C-20 proton of cholesterol. Chem. Commun. 1969:1423-1424.

Caspi, E., Zander, J. M., Greig, J. B., Mallory, F. B., Conner, R. L., and Landrey, J. R. 1968. Evidence for a nonoxidative cyclization of squalene in the biosynthesis of tetrahymanol. J. Amer. Chem. Soc. 90:3563-3564.

Celliers, P. G. 1961. Microbiological evaluation of the nutritive value of South African marine products with *Tetrahymena pyriformis* W. S. Afr. J. Agr. Sci. 4:191-204.

Cerroni, R. E. and Goldstein, N. O. 1960. Quantitative biochemical and microspectrophotometric Feulgen DNA analyses on *Tetrahymena* cells treated with iron. Exp. Cell Res. 20:258-261.

Cerroni, R. E. and Neff, R. J. 1959. Inhibition of the Feulgen reaction by ions. Exp. Cell Res. 16:465-470.

Cerroni, R. E. and Zeuthen, E. 1962. Asynchrony of nuclear incorporation of tritiated thymidine into *Tetrahymena* cells synchronized for division. C. R. Trav. Lab. Carlsberg 32:499-511.

Cerroni, R. E. and Zeuthen, E. 1962. Inhibition of macromolecular synthesis and of cell division in synchronized *Tetrahymena*. Exp. Cell Res. 26:604-605.

Chaix, P. and Baud, C. A. 1947. Étude de la lyse de *Glaucoma piriformis* (*Tetrahymena geleii*) par l'acide linoléique et quelques autres acides gras. Arch. Sci. Physiol. 1:3-9.

Chaix, P., Chauvet, J., and Fromageot, C. L. 1947. Sur la respiration du cilié *Tetrahymena geleii*. Antonie van Leeuwenhoek J. Microbiol. Serol. 12:145-155.

Chaix, P., Lacroix, L., and Fromageot, C. L. 1948. La toxicité des α-, β-, γ- et δ-hexachlorocyclohexanes vis-à-vis de *Glaucoma piriformis* (*Tetrahymena geleii*). Biochim. Biophys. Acta 2:57-63.

Chapman-Andresen, C. 1958. Pinocytosis of inorganic salts by *Amoeba proteus* (*Chaos diffluens*). C. R. Trav. Lab. Carlsberg 31:77-92.

Chapman-Andresen, C. 1967. Studies on endocytosis in amoebae. The distribution of pinocytically ingested dyes in relation to food vacuoles in *Chaos chaos*. I. Light microscopic observations. C. R. Trav. Lab. Carlsberg 36:161-187.

Chapman-Andresen, C. and Holtzer, H. 1960. The uptake of fluorescent albumin by pinocytosis in *Amoeba proteus*. J. Biophys. Biochem. Cytol. 8:288-291.

Chapman-Andresen, C. and Nilsson, J. R. 1968. On vacuole formation in *Tetrahymena pyriformis* GL. C. R. Trav. Lab. Carlsberg 36:405-432.

Chapman-Andresen, C. and Prescott, D. M. 1956. Studies on pinocytosis in the amoebae *Chaos chaos* and *Amoeba proteus*. C. R. Trav. Lab. Carlsberg 30:57-78.

Charret, R. 1969. L'ADN nucléolaire chez *Tetrahymena pyriformis*: Chronologie de sa réplication. Exp. Cell Res. 54:353-361.

Charret, R. 1972. Modifications des mitochondries et synthèse de l'ADN mitochondrial de *Tetrahymena pyriformis* après action du bromuré d'éthidium. J. Microsc. 14:279-298.

Charret, R. and André, J. 1968. La synthèse de l'ADN mitochondrial chez *Tetrahymena pyriformis*. Étude radiographique quantitative au microscope électronique. J. Cell Biol. 39:369-381.

Chasey, D. 1969. Observations on the central pair of microtubules from the cilia of *Tetrahymena pyriformis*. J. Cell Sci. 5:453-458.

Chasey, D. 1972. Further observations on the ultrastructure of cilia from *Tetrahymena pyriformis.* Exp. Cell Res. 74:471-479.

Chasey, D. 1972. Subunit arrangement in ciliary microtubules from *Tetrahymena pyriformis.* Exp. Cell Res. 74:140-146.

Chatton, E. and Brachon, S. 1935. Discrimination, chez deux infusoires du genre *Glaucoma,* entre système argentophile et infraciliature. C. R. Soc. Biol. 118:399-403.

Chatton, E. and Brachon, S. 1935. Les relations du chondriome avec l'infraciliature chez divers ciliés. Mitochondries ciliaires et parabasaux. C. R. Soc. Biol. 118:958-961.

Chatton, E. and Lwoff, A. 1935. La constitution primitive de la strie ciliaire des infusoires. La desmodexie. C. R. Soc. Biol. 118:1068-1072.

Chatton, E. and Lwoff, A. 1936. Techniques pour l'étude des protozoaires, spécialement de leurs structures superficielles (cinétome et argyrome). Bull. Soc. Fr. Microsc. 5:25-39.

Chatton, E., Lwoff, A., and Lwoff, M. 1931. La dualité figurée, substantielle et génétique des corps basaux des cils chez les infusoires; granule infraciliaire et corpuscule ciliare. C. R. Soc. Biol. 107:560-564.

Chatton, E., Lwoff, A., Lwoff, M., and Monod, J. L. 1931. La formation l'ébauche buccale postérieure chez les ciliés en division et ses relations de continuité topographique et génétique avec la bouche antérieure. C. R. Soc. Biol. 107:540-544.

Chatton, E. and Tellier, L. 1927. Sur les limites de résistance de quelques infusoires d'eau douce aux solutions de chlorures. C. R. Soc. Biol. 97:285-288.

Chatton, E. and Tellier, L. 1927. Sur les limites et les vitesses de retour à l'équilibre physiologique du *Glaucoma piriformis* porté dans les solutions hypertoniques de NaCl. C. R. Soc. Biol. 97:780-784.

Chatton, E. and Tellier, L. 1934. Comparaison de l'action de l'acide arsénieux avec celle du chlorure de sodium sur l'infusoire *Glaucoma piriformis.* Distinction entre une limite physique et une limite chimique de sa résistance à ces corps. C. R. Soc. Biol. 116:475-477.

Chatton, E. and Tellier, L. 1934. Sur les lésions de la chromatine produites par l'arsenic chez l'infusoire *Glaucoma piriformis.* C. R. Soc. Biol. 116:950-953.

Chi, J. C. H. and Suyama, Y. 1970. Comparative studies on mitochondrial and cytoplasmic ribosomes of *Tetrahymena pyriformis.* J. Mol. Biol. 53:531-556.

Child, F. M. 1959. The characterization of the cilia of *Tetrahymena pyriformis.* Exp. Cell Res. 18:258-267.

Child, F. M. 1961. Some aspects of the chemistry of cilia and flagella. Exp. Cell Res. 8(suppl.):47-53.

Child, F. M. 1963. The cilia of *Tetrahymena. In* J. Ludvík, J. Lom, and J. Vávra, eds., Progress in protozoology, Proceedings of the 1st International Congress on Protozoology, Prague, August 1961, Nakladatelstvi Československé Akademie, Prague, pp. 415-416.

Child, F. M. 1967. The chemistry of protozoan cilia and flagella. *In* M. Florkin and B. J. Scheer, eds., Chemical zoology, vol. 1, Protozoa (G. W. Kidder, ed.), Academic Press, New York, pp. 381-391.

Child, F. M. and Mazia, D. 1956. A method for the isolation of the parts of ciliates. Experientia 12:161-162.

Chou, S.-C., Conklin, D. A., Yamada, K., and Hokama, Y. 1971. Biochemical characterization of *Tetrahymena pyriformis* catalase. Int. J. Biochem. 2:705-710.

Chou, S.-C., Ramanathan, S., and Cutting, W. D. 1968. Quinacrine: Inhibition of synchronized cell division of *Tetrahymena.* Pharmacology 7:60-64.

Chou, S.-C., Ramanathan, S., Heu, P., and Conklin, K. A. 1971. Chlorpromazine effects on macromolecular syntheses in synchronized *Tetrahymena.* Pharmacology 6:1-8.

Chou, S.-C. and Scherbaum, O. H. 1963. Temperature-induced changes in phosphorus metabolism in synchronized *Tetrahymena.* Biochim. Biophys. Acta 71:221-224.

Chou, S.-C. and Scherbaum, O. II. 1965. Isolation and preliminary characteristics of two phosphorus-containing deoxysugars accumulating in division-blocked *Tetrahymena.* Exp. Cell Res. 39:346-354.

Chou, S.-C. and Scherbaum, O. H. 1965. Occurrence of acid soluble phosphorylated deoxysugars in division synchronized cells. Exp. Call Res. 40:217-223.

Chou, S.-C. and Scherbaum, O. H. 1967. Ethanolamine and phosphonic acid complexes in heat-treated *Tetrahymena* cells. Exp. Cell Res. 45:31-38.

Christensen, E. and Giese, A. C. 1956. Increased photoreversal of ultraviolet injury by flashing light. J. Gen. Physiol. 39:513-526.

Christensson, E. G. 1959. Changes in free amino acids and proteins during cell growth and synchronous division in mass cultures of *Tetrahymena pyriformis*. Acta Physiol. Scand. 45:339-349.

Christensson, E. G. 1962. Different RNA fractions during cell growth and synchronous division in mass cultures of *Tetrahymena pyriformis*. Acta Physiol. Scand. 54:1-8.

Christensson, E. G. 1967. Histones and basic proteins during cell growth and cell division in heat-synchronized mass cultures of *Tetrahymena pyriformis* GL. Ark. Zool. 19:297-308.

Christensson, E. G. 1968. Evidence of a synchronized RNA metabolism during cell growth and cell division in heat-treated mass cultures of *Tetrahymena pyriformis* GL. Z. Biol. 12:243-258.

Christensson, E. G. 1970. Studies of messenger RNA during the cell cycle in synchronized mass cultures of *Tetrahymena pyriformis* GL by hybridization. J. Protozool. 17:496-501.

Chua, A. S. and Ronkin, R. R. 1967. Acyl-CoA in aging populations of *Tetrahymena*. Comp. Biochem. Physiol. 21:425-429.

Cirillo, V. P. 1962. Mechanism of arabinose transport in *Tetrahymena pyriformis*. J. Bacteriol. 84:754-758.

Claff, C. L. 1940. A migration dilution apparatus for the sterilization of protozoa. Physiol. Zool. 13:334-341.

Clancy, C. F. 1968. The lethal effect of certain antimalarial drugs on *Tetrahymena pyriformis*. Amer. J. Trop. Med. Hyg. 17:359-363.

Claparède, E. and Lachmann, J. 1858-1861. Études sur les infusoires et les rhizopodes. Mém. Inst. Nat. Génévois 5:1-260; 6:261-482; 7:1-291.

Cleffmann, G. 1965. Die Schwellen der Hemmung der Nucleinsäuresynthese und der Teilung durch Actinomycin bei *Tetrahymena pyriformis*. Z. Zellforsch. Mikrosk. Anat. 67:343-350.

Cleffmann, G. 1966. Bildung zusätzlicher DNA nach Blockierung der Zellteilung von *Tetrahymena* durch Actinomycin. Z. Zellforsch. Mikrosk. Anat. 70:290-297.

Cleffmann, G. 1967. Temperaturabhängigkeit der Phasen der Teilungszyklus von *Tetrahymena pyriformis* HSM. Z. Zellforsch. Mikrosk. Anat. 79:599-602.

Cleffmann, G. 1968. Regulierung der DNA-Menge im Makronucleus von *Tetrahymena*. Exp. Cell Res. 50:193-207.

Cleffmann, G. 1969. Wachstumsparameter in normalen und durch Actinomycin verlängerten Zellzyklen von *Tetrahymena*. Z. Naturforsch. 24B:1624-1629.

Cline, S. G., 1966. Influence of ions and tonicity on RNA metabolism in *Tetrahymena pyriformis*. J. Cell Physiol. 68:157-164.

Cline, S. G. and Conner, R. L. 1966. Orthophosphate excretion as related to RNA metabolism in *Tetrahymena*. J. Cell Physiol. 68:149-156.

Cohen, A. I. 1956. The effect of enucleation on the DPN level of amoeba. J. Biophys. Biochem. Cytol. 2:15-21.

Cohen, A. I. 1957. Electron microscopic observations of *Amoeba proteus* in growth and inanition. J. Biophys. Biochem. Cytol. 3:859-865.

Cohen, A. I. 1959. Physiological and morphological observations on amoeba. Ann. N.Y. Acad. Sci. 78:609-622.

Coleman, J. R., Nilsson, J. R., Warner, R. R., and Batt, P. 1972. Qualitative and quantitative electron probe analysis of cytoplasmic granules in *Tetrahymena pyriformis*. Exp. Cell Res. 74:207-219.

Conklin, K. A. and Chou, S.-C. 1970. Antimalarials: Effect on *in vivo* and *in vitro* protein synthesis. Science 170:1213-1214.

Conklin, K. A. and Chou, S.-C. 1971. Protein synthesis in a cell-free preparation from *Tetrahymena pyriformis* GL. Comp. Biochem. Physiol. 40B:855-862.

Conklin, K. A. and Chou, S.-C. 1972. The effects of antimalarial drugs on uptake and incorporation of macromolecular precursors by *Tetrahymena pyriformis*. J. Pharmacol. Exp. Ther. 180:158-166.

Conklin, K., Chou, S.-C., and Heu, P. 1971. Quinine: Effect on *Tetrahymena pyriformis*. III. Energetics of isolated mitochondria in the presence of quinine and other antimalarial drugs. Biochem. Pharmacol. 20:1877-1882.

Conklin, K. A., Chou, S.-C., and Ramanathan, S. 1969. Quinine: Effect on *Tetrahymena pyriformis*. I. Inhibition of synchronized cell division and site of action. Pharmacology 2:247-256.

Conn, H. W. 1905. A preliminary report on the protozoa of the fresh waters of Connecticut. Bull. Conn. Geol. Nat. Hist. Survey, no. 2, Hartford Press. 69 pp.

Conn, H. W. and Edmondson, C. H. 1918. Flagellate and ciliate protozoa. *In* H. B. Ward and G. C. Whipple, eds., Fresh-water biology, John Wiley, New York, pp. 238-300.

Conner, R. L. 1957. Interaction of stigmasterol and 2,4-dinitrophenol in the growth of *Tetrahymena pyriformis*. Science 126:698.

Conner, R. L. 1959. Inhibition of growth of *Tetrahymena piriformis* by certain steroids. J. Gen. Microbiol. 21:180-185.

Conner, R. L. 1967. Transport phenomena in protozoa. *In* M. Florkin and B. J. Scheer, eds., Chemical zoology, vol. 1, Protozoa (G. W. Kidder, ed.), Academic Press, New York, pp. 309-350.

Conner, R. L. and Cline, S. G. 1964. Iron deficiency and the metabolism of *Tetrehymena pyriformis*. J. Protozool. 11:486-491.

Conner, R. L. and Cline, S. G. 1967. Some factors governing respiration, glucose metabolism and iodoacetate sensitivity in *Tetrahymena pyriformis*. J. Protozool. 14:22-26.

Conner, R. L., Goldberg, R., and Kornacker, M. S. 1961. The influence of hydrogen ion concentration and 2,4-dinitrophenol on orthophosphate accumulation in *Tetrahymena pyriformis*. J. Gen. Microbiol. 24:239-246.

Conner, R. L., Kornacker, M. S., and Goldberg, R. 1961. Influence of certain sterols and 2,4-dinitrophenol on phosphate accumulation and distribution in *Tetrahymena pyriformis*. J. Gen. Microbiol. 26: 437-442.

Conner, R. L., Landrey, J. R., Burns, C. H., and Mallory, F. B. 1968. Cholesterol inhibition of pentacyclic triterpenoid biosynthesis in *Tetrahymena pyriformis*. J. Protozool. 15:600-605.

Conner, R. L. and Linden, C. 1970. Purine and pyrimidine 5′-nucleotide catabolism in *Tetrahymena pyriformis* W. J. Protozool. 17:659-662.

Conner, R. L. and McDonald, L. A. 1964. The nature of the phosphatases associated with nucleotide metabolism in *Tetrahymena pyriformis*. J. Cell. Comp. Physiol. 64:257-263.

Conner, R. L. and Mallory, F. B. 1969. Biosynthesis of sterols and pentacyclic triterpenoids in *Tetrahymena pyriformis*. *In* K. Schubert, ed., Symposium über biochemische Aspekte der Steroidforschung, Akademie-Verlag, Berlin, pp. 5-16.

Conner, R. L., Mallory, F. B., Landrey, J. R., and Iyengar, W. 1969. The conversion of cholesterol to $\Delta^{5,7,22}$-cholestatrien-3β-ol by *Tetrahymena pyriformis*. J. Biol. Chem. 244:2325-2333.

Conner, R. L., Mallory, F. B., Landrey, J. R., Ferguson, K. A., Kaneshiro, E. S., and Ray, E. 1971. Ergosterol replacement of tetrahymanol in *Tetrahymena* membranes. Biochem. Biophys. Res. Commun. 44:995-1000.

Conner, R. L. and Nakatani, M. 1958. Stigmasterol antagonism of certain growth inhibitors for *Tetrahymena piriformis*. Arch. Biochem. Biophys. 74:175-181.

Conner, R. L. and Pruett, P. O. 1963. The influence of growth and suspension media on phosphate accumulation by *Tetrahymena pyriformis*. *In* J. Ludvík, J. Lom, and J. Vávra, eds., Progress in protozoology, Proceedings of the 1st International Congress on Protozoology, Prague, August 1961, Nakladatelstvi Československé Akademie, Prague, pp. 143-147.

Conner, R. L. and Ungar, F. 1964. The accumulation of cholesterol by *Tetrahymena pyriformis*. Exp. Cell Res. 36:134-144.

Connett, R. J. and Blum, J. J. 1971. Metabolic pathways in *Tetrahymena*: Distribution of carbon label by reactions of the tricarboxylic acid and glyoxylate cycles in normal and dimethylimipramine-treated cells. Biochemistry 10:3299-3309.

Connett, R. J., Wittels, B., and Blum, J. J. 1972. Metabolic pathways in *Tetrahymena*. II. Compartmentalization of acetylcoenzyme A and structure of the glycolytic and glyconeogenic pathways. J. Biol. Chem. 247:2657-2661.

Cook, D. E., Rangaraj, D. E., Best, N., and Wilkin, D. R. 1968. Conversion of glucose to glycogen in *Tetrahymena pyriformis*. Arch. Biochem. Biophys. 127:72-78.

Cooney, W. J. and Bradley, S. G. 1963. Action of cycloheximide on animal cells. Antimicrob. Agents Chemother. 1962:237-244.

Corbett, J. J. 1957. Changes in cell volume in relation to age of cultures and other factors in certain protozoa and bacteria. J. Cell. Comp. Physiol. 50:309-332.

Corbett, J. J. 1958. Factors influencing substrate utilization by *Tetrahymena pyriformis.* Exp. Cell Res. 15:512-521.

Corbett, J. J. 1970. Biochemical comparison of some classical *Tetrahymena pyriformis* and *Tetrahymena vorax* strains. J. Protozool. 17:181-182.

Corbett, J. J. and Sweeney, J. 1966. Antigenic differences among some classical *Tetrahymena pyriformis* and *Tetrahymena vorax* strains. J. Protozool. 13:359-366.

Corliss, J. O. 1952. Le cycle autogamique de *Tetrahymena rostrata.* C. R. Acad. Sci., Paris 235:399-402.

Corliss, J. O. 1952. Systematic status of the pure culture ciliate known as "*Tetrahymena geleii*" and "*Glaucoma piriformis.*" Science 116:188-119.

Corliss, J. O. 1952. Comparative studies on holotrichous ciliates in the *Colpidium-Glaucoma-Leucophrys-Tetrahymena* group. I. General considerations and history of strains in pure culture. Trans. Amer. Microsc. Soc. 71:159-184.

Corliss, J. O. 1953. Comparative studies on holotrichous ciliates in the *Colpidium-Glaucoma-Leucophrys-Tetrahymena* group. II. Morphology, life cycles, and systematic status of strains in pure culture. Parasitology 43:49-87.

Corliss, J. O. 1953. Silver impregnation of ciliated protozoa by the Chatton-Lwoff technic. Stain Technol. 28:97-100.

Corliss, J. O. 1953. Protozoa and systematics. Yale Sci. Mag. 28:14-17, 36, 38, 40.

Corliss, J. O. 1954. The literature on *Tetrahymena*: Its history, growth, and recent trends. J. Protozool. 1:156-169.

Corliss, J. O. 1954. The ciliated protozoan *Tetrahymena pyriformis* (alias "*Colpidium striatum*"). Carolina Tips 17:5.

Corliss, J. O. 1956. On the evolution and systematics of ciliated protozoa. Syst. Zool. 5:68-91, 121-140.

Corliss, J. O. 1958. The phylogenetic significance of the genus *Pseudomicrothorax* in the evolution of holotrichous ciliates. Acta Biol. Acad. Sci. Hung. 8:367-388.

Corliss, J. O. 1958. Proposed type genera for higher taxa within the sub-phylum Ciliophora (phylum Protozoa). Bull. Zool. Nomencl. 15:520-522.

Corliss, J. O. 1958. Order/class names: Problems in the selection of type genera. Bull. Zool. Nomencl. 15:1073-1079.

Corliss, J. O. 1960. *Tetrahymena chironomi* sp. nov., a ciliate from midge larvae, and the current status of facultative parasitism in the genus *Tetrahymena.* Parasitology 50:111-153.

Corliss, J. O. 1961. The ciliated protozoa: Characterization, classification, and guide to the literature. Pergamon Press, New York. 310 pp.

Corliss, J. O. 1961. Natural infection of tropical mosquitoes by ciliated protozoa of the genus *Tetrahymena.* Trans. Roy. Soc. Trop. Med. Hyg. 55:149-152.

Corliss, J. O. 1965. *Tetrahymena,* a ciliate genus of unusual importance in modern biological research. Acta Protozool. 3:1-20.

Corliss, J. O. 1965. L'autogamie et la sénescence du cilié hyménostome *Tetrahymena rostrata* (Kahl). Ann. Biol. 4:49-69.

Corliss, J. O. 1967. An aspect of morphogenesis in the ciliate Protozoa. J. Protozool. 14:1-8.

Corliss, J. O. 1967. Systematics of the phylum Protozoa. *In* M. Florkin and B. J. Scheer, eds., Chemical zoology, vol. 1, Protozoa (G. W. Kidder, ed.), Academic Press, New York, pp. 1-20.

Corliss, J. O. 1968. The value of ontogenetic data in reconstructing protozoan phylogenies. Trans. Amer. Microsc. Soc. 87:1-20.

Corliss, J. O. 1970. The comparative systematics of species comprising the hymenostome ciliate genus *Tetrahymena.* J. Protozool. 17:198-209.

Corliss, J. O. 1971. Type-specimens for the ten species of *Tetrahymena* (Protozoa: Ciliophora). Trans. Amer. Microsc. Soc. 90:243-252.

Corliss, J. O. 1971. Establishment of a new family (Glaucomidae n. fam.) in the holotrich hymenostome ciliate suborder Tetrahymenina, and description of a new genus (*Epenardia* n. g.) and a new species (*Glaucoma dragescui* n. sp.) contained therein. Trans. Amer. Microsc. Soc. 90:344-362.

Corliss, J. O. 1972. Current status of the International Collection of Ciliate Type-Specimens and guidelines for future contributors. Trans. Amer. Microsc. Soc. 91:221-235.

Corliss, J. O. 1972. *Tetrahymena* and some thoughts on the evolutionary origin of endoparasitism. Trans. Amer. Microsc. Soc. 91:566-573.

Corliss, J. O. and Dougherty, E. C. 1967. An appeal for stabilization of certain names in the protozoan family Tetrahymenidae (subphylum Ciliophora, order Hymenostomatida), with special reference to the generic name *Tetrahymena* Furgason, 1940. Bull. Zool. Nomencl. 24:155-185.

Corliss, J. O., Smith, A. C., and Foulkes, J. 1962. A species of *Tetrahymena* from the British garden slug *Milax budapestensis*. Nature 196:1008-1009.

Cox, D. 1970. Prolonged survival of *Tetrahymena* at 0-5 C in citrated, lecithinized, defined media. J. Protozool. 17:150-152.

Cox, D., Frank, O., Hutner, S. H., and Baker, H. 1968. Growth of *Tetrahymena* in carbohydrate-free high-glutamate media. J. Protozool. 15:713-716.

Cox, F. E. G. 1967. A quantitative method for measuring the uptake of carbon particles by *Tetrahymena pyriformis*. Trans. Amer. Microsc. Soc. 86:261-267.

Crane, F. L. 1962. Quinones in lipoprotein electron transport systems. Biochemistry 1:510-517.

Crockett, R. L., Dunham, P. B., and Rasmussen, L. 1965. Protein metabolism in *Tetrahymena pyriformis* cells dividing synchronously under starvation conditions. C. R. Trav. Lab. Carlsberg 34:451-486.

Crowther, J. A. 1926. The action of x-rays on *Colpidium colpoda*. Proc. Roy. Soc. 100:390-404.

Culbertson, J. R. 1966. Physical and chemical properties of cilia isolated from *Tetrahymena pyriformis*. J. Protozool. 13:397-406.

Culbertson, J. R. 1966. Cilia isolated from *Tetrahymena* after membrane stabilization by 1,5-difluoro-2,4-dinitrobenzene. Science 153:1390-1391.

Cummins, J. E. and Plaut, W. 1962. The effect of starvation on the nucleotide composition of the ribonucleic acid in *Amoeba proteus*. Biochim. Biophys. Acta 55:418-420.

Curds, C. R. 1969. An illustrated key to the British freshwater ciliated protozoa commonly found in activated sludge. Ministry of Technology, H. M. S. O., London. 90 pp.

Curds, C. R. and Cockburn, A. 1968. Studies on the growth and feeding of *Tetrahymena pyriformis* in axenic and monoxenic culture. J. Gen. Microbiol. 54:343-358.

Curds, C. R. and Cockburn, A. 1971. Continuous monoxenic culture of *Tetrahymena pyriformis*. J. Gen. Microbiol. 66:95-108.

Czapik, A. 1968. La famille Tetrahymenidae et son importance dans la systématique et l'évolution des ciliés. Acta Protozool. 5:315-357.

Dan, K. 1966. Behavior of sulfhydryl groups in synchronous division. *In* I. L. Cameron and G. M. Padilla, eds., Cell synchrony, Academic Press, New York, pp. 307-327.

Danforth, W. F. 1967. Respiratory metabolism. *In* T.-T. Chen, ed., Research in protozoology, vol. 1, Pergamon Press, New York, pp. 201-306.

Dass, C. M. S. 1950. Chromatin elimination in *Glaucoma pyriformis* (Ehrb.). Nature 165:693.

Debault, L. E. and Ringertz, N. R. 1967. A comparison of normal and cold-synchronized cell divisions in *Tetrahymena*. Exp. Cell Res. 45:509-518.

de Boissezon, P. 1930. Contribution à l'étude de la biologie et de l'histophysiologie de *Culex pipiens* L. Arch. Zool. Exp. Gén. 70:281-431.

de Castro, F. T. and Couceiro, A. 1955. Studies on vital staining of protozoa. Exp. Cell Res. 8:245-247.

de Duve, C. and Baudhuin, P. 1966. Peroxisomes (microbodies) and related particles. Physiol. Rev. 46:323-257.

de Duve, C. and Wattiaux, R. 1966. Functions of lysosomes. Ann. Rev. Physiol. 28:435-492.

Deflandre, G. 1938. Les corpuscules biréfringents des ciliés et des cryptomonadines. Bull. Soc. Fr. Microsc. 7:110-129.

Dekker, E. E., Schlesinger, M. J., and Coon, M. J. 1958. β-hydroxy-β-methylglutaryl coenzyme A deacylase. J. Biol. Chem. 233:434-438.

Den, H., Robinson, W. G., and Coon, M. J. 1959. Enzymatic conversion of β-hydroxyproprionate to malonic semialdehyde. J. Biol. Chem. 234:1666-1671.

Dennis, E. A. and Kennedy, E. P. 1970. Enzymatic synthesis and decarboxylation of phosphatidyl-serine in *Tetrahymena pyriformis*. J. Lipid Res. 11:394-403.

Dessi, P. 1964. Azioni della morfina sullo sviluppo in colture di elementi HeLa e di *tetrahimena* [sic] *pyriformis* W. Arch. Ital. Sci. Farmacol. 15:172-176.

DeVault, R. L., Schmitz, H., and Hooper, I. R. 1966. Demetric acid, a new cytotoxic agent. Antimicrob. Agents Chemother. 1965:796-800.

DeWaart, J. 1967. Detection of bacterial and fungal toxins in foods by means of biological test systems. Antonie van Leeuwenhoek J. Microbiol. Serol. 33:230.

Dewey, V. C. 1941. The nutrition of *Tetrahymena geleii* (Protozoa, Ciliata). Proc. Soc. Exp. Biol. Med. 46:482-484.

Dewey, V. C. 1944. Biochemical factors in the maximal growth of *Tetrahymena*. Biol. Bull. 87:107-120.

Dewey, V. C. 1967. Lipid composition, nutrition and metabolism. *In* M. Florkin and B. J. Scheer, eds., Chemical zoology, vol. 1, Protozoa (G. W. Kidder, ed.), Academic Press, New York, pp. 161-274.

Dewey, V. C., Heinrich, M. R., and Kidder, G. W. 1957. Evidence for the absence of the urea cycle in *Tetrahymena*. J. Protozool. 4:211-219.

Dewey, V. C., Heinrich, M. R., Markees, D. G., and Kidder, G. W. 1959. Multiple inhibition by 6-methylpurine. Biochem. Pharmacol. 3:173-180.

Dewey, V. C. and Kidder, G. W. 1952. Protogen and its relation to ketoacids in metabolism of *Tetrahymena*. Proc. Soc. Exp. Biol. Med. 80:302-306.

Dewey, V. C. and Kidder, G. W. 1953. Factors affecting the requirement of *Tetrahymena pyriformis* (*geleii*) for folic acid. J. Gen. Microbiol. 9:445-453.

Dewey, V. C. and Kidder, G. W. 1953. Further studies on effect of folic acid analogs on growth of *Tetrahymena*. Proc. Soc. Exp. Biol. Med. 83:586-588.

Dewey, V. C. and Kidder, G. W. 1954. Comparative activity of pantothenic acid and its conjugates for tetrahymenid ciliates. Proc. Soc. Exp. Biol. Med. 87:198-199.

Dewey, V. C. and Kidder, G. W. 1958. Amino acid antagonisms in *Tetrahymena*. Arch. Biochem. Biophys. 73:29-37.

Dewey, V. C. and Kidder, G. W. 1960. Antimetabolites of acetate in *Tetrahymena*. Arch. Biochem. Biophys. 88:78-82.

Dewey, V. C. and Kidder, G. W. 1960. The influence of folic acid, threonine and glycine on serine synthesis in *Tetrahymena*. J. Gen. Microbiol. 22:72-78.

Dewey, V. C. and Kidder, G. W. 1960. Serine synthesis in *Tetrahymena* from non-amino acid sources; compounds derived from serine. J. Gen. Microbiol. 22:79-92.

Dewey, V. C. and Kidder, G. W. 1960. The sites of action of azapurines in *Tetrahymena*. Z. Allg. Mikrobiol. 1:1-12.

Dewey, V. C. and Kidder, G. W. 1962. Biological activity of sterol glycosides. Biochem. Pharmacol. 11:53-56.

Dewey, V. C. and Kidder, G. W. 1964. Inhibition by puromycin and its reversal by peptides. Biochem. Pharmacol. 13:353-360.

Dewey, V. C. and Kidder, G. W. 1967. The use of Sephadex for the concentration of pteridines. J. Chromatogr. 31:326-336.

Dewey, V. C. and Kidder, G. W. 1972. Proline metabolism in *Tetrahymena*. J. Protozool. 19:50-53.

Dewey, V. C., Kidder, G. W., and Markees, D. G. 1959. Purine antagonists and growth of *Tetrahymena*. Proc. Soc. Exp. Biol. Med. 102:306-311.

Dewey, V. C., Kidder, G. W., and Parks, R. E., Jr. 1951. Growth inhibition in *Tetrahymena* by folic acid analogs. Proc. Soc. Exp. Biol. Med. 78:91-95.

Dewey, V. C., Kidder, G. W., and Parks, R. E., Jr. 1952. The effect of thiosemicarbazide and other urea analogues on the growth of *Tetrahymena geleii*. J. Gen. Microbiol. 6:181-186.

Dewey, V. C., Parks, R. E., Jr., and Kidder, G. W. 1950. Growth responses of *Tetrahymena geleii* to changes in the basal media. Arch. Biochem. 29:281-290.

Diamond, L. S. 1964. Freeze-preservation of protozoa. Cryobiology 1:95-102.

Dickie, N. and Liener, I. E. 1962. A study of the proteolytic system of *Tetrahymena pyriformis* W. I. Purification and partial characterization of the constituent proteinases. Biochim. Biophys. Acta 64:41-51.

Dickie, N. and Liener, I. E. 1962. A study of the proteolytic system of *Tetrahymena pyriformis* W. II. Substrate specificity of the constituent proteinases. Biochim. Biophys. Acta 64:52-59.

Diesterhaft, M. D., Hsieh, H. C., Elson, C., Sallach, H. J., and Shrago, E. 1972. Enzymatic regulation of the metabolism of phosphoenolpyruvate in *Tetrahymena pyriformis*. J. Biol. Chem. 247:2755-2762.

Dobrzańska, J. 1958. Investigations on ciliates living in lamellibranchiates of small water bodies. Bull. Acad. Pol. Sci. 6:113-118.

Dobrzańska, J. 1959. The occurrence of ciliates of the genus *Tetrahymena* Furgason, 1940 in freshwater mussels. Bull. Acad. Pol. Sci. 7:377-382.

Doroszewski, M. 1963. The response of the ciliate *Dileptus* and its fragments to the water shake. Acta Biol. Exp. (Warsaw) 23:3-10.

Dougherty, E. C. 1953. Problems of nomenclature for the growth of organisms of one species with and without associated organisms of other species. Parasitology 42:259-261.

Doyle, W. L. and Harding, J. P. 1937. Quantitative studies on the ciliate *Glaucoma*. Excretion of ammonia. J. Exp. Biol. 14:462-469.

Doyle, W. L. and Patterson, E. K. 1942. Origin of dipeptidase in a protozoon. Science 95:206.

Dragesco, J. and Njiné, T. 1971. Compléments à la connaissance des ciliés libres du Cameroun. Ann. Fac. Sci. Cameroun, no. 7-8:97-140.

Dryl, S. and Grebecki, A. 1966. Progress in the study of excitation and response in ciliates. Protoplasma 62:255-284.

Ducoff, H. S. 1956. Radiation-induced fission block in synchronized cultures of *Tetrahymena pyriformis* W. Exp. Cell Res. 11:218-220.

Ducoff, H. S., Williams, M., Syn, T. B., and Butler, B. D. 1964. Quantitative studies on metabolism in phagotrophic ciliates. I. Growth of *Tetrahymena pyriformis* W on viable and on killed bacteria. J. Protozool. 11:309-313.

Dugaiczyk, A. and Eiler, J. J. 1968. The inhibition of synchronous division of *Tetrahymena pyriformis* by actinomycin D. Curr. Mod. Biol. 1:333-336.

Dujardin, F. 1841. Histoire naturelle des zoophytes. Infusoires. Roret, Paris. 678 pp.

Dunham, L. J. 1937. Relationships between free-living thermal ciliates and ingested acid-fast bacteria. Amer. Rev. Tuberc. 36:553-566.

Dunham, P. B. 1964. The adaptation of *Tetrahymena* to a high NaCl environment. Biol. Bull. 126:373-390.

Dunham, P. B. and Child, F. M. 1961. Ion regulation in *Tetrahymena*. Biol. Bull. 121:129-140.

Dunham, P. B. and Stoner, L. C. 1969. Indentation of the pellicle of *Tetrahymena* at the contractile vacuole pore before systole. J. Cell Biol. 43:184-188.

Dunn, M. S. 1949. Determination of amino acids by microbiological assay. Physiol. Rev. 29:219-259.

Dunn, M. S. and Rockland, L. B. 1947. Biological value of proteins determined with *Tetrahymena geleii* H. Proc. Soc. Exp. Biol. Med. 64:377-379.

Ebringer, L. 1963. Some effects of antibiotics on protozoa. *In* J. Ludvík, J. Lom, and J. Vávra, eds., Progress in protozoology, Proceedings of the 1st International Congress on Protozoology, Prague, August 1961, Nakladatelstvi Československé Akademie, Prague, pp. 198-199.

Edmondson, C. H. 1906. The protozoa of Iowa. Proc. Davenport Acad. Sci. 11:1-124.

Edmondson, C. H. 1920. Protozoa of Devils Lake complex, North Dakota. Trans. Amer. Microsc. Soc. 39:182-198.

Edwin, E. E. and Green, J. G. 1960. Reversal by α-tocopherol and other substances of succinoxidase inhibition produced by a *Tetrahymena pyriformis* preparation. Arch. Biochem. Biophys. 87:337-338.

Ehrenberg, C. G. 1830. Organisation, Systematik und geographisches Verhaltniss der Infusionsthierch-en. Königlich Akademie Wissenschaften, Berlin, pp. 1-108.

Ehrenberg, C. G. 1832. Beiträge zur Kenntniss der Organization der Infusorien und ihrer geographischen Verbreitung besonders in Siberien. Abh. Akad. Wiss. Berlin 1830:1-88.

Ehrenberg, C. G. 1832. Uber die Entwicklung und Lebensdauer der Infusionsthiere; nebst ferneren Beiträgen zu einer Vergleichung ihrer organischen Systeme. Abh. Akad. Wiss. Berlin 1831:1-154.

Ehrenberg, C. G. 1838. Die Infusionsthierchen als vollkommene Organismen. Leopold Voss, Leipzig. 612 pp.

Eichel, H. J. 1954. Studies on the oxidation of succinic acid by cell-free homogenates of *Tetrahymena pyriformis* S and W. J. Biol. Chem. 206:159-169.

Eichel, H. J. 1956. Effects of atabrine and flavin mononucleotide on oxidation of succinic acid by *Tetrahymena* preparations. Biochim. Biophys. Acta 22:571-573.

Eichel, H. J. 1956. Purine-metabolizing enzymes of *Tetrahymena pyriformis*. J. Biol. Chem. 220:209-220.

Eichel, H. J. 1956. Respiratory enzyme studies in *Tetrahymena pyriformis*. II. Reduced diphosphopyridine nucleotide oxidase and reduced diphosphopyridine nucleotide cytochrome reductase. J. Biol. Chem. 222:121-136.

Eichel, H. J. 1956. Respiratory enzyme studies in *Tetrahymena pyriformis*. III. Some properties of re-duced diphosphopyridine nucleotide oxidase of cell-free homogenates. J. Biol. Chem. 222:137-144.

Eichel, H. J. 1959. A mammalian and protozoan electron transport inhibitor in *Tetrahymena pyrifor-mis*. Biochim. Biophys. Acta 34:589-591.

Eichel, H. J. 1959. Respiratory enzyme studies in *Tetrahymena pyriformis*. IV. Stabilization of elec-tron transport components. Biochem. J. 71:106-118.

Eichel, H. J. 1960. Electron-transport and phosphorylation inhibitor in *Tetrahymena* and evidence for its formation by a phospholipase. Biochim. Biophys. Acta 43:364-366.

Eichel, H. J. 1961. Respiratory enzyme studies in the ciliated protozoon, *Tetrahymena pyriformis*. Symp. Genet. Biol. Ital. 8:381-417.

Eichel, H. J. 1966. Oxidation of L-α-hydroxy acids by *Tetrahymena pyriformis*. Biochim. Biophys. Acta 128:183-186.

Eichel, H. J., Goldenberg, E. K., and Rem, L. T. 1964. Particle-bound DPN-linked lactate dehydro-genase in *Tetrahymena pyriformis* and its separation from the L-(+)-lactate oxidase. Biochim. Bio-phys. Acta 81:172-175.

Eichel, H. J. and Rem, L. T. 1959. Lactic oxidase of *Tetrahymena pyriformis*. Arch. Biochem. Biophys. 82:484-485.

Eichel, H. J. and Rem, L. T. 1959. Lactic oxidase of *Tetrahymena pyriformis*: The over-all reaction mechanism. Biochim. Biophys. Acta 35:571-573.

Eichel, H. J. and Rem, L. T. 1962. Respiratory enzyme studies in *Tetrahymena pyriformis*. V. Some properties of an L-lactic oxidase. J. Biol. Chem. 237:940-945.

Eichel, H. J. and Roth, J. S. 1953. The effect of x-irradiation on nuclease activity and respiration of *Tetrahymena geleii* W. Biol. Bull. 104:351-358.

Eidinoff, M. L. and Rich, M. A. 1959. Growth inhibition of a human tumor cell strain by 5-fluoro-2′-deoxyuridine: Time parameters for subsequent reversal by thimidine. Cancer Res. 19:521-527.

Eigen, E. and Shockman, B. D. 1963. The folic acid group. *In* F. Kavanagh, ed., Analytical microbiol-ogy, Academic Press, New York, pp. 431-488.

Eiler, J. J., Krezanoski, J. Z., and Lee, K. H. 1959. The biological action of cellular depressants and stimulants. I. The action of ethyl carbonate (urethane) on the endogenous respiration and the rate of cell division of *Tetrahymena pyriformis*. J. Amer. Pharm. Ass. 48:290-296.

Elliott, A. M. 1933. Isolation of *Colpidium striatum* Stokes in bacteria-free cultures and the relation of growth to pH of the medium. Biol. Bull. 65:45-56.

Elliott, A. M. 1935. Effect of carbohydrates on growth of *Colpidium*. Arch. Protistenk. 84:156-174.

Elliott, A. M. 1935. Effects of certain organic acids and protein derivatives on the growth of *Colpidium*. Arch. Protistenk. 84:225-231.

Elliott, A. M. 1935. The influence of pantothenic acid on the growth of protozoa. Biol. Bull. 68:82-92.

Elliott, A. M. 1936. Nutritional studies of protozoa. Proc. Minn. Acad. Sci. pp. 1-6.

Elliott, A. M. 1938. The influence of certain plant hormones on the growth of protozoa. Physiol. Zool. 11:31-39.

Elliott, A. M. 1939. The vitamin B complex and the growth of *Colpidium striatum*. Physiol. Zool. 12:363-373.

Elliott, A. M. 1939. A volumetric method for estimating population densities of protozoa. Trans. Amer. Microsc. Soc. 58:97-99.

Elliott, A. M. 1949. The amino acid requirement of *Tetrahymena geleii* (E). Physiol. Zool. 22:337-345.

Elliott, A. M. 1949. A photoelectric colorimeter for estimating protozoan population densities. Trans. Amer. Microsc. Soc. 68:228-233.

Elliott, A. M. 1950. The growth-factor requirements of *Tetrahymena geleii* E. Physiol. Zool. 23:85-91.

Elliott, A. M. 1959. Biology of *Tetrahymena*. Ann. Rev. Microbiol. 13:79-96.

Elliott, A. M. 1959. A quarter century exploring *Tetrahymena*. J. Protozool. 6:1-7.

Elliott, A. M. 1963. The fine structure of *Tetrahymena pyriformis* during mitosis. *In* L. Levine, ed., The cell in mitosis, Academic Press, New York, pp. 107-124.

Elliott, A. M. 1965. Primary lysosomes in *Tetrahymena pyriformis*. Science 149:640-641.

Elliott, A. M. 1970. The distribution of *Tetrahymena pyriformis*. J. Protozool. 17:162-168.

Elliott, A. M., Addison, M. A., and Carey, S. E. 1962. Distribution of *Tetrahymena pyriformis* in Europe. J. Protozool. 9:135-141.

Elliott, A. M. and Bak, I. J. 1964. The fate of mitochondria during aging in *Tetrahymena pyriformis*. J. Cell Biol. 20:113-129.

Elliott, A. M. and Bak, I. J. 1964. The contractile vacuole and related structures in *Tetrahymena pyriformis*. J. Protozool. 11:250-261.

Elliott, A. M., Brownell, L. E., and Gross, J. A. 1954. The use of *Tetrahymena* to evaluate the effects of gamma radiation on essential nutrilites. J. Protozool. 1:193-199.

Elliott, A. M. and Clark, G. M. 1956. The induction of haploidy in *Tetrahymena pyriformis* following x-irradiation. J. Protozool. 3:181-188.

Elliott, A. M. and Clark, G. M. 1957. Further cytological studies on haploid and diploid strains of *Tetrahymena pyriformis* with special reference to univalent spindles. Cytologia 22:355-359.

Elliott, A. M. and Clark, G. M. 1958. Genetic studies of the pyridoxine mutant in variety two of *Tetrahymena pyriformis*. J. Protozool. 5:235-240.

Elliott, A. M. and Clark, G. M. 1958. Genetic studies of the serine mutant in variety nine of *Tetrahymena pyriformis*. J. Protozool. 5:240-246.

Elliott, A. M. and Clemmons, G. L. 1966. An ultrastructural study of ingestion and digestion in *Tetrahymena pyriformis*. J. Protozool. 13:311-323.

Elliott, A. M. and Hayes, R. E. 1953. Mating types in *Tetrahymena*. Biol. Bull. 105:269-284.

Elliott, A. M. and Hayes, R. E. 1955. *Tetrahymena* from Mexico, Panama, and Colombia with special reference to sexuality. J. Protozool. 2:75-80.

Elliott, A. M. and Hogg, J. F. 1952. Culture variation in *Tetrahymena*. Physiol. Zool. 25:318-323.

Elliott, A. M. and Hunter, R. L. 1951. Phosphatase activity in *Tetrahymena*. Biol. Bull. 100:165-172.

Elliott, A. M. and Kennedy, J. R., Jr. 1962. The morphology and breeding system of variety 9, *Tetrahymena pyriformis*. Trans. Amer. Microsc. Soc. 81:300-308.

Elliott, A. M., Kennedy, J. R., Jr., and Bak, I. J. 1962. Macronuclear events in synchronously dividing *Tetrahymena pyriformis*. J. Cell Biol. 12:515-531.

Elliott, A. M. and Nanney, D. L. 1952. Conjugation in *Tetrahymena*. Science 116:33-34.

Elliott, A. M., Studier, M. A., and Work, J. A. 1966. *Tetrahymena pyriformis* from several Pacific Islands and Australia. J. Protozool. 11:370-378.

Elliott, A. M., Travis, D. M., and Work, J. A. 1966. An ultrastructural study of the effects of aeration and physical activity on aging in *Tetrahymena pyriformis*. J. Exp. Zool. 161:177-192.

Elliott, A. M. and Tremor, J. W. 1958. The fine structure of the pellicle in the contact area of conjugating *Tetrahymena pyriformis*. J. Biophys. Biochem. Cytol. 4:839-840.

Elliott, A. M. and Zieg, R. G. 1968. A Golgi apparatus associated with mating in *Tetrahymena pyriformis*. J. Cell Biol. 36:391-398.

Elson, C., Hartmann, H. A., Shug, A. L., and Shrago, E. 1970. Actomycin A: Stimulation of cell division and protein synthesis in *Tetrahymena pyriformis.* Science 168:385-386.

Engberg, J. and Pearlman, R. E. 1972. The amount of ribosomal RNA in *Tetrahymena pyriformis* in different physiological states. Eur. J. Biochem. 26:393-400.

Englehard, W. E. and McCashland, B. W. 1959. Antigenic relationships in seven strains of *Tetrahymena.* Appl. Microbiol. 7:41-45.

Ephrussi, B. and Lwoff, A. 1923. Sur la double périodicité cyclique de la zone de devision chez un cilié *Colpidium colpoda,* mise en évidence par la reversion de la scission. C. R. Acad. Sci., Paris 176:930-932.

Epstein, H. V. 1926. Infektion des Nervensystems von Fischen durch Infusorien. Russ. Arkh. Protist. 5:169-180. In Russian with German summary.

Epstein, S. S., Forsyth, J., Saporoschetz, I. B., and Mantel, N. 1966. An exploratory investigation on the inhibition of selected photosensitizers by agents of varying antioxidant activity. Radiat. Res. 28:322-335.

Epstein, S. S. and Niskanen, E. E. 1967. Effects of Tween 60 on benzo(*a*)pyrene uptake by *Tetrahymena pyriformis* and by isolated rat liver mitochondria. Exp. Cell Res. 46:211-234.

Epstein, S. S., Saporoschetz, I. B., and Hutner, S. H. 1967. Toxicity of antioxidants to *Tetrahymena pyriformis.* J. Protozool. 14:238-244.

Epstein, S. S., Saporoschetz, I. B., and Mantel, N. 1966. Interactions between antioxidant and photosensitizer in the photodynamic bioassay for antioxidants. Life Sci. 5:783-793.

Epstein, S. S., Saporoschetz, I. B., Small, M., Park, W., and Mantel, N. 1965. A simple bioassay for antioxidants based on protection of *Tetrahymena pyriformis* from the photodynamic toxicity of benzo(*a*)pyrene. Nature 208:655-658.

Erhardová, B. 1952. Experimental parasitism induced in free-living protozoa. Česk. Biol. 1:171-174. In Czechoslovakian with German summary.

Erwin, J. A. 1970. The reversal of temperature-induced cell-surface deformation in *Tetrahymena* by polyunsaturated fatty acids. Biochim. Biophys. Acta 202:21-34.

Erwin, J. A. and Bloch, K. 1963. Lipid metabolism of ciliated protozoa. J. Biol. Chem. 238:1618-1624.

Erwin, J. A. and Holz, G. G., Jr. 1962. Production of a vitamin B_{12} compound by tetrahymenids. J. Protozool. 9:211-214.

Erwin, J. A., Beach, D., and Holz, G. G., Jr. 1966. Effect of dietary cholesterol on unsaturated fatty acid biosynthesis in a ciliated protozoan. Biochim. Biophys. Acta 125:614-616.

Evans, F. R. and Corliss, J. O. 1964. Morphogenesis in the hymenostome ciliate *Pseudocohnilembus persalinus* and its taxonomic and phylogenetic implications. J. Protozool. 11:353-370.

Evans, R. J. and Bandemer, S. L. 1967. Nutritive values of some oil seed proteins. Cereal Chem. 44:417-426.

Evans, R. J. and Bandemer, S. L. 1967. Nutritive value of legume seed proteins. J. Agr. Food Chem. 15:439-445.

Everhart, L. P., Jr. 1972. Methods with *Tetrahymena. In* D. M. Prescott, ed., Methods in cell physiology, vol. 5, Academic Press, New York, pp. 220-288.

Everhart, L. P., Jr. and Ronkin, R. R. 1966. Changes in the lipids of cells from aging populations of *Tetrahymena pyriformis.* J. Protozool. 13:646-650.

Fabre-Domergue, P. L. M. 1888. Recherches anatomiques et physiologiques sur les infusoires ciliés. Masson et Cie, Paris. 140 pp.

Falk, H., Wunderlich, F., and Frank, W. W. 1968. Microtubular structures in macronuclei of synchronously dividing *Tetrahymena pyriformis.* J. Protozool. 15:776-780.

Fauré-Fremiet, E. 1906. Le *Glaucoma pyriformis* et l'organisation de la substance vivante. C. R. Ass. Anat. 8:120-127.

Fauré-Fremiet, E. 1910. Étude sur les mitochondries des protozoaires et des cellules sexuelles. Arch. Anat. Microsc. 11:457-648.

Fauré-Fremiet, E. 1911. Appareil nucléaire, chromidies, mitochondries. Arch. Protistenk. 21:186-208.

Fauré-Fremiet, E. 1912. Études cytologiques sur quelques infusoires des marais salants du Croisic. Arch. Anat. Microsc. 13:401-479.

Fauré-Fremiet, E. 1945. Symétrie et polarité chez les ciliés bi- ou multicomposites. Bull. Biol. Fr. Belg. 79:106-150.

Fauré-Fremiet, E. 1948. Doublets homopolaires et régulation morphogénétique chez le cilié *Leucophyrs patula.* Arch. Anat. Microsc. 37:183-203.

Fauré-Fremiet, E. 1948. Les mécanismes de la morphogenèse chez les ciliés. Folia Biotheoret. 3:25-58.

Fauré-Fremiet, E. 1949. Action du lithium sur la stomatogenèse chez les ciliés. J. Cyto-embryol. Belgonéerland., Gand, pp. 100-102.

Fauré-Fremiet, E. 1950. Morphologie comparée et systématique des ciliés. Bull. Soc. Zool. Fr. 75:109-122.

Fauré-Fremiet, E. 1953. Morphology of protozoa. Ann. Rev. Microbiol. 7:1-18.

Fauré-Fremiet, E. 1953. L'hypothèse de la sénescence et les cycles de réorganisation nucléaire chez les ciliés. Rev. Suisse Zool. 426-438.

Fauré-Fremiet, E. 1961. Documents et observations écologiques et pratiques sur la culture des infusoires ciliés. Hydrobiologia 18:300-320.

Fauré-Fremiet, E. 1967. Chemical aspects of ecology. *In* M. Florkin and B. J. Scheer, eds., Chemical zoology, vol. 1, Protozoa (G. W. Kidder, ed.), Academic Press, New York, pp. 21-54.

Fauré-Fremiet, E. 1967. La régénération chez les protozoaires. Bull. Soc. Zool. Fr. 92:249-272.

Fauré-Fremiet, E. and Mugard, H. 1949. Le dimorphisme de *Espejoia mucicola.* Hydrobiologia 1:379-393.

Fedorova, I. G. and Burnasheva, S. A. 1963. Electron microscopic investigation of the fine structure of the cilia of *Tetrahymena pyriformis.* Tsitologiya 5:689-691. In Russian with English summary.

Fennell, R. A. 1951. The relation between growth substances, cytochemical properties of *Tetrahymena geleii,* and lesion induction in the chorioallantois. J. Elisha Mitchell Sci. Soc. 67:219-229.

Fennell, R. A. and Degenhardt, E. F. 1957. Some factors affecting alkaline phosphatase activity in *Tetrahymena pyriformis* W. J. Protozool. 4:30-42.

Fennell, R. A. and Marzke, F. O. 1954. The relation between vitamins, inorganic salts and the histochemical characteristics of *Tetrahymena geleii* W. J. Morphol. 94:587-616.

Fennell, R. A. and Pastor, E. P. 1958. Some observations on the esterases of *Tetrahymena pyriformis* W. I. Evidence for the existence of aliesterases. J. Morphol. 103:187-201.

Ferguson, K. A., Conner, R. I., and Mallory, F. B. 1971. The effect of ergosterol on the fatty acid composition of *Tetrahymena pyriformis.* Arch. Biochem. Biophys. 144:448-450.

Ferguson, K. A., Conner, R. L., Mallory, F. B., and Mallory, C. W. 1972. α-Hydroxy fatty acids in sphingolipids of *Tetrahymena.* Biochim. Biophys. Acta 270:111-116.

Fernell, W. R. and Rosen, G. D. 1956. Microbiological evaluation of protein quality with *Tetrahymena pyriformis* W. I. Characteristics of growth of the organism and determination of relative nutritive values of intact proteins. Brit. J. Nutr. 10:143-156.

Fitt, P. S. 1966. Incorporation of ribonucleoside triphosphates into RNA by extracts of *Tetrahymena pyriformis* S. J. Protozool. 13:507-509.

Flavell, R. A. and Jones, I. G. 1970. Kinetic complexity of *Tetrahymena pyriformis* nuclear deoxyribonucleic acid. Biochem. J. 116:155-157.

Flavell, R. A. and Jones, I. G. 1970. Mitochondrial deoxyribonucleic acid from *Tetrahymena pyriformis* and its kinetic complexity. Biochem. J. 116:811-817.

Flavell, R. A. and Jones, I. G. 1971. DNA from isolated pellicles of *Tetrahymena.* J. Cell Sci. 9:719-726.

Flavin, M. and Engleman, M. 1953. Aminopurine interconversion in *Tetrahymena geleii:* Role of 8-azaguanine and hypoxanthine. J. Biol. Chem. 200:59-68.

Flavin, M. and Graff, S. 1951. Utilization of guanine for nucleic acid biosynthesis by *Tetrahymena geleii.* J. Biol. Chem. 191:55-61.

Flavin, M. and Graff, S. 1951. The utilization of adenine for nucleic acid biosynthesis by *Tetrahymena geleii.* J. Biol. Chem. 192:485-488.

Flickinger, C. J. 1965. The fine structure of the nuclei of *Tetrahymena pyriformis* throughout the cell cycle. J. Cell Biol. 27:519-529.

Foley, G. E., Eagle, H., Snell, E. E., Kidder, G. W., and Thayer, P. S. 1958. Studies on the use of *in vitro* procedures for the screening of potential antitumor agents: Comparison of activity in mammalian cell cultures and microbiological assays alone and in combination with experimental antitumor activity. Ann. N.Y. Acad. Sci. 76:952-960.

Foley, G. E., McCarthy, R. E., Binns, V. M., Snell, E. E., Guirard, B. M., Kidder, G. W., Dewey, V. C., and Thayer, P. S. 1958. A comparative study of the use of microorganisms in the screening of potential antitumor agents. Ann. N.Y. Acad. Sci. 76:413-438.

Forer, A., Nilsson, J. R., and Zeuthen, E. 1970. Studies on the oral apparatus of *Tetrahymena pyriformis* GL. C. R. Trav. Lab. Carlsberg 38:67-86.

Forrest, H. S. and Van Baalen, C. 1970. Microbiology of unconjugated pteridines. Ann. Rev. Microbiol. 24:91-108.

Forrest, I. S., Quesada, F., and Dietshman, G. L. 1964. Unicellular organisms as model systems for the mode of actions of phenothiazine and related drugs. Proc. West. Pharmacol. Soc. 7:42-44.

Francis, A. A. and Whitson, G. L. 1969. Ultraviolet-induced pyrimidine dimers in *Tetrahymena pyriformis*. II. *In vivo* photoreactivations. Biochim. Biophys. Acta 179:253-257.

Frank, O., Baker, H., Ziffer, H., Aaronson, S., Hutner, S. H., and Leevy, C. M. 1963. Metabolic deficiencies in protozoa induced by thalidomide. Science 139:110-111.

Franke, W. W. 1967. Zür Feinstruktur isolierter Kernmembranen aus tierischen Zellen. Z. Zellforsch. Mikrosk. Anat. 80:585-593.

Franke, W. W. and Eckert, W. A. 1971. Cytomembrane differentiation in a ciliate, *Tetrahymena pyriformis*. II. Bifacial cisternae and tubular formation. Z. Zellforsch. Mikrosk. Anat. 122:244-253.

Franke, W. W., Eckert, W. A., and Krien, S. 1971. Cytomembrane differentiation in a cilate, *Tetrahymena pyriformis*. I. Endoplasmic reticulum and dictyosomal equivalents. Z. Zellforsch. Mikrosk. Anat. 119:577-604.

Franke, W. W. and Kartenbeck, J. 1971. Outer mitochondrial membrane continuous with endoplasmic reticulum. Protoplasma 73:35-41.

Frankel, J. 1962. The effects of heat, cold, and *p*-fluorophenylalanine on morphogenesis in synchronized *Tetrahymena pyriformis* GL. C. R. Trav. Lab. Carlsberg 33:1-52.

Frankel, J. 1964. Cortical morphogenesis and synchronization in *Tetrahymena pyriformis* GL. Exp. Cell Res. 35:349-360.

Frankel, J. 1964. The effects of high temperatures on the pattern of oral development in *Tetrahymena pyriformis* GL. J. Exp. Zool. 155:403-436.

Frankel, J. 1964. Morphogenesis and division in chains of *Tetrahymena pyriformis* GL. J. Protozool. 11:514-526.

Frankel, J. 1965. The effect of nucleic acid antagonists on cell division and oral organelle development in *Tetrahymena pyriformis*. J. Exp. Zool. 159:113-148.

Frankel, J. 1967. Studies on the maintenance of development in *Tetrahymena pyriformis* GL-C. I. Analysis of the mechanism of resorption of developing oral structures. J. Exp. Zool. 164:435-460.

Frankel, J. 1967. Studies on the maintenance of development in *Tetrahymena pyriformis* GL-C. II. The relationship of protcin synthesis to cell division and oral organelle development. J. Cell Biol. 34:841-858.

Frankel, J. 1967. Critical phases of oral primordium development in *Tetrahymena pyriformis* GL-C: An analysis employing low temperature treatment. J. Protozool. 14:639-649.

Frankel, J. 1969. Participation of the undulating membrane in the formation of oral replacement primordia in *Tetrahymena pyriformis*. J. Protozool. 16:26-35.

Frankel, J. 1969. The relationship of protein synthesis to cell division and oral development in synchronized *Tetrahymena pyriformis* CL-C: An analysis employing cycloheximide. J. Cell. Physiol. 74:135-148.

Frankel, J. 1970. An analysis of the recovery of *Tetrahymena* from effects of cycloheximide. J. Cell. Physiol. 76:55-64.

Frankel, J. 1970. The synchronization of oral development without cell division in *Tetrahymena pyriformis* GL-C. J. Exp. Zool. 173:79-100.

Frankel, J. 1972. The stability of cortical phenotypes in continuously growing cultures of *Tetrahymena pyriformis*. J. Protozool. 19:648-654.

Frankel, J. and Heckmann, K. 1968. A simplified Chatton-Lwoff silver impregnation procedure for use in experimental studies with ciliates. Trans. Amer. Microsc. Soc. 87:317-321.

Friedberg, S. J. and Greene, R. C. 1968. The synthesis of glyceryl ether glycerol in *Tetrahymena pyriformis*. Biochim. Biophys. Acta 164:602-603.

Friedberg, S. J., Heifetz, A., and Greene, R. C. 1971. Loss of hydrogen from dihydroxyacetone phosphate during glyceryl ether synthesis. J. Biol. Chem. 256:5822-5827.

Friedberg, S. J., Heifetz, A., and Greene, R. C. 1972. Studies on the mechanism of *O*-alkyl lipid synthesis. Biochemistry 11:297-301.

Friedkin, M. 1953. Purine and pyrimidine metabolism in micro-organisms. J. Cell. Comp. Physiol. 41 (suppl.):261-282.

Fulton, C. 1971. Centrioles. *In* J. Reinert and H. Ursprung, eds., Origin and continuity of cell organelles, vol. 2, Results and problems in cell differentiation, (W. Beermann, J. Reinert, H. Ursprung, eds.), Springer, Berlin, pp. 170-221.

Furgason, W. H. 1940. The significant cytostomal pattern of the "*Glaucoma-Colpidium* group," and a proposed new genus and species, *Tetrahymena geleii*. Arch. Protistenk. 94:224-266.

Galigher, A. E. and Kozloff, E. N. 1971. Essentials of practical microtechnique, 2nd ed. Lea and Febiger, Philadelphia. 531 pp.

Gavin, R. H. 1965. The effects of heat and cold on cellular development in synchronized *Tetrahymena pyriformis* WH-6. J. Protozool. 12:307-318.

Gavin, R. H. and Frankel, J. 1966. The effects of mercaptoethanol on cellular development in *Tetrahymena pyriformis*. J. Exp. Zool. 161:63-82.

Gavin, R. H. and Frankel, J. 1969. Macromolecular syntheses, differentiation, and cell division in *Tetrahymena pyriformis* mating type I variety 1. J. Cell. Physiol. 74:123-134.

Geike, F. 1969. Zur Biosynthese der Phosphonoaminosäuren. II. Einbau von Glucose-U-^{14}C in Aminoäthylphosphonsäure durch zellfreie Präparationen von *Tetrahymena*. Naturwissenschaften 56:462.

Gellért, J. 1956. Ciliaten des sich unter dem Moosrasen auf Felsen gebildeten Humus. Acta Biol. Acad. Sci. Hung. 6:337-359.

Gellért, J. 1957. Ciliatenfauna im Humus einiger ungarischen Laub- und Nadelholzwälder. Ann. Inst. Biol. (Tihany) Hung. Acad. Sci. 24:11-34.

Genghof, D. S. 1949. The sulfur amino acid requirement of *Tetrahymena geleii*. Arch. Biochem. Biophys. 23:85-95.

Genghof, D. S. 1951. The specificity of the sulfur amino acid requirement of *Tetrahymena geleii*. Arch. Biochem. Biophys. 34:112-120.

Ghosh, E. 1925. On a new ciliate, *Balantidium knowlesii* sp. nov., a coelomic parasite in *Culicoides peregrinus*. Parasitology 17:189.

Gibbons, I. R. 1963. Studies of the protein components of cilia from *Tetrahymena pyriformis*. Proc. Natl. Acad. Sci. U.S. 50:1002-1010.

Gibbons, I. R. 1965. Chemical dissection of cilia. Arch. Biol. (Liège) 76:317-352.

Gibson, I. 1966. Chemical homologies between protozoa. J. Protozool. 13:650-653.

Gibson, I. 1970. Interacting genetic systems in *Paramecium*. Adv. Morphog. 8:159-208.

Gibson, I. and Martin, N. 1971. DNA amounts in the nuclei of *Paramecium aurelia* and *Tetrahymena pyriformis*. Chromosoma 35:374-382.

Giese, A. C. 1938. Differential susceptibility of a number of protozoans to ultra-violet radiations. J. Cell. Comp. Physiol. 12:129-138.

Giese, A. C. 1967. Effects of radiation upon protozoa. *In* T.T. Chen, Research in protozoology, vol. 2, Pergamon Press, New York, pp. 267-356.

Giese, A. C. and Alden, R. H. 1938. Cannibalism and giant formation in *Stylonychia*. J. Exp. Zool. 78:117-134.

Glaser, R. W. and Coria, N. A. 1930. Methods for pure culture of certain protozoa. J. Exp. Med. 51:787-806.

Glaser, R. W. and Coria, N. A. 1935. The culture and reactions of purified protozoa. Amer. J. Hyg. 21:111-120.

Glonek, T., Henderson, T., Hildebrand, R., and Myers, T. 1970. Biological phosphonates: Determination by phosphorus-31 nuclear magnetic resonance. Science 169:192-194.

Goldstein, L. 1958. Localization of nucleus-specific protein as shown by transplantation experiments in *Amoeba proteus*. Exp. Cell Res. 15:635-637.

Goldwasser, E. and Heinrikson, R. I. 1966. The biochemistry of pseudouridine. Progr. Nucleic Acid Res. Mol. Biol. 5:399-415.

Gordee, R. S. and Matthews, T. R. 1968. Evaluation of the *in vitro* and *in vivo* antifungal activity of pyrrolnitrin. Antimicrob. Agents Chemother. 1967:378-387.

Gorovsky, M. A. 1970. Studies on nuclear structure and function in *Tetrahymena pyriformis*. II. Isolation of macro- and micronuclei. J. Cell Biol. 47:619-630.

Gorovsky, M. A. 1970. Studies on nuclear structure and function in *Tetrahymena pyriformis*. III. Comparison of the histones of macro- and micronuclei by quantitative polyacrylamide gel electrophoresis. J. Cell Biol. 47:631-636.

Gorovsky, M. A. 1973. Macro- and micronuclei of *Tetrahymena pyriformis*: A model system for studying the structure and function of eukaryotic nuclei. J. Protozool. 20:19-25.

Gorovsky, M. A., Hattman, S., and Pleger, G. L. 1973. [^{6}N] methyl adenine in the nuclear DNA of a eucaryote *Tetrahymena pyriformis*. J. Cell Biol. 56:697-701.

Gorovsky, M. A. and Woodard, J. 1969. Studies on nuclear structure and function in *Tetrahymena pyriformis*. I. RNA synthesis in macro- and micronuclei. J. Cell Biol. 42:673-682.

Granátová, R. 1963. Experiments with artificial parasitism. Česk. Parasitol. 10:95-101. In Czechoslovakian with English summary.

Granick, S. 1954. Enzymatic conversion of δ-amino levulinic acid to porphobilinogen. Science 120:1105-1106.

Grass, F. S. 1972. An immobilization antigen in *Tetrahymena pyriformis* expressed under conditions of high salt stress. J. Protozool. 19:505-511.

Grass, F. S. 1972. Genetics of the St serotype system in *Tetrahymena pyriformis*, syngen 1. Genetics 70:521-536.

Grassé, P.-P. and de Boissezon, P. 1929. *Turchiniella culicis* n. g., n. sp. infusoire parasite de l'hemocoele d'un *Culex* adulte. Bull. Soc. Zool. Fr. 54:187-191.

Green, J., Price, S. A., and Gare, L. 1959. Tocopherols in micro-organisms. Nature 184:1339.

Grell, K. G. 1962. Morphologie und Fortpflanzung der Protozoen (einschliesslich Entwicklungsphysiologie und Genetik). Fortschr. Zool. 14:1-85.

Grell, K. G. 1968. Protozoologie, 2nd ed. Springer-Verlag, Berlin. 511 pp.

Griffin, J. L. 1960. An improved mass culture method for the large, free-living amoebae. Exp. Cell Res. 21:170-178.

Griffin, J. L. 1961. Identification of ameba crystals. I. Triuret in two crystal forms. Biochim. Biophys. Acta 47:433-439.

Griffin, J. L. 1973. Culture: Maintenance, large yields and problems of approaching axenic culture. *In* K. W. Jeon, ed., The Biology of *Amoeba*, Academic Press, New York, pp. 83-98.

Gross, D. and Tarver, II. 1955. Studies on ethionine. IV. The incorporation of ethionine into the proteins of *Tetrahymena*. J. Biol. Chem. 217:169-182.

Gross, J. A. 1955. The temperature dependence of the inhibition of growth of a protozoan by antibiotics. Biochim. Biophys. Acta 18:452-453.

Gross, J. A. 1955. A comparison of different criteria for determining the effects of antibiotics on *Tetrahymena pyriformis* E. J. Protozool. 2:42-47.

Gross, J. A. 1962. Cellular responses to thermal and photostress. II. Chlorotic euglenids and *Tetrahymena*. J. Protozool. 9:415-418.

Groupé, V. 1945. Effect of atabrine on *Tetrahymena geleii* (Protozoa, Ciliata). Proc. Soc. Exp. Biol. Med. 60:321-323.

Groupé, V., Herrmann, E. C., Jr., and Rauscher, F. J. 1955. Ingestion and destruction of influenza virus by a free living ciliate *Tetrahymena pyriformis*. Proc. Soc. Exp. Biol. Med. 88:479-482.

Groupé, V. and Pugh, L. H. 1952. Inactivation of influenza virus and of viral hemoglutinin by the ciliate *Tetrahymena geleii*. Science 115:307-308.

Gruchy, D. G. 1955. The breeding system and distribution of *Tetrahymena pyriformis*. J. Protozool.

2:178-185.

Guttman, H. N. and Friedman, W. 1963. Protozoa as pharmacological tools: The phenothiazine tranquilizers. Trans. N.Y. Acad. Sci. 26:75-89.

Hack, M. H., Yeager, R. G., and McCaffery, T. D. 1962. Comparative lipid biochemistry. II. Lipids of plant and animal flagellates, a non-motile alga, an ameba and a ciliate. Comp. Biochem. Physiol. 6:247-252.

Haessler, H. A. and Cunningham, L. 1957. A comparison of several deoxyribonucleases of type II. Exp. Cell Res. 13:304-311.

Haines, T. H. 1965. A microbial sulfolipid. I. Isolation and physiological studies. J. Protozool. 12:655-659.

Hall, R. H. 1938. The oxygen-consumption of *Colpidium campylum.* Biol. Bull. 75:395-408.

Hall, R. H. 1941. The effect of cyanide on oxygen consumption of *Colpidium campylum.* Physiol. Zool. 14:193-208.

Hall, R. P. 1937. Effects of manganese on the growth of *Euglena anabaena, Astasia* sp. and *Colpidium campylum.* Arch. Protistenk. 90:178-184.

Hall, R. P. 1939. Pimelic acid as a growth stimulant for *Colpidium campylum.* Arch. Protistenk. 92:315-319.

Hall, R. P. 1941. Vitamin deficiency as one explanation for inhibition of protozoan growth by conditioned medium. Proc. Soc. Exp. Biol. Med. 47:306-308.

Hall, R. P. 1941. Food requirements and other factors influencing growth of protozoa in pure cultures. *In* G. N. Calkins and F. M. Summers, eds., Protozoa in biological research, Columbia University Press, New York, pp. 475-516.

Hall, R. P. 1942. Incomplete proteins as nitrogen sources, and their relation to vitamin requirements in *Colpidium campylum.* Physiol. Zool. 15:95-107.

Hall, R. P. 1943. Growth factors for protozoa. Vitamins Hormones 1:249-268.

Hall, R. P. 1944. Comparative effects of certain vitamins on populations of *Glaucoma piriformis.* Physiol. Zool. 17:200-209.

Hall, R. P. 1953. Protozoology. Prentice-Hall, Englewood Cliffs, N.J. 682 pp.

Hall, R. P. 1954. Effects of certain metal ions on growth of *Tetrahymena pyriformis.* J. Protozool. 1:74-79.

Hall, R. P. 1954. Data on the metal requirements of *Tetrahymena pyriformis.* Trans. N.Y. Acad. Sci. 16:418-419.

Hall, R. P. 1965. Protozoan nutrition. Blaisdell, New York. 90 pp.

Hall, R. P. 1967. Nutrition and growth of protozoa. *In* T.-T. Chen, ed., Research in protozoology, vol. 1, Pergamon Press, New York, pp. 337-404.

Hall, R. P. and Cosgrove, W. B. 1944. The question of the synthesis of thiamin by the ciliate, *Glaucoma piriformis.* Biol. Bull. 86:31-40.

Hall, R. P. and Cosgrove, W. B. 1945. Mineral deficiency as a factor limiting growth of the ciliate, *Tetrahymena geleii.* Physiol. Zool. 18:425-430.

Hall, R. P. and Elliott, A. M. 1935. Growth of *Colpidium* in relation to certain incomplete proteins and amino acids. Arch. Protistenk. 85:443-450.

Hall, R. P., Johnson, D. F., and Loefer, J. B. 1935. A method for counting protozoa in the measurement of growth under experimental conditions. Trans. Amer. Microsc. Soc. 54:298-300.

Hall, R. P. and Loefer, J. B. 1940. Effects of culture filtrates and old medium on growth of the ciliate, *Colpidium campylum.* Proc. Soc. Exp. Biol. Med. 43:128-133.

Hall, R. P. and Shottenfeld, A. 1941. Maximal density and phases of death in populations of *Glaucoma piriformis.* Physiol. Zool. 14:384-393.

Hamana, K. and Iwai, K. 1971. Effects of steroid hormones on the growth of *Tetrahymena.* J. Biochem. 69:463-469.

Hamburger, K. 1962. Division delays induced by metabolic inhibitors in synchronized cells of *Tetrahymena pyriformis.* C. R. Trav. Lab. Carlsberg 32:359-370.

Hamburger, K. and Zeuthen, E. 1957. Synchronous divisions in *Tetrahymena pyriformis* as studied in an inorganic medium. Exp. Cell Res. 13:443-453.

Hamburger, K. and Zeuthen, E. 1960. Some characteristics of growth in normal and synchronized populations of *Tetrahymena pyriformis*. C. R. Trav. Lab. Carlsberg 32:1-18.

Hamburger, K. and Zeuthen, E. 1971. Respiratory response to dissolved food of starved, normal and division-synchronized *Tetrahymena* cells. C. R. Trav. Lab. Carlsberg 38:145-161.

Hamilton, L. 1953. Utilization of purines for nucleic acid synthesis in chrysomonads and other organisms. Ann. N.Y. Acad. Sci. 56:961-968.

Hamilton, L., Hutner, S. H., and Provasoli, L. 1952. The use of protozoa in analysis. Analyst 77:618-628.

Hanson, E. D. 1967. Protozoan development. *In* M. Florkin and B. J. Scheer, eds., Chemical zoology, vol. 1, Protozoa (G. W. Kidder, ed.), Academic Press, New York, pp. 395-539.

Hanson, R. W. and Barker, S. B. 1956. Some aspects of carbohydrate metabolism in *Tetrahymena*, a ciliated protozoan. J. Ala. Acad. Sci. 28:117-118.

Harding, J. P. 1937. Quantitative studies on the ciliate *Glaucoma*. I. The regulation of the size and the fission rate by the bacterial food supply. J. Exp. Biol. 14:422-430.

Harding, J. P. 1937. Quantitative studies on the ciliate *Glaucoma*. II. The effects of starvation. J. Exp. Biol. 14:431-439.

Harman, W. J. and Corliss, J. O. 1956. Isolation of earthworm setae by use of histophagous protozoa. Trans. Amer. Microsc. Soc. 75:332-333.

Harris, J. O. 1967. Non-adaptive growth responses of *Tetrahymena pyriformis* grown on single bacterial species. J. Protozool. 14:600-602.

Harrison, J. A. and Fowler, E. H. 1945. An antigen-antibody reaction with *Tetrahymena* which results in dystomy. Science 102:65-66.

Hartman, H. and Dowben, R. M. 1970. Polyribosomes from *Tetrahymena*. J. Cell Biol. 44:676-677.

Hartman, H. and Dowben, R. M. 1970. Polyribosome profiles from heat synchronized *Tetrahymena*. Biochem. Biophys. Res. Commun. 40:964-967.

Hauschka, T. S. and Doll, R. B. 1944. *Paraglaucoma* sp., a facultative parasite of coelenterates. J. Parasitol. 30:198-199.

Heinrich, M. R., Dewey, V. C., and Kidder, G. W. 1953. Utilization of guanine by *Tetrahymena geleii*. J. Amer. Chem. Soc. 75:1741-1742.

Heinrich, M. R., Dewey, V. C., and Kidder, G. W. 1957. The origin of thymine and cytosine in *Tetrahymena*. Biochim. Biophys. Acta 25:199-200.

Heinrich, M. R., Dewey, V. C., and Kidder, G. W. 1959. Chromatography of pteroylglutamic acid and related compounds on ion-exchange resins. J. Chromatogr. 2:85.

Heinrich, M. R., Dewey, V. C., Parks, R. E., Jr., and Kidder, G. W. 1952. The incorporation of 8-azaguanine into the nucleic acid of *Tetrahymena geleii*. J. Biol. Chem. 197:199-204.

Heinrikson, R. L. and Goldwasser, E. 1963. The enzymatic synthesis of 5-ribosyluridylic acid. J. Biol. Chem. 238:485-486.

Heinrikson, R. L. and Goldwasser, E. 1964. Studies on the biosynthesis of 5-ribosyluracil 5′-monophosphate in *Tetrahymena pyriformis*. J. Biol. Chem. 239:1177-1187.

Henshaw, R. E. 1961. Electron microscope observations of intracellular response to immobilization antibody in *Tetrahymena pyriformis*. Proc. Iowa Acad. Sci. 68:562-580.

Hetherington, A. 1932. The constant culture of *Stentor coeruleus*. Arch. Protistenk. 76:118-129.

Hetherington, A. 1933. The culture of some holotrichous ciliates. Arch. Protistenk. 80:255-280.

Hetherington, A. 1934. The sterilization of protozoa. Biol. Bull. 67:315-321.

Hetherington, A. 1934. The role of bacteria in the growth of *Colpidium colpoda*. Physiol. Zool. 7:618-641.

Hetherington, A. 1936. The precise control of growth in the pure culture of a ciliate, *Glaucoma pyriformis*. Biol. Bull. 70:426-440.

Hetherington, A. 1937. Controlled cultures of freshwater ciliates. *In* J. G. Needham, P. S. Galtsoff, F. E. Lutz, and P. S. Welch, eds., Culture methods for invertebrate animals, Comstock Publishing Co., Ithaca, New York, pp. 103-108.

Hill, D. L. 1972. The biochemistry and physiology of *Tetrahymena*. Academic Press, New York. 230 pp.

Hill, D. L. and Chambers, P. 1967. The purine and pyrimidine metabolism of *Tetrahymena pyriformis*. J. Cell. Physiol. 69:321-330.

Hill, D. L. and Chambers, P. 1967. The biosynthesis of proline by *Tetrahymena pyriformis*. Biochim. Biophys. Acta 148:435-447.

Hill, D. L., Judd, J., and van Eys, J. 1965. Search for a phosphagen in *Tetrahymena pyriformis*. Comp. Biochem. Physiol. 14:1-10.

Hill, D. L., Straight, S., and Allan, P. W. 1970. Use of *Tetrahymena pyriformis* to evaluate the effects of purine and pyrimidine analogs. J. Protozool. 17:619-623.

Hill, D. L. and van Eys, J. 1964. The source and nature of the non-protein nitrogen of *Tetrahymena pyriformis*. J. Tenn. Acad. Sci. 39:117-124.

Hill, D. L. and van Eys, J. 1965. The relationship between arginine, citrulline, and ornithine in *Tetrahymena pyriformis*. J. Protozool. 12:259-265.

Hjelm, K. K. 1970. A technique for cultivation of *Tetrahymena* in rotating bottles. Exp. Cell Res. 60:191-198.

Hjelm, K. K. and Zeuthen, E. 1967. Synchronous DNA synthesis following heat-synchronized cell division in *Tetrahymena*. Exp. Cell Res. 48:231-232.

Hjelm, K. K. and Zeuthen, E. 1967. Synchronous DNA-synthesis induced by synchronous cell division in *Tetrahymena*. C. R. Trav. Lab. Carlsberg 36:127-160.

Hofer, H. W., Pette, D., Schwab-Stey, H., and Schwab, D. 1972. Enzyme activity pattern and mitochondria-cytoplasmic relations in protozoa. Comparative study of *Tetrahymena pyriformis, Allogromia laticollaris,* and *Labyrinthula coenocystis,* J. Protozool. 19:532-537.

Hoffmann, E. J. 1965. The nucleic acids of basal bodies isolated from *Tetrahymena pyriformis.* J. Cell Biol. 25:217-228.

Hoffmann, E. K. and Kramhøft, B. 1969. A relationship between amino acid and sodium transport in *Tetrahymena pyriformis.* Exp. Cell Res. 56:265-268.

Hoffmann, E. K. and Rasmussen, L. 1972. Phenylalanine and methionine transport in *Tetrahymena pyriformis.* Characteristics of a concentrating, inducible transport system. Biochim. Biophys. Acta 266:206-216.

Hoffman, E. K., Rasmussen, L., and Zeuthen, E. 1970. Evidence for a common transport system for a group of amino acids in *Tetrahymena*. C. R. Trav. Lab. Carlsberg 38:133-143.

Hogg, J. F. 1963. Glyconeogenesis in protozoa (*Tetrahymena pyriformis*). Biochem. J. 89:88-89.

Hogg, J. F. 1969. Peroxisomes in *Tetrahymena* and their relation to gluconeogenesis. Ann. N.Y. Acad. Sci. 168:281-291.

Hogg, J. F. and Elliott, A. M. 1951. Comparative amino acid metabolism of *Tetrahymena geleii*. J. Biol. Chem. 192:131-139.

Hogg, J. F. and Kornberg, H. L. 1961. Role of the glyoxylate cycle in protozoal glyconeogenesis. Biochem. J. 81:17P.

Hogg, J. F. and Kornberg, H. L. 1963. The metabolism of C_2-compounds in microorganisms. 9. Role of the glyoxylate cycle in protozoal glyconeogenesis. Biochem. J. 86:462-468.

Holm, B. J. 1966. Stability of DNA in synchronized cultures of *Tetrahymena pyriformis* GL. Biochim. Biophys. Acta 119:647-649.

Holm, B. J. 1968. Changes in the amount of DNA in synchronized cultures of *Tetrahymena* cells. Exp. Cell Res. 53:18-36.

Holm, B. J. 1969. Inhibition of deoxyribonuclease activity in synchronized *Tetrahymena*. J. Protozool. 16:655-659.

Holm, B. J. 1970. Changes in ion content in heat-treated *Tetrahymena* cells. J. Protozool. 17:615-618.

Holmlund, C. E. 1971. On the mechanism of growth inhibition of *Tetrahymena pyriformis* by steroids. Biochim. Biophys. Acta 248:363-376.

Holmlund, C. E. and Bohonos, N. 1966. Growth inhibition of *Tetrahymena pyriformis* by 3-dialkyl-aminoethoxysteroids. Life Sci. 5:2133-2139.

Holsen, R. C., Jr. and Buhse, H. E., Jr. 1969. The effect of heat shocks on the morphogenetic events associated with macrostome formation in *Tetrahymena vorax*. *In* A. A. Strelkov, K. M. Sukhanova, and I. B. Raikov, eds., Progress in protozoology, 3rd International Congress on Protozoology, Leningrad, July 1969, Nauka, Leningrad, pp. 97-98.

Holter, H. and Doyle, W. L. 1938. Studies on enzymatic histochemistry. XXVIII. Enzymatic studies on protozoa. J. Cell. Comp. Physiol. 12:292-308.

Holz, G. G., Jr. 1960. Structural and functional changes in a generation in *Tetrahymena*. Biol. Bull. 118:84-95.

Holz, G. G., Jr. 1964. Nutrition and metabolism of ciliates. *In* S. H. Hutner, ed., Biochemistry and physiology of protozoa, vol. 3, Academic Press, New York, pp. 199-242.

Holz, G. G., Jr. 1966. Is *Tetrahymena* a plant? J. Protozool. 13:2-4.

Holz, G. G., Jr. and Corliss, J. O. 1956. *Tetrahymena setifera* n. sp., a member of the genus *Tetrahymena* with a caudal cilium. J. Protozool. 3:112-118.

Holz, G. G., Jr., Erwin, J. A., and Davis, R. J. 1959. Some physiological characteristics of the mating types and varieties of *Tetrahymena pyriformis*. J. Protozool. 6:149-156.

Holz, G. G., Jr., Erwin, J. A., Rosenbaum, N., and Aaronson, S. 1962. Triparanol inhibition of *Tetrahymena* and its prevention by lipids. Arch. Biochem. Biophys. 98:312-322.

Holz, G. G., Jr., Erwin, J. A., and Wagner, B. 1961. The sterol requirement of *Tetrahymena paravorax* RP. J. Protozool. 8:297-300.

Holz, G. G., Jr., Erwin, J. A., Wagner, B., and Rosenbaum, N. 1962. The nutrition of *Tetraymena setifera* HZ-1; sterol and alcohol requirements. J. Protozool. 9:359-363.

Holz, G. G., Jr., Rasmussen, L., and Zeuthen, E. 1963. Normal versus synchronized division in *Tetrahymena pyriformis*. A study with antimetabolites. C. R. Trav. Lab. Carlsberg 33:289-300.

Holz, G. G., Jr., Scherbaum, O. H., and Williams, N. E. 1957. The arrest of mitosis and stomatogenesis during temperature-induction of synchronous division in *Tetrahymena pyriformis*, mating type 1, variety 1. Exp. Cell Res. 13:618-621.

Holz, G. G., Jr., Wagner, B., Erwin, J. A., Britt, J. J., and Bloch, K. 1961. Sterol requirements of a ciliate *Tetrahymena corlissi* Th-X. I. A nutritional analysis of the sterol requirements of *T. corlissi* Th-X. II. Metabolism of tritiated lophenol in *T. corlissi* Th-X. Comp. Biochem. Physiol. 2:202-217.

Honigberg, B. M. and Mohn, F. A. 1973. An improved method for the isolation of highly polymerized native deoxyribonucleic acid from certain protozoa. J. Protozool. 20:146-150.

Hooper, F. F. and Elliott, A. M. 1953. Release of inorganic phosphorus from extracts of lake mud by protozoa. Trans. Amer. Microsc. Soc. 72:276-281.

Hopkins, J. M. and Watson, M. R. 1963. The cilia of *Tetrahymena pyriformis*. Isolation of ciliary segments. Exp. Cell Res. 32:187-189.

Horiguchi, M. 1972. Biosynthesis of 2-aminoethylphosphonic acid in cell-free preparations from *Tetrahymena*. Biochim. Biophys. Acta 261:102-113.

Horiguchi, M. and Kandatsu, M. 1960. Ciliatine: A new aminophosphonic acid contained in rumen ciliate protozoa. Studies on the reticulo-rumen digestion. Part XVII. Bull. Agr. Chem. Soc. Japan 24:565-570.

Horiguchi, M. and Kandatsu, M. 1964. Polymorphism of ciliatine (2-aminoethylphosphonic acid) from rumen ciliated protozoa and *Tetrahymena pyriformis* W. Agr. Biol. Chem. 28:408-410.

Horiguchi, M., Kittredge, J. S., and Roberts, E. 1968. Biosynthesis of 2-aminoethylphosphonic acid in *Tetrahymena*. Biochim. Biophys. Acta 165:164-166.

Horowitz, J. and Chargaff, E. 1959. Massive incorporation of 5-fluorouracil into a bacterial ribonucleic acid. Nature 184:1213-1214.

Huddleston, M. S. 1951. Effects of nutrients on *Tetrahymena geleii* W populations. Physiol. Zool. 24:141-155.

Huddleston, M. S., Cavalier, L., Eliassen, J., Kozak, M., and Maledon, A. 1964. The effect of proteose peptone concentration on cell size in *Tetrahymena pyriformis* W. Life Sci. 3:1181-1190.

Hull, R. W. 1961. Studies on suctorian protozoa: The mechanism of prey adherence. J. Protozool. 8:343-350.

Hull, R. W. 1961. Studies on suctorian protozoa: The mechanism of ingestion of prey cytoplasm. J. Protozool. 8:351-359.

Hull, R. W. and Morrissey, J. E. 1960. Metabolic efficiency of *Tetrahymena pyriformis,* strain WL as related to culture age. Trans. Amer. Microsc. Soc. 79:127-135.

Humphreys, W. J., Wodzicki, T. J., and Paulin, J. J. 1973. Fractographic studies of plastic-embedded cells by scanning electron microscopy. J. Cell Biol. 56:876-880.

Hutner, S. H. 1964. Protozoa as toxicological tools. J. Protozool. 11:1-6.

Hutner, S. H. 1964. Prospects in the industrial use of protozoa. Develop. Ind. Microbiol. 6:31-34.

Hutner, S. H., Baker, H., Frank, O., and Cox, D. 1972. Nutrition and metabolism in protozoa. *In* R. N. Fiennes, ed., Biology of nutrition, Pergamon Press, New York, pp. 85-177.

Hutner, S. H., Cury, A., and Baker, H. 1958. Microbiological assays. Anal. Chem. 30:849-867.

Hutner, S. H. and Holz, G. G., Jr. 1962. Lipid requirements of microorganisms. Ann. Rev. Microbiol. 16:189-204.

Hutner, S. H., Nathan, H. A., Aaronson, S., Baker, H., and Scher, S. 1958. General considerations in the use of microorganisms in screening antitumor agents. Ann. N.Y. Acad. Sci. 76:457-468.

Hwang, S. W., Davis, E. E., and Alexander, M. T. 1964. Freezing and viability of *Tetrahymena pyriformis* in dimethylsulfoxide. Science 144:64-65.

Hyman, L. H. 1940. The invertebrates: Protozoa through Ctenophora, voL 1. McGraw-Hill, New York. 726 pp.

Ikeda, M. and Watanabe, Y. 1965. Studies on protein-bound sulfhydryl groups in synchronous cell division of *Tetrahymena pyriformis.* Exp. Cell Res. 39:584-590.

Inoki, S. and Matsushiro, A. 1958. Antigenic variation in *Tetrahymena pyriformis.* Med. J. Osaka Univ. 8:763-770.

International Commission on Zoological Nomenclature. 1970. Opinion 915. *Tetrahymena* Furgason, 1940 (Ciliophora, Hymenostomatida): Preserved under the plenary powers and related matters. Bull. Zool. Nomencl. 27:33-38.

Irlina, I. S. 1960. Changes in the thermoresistance of some free-living protozoa under the effect of previous thermic conditions. Tsitologiya 2:227-234. In Russian.

Irlina, I. S. 1969. Ontogenetic changes in thermoresistance of *Tetrahymena pyriformis. In* A. A. Strelkov, K. M. Sukhanova, and I. B. Raikov, eds., Progress in protozoology, 3rd International Congress on Protozoology, Leningrad, July 1969, Nauka, Leningrad, pp. 143-144.

Ito, J., Lee, Y. C., and Scherbaum, O. H. 1968. Intranuclear microtubules in *Tetrahymena pyriformis,* GL. Exp. Cell Res. 53:85-93.

Iverson, R. M. 1958. Nuclear transfer studies on ultraviolet-irradiated *Amoeba proteus.* Exp. Cell Res. 15:268-270.

Iverson, R. M. and Giese, A. C. 1957. Nucleic acid content and ultraviolet susceptibility of *Tetrahymena pyriformis.* Exp. Cell Res. 13:213-223.

Iwai, K., Hamana, K., and Yabuki, H. 1971. Histone fraction of protozoa: Their correlation with mammalian histone. J. Biochem. (Tokyo) 68:597-601.

Iwai, K., Shiomi, H., Ando, T., and Mita, T. 1965. The isolation of histone from *Tetrahymena.* J. Biochem. (Tokyo) 58:312-314.

Iwata, H., Kariya, K., and Fujimoto, S. 1968. Variance of growth of *Tetrahymena* by the compounds which affect the adrenergic mechanism. Jap. J. Pharmacol. 18:329-330.

Iwata, H., Kariya, K., and Fujimoto, S. 1969. Effect of compounds affecting the adrenergic mechanism on cell growth and division of *Tetrahymena pyriformis* W. Jap. J. Pharmacol. 19:275-281.

Jacobson, K. B. and Prescott, D. M. 1964. The nucleotide pools for thymidine and cytidine in *Tetrahymena pyriformis.* Exp. Cell Res. 36:561-567.

Jahn, T. L. 1934. Problems of population growth in the protozoa. Cold Spring Harbor Symp. Quant. Biol. 2:167-180.

Jahn, T. L. 1936. The effect of aeration and lack of CO_2 on growth of bacteria-free cultures of protozoa. Proc. Soc. Exp. Biol. Med. 33:494-498.

Jahn, T. L. 1941. Respiratory metabolism. *In* G. N. Calkins and F. M. Summers, eds., Protozoa in biological research, Columbia University Press, New York, pp. 352-403.

Jahn, T. L. and Bovee, E. C. 1967. Motile behavior of protozoa. *In* T.-T. Chen, ed., Research in protozoology, vol. 1, Pergamon Press, New York, pp. 41-200.

Jahn, T. L., Bovee, E. C., Dauber, M., Winet, H., and Brown, M. 1965. Secretory activity of the oral apparatus of ciliates: Trails of adherent particles left by *Paramecium multimicronucleatum* and *Tetrahymena pyriformis.* Ann. N.Y. Acad. Sci. 118:912-920.

Jahn, T. L. and Jahn, F. F. 1949. How to know the protozoa. Wm. C. Brown, Dubuque, Iowa. 234 pp.

James, T. W. and Read, C. P. 1957. The effect of incubation temperature on the cell size of *Tetrahymena pyriformis.* Exp. Cell Res. 13:510-516.

Janakidevi, K., Dewey, V. C., and Kidder, G. W. 1966. Serotonin in protozoa. Arch. Biochem. Biophys. 113:758-759.

Janakidevi, K., Dewey, V. C., and Kidder, G. W. 1966. The biosynthesis of catecholamines in two genera of protozoa. J. Biol. Chem. 241:2576-2578.

Janda, V. and Jírovec, O. 1937. Ueber künstlich hervorgerufenen Parasitismus eines freilebenden Ciliaten *Galucoma piriformis* und Infektionsversuche mit *Euglena gracilis* und *Spirochaeta biflexa.* Věst. Čsl. Zool. Spol. 5:34-58.

Jankowski, A. W. 1962. A case of parasitism by the infusorian *Tetrahymena* in the turbellarian *Microstomum.* Vest. Leningrad Univ., no. 21, Ser. Biol., 4:153-155. In Russian with English summary.

Jankowski, A. W. 1964. Morphology and evolution of Ciliophora. IV. Sapropelebionts of the family Loxocephalidiae fam. nova, their taxonomy and evolutionary history. Acta Protozool. 2:33-58.

Jankowski, A. W. 1967. The boundaries and composition of the genera *Tetrahymena* and *Colpidium.* Zool. Zh. 46:17-23. In Russian with English summary.

Jankowski, A. W. 1967. A new system of ciliate protozoa (Ciliophora). Acad. Sci. USSR, Trav. Zool. Inst. 43:3-52. In Russian.

Jasmin, R. 1961. Apparation d'une pigmentation noire chez *Tetrahymena pyriformis* S après traitement par certains Tweens et autres substances. Influence de cette pigmentation sur la densité optique. Rev. Can. Biol. 20:813-818.

Jauker, F. 1970. Effects in *Tetrahymena pyriformis* adapting to a longtime treatment with actomycin. Cytobiologie 1:208-227.

Jauker, F. and Cleffmann, G. 1970. Oscillation of individual generation times in cell lines of *Tetrahymena pyriformis.* Exp. Cell Res. 62:477-480.

Jeffery, W. R. 1972. Evidence for a temporal incompatibility between DNA replication and division during the cell cycle of *Tetrahymena.* J. Cell Biol. 53:624-634.

Jeffery, W. R., Stuart, K. D., and Frankel, J. 1970. The relationship between deoxyribonucleic acid replication and cell division in heat synchronized *Tetrahymena.* J. Cell Biol. 46:533-543.

Jeffries, W. B. 1956. Studies on excystment in the hypotrichous ciliate *Pleurotricha lanceolata.* J. Protozool. 3:136-144.

Jerka-Dziadosz, M. and Frankel, J. 1969. An analysis of the formation of ciliary primordia in the hypotrich ciliate *Urostyla weissei.* J. Protozool. 16:612-637.

Jíra, J. 1949. The influence of sulfonamides on protozoa in vitro. Věst. Čsl. Zool. Spol. 13:191-215.

Jírovec, O. 1943. Über einige interessante physikalisch-chemische und biologische Eigenschaften von Germanin (Bayer 205) II. Biochem. Z. 315:69-82.

Jírovec, O. 1947. The free-living ciliate *Glaucoma piriformis* as test-objects for the toxicity of some pharmacological substances. Věst. Čsl. Zool. Spol. 11:206-217. In Czechoslovakian with English summary.

Jírovec, O. 1948. The influence of sulphonamides on protozoa. Biol. Listy 29:52-61.

Jírovec, O. 1949. Der Einfluss des Streptomycins und Patulins auf einige Protozoen. Experientia 5:74-77.

Jírovec, O. 1949. Influence of antibiotics on some protozoa. Věst. Čsl. Zool. Spol. 13:216-238. In Czechoslovakian with English summary.

Jírovec, O. 1950. Das Infusorium *Glaucoma piriformis* als Testobjekt in Pharmakologie und Physiologie. Pathol. Bakteriol. 13:129-138.

Jírovec, O. 1951. Die Wirkung einiger Antibiotica auf Protozoen. Pathol. Bakteriol. 14:653-666.

Jírovec, O. 1963. Protozoa as models in biological research. *In* J. Ludvík, J. Lom, and J. Vávra, eds., Progress in protozoology, Proceedings of the 1st International Congress on Protozoology, Prague, August 1961, Nakladatelstvi Československé Akademie, Prague, pp. 31-37.

Jírovec, O. and Sosna, M. 1954. Wirkung des Sonnenlichtes auf einige freilebende Protisten. Čsl. Biol. 3:108-113.

Johnson, D. F. 1935. Isolation of *Glaucoma ficaria* Kahl in bacteria-free cultures, and growth in relation to pH of the medium. Arch. Protistenk. 86:263-277.

Johnson, D. F. 1936. Growth of *Glaucoma ficaria* Kahl in culture with single species of other microorganisms. Arch. Protistenk. 86:359-378.

Johnson, I. S., Simpson, P. J., and Cline, J. C. 1962. Comparative studies with chemotherapeutic agents in biologically diverse *in vitro* cell systems. Cancer Res. 22:617-626.

Johnson, I. S., Vlantis, J., Mattas, B., and Wright, H. F. 1961. Anti-tumor principles derived from *Vinca rosea* Linn. II. Further studies of biological activities of vincaleukoblastine. 4th Canadian Cancer Conference, 1960, Academic Press, New York, pp. 339-353.

Johnson, W. H. 1941. Populations of ciliates. Amer. Nat. 75:438-457.

Johnson, W. H. 1941. Nutrition in the protozoa. Quart. Rev. Biol. 16:336-348.

Johnson, W. H. 1956. Nutrition of protozoa. Ann. Rev. Microbiol. 10:193-212.

Johnson, W. H. and Baker, E. G. S. 1943. Effects of certain B vitamins on populations of *Tetrahymena geleii*. Physiol. Zool. 16:172-185.

Jonah, M. and Erwin, J. A. 1971. The lipids of membranous cell organelles isolated from the ciliate, *Tetrahymena pyriformis*. Biochim. Biophys. Acta 231:80-92.

Jones, A. S. and Thompson, T. W. 1963. The deoxyribonucleic acids of some protozoa and a mould. J. Protozool. 10:91-93.

Jones, R. F. and Baker, H. G. 1946. Formation of aggregations of *Glaucoma pyriformis* Kahl by means of phenanthraquinone and other substances. Nature 157:554.

Jordan, L. E. 1932. An attempt to grow tubercle bacilli within the single celled organism, *Colpidium campylum*. J. Infect. Dis. 51:482-488.

Jergensmeyer, E. B. 1969. Serotype expression and transformation in *Tetrahymena pyriformis*. J. Protozool. 16:433-452.

Jukes, T. H. 1955. Assay of compounds with folic acid activity. Methods Biochem. Anal. 2:121-151.

Jukes, T. H. and Stokstad, E. L. R. 1948. Pteroylglutamic acid and related compounds. Physiol. Rev. 28:51-106.

Kahl, A. 1926. Neue und wenig bekannte Formen der holotrichen und heterotrichen Ciliaten. Arch. Protistenk. 55:197-438.

Kahl, A. 1930-1935. Urtiere oder Protozoa. I: Wimpertiere oder Ciliata (Infusoria), eine Bearbeitung der freilebenden und ectocommensalen Infusorien der Erde, unter Ausschluss der marinen Tintinnidae. *In* F. Dahl, ed., Die Tierwelt Deutschlands, G. Fischer, Jena, parts 18 (1930), 21 (1931), 25 (1932), 30 (1935), pp. 1-886.

Kamiya, T. 1959. Carbohydrate changes in the acid-soluble fractions of the ciliate protozoon *Tetrahymena geleii* W during the course of synchronous culture. J. Biochem. 46:1187-1192.

Kamiya, T. 1960. Acid-soluble phosphate compounds of the ciliate protozoon *Tetrahymena geleii* W. J. Biochem. 47:69-76.

Kamiya, T. and Takahashi, T. 1961. Mechanism of cell division. I. Changes in pyridine nucleotide and DPNH-cytochrome reductase in *Tetrahymena geleii* W during the course of synchronous culture. J. Biochem. 50:277-283.

Kandatsu, M. and Horiguchi, M. 1962. Occurrence of ciliatine (2-aminoethylphosphonic acid) in *Tetrahymena*. Agr. Biol. Chem. 26:721-722.

Kapoulas, V. M. 1969. The chromatographic separation of phosphonolipids from their phospholipid analogs. Biochim. Biophys. Acta 176:324-327.

Kapoulas, V. M. and Thompson, G. A., Jr. 1969. The formation of glyceryl ethers by cell-free *Tetrahymena* extracts. Biochim. Biophys. Acta 187:594-597.

Kapoulas, V. M., Thompson, G. A., Jr., and Hanahan, D. J. 1969. Metabolism of α-glyceryl ethers by *Tetrahymena pyriformis*. I. Characteristics of the *in vivo* degradation system. Biochim. Biophys. Acta 176:237-249.

Kapoulas, V. M., Thompson, G. A., Jr., and Hanahan, D. J. 1969. Metabolism of α-glyceryl ethers by *Tetrahymena pyriformis*. II. Properties of a cleavage system *in vitro*. Biochim. Biophys. Acta 176:250-264.

Kazubski, S. L. 1958. *Trichia lubomirskii* Slós. (Helicidae), a new host of *Tetrahymena limacis* (Warren, 1932) Kozloff, 1946 (Ciliata) and *Zonitoides nitidus* Müll. (Zonitidae), a new host of *T. rostrata* (Kahl, 1926) Corliss, 1952 in Poland. Bull. Acad. Pol. Sci. 6:247-252.

Keenan, R. W. 1972. Sphingolipid base phosphorylation by cell-free preparations from *Tetrahymena pyriformis*. Biochim. Biophys. Acta 270:383-396.

Keiding, J. and Westergaard, O. 1971. Induction of DNA polymerase activity in irradiated *Tetrahymena* cells. Exp. Cell Res. 64:317-332.

Keilin, D. 1921. On a new ciliate, *Lambornella stegomyiae*, n. g., n. sp., parasitic in the body cavity of the larvae of *Stegomyia scutellaris* Walker (Diptera, Nematocera, Culicidae). Parasitology 13:216-224.

Keilin, D. and Ryley, J. F. 1953. Haemoglobin in protozoa. Nature 172:451.

Kennedy, J. R., Jr. 1969. The role of microtubules in the cell cycle. *In* G. M. Padilla, G. L. Whitson, and I. L. Cameron, eds., The cell cycle: Gene-enzyme interactions, Academic Press, New York, pp. 227-248.

Kennedy, J. R., Jr., and Elliott, A. M. 1970. Cigarette smoke: The effect of residue on mitochondrial structure. Science 168:1097-1098.

Kennedy, J. R., Jr., and Zimmerman, A. M. 1970. The effects of high hydrostatic pressure on the microtubules of *Tetrahymena pyriformis*. J. Cell Biol. 47:568-576.

Kennedy, K. E. and Thompson, G. A., Jr. 1970. Phospholipids: Localization in surface membranes of *Tetrahymena*. Science 168:989-991.

Kent, W. S. 1880-1882. Manual of the infusoria. 3 vols., David Bogue, London. 913 pp.

Kidder, G. W. 1939. The effect of biologically conditioned medium on the growth rate and population yield of certain ciliated protozoa. Science 90:405-406.

Kidder, G. W. 1941. Growth studies on ciliates. V. The acceleration and inhibition of ciliate growth in biologically conditioned medium. Physiol. Zool. 14:209-226.

Kidder, G. W. 1941. Growth studies on ciliates. VII. Comparative growth characteristics of four species of sterile ciliates. Biol. Bull. 80:50-68.

Kidder, G. W. 1941. The technique and significance of control in protozoan culture. *In* G. N. Calkins and F. M. Summers, eds., Protozoa in biological research, Columbia University Press, New York, pp. 448-474.

Kidder, G. W. 1946. Studies on the biochemistry of *Tetrahymena*. VI. Folic acid as a growth factor for *Tetrahymena geleii* W. Arch. Biochem. 9:51-55.

Kidder, G. W. 1947. The nutrition of monocellular animal organisms. Ann. N.Y. Acad. Sci. 49:99-110.

Kidder, G. W. 1951. Nutrition and metabolism of protozoa. Ann. Rev. Microbiol. 5:139-156.

Kidder, G. W. 1952. Enzyme capacities and their relation to cell nutrition in animals. Int. Rev. Cytol. 1:27-34.

Kidder, G. W. 1953. The nutrition of invertebrate animals. *In* G. H. Bourne and G. W. Kidder, eds., Biochemistry and physiology of nutrition, vol. 2, Academic Press, New York, pp. 162-196.

Kidder, G. W. 1953. The nutrition of ciliated protozoa. *In* Symposium on nutrition and growth factors, Istituto Superiore di Sanita, Rome, September 1953, pp. 44-61.

Kidder, G. W. 1967. Nitrogen: Distribution, nutrition, and metabolism. *In* M. Florkin and B. J. Scheer, eds., Chemical zoology, vol. 1, Protozoa (G. W. Kidder, ed.), Academic Press, New York, pp. 93-159.

Kidder, G. W. and Dewey, V. C. 1942. The biosynthesis of thiamine by normally athiaminogenic micro-organisms. Growth 6:405-418.

Kidder, G. W. and Dewey, V. C. 1944. Thiamine and *Tetrahymena*. Biol. Bull. 87:121-133.

Kidder, G. W. and Dewey, V. C. 1945. Studies on the biochemistry of *Tetrahymena*. I. Amino acid requirements. Arch. Biochem. 6:425-432.

Kidder, G. W. and Dewey, V. C. 1945. Studies on the biochemistry of *Tetrahymena*. II. Factor three. Arch. Biochem. 6:433-437.

Kidder, G. W. and Dewey, V. C. 1945. Studies on the biochemistry of *Tetrahymena*. III. Strain differences. Physiol. Zool. 18:136-157.

Kidder, G. W. and Dewey, V. C. 1945. Studies on the biochemistry of *Tetrahymena*. IV. Amino acids and their relation to the biosynthesis of thiamin. Biol. Bull. 89:131-143.

Kidder, G. W. and Dewey, V. C. 1945. Studies on the biochemistry of *Tetrahymena*. V. The chemical nature of factors I and III. Arch. Biochem. 8:293-301.

Kidder, G. W. and Dewey, V. C. 1945. Studies on the biochemistry of *Tetrahymena*. VII. Riboflavin, pantothen, biotin, niacin and pyridoxine in the growth of *Tetrahymena geleii* W. Biol. Bull. 89:229-241.

Kidder, G. W. and Dewey, V. C. 1947. Studies on the biochemistry of *Tetrahymena*. IX. Folic acid components and conjugates. Proc. Natl. Acad. Sci. U.S. 33:95-102.

Kidder, G. W. and Dewey, V. C. 1947. Studies on the biochemistry of *Tetrahymena*. X. Quantitative response to essential amino acids. Proc. Natl. Acad. Sci. U.S. 33:347-356.

Kidder, G. W. and Dewey, V. C. 1948. Studies on the biochemistry of *Tetrahymena*. XIV. The activity of natural purines and pyrimidines. Proc. Natl. Acad. Sci. U.S. 34:566-574.

Kidder, G. W. and Dewey, V. C. 1948. Dietary factors in the utilization of homocystine. Proc. Natl. Acad. Sci. U.S. 34:81-88.

Kidder, G. W. and Dewey, V. C. 1949. Studies on the biochemistry of *Tetrahymena*. XI. Components of factor II of known chemical nature. Arch. Biochem. 20:433-443.

Kidder, G. W. and Dewey, V. C. 1949. Studies on the biochemistry of *Tetrahymena*. XII. Pyridoxine, pyridoxal, and pyridoxamine. Arch. Biochem. 21:58-65.

Kidder, G. W. and Dewey, V. C. 1949. Studies on the biochemistry of *Tetrahymena*. XIII. B vitamin requirements. Arch. Biochem. 21:66-73.

Kidder, G. W. and Dewey, V. C. 1949. The biological activity of substituted pyrimidines. J. Biol. Chem. 178:383-387.

Kidder, G. W. and Dewey, V. C. 1949. The biological activity of substituted purines. J. Biol. Chem. 179:181-187.

Kidder, G. W. and Dewey, V. C. 1951. The biochemistry of ciliates in pure culture. *In* A. Lwoff, ed., Biochemistry and physiology of protozoa, vol. 1, Academic Press, New York, pp. 323-400.

Kidder, G. W. and Dewey, V. C. 1955. Inhibition of *Tetrahymena* by a new tryptophan analog. Biochim. Biophys. Acta 17:288.

Kidder, G. W. and Dewey, V. C. 1957. Deazapurines as growth inhibitors. Arch. Biochem. Biophys. 66:486-492.

Kidder, G. W. and Dewey, V. C. 1961. Folic acid sparing by triamino pyrimidines. Biochem. Biophys. Res. Commun. 5:324-327.

Kidder, G. W. and Dewey, V. C. 1968. A new pteridine from *Tetrahymena*. J. Biol. Chem. 243:826-833.

Kidder, G. W., Dewey, V. C., Andrews, M. B., and Kidder, R. R. 1949. Tryptophan and nicotinamide in the nutrition of the animal microorganism, *Tetrahymena*. J. Nutr. 37:521-529.

Kidder, G. W., Dewey, V. C., and Heinrich, M. R. 1954. The effect of nonionic detergents on the growth of *Tetrahymena*. Exp. Cell Res. 7:256-264.

Kidder, G. W., Dewey, V. C., and Parks, R. E., Jr. 1950. Protogen and acetate in *Tetrahymena*. Arch. Biochem. 27:463-464.

Kidder, G. W., Dewey, V. C., and Parks, R. E., Jr. 1951. Studies on the inorganic requirements of *Tetrahymena*. Physiol. Zool. 24:69-75.

Kidder, G. W., Dewey, V. C., and Parks, R. E., Jr. 1951. Growth promotion in *Tetrahymena* by folic acid analogs. Proc. Soc. Exp. Biol. Med. 78:88-91.

Kidder, G. W., Dewey, V. C., and Parks, R. E., Jr. 1952. Effect of lowered essential metabolites on 8-azaguanine inhibition. J. Biol. Chem. 197:193-198.

Kidder, G. W., Dewey, V. C., Parks, R. E., Jr., and Heinrich, M. R. 1950. Further studies on the purine and pyrimidine metabolism of *Tetrahymena*. Proc. Natl. Acad. Sci. U.S. 36:431-439.

Kidder, G. W., Dewey, V. C., Parks, R. E., Jr., and Woodside, G. L. 1949. Purine metabolism in *Tetrahymena* and its relation to malignant cells in mice. Science 109:511-514.

Kidder, G. W., Dewey, V. C., Parks, R. E., Jr., and Woodside, G. L. 1951. Further evidence on the mode of action of 8-azaguanine (guanazolo) in tumor inhibition. Cancer Res. 11:204-211.

Kidder, G. W. and Fuller, R. C., III. 1946. The growth response of *Tetrahymena geleii* W to folic acid and to the *Streptococcus lactis* R factor. Science 104:160-161.

Kidder, G. W., Lilly, D. M., and Claff, C. L. 1940. Growth studies on ciliates. IV. The influence of food on the structure and growth of *Glaucoma vorax* sp. nov. Biol. Bull. 78:9-23.

Kidder, G. W., Stuart, C. A., McGann, V. G., and Dewey, V. C. 1945. Antigenic relationships in the genus *Tetrahymena*. Physiol. Zool. 18:415-425.

Kilsheimer, G. S. and Axelrod, B. 1958. Phylogenetic distribution of acid phosphatase inhibited by +-tartrate. Nature 182:1735-1736.

Kimball, R. F. 1964. Physiological genetics of the ciliates. *In* S. H. Hutner, ed., Biochemistry and physiology of protozoa, vol. 3, Academic Press, New York, pp. 243-275.

Kirby, H., Jr. 1941. Relationships between certain protozoa and other animals. *In* G. N. Calkins and F. M. Summers, eds., Protozoa in biological research, Columbia University Press, New York, pp. 890-1008.

Kitching, J. A. 1957. Effects of high hydrostatic pressures on the activity of flagellates and ciliates. J. Exp. Biol. 34:494-510.

Kitching, J. A. 1967. Contractile vacuoles, ionic regulation and excretion. *In* T.-T. Chen, ed., Research in protozoology, Pergamon Press, New York, pp. 307-337.

Kittredge, J. S. and Hughes, R. R. 1964. The occurrence of α-amino-β-phosphonopropionic acid in the zoanthid, *Zoanthus sociatus,* and the ciliate, *Tetrahymena pyriformis.* Biochemistry 3:991-996.

Kittredge, J. S. and Roberts, E. 1969. A carbon-phosphorus bond in nature. Science 164:37-42.

Klamer, B. and Fennell, R. A. 1963. Acid phosphatase activity during growth and synchronous division of *Tetrahymena pyriformis* W. Exp. Cell Res. 29:166-175.

Klein, B. M. 1926. Ergebnisse mit einer Silbermethode bei Ciliaten. Arch. Protistenk. 56:243-279.

Klein, B. M. 1927. Die Silberliniensysteme der Ciliaten. Ihr Verhalten während Teilung und Conjugation, neue Silberbilder, Nachträge. Arch. Protistenk. 58:55-142.

Klein, B. M. 1930. Das Silberliniensystem der Ciliaten. Weitere Ergebnisse, IV. Arch. Protistenk. 66:235-326.

Klein, B. M. 1934. Strukturelle und formative Reaktionen des Silberliniensystems. Ann. Protistol. 4:55-68.

Klein, B. M. 1942. Reaktionen des neuroformativen Systems bei Beute-Infusorien im Leibesinnern eines Raub-Infusors. Ann. Naturhist. Mus., Wien 52:54-65.

Klein, B. M. 1943. Das Silberlinien- oder neuroformative System der Ciliaten. Ann. Naturhist. Mus., Wien 53:156-336.

Klein, B. M. 1958. The "dry" silver method and its proper use. J. Protozool. 5:99-103.

Kline, A. P. 1943. The nitrogen compounds necessary for growth in *Colpidium striatum* Stokes, with special reference to the amino acids. Physiol. Zool. 16:405-417.

Knight, D. R. and McDougle, H. C. 1944. A protozoon of the genus *Tetrahymena* found in domestic fowl. Amer. J. Vet. Res. 5:113-116.

Kobayashi, S. 1965. Preparation and properties of mitochondria from the ciliated protozoan, *Tetrahymena*. J. Biochem. 58:447-457.

Koch, E. G. and Scherbaum, O. H. 1967. Some aspects of lipid metabolism in *Tetrahymena pyriformis* GL during environmental changes. Z. Allg. Mikrobiol. 7:349-361.

Koehler, L. D. and Fennell, R. A. 1964. Histochemistry and biochemistry of the polysaccharides, esterases, and dehydrogenases of *Tetrahymena pyriformis* (strain W). J. Morphol. 114:209-224.

Kovács, E. and Bucz, B. 1967. Propagation of mammalian viruses in protista. II. Isolation of complete virus from yeast and *Tetrahymena* experimentally infected with Picorna viral particles or their infectious RNA. Life Sci. 6:347-348.

Kovács, E., Bucz, B., and Kalompar, G. 1966. Propagation of mammalian viruses in protista. I. Visualization of fluorochrome labelled EMC virus in yeast and *Tetrahymena*. Life Sci. 5:2117-2126.

Kovács, E., Kalompar, G., and Bucz, B. 1967. Propagation of mammalian viruses in protista. III. Change in population densities, viability and multiplication of yeasts and *Tetrahymena* experimentally infected with encephalomyocarditis virus. Life Sci. 6:2359-2372.

Koyama, S. and Tsukada, H. 1970. Catalase pI-isozymes in *Tetrahymena pyriformis* (GL) and their changes during the cell multiplication and maturation. Tumor Res. 5:29-51.

Kozloff, E. N. 1946. The morphology and systematic position of a holotrichous ciliate parasitizing *Deroceras agreste* (L.). J. Morphol. 79:445-465.

Kozloff, E. N. 1956. Experimental infection of the gray garden slug, *Deroceras reticulatum* (Müller), by the holotrichous ciliate *Tetrahymena pyriformis* (Ehrenberg). J. Protozool. 3:17-19.

Kozloff, E. N. 1956. A comparison of the parasitic phase of *Tetrahymena limacis* (Warren) with clones in culture, with particular reference to variability in the number of primary ciliary meridians. J. Protozool. 3:20-28.

Kozloff, E. N. 1956. *Tetrahymena limacis* from the terrestrial pulmonate gastropods *Monadenia fidelis* and *Prophysaon andersoni*. J. Protozool. 3:204-208.

Kozloff, E. N. 1957. A species of *Tetrahymena* parasitic in the renal organ of the slug *Deroceras reticulatum*. J. Protozool. 4:75-79.

Kramhøft, B. 1970. Induced potassium uptake as a result of increased medium osmolarity in suspensions of *Tetrahymena* cells. C. R. Trav. Lab. Carlsberg 37:343-358.

Krenzanoski, J. Z. 1961. Effect of iproniazid on *Tetrahymena pyriformis*. J. Pharm. Sci. 50:421-424.

Kudo, R. R. 1966. Protozoology, 5th ed. Charles C Thomas, Springfield, Illinois. 1174 pp.

Kumar, A. 1969. Studies on ribosomal RNA from *Tetrahymena* by zone velocity sedimentation in sucrose gradients and base ratio analysis. Biochim. Biophys. Acta 186:326-331.

Kumar, A. 1970. Ribosome synthesis in *Tetrahymena pyriformis*. J. Cell Biol. 45:623-634.

Kuzmich, M. J. and Zimmerman, A. M. 1972. The action of mercaptoethanol on cell division in synchronized *Tetrahymena*. J. Protozool. 19:129-132.

Laird, M. 1959. Parasites of Singapore mosquitoes, with particular reference to the significance of larval epibionts as an index of habitat pollution. Ecology, 40:206-221.

Lamborn, W. A. 1921. A protozoon pathogenic to mosquito larvae. Parasitology 13:213-215.

Lane, N. and Padilla, G. M. 1971. Shifts in nuclear and cytoplasmic nucleic acid content in temperature-synchronized *Tetrahymena pyriformis* (HSM). J. Cell. Physiol. 77:93-102.

Lantos, T., Müller, M., Tőro, I., Druga, A., and Vargha, I. 1964. Activity of acid phosphatase and protease in *Tetrahymena pyriformis* GL of different states of nutrition. Acta Biol. Acad. Sci. Hung., Suppl. 6:22-29.

Lascelles, J. 1957. Synthesis of porphyrins by cell suspensions of *Tetrahymena vorax*: Effect of membranes of the vitamin B group. Biochem. J. 66:65-72.

Lawrie, N. R. 1935. Studies in the metabolism of protozoa. II. Some biochemical reactions occurring in the presence of the washed cells of *Glaucoma piriformis*. Biochem. J. 29:2297-2302.

Lawrie, N. R. 1937. Studies in the metabolism of protozoa. III. Some properties of a proteolytic extract obtained from *Glaucoma piriformis*. Biochem. J. 31:789-798.

Lazarus, L. H., Levy, M. R., and Scherbaum, O. H. 1964. Inhibition of synchronized cell division in *Tetrahymena* by actinomycin D. Exp. Cell Res. 36:672-676.

Lazarus, L. H. and Scherbaum, O. H. 1966. Effect of temperature on the activity of ribonuclease from *Tetrahymena pyriformis*. J. Cell. Physiol. 68:95-98.

Lazarus, L. H. and Scherbaum, O. H. 1967. Isolation and specificity of the intracellular ribonuclease from *Tetrahymena pyriformis*. Biochim. Biophys. Acta 142:368-384.

Lazarus, L. H. and Scherbaum, O. H. 1967. Activity of ribosomal phosphodiesterase in a protozoan. Nature 213:887-888.

Lazarus, L. H. and Scherbaum, O. H. 1968. Activity of ribonuclease, acid phosphatase and phosphodiesterase in *Tetrahymena pyriformis* during growth. J. Cell Biol. 36:415-418.

Leboy, P. S., Cline, S. G., and Conner, R. L. 1964. Phosphate, purines and pyrimidines as excretory products of *Tetrahymena*. J. Protozool. 11:217-222.

Lederberg, S. and Mazia, D. 1960. Protein synthesis and cell division in a pyrimidine-starved protozoan. Exp. Cell Res. 21:590-595.

Lee, D. 1969. Photolysis in a culture medium for *Tetrahymena pyriformis*. J. Cell. Physiol. 74:295-298.

Lee, D. 1970. The relative permeability of lysosomes from *Tetrahymena pyriformis* to carbohydrates, lactate and cryoprotective nonelectrolytes glyccrol and dimethylsulfoxide. Biochim. Biophys. Acta 211:550-554.

Lee, D. 1970. Variation in the latency of acid phosphatase and ribonuclease throughout the growth cycle of *Tetrahymena pyriformis*. J. Cell. Physiol. 76:17-22.

Lee, D. 1971. The stability to temperature of lysosomes in homogenates of *Tetrahymena pyriformis*. J. Protozool. 18:173-175.

Lee, D. 1971. The relative permeability of lysosomes from *Tetrahymena pyriformis* to some amino acids and dipeptides. Biochim. Biophys. Acta 225:108-112.

Lee, K.-H. 1959. Studies on cell growth and cell division. I. A temperature-controlled system for inducing synchronous cell division. J. Amcr. Pharm. Ass. 48:468-469.

Lee, K.-H. and Yuzuriha, Y. O. 1964. Studies on cell growth and cell division. III. Action of azaserine on cell division. J. Pharm. Sci. 53:290-293.

Lee, K.-H., Yuzuriha, Y. O., and Eiler, J. J. 1959. Studies on cell growth and cell division. II. Selective activity of chloramphenicol and azaserine on cell growth and cell division. J. Amer. Pharm. Ass. 48:470-473.

Lee, Y. C. and Byfield, J. E. 1970. Characteristics of *in vitro* ribonucleic acid synthesis by macronuclei of *Tetrahymena pyriformis*. Biochemistry 9:3947-3959.

Lee, Y. C. and Scherbaum, O. H. 1965. Isolation of macronuclei from the ciliate *Tetrahymena pyriformis* GL. Nature 208:1350-1351.

Lee, Y. C. and Scherbaum, O. H. 1966. Nucleohistone composition in stationary and division-synchronized *Tetrahymena* cultures. Biochemistry 5:2067-2075.

Lees, A. M. and Korn, E. D. 1966. Separation of positional isomers of long-chain unsaturated fatty acid methyl esters by thin layer chromatography and presence of mono-trans, di-cis isomers in commercial linolenic acid. Biochim. Biophys. Acta 116:403-406.

Lees, A. M. and Korn, E. D. 1966. Metabolism of unsaturated fatty acids in protozoa. Biochemistry 5:1475-1481.

Leeuwenhoek, A. van 1677. Observations communicated to the publisher by Mr. Antony van Leeuwenhoek in a Dutch letter of the 9th of October, 1676, here English'd: Concerning little animals by him observed in rain- well- sea- and snow-water; as also in water wherein pepper had lain infused. Philos. Trans. 12(133):821-831.

Leick, V. 1967. Growth rate dependency of protein and nucleic acid composition of *Tetrahymena pyriformis* and the control of synthesis of ribosomal and transfer RNA. C. R. Trav. Lab. Carlsberg 36:113-126.

Leick, V. 1968. Ratios between contents of DNA, RNA and protein in different micro-organisms as a function of maximal growth rate. Nature 217:1153-1155.

Leick, V. 1969. Formation of subribosomal particles in the macronuclei of *Tetrahymena pyriformis*. Eur. J. Biochem. 8:221-228.

Leick, V. and Andersen, S. B. 1970. Pools and turnover rates of nuclear ribosomal RNA in *Tetrahymena pyriformis*. Eur. J. Biochem. 14:460-461.

Leick, V., Engberg, J., and Emmerson, J. 1970. Nascent subribosomal particles in *Tetrahymena pyriformis*. Eur. J. Biochem. 13:238-246.

Leick, V. and Plesner, P. 1968. Formation of ribosomes in *Tetrahymena pyriformis*. Biochim. Biophys. Acta 169:398-408.

Leick, V. and Plesner, P. 1968. Precursors of ribosomal sub-units in *Tetrahymena pyriformis*. Biochim. Biophys. Acta 169:409-415.

Lengerová-Kučerová, A. 1950. Cell volume of *Glaucoma piriformis* as a function of hydrogen ion concentration in the medium. Věst. Čsl. Zool. Spol. 14:207-228.

Letts, P. J. and Zimmerman, A. M. 1970. Polypeptide synthesis with microsomes from pressure-treated *Tetrahymena*. J. Protozool. 17:593-596.

Levy, M. R. 1967. Interaction of acetate, glucose and growth conditions on glyconeogenesis and isocitrate lyase activity in *Tetrahymena*. Comp. Biochem. Physiol. 21:291-298.

Levy, M. R. 1967. Effects of inhibitors of RNA and protein synthesis on activation of glyconeogenesis in *Tetrahymena*. J. Cell. Physiol. 69:247-252.

Levy, M. R. 1970. Localization of acetyl-coenzyme A synthetase on peroxisomes in *Tetrahymena*. Biochem. Biophys. Res. Commun. 39:1-6.

Levy, M. R. 1972. Regulation of isocitrate metabolism in peroxisomes in *Tetrahymena pyriformis*. Arch. Biochem. Biophys. 152:463-471.

Levy, M. R. 1973. Decrease of peroxisomal enzymes in *Tetrahymena* and its prevention by actinomycin D and cyclohexamide. Biochim. Biophys. Acta. In press.

Levy, M. R. and Elliott, A. M. 1968. Biochemical and ultrastructural changes in *Tetrahymena pyriformis* during starvation. J. Protozool. 15:208-222.

Levy, M. R., Gollon, C. E., and Elliott, A. M. 1969. Effects of hyperthermia on *Tetrahymena pyriformis*. I. Localization of acid hydrolases and changes in cell ultrastructure. Exp. Cell Res. 55:295-305.

Levy, M. R. and Hunt, A. 1967. L-α-Hydroxy acid oxidase activity in *Tetrahymena*. Change with physiological state. J. Cell Biol. 34:911-915.

Levy, M. R. and Scherbaum, O. H. 1965. Induction of the glyoxylate cycle in *Tetrahymena*. Arch. Biochem. Biophys. 109:116-121.

Levy, M. R. and Scherbaum, O. H. 1965. Glyconeogenesis in growing and non-growing cultures of *Tetrahymena pyriformis*. J. Gen. Microbiol. 38:211-230.

Levy, M. R. and Wasmuth, J. J. 1970. Effects of carbohydrate on glycolytic and peroxisomal enzymes in *Tetrahymena*. Biochim. Biophys. Acta 201:205-214.

Liang, C. R. 1970. Enzymic synthesis of [^{14}C]-phosphoenolpyruvate from [^{14}C]-pyruvate for use as a precursor in biosynthesis of phosphonates. Bull. Inst. Chem. Acad. Sinica 18:70-73.

Liang, C. R. and Rosenberg, H. 1966. The biosynthesis of the phosphonic analogue of cephalin in *Tetrahymena*. Biochim. Biophys. Acta 125:548-562.

Liang, C. R. and Rosenberg, H. 1968. The biosynthesis of the carbon phosphorus bond in *Tetrahymena*. Biochim. Biophys. Acta 156:437-439.

Lilly, D. M. 1942. Nutritional and supplementary factors in the growth of carnivorous ciliates. Physiol. Zool. 15:146-167.

Lilly, D. M. 1953. The nutrition of carnivorous protozoa. Ann. N.Y. Acad. Sci. 56:910-920.

Lilly, D. M. 1967. Growth factors in protozoa. *In* M. Florkin and B. J. Scheer, eds., Chemical zoology, vol. 1, Protozoa (G. W. Kidder, ed.), Academic Press, New York, pp. 93-159.

Lilly, D. M. and Cevallos, W. H. 1956. Chemical supplements promoting growth in carnivorous ciliates. Trans. N.Y. Acad. Sci. 18:531-539.

Lilly, D. M. and Henry, S. M. 1956. Supplementary factors in the nutrition of *Euplotes*. J. Protozool. 3:200-203.

Lilly, D. M., Sterbenz, F. J., and Tarantola, V. 1953. Growth promotion in carnivorous protozoa by substituted purines. Proc. Soc. Exp. Biol. Med. 83:434-438.

Lilly, D. M. and Stillwell, R. H. 1965. Probiotics: Growth-promoting factors produced by microorganisms. Science 147:747-748.

Lindh, N. O. and Christensson, E. G. 1963. The carbohydrate metabolism during growth and division of *Tetrahymena pyriformis* GL. Ark. Zool. 15:163-180.

Lipmann, F., Kaplan, N. O., Novelli, G., Tuttle, L. C. and Guirard, B. 1947. Coenzyme for acetylation, a pantothenic acid derivative. J. Biol. Chem. 167:869-875.

Lloyd, D., Brightwell, R., Venables, S. E., Roach, G. I., and Turner, G. 1971. Subcellular fractionation of *Tetrahymena pyriformis* St by zonal centrifugation: Changes in activities and distribution of enzymes during the growth cycle and on starvation. J. Gen. Microbiol. 65:209-223.

Lloyd, D., Turner, G., Poole, R. K., Nicholl, W. G., and Roach, G. I. 1971. A hypothesis of nuclear-mitochondrial interactions for the control of mitochondrial biogenesis based on experiments with *Tetrahymena pyriformis*. Sub Cell. Biochem. 1:91-95.

Loefer, J. B. 1938. The utilization of dextrose by *Colpidium*, *Glaucoma*, *Chilomonas*, and *Chlorogonium* in bacteria-free cultures. J. Exp. Zool. 79:167-183.

Loefer, J. B. 1939. Acclimatization of fresh-water ciliates and flagellates to media of higher osmotic pressure. Physiol. Zool. 12:161-172.

Loefer, J. B. 1949. Penicillin and streptomycin sensitivity of some protozoa. Tex. J. Sci. 1:92-94.

Loefer, J. B. 1951. Chloromycetin and growth of certain protozoa. Physiol. Zool. 24:155-163.

Loefer, J. B. 1952. Some observations on the size of *Tetrahymena*. J. Morphol. 90:407-414.

Loefer, J. B. 1952. Growth inhibition of protozoa with bacitracin. Tex. J. Sci. 4:73-76.

Loefer, J. B. 1967. Criteria for the taxonomy of *Tetrahymena*. Bull. Natl. Inst. Sci. India, no. 34:34-47.

Loefer, J. B. and Hall, R. P. 1936. Effect of ethyl alcohol on the growth of eight protozoan species in bacteria-free cultures. Arch. Protistenk. 87:123-130.

Loefer, J. B. and Matney, T. S. 1952. Growth inhibition of free-living protozoa by actidione. Physiol. Zool. 25:272-276.

Loefer, J. B. and Mefferd, R. B., Jr. 1952. Concerning pattern formation by free-swimming microorganisms. Amer. Nat. 86:325-329.

Loefer, J. B. and Mefferd, R. B., Jr. 1952. Application of the most probable number method to determine heat sensitivity of protozoa. Biol. Bull. 103:364-368.

Loefer, J. B. and Owen, R. D. 1961. Characterization and distribution of "H" serotypes in 25°C cultures of *Tetrahymena pyriformis*, variety 1. J. Protozool. 8:387-391.

Loefer, J. B., Owen, R. D., and Christensen, E. 1958. Serological types among thirty-one strains of the ciliated protozoan, *Tetrahymena pyriformis*. J. Protozool. 5:209-217.

Loefer, J. B. and Scherbaum, O. H. 1961. Amino acid composition of protozoa. Comparative studies of *Tetrahymena*. J. Protozool. 8:184-192.

Loefer, J. B. and Scherbaum, O. H. 1962. Free amino acids in protozoa. *In* J. T. Holden, ed., Amino acid pools. Elsevier, Amsterdam, pp. 109-114.

Loefer, J. B. and Scherbaum, O. H. 1963. Free amino acids in Tetrahymenidae. J. Protozool. 10:275-279.

Loefer, J. B. and Scherbuam, O. H. 1963. Serological and biochemical factors relative to taxonomy of *Tetrahymena*. Syst. Zool. 12:175-177.

Loefer, J. B., Small, E. B., and Furgason, W. H. 1966. Range of variation in the somatic infraciliature and contractile vacuole pores of *Tetrahymena pyriformis*. J. Protozool. 13:90-103.

Lom, J. 1959. *Tetrahymena* infection in the earthworm. J. Parasitol. 45:320.

Lom, J., Corliss, J. O., and Noirot-Timothée, C. 1968. Observations on the ultrastructure of the buccal apparatus in thigmotrich ciliates and their bearing on thigmotrich-peritrich affinities. J. Protozool. 15:824-840.

Lövlie, A. 1963. Growth in mass and respiration rate during the cell cycle of *Tetrahymena pyriformis*. C. R. Trav. Lab. Carlsberg 33:377-413.

Lowe-Jinde, L. and Zimmerman, A. M. 1969. Heavy water and hydrostatic pressure effects on *Tetrahymena pyriformis*. J. Protozool. 16:226-230.

Lowe-Jinde, L. and Zimmerman, A. M. 1971. The incorporation of phenylalanine and uridine in *Tetrahymena*: A pressure study. J. Protozool. 18:20-23.

Lowy, B. and Leick, V. 1969. The synthesis of DNA in synchronized cultures of *Tetrahymena pyriformis* GL. Exp. Cell Res. 57:277-288.

Luckey, T. D. 1954. A single diet for all living organisms. Science 120:396-398.

Ludwig, W. 1928. Über den funktionellen Zusammenhang zwischen Populationschlichte, Nahrungsdichte und Teilungsräte bei Protisten und über die Zunahme der Bevölkerungsdichte überhaupt. Biol. Gen. 4:351-376.

Lundin, F. C. and West, L. S. 1963. The free-living protozoa of the upper peninsula of Michigan. Northern Michigan College Press, Marquette, Michigan. 175 pp.

Lutton, J. D. and McCashland, B. W. 1971. The effect of isoriboflavin and cyanide on growth and respiration in *Tetrahymena pyriformis*. Acta Protozool. 9:107-111.

Lwoff, A. 1923. Sur la nutrition des infusoires. C. R. Acad. Sci., Paris 176:928-930.

Lwoff, A. 1924. Infection expérimentale à *Glaucoma piriformis* chez *Galleria mellonella* (lepidoptère). C. R. Acad. Sci., Paris 178:1106-1108.

Lwoff, A. 1924. Le pouvoir de synthèse d'un protiste hétérotrope: *Glaucoma piriformis*. C. R. Soc.

Biol. 91:344-345.

Lwoff, A. 1925. Influence d'extraits de glandes et d'organes sur la vitesse de multiplication des infusoires. C. R. Soc. Biol. 93:1352-1354.

Lwoff, A. 1925. La nutrition des infusoires au dépens des substances dissoutes. C. R. Soc. Biol. 93:1272-1273.

Lwoff, A. 1929. Milieux de culture et d'entretien pour *Glaucoma piriformis* (cilié). C. R. Soc. Biol. 100:635-636.

Lwoff, A. 1932. Recherches biochimiques sur la nutrition des protozaires. Le pouvoir de synthèse. Masson et Cie, Paris. 158 pp.

Lwoff, A. 1943. L'évolution physiologique. Étude des pertes de fonctions chez le microorganismes. Hermann et Cie, Paris. 308 pp.

Lwoff, A. 1947. Some aspects of the problem of growth factors for protozoa. Ann. Rev. Microbiol. 1:101-114.

Lwoff, A. 1949. Kinetosomes and the development of ciliates. Growth, 9th Symp. Suppl. 13:61-91.

Lwoff, A. 1950. Problems of morphogenesis in ciliates. John Wiley, New York. 103 pp.

Lwoff, A. and Lwoff, M. 1937. L'aneurine, facteur de croissance pour le cilié *Glaucoma piriformis*. C. R. Soc. Biol. 126:644-646.

Lwoff, A. and Lwoff, M. 1938. La spécificité de l'aneurine, facteur de croissance pour le cilié *Glaucoma piriformis*. C. R. Soc. Biol. 127:1170-1172.

Lwoff, A. and Roukhelman, N. 1926. Variations de quelques formes d'azote dans une culture pure d'infusoires. C. R. Acad. Sci., Paris 183:156-158.

Lwoff, M. 1934. Sur la respiration du cilié *Glaucoma piriformis*. C. R. Soc. Biol. 115:237-241.

Lynch, V. H. and Calvin, M. 1952. Carbon dioxide fixation by microorganisms. J. Bacteriol. 63:525-531.

Lyttleton, J. W. 1963. A simple method of isolating ribosomes from *Tetrahymena pyriformis*. Exp. Cell Res. 31:385-389.

MacArthur, W. P. 1922. A holotrichous ciliate pathogenic to *Theobaldia annulata* Schrank. J. Roy. Army Med. Corps 38:83-92.

McCashland, B. W. 1955. Adaptation by *Tetrahymena pyriformis* to potassium cyanide. I. Adaptive reversal of cyanide inhibition of growth. J. Protozool. 2:97-100.

McCashland, B. W. 1956. Adaptation by *Tetrahymena pyriformis* to potassium cyanide. II. Adaptation against respiratory inhibition. J. Protozool. 3:131-135.

McCashland, B. W. 1963. Growth difference in progeny of single *Tetrahymena pyriformis*. Growth 27:39-45.

McCashland, B. W. and Andresen, W. F. 1963. The nature of cyanide adaptation in *Tetrahymena pyriformis* W. Growth 27:47-56.

McCashland, B. W. and Johnson, V. A. 1957. Variations in growth characteristics of seven strains of *Tetrahymena*. Growth 21:11-19.

McCashland, B. W. and Kronschnabel, J. M. 1962. Exogenous factors affecting respiration in *Tetrahymena pyriformis*. J. Protozool. 9:276-279.

McCashland, B. W., Marsh, W. R., and Kronschnabel, J. M. 1957. Variations in the inhibitory effect of potassium cyanide upon growth and respiration in seven strains of *Tetrahymena*. Growth 21:21-27.

McCashland, B. W. and Pace, D. M. 1952. The influence of potassium cyanide on growth and respiration in *Tetrahymena geleii* Furgason. Growth 16:75-83.

McCashland, B. W. and Steinacher, R. H. 1962. Metabolism changes in *Tetrahymena pyriformis* W adapted to potassium cyanide. Proc. Soc. Exp. Biol. Med. 111:789-793.

Macdonald, A. G. 1967. Delay in the cleavage of *Tetrahymena pyriformis* exposed to high hydrostatic pressure. J. Cell. Physiol. 70:127-129.

McDonald, B. B. 1958. Quantitative aspects of deoxyribose nucleic acid (DNA) metabolism in an amicronucleate strain of *Tetrahymena*. Biol. Bull. 114:71-94.

McDonald, B. B. 1962. Synthesis of deoxyribonucleic acid by micro- and macronuclei of *Tetrahymena pyriformis*. J. Cell Biol. 13:193-203.

McDonald, B. B. 1966. The exchange of RNA and protein during conjugation in *Tetrahymena*. J. Protozool. 13:277-285.

McKee, C. M., Dutcher, J. D., Groupé, V., and Moore, M. 1947. Antibacterial lipids from *Tetrahymena geleii*. Proc. Soc. Exp. Biol. Med. 65:326-332.

Mackenzie, T. B., Stone, G. E., and Prescott, D. M. 1966. The durations of G_1, S, and G_2 at different temperatures in *Tetrahymena pyriformis* HSM. J. Cell Biol. 31:633-635.

McKinney, R. E. 1956. Protozoa and activated sludge. Sewage Ind. Wastes 28:1219-1231.

Mackinnon, D. L. and Hawes, R. S. J. 1961. An introduction to the study of protozoa. Clarendon Press, Oxford. 506 pp.

McLaughlan, J. M., Shenoy, K. G., and Campbell, J. A. 1961. Some apparent drug-vitamin interrelationships in *Lactobacillus leichmannii* and *Tetrahymena pyriformis*. J. Amer. Pharm. Ass. 50:59-63.

Mager, J. 1960. Chloramphenicol and chlortetracycline inhibition of amino acid incorporation into proteins in a cell-free system from *Tetrahymena pyriformis*. Biochim. Biophys. Acta 38:150-152.

Mager, J. and Lipmann, F. 1958. Amino acid incorporation and the reversion of its initial phase with cell-free *Tetrahymena* preparations. Proc. Natl. Acad. Sci. U.S. 44:305-309.

Malecki, M., Vojtech, O., and Eiler, J. J. 1971. The effect of low partial pressure of oxygen and shaking on the rate of division in *Tetrahymena*. Curr. Mod. Biol. 3:291-298.

Mallory, F. B. and Conner, R. L. 1971. Dehydrogenation and dealkylation of various sterols by *Tetrahymena pyriformis*. Lipids 6:149-153.

Mallory, F. B., Conner, R. L., Landrey, J. R., and Lyengar, C. W. L. 1968. The biosynthesis of 7,22-bisdehydrocholesterol from cholesterol. Tetrahedron Lett., no. 58:6103-6106.

Mallory, F. B., Conner, R. L., Landrey, J. R., Zander, J. M., Greig, J. B., and Caspi, E. 1968. The biosynthesis of tetrahymanol from $(4R)$-mevalonic-4-^{3}H-2-^{14}C acid. J. Amer. Chem. Soc. 90:3564-3566.

Mallory, F. B., Gordon, J. T., and Conner, R. L. 1963. The isolation of a pentacyclic triterpenoid alcohol from a protozoan. J. Amer. Chem. Soc. 85:1362-1363.

Mandel, M. 1967. Nucleic acids of protozoa. *In* M. Florkin and B. J. Scheer, eds., Chemical zoology, vol. 1, Protozoa (G. W. Kidder, ed.), Academic Press, New York, pp. 541-572.

Manners, D. J. and Ryley, J. F. 1952. Studies on the metabolism of the protozoa. II. The glycogen of the ciliate *Tetrahymena pyriformis (Glaucoma piriformis)*. Biochem. J. 52:480-482.

Manwell, R. D. 1961. Introduction to protozoology. St. Martin's Press, New York. 642 pp.

Marburger, J. P. 1943. The production of growth substances by *Colpidium striatum* (Stokes). Physiol. Zool. 18:186-198.

Margolin, P., Loefer, J. B., and Owen, R. D. 1959. Immobilizing antigens of *Tetrahymena pyriformis*. J. Protozool. 6:207-215.

Mark, H. F., Imparato, A. M., Hutner, S. H., and Baker, H. 1963. Estimate of toxicity of radiopaque agents by means of a ciliate. Angiology 14:383-389.

Markees, D. G., Dewey, V. C., and Kidder, G. W. 1960. The inhibition of certain protozoa by diaminoalkoxypyridines. Arch. Biochem. Biophys. 86:179-184.

Markees, D. G., Dewey, V. C., and Kidder, G. W. 1968. The synthesis and biological activity of substituted 2,6-diaminopyridines. J. Med. Chem. 11:126-129.

Markees, D. G. and Kidder, G. W. 1956. The synthesis of 5-amino-7-hydroxy-1,3,4-imidazopyridine (1-deazaguanine) and related compounds. J. Amer. Chem. Soc. 78:4130-4135.

Martin, T. E. and Wool, I. G. 1969. Active hybrid 80 S particles formed from subunits of rat, rabbit and protozoan (*Tetrahymena pyriformis*) ribosomes. J. Mol. Biol. 43:151-161.

Mast, S. O. and Pace, D. M. 1946. The nature of the growth-substance produced by *Chilomonas paramecium*. Physiol. Zool. 19:223-235.

Matalon, R., Molho-Lacroix, L., and Cohen, M. 1950. Lyse du cilié *Glaucoma piriformis* par les alcools. Influence de la longueur de la chaine hydrocarbonée et de la structure. C. R. Acad. Sci., Paris 230:1542-1543.

Matalon, R., Molho-Lacroix, L., and Cohen, M. 1950. Lyse du cilié *Glaucoma piriformis* par les alcools et leurs dérivés polyoxyéthylés. C. R. Acad. Sci., Paris 230:1985-1987.

Maupas, E. 1883. Contributions à l'étude morphologique et anatomique des infusoires ciliés. Arch. Zool. Exp. Gén. (sér. 2) 1:427-664.

Maupas, E. 1886. Sur la multiplication de la *Leucophrys patula* Ehrb. C. R. Acad. Sci., Paris 103:1270-1273.

Maupas, E. 1887. Réponse à M. Balbiani à propos de la *Leucophrys patula.* C. R. Acad. Sci., Paris 104:308-310.

Maupas, E. 1888. Recherches expérimentales sur la multiplication des infusoires ciliés. Arch. Zool. Exp. Gén. (sér. 2) 6:165-277.

Maupas, E. 1889. La rajeunissement karyogamique chez les ciliés. Arch. Zool. Exp. Gén. (sér. 2) 7:149-517.

Mavrides, C. and D'Iorio, A. 1969. The regulation of tyrosine aminotransferase in *Tetrahymena pyriformis.* Biochem. Biophys. Res. Commun. 35:467-473.

Mazia, D. 1956. The life history of the cell. Amer. Sci. 44:1-32.

Mazia, D. and Zeuthen, E. 1966. Blockage and delay of cell division in synchronized populations of *Tetrahymena* by mercaptoethanol (monothioethylene glycol). C. R. Trav. Lab. Carlsberg 35:341-361.

Mefferd, R. B., Jr., and Loefer, J. B. 1953. Lethality of ultraviolet radiation for *Tetrahymena pyriformis.* Physiol. Zool. 26:341-344.

Mefferd, R. B., Jr., and Loefer, J. B. 1954. Inhibition of respiration in *Tetrahymena pyriformis* S by actidione. Physiol. Zool. 27:115-118.

Meister, H. 1970. Über die Modifizierbarkeit der DNS-Verteilung im Makronucleus bei *Stentor coeruleus* Ehrenberg. Arch. Protistenk. 112:314-342.

Metz, C. B. and Westfall, J. A. 1954. The fibrillar systems of ciliates as revealed by the electron microscope. II. *Tetrahymena.* Biol. Bull. 107:106-122.

Meyer, R. R., Boyd, C. R., Rein, D. C., and Keller, S. J. 1972. Effects of ethidium bromide on growth and morphology of *Tetrahymena pyriformis.* Exp. Cell Res. 70:233-237.

Michelson, E. H. 1971. Distribution and pathogenicity of *Tetrahymena limacis* in the slug *Deroceras reticulatum.* Parasitology 62:125-131.

Milkovitsch, G. 1929. Action du sérum humain sur un infusoire, *Glaucoma piriformis.* C. R. Soc. Biol. 100:417-420.

Milkovitsch, G. 1931. Action du sérum humain sur un infusoire. Bull. Biol. Fr. Belg. 65:103-113.

Miller, J. E. 1965. Biosynthesis of the benzoquinone rings of ubiquinone in *Tetrahymena pyriformis.* Biochem. Biophys. Res. Commun. 19:335-339.

Miller, O. L., Jr., and Stone, G. E. 1963. Fine structure of the oral area of *Tetrahymena patula.* J. Protozool. 10:280-288.

Miller, O. L., Jr., Stone, G. E., and Prescott, D. M. 1964. Autoradiography of soluble materials. J. Cell Biol. 23:654-658.

Millis, A. J. T. and Suyama, Y. 1972. Effects of chloramphenicol and cycloheximide on the biosynthesis of mitochondrial ribosomes in *Tetrahymena.* J. Biol. Chem. 247:4063-4073.

Ming Chu, I., Wheeler, M. A., and Holmlund, C. E. 1972. Fatty acid methyl and ethyl esters in *Tetrahymena pyriformis.* Biochim. Biophys. Acta 270:18-22.

Mita, T. 1965. Effects of actinomycin D on the RNA synthesis and the synchronous cell division of *Tetrahymena pyriformis* GL. Biochim. Biophys. Acta 103:182-185.

Mita, T., Kawazoe, Y., and Araki, M. 1969. Parallelism between carcinogenic activity of 4-nitroquinoline 1-oxide derivatives and their ability to induce unequal nuclear division in the synchronized division of *Tetrahymena.* Gann 60:155-160.

Mita, T. and Scherbaum, O. H. 1965. Chromatographic fractionation of DNA isolated from normally and synchronously dividing *Tetrahymena pyriformis* GL. J. Biochem. 58:130-136.

Mita, T., Shiomi, H., and Iwai, K. 1966. Isolation of nuclei from exponentially growing *Tetrahymena pyriformis.* Exp. Cell Res. 43:696-699.

Mita, T., Tokuzen, R., Fukuoka, F., and Nakahara, W. 1965. Effect of 4-nitroquinoline 1-oxide and related compounds on normally and synchronously dividing *Tetrahymena pyriformis* GL. Gann 56:293-299.

Mita, T., Tokuzen, R., Fukuoka, F., and Nakahara, W. 1967. Result of mating normal and carcinogen-induced pleomorphic clones of *Tetrahymena.* Gann 58:479-480.

Mita, T., Tokuzen, R., Fumiko, F., and Nakahara, W. 1966. Experimental production of the clone with atypical multiplication from synchronous cultures of *Tetrahymena pyriformis* GL treated with 4-nitroquinoline 1-oxide. Gann 57:273-278.

Mizuno, K., Ohkubo, Y., Yokoyama, S., Hamada, M., Maeda, K., and Umezawa, H. 1967. A new antibiotic, leucinamycin. J. Antibiot., Ser. A, 20:194-199.

Moikeha, S. N. and Chu, G. W. 1971. Dermatitis-producing alga *Lyngbya majuscula* Gomont in Hawaii. II. Biological properties of the toxic factor. J. Phycol. 7:8-13.

Molho, D. and Lacroix, L. 1949. Antibiotic action of some quinones. I. Behavior of *Glaucoma piriformis* in peptone medium in presence of vitamin K plus anti-vitamin K. Bull. Soc. Chem. Biol. 31:3141-3147.

Møller, K. M. and Prescott, D. M. 1955. Observations on the cytochromes of *Amoeba proteus, Chaos chaos* and *Tetrahymena geleii*. Exp. Cell Res. 9:375-377.

Moner, J. G. 1965. RNA synthesis and cell division in heat-synchronized populations of *Tetrahymena pyriformis*. J. Protozool. 12:505-509.

Moner, J. G. 1967. Temperature, RNA synthesis and cell division in heat-synchronized cells of *Tetrahymena*. Exp. Cell Res. 45:618-630.

Moner, J. G. 1972. The effects of temperature and heavy water on cell division in heat-synchronized cells of *Tetrahymena*. J. Protozool. 19:382-385.

Moner, J. G. and Berger, R. O. 1966. RNA synthesis and cell division in cold-synchronized cells of *Tetrahymena pyriformis*. J. Cell. Physiol. 67:217-224.

Monod, J. 1934. Independence du galvano-tropisme et de la densité du courant chez les infuroires ciliés. C. R. Acad. Sci., Paris 198:122-124.

Monod, J. 1935. Le taux de croissance en fonction de la concentration de l'ailment dans une population de *Glaucoma piriformis* en culture pure. C. R. Acad. Sci., Paris 201:1513-1515.

Moore, K. C. 1972. Pressure induced regression of oral apparatus microtubules in synchronized *Tetrahymena*. J. Ultrastr. Res. 41:499-518.

Morishita, I. 1968. Protozoa in activated sludge. Jap. J. Water Treatment Biol. 4(2):12-20. In Japanese with English summary.

Morishita, I. 1970. Studies on protozoa-populations in activated sludge of sewage and waste treatment plants. Jap. J. Protozool. 3:1-13.

Mouton, R. F. 1971. DNA repair inhibitors and carcinogenesis. Inhibition of post-UV irradiation growth in the dark of *Tetrahymena pyriformis* by caffeine and the oncogenic mycotoxin luteoskyrin. *In* E. Broda, ed., 1st European Biophysics Congress, vol. 2, Medical Akad., Vienna, pp. 241-250.

Mouton, R. F. and Fromageot, P. 1971. Inhibition of post-UV irradiation growth in the dark of *Tetrahymena pyriformis* by caffeine and the oncogenic mycotoxin luteoskyrin. FEBS Lett. 15:45-48.

Mučibabić, S. 1957. The growth of mixed populations of *Chilomonas paramecium* and *Tetrahymena patula*. Quart. J. Microsc. Sci. 98:251-263.

Mučibabić, S. 1957. The growth of mixed populations of *Chilomonas paramecium* and *Tetrahymena pyriformis*. J. Gen. Microbiol. 16:561-571.

Mugard, H. 1949. Contribution à l'étude des infusoires hyménostomes histiophages. Ann. Sci. Nat. Zool. Biol. Anim. (sér. 11) 10:171-268.

Mugard, H. 1951. Phosphatase alcaline chez les infusoires ciliés. Bull. Soc. Zool. Fr. 76:39-41.

Mulheirn, L. J., Aberhart, D. J., and Caspi, E. 1971. Dehydrogenation of sterols by the protozoan *Tetrahymena pyriformis*. J. Biol. Chem. 246:6556-6559.

Müller, M. 1967. Digestion. *In* M. Florkin and B. T. Scheer, eds., Chemical zoology, vol. 1, Protozoa (G. W. Kidder, ed.), Academic Press, New York, pp. 351-380.

Müller, M. 1969. Peroxisomes of protozoa. Ann. N.Y. Acad. Sci. 168:292-301.

Müller, M. 1971. Lysosomes in *Tetrahymena pyriformis*. II. Intracellular distribution of several acid hydrolases. Acta Biol. Acad. Sci. Hung. 22:179-186.

Müller, M. 1972. Secretion of acid hydrolases and its intracellular source in *Tetrahymena pyriformis*. J. Cell Biol. 52:478-487.

Müller, M., Baudhuin, P., and de Duve, C. 1966. Lysosomes in *Tetrahymena pyriformis*. I. Some properties and lysosomal localization of acid hydrolases. J. Cell. Physiol. 68:165-176.

Müller, M., Hogg, J. F., and de Duve, C. 1968. Distribution of tricarboxylic acid cycle and of glyoxylate cycle enzymes between mitochondria and peroxisomes in *Tetrahymena pyriformis.* J. Biol. Chem. 243:5385-5395.

Müller, M. and Röhlich, P. 1961. Studies on feeding and digestion in protozoa. II. Food vacuole cycle in *Tetrahymena corlissi.* Acta Microbiol. Acad. Sci. Hung. 10:297-305.

Müllcr, M., Röhlich, P., and Törő, I. 1965. Studies on feeding and digestion in protozoa. VII. Ingestion of polystyrene latex particles and its early effect on acid phosphatase in *Paramecium multimicronucleatum* and *Tetrahymena pyriformis.* J. Protozool. 12:27-34.

Müller, M., Röhlich, P., Tóth, J., and Törő, I. 1963. Fine structure and enzymatic activity of protozoan food vacuoles. *In* A. V. S. de Reuck and M. P. Cameron, eds., Ciba Foundation symposium on lysosomes, Churchill, London, pp. 201-216.

Müller, M., Törő, I., Röhlich, P., Tóth, J., and Tóth, G. 1961. Cytological studies on the feeding and digestion of a histophagous ciliate, *Tetrahymena corlissi.* Biol. Kozl. 9:55-62. In Hungarian.

Müller, M., Tóth, J., and Törő, I. 1960. Increase in esterase activity during intracellular digestion in a histophagus ciliate. Nature 187:65.

Müller, O. F. 1773. Vermium terrestrium et fluviatilium, seu animalium infusorium, helminthicorum et testaceorum, non marinorum, succincta historia. Heineck et Faber, Havniae et Lipsiae. 135 pp.

Müller, O. F. 1786. Animalcula infusoria fluviatilia et marina. N. Mölleri, Havniae et Lipsiae. 367 pp.

Munk, N. and Rosenberg, H. 1969. On the deposition and utilization of inorganic pyrophosphate in *Tetrahymena pyriformis.* Biochim. Biophys. Acta 177:629-640.

Munn, E. 1971. Fine structure of basal bodies (kinetosomes) and associated components of *Tetrahymena.* Tissue Cell 2:499-512.

Murdmatsu, M. 1970. Isolation of nuclei and nucleoli. *In* D. M. Prescott, ed., Methods in cell physiology, vol. 4, Academic Press, New York, pp. 195-230.

Murti, K. G. 1972. "Nucleoli" in micronucleus of *Tetrahymena.* Exp. Cell Res. 72:584-586.

Murti, K. G. and Prescott, D. M. 1970. Micronuclear ribonucleic acid in *Tetrahymena pyriformis.* J. Cell Biol. 47:460-467.

Muspratt, J. 1945. Observations on the larvae of tree-hole breeding Culicini (Diptera: Culicidae) and two of their parasites. J. Entomol. Soc. S. Afr. 8:13-20.

Muspratt, J. 1947. Notes on a ciliate protozoan, probably *Glaucoma pyriformis,* parasitic in culicine mosquito larvae. Parasitology 38:107-110.

Nachtwey, D. S. 1965. Division of synchronized *Tetrahymena pyriformis* after emacronucleation. C. R. Trav. Lab. Carlsberg 35:25-35.

Nachtwey, D. S. and Cameron, I. L. 1968. Cell cycle analysis. *In* D. M. Prescott, ed., Methods in cell physiology, vol. 3, Academic Press, New York, pp. 213-259.

Nachtwey, D. S. and Dickinson, W. J. 1967. Actinomycin D: Blockage of cell division of synchronized *Tetrahymena pyriformis.* Exp. Cell Res. 47:581-595.

Nachtwey, D. S. and Giese, A. C. 1968. Effects of ultraviolet irradiation and heat shocks on cell division in synchronized *Tetrahymena pyriformis.* Exp. Cell Res. 50:167-176.

Nakano, N. 1968. Inhibitory effects of fusarenon on multiplication of *Tetrahymena pyriformis.* Jap. J. Med. Sci. Biol. 21:351-357.

Nanney, D. L. 1953. Nucleo-cytoplasmic interaction during conjugation in *Tetrahymena.* Biol. Bull. 105:133-148.

Nanney, D. L. 1956. Caryonidal inheritance and nuclear differentiation. Amer. Nat. 90:291-307.

Nanney, D. L. 1957. Inbreeding degeneration in *Tetrahymena.* Genetics 42:137-146.

Nanney, D. L. 1957. The role of the cytoplasm in heredity. *In* W. D. McElroy and B. Glass, eds., The chemical basis of heredity, Johns Hopkins Press, Baltimore, pp. 134-164.

Nanney, D. L. 1959. Vegetative mutants and clonal senility in *Tetrahymena.* J. Protozool. 6:171-177.

Nanney, D. L. 1959. Nuclear differentiation in serotype determination in *Tetrahymena.* Science 130:1420-1421.

Nanney, D. L. 1960. Genetic factors affecting mating type frequencies in variety 1 of *Tetrahymena pyriformis*. Genetics 44:1173-1184.

Nanney, D. L. 1960. The relationship between the mating type and the H serotype systems in *Tetrahymena*. Genetics 45:1351-1358.

Nanney, D. L. 1960. Temperature effects on nuclear differentiation in variety 1 of *Tetrahymena pyriformis*. Physiol. Zool. 33:146-151.

Nanney, D. L. 1962. The relationship between the mating type and the N serotype systems in *Tetrahymena*. Genetics 45:1351-1358.

Nanney, D. L. 1962. Anomalous serotypes in *Tetrahymena*. J. Protozool. 9:485-487.

Nanney, D. L. 1963. Irregular genetic transmission in *Tetrahymena* crosses. Genetics 48:737-744.

Nanney, D. L. 1963. Aspects of mutual exclusion in *Tetrahymena*. *In* R. J. C. Harris, ed., Biological organization at the cellular and supercellular level, Academic Press, New York, pp. 91-109.

Nanney, D. L. 1963. The inheritance of H-L serotype differences at conjugation in *Tetrahymena*. J. Protozool. 10:152-155.

Nanney, D. L. 1963. Cytoplasmic inheritance in Protozoa. *In* W. J. Burdette, ed., Methodology in basic genetics, Holden-Day, San Francisco, pp. 353-378.

Nanney, D. L. 1964. Macronuclear differentiation and subnuclear assortment in ciliates. *In* M. Locke, ed., The role of chromosomes in development, Academic Press, New York, pp. 253-273.

Nanney, D. L. 1966. Corticotypes in *Tetrahymena pyriformis*. Amer. Nat. 100:303-318.

Nanney, D. L. 1966. Corticotype transmission in *Tetrahymena*. Genetics 54:955-968.

Nanney, D. L. 1966. Cortical integration in *Tetrahymena*: An exercise in cytogeometry. J. Exp. Zool. 161:307-318.

Nanney, D. L. 1966. Corticotypic technics in *Tetrahymena* taxonomy. J. Protozool. 13:483-490.

Nanney, D. L. 1967. Comparative corticotype analyses in *Tetrahymena*. J. Protozool. 14:690-697.

Nanney, D. L. 1967. Cortical slippage in *Tetrahymena*. J. Exp. Zool. 166:163-170.

Nanney, D. L. 1968. Patterns of cortical stability in *Tetrahymena*. J. Protozool. 15:109-112.

Nanney, D. L. 1968. Ciliate genetics: Patterns and programs of gene action. Ann. Rev. Genet. 2:121-140.

Nanney, D. L. 1968. Cortical patterns in cellular morphogenesis. Science 160:496-502.

Nanney, D. L. 1970. Temperature effects on cortical characteristics of *Tetrahymena*. J. Exp. Zool. 175:383-390.

Nanney, D. L. 1971. Cortical characteristics of strains of syngens 10, 11 and 12 of *Tetrahymena pyriformis*. J. Protozool. 18:33-37.

Nanney, D. L. 1971. The constancy of cortical units in *Tetrahymena* with varying numbers of rows. J. Exp. Zool. 178:177-181.

Nanney, D. L. 1971. The pattern of replication of cortical units in *Tetrahymena*. Develop. Biol. 26:296-305.

Nanney, D. L. 1972. Cytogeometric integration in the ciliate cortex. Ann. N.Y. Acad. Sci. 193:14-28.

Nanney, D. L. and Allen, S. L. 1959. Intranuclear co-ordination in *Tetrahymena*. Physiol. Zool. 32:221-229.

Nanney, D. L. and Caughey, P. A. 1953. Mating type determination in *Tetrahymena pyriformis*. Proc. Natl. Acad. Sci. U.S. 39:1057-1063.

Nanney, D. L. and Caughey, P. A. 1955. An unstable nuclear condition in *Tetrahymena pyriformis*. Genetics 40:388-398.

Nanney, D. L., Caughey, P. A., and Tefankjian, A. 1955. The genetic control of mating type potentialities in *Tetrahymena pyriformis*. Genetics 40:668-680.

Nanney, D. L. and Doerder, F. P. 1972. Transitory heterosis in numbers of basal bodies in *Tetrahymena pyriformis*. Genetics 72:227-238.

Nanney, D. L. and Dubert, J. M. 1960. The genetics of the H serotype system in variety 1 of *Tetrahymena pyriformis*. Genetics 45:1335-1349.

Nanney, D. L. and Nagel, M. J. 1964. Nuclear misbehavior in an aberrant inbred *Tetrahymena*. J. Protozool. 11:465-473.

Nanney, D. L., Nagel, M. J., and Touchberry, R. W. 1964. The timing of N antigenic differentiation in *Tetrahymena.* J. Exp. Zool. 155:25-42.

Nanney, D. L., Reeve, S. J., Nagel, M. J., and DePinto, S. 1963. H serotype differentiation in *Tetrahymena.* Genetics 48:803-813.

Nanney, D. L. and Rudzinska, M. A. 1960. Protozoa. *In* J. Brachet and A. E. Mirsky, eds., The cell, vol. 4, Academic Press, New York, pp. 109-150.

Nardone, R. M. and Blaszczynski, H. J. 1954. Growth effects induced in *Tetrahymena pyriformis* by streptomycin and its components. J. Exp. Zool. 125:119-125.

Nardone, R. M. and Wilber, C. G. 1950. Nitrogenous excretion in *Colpidium campylum.* Proc. Soc. Exp. Biol. Med. 75:559-561.

Nath, V. and Dutta, G. P. 1962. Cytochemistry of protozoa with special reference to the Golgi apparatus and the mitochondria. Int. Rev. Cytol. 13:323-366.

Nathan, H. A. and Friedman, W. 1962. Chlorpromazine affects permeability of resting cells of *Tetrahymena pyriformis.* Science 135:793-794.

Nelsen, E. M. 1970. Division delays and abnormal oral development produced by colchicine in *Tetrahymena.* J. Exp. Zool. 175:69-84.

Németh, G. 1972. Size and shape of cold- and heat-treated specimens of *Tetrahymena pyriformis* (GL). Acta Protozool. 9:323-327.

Németh, G. and Csík, L. 1961. Effect of penicillin on *Tetrahymena pyriformis,* strain GL. I. Multiplication and change of form. Acta Biol. Acad. Sci. Hung. 11:405-410.

Németh, G. and Csík, L. 1963. Effect of colchicine and ultraviolet irradiation on the size and shape of *Tetrahymena pyriformis* GL. Acta Biol. Acad. Sci. Hung. 13:217-222.

Nes, W. R., Malya, P. A. G., Mallory, F. B., Ferguson, K. A., Landrey, J. R., and Conner, R. L. 1971. Conformational analysis of the enzyme-substrate complex in the dehydrogenation of sterols by *Tetrahymena pyriformis.* J. Biol. Chem. 246:561-568.

Nettleton, R. M., Jr., Mefferd, R. B., Jr., and Loefer, J. B. 1953. Pattern formation in concentrated particulate suspensions. Amer. Nat. 87:117-118.

Newton, B. A. 1963. Protein and nucleic acid in protozoa (*Tetrahymena pyriformis*). Biochem. J. 89:87.

Nexø, B. A., Hamburger, K., and Zeuthen, E. 1972. Simplified microgasometry with gradient divers. C. R. Trav. Lab. Carlsberg 39:33-63.

Nicolette, J. A., Buhse, H. E., Jr., and Robin, M. S. 1971. The effect of 2-mercapto-1-(beta-4-pyridethyl) benzimidazole (MPB) on cell differentiation and RNA synthesis in the protozoon *Tetrahymena vorax.* J. Protozool. 18:87-89.

Nilsson, J. R. 1970. Macromolecular changes in *Tetrahymena pyriformis* GL during the cell cycle and in response to alterations in environmental condition. C. R. Trav. Lab. Carlsberg 37:285-300.

Nilsson, J. R. 1970. Suggestive structural evidence for macronuclear "subnuclei" in *Tetrahymena pyriformis* GL. J. Protozool. 17:539-548.

Nilsson, J. R. 1970. Cytolysomes in *Tetrahymena pyriformis* GL. I. Synchronized cells dividing in synchronized salt medium. C. R. Trav. Lab. Carlsberg 38:87-106.

Nilsson, J. R. 1970. Cytolysomes in *Tetrahymena pyriformis* GL. II. Reversible degeneration. C. R. Trav. Lab. Carlsberg 38:107-121.

Nilsson, J. R. 1972. Further studies on vacuole formation in *Tetrahymena pyriformis* GL. C. R. Trav. Lab. Carlsberg 39:83-110.

Nilsson, J. R. and Behnke, O. 1971. Studies on a surface coat of *Tetrahymena.* J. Ultrastruct. Res. 36:542-544.

Nilsson, J. R. and Leick, V. 1970. Nucleolar organization and ribosome formation in *Tetrahymena pyriformis* GL. Exp. Cell Res. 60:361-372.

Nilsson, J. R. and Williams, N. E. 1966. An electron microscope study of the oral apparatus of *Tetrahymena pyriformis.* C. R. Trav. Lab. Carlsberg 35:119-141.

Nishi, A. and Scherbaum, O. H. 1962. Levels of pyridine nucleotides and DPNH oxidase activity in growing and stationary protozoan cultures. Biochim. Biophys. Acta 65:411-418.

Nishi, A. and Scherbaum, O. H. 1962. Oxidative phosphorylation in synchronized cultures of *Tetrahymena pyriformis.* Biochim. Biophys. Acta 65:419-424.

Njiné, T. 1972. La transformation microstome-macrostome et macrostome-microstome chez *Tetrahymena paravorax* Corliss 1957. Ann. Fac. Sci. Cameroun, no. 10:69-84.

Noland, L. E. 1959. Ciliophora. *In* W. T. Edmondson, H. B. Ward, and G. C. Whipple, eds., Freshwater biology, 2nd ed., John Wiley, New York, pp. 265-297.

Noland, L. E. and Gojdics, M. 1967. Ecology of free-living protozoa. *In* T.-T. Chen, ed., Research in protozoology, vol. 2, Pergamon Press, New York, pp. 217-266.

Nordwig, A. 1969. Is collagen present in unicellular organisms? Hoppe-Seyler's Z. Physiol. Chem. 350:245-248.

Nozawa, K. 1944. Studies on the nucleo-plasmic ratio. III. The ratio in the ciliate *Glaucoma pyriformis*. Cytologia 13:360-368.

Nozawa, Y. and Thompson, G. A., Jr. 1971. Membrane formation in *Tetrahymena pyriformis*. II. Isolation and lipid analysis of cell fractions. J. Cell Biol. 49:712-721.

Nozawa, Y. and Thompson, G. A., Jr. 1971. Studies of membrane formation in *Tetrahymena pyriformis*. III. Lipid incorporation into various cellular membranes of logarithmic phase cultures. J. Cell Biol. 49:722-730.

Nozawa, Y. and Thompson, G. A., Jr. 1972. Studies of membrane formation in *Tetrahymena pyriformis*. V. Lipid incorporation into various cellular membranes of stationary phase cells, starving cells, and cells treated with metabolic inhibitors. Biochim. Biophys. Acta 282:93-104.

Nünn, J. F., Dixon, K. L., and Moore, J. R. 1968. Effect on halothane on *Tetrahymena pyriformis*. Brit. J. Anaesth. 40:145.

Organ, A. E., Bovee, E. C., and Jahn, T. L. 1972. The mechanism of the water explusion vesicle of the ciliate *Tetrahymena pyriformis*. J. Cell Biol. 55:644-652.

Orias, E. 1960. The genetic control of the lethal traits in variety 1, *Tetrahymena pyriformis*. J. Protozool. 7:64-69.

Orias, E. 1963. Mating type determination in variety 8, *Tetrahymena pyriformis*. Genetics 48:1509-1518.

Orias, E. and Rohlf, F. J. 1964. Population genetics of the mating type locus in *Tetrahymena pyriformis,* variety 8. Evolution 18:620-629.

Ormsbee, R. A. 1942. The normal growth and respiration of *Tetrahymena geleii*. Biol. Bull. 82:423-437.

Ormsbee, R. A. and Fisher, K. C. 1944. The effect of urethane on the consumption of oxygen and the rate of cell division in the ciliate *Tetrahymena geleii*. J. Gen. Physiol. 27:461-468.

Ottova, L. 1955. The dependence of the consumption of oxygen in the ciliate *Tetrahymena geleii* Furg. on some biological, chemical and physical factors. Věst. Čsl. Zool. Spol. 19:269-286.

Outka, D. E. 1961. Condition for mating and inheritance of mating type in variety seven of *Tetrahymena pyriformis*. J. Protozool. 8:179-184.

Pace, D. M. and Ireland, R. L. 1945. The effects of oxygen, carbon dioxide, and pressure on growth in *Chilomonas paramecium* and *Tetrahymena geleii* Furgason. J. Gen. Physiol. 28:547-557.

Pace, D. M. and Lyman, E. D. 1947. Oxygen consumption and carbon dioxide elimination in *Tetrahymena geleii* Furgason. Biol. Bull. 92:210-216.

Padilla, G. M. and Cameron, I. L. 1964. Synchronization of cell division in *Tetrahymena pyriformis* by a repetitive temperature cycle. J. Cell. Comp. Physiol. 64:303-307.

Padilla, G. M., Cameron, I. L., and Elrod, L. H. 1966. The physiology of repetitively synchronized *Tetrahymena*. *In* I. L. Cameron and G. M. Padilla, eds., Cell synchrony, Academic Press, New York, pp. 269-288.

Padilla, G. M. and James, T. W. 1964. Continuous synchronous cultures of protozoa. *In* D. M. Prescott, ed., Methods in cell physiology, vol. 1, Academic Press, New York, pp. 141-158.

Padilla, G. M. and Lane, N. M. 1971. Characteristics of macronuclear RNA in the cell cycle of synchronized *Tetrahymena*. *In* I. L. Cameron, G. M. Padilla, and A. M. Zimmerman, eds., Developmental aspects of the cell cycle, Academic Press, New York, pp. 23-39.

Painter, T. S. 1945. Chromatin diminution. Trans. Conn. Acad. Sci. 36:443-448.

Pannbacker, R. G. and Wright, B. E. 1967. Carbohydrate accumulation in the protist—A biochemical model of differentiation. *In* M. Florkin and B. T. Scheer, eds., Chemical zoology, vol. 1, Protozoa (G. W. Kidder, ed.), Academic Press, New York, pp. 573-616.

Park, O. 1929. The osmiophilic bodies of the protozoans, *Stentor* and *Leucophrys*. Trans. Amer. Microsc. Soc. 48:20-29.

Parks, R. E., Jr. 1955. Antimetabolite studies in *Tetrahymena* and tumors. *In* C. P. Rhoads, ed., Antimetabolites and cancer, Amer. Ass. Adv. Sci. Publ., Washington, D.C., pp. 175-197.

Parks, R. E., Jr., Kidder, G. W., and Dewey, V. C. 1952. Thiosemicarbazide toxicity in mice. Proc. Soc. Exp. Biol. Med. 79:287-289.

Parsons, J. A. 1965. Mitochondrial incorporation of tritiated thymidine in *Tetrahymena pyriformis*. J. Cell Biol. 25:641-646.

Parsons, J. A. and Rustad, R. C. 1968. The distribution of DNA among dividing mitochondria of *Tetrahymena pyriformis*. J. Cell Biol. 37:683-693.

Pastor, E. P. and Fennell, R. A. 1959. Some observations on the esterases of *Tetrahymena pyriformis* W. II. Some factors affecting aliesterase and cholinesterase activity. J. Morphol. 104:143-157.

Pearlman, R. E. and Westergaard, O. 1969. DNA polymerase activity from exponentially multiplying and division-synchronized *Tetrahymena*. C. R. Trav. Lab. Carlsberg 37:77-86.

Perlman, B. S. 1973. Basal body addition in ciliary rows of *Tetrahymena phyriformis*. J. Exptl. Zool. 184:365-368.

Penard, E. 1922. Études sur les infusoires d'eau douce. Georg, Geneva. 331 pp.

Peng, Y. M. and Elson, C. E. 1971. Effect of iron on lipid metabolism of *Tetrahymena pyriformis*. J. Nutr. 101:1177-1184.

Perlman, B. S. 1973. Temperature effects on maturity periods in *Tetrahymena pyriformis* syngen 1. J. Protozool. 20:106-107.

Peters, R. A. 1920. Nutrition of protozoa. 1. The growth of *Paramecium* in sterile culture medium. J. Physiol. 53:53-60.

Peters, R. A. 1920. Nutrition of protozoa. 2. The carbon and nitrogen compounds needed for *Paramecium* (*Colpidium*). J. Physiol. 54:260-266.

Peters, R. A. 1921. The substances needed for the growth of a pure culture of *Colpidium colpoda*. J. Physiol. 55:1-32.

Peters, R. A. 1927. Observations upon the oxygen consumption of *Colpidium colpoda*. J. Physiol. 69:ii-iii.

Peterson, R. E. 1942. Essential factors for the growth of the ciliate protozoan, *Colpidium campylum*. J. Biol. Chem. 146:537-545.

Phelps, A. 1935. Growth of protozoa in pure culture. I. Effect upon the growth curve of the age of the inoculum and of the amount of the inoculum. J. Exp. Zool. 70:109-130.

Phelps, A. 1936. Growth of protozoa in pure culture. II. Effect upon the growth curve of different concentrations of nutrient materials. J. Exp. Zool. 72:479-496.

Phelps, A. 1946. Growth of protozoa in pure culture. III. Effect of temperature upon the division rate. J. Exp. Zool. 102:277-292.

Phelps, A. 1949. Thermal adaptations in two strains of the ciliate, *Tetrahymena geleii*. Bull. Ecol. Soc. Amer. 30(4):50-59.

Phelps, A. 1959. Effect of visible light on the growth of *Tetrahymena pyriformis*. Ecology 40:512-513.

Phillips, R. B. 1967. Inheritance of T serotypes in *Tetrahymena*. Genetics 56:667-681.

Phillips, R. B. 1967. Differentiation of T serotype in *Tetrahymena*. Genetics 56:683-692.

Phillips, R. B. 1968. Mating-type alleles in Illinois strains of *Tetrahymena pyriformis*, syngen 1. Genet. Res. Camb. 11:211-214.

Phillips, R. B. 1969. Mating type inheritance in syngen 7 of *Tetrahymena pyriformis*: Intra- and interallelic interactions. Genetics 63:349-359.

Phillips, R. B. 1971. Inheritance of immobilization antigens in syngen 7 of *Tetrahymena pyriformis*: Evidence for a regulatory gene. Genetics 67:391-398.

Phillips, R. B. 1971. Induction of competence for mating in *Tetrahymena* by cell-free fluids. J. Protozool. 18:163-165.

Phillips, R. B. 1972. Similar times of differentiation of acid phosphatase heterozygotes in two syngens of *Tetrahymena pyriformis*. Develop. Biol. 29:65-72.

Pilcher, H. L. and Williams, H. H. 1954. Microbiological evolution of protein quality. II. Studies of the responses of *Tetrahymena pyriformis* W to intact proteins. J. Nutr. 53:589-599.

Pitelka, D. R. 1961. Fine structure of the silverline and fibrillar system of three tetrahymenid ciliates. J. Protozool. 8:75-89.

Pitelka, D. R. 1963. Electron-microscopic structure of protozoa. Pergamon Press, New York. 269 pp.

Pitelka, D. R. 1969. Fibrillar systems in protozoa. *In* T.-T. Chen, ed., Research in protozoology, vol. 3, Pergamon Press, New York, pp. 279-388.

Pitelka, D. R. and Child, F. M. 1964. The locomotor apparatus of ciliates and flagellates: Relations between structure and function. *In* S. H. Hutner, ed., Biochemistry and physiology of protozoa, vol. 3, Academic Press, New York, pp. 131-198.

Pitts, R. F. 1932. Effect of cyanide on respiration of the protozoan, *Colpidium campylum*. Proc. Soc. Exp. Biol. Med. 29:542-544.

Platt, J. R. 1961. "Bioconvection patterns" in cultures of free-swimming organisms. Science 133:1766-1767.

Plesner, P. 1958. The nucleoside triphosphate content of *Tetrahymena piriformis* during the division cycle in synchronously dividing mass culture. Biochim. Biophys. Acta 29:462-463.

Plesner, P. 1961. Changes in ribosome structure and function during synchronized cell division. Cold Spring Harbor Symp. Quant. Biol. 26:159-162.

Plesner, P. 1964. Nucleotide metabolism during synchronized cell division in *Tetrahymena pyriformis*. C. R. Trav. Lab. Carlsberg 34:1-76.

Plesner, P., Rasmussen, L., and Zeuthen, E. 1964. Technique used in the study of synchronous *Tetrahymena*. *In* E. Zeuthen, ed., Synchrony in cell division and growth, John Wiley, New York, pp. 543-563.

Poljansky, G. I. and Cheissin, E. M. 1965. Dogiel's General protozoology, rev. ed. Oxford University Press, London. 747 pp.

Pollard, W. O., Shorb, M. S., Lund, P., and Vasaitis, V. 1964. Effect of triparanol on synthesis of fatty acids by *Tetrahymena pyriformis*. Proc. Soc. Exp. Biol. Med. 116:539-543.

Poole, R. K., Nicholl, W. G., Howells, L., and Lloyd, D. 1971. The microsomal fraction from *Tetrahymena pyriformis* strain St: Characterization and subfractionation. J. Gen. Microbiol. 68:283-294.

Porter, P., Blum, J. J., and Elrod, H. 1972. Subcellular distribution of aspartate transaminase, alanine aminotransferase, glutamate dehydrogenase, and lactate dehydrogenase in *Tetrahymena*. J. Protozool. 19:375-378.

Preer, J. R., Jr. 1969. Genetics of the protozoa. *In* T.-T. Chen, ed., Research in protozoology, vol. 3, Pergamon Press, New York, pp. 129-278.

Prescott, D. M. 1956. Mass and clone culturing of *A. proteus* and *Chaos chaos*. C. R. Trav. Lab. Carlsberg 30:1-12.

Prescott, D. M. 1957. Change in the physiological state of a cell population as a function of culture growth and age (*Tetrahymena geleii*). Exp. Cell Res. 12:126-134.

Prescott, D. M. 1957. Relation between multiplication rate and temperature in *Tetrahymena pyriformis* strains HS and GL. J. Protozool. 4:252-256.

Prescott, D. M. 1958. The growth rate of *Tetrahymena geleii* HS under optimal conditions. Physiol. Zool. 31:111-117.

Prescott, D. M. 1959. Variations in the individual generation times of *Tetrahymena geleii* HS. Exp. Cell Res. 16:279-284.

Prescott, D. M. 1960. Relation between cell growth and cell division. IV. The synthesis of DNA, RNA and protein from division to division in *Tetrahymena*. Exp. Cell Res. 19:228-238.

Prescott, D. M. 1961. RNA synthesis in the nucleus and RNA transfer to the cytoplasm in *Tetrahymena pyriformis*. *In* T. W. Goodwin and O. Lindberg, eds., Biological structure and function,

Proceedings of the 1st IUB/IUBS International Symposium vol. 2, Academic Press, New York, New York, pp. 527-536.

Prescott, D. M. 1962. Synthetic processes in the cell nucleus. II. Nucleic acid and protein metabolism in the macronuclei of two ciliated protozoa. J. Histochem. Cytochem. 10:145-174.

Prescott, D. M. 1964. The normal cell cycle. *In* E. Zeuthen, ed., Synchrony in cell division and growth, John Wiley, New York, pp. 71-97.

Prescott, D. M. 1964. Cellular sites of RNA synthesis. Progr. Nucleic Acid Res. Mol. Biol. 33:33-58.

Prescott, D. M., Bollum, F. J., and Kluss, B. C. 1962. Is DNA polymerase a cytoplasmic enzyme? J. Cell Biol. 13:172-174.

Prescott, D. M., Bostock, C., Gamow, E., and Lauth, M. 1971. Characterization of rapidly labeled RNA in *Tetrahymena pyriformis*. Exp. Cell Res. 67:124-128.

Prescott, D. M. and Carrier, R. F. 1964. Experimental procedures and culture methods for *Euplotes eurystomus* and *Amoeba proteus*. *In* D. M. Prescott, ed., Methods of cell physiology, vol. 1, Academic Press, New York, pp. 85-95.

Prescott, D. M. and James, T. W. 1955. Culturing of *Amoeba proteus* on *Tetrahymena*. Exp. Cell Res. 8:256-258.

Prescott, D. M., Rao, M. V. N., Evenson, D. P., Stone, G. E., and Thrasher, J. D. 1966. Isolation of single nuclei and mass preparations of nuclei from several cell types. *In* D. M. Prescott, ed., Methods in cell physiology, vol. 2, Academic Press, New York, pp. 131-142.

Prescott, D. M. and Stone, G. E. 1967. Replication and function of the protozoan nucleus. *In* T.-T. Chen, ed., Research in protozoology, vol. 2, Pergamon Press, New York, pp. 117-146.

Price, K. E., Buck, R. E., Schlein, A., and Siminoff, P. 1962. A comparison of the *in vitro* susceptibility of HeLa and protozoan cells to antitumor antibiotics. Cancer Res. 22:885-891.

Provasoli, L., Balamuth, W., Becker, E. R., Corliss, J. O., Hall, R. P., Holz, G. G., Jr., Levine, N. D., Manwell, R. D., Nigrelli, R. F., Starr, R. C., and Trager, W. 1958. A catalogue of laboratory strains of free-living and parasitic protozoa (with sources from which they may be obtained and directions for their maintenance). J. Protozool. 5:1-38.

Prowazek, S. von 1909. Duplicidade morfolojica nos infuzorios ciliados (Formdimorphismus bei ciliaten Infusorien). Mem. Inst. Osw. Cruz 1:105-108.

Pruett, P. O., Conner, R. L., and Pruett, J. R. 1967. Orthophosphate flux across the membrane of *Tetrahymena pyriformis* W. J. Cell. Physiol. 70:217-224.

Pruthi, H. S. 1927. On the hydrogen-ion concentration of hay infusions, with special reference to its influence on the protozoan sequence. Brit. J. Exp. Biol. 4:292-300.

Przybylski, R. J. 1961. Electron microscope autoradiography. Exp. Cell Res. 24:181-184.

Pyne, C. 1968. Sur l'absence d'incorporation de la thymidine tritiée dans les cinétosomes de *Tetrahymena pyriformis* (cilié holotriche). C. R. Acad. Sci. Paris 267:755-757.

Raabe, Z. 1964. Zarys protozoologii. Państwowe Wydawnictwo Naukowe, Warsaw. 283 pp. In Polish.

Raff, E. C. and Blum, J. J. 1966. The effects of adenosine triphosphate and related compounds on some hydrodynamic properties of glycerinated cilia. J. Cell Biol. 31:445-453.

Raff, E. C. and Blum, J. J. 1969. Some properties of a model assay for cilia contractility. J. Cell Biol. 42:831-834.

Rahat, M., Judd, J., and van Eys, J. 1964. Oxidation of reduced diphosphopyridine nucleotide by *Tetrahymena pyriformis* preparations. J. Biol. Chem. 239:3537-3545.

Rahn, O. 1931. The order of death of organisms larger than bacteria. J. Gen. Physiol. 14:315-337.

Raikov, I. B. 1969. The macronucleus of ciliates. *In* T.-T. Chen, ed., Research in protozoology, vol. 3, Pergamon Press, New York, pp. 4-128.

Raikov, I. B. 1972. Nuclear phenomena during conjugation and autogamy in ciliates. *In* T.-T. Chen, ed., Research in protozoology, vol. 4, Pergamon Press, New York, pp. 147-289.

Rampton, V. W. 1962. Kinetosomes of *Tetrahymena*. Nature 195:195.

Randall, J. T. and Disbrey, C. 1965. Evidence for the presence of DNA at basal body sites in *Tetrahymena pyriformis*. Proc. Roy. Soc., B. 162:473-491.

Rannestad, J. and Williams, N. E. 1971. The synthesis of microtubules and other proteins of the oral apparatus in *Tetrahymena pyriformis*. J. Cell Biol. 50:709-720.

Rasmussen, L. 1963. Delayed divisions in *Tetrahymena* as induced by short-time exposures to anaerobiosis. C. R. Trav. Lab. Carlsberg 33:53-71.

Rasmussen, L. and Kludt, T. 1970. Particulate material as a prerequisite for rapid cell multiplication in *Tetrahymena* cultures. Exp. Cell Res. 59:457-463.

Rasmussen, L. and Modeweg-Hansen, L. 1973. Cell multiplication in *Tetrahymena* cultures after addition of particulate material. J. Cell Sci. 12:275-286.

Rasmussen, L. and Zeuthen, E. 1962. Cell division and protein synthesis in *Tetrahymena* as studied with *p*-fluorophenylalanine. C. R. Trav. Lab. Carlsberg 32:333-358.

Rasmussen, L. and Zeuthen, E. 1966. Cell division in *Tetrahymena* adapting to *p*-DL-fluorophenylalanine. Exp. Cell Res. 41:462-465.

Rasmussen, L. and Zeuthen, E. 1966. Amino acid antagonisms in *Tetrahymena*. C. R. Trav. Lab. Carlsberg 35:85-100.

Ray, C., Jr. 1956. Meiosis and nuclear behavior in *Tetrahymena pyriformis*. J. Protozool. 3:88-96.

Ray, C., Jr. 1956. Preparation of chromosomes of *Tetrahymena pyriformis* for photomicrography. Stain Technol. 31:271-274.

Ray, C., Jr., and Coleman, M. T. 1963. Detection of surface-antigens in *Tetrahymena pyriformis* by fluorescent antibodies. *In* J. Ludvík, J. Lom, and J. Vávra, eds., Progress in protozoology, Proceedings of the 1st International Congress on Protozoology, Prague, August 1961, Nakladatelstvi Československe Akademie, Prague, pp. 159-161.

Ray, H. and Frankel, J. 1969. Macromolecular synthesis, differentiation and cell division in *Tetrahymena pyriformis* mating type 1, variety 1. J. Cell. Physiol. 74:123-134.

Read, S. A. and Buetow, D. E. 1971. Non-continuous synthesis of DNA in *Tetrahymena pyriformis* W. Exp. Cell Res. 64:239-240.

Reeves, H., Papa, M., Seaman, G. R., and Ajil, S. 1961. Malate synthesis and isocitritase in *Tetrahymena pyriformis*. J. Bacteriol. 81:154-155.

Reid, R., Cox, D., Baker, H., and Frank, O. 1969. Phytosterols and other lipids as survival factors for *Tetrahymena* at 0-5°C. J. Protozool. 16:231-235.

Renaud, F. L., Rowe, A. J., and Gibbons, I. R. 1968. Some properties of the protein forming the outer fibers of cilia. J. Cell Biol. 36:79-90.

Rendina, G. and Coon, M. J. 1957. Enzymatic hydrolysis of the coenzyme A thiol esters of β-hydroxypropionic and β-hydroxyisobutyric acids. J. Biol. Chem. 225:523-534.

Rene, A. A. and Nardone, R. M. 1961. Protective and chemoresuscitative study of ultra-violet exposed *Tetrahymena pyriformis* strain WH. Exp. Cell Res. 23:549-554.

Reynolds, B. L. 1962. The modes of action of antibiotics. Austral. J. Sci. 25:243-249.

Reynolds, H. 1964. Potential analytical applications of *Tetrahymena pyriformis*. J. Wash. Acad. Sci. 54:99-108.

Reynolds, H. 1969. An apparent carbohydrate-amino acid interaction in *Tetrahymena pyriformis*. J. Protozool. 16:204-210.

Reynolds, H. 1970. Effect of type of carbohydrate on amino acid accumulation and utilization by *Tetrahymena*. J. Bacteriol. 104:719-725.

Reynolds, H. and Wragg, J. B. 1962. Effect of type of carbohydrate on growth and protein synthesis by *Tetrahymena pyriformis*. J. Protozool. 9:214-222.

Richards, O. W. 1941. The growth of the protozoa. *In* G. N. Calkins and F. M. Summers, eds., Protozoa in biological research, Columbia University Press, New York, pp. 517-564.

Ricketts, T. R. 1970. Effect on endocytosis upon acid phosphatase activity of *Tetrahymena pyriformis*. Protoplasma 71:127-137.

Ricketts, T. R. 1971. Endocytosis in *Tetrahymena pyriformis*. Selectivity of uptake of particles and the adaptive increase in cellular acid phosphatase activity. Exp. Cell Res. 66:49-58.

Ricketts, T. R. 1972. The interaction of particulate material and dissolved foodstuffs in food uptake by *Tetrahymena pyriformis*. Arch. Mikrobiol. 81:344-349.

Ricketts, T. R. 1972. The induction of endocytosis in starved *Tetrahymena pyriformis*. J. Protozool. 19:373-375.

Rifkin, J. L. 1973. The role of the contractile vacuole in the osmoregulation of *Tetrahymena pyriformis.* J. Protozool. 20:108-114.

Ringertz, N. R., Bolund, L., and Debault, L. E. 1967. Isolation and chemical composition of macronuclei from *Tetrahymena pyriformis.* Exp. Cell Res. 45:519-532.

Ris, H., Tolmach, L. J., Lajtha, L. G., Smith, C. L., Das, N. K., and Zeuthen, E. 1963. Differential sensitivity of the cell life cycle. J. Cell. Comp. Physiol., Suppl. 1, 62:141-156.

Risse, H. S. and Blum, J. J. 1972. Pathways of carbohydrate metabolism in *Tetrahymena.* Subcellular distribution and properties of hexokinase, aldolase and α-glycerophosphate dehydrogenase. Arch. Biochem. Biophys. 149:329-335.

Roberts, E., Simonsen, D. G., Horiguchi, M., and Kittredge, J. S. 1968. Transamination of aminoalkylphosphonic acids with alpha ketoglutarate. Science 159:886-888.

Robertson, M. 1939. A study of the reactions *in vitro* of certain ciliates belonging to the *Glaucoma-Colpidium* group to antibodies in the sera of rabbits immunized therewith. J. Pathol. Bacteriol. 48:305-322.

Robertson, M. 1939. An analysis of some of the antigenic properties of certain ciliates belonging to the *Glaucoma-Colpidium* group as shown in their response to immune serum. J. Pathol. Bacteriol. 48:323-338.

Robin, Y. and Viala, B. 1966. Sur la presence d'ATP: Arginine phosphotransferase chez *Tetrahymena pyriformis* W Cambridge. Comp. Biochem. Physiol. 18:405-413.

Robinson, W. G. and Coon, M. J. 1957. The purification and properties of β-hydroxyisobutyric dehydrogenase. J. Biol. Chem. 225:511-521.

Rockland, L. B. and Dunn, M. S. 1946. The microbiological determination of tryptophan in unhyrolyzed casein with *Tetrahymena geleii* H. Arch. Biochem. 11:541-543.

Rockland, L. B. and Dunn, M. S. 1949. Determination of the biological value of proteins with *Tetrahymena geleii.* Food Technol. 3:289-292.

Rockland, L. B. and Dunn, M. S. 1949. Growth studies on *Tetrahymena geleii.* J. Biol. Chem. 179:511-521.

Rogers, C. G. 1966. Effects of phenothiazines on growth, glucose uptake, and cell composition in *Tetrahymena pyriformis* W. Can. J. Biochem. 44:1493-1503.

Rogers, C. G. 1968. Uptake by ^{32}P-orthophosphate and incorporation into phospholipids in *Tetrahymena pyriformis* W exposed to phenothiazine derivatives. Can. J. Biochem. 46:331-340.

Rohdenburg, G. L. and Nagy, S. M. 1937. Growth stimulating and inhibiting substances in human urine. Amer. J. Cancer 29:66-67.

Rohdenburg, G. L. and Nagy, S. M. 1937. Cell division stimulating and inhibiting substances in tissues. Amer. J. Cancer 30:335-340.

Rohdenburg, G. L. and Nagy, S. M. 1937. Some further physiological effects of cell division stimulants and inhibitors. Amer. J. Cancer 30:512-516.

Ron, A. and Guttman, R. 1961. The effect of kinetin on synchronous cell division in *Tetrahymena pyriformis.* Exp. Cell Res. 25:176-178.

Rooney, D. W. and Eiler, J. J. 1967. Synchronization of *Tetrahymena* cell division by multiple hypoxic shocks. Exp. Cell Res. 48:649-652.

Rooney, D. W. and Eiler, J. J. 1969. Hypoxic synchronization of *Tetrahymena* cell division with an automatic apparatus. Exp. Cell Res. 54:49-52.

Rooney, D. W. and Eiler, J. J. 1969. Effects of division-synchronizing hypoxic and hyperthermic shocks upon *Tetrahymena* respiration and intracellular ATP concentration. J. Cell Biol. 41:145-153.

Roop, W. F., Tan, S. A., and Roop, B. L. 1971. Detection of aminoalkylphosphonates with a ninhydrin-molybdate chromogenic system. Anal. Biochem. 44:77-80.

Roque, M., de Puytorac, P., and Savoie, A. 1971. Caractéristiques morphologiques et biologiques de *Tetrahymena bergeri* sp. nov., cilié hyménostome tetrahyménien. Protistologica 6(1970):343-351.

Rosen, G. D. and Fernell, W. R. 1956. Microbiological evaluation of protein quality with *Tetrahymena pyriformis* W. II. Relative nutritive values of proteins in foodstuffs. Brit. J. Nutr. 10:156-169.

Rosenbaum, J. L. and Carlson, K. 1969. Cilia regeneration in *Tetrahymena* and its inhibition by colchicine. J. Cell Biol. 40:415-425.

Rosenbaum, J. L. and Holz, G. G., Jr. 1966. Amino acid activation in subcellular fractions of *Tetrahymena pyriformis*. J. Protozool. 13:115-123.

Rosenbaum, N., Erwin, J., Beach, D., and Holz, G. G., Jr. 1966. The induction of a phospholipid requirement and morphological abnormalities in *Tetrahymena pyriformis* by growth at supraoptimal temperature. J. Protozool. 13:535-546.

Rosenberg, H. 1964. Distribution and fate of 2-aminoethylphosphonic acid in *Tetrahymena*. Nature 203:299-300.

Rosenberg, H. 1966. The isolation and identification of "volutin" granules from *Tetrahymena*. Exp. Cell Res. 41:397-410.

Rosenberg, H. and Munk, N. 1969. Transport phenomena associated with the disappearance of pyrophosphate granules in *Tetrahymena pyriformis*. Biochim. Biophys. Acta 184:191-197.

Roth, J. S. 1954. Certain effects of 2-aminofluorene and Cl- and β-napthylamines on *Tetrahymena pyriformis*. Cancer Res. 14:346-351.

Roth, J. S. 1954. A possible function of intracellular ribonuclease. Nature 174:129.

Roth, J. S. 1956. Studies on the function of intracellular ribonuclease. II. The action of cobalt and nickel on *Tetrahymena pyriformis* W. Exp. Cell Res. 10:146-154.

Roth, J. S. 1959. Comparative studies on tissue ribonucleases. Ann. N.Y. Acad. Sci. 81:611-617.

Roth, J. S. 1962. Biochemical studies on irradiated protozoa. I. Effect of metabolites on the respiration of x-irradiated *Tetrahymena pyriformis*. J. Protozool. 9:142-146.

Roth, J. S. and Buccino, G. 1963. Biochemical studies on irradiated protozoa. III. Catalase activity in *Tetrahymena pyriformis* W. J. Protozool. 12:432-438.

Roth, J. S. and Eichel, H. J. 1955. The effect of x-radiation on enzyme systems of *Tetrahymena pyriformis*. Biol. Bull. 108:308-317.

Roth, J. S. and Eichel, H. J. 1961. Studies on the metabolism of L-phenylalanine by *Tetrahymena pyriformis* W. J. Protozool. 8:69-71.

Roth, J. S., Eichel, H. J., and Ginter, E. 1954. The oxidation of amino acids by *Tetrahymena pyriformis* W. Arch. Biochem. Biophys. 48:112-119.

Roth, L. E. and Minick, O. T. 1961. Electron microscopy of nuclear and cytoplasmic events during division of *Tetrahymena pyriformis* strains W and HAM 3. J. Protozool. 8:12-21.

Roux, J. 1899. Observations sur quelques infusoires ciliés des environs de Genève avec la description de nouvelles espèces. Rev. Suisse Zool. 6:557-636.

Roux, J. 1901. Faune infusorienne des eaux stagnantes des environs de Genève. Kündig, Geneva. 148 pp.

Rudick, M. J. and Cameron, I. L. 1972. Regulation of DNA synthesis and cell division in starved-refed synchronized *Tetrahymena pyriformis* HSM. Exp. Cell Res. 70:411-416.

Rudzinska, M. A. 1951. An abnormal type of development in the life cycle of a suctorian, *Tokophrya infusionum*. Trans. Amer. Microsc. Soc. 70:168-172.

Rudzinska, M. A. 1951. The influence of amount of food on the reproduction rate and longevity of a suctorian (*Tokophrya infusionum*). Science 113:10-11.

Rudzinska, M. A. 1952. Overfeeding and life span in *Tokophrya infusionum*. J. Gerontol. 7:544-548.

Rudzinska, M. A. 1953. Giant individuals and vigor of populations in *Tokophrya infusionum*. Ann. N.Y. Acad. Sci. 56:1087-1090.

Rudzinska, M. A. 1955. A simple method for paraffin and plastic embedding of protozoa. J. Protozool. 2:188-198.

Rudzinska, M. A. 1962. The use of a protozoan for studies on aging. III. Similarities between young and overfed and old normally fed *Tokophrya infusionum*: A light and electron microscope study. Gerontologia 6:206-226.

Rudzinska, M. A. 1970. The mechanism of food intake in *Tokophrya infusionum* and ultrastructural changes in food vacuoles during digestion. J. Protozool. 17:626-641.

Rudzinska, M. A. 1973. Do suctoria really feed by suction? BioScience 23:87-94.

Rudzinska, M. A. and Chambers, R. 1951. The activity of the contractile vacuole in a suctorian (*Tokophrya infusionum*). Biol. Bull. 100:49-58.

Rudzinska, M. A. and Granick, S. 1953. Protoporphyrin production of *Tetrahymena geleii.* Proc. Soc. Exp. Biol. Med. 83:525-526.

Rudzinska, M. A. and Porter, K. R. 1954. The fine structure of *Tokophrya infusionum* with emphasis on the feeding mechanism. Trans. N.Y. Acad. Sci. 16:408-411.

Ruffner, B., Jr., and Anderson, E. P. 1969. Adenosine triphosphate, uridine monophosphate, cytidine monophosphate phosphotransferase from *Tetrahymena pyriformis.* J. Biol. Chem. 244:5994-6002.

Rutter, W. J. and Groves, W. E. 1964. Coherence and variation in macromolecular structures in phylogeny. *In* C. A. Leone, ed., Taxonomic biochemistry and serology, Ronald Press, New York, pp. 417-434.

Ryley, J. F. 1952. Studies on the metabolism of the protozoa. 3. Metabolism of the ciliate *Tetrahymena pyriformis (Glaucoma pyriformis).* Biochem. J. 52:483-492.

Ryley, J. F. 1967. Carbohydrates and respiration. *In* M. Florkin and B. T. Scheer, eds., Chemical zoology, vol. 1, Protozoa (G. W. Kidder, ed.), Academic Press, New York, pp. 55-92.

Sanders, M. and Nathan, H. A. 1959. Protozoa as pharmacological tools: The antihistamines. J. Gen. Microbiol. 21:264-270.

Sandon, H. 1927. The composition and distribution of the protozoan fauna of the soil. Oliver and Boyd, Edinburgh. 237 pp.

Satir, B. 1967. Effect of actinomycin D on cultural growth phases and on the pattern of total RNA synthesis in *Tetrahymena.* Exp. Cell Res. 48:253-262.

Satir, B. and Dirksen, E. R. 1971. Nucleolar aging in *Tetrahymena* during the cultural growth cycle. J. Cell Biol. 48:143-154.

Satir, B. and Rosenbaum, J. L. 1965. The isolation and identification of kinetosome-rich fractions from *Tetrahymena pyriformis.* J. Protozool. 12:397-405.

Satir, B., Schooley, C., and Satir, P. 1973. Membrane fusion in a model system: Mucocyst secretion in *Tetrahymena.* J. Cell Biol. 56:153-176.

Satir, B. and Zeuthen, E. 1961. Cell cycle and the relationship of growth rate to reduced weight (RW) in the giant amoeba *Chaos chaos* L. C. R. Trav. Lab. Carlsberg 32:241-264.

Satir, P. and Satir, B. 1964. A model for ninefold symmetry in α-keratin and cilia. J. Theoret. Biol. 7:123-128.

Sato, H. and Saito, M. 1959. Morphological study on the macronuclear structure in interphase and pre-fission stage of *Tetrahymena geleii* W. Zool. Mag. (Tokyo) 68:209-214. In Japanese with English summary.

Sauberlich, H. E. and Baumen, C. A. 1948. A factor required for the growth of *Leuconostoe citrovorum.* J. Biol. Chem. 176:165.

Savoie, A. 1968. Les ciliés histophages en biologie cellulaire. Ann. Univ. Ferrara (N. S., Sect. III) 3:65-71.

Schensted, I. V. 1958. Model of subnuclear segregation in the macronucleus of ciliates. Amer. Nat. 92:161-170.

Scherbaum, O. H. 1956. Cell growth in normal and synchronously dividing mass culture of *Tetrahymena pyriformis.* Exp. Cell Res. 11:464-476.

Scherbaum, O. H. 1957. The application of a standard counting method in estimation of growth in normal and heat-treated cultures of *Tetrahymena pyriformis.* Acta Pathol. Microbiol. Scand. 40:7-12.

Scherbaum, O. H. 1957. Studies on the mechanism of synchronous cell division in *Tetrahymena pyriformis.* Exp. Cell Res. 13:11-23.

Scherbaum, O. H. 1957. The content and composition of nucleic acid in normal and synchronously dividing mass cultures of *Tetrahymena pyriformis.* Exp. Cell Res. 13:24-30.

Scherbaum, O. H. 1957. The division index and multiplication in a mass culture of *Tetrahymena* following inoculation. J. Protozool. 4:257-259.

Scherbaum, O. H. 1960. Possible sites of metabolic control during the induction of synchronous cell division. Ann. N.Y. Acad. Sci. 90:565-579.

Scherbaum, O. H. 1960. Synchronous division of micro-organisms. Ann. Rev. Microbiol. 14:283-310.

Scherbaum, O. H. 1962. A comparison of synchronized cell division in protozoa. J. Protozool. 9:61-64.

Scherbaum, O. H. 1963. Chemical prerequisities for cell division. *In* L. Levine, ed., The cell in mitosis, Academic Press, New York, pp. 125-157.

Scherbaum, O. H. 1964. Biochemical studies on synchronized *Tetrahymena*. *In* E. Zeuthen, ed., Synchrony in cell division and growth, John Wiley, New York, pp. 185-195.

Scherbaum, O. H. 1964. Comparison of synchronous and synchronized cell division. Exp. Cell Res. 33:89-98.

Scherbaum, O. H., Chou, S.-C., Seraydarian, K. H., and Byfield, J. E. 1962. The effect of temperature shifts on the intracellular level of nucleoside triphosphates in *Tetrahymena pyriformis*. Can. J. Microbiol. 8:753-760.

Scherbaum, O. H. and Jahn, T. L. 1964. Synchronization of *Tetrahymena* in large mass cultures. Exp. Cell Res. 33:99-104.

Scherbaum, O. H., James, T. W., and Jahn, T. L. 1959. The amino acid composition in relation to cell growth and cell division in synchronized culture of *Tetrahymena pyriformis*. J. Cell. Comp. Physiol. 53:110-137.

Scherbaum, O. H. and Levy, M. 1961. Some aspects of the carbohydrate metabolism in relation to cell growth and cell division. Pathol.-Biol. 9:514-517.

Scherbaum, O. H. and Loefer, J. B. 1964. Environmentally induced growth oscillations in protozoa. *In* S. H. Hutner, ed., Biochemistry and physiology of protozoa, vol. 3, Academic Press, New York, pp. 9-59.

Scherbaum, O. H., Louderback, A. L., and Jahn, T. L. 1958. The formation of subnuclear aggregates in normal and synchronized protozoan cells. Biol. Bull. 115:269-275.

Scherbaum, O. H., Louderback, A. L., and Jahn, T. L. 1959. DNA synthesis, phosphate content and growth in mass and volume in synchronously dividing cells. Exp. Cell Res. 18:150-166.

Scherbaum, O. H. and Rasch, G. 1957. Cell size distribution and single cell growth in *Tetrahymena pyriformis* GL. Acta Pathol. Microbiol. Scand. 51:161-182.

Scherbaum, O. H. and Zeuthen, E. 1954. Induction of synchronous cell division in mass cultures of *Tetrahymena pyriformis*. Exp. Cell Res. 6:221-227.

Scherbaum, O. H. and Zeuthen, E. 1955. Temperature-induced synchronous divisions in the ciliate protozoon *Tetrahymena pyriformis* growing in synthetic and proteose-peptone media. Exp. Cell Res., Suppl. 3:312-325.

Schewiakoff, W. 1889. Beiträge zur Kenntniss der holotrichen Ciliaten. Biblioth. Zool. 5:1-77.

Schewiakoff, W. 1896. The organization and systematics of the Infusoria Aspirotricha (Holotricha *auctorum*). Mém. Acad. Imp. Sci. St. Pétersbourg (sér. 8) 4:1-395. In Russian.

Schildkraut, J. L., Mandel, M., Levisohn, S., Smith-Sonneborn, J. E., and Marmur, J. 1962. Dexoyribonucleic acid base composition and taxonomy of some protozoa. Nature 196:795-796.

Schleicher, J. d'A. 1959. Paper chromatography analyses of amino acids in protozoa: Some aspects of the metabolism of aspartic acid. Catholic University of America Biology Studies, no. 53:1-73.

Schmid, P. 1967. Temperature adaptation of the growth and division process of *Tetrahymena pyriformis*. I. Adaptive phase. Exp. Cell Res. 45:460-470.

Schmid, P. 1967. Temperature adaptation of the growth and division process of *Tetrahymena pyriformis*. II. Relationship between cell growth and cell replication. Exp. Cell Res. 45:471-486.

Schneiderman, H. A., Gilbert, L. I., and Weinstein, M. J. 1960. Juvenile hormone activity in microorganisms and plants. Nature 188:1041-1042.

Schnitzer, R. J. 1955. Mechanism of the origin of resistance to drugs in protozoa. *In* M. G. Sevag, R. D. Reid, and O. E. Reynolds, eds., Origins of resistance to toxic agents, Academic Press, New York, pp. 16-81.

Schoenborn, H. W. 1949. Growth of protozoa in glassware cleaned with sulphuric acid-potassium bichromate solution. Physiol. Zool. 22:30-35.

Schouten, S. L. 1927. On digestion in protozoa. K. Akad. Wet. Amst. Proc., Sect. Sci. 30:619-623.

Schuster, G. L. and Vennes, J. W. 1960. Synchronous division of *Tetrahymena pyriformis* in a biphasic medium. Proc. N. Dak. Acad. Sci. 14:43-48.

Seaman, G. R. 1947. Penicillin as an agent for sterilization of protozoan cultures. Science. 106:327.

Seaman, G. R. 1949. The presence of the tricarboxylic acid cycle in the ciliate *Colpidium campylum.* Biol. Bull. 96:257-262.

Seaman, G. R. 1949. Relation between growth and lipid synthesis in *Colpidium campylum.* J. Cell. Comp. Physiol. 33:137-143.

Seaman, G. R. 1949. Cytochrome C, diphosphopyridine nucleotide, gluthathione, and adenosine triphosphatase content of the ciliate *Colpidium campylum.* J. Cell. Comp. Physiol. 33:441-443.

Seaman, G. R. 1950. Utilization of acetate by *Tetrahymena geleii* S. J. Biol. Chem. 186:97-104.

Seaman, G. R. 1950. The presence of sterols in *Tetrahymena geleii* S. J. Cell. Comp. Physiol. 36:129-131.

Seaman, G. R. 1951. Enzyme systems in *Tetrahymena geleii* S. I. Anaerobic dehydrogenases concerned with carbohydrate oxidation. J. Gen. Physiol. 34:775-783.

Seaman, G. R. 1951. Enzyme systems in *Tetrahymena geleii* S. III. Aerobic utilization of hexoses. J. Biol. Chem. 191:439-446.

Seaman, G. R. 1951. Studies on protozoan metabolism and their relation to general problems of physiology and biochemistry. Tex. Rep. Biol. Med. 9:171-179.

Seaman, G. R. 1951. Localization of acetylcholinesterase activity in the protozoan *Tetrahymena geleii.* S. Proc. Soc. Exp. Biol. Med. 76:169-170.

Seaman, G. R. 1952. Enzyme systems in *Tetrahymena geleii* S. IV. Combination of arsonoacetate with carboxyl affinity points on the succinic dehydrogenase. Arch. Biochim. Biophys. 35:132-139.

Seaman, G. R. 1952. Inhibition of the succinic dehydrogenase of *Tetrahymena geleii* S by a phosphono-substituted succinate analog. Arch. Biochem. Biophys. 39:241-243.

Seaman, G. R. 1952. The phosphagen of protozoa. Biochim. Biophys. Acta 9:693-696.

Seaman, G. R. 1952. Replacement of protogen by lipoic acid in the growth of *Tetrahymena.* Proc. Soc. Exp. Biol. Med. 79:158-159.

Seaman, G. R. 1952. Role of protogen in the oxidation of pyruvic acid. Proc. Soc. Exp. Biol. Med. 80:308-310.

Seaman, G. R. 1953. The metabolism of protogen in *Tetrahymena.* Ann. N.Y. Acad. Sci. 56:921-928.

Seaman, G. R. 1953. Effect of thioctic acid on the incorporation of carbon dioxide into pyruvate. J. Bacteriol. 65:744-745.

Seaman, G. R. 1953. Biological conversion of protogen B to protogen A. J. Biol. Chem. 200:813-818.

Seaman, G. R. 1953. Synthesis of vitamins by *Tetrahymena geleii* S. Physiol. Zool. 26:22-28.

Seaman, G. R. 1953. Role of thioctic acid in the transfer of acyl groups. Proc. Soc. Exp. Biol. Med. 82:184-189.

Seaman, G. R. 1954. Discussion: Removal of thioctic acid from enzyme proteins. Fed. Proc. 13:731-733.

Seaman, G. R. 1954. Enzyme systems in *Tetrahymena.* V. Comparison of succinic oxidase activity in different strains. Arch. Biochem. Biophys. 48:424-430.

Seaman, G. R. 1954. Participation of thioctic acid in the acetate-activating reaction. J. Amer. Chem. Soc. 76:1712.

Seaman, G. R. 1954. Pyruvate oxidation by extracts of *Tetrahymena pyriformis.* J. Gen. Microbiol. 11:300-306.

Seaman, G. R. 1954. Enzyme systems in *Tetrahymena pyriformis.* VI. Urea formation and breakdown. J. Protozool. 1:207-210.

Seaman, G. R. 1955. Metabolism of free-living ciliates. *In* S. H. Hutner and A. Lwoff, eds., Biochemistry and physiology of protozoa, vol. 2, Academic Press, New York, pp. 91-158.

Seaman, G. R. 1956. Succinate metabolism of hemoflagellates. Exp. Parasitol. 5:138-148.

Seaman, G. R. 1957. The metabolism of pyruvic oxime by extracts of *Tetrahymena pyriformis* S. Biochim. Biophys. Acta 26:313-317.

Seaman, G. R. 1957. Preparation and properties of the succinate-cleaving enzymes. J. Biol. Chem. 228:149-161.

Seaman, G. R. 1959. Cytological evidence for urease activity in *Tetrahymena.* J. Protozool. 6:331-333.

Seaman, G. R. 1960. Large-scale isolation of kinetosomes from the ciliated protozoan *Tetrahymena pyriformis.* Exp. Cell Res. 21:292-302.

Seaman, G. R. 1961. Acid phosphate activity associated with phagotrophy in the ciliate, *Tetrahymena*. J. Biophys. Biochem. Cytol. 9:243-245.

Seaman, G. R. 1961. Some aspects of phagotrophy in *Tetrahymena*. J. Protozool. 8:204-212.

Seaman, G. R. 1962. Protein synthesis by kinetosomes isolated from the protozoan, *Tetrahymena*. Biochim. Biophys. Acta 55:889-899.

Seaman, G. R. 1962. Two techniques applicable for the study of phagotrophy in ciliates. J. Protozool. 9:335.

Seaman, G. R. 1963. Metabolism of purines by extracts of *Tetrahymena*. J. Protozool. 10:87-91.

Seaman, G. R. 1970. Activity of the glyoxylate cycle in *Tetrahymena* after infection of the cockroach, *Periplaneta*. Exp. Parasitol. 27:15-21.

Seaman, G. R. and Clement, J. J. 1970. Tricarboxylic acid cycle enzymes in *Tetrahymena pyriformis* after infection of the cockroach, *Periplaneta americana*. J. Protozool. 17:287-290.

Seaman, G. R. and Houlihan, R. K. 1949. Effect of arsonoacetic, trans-1,2-cyclopentanedicarboxylic, and β-phosphonopropionic acids on enzyme systems in the ciliate, *Colpidium campylum*. Biol. Bull. 97:257-262.

Seaman, G. R. and Houlihan, R. K. 1950. Trans-1,2-cyclopentanedicarboxylic acid, a succinic acid analog affecting the permeability of the cell membrane. Arch. Biochem. 26:436-441.

Seaman, G. R. and Houlihan, R. K. 1951. Enzyme systems in *Tetrahymena geleii* S. II. Acetylcholinesterase activity. Its relation to motility of the organism and to coordinated ciliary action in general. J. Cell. Comp. Physiol. 37:309-321.

Seaman, G. R. and Naschke, M. D. 1954. The direct formation of acetylcoenzyme A from succinate. J. Amer. Chem. Soc. 76:5572.

Seaman, G. R. and Naschke, M. D. 1955. Removal of thioctic acid from enzymes. J. Biol. Chem. 213:705-711.

Seaman, G. R. and Naschke, M. D. 1955. Reversible cleavage of succinate by extracts of *Tetrahymena*. J. Biol. Chem. 217:1-12.

Seaman, G. R. and Reifel, R. M. 1963. Chemical composition and metabolism of protozoa. Ann. Rev. Microbiol. 17:451-472.

Seaman, G. R. and Roberts, N. L. 1968. Immunological response of male cockroaches to injection of *Tetrahymena pyriformis*. Science 161:1359-1361.

Seaman, G. R., Tosney, T., Berglund, R., and Goldberg, G. 1972. Infectivity and recovery of *Tetrahymena pyriformis* strain S from adult female cockroaches (*Periplaneta americana*). J. Protozool. 19:644-647.

Seamster, A. 1952. An unusual ciliate infection of the spinal fluid. Tex. J. Sci. 4:531-532.

Sedar, A. W. and Rudzinska, M. A. 1956. Mitochondria of protozoa. J. Biophys. Biochem. Cytol. 2 (suppl.):331-336.

Shaw, C. R. 1964. The use of genetic variation in the analysis of isozyme structure. Brookhaven Symp. Biol. 17:117-130.

Shaw, R. F. and Williams, N. E. 1963. Physiological properties of *Tetrahymena vorax*. J. Protozool. 10:486-491.

Shenoy, K. G. and McLaughlan, J. M. 1959. The effect of sulphonylurea drugs on pantothenic acid metabolism in *Tetrahymena pyriformis*. Can. J. Biochem. Physiol. 37:1388-1389.

Shepard, D. C. 1965. Production and elimination of excess DNA in ultraviolet-irradiated *Tetrahymena*. Exp. Cell Res. 38:570-579.

Shorb, M. S. 1963. The lipid composition of *Tetrahymena pyriformis* and *Trichomonas gallinae*. In J. Ludvík, J. Lom, and J. Vávra, eds., Progress in protozoology, Proceedings of the 1st International Congress on Protozoology, Prague, August 1961, Nakladatelstvi Československé Akademie, Prague, pp. 153-158.

Shorb, M. S., Dunlap, B. E., and Pollard, W. O. 1965. Effect of triparanol on synthesis of squalene and tetrahymanol by *Tetrahymena pyriformis*. Proc. Soc. Exp. Biol. Med. 118:1140-1145.

Shoup, G. D., Prescott, D. M., and Wykes, J. R. 1966. Thymidine triphosphate synthesis in *Tetrahymena*. I. Studies on thymidine kinase. J. Cell Biol. 31:295-318.

Shrago, E., Brech, W., and Templeton, K. 1967. Glyconeogenesis in *Tetrahymena pyriformis*. Relationship of enzyme adapation to the carbon pathway. J. Biol. Chem. 242:4060-4066.

Shrago, E. and Shug, A. L. 1966. Phosphoenolpyruvate carboxykinase from *Tetrahymena pyriformis.* Biochim. Biophys. Acta 122:376-378.

Shrago, E., Shug, A. L., and Ferguson, S. M. 1971. The effects of decreased aeration on the electron transport system and glyconeogenesis during growth of *Tetrahymena pyriformis.* Int. J. Biochem. 2:312-318.

Shug, A. L., Ferguson, S. M., and Shrago, E. 1968. Stimulation of growth and metabolism in *Tetrahymena pyriformis* by antimycin A. Biochem. Biophys. Res. Commun. 32:81-85.

Shug, A. L., Elson, C., and Shrago, E. 1969. Effect of iron on growth, cytochromes, glycogen and fatty acid of *Tetrahymena pyriformis.* J. Nutr. 99:379-386.

Shumway, W. 1940. A ciliate protozoon parasitic in the central nervous system of larval *Ambystoma.* Biol. Bull. 78:283-288.

Silvester, N. R. 1964. The cilia of *Tetrahymena pyriformis*: X-ray diffraction by the ciliary membrane. J. Mol. Biol. 8:11-19.

Simon, E. M. 1972. Freezing and storage in liquid nitrogen of axenically and monoxenically cultivated *Tetrahymena pyriformis.* Cryobiology 9:75-81.

Simon, E. M. and Hwang, S. 1967. *Tetrahymena*: Effect of freezing and subsequent thawing on breeding performance. Science 155:694-695.

Simpson, R. E. and Williams, N. E. 1970. The effects of pressure on cell division and oral morphogenesis in *Tetrahymena.* J. Exp. Zool. 175:85-98.

Sinclair, I. J. B. 1958. The role of complement in the immune reactions of *Paramecium aurelia* and *Tetrahymena pyriformis.* Immunology 1:291-299.

Singer, S. 1961. Some amino-folic acid interrelationships in *Tetrahymena pyriformis* H. J. Protozool. 8:265-271.

Singer, W. and Eiler, J. J. 1960. The biological action of cellular depressants and stimulants. V. The effect of phenylurethane on cellular synthesis by *Tetrahymena pyriformis* GL. J. Amer. Pharm. Ass. 49:669-673.

Singer, W., Lee, K. H., and Eiler, J. J. 1960. The biological action of cellular depressants and stimulants. IV. The effect of phenylurethane on the onset and the magnitude of synchronous cell division of *Tetrahymena pyriformis* GL. J. Amer. Pharm. Ass. 49:90-94.

Sipe, J. D. and Holmlund, C. E. 1972. A comparison of the effects of some hypocholesterolemic compounds on squalene metabolism in *Tetrahymena pyriformis* and rat liver. Biochim. Biophys. Acta 280:145-160.

Sládeček, V. 1963. Quantitative characteristic of protozoan communities in waste water. *In* J. Ludvík, J. Lom, and J. Vávra, eds., Progress in protozoology, Proceedings of the 1st International Congress on Protozoology, Prague, August 1961, Nakladatelstvi Československé Akademie, Prague, 338-340 pp.

Slater, J. V. 1952. The magnesium requirement of *Tetrahymena.* Physiol. Zool. 25:283-287.

Slater, J. V. 1952. The influence of cobalt on the growth of the protozoan, *Tetrahymena.* Physiol. Zool. 25:323-332.

Slater, J. V. 1952. Comparative biological activity of α-lipoic acid and protogen in the growth of *Tetrahymena.* Science 115:376-377.

Slater, J. V. 1953. Influence of pH on the growth of *Tetrahymena* in synthetic medium. J. Cell. Comp. Physiol. 41:519-522.

Slater, J. V. 1954. Temperature tolerance in *Tetrahymena.* Amer. Nat. 78:168-174.

Slater, J. V. 1955. Some observations on the cultivation and sterilization of protozoa. Trans. Amer. Microsc. Soc. 74:80-85.

Slater, J. V. 1957. Radiocobalt accumulation in *Tetrahymena.* Biol. Bull. 112:390-399.

Slater, J. V. and Tremor, J. W. 1962. Radioactive phosphorus accumulation and distribution in *Tetrahymena.* Biol. Bull. 122:298-309.

Sleigh, M. A. 1962. The biology of cilia and flagella. Pergamon Press, New York. 242 pp.

Sleigh, M. A. 1969. The physiology and biochemistry of cilia and flagella. *In* A. Lima-de-Faria, ed., Handbook of molecular cytology, John Wiley, New York, pp. 1243-1258.

Slotnick, I. J., Dougherty, M., and James, D. H. 1966. Vincristine inhibition of DNA synthesis in *Tetrahymena pyriformis.* Cancer Res. 26:673-675.

Small, E. B. 1967. The Scuticociliatida, a new order of the class Ciliatea (phylum Protozoa, subphylum Ciliophora). Trans. Amer. Microsc. Soc. 86:345-370.

Small, E. B. 1973. A study of ciliate protozoa from a small polluted stream in east-central Illinois. Amer. Zool. 13:225-230.

Small, E. B. and Marszalek, D. S. 1969. Scanning electron microscopy of fixed, frozen, and dried protozoa. Science 163:1064-1065.

Small, E. B., Marszalek, D. S., and Antipa, G. A. 1971. A survey of ciliate surface patterns and organelles as revealed with scanning electron microscopy. Trans. Amer. Microsc. Soc. 90:283-294.

Small, M., Mantel, N., and Epstein, S. S. 1967. The role of cell-uptake of polycyclic compounds in photodynamic injury of *Tetrahymena pyriformis.* Exp. Cell Res. 45:206-217.

Smith, J. D. and Law, J. H. 1970. Phosphonic acid metabolism in *Tetrahymena.* Biochemistry 9:2152-2157.

Smith, J. D. and Law, J. H. 1970. Phosphatidylcholine biosynthesis in *Tetrahymena pyriformis.* Biochim. Biophys. Acta 202:141-152.

Smith, J. D., Snyder, W. R., and Law, J. H. 1970. Phosphonolipids in *Tetrahymena* cilia. Biochem. Biophys. Res. Commun. 39:1163-1169.

Smith, R. M., Joslyn, D. A., Gruhzit, O. M., McLean, I. W., Jr., Penner, M. A., and Ehrlich, J. 1948. Chloromycetin: Biological studies. J. Bacteriol. 55:425-448.

Snell, E. E. and Broquist, H. P. 1949. On the probable identity of several unidentified growth factors. Arch. Biochem. 23:321-328.

Snyder, W. R. and Law, J. H. 1970. A quantitative determination of phosphonate phosphorus in naturally occurring aminophosphonates. Lipids 5:800-802.

Sonneborn, T. M. 1957. Breeding systems, reproductive methods, and species problems in protozoa. *In* E. Mayr, ed., The species problem, Amer. Ass. Adv. Sci. Publ., Washington, D.C., pp. 155-324.

Sonneborn, T. M. 1963. Does preformed cell structure play an essential role in cell heredity? *In* J. M. Allen, ed., The nature of biological diversity, McGraw-Hill, New York, pp. 165-219.

Speth, V. and Wunderlich, F. 1970. The macronuclear envelope of *Tetrahymena pyriformis* GL in different physiological states. III. Appearance of freeze-etched nuclear pore complexes. J. Cell Biol. 47:772-777.

Šrámek-Hušek, R. 1951. Vorläufiges Verzeichniss der Ciliaten aus Böhmen, Mähren und Schlesien bis zum Jahre 1950. I. Cas. Nar. Mus. Praze 121:140-148.

Šrámek-Hušek, R. 1954. Neue und wenig bekannte Ciliaten aus der Tschechoslowakei und ihre Stellung im Saprobiensystem. Arch. Protistenk. 100:249-267.

Stein, F. 1860. Vortrag über die bisher unbekannt gebliebene *Leucophrys patula* Ehrbg. und über zwei neue Infusoriengattungen *Gyrocorys* und *Lophomonas.* Sitz.-ber. K. Böhm. Ges. Wiss. Prag. 1860: 44-50.

Stein, F. 1867. Der Organismus der Infusionsthiere nach eigenen Forschungen in systematischer Reihenfolge bearbeitet. II. Engelmann, Leipzig. 355 pp.

Stephens, G. C. and Kerr, N. S. 1962. Uptake of phenylalanine by *Tetrahymena pyriformis.* Nature 194:1094-1095.

Sterbenz, F. J. and Lilly, D. M. 1956. Factors influencing abnormal growth in suctorial protozoa. Trans. N.Y. Acad. Sci. 18:522-530.

Stillwell, R. H. 1967. *Colpidium*-produced RNA as a growth stimulant for *Tetrahymena.* J. Protozool. 14:19-22.

Stokes, A. C. 1888. A preliminary contribution toward a history of the fresh-water infusoria of the United States. J. Trenton Nat. Hist. Soc. 1:71-344.

Stokstad, E. L. R., Hoffman, C. E., Regan, M. A., Fordham, D., and Jukes, T. H. 1949. Observations on an unknown growth factor essential for *Tetrahymena geleii.* Arch. Biochem. 20:75-82.

Stokstad, E. L. R., Seaman, G. R., and Hutner, S. H. 1956. Assay for thioctic acid. Methods Biochem. Anal. 3:23-47.

Stolk, A. 1959. A parasitic ciliate in the central nervous system of a larval newt. Naturwissenschaften 46:631.

Stolk, A. 1959. *Glaucoma* species in the central nervous system of the carp. Nature 184:1737.

Stolk, A. 1960. A parasitic ciliate in the central nervous system of a larval newt *Triturus taeniatus.* Proc. Acad. Sci. Amst. 63:71-78.

Stolk, A. 1960. *Glaucoma* sp. in the central nervous system of the carp *Cyprinus carpio* L. Proc. Acad. Sci. Amst. 63:79-86.

Stone, G. E. 1963. Polymorphic properties of *Tetrahymena patula* during growth in axenic culture. J. Protozool. 10:74-80.

Stone, G. E. 1968. Synchronized cell division in *Tetrahymena pyriformis* following inhibition with vinblastine. J. Cell Biol. 39:556-563.

Stone, G. E. 1968. Method for reversible inhibition of cell division in *Tetrahymena pyriformis* using vinblastine sulfate. *In* D. M. Prescott, ed., Methods in cell physiology, vol. 3, Academic Press, New York, pp. 161-170.

Stone, G. E. 1969. Change in histone DNA ratio during population growth in *Tetrahymena.* J. Cell Biol. 42:837-840.

Stone, G. E. and Cameron, I. L. 1964. Methods for using *Tetrahymena* in studies of the normal cell cycle. *In* D. M. Prescott, ed., Methods in cell physiology, vol. 1, Academic Press, New York, pp. 127-140.

Stone, G. E. and Miller, O. L., Jr. 1965. A stable mitochondrial DNA in *Tetrahymena pyriformis.* J. Exp. Zool. 159:33-37.

Stone, G. E., Miller, O. L., Jr., and Prescott, D. M. 1965. H^3-Thymidine derivative pools in relation to macronuclear DNA synthesis in *Tetrahymena pyriformis.* J. Cell Biol. 25:171-177.

Stone, G. E. and Prescott, D. M. 1964. Cell division and DNA synthesis in *Tetrahymena pyriformis* deprived of essential amino acids. J. Cell Biol. 21:275-281.

Stone, G. E. and Prescott, D. M. 1965. Amino acid deprivation and deoxyribonucleic acid synthesis in *Tetrahymena. In* C. P. Leblond and K. B. Warren, eds., The use of radioautography in investigating protein synthesis, Symposia of the International Society for Cell Biology, vol. 4, Academic Press, New York, pp. 95-106.

Stoner, L. and Dunham, P. B. 1970. Regulation of cellular osmolarity and volume in *Tetrahymena.* J. Exp. Biol. 53:391-399.

Stott, J. A. and Smith, H. 1966. Microbiological assay of protein quality with *Tetrahymena pyriformis* W. 4. Measurement of available lysine, methionine, arginine and histidine. Brit. J. Nutr. 20:663-673.

Stott, J. A., Smith, H., and Rosen, G. D. 1963. Microbiological evaluation of protein quality with *Tetrahymena pyriformis* W. 3. A simplified assay procedure. Brit. J. Nutr. 17:227-233.

Stout, J. D. 1954. The ecology, life history, and parasitism of *Tetrahymena* [*Paraglaucoma*] *rostrata* (Kahl) Corliss. J. Protozool. 1:211-215.

Stout, J. D. 1963. Some observations on the protozoa of some beechwood soils on the Chiltern Hills. J. Anim. Ecol. 32:281-287.

Stout, J. D. 1973. The relationship between protozoan populations and biological activity in soils. Amer. Zool. 13:193-201.

Stout, J. D. and Dutch, M. E. 1968. Rates of organic matter decomposition measured by Warburg respiratory experiments. Trans. 9th Int. Congr. Soil Sci., Adelaide, 3:203-209.

Stout, J. D. and Heal, O. W. 1967. Protozoa. *In* A. Burges and F. Raw, eds., Soil biology, Academic Press, New York, pp. 149-195.

Sueoka, N. 1961. Variation and heterogeneity of base composition of DNA: Compilation of old and new data. J. Mol. Biol. 3:31-40.

Sueoka, N. 1961. Compositional correlation between deoxyribonucleic acid and protein. Cold Spring Harbor Symp. Quant. Biol. 26:35-43.

Sugita, M. and Hori, T. 1971. Isolation of diacylglycerol-2-aminoethylphosphonate from *Tetrahymena pyriformis.* J. Biochem. (Tokyo) 69:1149-1150.

Suhama, M. and Yamataka, S. 1960. On the cortex of *Tetrahymena.* I. The comparison between the silver-plated structures and the fine structures through the electron microscope in *Tetrahymena vorax.* Zool. Mag. (Tokyo) 69:153-157. In Japanese with English summary.

Suhama, M. and Yamataka, S. 1960. On the cortex of *Tetrahymena.* II. The fine structures of cilia, kinetosomes and kinetodesmas in *Tetrahymena vorax.* Zool. Mag. (Tokyo) 69:158-162. In Japanese with English summary.

Sullivan, W. D. 1948. The effect of cholesterol on growth in *Colpidium campylum*. Trans. Amer. Microsc. Soc. 67:262-265.

Sullivan, W. D. 1950. Distribution of alkaline phosphatase in *Colpidium campylum*. Trans. Amer. Microsc. Soc. 69:267-271.

Sullivan, W. D. 1957. Identification of protozoa from the region of Point Barrow, Alaska. Trans. Amer. Microsc. Soc. 76:189-196.

Sullivan, W. D. 1959. The effect of ultraviolet radiation and the interaction with sulfur compounds on growth and division of *Tetrahymena*. Trans. Amer. Microsc. Soc. 78:181-193.

Sullivan, W. D. 1959. The sensitivity of interphase and dividing stages of *Tetrahymena pyriformis* W to ultraviolet light. III. Special reference to the protective effect of methionine. Trans. Amer. Microsc. Soc. 78:285-294.

Sullivan, W. D. 1960. The effect of ultraviolet radiation on a cysteine culture of *Tetrahymena pyriformis* W. Trans. Amer. Microsc. Soc. 79:392-396.

Sullivan, W. D. and McCormick, S. J. 1962. The protective effect of high concentration of cysteine against ultraviolet radiation in *Tetrahymena pyriformis* W. Trans. Amer. Microsc. Soc. 81:80-84.

Sullivan, W. D. and Rice, F. M. 1965. Autoradiographic study of the incorporation of H^3-thymidine by synchronously dividing *Tetrahymena pyriformis* GL. Trans. Amer. Microsc. Soc. 84:48-60.

Sullivan, W. D. and Snyder, R. L. 1962. The effect of x-radiation of fumarase activity during different stages of cell division of *Tetrahymena pyriformis* GL. Exp. Cell Res. 28:239-247.

Sullivan, W. D. and Sparks, J. T. 1961. The effect of ultraviolet radiation on succinic dehydrogenase activity during different stages of cell division. Exp. Cell Res. 23:436-442.

Summers, F. M. and Hughes, H. K. 1940. Experiments with *Colpidium campylum* in high frequency electric and magnetic fields. Physiol. Zool. 13:227-242.

Summers, L. G. 1963. Variation of cell and nuclear volume of *Tetrahymena pyriformis* with three parameters of growth: Age of culture, age of cell, and generation time. J. Protozool. 10:288-293.

Summers, L. G. 1963. The effects of nutrient deficiencies and analogs on the cell and nuclear volume of *Tetrahymena pyriformis*. J. Protozool. 10:293-297.

Summers, L. G., Bernstein, E., and James, T. W. 1957. A correlation between nuclear activity and the growth phase in cultures of protozoan cells. Exp. Cell Res. 13:436-437.

Suyama, Y. 1966. Mitochondrial deoxyribonucleic acid of *Tetrahymena*: Its partial physical characterization. Biochemistry 5:2214-2221.

Suyama, Y. 1967. The origins of mitochondrial ribonucleic acids in *Tetrahymena pyriformis*. Biochemistry 6:2829-2839.

Suyama, Y. and Eyer, J. 1967. Leucyl-RNA and leucyl-RNA synthetase in mitochondria of *Tetrahymena pyriformis*. Biochem. Biophys. Res. Commun. 28:746-751.

Suyama, Y. and Eyer, J. 1968. Ribonucleic acid synthesis in isolated mitochondria from *Tetrahymena*. J. Biol. Chem. 243:320-328.

Suyama, Y. and Miura, K. 1968. Size and structural variations of mitochondrial DNA. Proc. Natl. Acad. Sci. U.S. 60:235-242.

Suyama, Y. and Preer, J. R., Jr. 1965. Mitochondrial DNA from protozoa. Genetics 52:1051-1058.

Swader, L. L. 1957. Relations between individual size and growth conditions in populations of *Tetrahymena pyriformis*. Bios 28:241-249.

Swift, H. 1965. Nucleic acids of mitochondria and chloroplasts. Amer. Nat. 99:201-228.

Swift, H., Adams, B., and Larsen, K. 1964. Electron microscope cyto-chemistry of nucleic acids in *Drosophila* salivary glands and *Tetrahymena*. J. Roy. Microsc. Soc. 83:161-167.

Szabó, I. and Németh, G. 1961. Effect of penicillin on *Tetrahymena pyriformis*, strain GL. II. Changes in DNA/RNA ratio. Acta Biol. Acad. Sci. Hung. 12:187-190.

Szyszko, A., Prazak, B. L., Ehret, C. F., Eisler, W. J., Jr., and Wille, J. J., Jr. 1968. A multiunit sampling system and its use in the characterization of ultradian and infradian growth in *Tetrahymena*. J. Protozool. 15:781-785.

Taketomi, T. 1961. Phospholipids in *Tetrahymena pyriformis* W. Z. Allg. Mikrobiol. 1:331-340.

Tamari, M. 1971. Intracellular distribution of ciliatine (2-aminoethylphosphonic acid) in *Tetrahymena*. Agr. Biol. Chem. (Tokyo) 35:1799-1802.

Tamura, S. 1971. Properties of microtubule proteins in different organelles in *Tetrahymena pyriformis*. Exp. Cell Res. 68:169-179.

Tamura, S. 1971. Synthesis and assembly of microtubule proteins in *Tetrahymena pyriformis*. Exp. Cell Res. 68:180-185.

Tamura, S., Toyoshima, Y., and Watanabe, Y. 1966. Mechanism of temperature-induced synchrony in *Tetrahymena pyriformis*. Analysis of the leading cause of synchronization. Jap. J. Med. Sci. Biol. 19:85-96.

Tamura, S., Tsuruhara, T., and Watanabe, Y. 1969. Function of nuclear microtubules in macronuclear division of *Tetrahymena pyriformis*. Exp. Cell Res. 55:351-358.

Tanzer, C. 1941. Serological studies with free-living protista. J. Immunol. 42:291-312.

Tartar, V. 1967. Morphogenesis in protozoa. *In* T.-T. Chen, ed., Research in protozoology, vol. 2, Pergamon Press, New York, pp. 1-116.

Taylor, C. V., Thomas, J. O., and Brown, M. G. 1933. Studies on the protozoa. IV. Lethal effects of the x-radiation of a sterile culture medium for *Colpidium campylum*. Physiol. Zool. 6:467-492.

Teunisson, D. J. 1961. Microbiological assay of intact proteins using *Tetrahymena pyriformis* W. I. Survey of protein concentrates. Anal. Biochem. 2:405-420.

Teunisson, D. J. 1971. Elutriation and Coulter counts of *Tetrahymena pyriformis* grown in peanut and cottonseed meal media. Appl. Microbiol. 21:878-887.

Teunisson, D. J. and Robertson, J. A. 1967. Degradation of pure aflatoxins by *Tetrahymena pyriformis*. Appl. Microbiol. 15:1099-1103.

Thayer, P. S., Gordon, H. L., and Macdonald, M. 1971. *In vitro* growth inhibition by 3837 compounds tested for antitumor activity: Comparison of tumor cell culture and microbial assays. Cancer Chemother. Rep. 2:27-133.

Thompson, G. A., Jr. 1967. Studies of membrane formation in *Tetrahymena pyriformis*. I. Rates of phospholipid biosynthesis. Biochemistry 6:2015-2022.

Thompson, G. A., Jr. 1969. The metabolism of 2-aminoethylphosphonate lipids in *Tetrahymena pyriformis*. Biochim. Biophys. Acta 176:330-338.

Thompson, G. A., Jr. 1969. The properties of an enzyme system degrading endogenous phospholipids of *Tetrahymena pyriformis*. J. Protozool. 16:397-400.

Thompson, G. A., Jr. 1972. *Tetrahymena pyriformis* as a model system for membrane studies. J. Protozool. 19:231-236.

Thompson, G. A., Jr. 1972. Ether-linked lipids in protozoa. *In* F. Snyder, ed., Ether lipids: Chemistry and biology, Academic Press, New York, pp. 321-327.

Thompson, G. A., Jr., Bambery, R. J., and Nozawa, Y. 1971. Further studies of the lipid composition and biochemical properties of *Tetrahymena pyriformis* membrane systems. Biochemistry 10:4441-4447.

Thompson, G. A., Jr., Bambery, R. J., and Nozawa, Y. 1972. Environmentally produced alterations of the tetrahymanol:phospholipid ratio in *Tetrahymena pyriformis* membranes. Biochim. Biophys. Acta 260:630-638.

Thompson, G. A., Jr., and Nozawa, Y. 1972. Lipids of protozoa: Phospholipids and neutral lipids. Ann. Rev. Microbiol. 26:249-278.

Thompson, J. C., Jr. 1958. Experimental infections of various animals with strains of the genus *Tetrahymena*. J. Protozool. 5:203-205.

Thompson, J. C., Jr. 1958. *Tetrahymena rostrata* as a facultative parasite in the grey garden slug. Va. J. Sci. 9:315-318.

Thompson, J. C., Jr. 1960. A protozoan "fingerprint." Turtox News 38:258-259.

Thompson, J. C., Jr. 1963. The generic significance of the buccal infraciliature in the family Tetrahymenidae and a proposed new genus and species, *Paratetrahymena wassi*. Va. J. Sci. 14:126-135.

Thompson, W. D. and Wingo, W. J. 1959. Comparative enzymatic studies on cells of *Tetrahymena pyriformis* Y from cultures at two stages of growth: L-Glutamic acid dehydrogenase. J. Cell. Comp. Physiol. 53:413-420.

Thormar, H. 1959. Delayed division in *Tetrahymena pyriformis* induced by temperature changes. C. R. Trav. Lab. Carlsberg 31:207-225.

Thormar, H. 1962. Cell size of *Tetrahymena pyriformis* incubated at various temperatures. Exp. Cell Res. 27:585-586.

Thormar, H. 1962. Effect of termperature on the reproduction rate of *Tetrahymena pyriformis.* Exp. Call Res. 28:269-279.

Tingle, L. E., Pavlat, W. A., and Cameron, I. L. 1973. Sublethal cytotoxic effects of mercuric chloride on the ciliate *Tetrahymena pyriformis.* J. Protozool. 20:301-304.

Tittler, I. A. 1948. An investigation of the effects of carcinogens on *Tetrahymena geleii.* J. Exp. Zool. 108:309-325.

Tittler, I. A. and Belsky, M. M. 1951. Pterolglutamic acid activity of aminopterin in *Tetrahymena geleii.* Science 114:493.

Tittler, I. A., Belsky, M. M., and Hutner, S. H. 1952. The role of aneurine in the nutrition of *Tetrahymena geleii.* J. Gen. Microbiol. 6:85-89.

Tittler, I. A. and Bovell, C. 1954. Effect of sulfonamides on thiamine requirement of *Tetrahymena geleii* H. Proc. Soc. Exp. Biol. Med. 85:495-496.

Tokuyasu, K. and Scherbaum, O. H. 1965. Ultrastructure of mucocysts and pellicle in *Tetrahymena pyriformis.* J. Cell Biol. 27:67-81.

Treillard, M. and Lwoff, A. 1924. Sur un infusoire parasite de la cavité générale des larves de chironomes. Sa sexualité. C. R. Acad. Sci., Paris 178:1761-1764.

Tsuda, Y., Morimoto, A., Sano, T., Inubushi, Y., Mallory, F. B., and Gordon, J. T. 1965. The synthesis of tetrahymenol. Tetrahedron Lett., no. 19:1427-1431.

Ueno, Y. and Saheki, M. 1968. Inhibitory effects of luteoskyrin, a hepatotoxic pigment of *Penicillium islandicum* Sopp, and chemically related pigments on a protozoon, *Tetrahymena pyriformis* GL. Jap. J. Exp. Med. 38:157-164.

Ueno, Y. and Yamakawa, H. 1970. Antiprotozoal activity of scirpene mycotoxins of *Fusarium nivale* Fn 2 B. Jap. J. Exp. Med. 40:385-390.

van de Vijver, G. 1966. Studies on the metabolism of *Tetrahymena pyriformis* GL. I. Influence of substrates on the respiration rate. Enzymologia 31:363-381.

van de Vijver, G. 1966. Studies on the metabolism of *Tetrahymena pyriformis* GL. II. Some properties of the terminal oxidation. Enzymologia 31:382-391.

van de Vijver, G. 1967. Studies on the metabolism of *Tetrahymena pyriformis* GL. III. Effects of gamma-irradiation on the respiratory rate. Enzymologia 33:331-344.

van de Vijver, G. 1969. Studies on the metabolism of *Tetrahymena pyriformis* GL. IV. Irradiation effects on cell permeability. Enzymologia 36:371-374.

van Eys, J. and Warnock, L. G. 1963. The inhibition of motility of ciliates through methonium drugs. J. Protozool. 10:465-467.

Van Niel, C. B., Thomas, J. O., Ruben, S., and Kamen, M. D. 1942. Radioactive carbon as an indicator of carbon dioxide utilization. IX. The assimilation of carbon dioxide by protozoa. Proc. Natl. Acad. Sci. U.S. 28:157-161.

van Thoai, N. and Roche, J. 1964. Diversity of phosphagens. *In* C. A. Leone, ed., Taxonomic biochemistry and serology, Ronald Press, New York, pp. 347-362.

van Wagtendonk, W. J. 1955. The nutrition of ciliates. *In* S. H. Hutner and A. Lwoff, eds., Biochemistry and physiology of protozoa, vol. 2, Academic Press, New York, pp. 57-84.

Varkirtzi-Lemonias, C., Kidder, G. W., and Dewey, V. C. 1963. Ubiquinone in four genera of protozoa. Comp. Biochem. Physiol. 8:331-334.

Villadsen, I. S. and Zeuthen, E. 1970. Synchronization of DNA synthesis in *Tetrahymena* populations by temporary limitation of access to thymine compounds. Exp. Cell Res. 61:302-310.

Villela, G. G. 1963. Enzimas do metabolismo purinico em *Tetrahymena pyriformis.* Ann. Acad. Brazil Cienc. 35:453-457.

496 General Bibliography

Villela, G. G. 1964. El ciliada *Tetrahymena* en la investigacion bioquimica. Sobretiro Cienc. Mex. 23:89-98.

Villela, G. G. 1965. Dehydrogenases of purine metabolism in *Tetrahymena pyriformis*. Proc. Soc. Exp. Biol. Med. 118:834-838.

Villeneuve-Brachon, S. 1940. Recherches sur les ciliés hétérotriches. Arch. Zool. Exp. Gén. 82:1-180.

Viswanatha, T. and Liener, I. E. 1956. Utilization of native and denatured proteins by *Tetrahymena pyriformis* W. Arch. Biochem. Biophys. 56:222-229.

Viswanatha, T. and Liener, I. E. 1956. Isolation and properties of a proteinase from *Tetrahymena pyriformis* W. Arch. Biochem. Biophys. 61:410-421.

Volm, M. and Goerttler, K. 1969. Sodium-cyclamate in the "*in vitro*-test." Macromolecular synthesis and oxygen consumption. Arch. Toxikol. 24:304-308.

Volm, M., Wayss, K., and Schwartz, V. 1970. Wirkung von Lithium- und Rhodanidionen auf die Nukleinsäure- und Protein-Synthese bei *Tetrahymena pyriformis* GL. Wilhelm Roux' Arch. 165:125-131.

von Gelei, J. 1935. *Colpidium glaucomaeforme* n. sp. (Hymenostomata) und sein Neuronemensystem. Arch. Protistenk. 85:289-302.

Waithe, W. I. 1964. Glucose utilization by suspensions of *Tetrahymena pyriformis*. Exp. Cell Res. 35:100-107.

Walker, P. M. B. and Mitchison, J. M. 1957. DNA synthesis in two ciliates. Exp. Cell Res. 13:167-170.

Wang, G.-T. and Marquardt, W. C. 1966. Survival of *Tetrahymena pyriformis* and *Paramecium aurelia* following freezing. J. Protozool. 13:123-128.

Wand, M., Zeuthen, E., and Evans, E. A. 1967. Tritiated thymidine: Effect of decomposition by self-radiolysis on specificity as a tracer for DNA synthesis. Science 157:436-438.

Warnock, L. G. and van Eys, J. V. 1962. Normal carbohydrate metabolism in *Tetrahymena pyriformis*. J. Cell. Comp. Physiol. 60:53-60.

Warnock, L. G. and van Eys, J. V. 1963. Inhibition of growth of *Tetrahymena pyriformis* by oxamic acid. J. Bacteriol. 85:1179-1181.

Warnock, L. G. and van Eys, J. V. 1963. Observations of *Tetrahymena pyriformis* relating to the Pasteur effect. J. Cell. Comp. Physiol. 61:309-316.

Warren, R. 1932. On a ciliate protozoon inhabiting the liver of a slug. Ann. Natal Mus. 7:1-53.

Watanabe, Y. 1963. Some factors necessary to produce division conditions in *Tetrahymena pyriformis*. Jap. J. Med. Sci. Biol. 16:107-124.

Watanabe, Y. 1971. Mechanism of synchrony induction. I. Some features of synchronous rounding in *Tetrahymena pyriformis*. Exp. Cell Res. 68:431-436.

Watanabe, Y. 1971. Mechanism of synchrony induction. II. Synthesis and turnover of biomolecules during the induction phase of synchronous rounding in *Tetrahymena pyriformis*. Exp. Cell Res. 68:437-441.

Watanabe, Y. 1971. Mechanism of synchrony induction. III. Changes of water insoluble protein fraction involved in synchronous rounding in *Tetrahymena pyriformis*. Exp. Cell Res. 69:324-328.

Watanabe, Y. 1971. Mechanism of synchrony induction. IV. Accumulation of a special protein in the induction process of synchronous rounding in *Tetrahymena pyriformis*. Exp. Cell Res. 69:329-335.

Watanabe, Y. and Ikeda, M. 1965. Evidence for the synthesis of the "division protein" in *Tetrahymena pyriformis*. Exp. Cell Res. 38:432-434.

Watanabe, Y. and Ikeda, M. 1965. Isolation and characterization of division protein in *Tetrahymena pyriformis*. Exp. Cell Res. 39:443-452.

Watanabe, Y. and Ikeda, M. 1965. Further confirmation of "division protein" fraction in *Tetrahymena pyriformis*. Exp. Cell Res. 39:464-469.

Watson, J. M. 1946. On the coprophilic habits of a ciliate—*Glaucoma piriformis*. J. Trop. Med. Hyg. 49(3):44-46.

Watson, M. R., Alexander, J. B., and Silvester, N. R. 1964. The cilia of *Tetrahymena pyriformis*: Fractionation of isolated cilia. Exp. Cell Res. 33:112-129.

Watson, M. R. and Hopkins, J. M. 1962. Isolated cilia from *Tetrahymena pyriformis.* Exp. Cell Res. 28:280-295.

Watson, M. R., Hopkins, J. M., and Randall, J. T. 1961. Isolated cilia from *Tetrahymena pyriformis.* Exp. Cell Res. 23:629-631.

Weller, D. L., Raina, A., and Johnstone, D. B. 1968. Characterization of ribosomes of *Tetrahymena pyriformis.* Biochim. Biophys. Acta 157:558-565.

Wells, C. 1960. The response of *Tetrahymena pyriformis* to ionizing radiation: Strain specific radiosensitivities. J. Cell. Comp. Physiol. 55:207-219.

Wells, C. 1960. *Tetrahymena pyriformis,* strain HS: Micronucleate or amicronucleate? J. Protozool. 7:313-314.

Wells, C. 1960. Identification of free and bound amino acids in three strains of *Tetrahymena pyriformis* using paper chromatography. J. Protozool. 7:7-10.

Wells, C. 1961. Evidence for micronuclear function during vegetative growth and reproduction of the ciliate, *Tetrahymena pyriformis.* J. Protozool. 8:284-290.

Wells, C. 1965. Age associated nuclear anomalies in *Tetrahymena.* J. Protozool. 12:561-563.

Wenyon, C. M. 1926. Protozoology, vol. 2, Baillière, Tindall and Cox, London, pp. 779-1563.

West, R. A., Jr., Barbera, P. W., Kolar, J. R., and Murrell, C. B. 1962. The agar layer method for determining the activity of diverse materials against selected protozoa. J. Protozool. 9:65-73.

Westergaard, O. 1970. Separation of the two DNA polymerase fractions from *Tetrahymena* cells after excision-reparable damage to DNA. Biochim. Biophys. Acta 213:36-44.

Westergaard, O., Marcker, K. A., and Keiding, J. 1970. Induction of a mitochondrial DNA polymerase in *Tetrahymena.* Nature 227:708-710.

Westergaard, O. and Pearlman, R. E. 1969. DNA polymerase activity in methotrexate plus uridine treated *Tetrahymena.* Exp. Cell Res. 54:309-313.

White, F. R. 1962. Mannitol myleran. Cancer Chemother. Rep. 23:71-77.

Whitlow, K. J., D'Iorio, A., and Mavrides, C. 1972. Regulation of enzymes of tyrosine catabolism in *Tetrahymena pyriformis.* Biochim. Biophys. Acta 264:440-449.

Whitson, G. L., Francis, A. A., and Carrier, W. L. 1968. Ultraviolet-induced pyrimidine dimers in *Tetrahymena.* I. Removal in the dark. Biochim. Biophys. Acta 161:285-290.

Whitson, G. L., Green, J. G., Francis, A. A., and Willis, D. D. 1967. Cyclic changes in the alcohol-soluble carbohydrates in synchronized *Tetrahymena.* I. Fractionation. J. Cell. Physiol. 70:169-177.

Whitson, G. L. and Padilla, G. M. 1964. The effects of actinomycin D on stomatogenesis and cell division in temperature-synchronized *Tetrahymena pyriformis.* Exp. Cell Res. 36:667-671.

Whitson, G. L., Padilla, G. M., Canning, R. E., Cameron, I. L., Anderson, N. G., and Elrod, L. H. 1966. The isolation of oral structures from *Tetrahymena pyriformis* by low speed zonal centrifugation. Natl. Cancer Inst. Monogr. 21:317-321.

Whitson, G. L., Padilla, G. M., and Fisher, W. D. 1966. Cyclic changes in polyribosomes of synchronized *Tetrahymena pyriformis.* Exp. Cell Res. 42:438-446.

Whitson, G. L., Padilla, G. M., and Fisher, W. D. 1966. Morphogenetic and macromolecular aspects of synchronized *Tetrahymena.* In I. L. Cameron and G. M. Padilla, eds., Cell synchrony, Academic Press, New York, pp. 289-306.

Wichterman, R. 1957. Biological effects of radiations on protozoa. Bios 28:3-20.

Wilber, C. G. and Seaman, G. R. 1948. The lipids in *Colpidium campylum.* Biol. Bull. 94:29-32.

Wille, J. J., Jr. 1972. Physiological control of DNA nucleotide sequence redundancy in the eukaryote, *Tetrahymena pyriformis.* Biochem. Biophys. Res. Commun. 46:677-684.

Wille, J. J., Jr. 1972. Selective amplification of repetitive DNA in the eukaryote, *Tetrahymena pyriformis.* Biochem. Biophys. Res. Commun. 46:692-699.

Wille, J. J., Jr., Barnett, A., and Ehret, C. F. 1972. Participation of rare DNA templates in DNA/RNA hybridization in the eukaryote, *Tetrahymena pyriformis.* Biochem. Biophys. Res. Commun. 46:685-691.

Wille, J. J., Jr., and Ehret, C. F. 1968. Light synchronization of an endogenous circadian rhythm of cell division in *Tetrahymena.* J. Protozool. 15:785-789.

Wille, J. J., Jr., and Ehret, C. F. 1968. Circadian rhythm of pattern formation in populations of a free-swimming organism, *Tetrahymena.* J. Protozool. 15:789-792.

Williams, B. L., Goodwin, T. W., and Ryley, J. F. 1966. The sterol content of some protozoa. J. Protozool. 13:227-230.

Williams, N. E. 1960. The polymorphic life history of *Tetrahymena patula*. J. Protozool. 7:10-17.

Williams, N. E. 1961. Polymorphism in *Tetrahymena vorax*. J. Protozool. 8:403-410.

Williams, N. E. 1963. Structural development in synchronously dividing *Tetrahymena*. *In* E. Zeuthen, ed., Synchrony of cell division and growth, John Wiley, New York, pp. 159-175.

Williams, N. E. 1964. Induced division synchrony in *Tetrahymena vorax*. J. Protozool. 11:230-236.

Williams, N. E. 1964. Relations between temperature sensitivity and morphogenesis in *Tetrahymena pyriformis*. J. Protozool. 11:566-572.

Williams, N. E. and Frankel, J. 1973. Regulation of microtubules in *Tetrahymena*. I. Electron microscopy of oral replacement. J. Cell Biol. 56:441-457.

Williams, N. E. and Luft, J. H. 1968. Use of a nitrogen mustard derivative in fixation for electron microscopy and observations on the ultrastructure of *Tetrahymena*. J. Ultrastruct. Res. 25:271-292.

Williams, N. E., Michelsen, O., and Zeuthen, E. 1969. Synthesis of cortical proteins in *Tetrahymena*. J. Cell Sci. 5:143-162.

Williams, N. E. and Nelsen, E. M. 1973. Regulation of microtubules in *Tetrahymena*. II. Relation between turnover of microtubule proteins and microtubule dissociation and assembly during oral replacement. J. Cell Biol. 56:458-465.

Williams, N. E. and Scherbaum, O. H. 1959. Morphogenetic events in normal and synchronously dividing *Tetrahymena*. J. Embryol. Exp. Morphol. 7:241-256.

Williams, N. E. and Zeuthen, E. 1966. The development of oral fibers in relation to oral morphogenesis and induced division synchrony in *Tetrahymena*. C. R. Trav. Lab. Carlsberg 35:101-118.

Williams, R. J. 1943. The significance of the vitamin content of tissues. Vitamins Hormones 1:229-247.

Wilton, D. and Akhtar, M. 1970. The stereochemistry of hydrogen elimination during 7,8-double bond formation by *Tetrahymena pyriformis*. Biochem. J. 116:337-339.

Winet, H. and Jahn, T. L. 1972. Effect of CO_2 and NH_3 on bioconvection: Patterns in *Tetrahymena* cultures. Exp. Cell Res. 71:356-360.

Wingo, W. J. 1950. An inoculating device for use in protozoan cultures. Tex. Rep. Biol. Med. 8:410-411.

Wingo, W. J. and Anderson, N. L. 1951. Effect of pH on medium upon the growth rate of *Tetrahymena geleii* T-P. J. Exp. Zool. 116:571-576.

Wingo, W. J. and Browning, I. 1951. Measurement of swimming speed of *Tetrahymena geleii* by stroboscopic photomicrography. J. Exp. Zool. 117:441-449.

Wingo, W. J. and Cameron, L. F. 1952. Effect of thyroxin upon growth and oxygen uptake of *Tetrahymena geleii* (Y). Tex. Rep. Biol. Med. 10:1075-1083.

Wingo, W. J. and Cameron, L. F. 1954. Variations with age of culture in the ability of *Tetrahymena geleii* (Y) to oxidize and to incorporate methionine. Tex. Rep. Biol. Med. 12:438-443.

Wingo, W. J. and Lockingen, L. S. 1954. Secondary growth in *Tetrahymena geleii* (Y). Tex. Rep. Biol. Med. 12:196-199.

Wingo, W. J., Lockingen, L. S., and Cameron, L. F. 1955. Effect of thiamine on the growth, survival, senescence and morphology of *Tetrahymena pyriformis* (Y). Tex. Rep. Biol. Med. 13:578-586.

Winicur, S. 1967. Reactivation of ethanol-calcium-isolated cilia from *Tetrahymena pyriformis*. J. Cell Biol. 35:C7-C9.

Winicur, S. and Roth, J. S. 1965. Biochemical studies on irradiated protozoa. III. The deamination of deoxycytidine and deoxycytidylic acid by *Tetrahymena pyriformis* W. J. Protozool. 12:166-170.

Wolfe, J. A. 1967. Structural aspects of amitosis: A light and electron microscope study of the isolated macronuclei of *Paramecium aurelia* and *Tetrahymena pyriformis*. Chromosoma 23:59-79.

Wolfe, J. A. 1970. Structural analysis of basal bodies of the isolated oral apparatus of *Tetrahymena pyriformis*. J. Cell Sci. 6:679-700.

Woodard, J., Kaneshiro, E. S., and Gorovsky, M. A. 1972. Cytochemical studies on the problem of macronuclear subnuclei in *Tetrahymena*. Genetics 70:251-260.

Woodside, G. L., Kidder, G. W., Dewey, V. C., and Parks, R. E., Jr. 1953. The influence of 8-azaguanine on the mitotic rate and histological appearance of certain normal and neoplastic tissues. Cancer Res. 13:289-291.

Wu, C. and Hogg, J. R. 1952. The amino acid composition and nitrogen metabolism of *Tetrahymena geleii*. J. Biol. Chem. 198:753-764.

Wu, C. and Hogg, J. F. 1956. Free and nonprotein amino acids of *Tetrahymena pyriformis*. Arch. Biochem. Biophys. 62:70-77.

Wunderlich, F. 1969. The macronuclear envelope of *Tetrahymena pyriformis* GL in different physiological states. I. Quantitative structural data. Exp. Cell Res. 56:369-374.

Wunderlich, F. 1969. The macronuclear envelope of *Tetrahymena pyriformis* GL in different physiological states. II. Frequency of central granules in the pores. Z. Zellforsch. Mikrosk. Anat. 101:581-587.

Wunderlich, F. 1972. The macronuclear envelope of *Tetrahymena pyriformis* GL in different physiological states. V. Nuclear pore complexes—A controlling system in protein synthesis? J. Memb. Biol. 7:220-230.

Wunderlich, F. and Franke, W. W. 1968. Structure of macronuclear envelopes of *Tetrahymena pyriformis* in the stationary phase of growth. J. Cell Biol. 38:458-462.

Wunderlich, F. and Peyk, D. 1969. Antimitotic agents and macronuclear division of ciliates. II. Endogenous recovery from colchicine and colcemid—A new method of synchronization in *Tetrahymena pyriformis* GL. Exp. Cell Res. 57:142-144.

Wunderlich, F. and Peyk, D. 1969. Antimitotic agents and macronuclear division of ciliates. IV. Reaseffect of colchicine on synchronized *Tetrahymena pyriformis* GL. Experientia 25:1275-1280.

Wunderlich, F. and Speth, V. 1970. Antimotic agents and macronuclear division of ciliates. IV. Reassembly of microtubules in macronuclei of *Tetrahymena* adapting to colchicine. Protoplasma 70:139-152.

Wunderlich, F. and Speth, V. 1972. The macronuclear envelope of *Tetrahymena pyriformis* GL in different physiological states. IV. Structural and functional aspect of nuclear pore complexes. J. Microsc. 13:361-382.

Wunderlich, F. and Weirich, E. 1972. Antimototic agents and macronuclear division of ciliates. V. Cytokinesis of *Tetrahymena* synchronized by deuterium oxide. Exp. Cell Res. 72:568-571.

Wykes, J. and Prescott, D. M. 1968. The metabolism of pyrimidine compounds by *Tetrahymena pyriformis*. J. Cell. Physiol. 72:173-183.

Yamanaka, T., Nagata, Y., and Okunuki, K. 1968. Purification and properties of cytochrome c-553 and cytochrome b-560 from *Tetrahymena pyriformis*. J. Biochem. 63:753-760.

Yuyama, S. and Zimmerman, A. M. 1972. RNA synthesis in *Tetrahymena*. Exp. Cell Res. 71:193-203.

Zadylak, A. H., Ehret, C. F., and Wille, J. J., Jr. 1970. Electron microscopic autoradiography of DNA replication in *Tetrahymena*. Atomlight 71:1-4.

Zander, J. M. and Caspi, E. 1970. The stereospecificity of the cholesterol Δ^{22}-dehydrogenase of *Tetrahymena pyriformis,* and the origin of the C-22 protons of cholesterol. J. Biol. Chem. 245:1682-1687.

Zander, J. M., Caspi, E., Pandey, G. N., and Mitra, C. R. 1969. The presence of tetrahymanol in *Oleandra wallichii*. Phytochemistry 8:2265-2267.

Zander, J. M., Greig, J., and Caspi, E. 1970. Tetrahymanol biosynthesis. Studies *in vitro* on squalene cyclization. J. Biol. Chem. 245:1247-1254.

Zebrun, W., Corliss, J. O., and Lom, J. 1967. Electron microscopical observations on the mucocysts of the ciliate *Tetrahymena rostrata*. Trans. Amer. Microsc. Soc. 86:28-36.

Zeuthen, E. 1953. Growth as related to the cell cycle in single cell cultures of *Tetrahymena pyriformis*. J. Embryol. Exp. Morphol. 1:239-250.

Zeuthen, E. 1958. Artificial and induced periodicity in living cells. Adv. Biol. Med. Phys. 6:37-73.

Zeuthen, E. 1961. Cell division and protein synthesis. *In* T. W. Goodwin and O. Lindberg, eds., Biological structure and function, vol. 2, Proceedings of the 1st IUB/IUBS International Symposium, Academic Press, New York, pp. 537-548.

Zeuthen, E. 1961. Synchronized growth in *Tetrahymena* cells. *In* M. X. Zarrow, ed., Growth in living systems, a symposium, Basic Books, New York, pp. 135-158.

Zeuthen, E. 1963. Independent cycles of cell division and of DNA synthesis in *Tetrahymena*. *In* R. J. C. Harris, ed., Cell growth and cell division, vol. 2, Symposia of the International Society of Cell Biologists, Academic Press, New York, pp. 1-7.

Zeuthen E. 1963. Protein synthesis as a condition for division in synchronized *Tetrahymena*. *In* J. Ludvík, J. Lom, and J. Vávra, eds., Progress in protozoology, Proceedings of the 1st International Congress on Protozoology, Prague, August 1961, Nakladatelstvi Československé Akademie, Prague, pp. 218-221.

Zeuthen, E. 1964. The temperature-induced division synchrony in *Tetrahymena*. *In* E. Zeuthen, ed., Synchrony of cell division and growth, John Wiley, New York, pp. 99-156.

Zeuthen, E. 1968. Thymine starvation by inhibition of uptake and synthesis of thymine-compounds in *Tetrahymena*. Exp. Cell Res. 50:37-46.

Zeuthen, E. 1970. Independent synchronization of DNA synthesis and of cell division in the same culture of *Tetrahymena* cells. Exp. Cell Res. 61:311-325.

Zeuthen, E. 1971. Synchrony in *Tetrahymena* by heat shocks spaced a normal cell generation apart. Exp. Cell Res. 68:49-60.

Zeuthen, E. 1971. Synchronization of the *Tetrahymena* cell cycle. *In* D. M. Prescott, L. Goldstein, and E. McConkey, eds., Advances in cell biology, vol. 2, Appleton-Century-Crofts, New York, pp. 111-152.

Zeuthen, E. and Rasmussen, L. 1972. Synchronized cell division in protozoa. *In* T.-T. Chen, ed., Research in protozoology, vol. 4, Pergamon Press, New York, pp. 9-145.

Zeuthen, E. and Scherbaum, O. H. 1954. Synchronous divisions in mass cultures of the ciliate protozoon *Tetrahymena pyriformis* as induced by temperature changes. *In* J. A. Kitching, ed., Recent developments in cell physiology, Proceedings of the 7th Symposium of the Colston Research Society, London, Butterworths, London, pp. 141-156.

Zeuthen, E. and Williams, N. E. 1969. Division-limiting morphogenetic processes in *Tetrahymena*. *In* E. V. Cowdry and S. Seno, eds., Nucleic acid metabolism, cell differentiation and cancer growth, Pergamon Press, New York, pp. 203-216.

Zimmerman, A. M. 1969. Effects of high pressure on macromolecular synthesis in synchronized *Tetrahymena*. *In* G. M. Padilla, G. L. Whitson, and I. L. Cameron, eds., The cell cycle: Gene-enzyme interactions, Academic Press, New York, pp. 203-225.

Zingher, J. A. 1933. Beobachtungen und Fetteinschlüssen bei einigen Protozoen. Arch. Protistenk. 81:57-87.

Zotikor, A. 1971. Changes in the ultraviolet fluorescence of irradiated cells. Izv. Akad. Nauk SSSR, Ser. Biol. 1:144-146.

Subject Index

A

Acanthamoeba castellani, 96
Acetate, 93
Acetylcholine esterase, 424
Acetyl-CoA synthetase, 242, 246
B-N-Acetylglucosaminidase, 93
Acid phosphatase, 72, 74-76, 79, 80, 96, 144,
    149, 152, 154, 159, 231, 249-250, 252-253,
    330, 364
Actinomycin D, 132, 232, 291, 294-295, 391,
    394
Active transport, 166, 191-192
Adenase, 145
Adenosine monophosphate deaminase, 149
AMP pyrophosphorylase, 146
ATPase, 59, 67, 179, 391
Aedes, 31
Aerobacter spp., 20, 157
Aging, 273
Aging, starvation and, 252-253
Agrobacterium, 115
Alanine, 65, 90
Alcohol requirement for, 93
Aleic acid, 94
Alkaline phosphatase, 249

Allyl alcohol, 317
Amethopterin, 422
Amicronucleate, 4
D-Amino acid oxidase, 231
Amino acid requirements, 90-91
Amitosis, 309
Amoeba, 95, 173, 183
Amphotericin B, 422
Amylase, 93
Anarchic field, 17
Antibiotics, experimental, 413
Antibiotics, toxicity of, 421
Antihistamines, 425-426
Antimalarial agents, 426-427
Antimycin A, 245
Antioxidants, 418-419
Antipteridines, 426-427
Antitumor-agent screening, 421-423
Antivitamins, 425
Arginine, 90
Asexual reproduction, 260-261
Aspartic acid, 65, 90-91
Autogamy, 26, 261, 313, 383
Autophagic vacuoles, 75, 77, 80, 144, 159, 234,
    252-253, 385
Axenize, 414

8-Azaquanine, 126-127, 151, 421
Azaserine, 422

B

Balantidium, 4
Basal bodies, see Kinetosomes
Basal media, 416
Benzpyrene, 417-419
Biopterin, 423
Biotin, 91
Blepharisma, 79
Body shape and size, 16, 20, 22, 25, 28-29, 31-32, 34, 37, 42, 57
Brassicasterol, 93
-5-Bromodeoxyuridine, 137-138
Buccal area, 16, 35, 379, 381
Bufo, 30

C

Caffeine, 317, 420, 424
Campanella umbellaria, 72, 74-75
Candida albicans, 421
Capesterol, 93
Carbohydrate, fermentation of, 93
Carbohydrate, requirements of, 93
Carcinogenic hydrocarbons, photodynamic test for, 417-419
Carcinogens, effects of, 419-421
Cardiolipin, 113-114, 119
Caryonides, 315-316, 324, 326-327, 347
Cathepsin, 231
Cell cycle, 84, 128, 131, 133, 199-211, 229, 260, 386, 395
Cell cycle terminology, 200
Cell division, 69, 71, 157, 180, 217-221, 290-292, 300, 376, 378, 383, 387, 398-401
    inhibition of, 194
    rate of, 131
Cell membrane, 61, 63, 233, 235, 420
Cell volume, 188, 193, 208-210, 217-220
    regulation of, 182
Centromere, 264
Chaos chaos, 173
Characteristics of species, 6, 8, 13
Chemically defined media, 93
Chironomus plumosus, 22
Chlamydomonas, 118, 314
Chloramphenicol, 413, 422
Chlorella, 118
Chlorotetracycline, 422
Cholesterol, 94-95, 109-111, 158, 195

Chromatin, 84, 287, 289
Chromosomes, 18, 85, 260, 264, 269, 287
    number of, 264, 309-310, 336
Cigarette smoke residue, effect of, 74
Cilia, 67, 69, 118-119, 140, 264, 333, 378, 391
    caudal, 16, 19, 21, 23, 27, 29-30, 32, 37, 42
    chemical composition of, 65, 67, 390
    formation of, 70, 378, 389,
    membranelles of, 67
    meridians of, 15, 19, 21, 23, 27, 29-30, 32, 36, 41, 57, 274, 334, 364, 378
    metachrony of, 423
    proteins of, 67
Ciliary-kinetosome complex, 59, 67, 333
Citrulline, 90
Clionasterol, 93
Clonal analysis, 317
Cobalt—60, 317
Coenzyme A, 93
Colchicine, 69, 85, 291, 390, 396-397, 421-422
Cold synchrony, 133
Colpidium, 2-4, 11, 35, 39, 77, 124, 183, 383
Colpoda, 4, 47
Conjugation, 18, 261-269, 298-299, 308, 313-314, 383, 403
    cytological events of, 264-269
    immaturity period of, 276-281, 314, 324
    refractory period of, 276-281
Contractile vacuole, 80-81, 83, 165, 174, 176-179, 181, 183, 186, 387
    function of, 166, 183, 187-189
Contractile vacuole pores, 16, 20, 22, 25, 28-29, 31-32, 37, 42, 65, 80, 81, 83, 333-334, 365, 387, 389, 405, 407
Cortical architecture, 6, 59, 61, 402-407
Cortical development, 375-409
Corticotype, 6, 320, 333-336, 364-365, 403-404, 406
Cortisone, 422
Crescent stage, 264, 269
Crithidia, 417, 422-423
Culex, 31
Culture age, changes in ultrastructure and, 251-252
Curimostoma renalis, 50
Cycloheximide, 391, 393-394, 413, 422
Cysts, 17, 20, 26, 28, 30-33, 35, 39, 43, 313, 378, 382, 402
Cytidine, 91
Cytidine-deoxycytidine deaminase, 152
Cytidine kinase, 152
5'-Cytidine monophosphate deaminase, 152
Cytidine phosphorylase, 152

Cytochrome c reductases, 249
Cytological techniques, 5
Cytolysomes, 385
Cytopharyngeal area, 35, 39, 67, 69, 95, 191-192
Cytopharynx, role in water balance, 190
Cytophotometry, 289
Cytopyge, 96, 190, 333
Cytopharyngeal area, 35, 39, 67, 69, 95, 191-192
Cytopharynx, role in water balance, 190
Cytotaxis, 5, 335

D

Deltophylum, 11
Deoxyribonuclease, 140, 295
Deoxyribonucleic acid
    base composition of, 127-128, 359
    content of, 127, 132, 221, 237, 292-293, 308,
      347-448
    depolymerization of, 140
    division of, between two daughter cells, 127
    hybridization of, 295, 362
    inhibition of synthesis of, 422
    kinetosomal, 127, 359
    mitochondrial, 73-74, 127-128, 137-139, 295,
      359, 362, 420
    nuclear, 137, 155, 209, 359, 362
    repair process of, 420
    synthesis of macronuclear, 73, 200-202, 289-
      293, 300, 309, 380-381, 392, 427
    synthesis of micronuclear, 200, 309
Deoxyribonucleic acid polymerases, 140
Desmosterol, 93
Dextrin, 90, 93
Diaphorase, 249
Dichotomous key, 11-13
Diglyceride phosphocholine transferase, 115
2,4-Dinitrophenol, 118, 158, 167, 193, 398, 426
Diphosphopyridine nucleotide, 230
DPNH-cytochrome C reductase, 230
DPNH oxidase, 230, 249
Diplopterol, 106, 108-109
Division furrow, 377, 380-381, 386, 393
Division protein, 295
Drosophila, 30
Dry weight of, 222
Dynen, 67

E

Ecological, 6, 43
Elementary particles, 73
Endoplasmic reticulum, 70, 73-74, 76, 80, 132,
    234

Environmental factors, effects of, 227-257
Enzyme levels and culture age, 249-250
Enzymes and structural genes, 330
Enzymes, forms of, 330-333
Epigenetic systems, 308
Epincphrine, 423, 424
Epistylis anastatica, 74
Ergosterol, 93-94
Escherichia coli, 1, 421, 358
Ethidium bromide, 141
Euglena, 156, 420, 422
Euplotes, 267, 387, 402, 404
Evolution, 49-50, 366-369
Exconjugant clones, 315
Expulsion vesicle, see contractile vacuole

F

Facultatively free-living forms, 46
Fatty Acids
    biosynthesis, 101, 105, 158
    composition of, 100-102
    distribution of, 102
    metabolism of, 103
    ratio of saturate to unsaturated, 232
    temperature effects on metabolism of, 105-
      106
Filamentous reticulum, 379
5-Fluorodeoxyuridine, 155
p-Fluorophenylalanine, 394
5-Fluorouridine, 394
Folic acid, 90, 126
Food vacuoles, 69, 75, 77, 79, 80, 96, 109, 144,
    185, 190, 295, 381
    stages in digestion of, 77, 79
    stages in formation of, 77, 94-95, 191-193
Free living forms, 46
Freeze-etching, 63, 72, 84
Frontonia leucas, 72
Fructose, 93
Fructose diphosphatase, 247
Fucosterol, 93

G

Generation time
    distribution of, 202-204
    media and, 205-206
    optimal, 207-208
    osmolarity and, 206-207
    pH and, 204-205
    temperature and, 204
Genetics, 8, 19, 21, 27-28, 307-373

Genetics, population, 338-339
Genome transcription, 224
Genomic exclusion, 276, 313-314, 320, 336, 339
Genotypes, 320-324
Geographical distribution, 276, 281-284
Gibberellin, 426
Glaucoma, 3, 4, 8, 124, 142, 172, 183, 191, 383
Glucose, 90, 93, 114
Glucosidase, 93
Glutamate dehydrogenase, 249
Glutamate-oxalacetate transaminase, 249
Glutamic acid, 65, 90, 93
Glycerol, 93
Glycerophosphate, 93-94
Glyceryl ether metabolism, 115
Glycine, 90
Glycogen, content of, 83, 138, 237, 242-245, 247,
    252
Glycogen synthetase, 247-248
Glycolysis, 115
Glyconeogenesis, 106, 232, 243, 245-246, 253,
    424
Golgi apparatus, 69, 70, 72, 74
Golgi apparatus during mating, 75, 267, 269
Growth
    balanced, 221-222
    CO_2 and Ammonia effects on, 238
    cell division cycle and, 210-211
    cell size and, 217
    characteristics of, 199-224
    macromolecular synthesis and, 222-224
    Oxygen effects on, 235-238
    population cycle and, 211-215
    temperature and, 229
Guanase, 147
Guanine, 91
GMP pryophosphorylase, 148
Guanosine phosphorylase, 147, 151

H

Habitat, 18, 20, 22, 26, 28, 30-31, 36, 41, 43
Heat shock, treatment, 100, 118, 233, 261, 291,
    294-295, 381, 383, 385-386, 396, 399-400
Hexokinase, 241
Histidine, 90
Histones, 309
Histophagous species, 22-23, 30, 47
History, 2
Hompolar doublets, 17, 35
Hydrostatic pressure, 64, 73, 294-295, 398-399,
    406
Hypocholesteremics, 426

I

Immaturity, see conjugation
Immobilization antigens, see serotypes
Inbred strains, 312
Infractiliature, 6
Inorganic element requirements, 93
Inosine monophosphate pyrophosphorylase, 147
Insulin, 426
Ionic regulation, 165-167, 170, 173, 176, 178-
    179, 181-182, 192
Isocitrate dehydrogenase, 242, 246, 249, 424
Isocitrate lyase, 232, 245-247, 424
Isoleucine, 90
Isozymes, 330-333

K

α-Ketoglutarate oxidase, 249
Kineties, 63, 333, 386-387, 389, 396, 402-404
Kinetodesmal fibers, 61, 63, 333, 386, 389
Kinetosome, 59, 61, 63, 67, 69, 70, 72, 127, 264,
    287, 297-298, 333, 380, 385-386, 389, 390,
    396-398, 403
    chemical composition, 65, 390
    formation of, 378-379, 389

L

Lactate dehydrogenase, 240-242
Lactate oxidase, 230-231, 241-242, 246
Lactobacillus casei, 417, 421
Lanosterol, 94
Lebistes reticulatus, 30
Leptoglena, 4
Leucine, 90, 102, 104
Leuconostoc mesenteroides, 416
Leucophrys, 3, 4, 261, 403
Leucophyridium, 4
Life cycle characteristics, 6, 259-281, 290
Linkage relationships of genes, 336-338
Linolenic acids, 94
Lipids
    content of, 100
    distribution of, 118
    drug effects on metabolism of, 118
    metabolism of phosphorus containing, 114
    neutral, 106
    phosphorus containing, contents of, 112
    reserves and culture age, 250-251
    sterol effects on biosynthesis of, 109
    synthesis of, 118
Lophenol, 93

Lysine, 90
Lysolipids, 114
Lysophospholipase, 100
Lysosomes, 72, 75, 79-80, 142, 144, 230, 235,
    249-250, 253, 333
Lysosomes and rough endoplasmic reticulum, 75-
    76

M

Macronuclear genotype, 309
Macronucleus, 17, 83, 127, 132, 134-135, 159,
    200-202, 209, 261, 264, 269, 287, 292-294,
    308-309, 317, 339, 348
    chromatin extrusion and, 9, 17, 36, 41, 287-
        288, 292-293, 309
    division of, 73, 380
Macrostomes, 32-37, 42-43, 45, 378, 381, 383,
    385, 394, 396, 400-402
Malate dehydrogenase, 245-247
Malate synthase, 245-246
Maltose, 93
Mannose, 93
Master-slave, hypothesis, 350, 358
Mating
    factors found to influence, 261-264
    induction of, 313-314
Mating types, 18, 269, 273, 320, 324-327
Meiotic division, 264
Membrane
    synthesis during vacuole formation, 95
    transport through, 168
Membranellar connectives, 379
Mendelian inheritance, 307, 317-320
Metafolliculina, 94
Methionine, 65, 90, 104
24-Methylene cholesterol, 93
N-Methyltransferase, 115
Mevalonic acid, 94
Micronuclear division, 380
Micronuclear genotype, 309
Micronucleus, 4, 85, 127, 137, 200, 260, 264,
    269, 287, 290, 308-309, 312-313, 317, 339,
    420
Microsomal, 109, 143
Microstomes, 32, 34-37, 40, 42-43, 45, 378, 381,
    383, 400
    transformation of, 35, 39, 294
Microtubules, 59, 61, 63, 65, 67, 81, 260, 264,
    269, 378-379, 386, 389, 390, 395, 397-399
    contractile vacuole and, 80
    function of, 64, 69

micronucleus and, 67, 84-85, 287, 289
    oral area and, 68
    protein of, 390-392, 395
Mitochondria, 72-73, 104, 114-115, 119, 127,
    132, 140, 144, 159, 231, 234-235, 241, 243,
    251-253, 287, 289
    cell division and, 74
    degeneration of, 74
    intermitochondrial masses of, 73-74
    shape and position changes of, 74
Mitomycin, 422
Morphogenesis, 65
Mucocysts, 65, 76, 95, 267
    formation of, 72, 76
    function of, 76
Mutants, 274, 317
    nutritional types of, 278-280

N

Nassula, 389
Neomycin, 413, 422
Neospongesterol, 93
Niacin, 91
Nicotinamide adenine dinucleotides, 230
Nitrogen mustard, 422
Nitrogenous base requirements, 91
Nitrosoquanidine, 317
Nomenclature, 4
Norepinephrine, 423-424
Nuclear characteristics, 17, 20, 22, 26, 28, 30-31,
    36, 41, 43
Nuclear membrane, 70-72, 84, 135
Nuclear pores, 84
Nucleases, 91
Nucleic Acids
    cellular organelle content and, 132, 287-289
    hybridization of, 362-363
    kinetosomes and, 139, 297, 390
    metabolism of, 140
    ratios of, 287
Nucleoli, 84, 132, 134-135, 137, 159, 211, 234,
    252-253, 291, 294
Nucleoside-deoxynucleosidomonophosphate, 154
Nucleoside hydrolase, 148
Nucleoside phosphorylase, 146
Nucleosides, metabolism of, 144
Nucleosidodiphosphate kinases, 154
3' Nucleotidase, 144, 154
Nucleotides of adenine and uracil, 67
Nutrient entry, 94
Nutritional requirements, 89, 273

O

Ochromonas, 118, 413, 417, 422-423
Oleic acids, 94
Oral apparatus, 67, 95, 267, 269, 377-378, 389-
    391, 395-396, 398, 405, 426
    isolation of, 69-70, 391
    membranelles of, 69
    oral ridges of, 67-69
Oral development, 69-70, 376-382, 391-399
    duration of stages of, 380
    oral structure regression and, 381
    protein synthesis and, 393-394
Oral primordium, 386-387, 389, 391, 392, 396
    resorption of, 398
Oral replacement, 35, 43, 376, 382-386, 394-395,
    399-400, 406
Organic solutes, accumulation of, 191
Ornithine, 90
Osmolarity, 206
Osmoregulation, 165-166, 182-184, 187
Oxidative phosphorylation, 230, 427

P

Palmitic acids, 94
Panthetheine, 93
Pantothenate, 93
Paraglaucoma, 4
Paramecium, 18, 61, 63, 79, 80, 83, 95, 148, 166,
    175, 187, 190, 192, 210, 260, 267, 269, 276,
    297-298, 310, 327, 329, 356, 364, 366, 368,
    387, 389-390, 402, 418, 422
Parasitism, 22, 45-47
    facultative, 18-19, 26, 44, 46, 76
Parasomal sac, 65, 192
Paratetrahymena, 11
Penicillium islandicum, 420
Peroxisomes, 76, 231, 234, 241-242, 245-246
pH and culture age, 251
Phagosomes, 144
Phenotypes, 320-324
Phenotypic diversity between syngens, 359-366
Phenotypic diversity in clones, 339-358
Phenylalanine, 90
Phosphatidylcholine, 112-113, 115
Phosphatidylethanolamine, 110, 112-115
Phosphatidylinositol, 114
Phosphodiesterase, 143-144
Phosphoenolpyruvate carboxykinase, 245-246
Phosphofructokinase, 240
Phosphoglucomutase, 248
Phosphoglycerides, 94

Phospholipase, 100, 250, 253
Phospholipids, 112-113, 115, 118
    biosynthesis of, 110
    metabolism of, 115
Phylogeny, 47-48
Physiological stress, ultrastructure and, 73
Physostigmine, 423
Pinocytic vacuoles, 80, 95
Platynematum, 11
Polymorphism, 25
Polymyxin, 422
Polyribosomes, 72, 133, 223, 295
Poriferasterol, 93
Probasal bodies, 297, 390, 396
Progesterone, 118, 426
Proline, 65, 90, 93
Pronuclei, 269
Proteases, 253
Protein
    assay for biological value of, 415
    content of, 132, 237
    synthesis of, 118, 139, 232, 294, 392, 427
Protobalantidium, 4
Pseudacris, 30
Pseudomonas, 118
Pseudotriton, 30
Pseudouridine synthetase, 153
Purine
    analogs of, 126-127
    metabolism of, 145-151
    nutritional requirements, 92, 124-125
Purine kinases, 148
Puromycin, 156, 393, 413
Pyridoxal, 91
Pyridoxamine, 91
Pyridoxine, 91
Pyrimidine
    analogs of, 126-127
    metabolism of, 151-156
    nutritional requirements, 92, 125
    ring biosynthesis, 125
Pyrimidine nucleoside knases, 126
Pyruvate kinase, 231, 240, 242

R

Rana, 30
Replicons, 348, 350
Reproductive cyst, 34
Reserpine, 424
Respiration, 74, 157, 229-230, 237, 248, 252,
    426
Respiration, culture age and, 248-249

Respiratory quotient, 248
Riboflavin, 93
Riboflavin monophosphate, 93
Ribonuclease, 84, 143-144
Ribonucleic acid, 126, 141, 155
    base ratios of, 131, 134, 157
    content of, 128, 131-132, 157
    depolymerization of, 141
    kinetosomes and, 140, 294
    metabolism of, 156-160
    mitochondria and, 139
    synthesis of, 118, 132, 156-160, 232, 293-
        295, 309, 392, 394, 426-427
Ribonucleic acid polymerase, 132, 156, 294
Ribosomes, 71-73, 75, 84, 132-136, 139, 158,
    211, 269, 294, 296, 299
    assembling of, 135
    biosynthesis of, 136
    formation of rRNA of, 134-135
    mitochondrial, 134
    protein of, 134
    ribonucleic acids of, 134, 136

S

Saccharomyce carlsbergensis, 91, 417, 421-422
Scenedesmus basilensis, 422
Selfing, 260, 327
Serine, 90, 93, 114-115
Serotonin, 423
Serotype, 273, 319-320, 327-330, 402
Sexuality, 18, 20, 22, 28, 30-31, 36, 41, 43
B-Sitosterol, 93, 413
Somatic cortex, development of, 386-389
Species, 3
Sphingolipids, 113-114
Spindle fibers, 85, 261
Squalene, 94
Stabilization point, 392, 394, 396, 398-399
Starch, 93
Stearic acids, 94
Stegochilum, 11
Stentor, 192, 298, 383
Sterol
    nutritional requirements, 93, 109
    transformation of, 110
Stigmasterol, 93
Stomatin, 35, 383, 400-401
Stomatogenesis, 17, 20, 22, 26, 28, 30-31, 35, 39,
    43, 333, 377-379, 381, 385, 399
Storage products, 83
Streptococcus faecalis, 417
Streptomyces, 118

Streptomycin, 422
Stylonychia mytilus, 348
Subnuclei, 132, 287, 289, 292, 348-349, 356,
    358
Succinate dehydrogenase, 230-231, 247, 249
Succinoxidase, 230, 249
Sulfolipid, 118
Sulfonamides, 426-427
Synchrony, for cell division, 72-73, 85, 100, 102,
    118, 180, 217-221, 294-295, 380-381, 385-
    386, 393, 400-401, 420, 426-427
Synclones, 315-316, 324
Syngens, 310, 324
    characters of, 274-281
    definition of, 261, 272
    nutritional requirements, 273
Synonyms, 4, 15, 19, 21, 23, 27, 29-30, 32, 36,
    41

T

Taxonomy, 3
Temperature
    morphology and ultrastructure, 233-234
    physiology and biochemistry, 229-233
    tolerance to, 276, 281
Testosterone, 426
Tetracyclines, 413
Tetrahymanol, 94, 106-109, 118, 232, 426
Tetrahymena, species of, 7, 9
    T. bergeri, 7, 9, 26, 381, 384
    T. chironomi, 3, 8-9, 12, 21-22, 44, 46-47, 49,
        76
    T. corlissi, 3, 8-9, 11, 13, 21, 29-30, 32, 45-47,
        77, 91, 93-94, 102, 103, 108, 413, 416
    T. faurei, 7, 8
    T. geleii, 4, 74
    T. glaucomaeformis, 7, 8
    T. limacis, 3, 8-9, 13, 27-28, 44-45, 47, 49,
        76, 93-94, 108, 402
    T. parasitica, 7, 8
    T. paravorax, 3, 7, 9, 12, 21, 32, 41-43, 45-46,
        50, 93, 102-104, 108, 124
    T. patula, 3-4, 9, 12-13, 16, 32, 34-37, 39, 43,
        45-46, 67, 93, 94, 108, 378, 381-385, 399-
        401, 403
    T. rostrata, 3, 9, 13, 23, 26-28, 32, 37, 44, 46-
        47, 49, 76, 95, 142, 313, 402
    T. setifera, 3, 7, 9, 13, 19, 21-22, 44, 45-46,
        49, 93, 96, 102-103, 108-110, 124, 126,
        158, 413
    T. stegomyiae, 3, 7, 9, 11, 13, 30, 32, 45, 47

T. vorax, 9, 12-13, 35-37, 39, 41, 43, 45-46, 49, 108, 124, 142, 294, 378, 381, 383, 385, 394, 396, 399-402, 404
Theophylline, 248, 424
Thioctic acid, 91, 93
Threonine, 90
Thymidine kinase, 154
Thymidine phosphorylase, 154
Thymine, 91
Thyroxine, 424, 426
Tomites, 378
Tranquilizers, 425-426
Transformation, polymorphic, 401
Triglyceride synthesis, 106
Triparanol, 118
Triphosphopyridine nucleotide, 230
Triton X-100, 332
Tryptophan, 90
Turchiniella, 4
Tween, 100, 418, 421, 426
Type material, 5, 15, 19, 21, 23, 27, 29, 30, 32, 36, 41
Type species, 13, 44
Tyrosine, 90
Tyrosine aminotransferase, 247

U

Ubiquinone, 106-107, 109
Ultraviolet irradiation, 291, 317
Uracil, 91
Uricase, 148
UMP pyrophosphorylase, 152
Uridine diphosphoglucose pyrophosphorylase, 248

Uridine kinase, 154
Uridine phosphorylase, 154
Undulating membrane, 35, 67-69, 380-381, 384

V

Vacuoles, see specific type
Valine, 90
Variety, definition of, 261
Vinblastine, 291
Vinca, 422
Vincristine, 422
Virus multiplication, 428
Vitamins
    assays for, 417
    B-complex, 91
    growth requirements of, 92

W

Wild strains, 310-312
Wild type, 274

X

Xanthine monophosphate, 148
Xanthine oxidase, 147-148
Xanthosine phosphorylase, 149
X-rays, 317

Z

Zygotic nucleus, 308
Zymosterol, 93